Donald E. Savage and Donald E. Russell
MAMMALIAN PALEOFAUNAS OF THE WORLD, 1983
ISBN 0-201-064944

Page 261 -- Please note relocation of areas on map in Fig. 6-25.

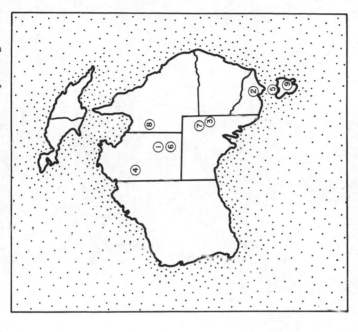

FIGURE 6-25 *Australian Miocene Localities and Stratigraphic Units.*

1　Alcoota—Late Miocene—Northern Territory—Waite Formation

2　Beaumaris (Port Phillip Bay)—Late Miocene—Victoria—Cheltenham sands

3　Billeroo Creek　(See Frome Embayment.)

4　Bullock Creek—Middle Miocene—Northern Territory—Camfield Beds

5　Fossil Bluff (Wynyard)—Early Miocene—Tasmania—Fossil Bluff Sandstone

3　Frome Embayment (Lake Namba)—Middle Miocene—South Australia—Namba Formation

3　Ian's Prospect　(See Frome Embayment.)

6　Kangaroo Well—Middle Miocene—Northern Territory

7　Kutjamarpu fauna—Middle Miocene—South Australia—Wipajiri Formation

7　Lake Palankarinna West—Ngapakaldi fauna—Middle Miocene—South Australia—Etadunna Formation

3　Lake Pinpa Southwest　(See Frome Embayment.)

7　Ngapakaldi fauna　(See Lake Palankarinna West.)

3　Pinpa fauna　(See Frome Embayment.)

2　Port Phillip Bay　(See Beaumaris.)

8　Riversleigh—Middle Miocene—Queensland—Carl Creek Limestone

3　South Prospect B　(See Frome Embayment.)

3　Tom O's Quarry　(See Frome Embayment.)

3　Tarkarooloo　(See Frome Embayment.)

5　Wynyard　(See Fossil Bluff.)

Australian Oligocene.

9　Geilston Bay—Late? Oligocene—Tasmania—Geilston Travertine

ADDISON-WESLEY PUBLISHING COMPANY
Advanced Book Program/World Science Division
Reading, Massachusetts
ISBN for Erratum Sheet:　0-201-06496-0

Mammalian Paleofaunas of the World

Mammalian Paleofaunas
of the World

Donald E. Savage 1917–

University of California,
Berkeley

Donald E. Russell

Muséum National d'Histoire Naturelle,
Paris

1983
ADDISON-WESLEY PUBLISHING COMPANY
Advanced Book Program/World Science Division
Reading, Massachusetts
LONDON · AMSTERDAM · DON MILLS, ONTARIO · SYDNEY · TOKYO

Front cover: Synthetoceras tricornatus Stirton, 1932E

Ten-million-year-old ruminant artiodactyl from Clarendon, Texas

Drawn by Mr. David Foster, Department of Paleontology University of California

Library of Congress Cataloging in Publication Data

Savage, Donald Elvin, 1917–
    Mammalian paleofaunas of the world.

   Bibliography: p.
   Includes indexes.
    1. Mammals, Fossil. I. Russell, Donald E.   II. Title.
QE881.S28   1983   569   82-13764
ISBN 0-201-06494-4
ABCDEFGHIJ-MA-89876543

Professor R. A. Stirton, about 1951
(*Drawn by Dr. Dawn Adams, Department of Paleontology,*
University of California, Berkeley)

Dedication

The late Professor R. A. Stirton, chairman of the Department of Paleontology and director of the Museum of Paleontology at the University of California from 1949 until 1966, initiated a series of study guides and lectures outlining the history of mammalian faunas of the world. This book is an elaboration of Professor Stirton's work, and we dedicate the following pages to him.

Contents

List of Figures

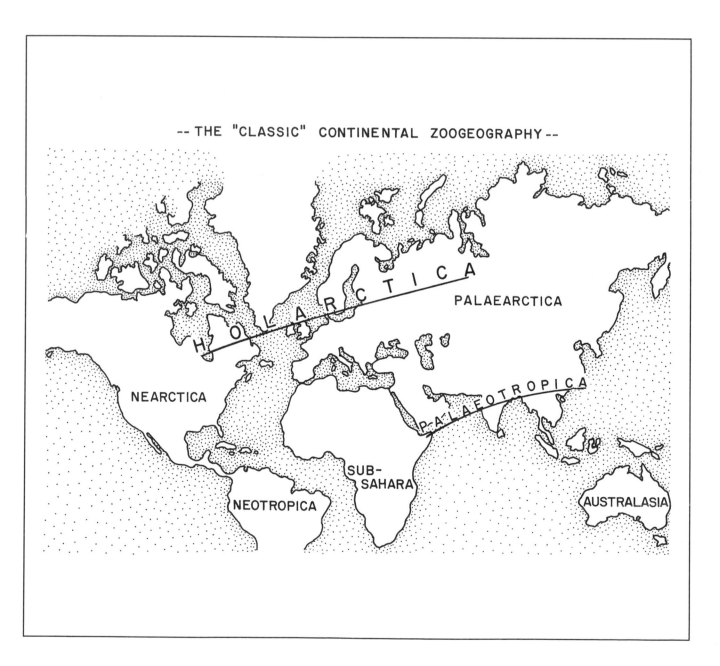

-- THE "CLASSIC" CONTINENTAL ZOOGEOGRAPHY --

HOLARCTICA

PALAEARCTICA

NEARCTICA

PALAEOTROPICA

SUB-SAHARA

NEOTROPICA

AUSTRALASIA

Preface

"Here, then, are the names that I propose to give to the five groups that I have thought best to establish in the Secondary Terrains . . . The name Mastozootic, applied to the fourth group, will recall that in the midst of these terrains there are found the bones of mammals, the study of which has given birth to the beautiful work which has, as it were, created geology among us."

J. J. d'Omalius d'Halloy, 1822. Observations sur un essai de carte géologique de la France, des Pays-Bas et des contrées voisines. *Annales des mines*, vol. 7, pp. 373–374 (Cited from translations by Wilmarth, 1925. *U.S. Geol. Surv. Bull.* 769:56–57.)

Mammalian Paleofaunas of the World presents an abbreviated view of the mammalian faunas of the world through 200 million years. We hope and intend that our emphasis on data and our extensive bibliographies covering the pertinent literature through 1982 will serve as a platform from which advanced students of fossil mammals and research workers can progress with investigations in their specialties. Those primarily concerned with zoological or geological or anthropological sciences, and also other disciplines contributing to knowledge about organic evolution, should find the data and summaries of this book applicable to hypotheses about mammalian evolution, biochorology, and geochronology. Those more casually interested in the history of mammals can more closely appreciate the mass of paleontologic facts and procedures supporting interpretations in the grand synthesis of historical paleontology.

We have tried to compile and organize diverse paleontologic, stratigraphic, and geographic facts that have not been completely synthesized since 1910, when H. F. Osborn published the *Age of Mammals*. Osborn's extraordinary volume, with its priceless photographic plates, cannot be reproduced or used as a model for a similar book at the present time; the price of such a book would be staggering for most students and researchers. Accordingly, our history of mammalian paleofaunas uses only pen-and-ink illustrations.

W. D. Matthew's *Climate and Evolution* (1939), while dealing with fossil mammals on a worldwide scale, emphasized hypotheses regarding the origins and dispersals of higher vertebrates. E. Thenius (1959) meticulously compiled a text-outline on Tertiary vertebrate faunas. Although an enormous amount of information has been published on this subject since 1959, Thenius's publication has been an invaluable reference for us. Lillegraven, Kielan-Jaworowska, and Clemens (1979) have given us an extremely valuable reference on Mesozoic mammals, detailing the first two-thirds of mammalian history. We therefore concentrate on the final and most conspicuous third, the astoundingly intricate and numerous details of Cenozoic history.

We do not review the ever-present problems and intrigue of conceiving phylogeny and proposing taxonomic classification among the Mammalia. Paleontologic phylogenies are being supplemented, corrected, or otherwise improved (and sometimes completely disproved) by data from living organisms: morphological, embryological, physiological (serological, immunological), genetic, and behavioral information. Many scholars are now engaged in formulating "a series of precise statements of phylogenetic affinity" to produce "testable hypotheses" (Delson, Eldredge, and Tattersall, 1977, p. 264). See, for example, Schaeffer, Hecht, and Eldredge (1972), Cracraft (1974), McKenna (1975), Hecht (1976), and Szalay (1978), and references cited by these authors. Inevitably, however, our arrangement of taxa in faunal lists indicates our general preference in classification. Simpson's 1945 classification of the Mammalia (Simpson, 1945I) continues to be the single most valuable reference for our purposes, and we hope that the revised classification and listing of mammalian taxa that is under way at the American Museum of Natural History by M. C. McKenna will be forthcoming soon.

We remain primarily concerned with fossil mammals and beg forgiveness from those who emphasize studies of the equally important fossils of fishes, amphibians, reptiles, and birds for slighting these classes. Together with fossil plants and fossils of certain invertebrate animals and protistans, these other groups of bony animals may indicate past climates and environments better than the mammals do.

We work also with aggregates of fossils. These aggregates may represent associations of living animals (communities) or fortuitous "death assemblages," or a varying percentage of living association plus death assemblage. Thus, these aggregates may faithfully or imperfectly represent the relations of the animals with their environment. Physical-chemical, geological, and biological associations may be interpreted partly by comparing fossils with the most similar present-day organisms (substantive uniformitarianism). Efremov (1940), Behrensmeyer (1975), Hanson (1980), Bown and Krause (1981), and others show us how important the detailed studies of the general entombment phenomena in sediments are. This research area, *taphonomy*, can, as Behrensmeyer claims, lead to a better knowledge of the character and evolution of vertebrate communities through time (Behrensmeyer, 1975, p. 474).

From the maxim that environments undoubtedly had dimension and configuration, we proceed to paleobiogeography—the study of areal distributions (chorology) of ancient life—which is the background information for biogeography. Thus biogeography must be historical in depth, as G. G. Simpson and others have emphasized.

O. Reig (1981) summarized his review of G. G. Simpson's book *Splendid Isolation. The Curious History of South American Mammals* (Simpson, 1980) with the statement: "Perhaps this book marks the end of an epoch in biogeographical thought. Let us hope that new generations will be inspired by it to build up the foundations of a new biogeographical science more theoretical than empirical, more mathematical than discriptive, and more integrative than merely paleontological in its factual foundations."

Since our book will be subjected to the same type of comment from the "new" generation, we reply to Reig: A "new biogeographical science" of theories, mathematics, and integrations built on theory may be impressive in its printed version and indeed may inspire future generations of workers, but Reig must agree that if these theories and the like are not built upon meticulous accumulation and organization of paleontological facts, chaos can ensue. Thus we endeavor to justify the "data base" that is the essence of our book.

The glossaries of fossil-mammal localities and lithostratigraphic units and the bibliographies of this book were compiled by means of card files developed over 25 years. Undoubtedly data lists like these lend themselves ideally to computerized inventory. However, because of the maturity of the "old-fashioned" card-file system, we chose not to expend the extra time and money for conversion to a computer-controlled system. Our data, as herein printed, can be converted, corrected, and updated with electronic control by a future worker so inclined.

The fossil record of mammals covers all continents and many islands. It represents an interval of about 200 million years. But it is a spotty record. Much work remains to be accomplished. During this year, more than 2000 scholars are studying the faunal history of mammals as it is related to and influenced by the great geologic and geographic changes of continents and seas. Our book cannot be completely abreast of the growing knowledge of earth and life history, and we apologize to researchers whose ideas and new taxa we have missed. We hope, however, that you who read this will find here a relatively modern guide to the state of knowledge about succession of the world's mammalian faunas.

Acknowledgments

We have benefited for many years from the advice, criticism, and information given us by our peers in North America, Europe, Asia, Africa, South America, and Australia. These colleagues are indicated in the citations and bibliographies.

Certain friends and certain former teachers who are not vertebrate paleontologists have given us special help and consultation. These people are Garniss H. Curtis, the late Edwin C. Allison, Dwight W. Taylor, William Glen, Daniel I. Axelrod, Jack A. Wolfe, and Howard Schorn.

We give special thanks to Judith A. Bacskai and to George V. Shkurkin, compilers of the bibliographies of fossil vertebrates. Their guidance for our reference searches and their translations of critical excerpts from the literature printed in the languages of eastern Europe and of Asia have been priceless.

The senior author is deeply grateful to Khin Than Myint for constant help and encouragement.

Most of the illustrations of this book except reproductions of previously published figures were done by Dr. Dawn Adams.

Mammalian Paleofaunas
of the
World

CHAPTER ONE

Introduction

BIASES AND BASES

Whose Paleontologic Taxa Are Better? Two competent professionals of the same generation of mammalian paleontologists study the same suite of fossils, noting essentially the same ensemble of anatomical characters. Both professionals use the accepted systems of data management for analyzing possible taxonomic relations. Professional *A* concludes that the suite represents one genus and two species of once-living animals, and Professional *B* concludes that three genera and six species are represented. In these analyses, *A* becomes the "lumper," *B* the "splitter." Whose taxa are better? Whose taxa should be used in a faunal list based on these fossils?

This is the age-old and eternal problem for colleagues specializing in certain groups who wish to compile a faunal list. Expectedly, this problem has been one of our greatest frustrations in compiling our book. While we cannot deny subjectivity and bias for the decisions made here, we do assert that our decisions were made with malice toward no one. In general, with the approximate conditions as listed, we have tended to follow the splitter. Our rationale is that the split list of taxa gives the nonspecialized reader a better comprehension of the possible geographic and temporal heterogeneity portrayed by the fossils. Inevitably, our compilation is not internally consistent in taxonomic judgment, for we usually follow the latest reviser of a group

unfamiliar to us. Even more subjectively, we have considered ourselves competent to judge the "validity" of paleontological taxa in several orders of land mammals without conferring with colleagues also working with those groups.

At times we have tried to retreat from a continuing dispute about suitable name and content of a family of mammals by using only an all-encompassing superfamily name and thus submerging the issue. Our consistent use of the superfamily names Ischyromyoidea and Erinaceoidea are examples of such retreat.

What do authors mean when they publish appellations such as *Mesodma* sp. or *Mustela* sp.? Is the particular sample of *Mesodma* (or of *Mustela*) unidentifiable for species or does it show a peculiar array of anatomical characters that the author intends to describe and name as a new taxon when time is available? The reader is confronted with a great ambiguity.

Most examples of *sp.* in the literature on fossil mammals seem to mean *species unidentifiable,* but we know also of many examples in which the author thought a new taxon was represented. We urge that future authors clarify, writing for example, *Mesodma*, n. sp. (or sp. nov.), versus *Mesodma*, sp. indet. (or sp. unident.).

Proposed infraspecies categories such as subspecies, races, "advanced or progressive," or "primitive stage" are not given in our faunal lists. Although such designations may be critical for interpreting local zonation or for ecology, we believe they are beyond the purview of this book.

Another of our frustrations in reviewing the literature on fossil mammals has been the perniciously infrequent citing of the author and date of a species being discussed. This mundane task has obviously been too much of a nuisance for our past and current generation of ''mega''-thinkers and ''mega''-speculators, who have been concerned with the true courses of evolution, the niceties of phylogeny and systematics, the intricacies of paleoecology, the efficacies of proposing new chronostratigraphic and geochronologic systems/boundaries/correlations, and the profundities of paleobiogeography as related to the new theories of plate tectonics on our dynamic earth. One of our greatest disappointments has been that the ''Romer Bibliographies'' of fossil vertebrates for the early non-American literature (Romer, Wright, Edinger, and Van Frank, 1962) could not have been expanded to include the cross-referencing and the systematic indices that had been instituted in the ''Hay-Camp-Gregory'' bibliographies, despite the great amount of time and money that such an enterprise would have required. Be this as it may, in hoping to help with the remedy for this malady, we have tracked down the author-date information whenever possible in the time available to us. These data are given in our taxonomic-faunal lists, and we have used the system and organization of the Hay-Camp-Gregory bibliographies of fossil vertebrates: Hay (1902), Hay (1929, 1930), Camp and VanderHoof (1940), Camp, Taylor, and Welles (1942), Camp, Welles, and Green (1949), Camp, Welles, and Green (1953), Camp and Allison (1961), Camp, Allison, and Nichols (1964), Camp, Allison, Nichols, and McGinnis (1968), Camp, Nichols, Brajnikov, Fulton, and Bacskai (1972), Gregory, Bacskai, Brajnikov, and Munthe (1973), Gregory, Bacskai, and Shkurkin (1978), and Gregory, Bacskai, Shkurkin, and Bryant (1981). For example:

1. *Thinohyus nanus* Marsh, 1894L   —Stirton and Woodburne (1965) (One can find the complete bibliographic citation for this species by looking in Hay (1902) under the heading *Marsh, O. C., 1894L*. Further, we have noted that Stirton and Woodburne (1965) have more recently substantiated Marsh's species and its generic assignment. See complete bibliographic citation for Stirton and Woodburne in Camp et al. (1972).)

2. *Aepinacodon rostratus* (Scott, 1894E)   —Macdonald (1956C) (One can find the complete bibliographic citation for this species by looking in Hay (1902) under the heading *Scott, W. B., 1894E*. Further, we have noted that Macdonald (1956C) more recently reviewed Scott's species; he assigned it to *Aepinacodon* Troxell (1921B). See complete bibliographic citation for Macdonald in Camp, Allison, and Nichols (1964).)

The boundaries of land-mammal biogeographic regions seldom coincide with present borders of continents except for South America while it was essentially an island and for the Australia–New Guinea–Tasmania complex. We know or can often surmise also that the ''fuzzy'' borders of land-mammal regions shifted drastically as the expanse of seas relative to lands changed. Often, for example, the fauna of North Africa or of Asian Turkey (Anatolia) may be more similar to the contemporaneous fauna of Europe than to the fauna of sub-Saharan Africa or eastern Asia. We shall try to keep you aware of these dynamic qualities of paleobiogeography as we discuss, for example, the early Miocene fauna of the Asian continent or the African continent, or the like.

ABOUT STRATIGRAPHY AND PALEONTOLOGIC CHRONOLOGY

Thousands of pages have been written on how to think and how to proceed when striving for tenable and useful conclusions in stratigraphy and geochronology. The dating and correlation of rocks from place to place invites extensive verbiage, oratory, and debate in the earth sciences. Inevitably, each writer or speaker in this area accepts, or should accept, the admonishment of G. G. Simpson (1961): ''The most fundamental reason for disagreement in science is, however, the inherent impossibility of complete certainty.''

We shall not belabor you with reviews of the historical development of terms and procedures in stratigraphy and geochronology; many good textbooks and monographs do this and are among the selected bibliography at the end of this chapter. Rather, with the desire to be utilitarian, we present a few brief definitions and concepts for terms used in this book.

Biotic-Paleontologic Terms

fauna: We use *fauna* in its broadest sense. Fauna is the total animal content of something, including the animal content as represented by fossils. *Examples:* the fauna of Sicily, the mammalian fauna of the Sespe Formation, the Miocene fauna, the Cenozoic molluscan fauna, the fauna on a dog's back.

local fauna: Following the proposal of H. Wood et al. (1941), this expression is used by some writers in reference to the localized fossil sample of a fauna. There are no exact limits—areally, stratigraphically, or ecologically—but the sample is usually from a thickness of a meter or less of strata in one quarry or in a relatively small district. As would be expected, one writer's ''local fauna'' is another writer's ''fauna.'' The proper name of a local fauna is, by convention, also the name of the sample locality. *Example: Candelaria local fauna* from the Candelaria locality (J. Wilson et al., 1968).

faunule and **paleofaunule:** Fenton and Fenton (1928) defined the term *faunule* as an assemblage of fossils representing, interpretively, an ecologically restricted society of animals, one of the components of the larger and more complex fauna that is represented. Sometimes we shall use *paleofaunule* merely to emphasize that the assemblage being studied is an aggregate of fossils. *Example:* the pond-bank faunule, the paleofaunule of the *X* Member of the *Y* Formation, the *Globigerinoides allisoni* paleofaunule.

Lithostratigraphic (Rock-Stratigraphic) Terms

formation: Used in accordance with the recommendations of Schenck and Müller (1941), the stratigraphic code of the American Commission of Stratigraphic Nomenclature (1970), the stratigraphic codes of many other countries, and the *International Guide to Stratigraphic Classification,* ed. H. Hedberg (1976). A formation is a mappable lithogenetic unit. It is recognized basically for mapping and is characterized by discrete lithology—either homogeneous or heterogeneous compared with adjacent formations—and by a traceable stratigraphic position.

member: A formally designated and mappable subdivision of a formation. The term may be useful for describing and interpreting the geologic history of an area even though the member may not be coextensive with the formation. *Example:* the Laney Member of the Green River Formation.

group: A group is composed of two or more stratigraphically contiguous formations, clustered in this lithostratigraphic category because of their general geographic and lithogenetic affinities with one another. Group designation may be especially useful for mapping and for explaining episodes of geologic history in areas in which component formations become indistinct lithologically. *Example:*

WEST ←----------------------------→*EAST*		
(Berkeley Hills, California)		(Lafayette, California)
Bald Peak Basalt		
Siesta Formation	Contra Costa Group	Contra Costa Group, undifferentiated
Moraga Formation		
Orinda Formation		

Paleontostratigraphic (Biostratigraphic) Term

zonule: A purely paleontologic-stratigraphic unit, comprising the faunule together with the strata, or stratum, or part of a stratum in which the faunule is entombed. A zonule may be found at different stratigraphic levels in a given district, corresponding to possible disappearances and subsequent reappearances of the environment in which the living faunule thrived. Such a phenomenon may be the result of back-and-forth travel of a shifting river channel, shifting vegetation, or other environmental shifts.

Chronostratigraphic (Time-Stratigraphic) Terms

chronozone: (paleontologic chronozone = "Oppelzone") A stratal interval identified by the *congregation* (W. Berry, 1966,

1968) and bounded above and below by "horizons" that are isochronous so far as paleontological record can determine. The lower limit of a given chronozone, then, is probably later than the underlying chronozone in any place that these two chronozones can be recognized. This paleontological chronozone equals "zone" or "assemblage zone" or "faunal zone" or "faunenzone" or "faunizone" or "fossil zone" or "index zone" of divers earlier workers, depending on the thought and application of each worker.

Congregation and *chronozone* apply especially to the chronological ordering of stratified marine sedimentary rocks in one basin of deposition or in one faunal province. Admirably, there is endeavor here to make stratigraphic correlation more accurate, with hope of avoiding the uncertainties of correlation engendered by repeated or time-transgressive (*diachronous* of Shaw, 1964) paleontological and lithological facies. The recognition of a "best" horizon (boundary) tokened by "appearances of certain new species" or "certain other new species" (Berry, 1968) is the subjective link in the chain of Oppelian methodology, wherein bias or seemingly practical convenience may have been the deciding factor. In practice (Kleinpell, 1938, p. 122, typification of the Luisian Stage; Mallory, 1959, p. 19, problems of control, and p. 30, typification of the Bulitian Stage), many of the boundaries of the paleontological chronozones have been set at the available lithologic-formational boundaries and are thus at depositional-ecologic boundaries also. Such boundaries cannot be assumed to be the total stratigraphic range of the congregation.

A species that has abundant fossil representation in a chronozone but is not necessarily confined to the chronozone provides the proper name for the chronozone. For example: the *Uvigerina gallowayi* chronozone of the Zemorrian Stage (Kleinpell, 1938, p. 110). Thus the Oppelian method for chronostratigraphic subdivision and correlation does not actually conquer the basic problem resulting from the possibility that lithofacies and paleontologic facies may change in age from place to place—the "diachronous homotaxial biostratigraphic or lithostratigraphic sequences" of Shaw (1964)—even though chronozones and stages have been defined as having isochronous limits. But at least in the marine sedimentary milieu, Oppelian zonation may be the method most closely approaching maximum use of the chronologic resolution that is available from the fossil record.

stage: A succession of strata or a stratigraphic interval with characterizing fossil aggregate delineated in "a continuous fossiliferous section exposing also fossiliferous sections of both subjacent and superjacent Stages" (Kleinpell, 1938, p. 103). Traditionally the stage has been extended geographically on the basis of its characterizing fossils. The top as well as the bottom of a stage is proposed to be the same age everywhere the stage is recognized so far as can be judged from the paleontology, and hence the stage is conceived as a chronostratigraphic unit. Radiometric dating is now providing an intensive criticism for previously proposed year-age estimates for stages and their boundaries. "Magnetostratigraphy" is giving tests for correlations of zones and stages that were defined by fossil content. The proper name of a stage is based on a geographic name in the district of the stratotype, for example, *Relizian Stage* from

Reliz Canyon, Monterey County, California (Kleinpell, 1938, p. 117).

Golden Spikes

Paleontologic datum plane, global datum plane, datum level, datum event, evolutionary first occurrence of a phyletic event, first-appearance datum (FAD), last-appearance datum (LAD), niveau répère, ... See Berggren and Van Couvering (1974), Berggren et al. (1980), Murphy (1977), Woodburne (1977), Jaeger and Hartenberger (1975), and Thaler (1972). In our opinion, all the above terms, though not completely synonymous, relate conceptually to the pioneer paleontostratigraphic and paleontogeochronologic notion of the *index taxon,* or the *guide fossil* (cf. Grabau, 1924). All are associated with the idea that a taxon (usually a species or a genus) may have widespread and isochronous appearance, distribution, and stratigraphic range—in short, that such a taxon, especially in its lowest stratal recovery, can be a "golden spike" to use in nailing down an accurate and precise geochronologic dating and chronostratigraphic correlation.

Although examples such as the *"Hipparion* datum," the "first-appearance-of-Rodentia datum," the *"Hyracotherium* datum," the *"Orbulina* global datum," the "Coderet niveau répère," FAD, and LAD are in the public eye and used frequently in this decade, we believe that these concepts overreach the real resolution inherent to the paleontological discipline in chronostratigraphy and geochronology. See Savage and Russell (1977) for further appraisal of these concepts.

Geochronologic Terms

Each geochronologic unit tokens the duration of a corresponding and first-described chronostratigraphic unit. Each geochronologic unit bears the same proper name as its corresponding chronostratigraphic unit. Kauffman (1970), among others, has objected to this duality. He opposes formal recognition and classification of the chronostratigraphic hierarchy, but we prefer to continue with this twinned system as above outlined. See discussion of the semantics of this problem in Savage (1975).

Each geochronologic unit is, by definition and belief, discrete. Again, however, it is axiomatic that adjacent units may overlap, since the paleontological discipline in geochronology may not always provide the hoped-for resolution. Common geochronologic terms follow.

age: Example: the Relizian Age, duration of the Relizian Stage.
epoch: Example: the Miocene Epoch, duration of the Miocene Series.

period: Example: the Cretaceous Period, duration of the Cretaceous System.
land-mammal age: Used in reference particularly to the "North American Provincial Ages" of H. E. Wood et al. (1941) and to geochronologic ("biochronologic") units of equivalent concept that have been proposed and are used in other parts of the world. In light of customary usage of the term *province* in the literature of biogeography and paleontology, we think that the entire North American area is larger than a faunal province; so we prefer to use Land-Mammal Age for North America (or Nearctica) or for other continental areas. The published land-mammal ages of the Cenozoic of North America, Europe, South America, and Africa are shown in Figure 1–1.

Paleontological sampling remains incomplete relative to the intricate continuum of change of the world's natural "paleophenomena." These phenomena include morphologic evolution of organisms, whether this evolution has been punctuated or smoothly gradual in its finer details. Paleontological sampling also remains incomplete relative to the gradualness of the dispersal of organisms and relative to the gradualness of change in taxonomic composition of living associations of organisms as their environment (climate and other environmental vectors) changes. As a result, two local faunas will seldom if ever have exactly the same roster of genera and species, and probably there will always be dispute about the precise age of a particular local fauna as it is compared taxonomically with other local faunas; thus, the "Lafayette" local fauna may be latest Barstovian to some workers but earliest Clarendonian to others. Aside from these boundary and subdivision disputes, however, the land-mammal ages—and correlated or tied-in marine mammal ages—appear to be discrete and sequential, as shown in Figure 1-1.

Assigning many of the recognized mammal ages to the classic epochs, that is, Eocene, Miocene, and so on, introduces a different and stupendous hierarchy of problems and controversies that have been with us for more than a hundred years. These problems and controversies stem from the vagueness of the initial definitions and usages of the epochs as they are viewed by present-day techniques and procedures in stratigraphy and geochronology. The controversies regarding placement of ages relative to epochs are symbolized by the slanted dashed lines that are placed between the names of the epochs. Figure 1-1 tells us, therefore, that faunas assigned to early Agenian or early Arikareean land-mammal ages may be termed late Oligocene by some workers but early Miocene by others. Likewise, deliberate omission of boundary lines and horizontal correlation lines on Table 1 is our device for showing the basic lack of resolution for delimiting mammal ages and for their intercontinental correlation. Our purpose in recognizing certain mammal-age names and proposing new names is not to promulgate one concept at the expense of another. If a mammalian fauna can be shown to be an integral part of a marine stage stratotype, as is frequently the case in the Paris Basin, no separate mammal-age name is needed. The decisions we have made about limits express expediency more than firm conviction.

Mammalia and Geochronology

For our purposes, a mammal is a vertebrate animal in which the squamosal bone of the cranium and the dentary bone of the mandible form the essential part of the jaw-hinging

CENOZOIC LAND MAMMAL AGES

EPOCHS	EUROPE [1]	EUROPE [2]	NORTH AMERICA	SOUTH AMERICA	AFRICA
PLEISTOCENE	STEINHEIMIAN	STEINHEIMIAN	RANCHOLABREAN	LUJANIAN	
	BIHARIAN	BIHARIAN	IRVINGTONIAN	ENSENADAN	
				UQUIAN	
PLIOCENE	VILLANYIAN VILLAFRANCHIAN CSARNOTAN	VILLAFRANCHIAN	BLANCAN	CHAPADMALALAN	MAKAPANIAN
	RUSCINIAN	RUSCINIAN		HUAYQUERIAN	LANGEBAANIAN
	TUROLIAN	TUROLIAN	HEMPHILLIAN		LOTHAGAMIAN
	VALLESIAN	VALLESIAN	CLARENDONIAN	CHASICOAN	NGORORAN
MIOCENE	ASTARACIAN	ASTARACIAN	BARSTOVIAN	FRIASIAN	TERNANIAN
	ORLEANIAN	ORLEANIAN	HEMINGFORDIAN		RUSINGAN
	AGENIAN	AGENIAN	ARIKAREEAN	SANTACRUCIAN	
	ARVERNIAN	Late OLIGOCENE	WHITNEYAN	COLHUEHUAPIAN	
OLIGOCENE		Middle OLIGOCENE	ORELLAN		FAYUMIAN
	SUEVIAN	Early OLIGOCENE	CHADRONIAN	DESEADAN	
				DIVISADERAN	
	HEADONIAN	HEADONIAN	DUCHESNEAN		
		ROBIACIAN	UINTAN	MUSTERSAN	
EOCENE	RHENANIAN	LUTETIAN			
		CUISIAN	BRIDGERIAN WASATCHIAN	CASAMAYORAN	
	NEUSTRIAN	SPARNACIAN	CLARKFORKIAN		
	Cernay	CERNAYSIAN		RIOCHICAN	
			TIFFANIAN		
PALEOCENE	Mons	"DANO-MONTIAN"	TORREJONIAN		
			PUERCAN		

[1] From Fahlbusch (1976)
[2] Age names that we use.

FIGURE 1-1 *Cenozoic Land-Mammal Ages of Europe, North America, South America, and Africa.*

mechanism. Diphyodont dentitions or complexly cuspate cheek teeth with two or more roots, or both, are the general basis for assigning many fossils to the Mammalia, especially the Mesozoic and earlier Cenozoic fossils. It is our thesis, and indeed the framework of this entire book, that many mammals, marine as well as nonmarine, contributed fossils to the record that are admirable tools for geochronology (relative dating and correlation). Let us examine the attributes of mammals supporting this assertion.

1. *Many genera of mammals* (consensus) *and probably many species* (our opinion) *have and have had trans-* *oceanic and transworld geographic distribution in the Northern Hemisphere.* Fossils of these taxa may be found around the world and are thus available for geochronologic correlation. Among the extant species of Carnivora and Cetacea, *Felis (Puma) concolor* is found from British Columbia to Patagonia, and *Mustela erminea*, the ermine, and *Gulo gulo*, the wolverine, are circumboreal species. Even the tiny *Reithrodontomys*, harvest mouse, ranges from southwestern Canada to Colombia and Ecuador, and its species *R. megalotis* extends from southwestern Canada to Oaxaca, Mexico. In the marine sphere, *Physeter catodon*, sperm whale, is found in tropical to polar waters; *Orcinus orca*, killer

whale, lives in all oceans. *Delphinus delphis,* common dolphin, is worldwide in warm and temperate seas but found also at times in cooler seas. *Zalophus californianus,* California sea lion, ranges from the west coast of North America to the Galapagos and into the southern part of the Sea of Japan. These data are from Walker (1968).

Three of the 13 genera of land mammals known from the late Jurassic Purbeck Formation of southern England are known also from the Morrison Formation at Como Bluff, Wyoming. Thirty to 35 percent of the 140 or more nonvolant nonmarine mammalian genera of the North American Pleistocene are also known from Pleistocene rocks of greater Europe, and a comparable percentage is known from Pleistocene rocks in eastern Asia. An almost incredible 50–60 percent of the genera of mammals from the Sparnacian Stage (Lower Eocene) of the Paris Basin, France, are found also in the Wasatch strata of Wyoming. Thus genera of mammals are potentially a valuable asset to Mesozoic and Cenozoic correlations between continents and across seaways, especially in the Northern Hemisphere.

The continental areas of the Southern Hemisphere, however, have been either partly (Africa, South America) or completely (Australasia) unavailable to land-mammal dispersal through the Cenozoic. Fossils of marine mammals, plants, invertebrates, and micro-organisms are used for relative dating and for paleontologic correlation of strata of these continents with strata on other continents. Cenozoic land-mammal ages are being recognized in each of these southern continents, however, especially in South America, and are useful within each continent.

2. *The taxa of mammals that we use in correlation are noteworthy for their relatively and actually restricted geochronologic range.* Colleagues may point to this statement as raw pragmatism, based on little more than indoctrination. But is it? Let us look at the known ranges of some of the Cenozoic mammalian genera (Table 1). Here we see a spectrum of mammalian genera, including some with longest-known ranges as well as some with shortest-known ranges. We have concentrated on plotting genera with transoceanic distribution and the year dates are approximate boundaries for epochs (e.g., the Paleocene-Eocene boundary is placed at about 54 million years before the present. Henceforth this time unit may be abbreviated "m.y.b.p."). Many fossil-mammal specialists will hold that we have overextended the time range of some of the genera.

Although insurmountable biological, genetic, and taxonomic uncertainties are involved in any comparison between mammalian genera and genera in other classes of organisms, Table 1 supports our assertion that mammalian genera are relatively and actually short-ranging in time. This is not a new idea. Sir Charles Lyell (1833, p. 253) noted, "We have more than once adverted to the fact that extinct mammalia are often found associated with assemblages of *recent* shells, a fact from which we have inferred the inferior duration of species in mammalia as compared to testacea."

We have used genera in our table simply because vertebrate paleontologists agree about identification only at this level of the taxonomic hierarchy.

3. *The genera and species of mammals disperse rapidly.* They walk, run, climb, swim, and fly. They are free-moving. Relative vagility of different classes of organisms is as difficult to assess as relative duration in geologic time. In asserting the mobility of marine invertebrates, Woodford (1965, p. 96) said, "Modern marine clams and snails have spread through scores or even hundreds of miles of shallow water in a few decades (Elton, 1958). If the ammonites got around as quickly as the clams and snails do now, which seems likely, the spread of a new European zonal fauna must have been geologically instantaneous." Erle Kauffman has reemphasized to us (written communication, 1974) that each generation of molluscs, echinoderms, worms, and the like in the marine realm normally disperses hundreds or thousands of miles by planktonic larval drift. We affirm that certain populations and individuals of marine and land mammals and birds are known to travel thousands of miles in migration over their home range and in permanent extension of their geographic range. Some species of mammals evidently have dispersed across a continental or ocean-basin surface in a few thousand years.

Tyndale-Biscoe (1973, pp. 225–226) summarizes the dispersal record of the opossum, *Didelphis marsupialis,* which is known from the United States, Mexico, Central America, and South America. This species was introduced in California during the period 1870–1915. It spread north into British Columbia and eastward to an elevation of 1500 meters (m) by 1958 and is now established along the Pacific Coast from southern Canada to northern Mexico. It was recorded to have spread northward at a rate of 50 kilometers (km) per year during a 26-year interval, although the influence of man's activities on this dispersal is not completely certain. Thus, as a conservative projection, if there were no ecological barriers or other restraints, *D. marsupialis* might extend its range 20,000 km (to Europe and Africa via Asia) in 1000–2000 years. This would be an insignificant interval compared with currently recognized durations of chronozones or the like. Such evidence strongly suggests that a species of mammal, known from far-flung provinces of the world, dispersed to these provinces too fast for the speed to be calibrated by present-day methods of geochronology.

Kurtén (1957) presented intriguing conclusions on the rate of dispersal of mammalian species in the faunas of the Pliocene, Pleistocene, and Recent of Europe and China. He decided (p. 217) from the data amassed that "an unchecked spread of some 1000 kilometers in a century would seem a moderate estimate for most larger mammals." Thus, following his assertion and assuming no special restraints,

TABLE 1 *Temporal Ranges of Certain Genera of Land Mammals in Millions of Years*

100	65	54	35	24	5	R	Genera
		XXXXX					*Plesiadapis*
		XXXXX					*Esthonyx*
					XXXXX		*Ochotona*
				XXXXXXXXX			*Hypolagus*
		XXXXX					*Paramys*
				XXXXXXXXXXXXX			*Sciurus*
					XXXX		*Castor*
				XXXX			*Monosaulax*
				XXXX			*Dipoides*
						XX	*Microtus*
		XXXX					*Proviverra*
			XXXX				*Pterodon*
			XXXXXXXX				*Hyaenodon*
		XXXXXXXXX					*Viverravus*
		XXXXXXXXX					*Miacis*
					XXXXXX		*Canis*
				XXXXX			*Hemicyon*
				XXXXXX			*Amphicyon*
					XXX		*Agriotherium*
					XXX		*Indarctos*
				XXXXXX			*Ursavus*
						XXX	*Ursus*
			XXXXX				*Palaeogale*
					XXX		*Plesiogulo*
				XXXXXXXXXX			*Mustela*
					XXXX		*Eomellivora*
				XXXXXXXXX			*Lutra*
				XXXXXXX			*Pseudaelurus*
					XXXXX		*Felis, s.l.*
					XXXX		*Machairodus*
		XXXXXXX					*Phenacodus*
		XXXXX					*Coryphodon*
				XXXXXXX			*Gomphotherium*
						XXX	*Mammuthus*
		XXX					*Hyracotherium*
				XXXXXX			*Anchitherium*
					XXXXXXX		*Hipparion, s.l.*
					XXXX		*Equus, s.l.*
		XXXXXXXXXXXXXXXX					*Phenacolemur*
		XXXX					*Prototomus*
		XXXX					*Heptodon*
		XXXX					*Hyrachyus*
			XXXXXXXXX				*Protapirus*
					XXXX		*Cervus*
		XXXXX					*Dissacus*
		XXXXX					*Palaeosinopa*
		XXX					*Pachyaena*
		XXX					*Arctocyon*
		XXXX					*Palaeonictis*
		XXXX					*Oxyaena*
						XXX	*Mimomys*
		XXX					*Pelycodus*
						XX	*Bison*
						XX	*Homo*

7

a mammalian species might extend its geographic range from the Bering Straits to western Europe in about 1000 years. Combined evidence indicates that all conceivable barriers to mammal dispersal over vast distances on land were less effective during the first 175 million years of mammalian history than during the Pleistocene and Recent. Thus we can agree with Kurtén and other workers that the Tertiary species of mammals, and perhaps the Mesozoic species as well, dispersed as quickly as Pleistocene and Recent species or more quickly and were as far-ranging.

BIBLIOGRAPHY

ALBRITTON, C.C., JR., ed. 1963–1964. *The Fabric of Geology.* Stanford, Calif.: Freeman, Cooper & Co.

BEHRENSMEYER, A.K. 1975. The taphonomy and paleoecology of Plio-Pleistocene assemblages east of Lake Rudolf, Kenya. *Mus. Comp. Zool. Bull.* (Harvard Coll.) 146:473–578.

BERGGREN, W.A., BURCKLE, L.H., CITA, M.B., COOKE, H.B.S., FUNNELL, B.M., GARTNER, S., HAYS, J.D., KENNETT, J.P., OPDYKE, N.D., PASTOURET, L., SHACKLETON, N.J., and TAKAYANAGI, Y. 1980. Towards a Quaternary time scale. *Quat. Research* 13:277–302.

BERGGREN, W.A., and VAN COUVERING, J.A. 1974. The Late Neogene: Biostratigraphy, geochronology and paleoclimatology of the last 15 million years in marine and continental sequences. *Palaeogeogr., Palaeoecol., Palaeoclimatol.* 16:1–216.

BERRY, W.B.N. 1966. Zone and zones—with exemplification from the Ordovician. *Bull. Am. Assoc. Pet. Geol.* 50:1487–1500.

BERRY, W.B.N. 1968. *Growth of a Prehistoric Time Scale.* San Francisco: W.H. Freeman. 158 pp.

CAMP, C.L., and ALLISON, H.J. 1961. *Bibliography of Fossil Vertebrates, 1949–1953.* Geol. Soc. Am., Mem., 84.

CAMP, C.L., ALLISON, H.J., and NICHOLS, R.H. 1964. *Bibliography of Fossil Vertebrates, 1954–1958.* Geol. Soc. Am., Mem., 92.

CAMP, C.L., ALLISON, H.J., NICHOLS, R.H., and McGINNIS, H. 1968. *Bibliography of Fossil Vertebrates, 1959–1963.* Geol. Soc. Am., Mem., 117.

CAMP, C.L., NICHOLS, R.H., BRAJNIKOV, B., FULTON, C., and BACSKAI, J.A. 1972. *Bibliography of Fossil Vertebrates, 1964–1968.* Geol. Soc. Am., Mem., 134.

CAMP, C.L., TAYLOR, D.N., and WELLES, S.P. 1942. *Bibliography of Fossil Vertebrates, 1934–1938.* Geol. Soc. Am., Spec. Pap., 42.

CAMP, C.L., and VANDERHOOF, V.L. 1940. *Bibliography of Fossil Vertebrates, 1928–1933.* Geol. Soc. Am., Spec. Pap., 27.

CAMP, C.L., WELLES, S.P., and GREEN, M. 1949. *Bibliography of Fossil Vertebrates, 1939–1943.* Geol. Soc. Am., Mem., 37.

CAMP, C.L., WELLES, S.P., and GREEN, M. 1953. *Bibliography of Fossil Vertebrates, 1944–1948.* Geol. Soc. Am., Mem., 57.

CRACRAFT, J. 1974. Phylogenetic models and classification. *Syst. Zool.* 23:71–90.

DELSON, E., ELDREDGE, N., and TATTERSALL, I. 1977. Reconstruction of hominid phylogeny: A testable framework based on cladistic analysis. *J. Human Evol.* 6:263–278.

EFREMOV, J.A. 1940. Taphonomy: A new branch of paleontology. *Pan.-Am. Geol.* 74:81–93.

FAHLBUSCH, V. 1976. Report on the International Symposium on mammalian stratigraphy of the European Tertiary. *Newsl. Stratigr.* 5:160–167.

FENTON, C.L., and FENTON, M.A. 1928. Ecologic interpretations of some biostratigraphic terms. *Am. Midl. Nat.* 11:1–23.

GRABAU, A.W.A. 1924. *Principles of Stratigraphy,* 2d ed. New York: A.G. Seiler. 1185 pp.

GREGORY, J.T., BACSKAI, J.A., BRAJNIKOV, B., and MUNTHE, K. 1973. *Bibliography of Fossil Vertebrates, 1969–1972.* Geol. Soc. Am., Mem., 141.

GREGORY, J.T., BACSKAI, J.A., and SHKURKIN, G.V. 1978. *Bibliography of Fossil Vertebrates, 1978.* Am. Geol. Inst., Soc. Vertebr. Paleo.

GREGORY, J.T., BACSKAI, J.A., SHKURKIN, G.V. and BRYANT, L.A. 1981. *Bibliography of Fossil Vertebrates, 1979.* Am. Geol. Inst., Soc. Vertebr. Paleo.

HANSON, C.B. 1980. Fluvial taphonomic processes: Models and experiments. In *Fossils in the Making,* ed. A.K. Behrensmeyer and A.P. Hill, pp. 156–181. Chicago: Univ. of Chicago Press.

HAY, O.P. 1902. *Bibliography and Catalogue of the Fossil Vertebrata of North America.* U.S. Geol. Surv.

HAY, O.P. 1929 and 1930. *Second Bibliography and Catalogue of the Fossil Vertebrata of North America.* Carnegie Inst. Wash. Publ., Vols. 1 and 2.

HAY, W.W. 1972. Probablistic stratigraphy. *Eclog. Geol. Helv.* 65:255–266.

HECHT, M.K. 1976. Phylogenetic inference and methodology as applied to the vertebrate record. In *Evolutionary Biology,* ed. M.K. Hecht, W.C. Steere, and R. Wallace, pp. 335–363. New York: Plenum Pub.

HEDBERG, H., ed. 1976. *International Stratigraphic Guide.* New York: John Wiley & Sons. 200 pp.

JAEGER, J.-J., and HARTENBERGER, J.-L. 1975. *Pour l'utilisation systematique de niveaux-repères en biochronologie mammalienne.* 3e Réunion anuelle des sciences de la terre, Montpellier, p. 201.

KAUFFMAN, E.G. 1970. Population systematics, radiometrics and zonation—a new biostratigraphy. *Proc. N.A. Paleontol. Convention,* Pt. F. Pp. 612–666.

KLEINPELL, R.M. 1938. *Miocene Stratigraphy of California.* Tulsa: Am. Assoc. Pet. Geol. 450 pp.

KURTÉN, B. 1957. Mammal migrations, Cenozoic stratigraphy, and the age of Peking man and the australopithecines. *J. Paleontol.* 31:215–227.

LILLEGRAVEN, J.A., KIELAN-JAWOROWSKA, Z., and CLEMENS, W.A., JR., eds. *Mesozoic Mammals: The First Two-Thirds of Mammalian History.* Berkeley: Univ. of Calif. Press.

LYELL, C. 1833. *Principles of Geology,* Vol. 3. London: John Murray. 398 pp.

McKENNA, M.C., 1975. Toward a phylogenetic classification of the Mammalia. In *Phylogeny of the Primates,* ed. W.P. Luckett and F.S. Szalay, pp. 21–46. New York: Plenum Pub.

MALLORY, V.S. 1959. *Lower Tertiary Biostratigraphy of the California Coast Ranges.* Am. Assoc. Pet. Geol. 416 pp.

MARSHALL, L.G., PASCUAL, R., DRAKE, R.E., and CURTIS, G.H. 1977. South American continental geochronology—a preliminary radiometric time scale for middle to late Tertiary mammal-bearing horizons, Patagonia, southern Argentina. *Science* 195:1325–1328.

MATTHEW, W.D. 1939. *Climate and Evolution,* 2d ed. N.Y. Acad. Sci. Spec. Publ., vol. 1. Pp. 1–223.

MURPHY, M.A. 1977. On chronostratigraphic units. *J. Paleontol.* 51:213–219.

OPPEL, A. 1856–1858. *Die Juraformation Englands, Frankreichs und des Sudwestlichen Deutschlands, . . . Abdr. Wurtt. Naturw. Jahresh.* 12–14.

OSBORN, H.F. 1910. *The Age of Mammals in Europe, Asia and North America.* New York: MacMillan Co. 635 pp.

ROMER, A.S., WRIGHT, N.E., EDINGER, T., and VAN FRANK, R. 1962. *Bibliography of Fossil Vertebrates exclusive of North America, 1509–1927.* Vols. 1 and 2. Geol. Soc. Am., Mem., 87.

SAVAGE, D.E. 1951C. Late Cenozoic vertebrates of the San Francisco Bay region. *Bull. Dept. Geol. Sci. Univ. Calif.* 28:215–314.

SAVAGE, D.E. 1975. Cenozoic—the primate episode. In *Approaches to Primate Paleobiology,* ed. F.S. Szalay pp. 2–27. Basel: S. Karger.

SAVAGE, D.E., and RUSSELL, D.E. 1977. Comments on mammalian paleontologic stratigraphy and geochronology; Eocene stages and mammal ages of Europe and North America. *Geobios, Mém. Spéc.,* 1:47–56.

SCHAEFFER, B., HECHT, M.K., and ELDREDGE, N. 1972. Phylogeny and paleontology. In *Evolutionary Biology,* ed. T. Dobzhansky et al., pp. 31–46. New York: Appleton-Century-Crofts.

SCHENCK, H.G., and MULLER, S.W. 1941. Stratigraphic terminology. *Bull. Geol. Soc. Am.* 52:1419–1426.

SCHIEBOUT, J.A. An overview of the terrestrial early Tertiary of southern North America—fossil sites and paleopedology. Tulane, *Studies Geol. Paleontol.* 15:75–94.

SHAW, A.B. 1964. *Time in Stratigraphy.* New York: McGraw-Hill. 365 pp.

SIMPSON, G.G. 1945I. The principles of classification and a classification of mammals. *Bull. Am. Mus. Nat. Hist.* 85:1–350.

SIMPSON, G.G. 1961. Notes on the nature of science by a biologist. In *Notes on the Nature of Science,* pp. 7–12. New York: Harcourt, Brace & World.

SZALAY, F.S. 1978. Phylogenetic relationships and a classification of the eutherian Mammalia. In *Major Patterns in Vertebrate Evolution,* pp. 315–374. NATO Advanced Study Institute. Ser. A, vol. 14. New York: Plenum Pub.

THALER, L. 1972. Datation, zonation et Mammifères. In *Colloque sur l'Eocène,* vol. 2, pp. 411–424. *Fr., Bur. Rech. Geol. Minières Mem.* 2.

THENIUS, E. 1959. Tertiär. 2, Pt. Wirbeltierfaunen. *Handb. Strat. Geol.* 3:1–328.

TYNDALE-BISCOE, H. 1973. *Life of Marsupials.* London: Edward Arnold.

WALKER, E.P. 1968. *Mammals of the World.* 2d. ed. Baltimore: Johns Hopkins Press.

WILSON, J.A., TWISS, P.C., DEFORD, R.K., and CLABAUGH, S.E. 1968. Stratigraphic succession, potassium-argon dates, and vertebrate faunas, Vieja Group, Rim Rock country, Trans-Pecos Texas. *Am. J. Sci.* 266:590–604.

WOOD, H.E., CHANEY, R.W., CLARK, J., COLBERT, E.H., JEPSEN, G.L., REESIDE, J.B. JR., and STOCK, C. 1941. Nomenclature and correlation of the North American continental Tertiary. *Bull. Geol. Soc. Am.* 52:1–48.

WOODBURNE, M.O. 1977. Definition and characterization in mammalian chronostratigraphy. *J. Paleontol.* 51:220–234.

WOODFORD, A.O. 1965. *Historical Geology.* San Francisco: W.H. Freeman & Co.

CHAPTER TWO

Mesozoic Mammalian Faunas

Mammals about 200 to 65 Million Years Ago

The excellent résumé of the Mesozoic mammals edited by Lillegraven, Kielan-Jaworowska, and Clemens (1979) makes an elaborate treatment of this subject unnecessary for the next decade. Each student interested in the earliest mammals will want to get a copy of Lillegraven et al. for reference.

In the following pages we present only a brief outline of the Mesozoic, based on Lillegraven et al. and supplemented by a few data that have come to light in 1980 and 1981. The importance of Mesozoic mammals cannot be overemphasized. They represent more than two-thirds of the total history of the Mammalia. At the beginning of this history, about 200 million years ago, little land animals recognized as mammals because of the especially strong function of squamosal and dentary bones in their masticatory apparatus and because of their peculiar dental anatomy were dispersed from a highly conjectural theater of origin into most, if not all, the large continental areas: Eurasia (including peninsular India), Africa, North America, and possibly South America. The Mesozoic mammalian record in Australia, in Antarctica, and in the Arctic is nonexistent at present; we speculate that because of the known worldwide distribution, however, the subaerial (nonmarine) portions of these latter regions also may have supported Mesozoic mammals. What happened immediately subsequent to this pangaeic distribution of earlier mammals is uncertain because the Mesozoic records are widely scattered and frustratingly meager before the Late Cretaceous.

Since different groups of mammals first appeared, they have adapted to two basic modes of eating; there are animal eaters and plant eaters, as shown in Figure 2-1. In the Late Cretaceous, about 75 to 65 million years ago, small and middle-sized mammals were more abundant than large ones, according to fossil findings. They showed adaptive radiation into insectivorous-carnivorous, herbivorous, and omnivorous methods of eating. Mammalian large herbivores (bear-sized and larger), aquatic herbivores and carnivores (those adapted for constant existence and locomotion in fresh water or in seas), and flyers are still missing from the Mesozoic records. These mammalian adaptees probably were not yet developed as late as 65 million years ago. Nevertheless, the basic evolutionary differentiations leading to or included in the main modern subgroups of mammals, such as marsupials *s.l.* (Metatheria) and ungulates *s.l.*, were extant before the end of the Mesozoic. Also, insectivorous and carnivorous groups—cimolestans and leptictans—probably not ancestral to the modern Insectivora and Carnivora, were abundant before the end of the Mesozoic. We are not ready to accept the one-tooth record of *Purgatorius* from the Hell Creek Formation of Montana as valid documenta-

MAMMALIAN DIETARY ADAPTATIONS: LATEST TRIASSIC INTO RECENT

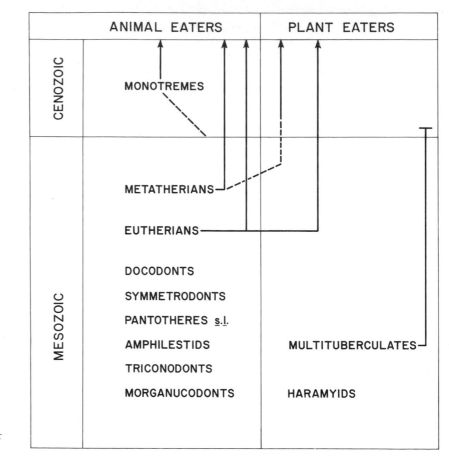

FIGURE 2-1 *Mammalian Dietary Adaptations: Latest Triassic into Recent.*

tion of latest Mesozoic primates, although the evolutionary trend toward the great Cenozoic radiation of primates must have been under way by late Cretaceous time.

The outline of Mesozoic mammal faunal history that follows is taken from Clemens et al. (1979) with only slight modification of organization and age designations. See Clemens et al. (1979) for maps of the localities and for a comprehensive bibliography of Mesozoic mammals' sites. Glossaries of locality names and of producing lithostratigraphic units are included in our chapter glossaries as a supplement to Clemens et al.

Succession of Mammalian Faunas of the Mesozoic

1. Late Triassic (about 200 m.y. ago). Equivalent of "Keuper" and Norian stages of Europe.

   Bones or teeth or trackways in East Germany (Halberstadt), Lesotho (South Africa, Stormberg Group), Kirghiz SSR (North Turkestan Range), North Carolina (New Egypt Coal Mine), Texas (Dockum Group), and Santa Maria, Brazil, indicate that haramiyids, morganucodontids?, and mammals *incertae*

sedis (= uncertain classification within the Mammalia) were extant and probably widespread during late Triassic time.

2. Latest Triassic into early Jurassic (about 195 to 190 m.y. ago). Equivalent of Rhaetian into "Liassic" (Hettangian) stages of Europe.

   Remains of mammals from the Rhaetic bone beds of southern Germany, northern Switzerland, and eastern France, from the Rhaeto-Liassic fissure fills of Wales and southwest England, from the Red Beds and Cave Sandstone of the Stormberg Group in southern Africa, from the Lufeng series in Yunnan, and from the Kayenta Formation of the Colorado Plateau, southwestern United States—together with suggestions of mammals approximately this age in southern South America—indicate that haramiyid mammals, morganucodontids, and other mammals, triconodonts, and symmetrodonts (Pantotheria *s.l.*) were thriving during the latest Triassic–earliest Jurassic passage. Of these mammalian groups, the morganucodontids appear to have occupied Eurasia, Africa, and North America, and perhaps were represented in South America. These sparse records of early mammals complement the geologic interpretation that the big continental plates formed a Pangaea at that time, probably providing corridors for dispersal of land vertebrates. Although the late and latest Triassic appears to have been an interval of rigorous and varying climates in many parts of the world, the "world travels" of small land vertebrates (especially certain mammal-like reptiles and mammals) seem to have been unhampered.

3. Early Jurassic (about 190–170 m.y. ago). Equivalent? of the "Liassic" of Europe.

   Datta, Yadagiri, and Rao (1978) have reported haramiyids or multituberculates and triconodonts or therians from the Kota Formation, Andhra Pradesh, India. And Datta (1981. Zool. J. Linn. Soc. London, *73:*307–312) has described *Kotatherium* (a symmetrodont) from here. Continuing studies apparently will support the thesis that the Indian lithospheric plate was a connected part of terrestrial Pangaea early in the age of mammals.

4. Middle Jurassic (about 165–155 m.y. ago). Bathonian Stage in the British Isles.

   The Stonesfield "Slate," Forest "Marble" (Freeman, 1979), "Monster" Bed of the Hampen Marly Beds of England, and the Ostracod Limestones of the Great Estuarine "Series" on the Isle of Skye of Scotland have provided record of triconodonts, eupantotheres, haramiyids or multituberculates, morganucodontids, docodonts, and symmetrodonts. Jaws from Stonesfield first made earth scientists realize that mammals existed during the Mesozoic Era (Buckland, 1824). All the British mid-Jurassic habitats were evidently adjacent to coastlines, for the mammals are found in dominantly marine lithostratigraphic units.

5. "Medial to Late Jurassic." From Nanjiang, northeast Sichuan, Chow and T. Rich have described *Shuotherium dongi*, a "pseudo-tribosphenic therian," and Chow reports (verbal communication, 1981) a jaw of cf. *Palaeoxonodon* Freeman, 1979.

   Chow and T. Rich are describing a triconodont, *Klamelia*, from Laoshangou, Xinjiang. It is late Jurassic.

6. Late Jurassic, or earliest Cretaceous (about 140–135 m.y. ago). Tithonian or Berriasian stages and equivalents in Europe, Africa, and North America.

   The Durlston Bay district (Purbeck beds) in southern England, Guimarota in Portugal, Morrison Formation sites in the Rocky Mountain province of the United States, and Tendaguru in Tanzania—especially the Morrison and Guimarota—have provided some of the most diversified records of Mesozoic mammals: multituberculates, docodonts, symmetrodonts, triconodonts, and eupantotheres. As of 1982, the woefully incomplete eupantothere jaw fragment from the dinosaur quarries at Tendaguru is the only post-"Rhaetic" Mesozoic mammal for the entire continent of Africa.

7. Late? Jurassic footprints in Patagonia.

   Casamiquela (1964, 1975) thought these footprints were made by mammals.

8. Early Cretaceous (about 130–120 m.y. ago). Valanginian, Hauterivian, and? Berriasian Stages and equivalents of Europe and correlatives in Asia and North America.

   The Weald area of southeastern England (Hastings and Wadhurst beds) has yielded multituberculates, symmetrodonts, and eupantotheres. Strata in the area near Fuhsien, northeastern China, and in the Guchin Depression of Mongolia have produced "Lower Cretaceous" multituberculates, symmetrodonts, triconodonts, eupantotheres, "Therians of metathere-euthere grade," and eutheria. The Cloverly Formation of Montana has produced triconodont skeletons that are being described at Harvard University. It now appears, from current geochronologic datings and taxonomic identifications, that the basic dichotomy of therian mammals metatherians divergent from eutherians—was under way in the early Cretaceous interval.

9. Early to Middle Cretaceous (within the interval of about 120–100 m.y. ago).

   Strata near Galve and Uña in central Spain show an association of fishes, crocodilians, dinosaurs, multituberculates, and eupantotheres.

10. Middle Cretaceous (about 105–100 m.y. ago). Correlated with the Albian Stage of Europe.

    Fossils from the Antlers or Paluxy sands in the Trinity Group of north-central Texas show that triconodonts, multituberculates, symmetrodonts, and "therians of metathere-euthere grade" (some of the last group having been designated as eutheres, others as metatheres, by Slaughter, 1971) were thriving at that time in habitats adjacent to the coast of the ancestral Gulf of Mexico. These mammals were associated with marine (and nonmarine?) fishes and with turtles, crocodilians, and dinosaurs.

11. Earlier Late Cretaceous (about 95–90 m.y. ago). Possible correlatives of the Cenomanian or Turonian stages, or both, of Europe.

    Mammal bones from the Frontier Formation of southeastern Wyoming and teeth from the Woodbine Sandstone of north-central Texas give only a tantalizing hint of the diversity of mammals through this interval. They represent multituberculates, "therians," and mammals, *inc. sed.* We hope that the paleontologists in Wyoming and in Texas will spare no effort to get better records from these formations.

12. Medial Late Cretaceous of North America (about 85–70 m.y. ago). Aquilan and Judithian mammal ages of L. Russell (1975), correlated with Santonian? and Campanian Stages of Europe.

Fossils from the Milk River and Oldman formations of Alberta, from the Judith River Formation and Mesa Verde Group of Montana and Wyoming, and from the ''El Gallo'' Formation of Baja California are the earliest in the relatively rich and complete fossil-mammal succession through the Late Cretaceous of North America. Represented in this composite are triconodonts (Milk River), therians *inc. sed.* (Milk River), symmetrodonts (Milk River), multituberculates, marsupials and eutherians, associated with nonmarine molluscs, fishes, amphibians, small reptiles, turtles, crocodilians, and dinosaurs. These associations evidently lived in lowland wooded and swampy areas bordering streams and ponds.

13. Medial Late to Late Cretaceous in Asia (about 75–70 m.y. ago). Correlated with Santonian-Campanian Stages in Europe.

The famous Djadokhta Formation at Bayn Dzak, Mongolia, and correlative formations and localities in the same country have yielded complete skulls and postcranial parts of multituberculates, eutherians, and ''therians of metathere-euthere grade.'' The lithologies of the containing rocks and the combined fauna, including dinosaurs (with fossilized eggs), lizards, crocodiles, and invertebrates, indicate unique, high-elevation, inland environments where windblown sand was deposited in river floodplains or in ponds adjacent to well-vegetated terrain. This milieu differs notably from most of the other Mesozoic mammal sites.

A fragmentary low-crowned lower molar of an uncertain order of eutherian mammals has been recovered from the Eutaw Formation (= ''Santonian'') at Vinton Bluff, Clay County, Mississippi. (Emry, A.J., Archibald, J.D., and Smith, C.C. 1981. A mammalian molar from the Late Cretaceous of northern Mississippi. *J. Paleontol.* 55:953–956.)

14. Late to Latest Cretaceous (about 70–65 m.y. ago). Correlatives of Maestrichtian or late Senonian Stages in Europe, Edmontonian, and Lancian mammal ages of L. Russell (1975) in North America.

Mammalian faunas from the St. Mary River and Edmonton-Scollard formations of Alberta and from the Fox Hills, Hell Creek, Lance, Laramie, Fruitland-Kirtland, North Horn, and other formations in the United States give strong testimony for increased diversity and great numbers of multituberculates, metatherians, and eutherians. Miscellaneous mammal localities with meager or undescribed samples from Mongolia, southern France, and Portugal are referred to this interval. The now very large samples of latest Cretaceous vertebrates from North America include hadrosaurian, carnosaurian, ceratopsian, ankylosaurian, and sauropodian dinosaurs, crocodilians, turtles, and many other smaller reptiles, together with diversified fishes, amphibians, and birds. The reptiles, especially, indicate a northern province (Alberta, Montana, Wyoming) distinguishable from a southern province (New Mexico, Texas). And the mammalian fossils in the northern province show penecontemporaneous biofacies; the *Lance–Hell Creek–Scollard facies*, dominated by multituberculates and metatherians, contrasts with the *Bug Creek facies* (based on the astoundingly rich sample from Bug Creek, Montana), dominated by multituberculates more closely related to Paleocene taxa and ''indexed'' by the abundant arctocyonian *Protungulatum*, which is believed to be ancestral to the great Cenozoic ungulate radiation.

15. A condylarthlike jaw fragment (*Perutherium*) together with didelphid teeth, charophytes and dinosaur? eggshell fragments from Laguna Umayo, Peru, suggest Cretaceous? metatherians and evolved eutherians on that continent. Paleontologists hope that work under way in Argentina will uncover more complete small-vertebrate records in the Cretaceous dinosaur beds of that country.

MAMMAL-BEARING LITHOSTRATIGRAPHIC UNITS—MESOZOIC

Alag Tag beds—late Cretaceous—Mongolia

Arlington Sandstone Member, Woodbine Formation—early Late Cretaceous—Texas

Chongqing (Chungking) Group—middle to late Jurassic —Sichuan

''Cliff End bone bed,'' Wadhurst Clay—early Cretaceous —England

Djadokhta Formation—late Cretaceous—Mongolia

Dockum Group—late Triassic—Texas

Dzun-Bain Formation—early Cretaceous—Mongolia

Edmonton Group—Edmontonian and Lancian, late Cretaceous —Alberta

''El Gallo'' Formation—=Campanian?, late Cretaceous—Baja California

Forest Marble—Bathonian, mid-Jurassic—England

Fox Hills Formation—=Maestrichtian, Lancian—South Dakota, Wyoming

Frenchman Formation—late Cretaceous—Saskatchewan

Frontier Formation   See Wall Creek Member.

Fruitland Formation—Edmontonian, late Cretaceous—New Mexico

Great Estuarine ''Series''—Bathonian, mid-Jurassic—Isle of Skye, Scotland (See also Ostracod Limestones.)

Great Oolite ''Series''—Bathonian, mid-Jurassic—England

Grinstead Clay, Hastings Beds—early Cretaceous—England

Hampen Marly Beds, ''Monster Bed''—Bathonian, mid-Jurassic—England

Hastings Beds—early Cretaceous—England

Hell Creek Formation—Lancian, late Cretaceous—Montana, Dakotas, Wyoming

Iron Lightning Member, Fox Hills Formation—Lancian = Maestrichtian—South Dakota

Javelina Formation—Lancian—Texas

Judith River Formation—Judithian, late Cretaceous—Montana

Kayenta Formation— = ''Rhaeto-Liassic,'' latest Triassic into earliest Jurassic—Arizona

Keuper ''Stage''—late Triassic—Germany

Kirtland Formation—Edmontonian, late Cretaceous—New Mexico

Kota Formation—early Jurassic—India

Lance Formation—Lancian, late Cretaceous—Wyoming

Laramie Formation—Lancian, late Cretaceous—Colorado

Lower Building Stone, Lulworth-Purbeck Beds—late Jurassic—England

Lufeng ''Series''—latest Triassic or earliest Jurassic—Yunnan

Lulworth Beds (Middle Purbeck)—late Jurassic—England

Mammal Bed, Lulworth Beds—late Jurassic—England

Milk River Formation, upper—Aquilan = early Campanian, late Cretaceous—Alberta

''Monster Bed'' (Hampen Marly Beds)—Bathonian, mid-Jurassic—England

Morrison Formation—late Jurassic (and? earliest Cretaceous?)—Colorado, Wyoming, Oklahoma

Navajo Formation—''Rhaeto-Liassic,'' latest Triassic? into early Jurassic—Arizona, Utah

Newark Canyon Formation—late Cretaceous—Nevada

North Horn Formation, lower—Lancian, late Cretaceous—Utah

Ojo Alamo Formation—Lancian, late Cretaceous—New Mexico

Oldman Formation—Judithian, late Cretaceous—Alberta

Ostracod Limestones, Great Estuarine ''Series''—Bathonian, mid-Jurassic—Isle of Skye, Scotland

Paskapoo Formation, lower—Lancian, late Cretaceous—Alberta

''*Plateosaurus*'' beds—Keuper, late Triassic—Germany

Purbeck Beds (Middle)—late Jurassic—England

St. Mary River Formation—Judithian, late Cretaceous—Alberta

Salt Wash Member, Morrison Formation—late Jurassic—Colorado

Santa Maria Formation—''Ischigualastian,'' middle to late Triassic—Brazil

Scollard Member, Paskapoo Formation—Lancian, late Cretaceous—Alberta

Shaximiao Formation—middle to late Jurassic—Sichuan

Shishugou Formation—late Jurassic—Xinjiang

Stonesfield ''Slate''—Bathonian, mid-Jurassic—England

Stormberg Group— = ''Norian to Rhaetian,'' late Triassic—Lesotho, South Africa

''Telham pebble bed,'' Wadhurst Clay—early Cretaceous—England

Toogreeg beds—late Cretaceous—Mongolia

Vilquechico Formation—late? Cretaceous?—Peru

Wadhurst Beds—early Cretaceous—England

Wall Creek Member, Frontier Formation—early Late Cretaceous—Wyoming

Wealden Beds—early Cretaceous—England

Woodbine Formation (Arlington Sandstone Member)—early Late Cretaceous—Texas

Mesozoic Mammal Localities

Albany County Cretaceous—early Late Cretaceous—Wyoming

Altan Ula—late Cretaceous—Mongolia

Ankylosaur Point—Judithian, late Cretaceous—Montana

Armstrong-Ziegler A, B, and C—Edmontonian, late Cretaceous—New Mexico

Atumcolla   (See Laguna Umayo.)

Aveiro North—late Cretaceous—Portugal

Barun Goyot Fm. sites—mid? Campanian, late Cretaceous—Mongolia

Barwin Quarry—Judithian, late Cretaceous—Wyoming

''Basutoland'' = Lesotho sites—''Rhaetian''—South Africa

Baybolat Well (Karasyor Farm)—late Cretaceous—Kazakhstan

Bayn Dzak (Bain Dzak, Djadokhta)—late? Santonian, late Cretaceous—Mongolia

''Bear River''—late Cretaceous—Wyoming

''Belly River''—late Cretaceous—Alberta

Birch Creek—Judithian, late Cretaceous—Montana

Birmingham—late Cretaceous (bird)—New Jersey

Black Buttes Station North—late Cretaceous—Wyoming

Bone Cabin Draw—late Jurassic—Wyoming

Bug Creek sites—Lancian, late Cretaceous—Montana

Bugeen Tsav beds sites—late Cretaceous—Mongolia

Butler Farm—Albian, mid-Cretaceous—Texas

Carpenter—Weld County—Lancian, late Cretaceous—Colorado

Champ-Garimond   late Senonian, late Cretaceous—France

Chatham coal fields   (See New Egypt.)

Chatzuyao   (See Sakusiyo.)

Cheyenne River North—Lancian, late Cretaceous—Wyoming

Chris's bonebed—Lancian—Montana

Clambank Hollow—Judithian—Montana

Clayball Hill—Judithian—Montana

Cliff End—Valanginian, early Cretaceous—England

Como Bluff, Quarry 9—late Jurassic—Wyoming

Cope-Wortman sites—Lancian, late Cretaceous—South Dakota

Cottonwood Creek ''Lance''—Lancian, late Cretaceous—Wyoming

Crooked Creek—Lancian, late Cretaceous—Montana

Crosby County— = ''Keuper,'' late Triassic—Texas

Degerloch bonebed site—Rhaetian, latest Triassic—S.W. Germany

Dry Mesa Quarry—late Jurassic—Colorado

Duchy Quarry—Rhaeto-Liassic, latest Triassic or earliest Jurassic—Wales

''Dumbbell Hill Lance''—Lancian, late Cretaceous—Wyoming

Durlston Bay–Isle of Purbeck—late Jurassic—England

Dzhalagash Rayon (Baybolat Well)—late Cretaceous—Kazakhstan

Edmonton Formation sites—Edmontonian, late Cretaceous—Alberta

Ekalaka South—Lancian, late Cretaceous—Montana

El Gallo Formation sites   (See Rosario.)

Estancia Laguna Manatiales–M.—late Jurassic—Patagonia

Eureka County—late Cretaceous—Nevada

Eureka Quarry—Lancian, late Cretaceous—South Dakota

Eweeny Quarry—Rhaeto-Liassic, latest Triassic to earliest Jurassic—Wales

Forestburg—Greenwood Canyon—Albian, mid-Cretaceous—Texas

Forsyth-Snow Creek—Lancian, late Cretaceous—Montana

Fort Peck Reservoir sites—Lancian, late Cretaceous—Montana

Freezout Hills—late Jurassic—Wyoming

Frenchman 1—late Cretaceous—Saskatchewan

Frontier Airlines Operations Building—early Late Cretaceous—Texas

Fruita Bowl—late Jurassic—Colorado

Gaisbrunnen bonebed site—Rhaetian, latest Triassic—Germany

Galve—early Cretaceous—Spain

Garden Park—late Jurassic—Colorado

Garfield County, Hell Creek sites—Lancian, late Cretaceous —Montana

Grand River—Lancian, late Cretaceous—South Dakota

Greenwood Canyon (Forestburg)— = Albian, mid-Cretaceous —Texas

Guimarota coal pit—lower? Kimmeridgian, late Jurassic —Portugal

Halberstadt—"Keuper," late Triassic—Germany

Hallau—Rhaetian, latest Triassic—Switzerland

Hei Koa Pen—"Rhaetian," latest Triassic or earliest Jurassic —Yunnan

Hilda West—Judithian, late Cretaceous—Alberta

Holwell—Rhaetian, latest Triassic or earliest Jurassic—England

Hop Brook— = Maestrichtian?, Lancian, late Cretaceous—New Jersey

Hsinchiu (Hsinch'inyao, Xinqiuyao) coal mine—early or mid-Cretaceous—Liaoning, China

Hunter Wash—Edmontonian, late Cretaceous—New Mexico

In Beceten— = "Senonian," late Cretaceous (no mammals to date)—Niger

Iron Lightning Village—Lancian, late Cretaceous—South Dakota

Isle of Wight–Weald—early Cretaceous—England

Joe Painter Quarry—Lancian, late Cretaceous—Wyoming

Ken's Saddle—Lancian, late Cretaceous—Montana

Khaychin–Ula I—late Cretaceous—Mongolia

Khermeen Tsav (Khermin Tsav, Khermintsav)—late Cretaceous—Mongolia

Khovboor (Khovbur, Khobur)—early to mid-Cretaceous —Mongolia

Khulsan (Nemegt Basin)—late Cretaceous—Mongolia

Kirby Creek "Lance"—Lancian, late Cretaceous—Wyoming

Kzyl–Orda Oblast (See Baybolat Well.)

Lady Brand—"Rhaetian," latest Triassic?—Orange Free State, South Africa

Laguna Umayo—late Cretaceous?—Peru

Lambert-Carter County—Lancian, late Cretaceous—Montana

Lance Creek sites—Lancian, late Cretaceous—Wyoming

Laoshangou, Zunggar Basin—late Jurassic—Xinjiang

Leiria (See Guimarota.)

Los Menucos—late Triassic? (mammal tracks?)—Patagonia

Lufeng Northeast—latest Triassic or earliest Jurassic—Yunnan

Lundbreck North—Edmontonian, late Cretaceous—Alberta

Madygen See Shurab.

Mafeteng—"Rhaetian," latest Triassic or earliest Jurassic —Lesotho

Makela-French sites—Judithian, late Cretaceous—Alberta

Mammal Hill—Lancian—Montana

Manyberries—Judithian, late Cretaceous—Alberta

McCone County Hell Creek sites—Lancian, late Cretaceous —Montana

Medicine Hat North—Judithian, late Cretaceous—Alberta

Milk River Formation sites—Aquilan, late Cretaceous—Alberta

Moreau River Headwaters—late Cretaceous—South Dakota

Nanjiang—middle to late Jurassic—Sichuan

Nemegt Basin— = late? Campanian?, late Cretaceous—Mongolia

New Egypt coal mine—"Rhaetian," latest Triassic?—North Carolina

North Nemegt— = late Campanian?, late Cretaceous—Mongolia

Ojo Alamo Formation sites—Lancian, late Cretaceous—New Mexico

Old Cement Works Quarry, Kirtlington—Bathonian, mid-Jurassic—England

Oldman Formation sites—Judithian, late Cretaceous—Alberta

Olgahain bonebed site—Rhaetian, latest Triassic?—Germany

Paddockhurst Park—Valanginian, early Cretaceous—England

Pant Quarry—"Rhaeto-Liassic"–latest Triassic into earliest Jurassic—Wales

Pemberton Marl Company   (See Birmingham, New Jersey.)

Phoebus Landing—late Cretaceous (no mammals to date)—North Carolina

Pokane—"Rhaetian," latest Triassic?—Lesotho

Polecat Bence "Lance" site—Lancian, late Cretaceous —Wyoming

Polyglyphanodon Quarry—Lancian, late Cretaceous—Utah

Pontalun or Pont Quarry—"Rhaeto-Liassic," latest Triassic into earliest Jurassic—Wales

Porto das Barcas—"late Kimmeridgian," late Jurassic—Portugal

Porto Pinheiro—"Kimmeridgian," late Jurassic—Portugal

Powderville P.O., South—Lancian, late Cretaceous—Montana

Purbeck   (See Durlston Bay, etc.)

Quarry 9, Como Bluff   (See Como Bluff.)

Red Lodge—Lancian, late Cretaceous—Montana

Red Owl North—Lancian, late Cretaceous—South Dakota

Rosario— = "Campanian?," late Cretaceous—Baja California

Saint-Nicolas-de-Port—Rhaetian, latest Triassic?—France

Sakusiyo coal field (Chatzuyao, Zhaziyao)—early Cretaceous —Laoning, China

Santa Maria—late Triassic? (mammal?)—Brazil

Scabby Butte—Edmontonian, late Cretaceous—Alberta

Schech Auwed—late Cretaceous (mammal?)—Egypt

Schlösslesmühle bonebed site—Rhaetian, latest Triassic? —Germany

Shabarakh Usu   (See Bayn Dzak.)

Shurab lignite field—late Triassic?—Khirgiz SSR

Skye, Isle of—Bathonian, mid-Jurassic—Scotland

Snow Creek—Lancian, late Cretaceous—Montana

Southeast Nemegt—late Cretaceous—Mongolia

South Saskatchewan River   (See Hilda West.)

Steinenbronn—Rhaetian, latest Triassic?—Germany

Steveville sites—Judithian, late Cretaceous—Alberta

Stonesfield—Bathonian, mid-Jurassic—England

Swindon–Town Gardens Quarry—late Jurassic—England

Ta Ti—"Rhaetian," latest Triassic?—Yunnan

Tendaguru— = "Kimmeridgian," late Jurassic—Tanzania

Tighe Farm—Valanginian, early Cretaceous—England

Tolman Ferry district—late Cretaceous—Alberta

Toogreeg (Toogreegeen Shireh)—late Cretaceous—Mongolia

Town Gardens Quarry   (See Swindon.)

Trochu East—Lancian, late Cretaceous—Alberta

Tsagan Khushu—late Cretaceous—Mongolia

Tsondolein-Khuduk—late Cretaceous—Gansu, China

Tuba City district— = "Rhaetian," latest Triassic or earliest Jurassic—Arizona

UALP 75137—Edmontonian, late Cretaceous—New Mexico

Ugab–Huab Rivers—"Rhaetian," latest Triassic?—South Africa?, Namibia?

Una—early to mid-Cretaceous—Spain

Univ. Wyoming "Lance" sites–Black Buttes–Bitter Creek district—Lancian, late Cretaceous—Wyoming

Upper White Beds sites (See Khermeen Tsav.)

Verdigris Coulee—Aquilan, late Cretaceous—Alberta

Watchet—"Rhaetian," latest Triassic?—England

Wealden district (See Cliff End, etc.)

West Birch Creek—Judithian, late Cretaceous—Montana

West (Watton) Cliff—Bathonian, mid-Jurassic—England

Woodeaten Quarry—Bathonian, mid-Jurassic—England

Württemberg district—Rhaetian, latest Triassic?—Germany

Xinqiuyao (See Hsinch'inyao.)

Yang T'sao Ti (Yancaodi)—"Rhaetian," latest Triassic? —Yunnan

Yanmanpalli area—early Jurassic—India

Zhaziyao (See Sakusiyo.)

Zunggar Basin (Klameli district)—late Jurassic—Xinjiang

BIBLIOGRAPHY

BUCKLAND, W. 1824. Notice on *Megalosaurus. Trans. Geol. Soc. London* (2)1:390–396.

BUTLER, P. 1978. A new interpretation of the mammalian teeth of tribosphenic pattern from the Albian of Texas. *Breviora* 446:1–27.

CASAMIQUELA, R.M. 1964. *Estudios icnologicos. Problems y metodos de la icnologia con aplication al estudio de pisadas mesozoicas (Reptilia, Mammalia) de la Patagonia.* (Ministerio de Asuntos Sociales de la Provincia de Rio Negro, Argentina.), Buenos Aires: Imprenta del Colegio Industrial, Pio 9.

CASAMIQUELA, R.M. 1975. Nuevo material y reinterpretacion de las icnitas mesozoicas (neotriasicas) de Los Menucos, Provincia de Rio Negro (Patagonia). *Actas del I Congreso Argentino de Paleontologia y Biostratigrafia,* Tucuman 1:555–580.

CLEMENS, W.A., JR. 1980. Rhaeto-Liassic mammals from Switzerland and West Germany. *Zitteliana, Abh. Bayer. Statts. Palaeont. Hist. Geol.* 5:51–92.

CLEMENS, W.A., JR., LILLEGRAVEN, J.A., LINDSAY, E.H., and SIMPSON, G.G. 1979. Where, When and What—A survey of known Mesozoic mammal distribution. In *Mesozoic Mammals: The First Two-Thirds of Mammalian History,* pp. 7–59. Berkeley: Univ. of Calif. Press.

DATTA, P.M., YADAGIRI, P., and RAO, B.R.J. 1978. Discovery of early Jurassic micromammals from upper Gondwana sequence of Pranhita Godivari Valley. *J. India. Geol. Soc. India* 19:64–68.

FOX, R.C. 1978. Upper Cretaceous terrestrial vertebrate stratigraphy of the Gobi Desert (Mongolian People's Republic) and western North America. *Geol. Assoc. Canada, Spec. Pap.* 18:577–594.

FOX, R.C. 1980A. Mammals from the Upper Cretaceous Oldman Formation, Alberta. 4. *Meniscoessus* Cope (Multituberculata). *Can. J. Earth Sci.* 17:1480–1488.

FOX, R.C. 1980B. *Picopsis pattersoni,* n. gen. and sp., an unusual therian from the Upper Cretaceous of Alberta, and the classification of primitive tribosphenic mammals. *Can. J. Earth Sci.* 17:1489–1498.

FOX, R.C. 1981. Mammals from the Upper Cretaceous Oldman Formation, Alberta. 5. *Eodelphis* Matthew, and the evolution of the Stagodontidae (Marsupialia), *Can. J. Earth Sci.* 18:350–365.

FREEMAN, E.F. 1979. A middle Jurassic mammal bed from Oxfordshire. *Palaeontol.* 22:135–166.

JOHNSTON, P.A. 1980. First record of Mesozoic mammals from Saskatchewan. *Can. J. Earth Sci.* 17:512–519.

KIELAN-JAWOROWSKA, Z., and DASHZEVEG, D. 1978. New Late Cretaceous mammal locality in Mongolia and a description of a new multituberculate. *Acta Palaeontol. Pol.* 23:115–128.

KIELAN-JAWOROWSKA, Z., and SLOAN, R.E. 1979. *Catopsalis* (Multituberculata) from Asia and North America and the problem of taeniolabidid dispersal in the Late Cretaceous. *Acta Palaeontol. Pol.* 24:187–197.

LUPTON, C., GABRIEL, D., and WEST, R.M. 1980. Paleobiology and depositional setting of a Late Cretaceous vertebrate locality, Hell Creek Formation, McCone County, Montana. *Univ. Wyo., Contrib. Geol.* 18:117–126.

RUSSELL, L.S. 1975. Mammalian faunal succession in the Cretaceous system of western North America. *Geol. Assoc. Canada, Spec. Pap.* 13:137–161.

SLAUGHTER, B. 1971. Mid-Cretaceous (Albian) therians of the Butler Farm local fauna, Texas. In *Early Mammals,* ed. D.M. Kermack and K.A. Kermack, pp. 131–143. *J. Linn. Soc. Zool.* 50, suppl. 1.

SLAUGHTER, B. 1981. The Trinity therians (Albian, Mid-Cretaceous) as marsupials and placentals. *J. Paleontol.* 55:682–683.

CHAPTER THREE

Paleocene Mammalian Faunas

(*Mammals about 65 to 54 Million Years Ago*)

CENOZOIC, TERTIARY, PALEOCENE

Cenozoic

Our primary emphasis in this book is the mammalian faunal succession of the Cenozoic (Kainozoic) Era—the last 65 million years of the history of life on this planet. Therefore let us first review briefly the origin and the concept of Cenozoic and its recognized main subdivisions.

Kainozoic was used by J. Phillips in 1840 (*Penny Cyclopaedia*, vol. 17, pp. 153–154): "As many systems or combinations of organic forms as are clearly traceable in the stratified crust of the globe, so many corresponding terms (as Paleozoic, Mesozoic, Kainozoic, etc.) may be made, nor will these necessarily require change upon every new discovery."

Nowadays, *Cenozoic* connotes a chronostratigraphic and geochronologic unit ("*Terrain*" and *Era*) comprising the rocks and fossils and the interval of events they represent in earth and life history, extending from about 65 million years ago to the present. We follow the convention of dividing Cenozoic into the *Tertiary* and *Quaternary* systems-periods. *Paleogene* (= "Nummulitic"), which consists of the Paleocene, Eocene, and Oligocene series-epochs, and *Neogene*, which is the post-Oligocene Cenozoic, are used for synthesis in descriptions at times in this book.

Tertiary

According to Arduino (1760; from a translation in Wilmarth, 1925):

> The tertiary mountains, or rather hills, are formed of a succession of strata of hard limestone, of consolidated or unconsolidated sand and gravel, and of rock, and vitrified earth (but different from that of the primary mountains) and of earth of various colors.
> These especially ought to be called tertiary; not only because they are seen in superposition on the slopes of the secondaries where the same hills terminate, but also because the greater part of their materials are shells, fragments, and comminuted sea shells; and fragments, flints, sand and muds derived. . . .

We use Paleocene, Eocene, Oligocene, Miocene, and Pliocene as the principal subdivisions of the Tertiary.

ABOUT 65 TO 54 MILLION YEARS AGO

Paleocene

Schimper was a paleobotanist. From his 1874 work we have the subtitle:

VIII. EPOQUE TERTIAIRE.
I. Période paléocène
Sables de Bracheux, Travertins Anciens de Sézanne, Lignites et Grès
du Soissonais (Suessonien).

Schimper was impressed that the *Paleocene* flora from the stratigraphic units mentioned in the subtitle of his work contained many plants now characteristic of the Northern Hemisphere, and partly characteristic of the modern European flora; they were in contrast to the Cretaceous floras, wherein Southern Hemisphere plants predominated.

It must be remembered that the Paleocene is based on stratigraphic units and fossils lying in the general type area of the previously named Eocene Series-Epoch (Lyell, 1833). So Paleocene, in a sense, was carved out of the bulk-Eocene.

Pomerol (1969) and Schorn (1971) independently have emphasized that Schimper included the Suessonien Stage of d'Orbigny (1852) in the list of stratigraphic units typifying Paleocene. "Suessonien" includes the Thanetian, Sparnacian, and Cuisian stages (or substages) of the currently recognized succession in the Paris Basin. Schorn notes that Schimper's paleobotanical conclusions must have been based on the earlier monograph of Watelet (1866, *Plantes Fossiles du Bassin de Paris.*) Watelet's fossils came from strata that would now be assigned to the Thanetian and Sparnacian, *and also* from the Belleu Sandstone at Soissons, France. The Belleu overlies the Lignites du Soissonais and is placed in the Cuisian Stage (or Substage) by French stratigraphers. The Cuisian is the later part of Early Eocene in the geochronology of most present-day workers! This illustrates the vagueness and confusion one encounters frequently as the attempt is made to unravel the original definitions and concepts on which most of the large chronostratigraphic-geochronologic subdivisions of the Cenozoic were based. Thus, the various units that would be included in the Paleocene on basis of strict adherence to Schimperian typology are as shown in the accompanying outline.

Cuisian Stage or Substage	*Nummulites planulatus* zone and beds, Belleu Sandstone, Ypresian Stage (upper), Argile de Flandres, and other units.
Sparnacian Stage or Substage	Lignites du Soissonais, l'Argile Plastique, Ilerdian Stage, Ypresian Stage (lower), and other units.
	Landenian Stage
Thanetian Stage	Sables de Bracheux, Sables de Châlons-sur-Vesle, Travertins de Sézanne, Thanet Sand, and other units.

Anyone following Schimper's priority without modification would be forced to put most of the Lower Eocene of current workers in the Paleocene. But we hold, in accord with many of the stratigraphers of the Paris Basin and with the current consensus of vertebrate paleontologists, that the best adjustment of Schimper's loose typology is to place the top of the Paleocene at the top of strata referable to the Thanetian Stage and its equivalents in the Paris Basin.

Paleontologists and geologists working with deep-water marine deposits and paleontologic stratigraphy based on the planktonic microfossils of that milieu are placing the top of the Paleocene higher than we do. They place it at the approximate top of the Sparnacian (King, 1981).

The Bracheux Sand and the Travertine of Sézanne, together with the sands of Châlons-sur-Vesle, the Rilly limestone, and various lithostratigraphic equivalents recognized by Blondeau, Cavelier, Feugueur, and Pomerol, or by Mégnier (1980) are typical Paleocene and are the local representatives of the Thanetian Stage (type in England) or the lower part of the Landenian Stage (type in Belgium) in the Paris Basin. The top of the uppermost of these units is taken as the top of the type Paleocene.

Next we must decide, How far down and how much earlier is the *beginning* of the Paleocene? Deeply embedded in this question is the problem of the system assignment of the marine Danian Stage of Denmark and France and the Montian Stage of Belgium.

DANO-MONTIAN

The Danian Stage

E. Desor (1847) identified the fossiliferous stratal units in the section south of Copenhagen, Denmark as here listed.

- Coralline limestone or chalk
- Faxoë limestone
- Fish clay
- White chalk (characteristic Cretaceous)

He concluded (our translation):

> Mr. Desor envisages as a consequence the Faxoë limestone, the coralline chalk, and the pisolitic stratum of Laversine and of Vigny [in France] as a particular stage of the chalk [Cretaceous], the most recent of all, such as Mr. Elie de Beaumont had proposed, but it should not include with it the terrains with nummulites, which he envisages as being a more recent epoch. Mr. Desor proposes to call this stage *danian terrain*, because it is especially well developed in the islands of Denmark. Likewise as Mr. Graves had thought, it is probable that it [the Danian] should be related later to the Maestricht terrain.

Note that Desor thought the Danian to be probably "related" to the Maestrichtian which is late Cretaceous. Thus the "Danian controversy" got under way in 1846.

Rosenkrantz (1940) has been able to demonstrate the complex nature of the Maestrichtian-Danian boundary and described it in the way geologists today agree to interpret it. Loeblich and Tappan (1957) noted, like various previous workers, that the Danian has no ammonites, no belemnites, and no *Inoceramus;* these are highly characteristic groups in the usual marine Cretaceous rocks. They concluded that the Danian is a local lithologic and faunal facies of the Lower Paleocene that is not represented in many places in Europe since the seas regressed or perhaps this lithofacies was not deposited. Jeletsky (1962) and many other workers have concluded that such characteristic Mesozoic groups as ammonites, true belemnites, "Cretaceous" planktonic foraminifers, mosasaurs, plesiosaurs, pterodactyls, and dinosaurs are found only in rocks correlative with the strata underlying Danian; hence, Danian belongs in the Cenozoic, despite its referral to the Chalk by its author, Mr. Desor. This is one of the many instances in stratigraphy in which "custom and usage" and later concept have replaced the prior and "legal" typology for a chronostratigraphic-geochronologic name. Articles in a two-volume publication on the "boundary event," the result of a symposium held in Copenhagen in 1979, Birkelund and Bromley, editors, give voluminous documentation of the present state of knowledge within the various disciplines concerning the Danian.

Oriented by consensus among paleontologists and marine stratigraphers about the Cretaceous-Tertiary boundary, L. Alvarez, W. Alvarez, Asaro, and Michel (1980 and later) have analyzed a "boundary clay" in marine successions in Italy, Denmark, New Zealand, Spain, and North Africa. They find that this clay bears a relatively enormous amount of the platinum metal iridium, compared with its concentration in other rocks of the earth's crust that they have examined. More recently (Alvarez and Alvarez, personal commun., 1981), they found an iridium concentration comparable to that found in Italy just above Late Cretaceous dinosaur bones and at the Hell Creek Formation (late Cretaceous)–Tullock Formation (early Paleocene) boundary in eastern Montana. From these data, they propose an extraterrestrial origin for this "end-of-Cretaceous" iridium concentration. And they believe that the concentration was effected by the collision with a 10-km ($\pm$4 km) asteroid whose dusty disturbance darkened the earth's atmosphere long enough to halt many photosynthetic food chains and cause mass extinction of many kinds of organisms in sea and on land. They are convinced that an array of physical and paleontological phenomena calls for us to conclude that one of the most abrupt, cataclysmic, brief, and life-changing events in the history of the earth happened about 65 million years ago, just before the beginning of the Danian Age. They do not try to resolve the paleontologic-paleoecologic problem of the *filtered* extinction at that time—some, but not all, of the small organisms and some, but not all, of the large organisms were exterminated—on land, in the sea, and in the air.

The Montian Stage-Age

J. Delwaque (1868) proposed a *Système montien* on the basis of the Mons Limestone at Mons, Belgium. Later workers have included the underlying Tufeau de Ciply in Delwaque's système. The history of controversy about the system-period assignment of the Montian is summarized by Blondeau, Cavelier, Feugueur, and Pomerol (1965). At present, on the basis of foraminifers, molluscs, and ostracods (Marie, 1964; Chavan, 1948, 1950; Damotte and Feugueur, 1963), the classic sites in the Paris Basin such as Laversines and Vigny, specified as Danian by Desor, are correlated also with either the Tufeau de Ciply or the Mons Limestone. In other words, they are correlated also with the Montian Stage. Thus we conclude that the Montian is the temporal equivalent of at least part of the Danian. And we shall treat this Dano-Montian as the lowest chronostratigraphic and the earliest geochronologic subdivision of the Tertiary in Western Europe.

Dano-Montian

Godfriaux and Thaler (1972), Thaler (1977), and Vianey-Liaud (1979) are identifying land mammals from the nonmarine beds of the Montian Stage at Hainin (Hainaut), Belgium. Their faunal list comprises the following genera.

Composite Dano-Montian Mammalian Fauna

MULTITUBERCULATA
  Ptilodontoidea, fam. indet.
    Boffius splendidus Vianey-Liaud, 1979
MULTITUBERCULATA, *Inc. Sed.*
    Hainina belgica Vianey-Liaud, 1979
    H. godfriauxi Vianey-Liaud, 1979
PANTOLESTA
  Pentacodontidae
    cf. *Aphronorus*
APATOTHERIA
  Apatemyidae, gen. et sp. unident.
CARNIVORA? ARCTOCYONIA?
  possibly miacids or paroxyclaenids
INSECTIVORA
  Erinaceoidea
    erinaceomorphs
    adapisoricids
  Soricoidea?
    cf. *Leptacodon tener*

These identifications suggest that the Hainin local fauna has endemic elements mixed with taxa of North American Torrejonian affinity. There may have been a significant dispersal of land vertebrates between Eurasia and North America during earlier Paleocene time.

EARLY PALEOCENE OF ASIA

From Khaïchin Ula II, Buginstav Basin, Mongolia: *Buginbaator transaltaiensis* (multituberculate) is believed to be earlier Paleocene (Trofimov, 1975).

Localities in the Nanxiong (Nanhsiung) Basin of Guangdong (Kwantung), the Chaling Basin of Hunan, the Tantou Basin of Henan, and the Qianshan Basin of Anhui have produced mammals tentatively assigned to the Middle Paleocene. They are believed to be earlier than the Gashato fauna of Mongolia. Possibly all these samples may be no earlier than the early Tiffanian mammal age of North America (beginning of later Paleocene), but present correlations are uncertain, and some could be considerably older.

Shanghu Fauna in Lower Strata of the Shanghu Formation, Nanxiong Basin, Guangdong —Chow, Chang, Wang, and Ting (1973)

ANAGALIDA
  Linnania lofoensis Chow, Chang, Wang, and Ting, 1973
PANTODONTA
  Bemalambda nanhsiungensis Chow, Chang, Wang, and Ting, 1973
  B. pachyosteus Chow, Chang, Wang, and Ting, 1973
  B. crassa Chow, Chang, Wang, and Ting, 1973
CARNIVORA
  Miacidae
  Pappictidops acies Wang, 1978
  P. obtusus Wang, 1978
TILLODONTIA
  Lofochaius brachyodus Chow, Chang, Wang, and Ting, 1973
TILLODONTIA?
  Dysnoetodon minuta Zhang, 1980
ACREODI
  Mesonychidae
  Yantanglestes feiganensis (Chow, Chang, Wang, and Ting, 1973)   —Ideker and Yan, 1980
  Dissacusium shanghoensis (Chow, Chang, Wang, and Ting, 1973)
  Hukoutherium ambiguum Chow, Chang, Wang, and Ting, 1973
CONDYLARTHRA
  Periptychidae?
  Ectoconus?
  Hyopsodontidae
  Yuodon protoselenoides Chow, Chang, Wang, and Ting, 1973
  Palasiodon siurenensis Chow, Chang, Wang, and Ting, 1973

Fauna from the Chaling Basin (Middle Paleocene), Hunan   —Gao (1975)

ANAGALIDA
  Stenanagale xiangensis Wang, 1975
PANTODONTA
  Hypsilolambda chalingensis Wang, 1975
  H. impensa Wang, 1975
  Bemalambda nanhsiungensis Chow, Chang, Wang, and Ting, 1973

ACREODI
  Mesonychidae
  Yantanglestes rotundus (Wang, 1975)   —Ideker and Yan (1980)
TILLODONTIA
  Meiostylodon zaoshiensis Wang, 1975

Fauna from Qiansham (Chianshan) Basin (lower strata), Wanghudun Formation

ANAGALIDA
  Huaiyangale chianshanensis Q. Xu, 1976A
  Wanogale hodungensis Q. Xu, 1976B
  Chianshania gianghuaiensis Q. Xu, 1976B
  Anaptogale wanghoensis Q. Xu, 1976B
  Diacronus anhuiensis Q. Xu, 1976B
  D. wanghuensis Q. Xu, 1976B
  Anchilestes impolitus Qiu (Chiu) and Li, 1977
  Anictops tabiepedis Qiu (Chiu), 1977
  Paranictops majuscula Qiu, 1977
  Heomys   —Li (1977)
  Mimotona wana Li, 1977
  Cartictops canina Ding and Tong, 1979
CARNIVORA
  Miacidae
  Pappictidops orientalis Qui and Li, 1977
CONDYLARTHRA
  Hyopsodontidae?
  Decoredon elongatus Q. Xu, 1977
Order?
  Obtususdon hanhuaensis Q. Xu, 1977
Order?
  Didymoconidae
  Zeuctherium niteles Tang and Yan, 1976
PANTODONTA
  Bemalambda sp.
  Harpyodus euros Qiu and Li, 1977
  Altilambda pactus Chow and Wang, 1978
  A. tenuis Chow and Wang, 1978
ACREODI
  Mesonychidae
  Yantanglestes conexus (Yan and Tang, 1976) —Ideker and Yan (1980)

Fauna from Chijiang Basin, Jiangxi

PANTODONTA
  Bemalambda shizikouensis Wang and Ding, 1979

We await the results of current work by Chinese colleagues regarding the exact positioning stratigraphically of the taxa and regarding possible intercontinental correlations. The early Paleocene Asian fauna, as currently constituted, appears to be composed of two elements: *endemic taxa* and *taxa of North American affinity*. The latter element may be only an artifact of the lack of mammalian faunas of this age from other parts of the world.

In the above small sample of 30 genera, 21 (70 percent) have not been found in the succeeding later Paleocene fauna of China and Mongolia, according to data available in 1981.

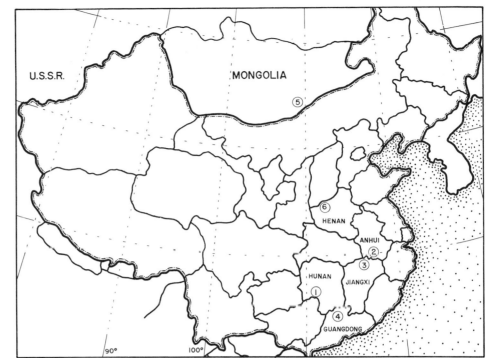

FIGURE 3-1 *Earlier Paleocene
Localities (Districts) in Asia.*
  1 Chaling Basin—Hunan
  *2 Chianshan (Qianshan) Basin—
     Anhui*
  3 Chijiang Basin—Jiangxi
  *4 Datang-Nanxiong Basin
     (Shanghu)—Guangdong*
  5 Khaïchin Ula II—Mongolia
  6 Tantou Basin—Henan

EARLY PALEOCENE OF AFRICA

Cappetta, Jaeger, Sabatier, Sudre, and Vianey-Liaud (1978)
have announced the discovery of eutherian mammals in the
southern part of the Ouarzazate Basin of Morocco. These
are associated with selachian and batoid elasmobranchs,
dipnoian, pycnodont, and teleost (characid) fishes, turtles,
crocodilians, and other reptiles. This paleofauna is in a calcareous stratum dated Montian (= earlier Paleocene) on the
data from the selachians and from fossils of invertebrates in
underlying and overlying beds. According to Cappetta and
others (1978), the mammalian fauna includes palaeoryctids

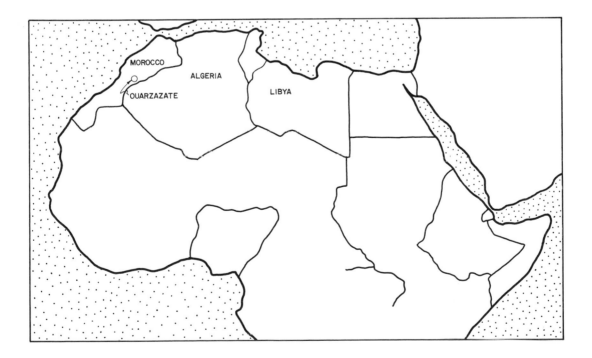

FIGURE 3-2 *Early Paleocene Locality in Morocco.*

(Cimolesta), Carnivora or Creodonta, provivierrine? hyaenodontids (Creodonta), and Miacidae? (Carnivora). Cappetta and his coauthors believe that this curious assemblage of small carnivorous land mammals indicates great antiquity for such forms in Africa; but many uncertainties remain about age and identification of the specimens.

PALEOCENE OF NORTH AMERICA

Nonmarine Paleocene sedimentary rocks of North America were deposited in stream channels, on floodplains, and in swamps and lakes. The areas of deposition were usually intermontane basins. In many places in the western region of this continent, within short distances, there is a striking change from the fine-grained texture of vertebrate-bearing claystones, siltstones, mudstones, and fine-grained sandstones to coarse, pebble, or cobble conglomerates. During this interval, districts in the Rocky Mountain region witnessed one or more of the orogenic movements that have been called the Laramide Revolution.

Areas of outcrop of most North American continental Paleocene formations are referred to customarily as *basins*. These basins are structural depressions, and most are also topographic low areas although they may be 1000 m or more above sea level at the present time. They were centers

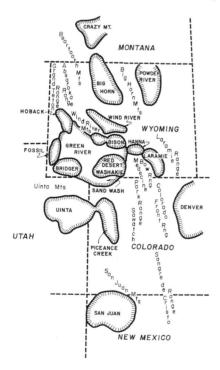

FIGURE 3-3 *Paleogene Basins of Nonmarine Deposition in the Rocky Mountain Region, United States.*

of deposition during Paleocene and Eocene times, and some had relatively continuous sedimentation from the Late Cretaceous until early in the Oligocene. They are famous names in the history of vertebrate-fossil collecting in the United States, for example, San Juan Basin, Uinta Basin, Green River Basin, Wind River Basin, Washakie Basin, and Bighorn Basin.

The following land-mammal ages were recognized as provincial time terms and as subdivisions of the Paleocene in North America by H.E. Wood et al. (1941):

· Clarkforkian
· Tiffanian
· Torrejonian
· Dragonian
· Puercan

Dragonian was proposed by H.E. Wood and others (1941), but we believe that it is best used as early Torrejonian. Van Valen (1978, p. 68) presented *Mantuan* as a pre-Puercan land-mammal age. Without denying that the Mantua local fauna may be earlier than most Puercan samples, we think Mantuan is best treated as early Puercan.

PUERCAN

Puercan was proposed as a provincial time term "based on the Puerco formation of the San Juan Basin, northwestern New Mexico, type locality, Rio Puerco area, most typical and only fossiliferous exposures, the escarpment running from northwest of Ojo Alamo about 25 miles to Arroyo Eduardo, east of Kimbetoh" (H. Wood et al., 1941). According to present-day stratigraphic nomenclature and concept, the Puercan was based on an aggregate of mammalian fossils, sometimes termed Puerco fauna, from the lower part of the Nacimiento Formation of Keyes (1906). See Simpson (1959) and D. Russell (1967) for the history of use of the name *Puerco*.

The stratigraphic range of the type fauna appears to be no more than about 50 m. Mammalian fossils from this range will here be regarded as one fauna, although Van Valen (1978) recognized a lower, *Hemithlaeus,* "facies" and an upper, *Taeniolabis,* "facies." A lower, *Ectoconus* or *Hemithlaeus,* "zone" and an upper, *Taeniolabis,* "zone" also have been proposed. The rocks producing this fauna are generally somber gray to brownish gray or reddish brown mudstones, and these intercalate with light brownish gray to almost white friable sandstone beds. Only a few good quarry sites have been located in the extensive desert badlands and canyons exposing the Nacimiento Formation. The fossiliferous mudstones are believed to be river floodplain deposits, and the associated *Lepisosteus* (gar), turtles, and crocodiles substantiate this interpretation.

MULTITUBERCULATA
Ptilodontidae
Kimbetohia campi (Granger, Gregory, and Colbert in
Matthew, 1937) —Simpson (1936E)
Ptilodus tsosiensis Sloan, 1981
Eucosmodontidae
Eucosmodon americanus (Cope, 1885M)
Taeniolabididae
Catopsalis foliatus Cope, 1882U
Taeniolabis taoensis (Cope, 1882AA)
MARSUPIALIA
Didelphidae
Peradectes Matthew and Granger, 1921A
CIMOLESTA
Palaeoryctidae
Cimolestes simpsoni (Reynolds, 1936)
TAENIODONTA
Stylinodontidae
Onychodectes tisonensis Cope, 1888CC
O. rarus Osborn and Earle, 1895A
Wortmania otariidens (Cope, 1885L)
CARNIVORA
Miacidae
cf. *Ictidopappus* Simpson, 1935G
ARCTOCYONIA
Arctocyonidae
Oxyclaenus cuspidatus (Cope, 1884K)
O. simplex (Cope, 1884K)
Loxolophus hyattianus (Cope, 1883J)
L. kimbetovius (Matthew, 1937)
L. pentacus (Cope, 1888CC)
Desmatoclaenus protogonioides (Cope, 1882EE)
Eoconodon gaudrianus (Cope, 1888CC)
E. heilprinianus (Cope, 1882E)
CONDYLARTHRA
Periptychidae
Hemithlaeus kowalevskianus Cope, 1882D
Gillisonchus gillianus (Cope, 1882C)—Rigby (1981)
Conacodon entoconus (Cope, 1882AA)
Periptychus coarctatus (Cope, 1883Q)
Oxyacodon agapetillus (Cope, 1884K)
O. apiculatus Osborn and Earle, 1895A
O.? cophater (Cope, 1884K)
Mioclaenidae
Choeroclaenus turgidunculus Simpson, 1937G
Promioclaenus priscus (Cope, 1888CC)
P. vanderhoofi (Simpson, 1936E)
P. wilsoni Van Valen, 1978
Hyopsodontidae
Haplaletes andakupensis Van Valen, 1978

This type Puerco fauna has been recognized by many workers to be a biased fossil sample of principally the middle-sized and larger, ground-dwelling mammals of the time. It and correlative faunas from Utah, Colorado, Wyoming, and Montana furnish the following composite Puercan mammalian fauna.[1]

Composite Puercan Mammalian Fauna

MULTITUBERCULATA
Neoplagiaulacidae
Parectypodus vanvaleni Sloan, 1981
Mesodma formosa (Marsh, 1889F) —Clemens
(1963C)
M. ambigua Jepsen, 1940
Neoplagiaulax macintyrei Sloan, 1981
Cimexomys
Ptilodontidae
Kimbetohia campi (Granger, Gregory and Colbert in
Matthew, 1937) —Simpson (1936E)
Eucosmodontidae
Eucosmodon americanus (Cope, 1885M)
E. gratus Jepsen, 1940
Stygimys sp.
Taeniolabididae
Catopsalis? foliatus Cope, 1882U
Taeniolabis taoensis (Cope, 1882AA)
MARSUPIALIA
Didelphidae
Peradectes
LEPTICTIDA
Leptictidae
Prodiacodon crustulum Novacek, 1977
CIMOLESTA
Palaeoryctidae
Cimolestes simpsoni (Reynolds, 1936)
cf. *Gelastops*
cf. *Acmeodon*
cf. *Palaeoryctes*
Procerberus plutonis Van Valen, 1978
TAENIODONTA
Stylinodontidae
Onychodectes tisonensis Cope, 1888CC
O. rarus Osborn and Earle, 1895A
Wortmania otariidens (Cope, 1885L)
APATOTHERIA
Apatemyidae
cf. *Unuchinia*
CARNIVORA
Miacidae
cf. *Ictidopappus*
INSECTIVORA
Soricoidea
Leptacodon proserpinae Van Valen, 1978
DERMOPTERA
Plagiomenidae
Elpidophorus?
PRIMATES
Plesiadapoidea
Purgatorius unio Van Valen and Sloan, 1965
ARCTOCYONIA
Arctocyonidae
Protungulatum sloani Van Valen, 1978
Oxyclaenus cuspidatus (Cope, 1884K)

O. simplex (Cope, 1884K)
Loxolophus hyattianus (Cope, 1885J)
L. nordicus (Jepsen, 1930C)
L. kimbetovius (Matthew, 1937)
L. pentacus (Cope, 1888CC)
Oxyprimus galadrielae Van Valen, 1978
O. putorius Van Valen, 1978
O. albertensis (Fox, 1968C) Van Valen, 1978
Desmatoclaenus protogonioides (Cope, 1882EE)
D. dianae Van Valen, 1978
D. hermaeus Gazin, 1941B
D. paracreodus Gazin, 1941B
Eoconodon gaudrianus (Cope, 1888CC)
E. heilprinianus (Cope, 1882E)
E. coryphaeus (Cope, 1885J)
E. nidhoggi Van Valen, 1978
E. copanus Van Valen, 1978
Mimotricentes subtrigonus (Cope, 1881H)
M. mirielae Van Valen, 1978
Thangorodrim thalion Van Valen, 1978
Ragnarok wovokae Van Valen, 1978
Platymastus palantir Van Valen, 1978
Baioconodon denverensis Gazin, 1941A
B. antiquus (Simpson, 1936E)   —Van Valen (1978)
Goniacodon hiawathae Van Valen, 1978
Chriacus calenancus Van Valen, 1978
CONDYLARTHRA
  Periptychidae
    Hemithlaeus kowalevskianus Cope, 1882DD
    Anisonchus athelas Van Valen, 1978
    A. eowynae Van Valen, 1978
    A. oligistus Van Valen, 1978
    Gillisonchus gillianus (Cope, 1882C)—Rigby (1981)
    Conacodon entoconus (Cope, 1882AA)
    Maiorana noctiluca Van Valen, 1978
    Fimbrethil ambaronae Van Valen, 1978
    Tinuviel eurydice Van Valen, 1978
    Ectoconus ditrigonus (Cope, 1882C)
    E. symbolus Gazin, 1941B
    Periptychus coarctatus Cope, 1883Q
    Carsioptychus hamaxitus Gazin, 1941B
    Haploconus elachistus Gazin, 1941B
    Oxyacodon agapetillus (Cope, 1884K)
    O. apiculatus Osborn and Earle, 1895A
    O. josephi Van Valen, 1978
    O. marshater Van Valen, 1978
    Oxyacodon? cophater (Cope, 1884K)
    Mimatuta makpialutae Van Valen, 1978
    M. minuial Van Valen, 1978
    Earendil undomiel Van Valen, 1978
  Mioclaenidae
    Choeroclaenus turgidunculus (Simpson, 1937G)
    Promioclaenus priscus (Cope, 1888CC)
    P. wilsoni Van Valen, 1978
    P. vanderhoofi (Simpson, 1936E)
    Bomburia prisca (Matthew, 1937)   —Van Valen, 1978

Protoselene bombadili Van Valen, 1978
Ellipsodon witkoi Van Valen, 1978
Hyopsodontidae
  Haplaletes andakupensis Van Valen, 1978

We have referred to a composite of genera from the Puercan as the *Taeniolabis-Ectoconus-Oxyclaenus-Purgatorius* fauna. *Earlier* Puercan, *Middle* Puercan, and *Later* Puercan have been applied to various faunas, depending on an author's concept of the phyletic position of certain species of condylarths, arctocyonians, or other groups. The list of taxa of small mammals from the Puercan that have been collected from localities such as Purgatory Hill and Garbani during the past ten years has been greatly enlarged, but newer specimens are largely undescribed.

A few of the Puercan genera such as *Mesodma* (and possibly other multituberculates), *Cimolestes, Procerberus,* and *Protungulatum* (and probably other arctocyonians) are known from North American late Cretaceous stocks. But lacking the necessary samples from the earlier strata outside North America, we shall not speculate about the phyletic origin or the geographic origin of the bulk of the Puercan mammalian fauna.

Of the 55 genera of mammals here recognized in the Puercan composite fauna, extending through possibly 3 million years, 9 are shared with the earlier, Lancian (latest Cretaceous), age and 28 (51 percent) are not known in the succeeding Torrejonian fauna of North America. Forty-six of the Puercan genera (84 percent) have earliest record in Puercan, and approximately 7 of the recognized families of mammals of the latest Cretaceous do not survive into the Puercan. All Puercan families carry on into the succeeding, Torrejonian, mammal age.

TORREJONIAN

The Torrejonian land-mammal age was proposed as a provincial time term, "based on the Torrejon formation of the San Juan Basin, New Mexico, type locality, the heads of Arroyo Torrejon; typical area runs from there northwest to Ojo Alamo, with additional poorer localities scattered to the north almost to the Colorado line" (H. Wood et al., 1941). Like the Puercan, and in keeping with present-day stratigraphic nomenclature and concept, the Torrejonian is actually based on an aggregate of fossils from the districts specified above and obtained from the upper part of the Nacimiento Formation (Keyes, 1906). This upper part, about 250 m thick and beautifully exposed in Kutz Canyon, south of Bloomfield, New Mexico, is generally lighter in color than the lower part, but there is no useful lithologic distinction between the two, and the upper part appears to be a continuation of the depositional environment of the lower part.

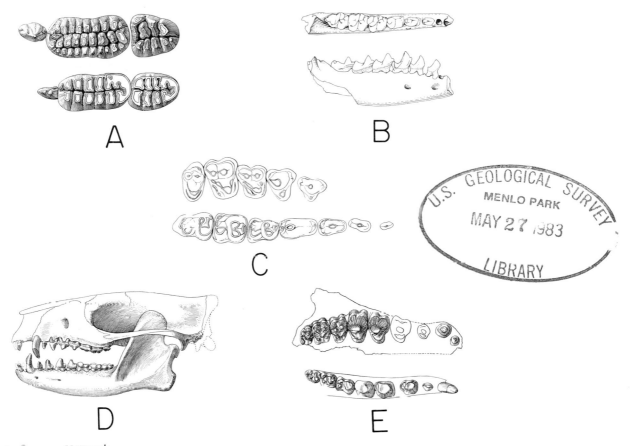

FIGURE 3-4 *Puercan Mammals.*

A *Taeniolabis taoensis* (Cope, 1882AA). *Figure 74 in Matthew (1937).* ×1 approx P^1–M^2 and P_4–M_2.

B *Purgatorius unio Van Valen and Sloan, 1965. Figure 1 in Clemens (1974). ×3, approx.*

C *Eoconodon heilprinianus* (Cope, 1882E). *Figures 1 and 2 in Matthew (1937). ×1/2.*

D *Onychodectes tisonensis Cope, 1888CC. Figure 58 in Matthew (1937). × 1/2.*

E *Periptychus coarctatus Cope, 1883Q. Figure 25 in Matthew (1937). × 1/2.*

(Figures A, C, D, and E published with permission of the American Philosophical Society, Philadelphia, Pennsylvania.
Figure B, Copyright © 1974 by the American Association for the Advancement of Science.)

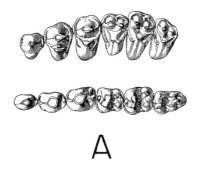

FIGURE 3-5 *Puercan Mammals.*

A *Hemithlaeus kowalevskianus Cope, 1882D. Figure 32 in Matthew (1937).* ×1½ approx.

B *Bombura prisca (Matthew, 1937). Figure 54 in Matthew (1937). 1½ approx.*

(Figures published with permission of the American Philosophical Society, Philadelphia, Pennsylvania.)

FIGURE 3-6 *Representative Puercan Localities and Stratigraphic Units.*

1 *Barrel Springs Arroyo—Nacimiento Formation, lower*
2 *Corral Bluffs—Dawson Arkose*
3 *Dawson—Dawson Arkose*
4 *Flagstaff Peak—North Horn Formation, upper*
5 *Garbani—Tullock Formation*
6 *Jimmy Camp Creek—Dawson Arkose*
7 *Mantua—Mantua lentil, Polecat Bench Formation*
8 *McKenzie—Fort Union Group*
9 *Pine Cree Park—Ravenscrag Formation*
10 *Purgatory Hill—Tullock Formation*
11 *South Table Mountain—Denver Formation, upper*
12 *Bitonitsoseh Wash or Arroyo (including Kimbetoh East and Mammelon Hill)—Nacimiento Formation, lower*
13 *Wagonroad—North Horn Formation, upper*
14 *Leidy Quarry—Polecat Bench Formation*

Composite Torrejonian Mammalian Fauna[2]

MULTITUBERCULATA
Neoplagiaulacidae
 Mesodma, n. sp.   —Rose (1981)
 Mimetodon silberlingi (Simpson, 1935G)
 M. krausei Sloan, 1981
 M. trovessartianus Jepsen, 1940   —[*Parectypodus*, acc. Tsentas (1981)]
 Neoplagiaulax grangeri (Simpson, 1935G)
 N. macrotomeus (R. Wilson, 1956C)
 N. cf. *hunteri* (Simpson, 1936F)   —Rigby (1980)
 Ectypodus sylviae Rigby, 1980
 E. szalayi Sloan, 1981
 Parectypodus sinclairi (Simpson, 1935G)
 P. clemensi Sloan, 1981
 Xanclomys mcgrewi Rigby, 1980
Ptilodontidae
 Ptilodus mediaevus Cope, 1881V
 P. ferronensis Gazin, 1941B
 P. montanus Douglass, 1908D
 P. douglassi Simpson, 1937G   —Krause (1977)
 P. wyomingensis Jepsen, 1940
Cimolodontidae
 Anconodon gidleyi (Simpson, 1935G)
 A. russelli (Simpson, 1935G)
Eucosmodontidae
 Eucosmodon molestus (Cope, 1869I)
 Stygimys jepseni (Simpson, 1935G)
 S. teilhardi (Granger and Simpson, 1929)
 Xironomys swainae Rigby, 1980

Taeniolabididae
 Catopsalis fissidens (Cope, 1884K)
 C. utahensis Gazin, 1939A
LEPTICTIDA
Leptictidae
 leptictid, n. sp.   —Rose (1981)
 Prodiacodon puercensis (Matthew and Granger, 1918H)
 P. concordiarcensis Simpson, 1935G
 P. furor Novacek, 1977
 "*Diacodon*" spp.   —Rose (1981)
 Myrmecoboides montanensis Gidley, 1915A
 Palaeictops?   —Rose (1981)
CIMOLESTA
Palaeoryctidae
 Palaeoryctes puercensis (Matthew, 1913A)
 Gelastops parcus Simpson, 1935G
 G. joni Rigby, 1980
 Stilpnodon simplicidens Simpson, 1935G
 Acmeodon secans Matthew and Granger, 1921A
 A. hyoni Rigby, 1980
 Avunculus didelphodonti Van Valen, 1966A
TAENIODONTA
Stylinodontidae
 Psittacotherium aspasiae Cope, 1882E
 P. multifragum Cope, 1882P
 Conoryctes comma Cope, 1881H
 Conoryctella dragonensis Gazin, 1939A
 Huerfanodon torrejonius Schoch and Lucas, 1981
 H. polecatensis Schoch and Lucas, 1981
PANTODONTA

Pantolambdidae
  Pantolambda bathmodon Cope, 1882V
  P. cavirictus Cope, 1883Q
  P. intermedius Simpson, 1935G
  Caenolambda jepseni Simons, 1960C
Titanoideidae   —Simons (1960C)
  Titanoides simpsoni Simons, 1960C
PANTOLESTA
  Pantolestidae
    Palaeotomus milleri Rigby, 1980
    Propalaeosinopa diluculi (Simpson, 1935G)
    P. thompsoni (Simpson, 1936F)
    Pantomimus leari Van Valen, 1957B
    Leptonysson basiliscus Van Valen, 1967B
  Pentacodontidae
    Pentacodon inversus (Cope, 1888EE)
    P. occultus Matthew, 1937
    Coriphagus encinensis (Matthew and Granger, 1921A)
            —Simpson (1937G)
    C. montanus Douglass, 1908D
    Aphronorus simpsoni Gasin, 1938B
    A. fraudator Simpson, 1935G
    A. orieli Gazin, 1969A
APATOTHERIA
  Apatemyidae
    Jepsenella praepropera Simpson, 1940C
    Unuchinia asaphes (Simpson, 1936F)   —Simpson
    (1937F)
CREODONTA?   —Rigby (1980)
  Hyaenodontidae?   —Rigby (1980)
    Prolimnocyon macfaddeni Rigby, 1980
CARNIVORA
  Miacidae
    Protictis (Protictis) haydenianus MacIntyre, 1966
    P., n. sp.   —Rose (1981)
    P. (Simpsonictis) tenuis MacIntyre, 1966
    P. (Simpsonictis) jaynanneae Rigby, 1980
    P. (Bryanictis) microlestes (Simpson, 1935G)
            —MacIntyre (1966)
    P. (B.) vanvaleni MacIntyre, 1966
    Ictidopappus mustelinus MacIntyre, 1966
INSECTIVORA? SCANDENTIA? PROTEUTHERIA?
  Mixodectidae
    Mixodectes pungens Cope, 1883M
    M. malaris (Cope, 1884K)
    Eudaemonema cuspidata Simpson, 1935G
    Dracontolestes aphantus Gazin, 1941B
INSECTIVORA
  Erinaceoidea
    Mckennatherium ladae (Simpson, 1935G)   —Van
    Valen, 1965F
    M. martinezi Rigby, 1980
    M. fredericki Rigby, 1980
  Soricoidea
    Leptacodon? munusculum Simpson, 1935G
DERMOPTERA
  Plagiomenidae
    Elpidophorus minor Simpson, 1937G
PRIMATES
  Plesiadapidae
    Pronothodectes matthewi Gidley, 1923A
    P. jepi Gingerich, 1975B

Carpolestidae
  Elphidotarsius florencae Gidley, 1923A
  E. russelli Krause, 1978
Picrodontidae
  Picrodus silberlingi Douglass, 1908D
  Draconodus apertus Tomida, 1982
Paromomyidae
  Paromomys maturus Gidley, 1923A
  P. depressidens Gidley, 1923A
  Plesiolestes problematicus Jepsen, 1930C
  Palaechthon alticuspis Gidley, 1923A
  P. nacimienti R. Wilson and Szalay, 1972A
  Palenochtha minor (Gidley, 1923A)
  P. weissae Rigby, 1980
  Torrejonia wilsoni Gazin, 1968A
  Talpohenach torrejonius Kay and Cartmill, 1979
ARCTOCYONIA
  Arctocyonidae
    Oxyclaenus pearcei Gazin, 1941B
    Loxolophus spiekeri (Gazin, 1938B)
    L. criswelli Rigby, 1980
    Arctocyon ferox (Cope, 1883M)   —Van Valen (1978)
    A. montanensis (Gidley, 1919B)   —Van Valen (1978)
    Colpoclaenus procyonoides (Matthew, 1937)   —Van
    Valen (1978); "*Neoclaenodon*," acc. Tsentas (1981)
    C. silberlingi (Gidley, 1919B)   —Van Valen (1978)
    Chriacus pelvidens (Cope, 1881Z)
    C. baldwini (Cope, 1882C)
    C. katrinae Van Valen, 1978
    C. elassus (Gazin, 1941B)
    C. crassicollidens (Cope, 1884K)
    Mimotricentes subtrigonus (Cope, 1881H)   —Van
    Valen (1978)
    M. latidens (Gidley) Simpson, 1935G
    Deltatherium fundaminis Cope, 1881N
    Deuterogonodon montanus Simpson, 1937G
    D. noletil Van Valen, 1978
    Metachriacus punitor Simpson, 1935G   —[retained
    as *Tricentes* by Rose (1981)]
    M. provocator Simpson, 1935G   —[retained as *Tri-
    centes* by Rose (1981)]
    Spanoxyodon latrunculus Simpson, 1935G   —[may
    be *Tricentes punitor,* acc. Rose (1981)]
    Triisodon quivirensis Cope, 1881Z
    T. antiquus (Cope, 1882E)   —Taylor (1981)
    T. crassicuspis (Cope, 1882E)   —Tsentas (1981)
    Goniacodon levisanus Cope, 1888CC
    Desmatoclaenus cf. *paracreodus* Gazin, 1941B
    Prothryptacodon ambiguus (Van Valen, 1967)
            —Van Valen (1978)
    P. furens Simpson, 1935G
    P. hilli (Rigby, 1980)
    Stelocyon arctylos Gingerich, 1978
ACREODI
  Mesonychidae
    Dissacus navajovius (Cope, 1881Z)
    Ankalagon saurognathus (Wortman in Matthew,
    1897C)   —Van Valen (1978)
    Microclaenodon assurgens (Cope, 1884K)

CONDYLARTHRA
  Periptychidae
    Anisonchus sectorius (Cope, 1881H)
    A. dracus Gazin, 1939A
    A. onostus Gazin, 1939A
    A. fortunatus Simpson, 1932H
    A. willeyi Rigby, 1980
    Haploconus angustus (Cope, 1881U)
    H. corniculatus Cope, 1888CC
    H. inopinatus Gazin, 1939A
    Periptychus carinidens Cope, 1881U
    P. gilmorei Gazin, 1939A
    Oxyacodon tecumsae Van Valen, 1978
  Mioclaenidae
    Mioclaenus turgidus Cope, 1881U
    Litaletes disjunctus Simpson, 1935G
    L. sternbergi (Gazin, 1939A)

    L. mantiensis (Gazin, 1939A)
    L. ondolinde Van Valen, 1978
    Promioclaenus lemuroides (Matthew, 1897E)
    P. acolytus (Cope, 1882C)
    P. aquilonius (Simpson, 1935G)   —Rose (1981)
    Ellipsodon inaequidens (Cope, 1884K)
    E. yotankae Van Valen, 1978
    E. grangeri R. Wilson, 1956A
    Protoselene opisthacus (Cope, 1882EE)
    P. griphus (Gazin, 1929A)
  Hyopsodontidae
    Litomylus osceolae Van Valen, 1978
    L. dissentaneus Simpson 1935G
    L. perissus (Gazin, 1941B)
    L. aequidens (Matthew, 1937)   —[*Promioclaenus,*
      acc. Tsentas (1981)]
    Haplaletes disceptatrix Simpson, 1935G
  Phenacodontidae
    Tetraclaenodon puercensis (Cope, 1881H)
    T. pliciferus (Cope, 1884K?)

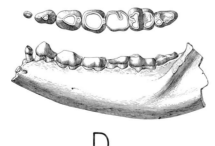

A

B

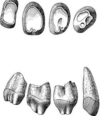

C

D

FIGURE 3-7 *Torrejonian Mammals.*
  A *Ptilodus mediaevus Cope, 1881V. Figure 77 in
    Matthew (1937). ×3. P¹–M³.*
  B *Psittacotherium multifragum Cope, 1882P. Figure
    62 in Matthew (1937). ×1/4.*
  C *Psittacotherium multifragum Cope, 1882P. Figure
    62 in Matthew (1937). ×1/2. P¹–M³.*
  D *Mioclaenus turgidus Cope, 1881U. Figure 50 in
    Matthew (1937). ×1. Palate and lower jaw.*
  (Figures published with permission of the American Philosophical Society, Philadelphia, Pennsylvania.)

We have referred to the composite of genera of Torrejonian age as the *Pantolambda-Tetraclaenodon-Pronothodectes* fauna.

The type fauna of the Torrejonian is supplemented by correlative faunas from the "Ft. Union Formation" and the Polecat Bench Formation in Wyoming and the Lebo Formation in Montana. Gidley and Silberling Quarries in the Lebo produced a large assortment of arboreal and woodland-dwelling small, insectivorous, carnivorous, and frugivorous mammals. For many years, six highly divergent genera from the Gidley Quarry were the earliest record of primates. More recently, the record of primates has been extended downward to the approximate Paleocene-Cretaceous boundary (Van Valen and Sloan, 1965; Clemens, 1974).

On the basis of the associated fossil plants as well as the mammalian fauna, Simpson (1937) concluded that during Torrejonian time in Montana there was a well-developed arboreal habitat with abundant food for browsing and frugivorous mammals. He thought that the more open plains habitat might have been restricted or absent. Simpson's conclusions are supported by later work in the area.

The chances of finding fossils of mammals in the cores of deep wells are almost negligible; yet oddly enough, two of the four North American specimens known to have been found in this manner are of Torrejonian age; Simpson (1932) and Jepsen (1940, p. 243) have discussed a jaw of *Anisonchus* (periptychid condylarth) from the core of a deep well in Louisiana and in a bed of probable marine deposition. More recently, a jaw of an arctocyonid of probable

Torrejonian age was found 720 ft below the earth's surface in Alberta (Fox, 1968).

Torrejonian, as above recognized, may have lasted about 2 million years.

About 88 genera in 29 families constitute the composite Torrejonian mammalian fauna. Twenty-seven of these genera (30 percent) and 16 families (55 percent) are found also in the preceding Puercan. Forty-five (51 percent) of the genera are not known in later faunas of North America, but all the Torrejonian families carry on into later faunas.

CERNAYSIAN

As previously indicated in this chapter, the principal stratal, floral, and faunal units of the type Paleocene in the Paris Basin (Bracheux Sand and its fauna, Travertine of Sézanne and its flora and fauna, and other marine and nonmarine units that have been assigned to the Thanetian Stage or to the lower part of the Landenian Stage) also represent the upper and later part of the Paleocene. Seaways occupied only restricted areas in northwestern and northern Europe during this time. Probably a more extensive seaway occupied the Mediterranean region (western Tethyan Sea). Much of central and northeastern France and adjacent countries to the northeast and east were land.

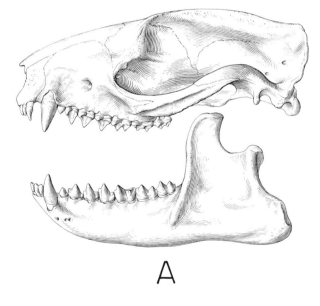

FIGURE 3-8 *Torrejonian Mammals.*

A *Pantolambda bathmodon Cope, 1882V. Figure 38 in Matthew (1937). ×1/2. Skull and lower jaw.*

B *Pantolambda cavirictis Cope, 1883Q. Figure 43 in Matthew (1937). ×1/2. P⁴–M³.*

C *Tetraclaenodon puercensis (Cope, 1881H). Figure 44 in Matthew (1937). × 3/4. P²–M³ and P₃–M₃.*

(*Figures published with permission of the American Philosophical Society, Philadelphia, Pennsylvania.*)

FIGURE 3-9 *Representative Torrejonian Localities and Stratigraphic Units.*

1 *Arroyo Torrejon (east branch), Big Pocket, Little Pocket (Nageezi)—Nacimiento Formation, upper*
2 *Balzac West—Paskapoo Formation*
3 *Belfry Southeast = Cub Creek—Polecat Bench or Lebo Formation*
4 *Caddo Parish–Junior Oil Company, Beard No. 1 Well—Midway Group*
5 *Carter—Fort Union Group*
6 *Contact—Fort Union Group*
7 *Dragon or Dragon Canyon—North Horn Formation, upper*
8 *Ekalaka = Medicine Rocks 1—Tongue River Formation*
9 *Heart Butte = type Fort Union = Tongue River sites—Fort Union Group*
10 *Gidley Quarry, Silberling Quarry—Lebo Formation*
11 *Jenkins Mountain (=Shotgun 2?)—Shotgun Formation*
12 *Laudate—Goler Formation*
13 *Little Muddy Creek—Evanston Formation*
14 *Rock Bench Quarry—Polecat Bench Formation*
15 *Swain Quarry—Fort Union Group*
16 *Western Tornillo Flat—Black Peaks Formation*
17 *Calgary 2E—Porcupine Hills Formation*
18 *Cochrane 11—Porcupine Hills Formation*

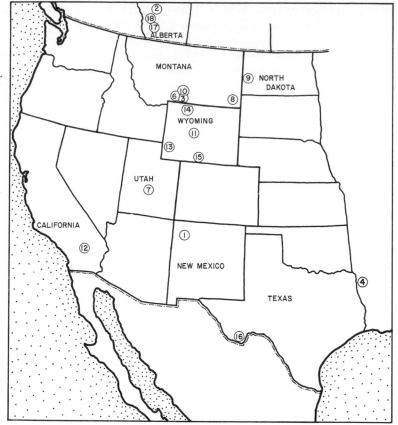

A tabulation of the percentage of species of plants in the later Paleocene floras that have leaves with entire margins (Menat in southern France, Vervins, Sézanne, and Ostricourt in northern France, Gelinden in Belgium) has led Schorn to conclude (personal commun., 1979) that the later Paleocene climate of this area ranged from cool-temperate to warm-tropical. Evidently there were extensive coastal lowland forests. In and around these forests thrived a vertebrate fauna of bony freshwater fishes, frogs, salamanders, champsosaurs, crocodilians, lizards, turtles, large ground birds, and diversified small to middle-sized mammals. This is the *Cernaysian* fauna and mammal age of Europe.

Cernaysian Mammal Age

Cernaysian was named from the Cernay fauna at Cernay-lès-Reims and Mont de Berru, northeastern Paris Basin.

Victor Lemoine (1880, p. 345; 1889, p. 265) termed this assemblage la faune cernaisienne, or Faune cernaysienne.

The two principal quarry districts for Cernay fossils lie on the slopes of Mont de Berru, east of the city of Reims. The Lemoine quarries are near the village of Cernay-lès-Reims. They are named after Victor Lemoine, medical doctor of Reims, who in the late 1800s did most of the early work on the Paleocene and Eocene vertebrates of northeastern France. Here is exposed the "conglomérat de Cernay," a sandy-shelly bed, rich in calcareous tubules of teredines. It is 0–2 m thick and has abundant fossil bones. This "conglomérat" lies upon pockets of white quartz sand, the Rilly Sand, which in turn, lie on the eroded surface of Cretaceous chalk. The Cernay conglomerat is overlaid here by greenish blue sticky clay, the Argile Plastique, which is assigned to the Sparnacian Stage, or Substage.

The Mouras sand quarry is about 1 km farther east, on the east slope of Mont de Berru. In this quarry, brownish

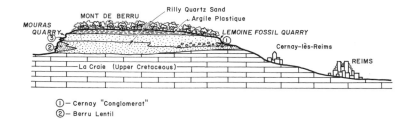

FIGURE 3-10 *Stratigraphic Profile of the Mont de Berru, France.*

① — Cernay "Conglomerat"
② — Berru Lentil
③ — Mouras Sandy Clays

argillaceous sand, often a sandy clay 10–30 cm thick lying on the Rilly Sand, has yielded beautiful jaws, skulls, and skeletal parts of *Arctocyon, Plesiadapis, Pleuraspidotherium,* and other genera. It is especially rich in bones of champsosaurs, the aquatic choristoderan reptiles also well known in the Upper Cretaceous and Paleocene of North America.

The Lemoine quarry and the Berru lens (a fossiliferous concentration in the Rilly Sand at the Mouras Quarry) are now believed to be slightly earlier than the Mouras local fauna on the basis of interpreted stage of evolution on some of the species (D. Russell, Louis, and Poirier, 1966).

Type Cernaysian Mammalian Fauna
(From the Lemoine quarry, Berru lens, and Mouras quarry.)

MULTITUBERCULATA

Neoplagiaulacidae

Neoplagiaulax eocaenus (Lemoine, 1880)

N. copei Lemoine, 1885

gen. nov.

Eucosmodontidae

Liotomus marshi (Lemoine, 1882)

FAMILY?

Hainina godfriauxi Vianey-Liaud, 1979

LEPTICTIDA

Leptictidae

cf. *Palaeictops levei* (D. Russell, Louis and Poirier, 1966B)

Leptictidae?

Adapisoriculus minimus Lemoine, 1883

CIMOLESTA

Palaeoryctidae

Aboletylestes

PANTOLESTA

Pantolestidae

Pagonomus dionysi D. Russell, 1964

APATOTHERIA?

Apatemyidae?

Jepsenella?

INSECTIVORA? SCANDENTIA?

Mixodectidae

Remiculus deutschi D. Russell, 1964

INSECTIVORA

Erinaceoidea

Adapisorex gaudryi Lemoine, 1883

Soricoidea

cf. *Leptacodon*

PRIMATES?

Uintasoricidae

Berruvius lasseroni D. Russell, 1964

B. gingerichi D. Russell, 1981

PRIMATES

Plesiadapidae

Plesiadapis tricuspidens Gervais, 1877

P. remensis Lemoine, 1887

Chiromyoides campanicus Stehlin, 1916

ARCTOCYONIA

Arctocyonidae

Arctocyon primaevus Blainville, 1841

Arctocyonides trouessarti Lemoine, 1891

A. arenae D. Russell, 1964

Landenodon phelizoni D. Russell, 1980C

L. lavocati D. Russell, 1980C

ACREODI

Mesonychidae

Dissacus europaeus Lemoine, 1891

CONDYLARTHRA

Tricuspiodontidae

Tricuspiodon magistrae D. Russell, 1964

T. rütimeyeri Lemoine, 1891

T. sobrinus D. Russell, 1980

Hyopsodontidae

Dipavali petri (D. Russell, 1964)

Paschatherium

Microhyus

"*Adunator*" *lehmani* (D. Russell, 1964)

Louisina mirabilis D. Russell, 1964

gen. nov.   —D. Russell (1964)

Meniscotheriidae   —(Mioclaenidae of Van Valen, 1978)

Pleuraspidotherium aumonieri Lemoine, 1878

P. remense Lemoine, 1878

Orthaspidotherium edwardsi Lemoine, 1885

Arctocyon, an animal the size of a small bear, is the largest mammal in the Cernay fauna. None of the large herbivorous ungulates (pantodonts and uintatheres) of the late Paleocene of Nearctica and Asia Palaearctica is now known from the Paleocene of Europe. The Cernay must be an unbalanced sample of a fauna that thrived in a warm-humid coastal lowland. We know from other faunas that large herbivores are not and were not restricted from such a habitat and so must assume that the large mammals had not reached the Cernay area by late Paleocene time. Because of the intensive collecting that has been accomplished here, we believe that if the large mammals had been present, their fossils would have been discovered.

The Cernay fauna shows some taxonomic resemblance to the Nearctic fauna. *Plesiadapis, Arctocyon, Dissacus, Parectypodus,* and *Neoplagiaulax* are common to both, but this resemblance in no way compares with the very great similarity between the Early Eocene faunas of the two continents and may have been mostly an inherited vestige from preceding faunas. D. Russell (1964) has concluded that *Plesiadapis tricuspidens* of Cernay is derivable from Nearctic forms rather than from a species in the slightly earlier Walbeck local fauna of Germany. There is much to learn about intercontinental dispersal of land vertebrates 60 million years ago.

The La Fère locality (Aisne) has produced a skull of *Arctocyon primaevus* (type of the genotypic species), turtle, and crocodilian from glauconitic calcareous sandstone in the zone of *Pholadomya konincki.* This level, according to Paris Basin stratigraphers, is lower than the beds containing the Cernay fauna, but there is probably no great age difference.

Walbeck Local Fauna, Central Germany. The Walbeck locality was fissure fills in Triassic beds in a large quarry about 85 km west of Berlin in the Allier Valley (Kühn, 1940; Weigelt, 1939, 1940, 1942, 1947; D. Russell, 1964). On the basis of the evolutionary grade of species, D. Russell believes that Walbeck is earlier than Cernay. Walbeck may be a temporal equivalent of the early Tiffanian in Nearctica, and it probably corresponds to the earliest Thanetian of the marine districts of northwestern Europe. Associated animals include crocodilians, lizards, salamanders, and birds.

Walbeck Mammalian Fauna

LEPTICTIDA
    Leptictidae
        Adunator lehmani D. Russell, 1964
        Diaphyodectes prolatus D. Russell, 1964
    Leptictidae?
        Adapisoriculus? germanicus D. Russell, 1964
CIMOLESTA
    Palaeoryctidae

        Aboletylestes hypselus D. Russell, 1964
PANTOLESTA
    Pantolestidae
        Pagonomus   —D. Russell (1964)
INSECTIVORA
    Erinaceoidea
        Adapisorex abundans D. Russell, 1964*
PRIMATES
    Plesiadapidae
        Plesiadapis walbeckensis D. Russell, 1964
        Saxonella crepaturae D. Russell, 1964
ARCTOCYONIA
    Arctocyonidae
        Arctocyon matthesi D. Russell, 1964*
        Arctocyonides weigelti D. Russell, 1964**
        Mentoclaenodon walbeckensis Weigelt, 1960
CONDYLARTHRA
    Tricuspiodontidae
        Paratricuspiodon krumbiegeli D. Russell, 1964
    Hyopsodontidae
        Louisina atavella D. Russell, 1964

---

* Abundant fossils
** Very abundant fossils

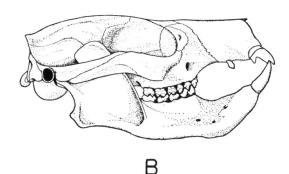

FIGURE 3-11 *Cernaysian Mammals.*
   A *Arctocyon primaevus de Blainville, 1841. Cranium, reconstructed. × 3/7 approximately. Figure 22 in D. Russell (1964).*
   B *Plesiadapis tricuspidens Gervais, 1877. Cranium, reconstructed. × 9/16, approximately. Figure 13 in D. Russell (1964).*
(All figures published with permission of the Muséum National d'Histoire Naturelle, Paris.)

You can note immediately that the Walbeck mammalian fauna lacks the multituberculates and the meniscotheres that are so abundant in the Cernay fauna. It also lacks the large herbivorous mammals. Walbeck appears to be a particularly and peculiarly unbalanced fissure-fill assemblage.

Of the 29 genera of mammals here recognized in the small composite Cernaysian fauna, extending through an interval of possibly 4 million years, 23 (80 percent) are not known in the succeeding, early Eocene (Sparnacian), fauna of Europe. In this succeeding fauna, less than 10 of its 61 known genera may have been derived by evolution in situ from Cernaysian stocks. (We shall consider this startling phenomenon in Chapter 4 relative to the Sparnacian.)

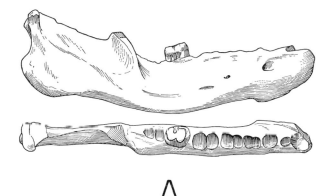
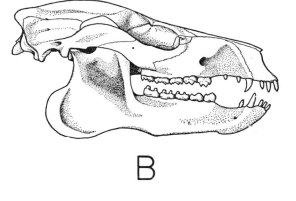

FIGURE 3-12 Cernaysian Mammals.

A *Tricuspiodon magistrae D. Russell, 1964. Figure 51 in D. Russell (1964).* × 2/3 approx.

B *Pleuraspidotherium aumonieri Lemoine, 1878. Figure 53 in D. Russell (1964).* × 5/7 approx.

(*Figures published with permission of the Muséum National d'Histoire Naturelle, Paris, France.*)

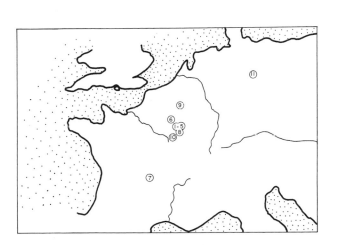
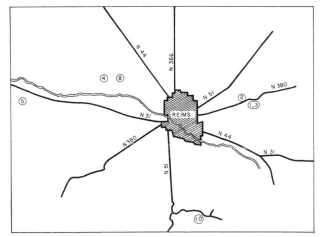

FIGURE 3-13

1 *Berru lens—Rilly Sand*
2 *Cernay-Lemoine Quarry—Conglomérat de Cernay*
3 *Cernay-Mouras Quarry—Mouras sandy mudstone, sands and clays*
4 *Chenay—Sables de Châlons-sur-Vesle (Bracheux Sand)*
5 *Jonchery—Sables de Châlons-sur-Vesle (Bracheux Sand)*
6 *La Fère—glauconitic calcareous sandstone*
7 *Menat—Lignites de Menat*
8 *Merfy—Sables de Châlons-sur-Vesle (Bracheux Sand)*
9 *Mesvin—Sables de Châlons-sur-Vesle (Bracheux Sand)*
10 *Rilly-la-Montagne—Calcaire de Rilly*
11 *Walbeck—fissure fills*

LATER PALEOCENE OF ASIA

Sites in the Gobi Desert of Mongolia and in China introduce us to the mammalian fauna approximately contemporaneous with the Cernaysian of Europe. Principal collections in these areas were first made by expeditions from the American Museum of Natural History. Polish-Mongolian expeditions in the 1960s and expeditions from the USSR in the 1960s and 1970s have made large new collections in the Mongolian region. Paleontologists from Beijing have been uncovering exciting new mammals from middle and late Paleocene strata in China in recent years.

The Gashato fauna of Mongolia and correlative faunas in northwestern China have been described by Matthew and Granger (1925), Matthew, Granger, and Simpson (1929), Flerov (1952, 1957), Sulimski (1969), Kielan-Jaworowska (1969), and Szalay and McKenna (1971). We add taxa listed in more recent publications by Chinese paleontologists to make the following composite Asian later Paleocene mammalian fauna.

Composite Asian Later Paleocene Mammalian Fauna

MULTITUBERCULATA
  Cimolomyidae
    Sphenopsalis nobilis Matthew, Granger, and Simpson, 1928
  Taeniolabididae (including Lambdopsalidae of Chow and Qi, 1978)
    Prionessus lucifer Matthew and Granger, 1925
    Lambdopsalis bulla Chow and Qi, 1978
EDENTATA
  Ernanodontidae
    Ernanodon antelios Ding, 1979
ANAGALIDA
  Zalambdalestidae
    Praolestes nanus Matthew, Granger, and Simpson, 1929
  Eurymylidae
    Eurymylus laticeps Matthew and Granger, 1925
    Mimotona wana Li, 1977
    M. borealis Chow and Qi, 1978
    M. robusta Li, 1977
    Heomys orientalis Li, 1977
    Gomphos elkema Shevyreva, 1975
  Pseudictopidae
    Pseudictops lophiodon Matthew, Granger and Simpson, 1929
    P. chaii Tong, 1979
    P.? tenuis Ding and Zhang, 1979
    Allictops inserrata Qiu (Chiu)
  Anagalidae
    Kashanagale zofiae Szalay and McKenna, 1971
    K.? sp.   —Szalay and McKenna (1971)
    Huaiyangale? leura Ding and Tong, 1979
    Haltictops mirabilis Ding and Tong, 1979
    H. meilingensis Ding and Tong, 1979

    Hsiuannania tabiensis Q. Xu, 1976
    H. maguensis Q. Xu, 1976
    H. minor Ding and Zhang, 1979
ANAGALIDA?
    Petrolemur brevirostre Tong, 1979A
CIMOLESTA?
  Palaeoryctidae? or Deltatheridiidae?
    Hyracolestes ermineus Matthew and Granger, 1925
    Sarcodon pygmaeus Matthew and Granger, 1925
    (includes *Opisthopsalis vetus* Matthew, Granger, and Simpson, 1929)
PANTODONTA
  Pantolambdodontidae
    Archaeolambda planicanina Flerov, 1952
    A. tabiensis Huang, 1977
    A. trofimovi Flerov and Dashzeveg, 1975
    A. dayuensis Tong, 1979
    A. yangtzeensis Huang, 1978
    Dilambda speciosa Tong, 1978
    Nanlingilambda chijiangensis Tong, 1979B
  Bemalambdidae
    Bemalambda nanhsiungensis Chow, Chang, Wang, and Ting, 1973
  Pastoralodontidae
    Pastoralodon lacustris Chow and Qi, 1978
    Convallisodon convexus Chow and Qi, 1978
    C. haliutensis Chow and Qi, 1978
    Altilambda pactus Chow and Wang, 1977
  Phenacolophidae
    Phenacolophus fallax Matthew and Granger, 1925
    Tienshanilophus subashiensis Tong, 1978
    T. lianmuqinensis Tong, 1978
    T. shengjinkouensis Tong, 1978
    Ganolophus lanikenensis Zhang, 1979
    Yuelophus validus Zhang, 1978
    Minchenella grandis (Zhang, 1978)   —Zhang (1980)
  Harpyodidae
    Harpyodus decorus Wang, 1979
DINOCERATA
  Uintatheriidae
    Prodinoceras martyr Matthew, Granger, and Simpson, 1929
    P. turfanensis Chow, 1960
    P. diconicus Tong, 1978
    Mongolotherium plantigradum Flerov, 1952
    M. efremovi Flerov, 1957
    Jiaoluotherium turfanensis Tong, 1978
    Houyanotherium primigenum Tong, 1978
    H. simplum Tong, 1978
NOTOUNGULATA
  Arctostylopidae
    Palaeostylops iturus Matthew and Granger, 1925
    P. macrodon Matthew, Granger, and Simpson, 1929
    Sinostylops promissus Tang and Yan, 1976
    S. progressus Tang and Yan, 1976
    Asiostylops spanios Zheng, 1979
    Allostylops periconotus Zheng, 1979
ORDER?
  Didymoconidae
    Archaeoryctes notialis Zheng, 1979
TILLODONTIA?
    Dysnoetodon minuta Zhang, 1980

ORDER?
> *Wanotherium xuanchengensis* Tang and Yan, 1976

ACREODI
> Mesonychidae
>> *Dissacus indigenous* Dashzeveg, 1976
>> *D. magushanensis* Yan and Tang, 1976
>> *Pachyaena nemegetica* Dashzeveg, 1976
>> *P.? sp.*   —Wang (1976)
>> *Plagiocristodon serratus* Chow and Qi, 1978
>> *Yantanglestes datangensis* (Wang, 1976)   —Ideker and Yan (1980)
>> *Jiangxia chaotoensis* Zhang, Zheng, and Ding, 1979
>> *Hapalodectes? sp.*   —Zhang, Zheng, and Ding (1979)

CONDYLARTHRA
> Periptychidae
>> *Pseudanisonchus antelios* Zhang, Zheng, and Ding, 1979
> Hyopsodontidae   —Zhang, Zheng, and Ding (1979)

ORDER?
> *Obtususdon hanhuaensis* Q. Xu, 1977

RODENTIA, indet.

Dashzeveg and McKenna (1977) and Rose (1980) have suggested that the paleofauna of the White Beds of the Naran Bulak Formation in the Nemegt Basin of Mongolia, which Dashzeveg and McKenna assigned to the early Eocene, may have lived at the same time as the Gashato paleofauna. This may be true, for many fossil species are common to the two faunas. However, *Hyopsodus*, a tapiroid perissodactyl, *Altanius* (the omomyid primate), *Pachyaena, Coryphodon*, and a hyaenodontid creodont are known from the Naran Bulak and not from the Gashato, and we believe these animals may have been post-Gashato arrivals

in this part of Asia. Reintensified collecting in the Gashato district may prove that we are wrong.

This Asian later Paleocene mammalian fauna is an intriguing mélange of allochthonous and endemic taxa. The multituberculates plus the anagalidoid-pseudictopid complex (endemic) are coupled with taxa presumed to be of Nearctic and Neotropical affinity, such as pantodonts, uintatheres, mesonychids, notoungulates, and the edentate.

Among the many phyletic and paleogeographic mysteries this fauna generates, the notoungulates are one of the most exciting. No qualified student of fossil mammals, so far as we know, believes that the splendid jaws of *Palaeostylops, Sinostylops, Asiostylops*, and *Allostylops* from China and Mongolia could possibly be so completely similar to Neotropical taxa as a result of convergent evolution from dissimilar ancestors. The only other Paleogene notoungulate known outside South America is *Arctostylops* from the late and latest Paleocene (Tiffanian and Clarkforkian) of North America (Wyoming). How and exactly when did this group of notoungulates disperse into the Palaearctic, Nearctic, and Neotropical regions? Where did they originate? Certainly not all over the world in an "instant," in our opinion. We believe that the dispersal must have been within a relatively short interval and that this interval began no earlier than mid-Paleocene (ca. 60 m.y. ago) and terminated during the earlier part of the Late Paleocene (ca. 58 m.y. ago).

Jepsen and Woodburne (1969) implied that notoungulates might actually have originated in a largely unknown

FIGURE 3-14 *Later Paleocene Localities and Districts of Asia.*

1. *Chianshan (Qianshan) Basin, Xuanchuang Basin, Lai'an—Anhui*
2. *Chijiang Basin—Jiangxi*
3. *Dazhang Formation site, Tantou Basin—Henan*
4. *Datang, Nanxiong Basin, Jintang Village, Shanghu Formation sites—Guangdong*
5. *Gashato—Mongolia*
6. *Gonghudong-Nomogen, Haliut-Nomogen, Nomogen Formation sites—Nei Mongol Autonomous Region*
7. *Khaïchin Ula II—Mongolia*
8. *Turfan Basin—Xinjiang-Uighur Autonomous Region*

fauna during an unrepresented interval in the later Paleocene of Nearctica, thereafter dispersing to Palaearctica and Neotropica. Zheng (personal commun., 1980), who is restudying the Chinese Paleocene notoungulates, believes that the Asian representatives of this order may predate South American records. He believes that notoungulates could have originated in eastern Asia.

Did a small but ecologically adaptable succession of troops of notoungulates survive the rigors of a 20,000-km, million-year, multigenerational trek from the homeland in equable South America to an outpost in eastern Asia, and was Wyoming a cul de sac "rest stop" for some of the troops? Or was this trek from China to South America?

Equally intriguing and challenging to the imagination is the recently published record of a xenarthrous edentate from the Datang (late Paleocene) Formation of Guangdong, *Ernanodon* Ding (1979). The skull and skeleton on which this genus is based are the most complete specimens of an animal referable to the Edentata from the Paleocene of the entire world. The Late Paleocene (Riochican) record of edentates from South America consists only of a few scattered scutes, teeth, and bones, and no animal of this character is known from North America. Ding's text description, together with the photographs of the skull and humerus, strongly supports her assertion that *Ernanodon* is not convergently "edentate" and xenarthrous. If a more detailed description and further comparisons support Ding's conclusion, we must completely reprogram our thinking and speculation about the theater of origin and early dispersals of the Edentata.

In the relatively small sample of 50 genera constituting the composite later Paleocene mammalian fauna of China and Mongolia, only 9 genera are shared with the earlier Paleocene fauna of the same area. We believe, however, that most of the new taxa of the later fauna—except possibly the notoungulates, uintatheres, condylarths?, and perissodactyls?—could have originated from stocks known in the earlier fauna of the same area.

TIFFANIAN

The Tiffanian land-mammal age was proposed as a provincial time term, "based on the Tiffany local fauna, often used in a more extended sense as a faunal level, northern rim of San Juan Basin, southwestern Colorado, typical area, Mason pocket, . . ." (H. Wood et al., 1941). Later work and study in this type area (Simpson, 1948) determined that the Mason pocket, the Tiffany fauna, and the "Tiffany beds" are contained in the San Jose Formation. Most of the small mammals and most of the taxa in this type fauna were found

in the Mason pocket concentration—thought to be a fissure fill comprising about a cubic meter of matrix. Larger mammals of this fauna were collected at the same general level in the formation as the beds were traced laterally for 15 km.

Mason Pocket, Type Tiffany Mammalian Fauna

MULTITUBERCULATA
  Neoplagiaulacidae
    Ectypodus musculus Matthew and Granger, 1921
MARSUPIALIA
  Didelphidae
    Peradectes elegans Matthew and Granger, 1921
PANTODONTA
  Barylambdidae
    Ignatiolambda barnesi Simons, 1960C
APATOTHERIA
  Apatemyidae
    Labidolemur soricoides Matthew and Granger, 1921
CREODONTA
  Oxyaenidae, gen. indet.
INSECTIVORA
  Erinaceoidea?
    Xenacodon mutilatus Matthew and Granger, 1921
  Soricoidea
    Leptacodon tener Matthew and Granger, 1921
PRIMATES?
  Uintasoricidae?
    Navajovius kohlhaasae Matthew and Granger, 1921
PRIMATES
  Plesiadapidae
    Nannodectes gidleyi (Matthew, 1917E)—Gingerich, 1976
  Carpolestidae
    Carpodaptes aulacodon Matthew and Granger, 1921
  Picrodontidae
    Zanycteris paleocena Matthew, 1917B
  Paromomyidae
    Phenacolemur (Ignacius) frugivorus (Matthew and Granger, 1921)
ARCTOCYONIA
  Arctocyonidae
    Chriacus
    Thryptacodon australis Simpson, 1935E
    Platymastus mellon Van Valen, 1978
ACREODI
  Mesonychidae
    Dissacus?
CONDYLARTHRA
  Periptychidae
    Periptychus superstes Simpson, 1935E
  Phenacodontidae
    Phenacodus grangeri Simpson, 1935E
    P. matthewi Simpson, 1935E
    P. gidleyi Simpson, 1935E

The Mason Pocket aggregate is not a large sample but contains some of the best-preserved and most complete specimens known for many of the Tiffanian taxa.

MULTITUBERCULATA
  Neoplagiaulacidae
    Ectypodus musculus Matthew and Granger, 1921
    E. cochranensis L. Russell, 1967A
    E. powelli Jepsen, 1940
    Parectypodus laytoni (Jepsen, 1940)   —Krause (1977)
    P. sinclairi (Simpson, 1935G)
    P. sloani Schiebout, 1974
    Neoplagiaulax hazeni (Jepsen, 1940)   —Krause (1977)
    N. hunteri (Simpson, 1936F)
    N. fractus (Dorr, 1952B)   —McKenna (1980B)
    N. nanophus Holtzman, 1978
    N. douglassi (Simpson, 1935G)   —Schiebout (1974)
    Mesodma   —Schiebout (1974)
    Mimetodon churchilli Jepsen, 1940
    M. silberlingi (Simpson, 1935G)   —Schiebout (1974)
  Ptilodontidae
    Ptilodus montanus Douglass, 1908D
    P. wyomingensis Jepsen, 1940
    P. kummae Krause, 1977
    P. mediaevus Cope, 1881V   —Schiebout (1974)
    Prochetodon cavus Jepsen, 1940
  Cimolodontidae
    Anconodon sp.
  Eucosmodontidae
    Microcosmodon conus Jepsen, 1930C
    M. woodi Holtzman and Wolberg, 1977
    Pentacosmodon pronus Jepsen, 1940
    Neoliotomus conventus Jepsen, 1930B
  Taeniolabididae
    Catopsalis calgariensis L. Russell, 1926A
MARSUPIALIA
  Didelphidae
    Peradectes elegans Matthew and Granger, 1921
    P. pauli Gazin, 1956B
    Peratherium? sp.   —Rose (1981)
LEPTICTIDA
  Leptictidae
    Prodiacodon pearcei (Gazin, 1956B)
    P. cf. *concordiarcensis* Simpson, 1935G
    Myrmecoboides, prob. n. sp.   —Rose (1981)
    Palaeictops?
CIMOLESTA
  Palaeoryctidae
    Palaeoryctes cf. *punctatus* Van Valen, 1966A
    Gelastops
    Pararyctes pattersoni Van Valen, 1966A
TAENIODONTA
  Stylinodontidae
    cf. *Lampadophorus*   —Rose (1981)
    Lampadophorus expectatus Patterson, 1949A
PANTODONTA
  Pantolambdidae   —Simons (1960C)
    Caenolambda pattersoni Gazin, 1956B
  Titanoideidae
    Titanoides primaevus Gidley, 1917A
    T. gidleyi Jepsen, 1930C

    T. zeuxis Simpson, 1937B
    T. majus Simons, 1960C
  Barylambdidae
    Barylambda jackwilsoni Schiebout, 1974
    Haplolambda quinni Patterson, 1939B
    Leptolambda schmidti Patterson and Simons, 1958
    Ignatiolambda barnesi Simons, 1960C
  Cyriacotheriidae   —Rose and Krause (1982)
    Cyriacotherium argyreum Rose and Krause (1982)
PANTOLESTA
  Pantolestidae
    Propalaeosinopa diluculi (Simpson, 1935G)
    P. thomsoni (Simpson, 1936F)   —McKenna (1980B)
    P. albertensis Simpson, 1927B
    Palaeosinopa simpsoni Van Valen, 1967B
    P. dorri Gingerich, 1980
    Bisonalveus browni Gazin, 1956B
    Niphredil radagasti Van Valen, 1978
    Paleotomus senior (Simpson, 1937)
APATOTHERIA
  Apatemyidae
    Labidolemur soricoides Matthew and Granger, 1921
    Apatemys bellus Marsh, 1872L
    Unuchinia dysmathes Holtzman, 1978
    Jepsenella   —Schiebout (1974)
CREODONTA
  Oxyaenidae
    Tytthaena parrisi Gingerich, 1980
    cf. *Oxyaena?*   —Rose (1981)
    Dipsalodon? matthewi Jepsen, 1930C   —Rose (1981)
    Dipsalodon churchillorum Rose, 1981
CARNIVORA
  Miacidae
    Protictis (Protictis) paralus Holtzman, 1978
    P. (P.) cf. *haydenianus* (Cope, 1882C)   —Rose (1981)
    Protictis cf. *tenuis* MacIntyre, 1966
    Protictis (Bryanictis) cf. *microlestes* MacIntyre, 1966
    Didymictis protenus (Cope, 1874O) (age?)   —McKenna (1980B)
    Viverravus, n. sp.   —Rose (1981)
INSECTIVORA?, SCANDENTIA?, PROTEUTHERIA?
  Mixodectidae
    Eudaemonema   —[Shotgun loc. = late Torrejonian, acc. Rose, 1981]
    Mixodectes?   —McKenna (1980B)
INSECTIVORA
    Inc. Sed. Cf. *Apternodus* spp.   —Jepsen and Woodburne (1969); Rose (1981)
  Erinaceoidea
    Litolestes notissimus Simpson, 1936F
    L. ignotus Jepsen, 1930C
    L. lacunatus Gazin, 1956B
    L., n. sp.   —Rose (1981)
    "*Diacodon*" *minutus* (Jepsen, 1930C)
    Mckennatherium cf. *ladae* (Simpson, 1935G)   —McKenna (1980B)
  Soricoidea
    Leptacodon tener Matthew and Granger, 1921

L. packi Jepsen, 1930C

L. munusculum Simpson, 1935G

DERMOPTERA

Plagiomenidae

Elpidophorus elegans Simpson, 1927B   —Rose
(1975A)

E. patratus Simpson, 1935G

PRIMATES?

Uintasoricidae

Navajovius kohlhaasae Matthew and Granger, 1921

Micromomys silvercouleei Szalay, 1973

M. vossae Krause, 1978

PRIMATES

Plesiadapidae

Pronothodectes cf. *matthewi* Gidley, 1923A

Plesiadapis fodinatus Jepsen, 1930C

P. dubius (Matthew, 1915F)

P. praecursor Gingerich, 1975B

P. churchilli Gingerich, 1975B

P. anceps Simpson, 1936F

P. rex (Gidley, 1923A)

P. simonsi Gingerich, 1975B

P. (Nannodectes) gidleyi (Matthew, 1917C)

P. (N.) gazini Gingerich, 1975B

P. (N.) simpsoni (Gazin, 1956B)

P. (N.) intermedius (Gazin, 1971A)

Chiromyoides minor Gingerich, 1975B

C. potior Gingerich, 1975B

C. caesor Gingerich, 1973

Carpolestidae   —Rose (1975B)

Carpodaptes aulacodon Matthew and Granger, 1921

C. hazelae Simpson, 1936F

C. hobackensis Dorr, 1952A

C. cygneus (L. Russell, 1967A)

C. jepseni Rose, 1975B

Carpolestes dubius Jepsen, 1930C

Picrodontidae

Picrodus cf. *silberlingi* Douglass, 1908D   —Rose (1981)

Zanycteris paleocena Matthew, 1917B

Paromomyidae

Paromomys? depressidens Gidley, 1923A?
—McKenna (1980B)

Paromomys? sp.

Phenacolemur (Phenacolemur) pagei Jepsen, 1930C

P. (Ignacius) frugivorus (Matthew and Granger, 1921)

Torrejonia? sirokyi (Szalay, 1973)

ARCTOCYONIA

Arctocyonidae

Arctocyon ferox (Cope, 1883A)

A. mumak Van Valen, 1978

A. cf. *montanensis* (Gidley, 1919B)

Mimotricentes subtrigonus (Cope, 1881H)
—McKenna (1980B)

Mentoclaenodon acrogenius (Gazin, 1956B)

Neoclaenodon sp.

Chriacus fremontensis (Gazin, 1956B)

C. truncatus (Cope, 1884K)   —Schiebout (1974)

C. metocometi Van Valen, 1978

C. oconostotae Van Valen, 1978

C. orthogonius L. Russell, 1929A

Colpoclaenus keeferi Patterson and McGrew, 1962

Desmatoclaenus mearae Van Valen, 1978

Thryptacodon australis Simpson, 1935E

T. antiquus Matthew, 1915D   —McKenna
(1980B)

Lambertocyon eximius Gingerich, 1979

Deltatherium durini Van Valen, 1978

Platymastus mellon Van Valen, 1978

Deuterogonodon   —Schiebout (1974)

Anacodon? nexus   —Rose (1981)

TILLODONTIA

Esthonyx sp. (early Clarkforkian?)   —McKenna
(1980B)

DINOCERATA

Uintatheriidae

Bathyopsoides harrisorum Patterson, 1939B

Probathyopsis newbilli Patterson, 1939B

ACREODI

Mesonychidae

Dissacus cf. *praenuntius* Matthew, 1915D

D. cf. *navajovius* (Cope, 1881M)   —Rose (1981)

NOTOUNGULATA

Arctostylopidae

Arctostylops cf. *steini* Matthew, 1915D

CONDYLARTHRA

Periptychidae

Periptychus superstes Simpson, 1935A

Mioclaenidae

Promioclaenus pipiringosi Gazin, 1956B

P. cf. *aquilonius*   —McKenna (1980B)

P. acolytus (Cope, 1882C)   —Schiebout (1974)

Ellipsodon

cf. *Litaletes disjunctus* Simpson, 1935G

Protoselene? novissimus Gazin, 1956B

Protoselene opisthacus (Cope, 1882EE)
—Schiebout (1974)

Mioclaenidae?

Phenacodaptes sabulosus Jepsen, 1930C
— [=*Apheliscus sabulosus*, acc. McKenna (1980B)]

Hyopsodontidae

Haplaletes pelicatus Gazin, 1956B

H. serior Gazin, 1956B

H. diminutivus Dorr, 1952A

H. disceptatrix Simpson, 1935G

Litomylus dissentaneus Simpson, 1935G

L.? alphamon Van Valen, 1978

L. ishami Gazin, 1956A

Phenacodontidae

Phenacodus primaevus Cope, 1873D

P. matthewi Simpson, 1935E

P. cf. *brachypternus* Cope, 1882E   —McKenna
(1980B)

P. gidleyi Simpson, 1935E

P. vortmani (Cope, 1880N) (early Clarkforkian?)
—McKenna (1980B)

P. bisonensis Gazin, 1956B

P. grangeri Simpson, 1935E   —Schiebout (1974)

Meniscotheriidae?

Ectocion wyomingensis (Gazin, 1956B)

E. cf. *montanensis* Simpson, 1935G   —Schiebout
(1974)

Prosthecion major Patterson and West, 1973
PHOLIDOTA?
  Epoicotheriidae
    Amelotabes simpsoni Rose, 1978
  Metacheiromyidae
    Propalaeanodon schaffi Rose, 1979
    We have referred to the composite of genera from the Tiffanian as the *Haplaletes-Plesiadapsis-Carpodaptes* fauna.

Producing Tiffanian sites have a great latitudinal range, which extends from the Swan Hills of Alberta (54°30′N) to sites in the Black Peaks Formation of the Big Bend National Park in Texas (29°30′N). The composite Tiffanian fauna is noteworthy for its abundance and diversity of multituberculates, insectivorans *s.l.*, primates, arctocyonians, and condylarths. The Plateau Valley fauna from western Colorado demonstrates that large mammals, especially pantodonts and uintatheres, were abundant and diversified in some districts, probably in the open woodland and savanna habitats.

We estimate that Tiffanian, as above recognized, lasted about 4 million years.

About 94 genera in 38 families make up the composite Tiffanian mammalian fauna. Forty-three of these genera (46 percent) and 27 of the families (71 percent) were holdovers from the Torrejonian. Fifty-three of the genera (56 percent) and 7 families (Cimolodontidae, Taeniolabididae, Pantolambdidae, Titanoideidae, Mixodectidae, Picrodontidae, and Periptychidae = 18 percent) are not known among later faunas in North America.

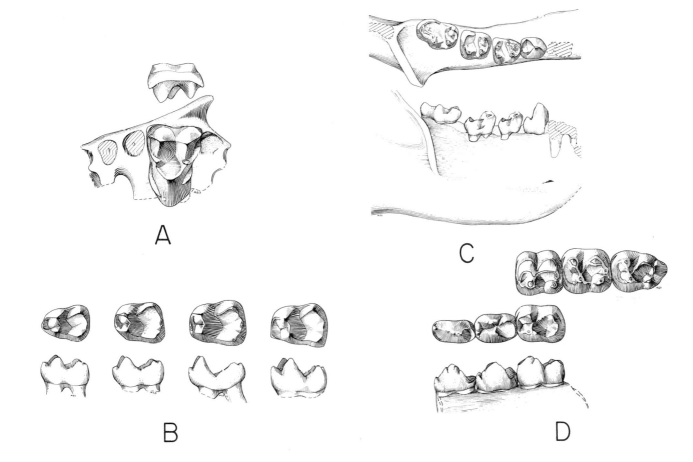

FIGURE 3-15 *Tiffanian Mammals.*
  A Navajovius kohlhaasae Matthew and Granger, 1921; Schiebout (1974). Figure 15 in Schiebout (1974). X10. Maxillary with M².
  B Navajovius kohlhaasae Matthew and Granger, 1921; Schiebout (1974). Figure 15 in Schiebout (1974). X10. P₄, M₁, M₂ and M₃.
  C Plesiadapis (Nannodectes) gidleyi (Matthew, 1917C); Schiebout (1974). Figure 17 in Schiebout (1974). X2. Lower jaw with P₄ — M₃.
  D Phenacodus grangeri Simpson, 1935E; Schiebout (1974). Figure 23 in Schiebout (1974). X1. M₁ — M₃ and P₃ — M₁.

(*Published with permission of the Texas Memorial Museum, The University of Texas, Austin, Texas. The original drawings were made by Dr. Margaret S. Stevens, Lamar Tech, Beaumont, Texas.*)

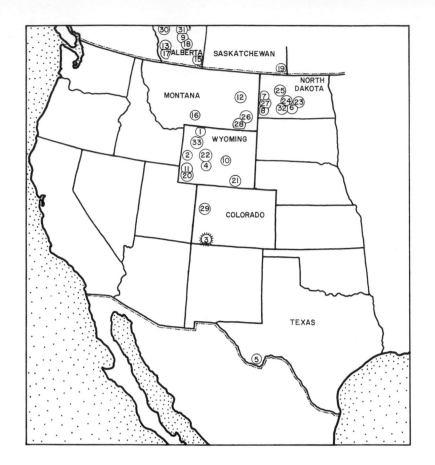

FIGURE 3-16 *Representative Tiffanian Localities and Stratigraphic Units.*

1 *Airport, Brice Canyon, Cedar Point Quarry, Croc Tooth Quarry, Divide Quarry, Fossil Hollow, Fritz Quarry, Highway Blowout, Horse Pasture, Jepsen Valley, Long Draw Quarry, Love Quarry, Lower Sand Draw, Middle Sand Draw, Princeton Quarry, Schaff Quarry, Silver Coulee beds sites, Storm Quarry, Sunday Locality, UM-SC 165—Polecat Bench Formation, upper.*

2 *Battle Mountain, Dell Creek—Hoback Formation*

3 *Bayfield, Mason Pocket, Tiffany beds sites—San Jose Formation*

4 *Bison Basin sites, Ledge, Saddle, Saddle Annex, Titanoides Locality, West End—Fort Union Group*

5 *Black Peaks Formation sites, Joe's Bonebed, Ray's Bonebed and Annex, Tornillo Flat, Western Tornillo Flat—Black Peaks Formation*

6 *Brisbane—Tongue River Formation*

7 *Buford Northeast—Fort Union Group*

8 *Bullion Butte—Fort Union Group*

9 *Canyon Ski Quarry—Paskapoo Formation*

10 *Cedar Ridge West, Malcolm's Locality—Shotgun Formation of Fort Union Group*

11 *Chappo 1 and 17—Chappo Member of Wasatch Formation*

12 *Circle—Tongue River Formation*

13 *Cochrane 1 and 2—Paskapoo Formation*

15 *Cypress Hills Paleocene, Police Point—Ravenscrag Formation*

16 *Douglass Quarry, Melville 13, Princeton Locality 11, Scarritt Quarry—Melville Formation*

17 *Elbow or Elbow River—Paskapoo Formation*

18 *Erickson's Landing—Paskapoo Formation*

19 *Estevan or Roche Percée—Ravenscrag Formation*

20 *Fossil Basin, Fossil Station, Twin Creeks Fork or Twin Creek—Evanston Formation*

21 *Hanna Basin—Hanna Formation*

22 *Jenkins Mountain, Keefer Hill, Shotgun Butte 1, New Anthill—Shotgun Formation of Fort Union Group*

23 *Judson—Tongue River Formation*

24 *Lloyd and Hares sites—Fort Union Group*

25 *Donnybrook—Tongue River Formation*

26 *Medicine Rocks II—Fort Union Group*

27 *McKenzie County sites—Sentinel Butte Formation, Fort Union Group*

28 *Olive—Tongue River Formation, Fort Union Group*

29 *Plateau Valley beds sites—Debeque Formation*

30 *Saunders Creek—Saunders Formation*

31 *Swan Hills—Paskapoo Formation*

32 *Wannagan Creek Quarry—Tongue River Formation*

33 *Love Quarry, "Low" and "Rohrer" localities—unnamed formation*

The Clarkforkian land-mammal age was formally proposed by H. Wood and others (1941) as a provincial time term, "based on the Clark Fork member (and faunal zone) of the Polecat Bench formation, type locality, scarp forming divide between Bighorn and Clark Fork basins and exposures near its base, Park County, Wyoming." First report of fossil mammals and a possibly distinct fauna from this general stratigraphic interval was by Sinclair and Granger (1912, pp. 59–60) and *Ralston beds,* or *formation,* was applied tentatively. Granger (1914) substituted *Clark Fork* beds for Ralston, and later Jepsen (1940, pp. 237–238), on the basis of his lifelong studies in this Clark's Fork River–Sand Coulee area, defined Clark Fork beds as the uppermost 500 ft of his 3500-ft-thick Polecat Bench Formation.

At the present time, K. Rose, P.D. Gingerich, and associates are culminating an intensive reinvestigation of the stratigraphy and paleontology of the Clark Fork: Gingerich and Rose (1977), Rose (1980), and especially Rose (1981). Their work corroborates Jepsen's previous conclusion that there is a discrete and mappable Clark Fork stratigraphic interval containing a significant vertebrate paleofauna. They note that the containing beds are 2100 ft thick with the lower 1200 ft in the Polecat Bench Formation and the upper 900 ft in the overlying red banded beds of the Willwood Formation. Thus, in the northern Bighorn Basin, and also along the western margin of Green River Basin in southwestern Wyoming (sites in the Chappo Member of the Wasatch Formation; Gazin, 1956A, Oriel, 1962), "Wasatch" lithology began forming in pre-Wasatchian time.

The list of mammalian species of the Clarkforkian, which follows, is based on Rose (1981), with slight modification of his classification.

Composite Clarkforkian Mammalian Fauna

MULTITUBERCULATA
  Neoplagiaulacidae
    Ectypodus powelli Jepsen, 1940
    Parectypodus laytoni Jepsen, 1940
  Ptilodontidae
    Prochetodon cf. *cavus* Jepsen, 1940
  Eucosmodontidae
    Neoliotomus conventus Jepsen, 1930B
    Microcosmodon rosei Krause, 1980
MARSUPIALIA
  Didelphidae
    Peradectes cf. *chesteri* (Gazin, 1952)
    Mimoperadectes? labrus Bown and Rose, 1979
LEPTICTIDA
  Leptictidae
    cf. *Prodiacodon tauricinerei* (Jepsen, 1930)
CIMOLESTA
  Palaeoryctidae
    Palaeoryctes punctatus Van Valen, 1966A
PANTODONTA
  Barylambdidae
    Barylambda faberi Patterson, 1933
    Leptolambda schmidti Patterson and Simons, 1958
  Coryphodontidae
    Coryphodon
  Cyriacotheriidae   —Rose and Krause (1982)
    Cyriacotherium psamminum Rose and Krause (1982)
PANTOLESTA
  Pantolestidae
    Palaeosinopa didelphoides (Cope, 1881D)
      —McKenna (1980B)
    cf. *P. lutreola* Matthew, 1918H
  Pentacodontidae
    Protentomodon ursirivalis Simpson, 1928B
    cf. *Aphronorus*   —Rose (1981)
APATOTHERIA
  Apatemyidae
    Apatemys kayi (Jepsen, 1974)   —Rose (1981)
TAENIODONTA
  Stylinodontidae
    Lampadophorus lobdelli Patterson, 1949A
    cf. *Ectoganus*   —McKenna (1980B)
CREODONTA
  Oxyaenidae
    Oxyaena aequidens Matthew, 1915D
    O. transiens Matthew, 1915D
    O. platypus (Matthew, 1915D)
    O.? lichna Rose, 1981
    Dipsalodon matthewi Jepsen, 1930C
    D. churchillorum Rose, 1981
    Palaeonictis peloria Rose, 1981
  Hyaenodontidae
    cf. *Prolimnocyon atavus* Matthew, 1915D
      —McKenna (1980B)
CARNIVORA
  Miacidae
    Didymictis protenus (Cope, 1874O)
    D., n. sp.   —Rose (1981)
    Uintacyon rudis Matthew, 1915D
    Viverravus politus Matthew, 1915D
    V. acutus Matthew, 1915D
    V., n. sp.   —Rose (1981)
INSECTIVORA
  Inc. Sed. Cf. *"Diacodon" minutus* (Jepsen, 1930)
    —Rose (1981)
  Erinaceoidea
    genera unident. (Washakie Basin)
    Leipsanolestes siegfriedti Simpson, 1928A
    Litolestes
  Soricoidea
    cf. *Pontifactor bestiola* West, 1974
    cf. *Leptacodon packi* Jepsen, 1930
    cf. *Plagioctenodon krausae* Bown, 1979
DERMOPTERA
    Planetetherium mirabile Simpson, 1928A
    Plagiomene accola Rose, 1981
    Worlandia inusitata Bown and Rose, 1979
PRIMATES?
  Uintasoricidae

Niptomomys doreenae McKenna, 1960H
cf. *Navajovius* Matthew and Granger, 1921
Tinimomys graybulliensis Szalay, 1974
Microsyopidae
Microsyops simplicidens Rose, 1981
PRIMATES
Plesiadapidae
Plesiadapis cookei Jepsen, 1930C
P. dubius (Matthew, 1915F)
P. gingerichi Rose, 1981
Chiromyoides major Gingerich, 1975B
C. potior Gingerich, 1975B
Carpolestidae
Carpolestes nigridens Simpson, 1928A
Paromomyidae
Phenacolemur praecox Matthew, 1915F
P. cf. *simonsi* Bown and Rose, 1976
P. pagei Jepsen, 1930C
P. (Ignacius) graybullianus Bown and Rose, 1976
ARCTOCYONIA
Arctocyonidae
Thryptacodon cf. *antiquus* Matthew, 1915D
T. pseudarctos Simpson, 1928A
Anacodon? nexus   —Rose (1981)
"Chriacus"?   —Rose (1981)
Lambertocyon ischyrus Gingerich, 1979
cf. *Tricentes*   —Rose (1981)
TILLODONTIA
Esthonychidae
Esthonyx xenicus Gingerich and Gunnell, 1979
E. ancylion Gingerich and Gunnell, 1979
E. grangeri Simpson, 1937D
DINOCERATA
Uintatheriidae
Probathyopsis praecursor Simpson, 1929F
ACREODI
Mesonychidae
Dissacus praenuntius Matthew, 1915D
NOTOUNGULATA
Arctostylopidae
Arctostylops steini Matthew, 1915F
CONDYLARTHRA
Hyopsodontidae?
Phenacodaptes?
Apheliscus nitidus Simpson, 1937D
Hyopsodontidae
Haplomylus simpsoni Rose, 1981
Aletodon gunnelli Gingerich, 1977
Hyopsodus sp.   —Rose (1981)
Phenacodontidae
Phenacodus primaevus Cope, 1873D
P. vortmani (Cope, 1880N)
P.? almiensis Gazin, 1942A   —Rose (1981)
Meniscotheriidae
Ectocion osbornianus (Cope, 1882E)
E. parvus Granger, 1915A
E.? semicingulatum (L. Russell, 1929A)
Meniscotherium priscum Granger, 1915A

PERISSODACTYLA
Equidae
Hyracotherium seekinsi Morris, 1968B   (age probably
Wasatchian)
PHOLIDOTA?
Metacheiromyidae
Palaeanodon parvulus Matthew, 1918H   —(species
questioned by Rose, 1981)
RODENTIA
Ischyromyoidea
Paramys atavus Jepsen, 1937
P. annectens Rose, 1981
P. cf. *excavatus* Loomis, 1907
Franimys amherstensis A. Wood, 1962B
We refer to the composite of genera from the Clarkforkian as
the *Ectocion-Plesiadapis-Carpolestes-Paramys* fauna.

As R. Wood noted (1967), the original "Clark Fork mammalian fauna," which was collected in the early 1900s, was transitional between the fauna of the underlying, classic, Late Paleocene Tiffanian Silver Coulee beds of the Polecat Bench Formation and the fauna of the overlying, classic, Early Eocene Graybullian (=early Wasatchian) "Gray Bull beds" (lower part of the Willwood Formation of Van Houten, 1944), since many of these "Clark Fork" specimens were actually collected from strata now known to be Tiffanian or Wasatchian respectively. However, meticulous collecting from the restricted Clark Fork interval by Jepsen during the later years of his field activities, and by Gingerich and Rose and associates during the last decade, shows that the restricted Clarkforkian fauna is indeed transitional taxonomically between Tiffanian and Wasatchian. Gingerich and Rose (1977) note that Clarkforkian is the earliest North American record of the following taxa: *Aletodon, Arctostylops* (now known from Tiffanian, Harvard specimen), *Probathyopsis,* Rodentia, *Plagiomene, Haplomylus, Apheliscus, Esthonyx,* Miacinae, *Oxyaena, Coryphodon,* and *Hyracotherium* (age?). They conclude that the above taxa indicate an important immigration of land mammals at about the beginning of Clarkforkian time. The apparently sudden appearance of the rodents (perhaps also *Esthonyx, Oxyaena,* and *Coryphodon*) lends strength to this conclusion.

A secondary consideration about Clarkforkian is whether it should be termed latest Paleocene or earliest Eocene or latest Paleocene *and* earliest Eocene. Gingerich and Rose (1977) contended that the Sparnacian locality of Meudon—a small sample collected about 1900 from a pocket in this suburb of Paris—includes the first representatives of a more modern fauna with rodents, perissodactyls, oxyaenoid creodonts, and *Coryphodon:* "Thus the basic difference between the Paleocene and Eocene mammal faunas of the Paris Basin (where the Paleocene-Eocene boundary is defined) is the appearance of these four groups at the beginning of the Eocene." Since rodents, oxyaenids, and *Coryphodon* appear in the Clarkforkian also, and because *Plesia-*

dapsis russelli from Meudon is similar to *P. cookei* of the Clark's Fork, these authors believed Clarkforkian to be properly correlated with the Sparnacian of the Paris Basin and that Clarkforkian should be regarded as earliest Eocene.

Subsequently, Rose (1981) has concluded that the Clarkforkian chronostratigraphic interval is divisible into three zones. His *Plesiadapis gingerichi* Zone—which he believes to bridge the Tiffanian-Clarkforkian boundary and to be early Clarkforkian = latest Paleocene in only its upper moiety—is based on the range-zone (local) of *P. gingerichi*. The Clarkforkian part of his *P. gingerichi* zone is recognized by the existence of *P. gingerichi* or *Paramys, Coryphodon, Esthonyx,* and *Haplomylus* or a combination (Rose 1981, pp. 11, 27). His overlying *Plesiadapis cookei* Zone, "middle Clarkforkian and earliest Eocene," is equivalent to the stratigraphic range of *P. cookei* in the Clark's Fork

Basin. His uppermost *Phenacodus-Ectocion* Zone has its lower boundary at the "last appearance of *P. cookei*" and "is further recognized by the evolutionary first occurrence of *Esthonyx grangeri* and *Phenacodus praecox* (which make their appearance during but not at the beginning of the zone)."

Rose judiciously adds (p. 28): "Whether the zones distinguished in the Clarkforkian of the Clark's Fork Basin will have broad geographic applicability must await more thorough collecting elsewhere."

This rodent-*Coryphodon-Esthonyx*-etc. fauna, characterizing the Clarkforkian and Wasatchian of North America and the Sparnacian of Europe, produces a noteworthy

FIGURE 3-17 *Clarkforkian Localities and Stratigraphic Units.*

1   *Bear Creek or Eagle Coal Mine—Polecat Bench Formation, upper*
2   *Big Multi—Fort Union Gap*
3   *Buckman Hollow or La Barge Creek—Chappo Member of Wasatch Formation*
4   *Chappo 12—Chappo Member of Wasatch Formation*
5   *Clark's Fork beds sites, Cleopatra Reservoir, Little Sand Coulee, Paint Creek, UM-SC70, UM-SC71, UM-SC107, UM-SC19, UM-SC62, UM-SC136, UM-SC183, UM-SC74, UM-SC110, UM-SC195, UM-SC188, UM-SC196 (Rough Gulch), UM-*

SC190 (Granger Mt.), vicinity of UM-SC102 (Big Sand Coulee, Upper Sand Draw—Polecat Bench Formation, upper and Willwood Formation, lower
6   *Hell's Half Acre—Atwell Gulch Member of Debeque Formation*
7   *Mt. Leidy Highlands or Togwotee or Togwotee Pass—Formation not named.*
8   *Punta Prieta (age?)—Tepetate? Formation*
9   *Reiss; Polecat Bench Formation, upper?*
10 *South Wall (137-meter Level)—Black Peaks Formation*

taxonomic turnover relative to the preceding mammalian faunas of North America and Europe. Consensus holds that it was a rapidly dispersing fauna (McKenna, 1975; D. Savage, 1971). Sloan (1970) has hypothesized that this fauna originated in Mexico and Central America, spreading across North America and into Eurasia. Dispersal lag time is conjectural, but certain families, subfamilies, and genera of the Clarkforkian might appear significantly later in the Sparnacian. We have discussed this possibility previously (Savage and Russell, 1977).

All dispute regarding the assignment of part or all of the Clarkforkian to the Eocene becomes merely a tempest in a teapot if the current consensus from marine paleoplanktonic stratigraphy becomes the standard. For in that case, Sparnacian and its correlatives (= P6 and NP9), equaling probably *most* of the Wasatchian (!), are included in the Paleocene.

Of the 63 genera and 35 families of mammals here recognized in the Clarkforkian composite fauna, extending through possibly 2 million years, 36 (57 percent) of the genera and 24 (69 percent) of the families are shared with the preceding Tiffanian. Sixteen genera (25 percent) and 5 families (Barylambdidae, Cyriacotheriidae, Pentacodontidae, Plesiadapidae, and Arctostylopidae = 14 percent) are not known in the succeeding Wasatchian fauna in North America.

RIOCHICAN

Riochican is based on the composite Rio Chico fauna from the Rio Chico Formation, a succession of mudstone and sandstone beds exposed near the coast of the Gulf of San Jorge, Chubut Province, Argentina (Simpson, 1940; Marshall et al., 1977). The Rio Chico Formation overlies the dinosaur-bearing Chubut (or Neuquen) Group; in certain districts, it overlies the marine Salamanca Formation. The Salamanca is correlated with the Dano-Montian of Europe on the basis of its foraminifers (Loeblich and Tappan, 1957). Thus, the Rio Chico is not earlier than late Paleocene. The mammalian fauna from the Rio Chico is peculiarly endemic and offers no direct means for intercontinental correlation.

At Saõ Jose de Itaborai, inland from Rio de Janeiro, Brazil, Paula Couto (1950, 1952a, 1952b, 1952c, 1952e, 1954, 1958) and others found a local fauna of small and medium-sized vertebrates in marl fills of cavernous pre-Riochican Paleocene limestone, exposed in a limestone quarry. The fossils are beautifully preserved. In addition to mammals, there are abundant remains of birds, sebecoid crocodilians, fish, turtles, amphibians, and lizards. We combine the mammalian taxa from Itaborai with those from the Rio Chico in the following generic list for the Riochican, basing identifications on Simpson (1967) and Paula Couto (ibid.).

Composite Riochican Mammalian Fauna

MARSUPIALIA
  Didelphoidea
    Protodidelphis vanzolinii Paula Couto, 1952C
    Didelphopsis cabrerai Paula Couto, 1952C
    Ischyrodidelphis castellanosi Paula Couto, 1952C
    Xenodelphis doelloi Paula Couto, 1962B
    Minusculodelphis minimus Paula Couto, 1970A
    Schaefferia fluminensis Paula Couto, 1952C
    Mirandotherium alipioi Paula Couto, 1952E
    Guggenheimia brasiliensis Paula Couto, 1952C
    Derorhynchus singularis Paula Couto, 1952C
    Gaylordia macrocynodonta Paula Couto, 1952C
    Monodelphopsis travassoi Paula Couto, 1952C
    Eobrasilia coutoi Paula Couto, 1962B
    Marmosopsis juradoi Paula Couto, 1962B
  Borhyaenidae
    Patene simpsoni Paula Couto, 1952B
    Nemolestes?   —Simpson (1948E)
  Polydolopidae
    Polydolops rothi Simpson, 1936G
    P. winecage Simpson, 1935B
    P.? kamedtsen Simpson, 1935B
    Epidolops ameghinoi Paula Couto, 1952B
    E. gracilis Paula Couto, 1952B
      Seumadia yapa Simpson, 1935B
  Inc. sed.
    Gashternia ctalehor Simpson, 1935B
EDENTATA
  Dasypodidae
    Utaetus?   —Simpson (1935B)
  Glyptodontidae—McKenna (personal commun., 1979)
LITOPTERNA
  Proterotheriidae
    Anisolambda prodromus Paula Couto, 1952B
    Josephleidya sp.
    Wainka tshotshe Simpson, 1935B
    Ricardolydekkeria?
  Macraucheniidae
    Victorlemoinea prototypica Paula Couto, 1952A
NOTOUNGULATA
  Henricosborniidae
    Henricosbornia waitehor Simpson, 1935B
    H.? lophodonta (Ameghino, 1901)   —Simpson (1948E)
    Othnielmarshia?
    Peripantostylops? orehor Simpson, 1935B
    Simpsonotus praecursor Pascual, Vucetich, and Fernandez, 1978
    S. major Pascual, Vucetich and Fernandez, 1978
  Inc. Sed.
    Brandmayria simpsoni Cabrera, 1935C
  Notostylopidae
    Seudenius cteronc Simpson, 1935B   (family?)
    Homalostylops? atavus Paula Couto, 1954B

Oldfieldthomasiidae
    Kibenikhora get Simpson, 1935B
    Colbertia magellanica Paula Couto, 1952D
    Camargomendesia pristina Paula Couto, 1978
    Itaboraitherium atavum Paula Couto, 1978
Isotemnidae
    Isotemnus haugi (Roth, 1902)
    I.? ctalego Simpson, 1935B
Archaeohyracidae
    Eohyrax?   —Simpson (1967A)
Interatheriidae
    Notopithecus?
    Transpithecus?
TRIGONOSTYLOPOIDEA
Trigonostylopidae
    "Trigonostylops" apthomasi (Price and Paula Couto, 1950)
    Shecenia ctirneru Simpson, 1935B
XENUNGULATA   [DINOCERATA?   —McKenna
    (1981)]
Carodniidae
    Carodnia feruglioi Simpson, 1935B
    C. vieirai Paula Couto, 1952A
CONDYLARTHRA
Didolodontidae
    Ernestokokenia yirunhor Simpson, 1935B
    E. protocenica Paula Couto, 1952A
    E. parayirinhor Paula Couto, 1952A
    E. chaishoer Simpson, 1935B   (age?)
    Lamegoia conodonta Paula Couto, 1952A
    Asmithwoodwardia scotti Paula Couto, 1952A

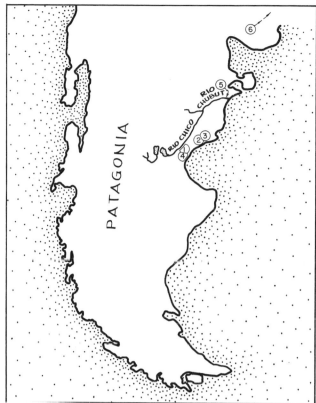

FIGURE 3-18   *Riochican Localities and Stratigraphic Units.*
  1 Cañadón Hondo—Rio Chico Formation
  2 Cerro Redondo—Rio Chico Formation
  3 Las Violetas—Rio Chico Formation
  4 Pico Salamanca—Rio Chico Formation
  5 Trelew-Gaiman region—Rio Chico Formation
  6 Itaborai, Brazil—Itaborai? Formation (fissure marl)

The composite Rio Chico fauna signifies a prior long or alternatively "explosive" interval of endemic mammalian evolution in South America. Simpson (1950 and other publications) has concluded that a few basic stocks of land mammals "filtered" into the Neotropical region from the Nearctic region, possibly by island-hopping across narrow seaways in the present-day Central American or Caribbean area, or both, about the end of Mesozoic time. These stocks consisted of didelphoid marsupials, predasypodoid edentates, and phenacodontoid condylarths (or perhaps arctocyonians). Then in the relatively brief interval of latest? Cretaceous and earlier Paleocene (ca. 70 to 60 m.y.b.p.) there was an adaptive radiation of land mammals into the large- and small-animal insectivore, carnivore, herbivore, omnivore, arboreal, terrestrial-ambulatory, and other niches that are demonstrated by the diversified Riochican taxa.

Reig (1981) and others, however, believe that the diversity seen in the Riochican fauna may be the product of a long Mesozoic mammalian history in South America.

Of the 46 genera of mammals here recognized in the composite Riochican fauna, extending through possibly 5 million years, 31 (about 67 percent) are not known in the succeeding Casamayoran (="early" Eocene) fauna in South America. In this succeeding fauna, all its 66 genera may have been derived by evolution in situ from Riochican stocks.

STRATIGRAPHIC DEMONSTRATIONS

Europe. There is no demonstration in Europe that strata with Cernaysian mammals directly overlie strata with Dano-Montian mammals. Dano-Montian underlies Landenian (=Thanetian plus Sparnacian) in Belgium, but to date, land mammals have not been obtained from the lower Landenian.

Asia. Qiu and others (1977) have documented a stratigraphic succession of Paleocene land-vertebrate localities in the Qianshan and Xuanchan Basins of Anhui. Here, there is evidently superposition of faunas representing a sequence through most of middle and later Paleocene time in China.

Africa. Sigé (personal commun., 1980) reports that Ypresian (=Sparnacian or Cuisian) land-micromammals are being found in marine strata overlying Thanetian marine strata in Morocco. Cappetta and others (1978) have previously reported mammals in the underlying Montian marine limestones of this same basin. Dating here has been furnished by the record of sharks and marine invertebrates.

North America. Demonstration of the superposition of Torrejonian mammals over Puercan mammals in the San Juan Basin of New Mexico is founded on short-distance, intrabasin stratigraphic correlation. Magnetic polarity intervals and the paleontologic stratigraphy relating to Puercan and to the Cretaceous-Tertiary boundary have been the chief concern for workers in this district to date. Figure 3-19, which shows these phenomena, is taken from R.F. Butler, E.H. Lindsay, and L.L. Jacobs, 1977, Magnetostratigraphy of the Cretaceous-Tertiary boundary in the San Juan Basin, New Mexico, *Nature* 267:318–323.

The Sand Coulee–Polecat Bench district of northwestern Wyoming demonstrates the only uninterrupted succession of Puercan-Torrejonian-Tiffanian-Clarkforkian-Wasatchian mammals. Figure 3-20, showing this sequence, is taken from Butler, Lindsay, and Gingerich (1980).

South America. In Patagonia, the Rio Chico Formation and its fauna overlie the marine Salamanca Formation in some districts. The Salamanca has been correlated with Danian by Loeblich and Tappan (1957) and other marine stratigraphers.

DATING AND CORRELATION

Year-age dating of and within the nonmarine Paleocene is supported by only a few radiometric determinations from associated volcanic rocks. Evernden and others (1964) gave a K-Ar date of 66.4 m.y.b.p. (corrected constants) for plagioclase from dacitic pumice cobbles at about the level of Puercan mammals in the Denver Formation, Colorado. A basalt at the top of this formation was 60.2 m.y. So far as we are aware, the remainder of year-age estimates for Paleocene rocks and paleobiota are based on glauconite determinations in marine strata (Hardenbol and Berggren, 1978)

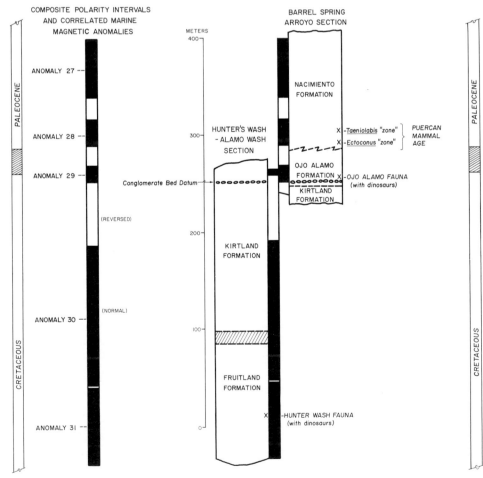

FIGURE 3-19 *The Puercan Mammal Age, the Cretaceous-Paleocene Boundary "Zone", Polarity Intervals and Correlated Marine Magnetic Anomalies in the San Juan Basin, New Mexico.*
(*Data from R.F. Butler, E.H. Lindsay, and L.L. Jacobs, 1977, Magnetostratigraphy of the Cretaceous-Tertiary boundary in the San Juan Basin, New Mexico. Nature 267:318–323.*)

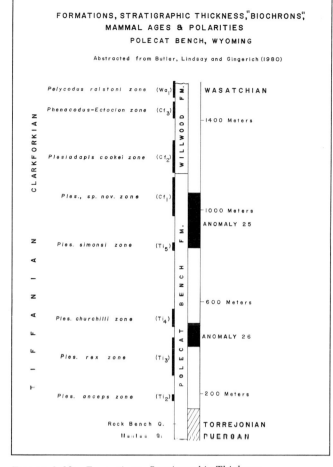

FORMATIONS, STRATIGRAPHIC THICKNESS,"BIOCHRONS",
MAMMAL AGES & POLARITIES
POLECAT BENCH, WYOMING

Abstracted from Butler, Lindsay and Gingerich (1980)

FIGURE 3-20 *Formations, Stratigraphic Thickness, "Biochrons," Mammal Ages, and Polarities in the Polecat Bench District, Wyoming.*
(Abstracted from R.F. Butler, E.H. Lindsay, and P.D. Gingerich. Magnetic polarity stratigraphy and Paleocene-Eocene biostratigraphy of Polecat Bench, northwestern Wyoming, in Early Cenozoic Paleontology and Stratigraphy of the Bighorn Basin, Wyoming. 1880–1980, *ed. Gingerich, P.D., pp. 95–98. Univ. Mich. Pap. on Paleontol. 24.)*

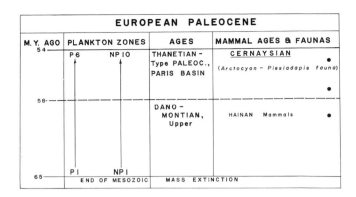

FIGURE 3-21 *Correlations within the European Paleocene. Black Circles Represent Recognized Subdivisions in a Mammal Age.*

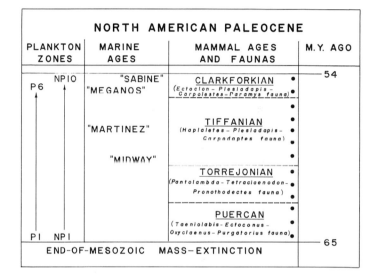

FIGURE 3-22 *Correlations within the North American Paleocene. Black Circles Represent Recognized Subdivisions in a Mammal Age.*

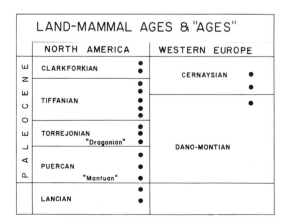

FIGURE 3-23 *Correlation between European and North American Paleocene Land-Mammal Ages. Black Circles Represent Recognized Subdivisions.*

or they are prorated on the basis of being earlier (stratigraphically lower) than dated early Eocene rocks (ca. 54 m.y.b.p.) or later (stratigraphically higher) than dated latest Cretaceous rocks (ca. 65–67 m.y.). Episodes in the magnetic "time scale" of the Paleocene have been assigned year dates from measurements of sea-floor anomalies with assumed constant rate of sea-floor spreading; see, for example, LaBrecque, Kent, and Cande (1977) and Lowrie and Alvarez (1981). The date of 65 m.y. for the Cretaceous-Paleocene boundary that we have adopted and used in this book is likely to be the minimum estimate, according to current publications.

Our placement of the marine subdivisions on the accompanying correlation charts for Europe and North America (Figures 3-21 and 3-22) follows King (1981) and Hardenbol and Berggren (1978).

49

BIOGEOGRAPHY AND THE MILIEU

The relation between land and sea was notably different about 60 million years ago, in mid-Paleocene time. Our Figure 3-24 is a diagram of the chief aspects of this difference, which are important in faunal comparisons. This is a crude but ''digestible'' presentation of the myriad publications and maps on Tertiary plate tectonics and paleogeography that have been issued during the past 15 years.

A small seaway (or perhaps several) appears to have isolated northwestern Europe (England and part of the present Scandinavian region) from central and southeastern Europe. An epicontinental sea projected from the Arctic Basin southward, separating European USSR from Asian USSR, and it may have connected with an east-west–trending Tethyan seaway that formed a marine barrier between the terrestrial African-Arabian plate and the terrestrial Eurasian plate. This Tethyan seaway extended across what is now sub-Himalayan India, on through southern Burma and to the southeast. Australia was evidently not yet a complete island continent, and the cojoined Antarctica may or may not have been subaerially connected to South America. The ''Salamancan'' Sea lapped farther inland than it does at present on Argentina's Atlantic side, and South America was separated from Central and North America. A belt of the present-day Gulf and Atlantic Coast of the United States was inundated by the sea, and a sea extended northward through the Mississippi River embayment as far as present-day Canada. Also, large districts along the West Coast of the United States were covered by Paleocene sea. North American land may have been connected to Asian land in the Alaska-Bering region. Oceanic currents moved along different routes 60 million years ago, causing global equability of climate, even in the Arctic and Antarctic latitudes, in contrast to present climatic zonations.

With this sketchy picture of the mid-Paleocene geography of the world, we can develop a scenario for the probable milieu of the land mammals of that day.

Wolfe (1980) summarized his conclusions regarding the climates of the earlier Tertiary as follows:

> During the Paleocene and Eocene, climates were characterized by a low mean annual range of temperature (a maximum of 10–15 °C), a moderate to high mean annual temperature (10–20 °C), and abundant precipitation; strong broad-leaved evergreen vegetation extended to almost lat. 60 °N during the Paleocene and to well above 61 °N during the Eocene. Poleward of the broad-leaved evergreen forests were forests that were broad-leaved deciduous; these deciduous forests, however, were unlike extant broad-leaved deciduous forests in general floristic composition and physiognomy. Coniferous forests probably occupied the northernmost latitudes.

Paleocene strata of the Western interior of North America contain numerous localities in which fossils of ferns, palms, conifers, angiosperms, ostracods, diatoms, insects, molluscs, fishes, amphibians, turtles, lizards, crocodilians, and primitive mammals can be found. Paleofloras are known from sites in southern California and

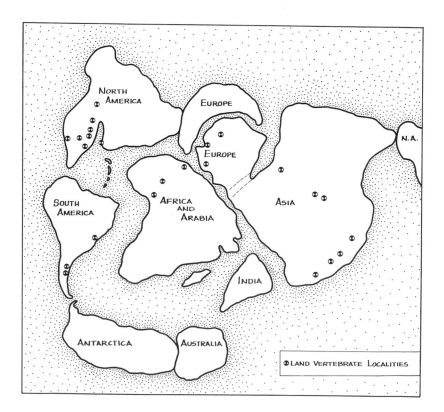

FIGURE 3-24 *Sketch of the World About 60 Million Years Ago.*

northern New Mexico northward to Alaska (Knowlton, 1914, 1922; Bell, 1949; Pabst, 1968; E.W. Berry, 1924; Dorf, 1940, 1942; Shoemaker, 1966; R.W. Brown, 1943, 1962; Anderson, 1960; Hall and Norton, 1967; Norton and Hall, 1969; and Tschudy, 1969).

There were forests and savannas thriving in relatively humid, frost free "subtropical" climate in higher latitudes. Much of the Paleocene lowland of the western United States was probably a "subtropical" savanna, comparable in its vegetation to parts of present-day Sinaloa, western Mexico (Axelrod, 1950). Wolfe and Hopkins (1967), analyzing percentage of plant species with leaves indicating warm-humid climate—that is, leaves having entire, smooth margins, large size, and acuminate distal points ("drip points")—concluded:

> No known Paleocene flora in North America indicates a climate cooler than warm-temperate, and subtropical climate existed at least as far north as latitude 62° in Alaska during part, if not all, of the Paleocene.

Sloan (1970, p. 428) also summarized the data from the paleobotanical literature as

> suggesting a climatic trend from an equable subtropical climate during the late Cretaceous to a warm temperate, less equable climate during the Paleocene . . . followed by a return to an equable subtropical climate during the early Eocene.

Hickey (1980) concluded that the Tiffanian deciduous broad-leaved forests of the Clark's Fork Basin in northwestern Wyoming were distinct from "those before and after in being relatively low in diversity and low in equitability or 'evenness.' Mean annual temperature in the Tiffanian was about 10 °C, compared with 13.5 °C in the Clarkforkian."

Hickey (1980) further concluded that the Clarkforkian flora of the Clark's Fork area comprised forests of mixed broad-leaved deciduous and broad-leaved evergreen trees, in keeping with a mean annual temperature of 13.5 °C.

The combined evidence indicates that during Paleocene time, from about 65 to about 54 million years ago, on land in North America, Eurasia, and (probably) South America, vertebrates thrived and died near streams and lakes in well-watered woodland and savanna country. And in these districts the climate was uniformly warm through the year, with no pronounced seasonality, no arid (cf. desert) districts, and little or no freezing temperature. An archaic land-mammal fauna lived in this pleasant milieu. This fauna was dominated, in taxonomic diversity and in probable numbers of individuals, by multituberculates (small herbivores), anagalidans (small omnivores? in Asia only), pantodonts (large herbivores), arctocyonians and condylarths (small to large omnivores and herbivores), notoungulates (small to medium herbivores in South America and in Asia), and mesonychids (medium omnivores). From 60–54 million years ago, specialized small primates, not demonstrably ancestral to anthropoids, and small insectivorans, together with small insectivora-like mammals such as leptictidans, apatotheres, microsyopids, palaeoryctids, and pantolestids, were abundant in the forests of the Northern Hemisphere.

Predictably smaller numbers of miacid carnivorans, carnivorous creodonts (Northern Hemisphere), and borhyaenid marsupials (South America) preyed on the great biomass of herbivorous and insectivorous mammals. We must not forget, however, that predator reptiles and birds also reduced the mammal populations during that time, just as they do today.

ENDNOTES

1. Although Van Valen often omits adequate citations, descriptions, and comparisons for synonymies, for changed family assignment of genera, for the proposal of new taxa, and the like, we have chosen to follow most of his revisions (1978) because of his extensive studies of the collections of earlier-Paleocene mammals through the past 15 years.

2. See note 1. Also, we do not wish to, but inevitably do, indicate a preference for one of the opposing arrangements (and its implied phyletic relations) for the higher categories—for example, Orders—of the Mammalia. Szalay (1977), for example, allies the Pantodonta with the Arctocyonidae on the basis of structure of the tarsi, rejecting McKenna's (1975) interpretation of close relation between pantodonts and the Cretaceous *Cimolestes*. Because of the anatomy of astragali and calcanea referred to the late Cretaceous *Procerberus, Gypsonictops,* and *Cimolestes,* Szalay aligns these genera with the Tertiary Leptictidae, Pantolestidae, Taeniodontidae, and possibly Microsyopidae, in his new order Leptictimorpha under the cohort Glires; this cohort also includes rodents and lagomorphs. And these conclusions strongly contrast also with McKenna's treatment of these groups. Our principal concern is to list the lower taxonomic categories in the same sequence in each fauna, to allow quick faunal-taxonomic comparisons by the reader.
McKenna, M.C. 1975. Toward a phylogenetic classification of the Mammalia. In *Phylogeny of the Primates,* ed. W.P. Luckett and F.S. Szalay, pp. 21–46. New York: Plenum Press.
Szalay, F.W. 1977. Phylogenetic relationships and a classification of the eutherian Mammalia. In *Major Patterns in Vertebrate Evolution,* ed. M.K. Hecht, P.C. Goody, and B.M. Hecht, pp. 315–374. New York: Plenum Press.

PALEOCENE MAMMAL-BEARING LITHOSTRATIGRAPHIC UNITS

We remind you that the age given for these mammal-bearing units and potentially mammal-bearing units is the age of the fossil mammals found in each unit. The whole unit may not be of the age given.

Bajo de la Palangana Sandstone—Riochican—Argentina

Bara Formation (crocodiles and turtles)—Paleocene—Pakistan

Berru lens—Cernaysian—France

Black Peaks Formation—Torrejonian and Tiffanian—Texas

Bracheux Sands (Sables de Bracheux)—Thanetian and Cernaysian—France

Calcaire de Rilly—Cernaysian—France

Cañadon Honda Sandstone—Riochican—Argentina

Carodnia zone—Riochican—Argentina

Cernay Conglomerate (Conglomérat de Cernay)—Cernaysian—France

Châlons-sur-Vesle Sands (Sables de Châlons-sur-Vesle)—Thanetian and Cernaysian—France

Chappo Member, Wasatch Formation—Tiffanian into Wasatchian—Wyoming

Chijiang (Chihkiang, Chihjiang) Formation—Late Paleocene—Jiangxi

"Clark Fork" beds (=upper "Silver Coulee" beds = upper Polecat Bench and lower Willwood Formations)—Clarkforkian—Wyoming

Datang (Datangxu) Member, Nongshan Formation—Late Paleocene—Guangdong

Dawson Arkose—Puercan—Colorado

Dazhang (Dashang) Formation—Late Paleocene—Henan

Debeque Formation—Tiffanian into Wasatchian—Colorado

Denver Formation, upper—Puercan—Colorado (includes South Table Mountain volcanic ash)

Doumu Formation—Late Paleocene—Anhui

Ernestokokenia zone—Riochican—Argentina

Evanston Formation, in part—Tiffanian—Wyoming

Fort Union Formation or Group—Puercan through Clarkforkian—Montana, North Dakota and Wyoming

Gaoyugou Formation—Middle Paleocene—Henan

Gashato Formation—Late Paleocene—Mongolia

Gelinden, marnes de—Montian—Belgium (birds and plants)

Goler Formation, in part—Torrejonian—California

Hanna Formation—Tiffanian—Wyoming

Hoback Formation—Tiffanian—Wyoming

Kibenikhoria zone—Riochican—Argentina

La Fère, Tuffeau de—Cernaysian—France

Landen Formation—Cernaysian and Sparnacian—Belgium

Lannikeng Member, Chijiang Formation—Late Paleocene—Jiangxi

Lebo Formation—Torrejonian—Montana

Lignites de Menat—Cernaysian—Belgium

Lincent, Tuffeau de—Cernaysian equivalent—Belgium

Livingston Formation—Tiffanian and Clarkforkian—Montana

Lofochai "series"—Middle? and Late Paleocene—Guangdong

Mantua lentil, Polecat Bench Formation—Puercan—Wyoming

Mason pocket, "Tiffany" beds, San Jose Formation—Tiffanian—Colorado

Mealla Formation—Riochican—Argentina

Melville Formation—Tiffanian—Montana

Menat, lignites de—Cernaysian—France

Midway Group, in part—Torrejonian—Louisiana

Montian, nonmarine (type Montian, in part)—Montian—Belgium

Mouras sandy clays, etc.—Cernaysian—France

Nacimiento Formation—Puercan and Torrejonian—New Mexico

Nao-mu-gen (see Nomogen)

Naran Bulak (lower?) Formation—Late Paleocene—Mongolia

Nomogen Formation—Late Paleocene—Nei Mongol (Inner Mongolia, China)

Nongshan (Nonshan) Formation—Late Paleocene—Guangdong

North Horn Formation, upper—Puercan and early Torrejonian—Utah

Nungshan Formation, Lofochai Group—=Nongshan Formation—Late Paleocene—Guangdong

Orp-le-Grand, sables de—Montian—Belgium

Ostricourt, sable de, in part—=Cernaysian—Belgium, France

Paskapoo Formation—Torrejonian and Tiffanian—Alberta

Polecat Bench Formation—Puercan into Clarkforkian—Wyoming

Porcupine Hills Formation—Torrejonian—Alberta

"Puerco" Formation (=Nacimiento Formation, lower)—Puercan—New Mexico

Ravenscrag Formation—Puercan and Tiffanian—Saskatchewan

Rilly marls and limestone (Calcaire de Rilly)—Cernaysian—France

Rio Chico Formation—Riochican—Argentina

"Rock Bench Quarry beds" (Polecat Bench Formation)—Torrejonian—Wyoming

Sables de Bracheux (stratotype Paleocene)—Thanetian (=Cernaysian)—France

Sables de Châlons-sur-Vesle (lateral variant of Sables de Bracheux)—Thanetian (=Cernaysian)—France

San Jose Formation—Tiffanian and Wasatchian—New Mexico and Colorado

Saunders Formation—Tiffanian—Alberta

Sentinel Butte Formation, Fort Union Group—Tiffanian—North Dakota

Shanghu Formation—Early? and Middle Paleocene—Guangdong

Shizikou Formation—Middle Paleocene—Jiangxi

Shotgun Formation—Torrejonian and Tiffanian—Wyoming

Shuangtasi Group—Late Paleocene—Anhui

"Silver Coulee beds" (Polecat Bench Formation, upper)—Tiffanian into Clarkforkian—Wyoming

South Table Mountain ash (Denver Formation, upper)—Puercan—Colorado

Taizicun Formation—Late Paleocene—Xinjiang

Tepetate Formation (questionable formational assignment of fossils)—Clarkforkian?—Baja California, Mexico

"Tiffany beds" (San Jose Formation)—Tiffanian—Colorado

Tongue River Formation—Torrejonian and Tiffanian—Montana and North Dakota

"Torrejon" beds (Nacimiento Formation, upper)—Torrejonian—New Mexico

Toumu (Doumu) Formation—Late Paleocene—Anhui

Travertin de Sézanne (with nonmammalian Cernay fauna)—Stratotype Paleocene—Cernaysian—France

Tsaoshin Formation—Middle Paleocene—Hunan

Tullock Formation—Puercan—Montana

Walbeck fissure fills—Cernaysian—Germany

Wanghutun (Wanghudun) Formation—Early? and Middle Paleocene—Anhui

Wangwu Member, Chijiang Formation—Late Paleocene—Jiangxi

Willwood Formation, lower—Clarkforkian and Wasatchian—Wyoming

Zaoshi Formation—Middle Paleocene—Hunan

Zhangshanji Formation—Late Paleocene—Anhui

Zhuguikeng Member, Nongshan Formation—Late Paleocene—Guangdong

PALEOCENE LOCALITIES

Airport—Tiffanian—Wyoming
Arroyo Torrejon (or Torreon), East Branch, district
   —Torrejonian, New Mexico
Bab's Basin—Torrejonian—New Mexico
Bajo de la Palangana—Riochican—New Mexico
Balzac West—Torrejonian?—Alberta
Bara Formation sites (crocodiles and turtles only)
   —Paleocene—Pakistan
Barrel Springs Arroyo—Puercan—New Mexico
Battle Mountain—Tiffanian—Wyoming
Bayfield—Tiffanian—Colorado
Bear Creek (Eagle Coal Mine)—Clarkforkian—Montana
Belfry Southeast—Torrejonian—Montana
Big Multi—Clarkforkian—Wyoming
Big Pocket—Torrejonian—New Mexico
Bison Basin (Ledge, Saddle, Titanoides, West End)
   —Tiffanian—Wyoming
Bitonitsoseh Wash or Arroyo—Puercan—New Mexico
Black Peaks Formation sites (Joe's Bonebed, Ray's Bonebed,
   Ray's Annex, Schiebout-Reeves Quarry, etc.)—Torrejonian?,
   Tiffanian, and Clarkforkian—Texas
Brisbane—Tiffanian—North Dakota
Brisbane West—Tiffanian—North Dakota
Buckman Hollow (La Barge Creek)—Clarkforkian—Wyoming
Buford Northeast—Tiffanian—North Dakota
Caddo Parish (Junior Oil Co., Beard No. 1)—Torrejonian
   —Louisiana
Calgary 2E—Torrejonian?—Alberta
Cañadon Hondo—Riochican and Casamayoran—Argentina
Canyon Ski Quarry—Tiffanian—Alberta
Cedar Point Quarry—Tiffanian—Wyoming
Cedar Ridge West—Tiffanian—Wyoming
Cernay-lès-Reims (Lemoine quarry, Mouras quarry)
   —Cernaysian—France
Cerro Redondo—Riochican—Argentina
Chaling Basin—Middle Paleocene—Hunan
Châlons-sur-Vesle Sands sites (Chenay, Merfy, Jonchery)
   —Cernaysian—France
Chappo 1 and 17—Tiffanian—Wyoming
Chappo 12—Clarkforkian—Wyoming
Chenay—Cernaysian—France
Chianshan (See Qiangshan or Chienshan)—Early?, Middle, and
   Late Paleocene—Anhui
Chijiang (Chikiang) Basin—Middle and Late Paleocene—Jiangxi
Circle—Tiffanian—Montana
Clark Fork sites (general)—Clarkforkian (and older and
   younger)—Wyoming
Cleopatra Reservoir–Foster Gulch—Clarkforkian—Wyoming
Cochrane 1 and 2—Tiffanian—Alberta
Cochrane 11—Torrejonian?—Alberta
Corral Bluff or Bluffs—Puercan—Colorado
Crazy Mountain Field ("Fish Creek")—Torrejonian and
   Tiffanian—Montana (see Gidley and Silberling quarries, Scar-
   ritt quarry, etc.)
Croc Tooth Quarry—Tiffanian—Wyoming
Cub Creek—Torrejonian—Montana
Cypress Hills Paleocene (See Police Point.)
Datang Commune and Nanxiong Basin—Early? and Middle and
   Late Paleocene—Guangdong
Dawson—Puercan—Colorado

Dazhang Formation site—Late Paleocene—Henan
Dell Creek—Tiffanian—Wyoming
Divide Quarry—Tiffanian—Wyoming
Donnybrook—Tiffanian—North Dakota
Douglass Quarry—Tiffanian—Montana
Dragon or Dragon Canyon—Torrejonian—Utah
Eagle Coal Mine (Bear Creek)—Clarkforkian—Montana
Ekalaka (Medicine Rocks)—Torrejonian—Montana
Elbow or Elbow River—Tiffanian—Alberta
Erickson's Landing—Tiffanian—Alberta
Erquelinnes, lower—Cernaysian champsosaur—Belgium
Estevan (Roche Percée)—Tiffanian—Saskatchewan
Flagstaff Peak—Puercan—Utah
Fossil Basin (See Fossil Station.)
Fossil Hollow—Tiffanian—Wyoming
Fossil Station—Tiffanian—Wyoming
Foster Gulch Oil Well No. 1—Clarkforkian—Wyoming
Fritz Quarry—"Tiffanian-Clarkforkian boundary" (Rose,
   1975)—Wyoming
Gaiman, lower (Pan de Azucar)—Riochican—Argentina
Garbani—Puercan—Montana
Gashato—Late Paleocene—Mongolia
Gelinden—Montian birds and plants—Belgium
Gidley Quarry—Torrejonian—Montana
Gonghudong (see Nomogen)—Late Paleocene—Nei Mongol
Hainin—Montian—Belgium
Haliut (See Nomogen)—Late Paleocene—Nei Mongol
Hanna Basin—Tiffanian?—Wyoming
Heart Butte—Torrejonian?—North Dakota
Hell's Half Acre (Debeque South-Southeast)—Clarkforkian?
   —Colorado
Highway Blowout—Tiffanian—Wyoming
Hoback Basin, general—Torrejonian into Wasatchian
   —Wyoming
Hsuankuang (Kuangte Basin)—Late Paleocene—Anhui
Huaining—Late Paleocene—Anhui
Itaborai (São Jose de Itaborai)—Riochican—Brazil
Jenkins Mountain (Shotgun)—late Torrejonian or early
   Tiffanian—Wyoming
Jep-Woodburne—Clarkforkian?—Wyoming
Jimmy Camp Creek—Puercan—Colorado
Jintang Village (Nanxiong Basin)—Early? into Late Paleocene
   —Guangdong
Joe's Bonebed—Tiffanian—Texas
Jonchery—Cernaysian—France
Judson—Tiffanian—North Dakota
Junior Oil Co., Beard No. 1 Well (Caddo Parish)
   —Torrejonian—Louisiana
Keefer Hill (Jenkins Mountain?, Shotgun)—Tiffanian
   —Wyoming
Khaichin (Khaychin) Ula II—Early Paleocene—Mongolia
Krebb de Sessao (littoral vertebrates; no mammals yet)
   —Paleocene—Niger
La Barge Creek (Buckman Hollow)—Clarkforkian—Wyoming
La Fère—Cernaysian—France
Lai'an—Late Paleocene—Anhui
Laolingbei (Dayu)—Late Paleocene—Jiangxi
Larry's Beaucoup—Torrejonian—New Mexico
Las Violetas—Riochican—Argentina

Laudate—Torrejonian—California
Ledge—Tiffanian—Wyoming
Leidy Quarry—Puercan—Wyoming
Lemoine Quarry—Cernaysian—France
Little Muddy Creek—Torrejonian—Wyoming
Little Pocket (Nageezi)—Torrejonian—New Mexico
Little Sand Coulee—Clarkforkian—Wyoming
Livingston Formation site—Tiffanian-Clarkforkian—Montana
Lloyd and Hares sites—Tiffanian—North Dakota
Long Draw Quarry—Tiffanian—Wyoming
Love Quarry—Tiffanian—Wyoming
Lower Sand Draw—Tiffanian—Wyoming
"Low Locality"—late Tiffanian or Clarkforkian—Wyoming
Malcolm's Locality—Tiffanian—Wyoming
Mantua—Puercan—Wyoming
Maret I—Cernaysian—Belgium
Maret II—"pre-Cernaysian"—Belgium
Mason Pocket—Tiffanian—Colorado
McKenzie County sites—Tiffanian—North Dakota
Medicine Rocks I—Torrejonian—Montana
Medicine Rocks II—Tiffanian—Montana
Melville 13—Tiffanian—Montana
Menat—Cernaysian—France
Merfy—Cernaysian—France
Mesvin—"pre-Cernaysian" bird—Belgium
Middle Sand Draw—Tiffanian—Wyoming
Mont de Berru (Lemoine and Mouras Quarries)—Cernaysian
—France
Mt. Leidy Highlands (Togwotee)—Clarkforkian and
Wasatchian—Wyoming
Mouras Quarry—Cernaysian—France
Nanxiong (Nanhsiung, Nanyung, Namyung) Basin—Early? into
Late Paleocene—Guangdong
Naran Bulak—lower fauna is Late Paleocene—Mongolia
Nemegetu Basin (Naran Bulak)—Late Paleocene and Early
Eocene—Mongolia
New Anthill—Tiffanian—Wyoming
Nomogen Commune (Haliut and Gonghudong)—Late
Paleocene—Nei Mongol
Olive—Tiffanian—Montana
Ouarzazate Basin—Early Paleocene—Morocco
Paint Creek—Clarkforkian—Wyoming
Pico Salamanca—Riochican and Casamayoran—Argentina
Pine Cree Park—Puercan—Saskatchewan
Plateau Valley beds sites—Tiffanian into Wasatchian—Colorado
Police Point (Cypress Hills Paleocene)—Tiffanian—Alberta
Princeton Loc. 11—Tiffanian—Montana
Princeton Quarry—Tiffanian—Wyoming
Punta Prieta—Clarkforkian?, Wasatchian?—Baja California,
Mexico
Purdy Basin   (See Mt. Leidy Highlands–Togwotee Pass.)
Purgatory Hill—Puercan—Montana
Qianshan (Chianshan, Chienshan) Basin—Early? into Late
Paleocene—Anhui
Ray's Bonebed and Annex—Tiffanian—Texas
Ries Loc.—Clarkforkian—Wyoming
Rilly-la-Montagne—Cernaysian—France
Riverdale—Tiffanian—North Dakota

Roche Percée (Estevan)—Tiffanian—Saskatchewan
Rock Bench Quarry—Torrejonian—Wyoming
"Rohrer" Locality—late Tiffanian or early Clarkforkian
—Wyoming
Rough Gulch—Clarkforkian—Wyoming
Ruby (Plateau Valley beds)—Tiffanian?—Colorado
Saddle (Bison Basin)—Tiffanian—Wyoming
Saō Jose de Itaborai—Riochican—Brazil
Saunders Creek—Tiffanian—Alberta
Scarritt Quarry—Tiffanian—Montana
Schaff Quarry—Tiffanian—Wyoming
Shanghu Formation sites (Nanxiong Basin)—Middle?
Paleocene—Guangdong
Shimen Basin—Paleocene-Shaanxi
Shotgun Butte 1—Tiffanian—Wyoming
Silberling Quarry—Torrejonian—Montana
Silver Coulee beds sites—Tiffanian and Clarkforkian—Wyoming
South Table Mountain—Puercan—Colorado
South Wall (132-Meter Level)—Clarkforkian?—Texas
Storm Quarry—Tiffanian-Clarkforkian—Wyoming
Swain Quarry—Torrejonian—Wyoming
Swan Hills—Tiffanian—Alberta
Tantou Basin (Gaoyugou Fm. site, Middle Paleocene; Dazhang
Fm. site, Late Paleocene)—Henan
Taylor Mound—Torrejonian—New Mexico
Tiffany beds sites, general—Tiffanian—Colorado
Titanoides (Bison Basin)—Tiffanian—Wyoming
Togwotee, or Togwotee Pass—Clarkforkian and Wasatchian
—Wyoming
Tornillo Flat—Tiffanian—Texas
Trelew-Gaiman region—Riochican and Colhuehuapiàn
—Argentina
Tres Cruces-Mina Aguilar—Riochican—Argentina
Turpan Basin (Taizicun)—Cretaceous and Late Paleocene
—Xinjiang
Twin Creeks Fork—Tiffanian—Wyoming
Ulan (Oulan) Bulak—Late Paleocene—Mongolia
UM-Sub-Wy 7—Clarkforkian—Wyoming
UM-Sub-Wy 10 and 20—Clarkforkian—Wyoming
UM SC-188—Clarkforkian—Wyoming
Upper Sand Draw—Clarkforkian—Wyoming
Vertain (lower)—Cernaysian champsosaurs—France
Wagonroad—Puercan—Utah
Walbeck—Cernaysian—Germany
Wannagan Creek Quarry—Tiffanian—North Dakota
West End (Bison Basin)—Tiffanian—Wyoming
Western Tornillo Flat—Torrejonian or early Tiffanian—Texas
Williams County—Tiffanian—North Dakota
Xuanchang Basin—Late Paleocene—Anhui
Zhulinshan (Dayu)—Late Paleocene—Jiangxi

BIBLIOGRAPHY

ANDERSON, R.Y. 1960. Cretaceous-Tertiary palynology, eastern
side of the San Juan Basin, New Mexico. *N.M. State Bur. Mines
Miner. Resour. Mem.* 6:1–58.
ARDUINO, G. 1760. *Nuova raccolta di opuscoli scientifici e filo-
logici del padre abate Angiolo Calogiera*, 6:clxii–clxiii, Ven-
ice.

AXELROD, D.I. 1950. Evolution of desert vegetation in western North America. *Carnegie Inst. Wash. Publ.* 590:215–306.

BELL, W.A. 1949. Uppermost Cretaceous and Paleocene floras of western Alberta, Canada. *Bull. Can. Geol. Surv.* 13:1–231.

BERGGREN, W.A. 1962. Some planktonic Foraminifera from the Maestrichtian and type Danian stages of southern Scandinavia. *Stockh. Contrib. Geol.* 9:1–106.

BERGGREN, W.A. 1964. The Maestrichtian, Danian and Montian stages and the Cretaceous/Tertiary boundary. *Stockh. Contrib. Geol.* 11:103–176.

BERGGREN, W.A. 1965. Paleocene—A micropaleontologist's point of view. *Bull. Am. Assoc. Pet. Geol.* 49:1473–1484.

BERRY, E.W. 1917. Rilly, a fossil lake. *Sci. Monthly* 5:175–185.

BERRY, E.W. 1924. American Tertiary terrestrial plants and their interdigitation with marine deposits. *Bull. Geol. Soc. Am.* 35:767–784.

BIRKELUND, T., and BROMLEY, R.G., eds. 1979. *Cretaceous-Tertiary Boundary Events: Symposium.* Vol. 1, *The Maestrichtian and Danian of Denmark,* 210 pp. Vol. 2, *Proceedings,* 250 pp. Denmark: Univ. of Copenhagen.

BIRYUKOV, M.D., and KOSTENKO, N.N. 1965. (Concerning the "Obayly" fauna of the Zaysan Depression.) *Vestn. Akad. Nauk. Kaz.* 21:75–77.

BLONDEAU, A., CAVELIER, C., FEUGUEUR, L., and POMEROL, C. 1965. Stratigraphie du Paléogène du bassin de Paris en relation avec les bassins avoisinants. *Bull. Soc. Géol. Fr.* 7:200–221.

BONHOMME, M., ODIN, G.S., and POMEROL, C. 1968. Age de formations glauconieuses de l'Albien et de l'Eocène du Bassin de Paris. Colloque sur l'Eocène. *Fr., Bur. Rech. Géol. Minières Mém.* 58:339–346.

BROWN, R.W. 1943. Cretaceous-Tertiary boundary in the Denver Basin, Colorado. *Bull. Géol. Soc. Am.* 54:65–86.

BROWN, R.W. 1962. Paleocene flora of the Rocky Mountains and Great Plains. *U.S. Geol. Surv. Prof. Pap.* 375:1–119.

CAPPETTA, H., JAEGER, J.-J., SABATIER, M., SUDRE, J., and VIANEY-LIAUD, M. 1978. Découverte dans le Paléocène du Maroc des plus anciens mammifères euthériens d'Afrique. *Géobios,* 11 (2):257–263.

CHAVAN, A. 1948. L'âge des principaux gisements du calcaire pisolithique. *Bull. Soc. Géol. F.* 18:565–574.

CHAVAN, A. 1950. Mise au point sur la question Danien-Paléocène. *C. R. Soc. Géol. Fr.* 1950:110.

CHOW (ZHOU), M. 1960. *Prodinoceras* and a summary of mammalian fossils of Sinkiang. *Vertebr. PalAsiatica* 4:99–102.

CHOW (ZHOU), M., CHANG, Y., WANG, B., and TING, S. 1973. New mammalian genera and species from the Paleocene of Nanhsiung, north Kwangtung. *Vertebr. PalAsiatica* 11:31–35.

CHOW, M., and QI, T. 1978. Paleocene mammalian fossils from Nomogen Formation of Inner Mongolia. *Vertebr. PalAsiatica* 16:77–85.

CLEMENS, W.A., JR. 1974. *Purgatorius,* an early paromomyid primate (Mammalia). *Science* 184:903–905.

CURRY, D., GULNICK, M., and POMEROL, C. 1969. Le Paléocène et l'Eocène dans les bassins de Paris, de Belgique et d'Angleterre. Colloque sur l'Eocène. *F., Bur. Rech. Géol. Minières Mem.* 69:361–370.

DAMOTTE, R., and FEUGUEUR, L. 1963. L'âge du calcaire de Vigny (S.-et-O.) à partir de données paléontologiques nouvelles. *C. R. Acad. Sci.,* Paris, Sér. D., 256:3864–3866.

DESOR, E. 1847. Sur le terrain danien, nouvel étage de la craie. *Bull. Soc. Géol. Fr.* 4:179–182.

DING, S. 1979. A new edentate from the Paleocene of Guangdong. *Vertebr. PalAsiatica* 17:57–64.

DING, S., and TONG, Y. 1979. Some Paleocene anagalids from Nanxiong, Guangdong. *Vertebr. PalAsiatica* 17:137–145.

DING, S., and ZHANG, Y. 1979. Insectivore and anagalid fossils from the Chijiang Basin, Jiangxi Province. In *Mesozoic and Cenozoic Red Beds of South China,* pp. 354–359. Conf. at Nanxiong, Guangdong. Acad. Sin. I.V.P.P. and Nanjing Inst. Geol. Paleontol.

DORF, E. 1940. Relationship between floras of type Lance and Fort Union Formations. *Bull. Geol. Soc. Am.* 51:213–236.

DORF, E. 1942. Application of paleobotany to the Cretaceous-Tertiary boundary problem. *Trans. N.Y. Acad. Sci.* 4:73–78.

DORR, J.A., JR. 1977. Partial skull of *Palaeosinopa simpsoni* (Mammalia, Insectivora), latest Paleocene Hoback Formation, central western Wyoming, with some general remarks on the family Pantolestidae. *Univ. Mich., Mus. Paleontol. Contrib.* 24:281–307.

EL-NAGGAR, Z.R. 1967. Remarques sur les divisions du Paléocène resultant d'études dans les localités-types en Europe occidentale. *Rev. Micropaléontol.* 10:215–216.

EVERNDEN, J.F., SAVAGE, D.E., CURTIS, G.H., and JAMES, G.T. 1964. Potassium-argon dates and the Cenozoic mammalian chronology of North America. *Am. J. Sci.* 262:145–198.

FARCHAD, H. 1936. Etude du Thanétien du Bassin de Paris. *Soc. Géol. Fr. Mém.* 30:1–101.

FLEROV, K.K. (FLEROW, C.C.). 1957. A new coryphodont from Mongolia, and on evolution and distribution of Pantodonta. *Vertebr. PalAsiatica* 1:73–81.

FOX, R.C. 1968. A new Paleocene mammal (Condylarthra: Arctocyonidae) from a well in Alberta, Canada. *J. Mammal.* 49:661–664.

GAO, H. 1975. Paleocene mammal-bearing beds of Chaling Basin, Hunan. *Vertebr. PalAsiatica* 13:89–96.

GAZIN, C.L. 1941A. Paleocene mammals from the Denver Basin, Colorado. *J. Wash. Acad. Sci.* 31:289–295.

GAZIN, C.L. 1941B. The mammalian faunas of the Paleocene of central Utah, with notes on the geology. *Proc. U.S. Natl. Mus.* 91:1–53.

GAZIN, C.L. 1953. The Tillodontia: An early Tertiary order of Mammals. *Smithson. Misc. Collect.* 121:1–110.

GAZIN, C.L. 1956A. Paleocene mammalian faunas of the Bison Basin in south-central Wyoming. *Smithson. Misc. Collect.* 131:1–57 (no. 6).

GAZIN, C.L. 1956B. The Upper Paleocene Mammalia from the Almy Formation in western Wyoming. *Smithson. Misc. Collect.* 131:1–18 (no. 7).

GAZIN, C.L. 1968. A new primate from the Torrejon Middle Paleocene of the San Juan Basin, New Mexico. *Proc. Biol. Soc. Wash.* 81:629–634.

GAZIN, C.L. 1969. A new occurrence of Paleocene mammals in the Evanston Formation, southwestern Wyoming. *Smithson. Contrib. Paleobiol.* 2:1–16.

GINGERICH, P.D. 1969. New North American Plesiadapidae (Mammalia, Primates) and a biostratigraphic zonation of the middle and upper Paleocene. *Univ. Mich., Mus. Paleontol. Contrib.* 24:135–148.

GINGERICH, P.D. 1976. Cranial anatomy and evolution of early

Tertiary Plesiadapidae (Mammalia, Primates). *Univ. Mich., Mus. Paleontol. Pap. Paleontol.* 15:1–140.

GINGERICH, P.D. 1977. *Aletodon gunnelli*, a new Clarkforkian hyopsodontid (Mammalia, Condylarthra) from the early Eocene of Wyoming. *Univ. Mich., Mus. Paleontol. Contrib.* 24:237–244.

GINGERICH, P.D. 1978. New Condylarthra (Mammalia) from the Paleocene and early Eocene of North America. *Univ. Mich., Mus. Paleontol. Contrib.* 25:1–9.

GINGERICH, P.D. 1979. *Lambertocyon eximius*, a new arctocyonid (Mammalia, Condylarthra) from the late Paleocene of western North America. *J. Paleontol.* 53:524–529.

GINGERICH, P.D. 1980. *Tytthaena parrisi*, oldest known oxyaenid (Mammalia, Creodonta) from the late Paleocene of western North America. *J. Paleontol.* 54:570–576.

GINGERICH, P.D., and DORR, J.A. 1979. Mandible of *Chiromyoides minor* (Mammalia, Primates) from the Upper Paleocene Chappo Member of the Wasatch Formation, Wyoming. *J. Paleontol.* 53:550–552.

GINGERICH, P.D., and GUNNELL, G.F. 1979. Systematics and evolution of the genus *Esthonyx* (Mammalia, Tillodontia) in the early Eocene of North America. *Univ. Mich., Mus. Paleontol. Contrib.* 25:125–131.

GINGERICH, P.D., and ROSE, K.D. 1977. Preliminary report on the American Clark Fork mammal fauna, and its correlation with similar faunas in Europe and Asia. *Géobios, Mém. Spéc.* 1:39–45.

GODFRIAUX, J., and ROBASZYNSKI, F. 1974. Le Montien continental et le Dano-Montien marin des sondages de Hainin (Hainaut, Belgique). *Ann. Soc. Géol. Belg.* 97:185–200.

GODFRIAUX, I., and THALER, L. 1972. Note sur la découverte de dents de mammifères dans le Montien continental du Hainaut (Belgique). *Bull. Acad. R. Belg. Cl. Sci.* 58:536–541.

GRANGER, W. 1914. On the names of the Lower Eocene faunal horizons of Wyoming and New Mexico. *Bull. Am. Mus. Nat. Hist.* 33:121–131.

GRUAS-CAVAGNETTO, C. 1968. Etude palynologique des divers gisements du Sparnacien du bassin de Paris. *Soc. Géol. Fr. Mém.* 110:1–145.

HALL, J.W., and NORTON, N.J. 1967. Palynological evidence of floristic change across the Cretaceous-Tertiary boundary in eastern Montana (USA). *Palaeogeogr., Palaeoclimatol., Palaeoecol.* 3:131–141.

HARDENBOL, J., and BERGGREN, W.A. 1978. A new Paleogene time scale. In *Contribution to the Geological Time Scale*, ed. G.V. Cohee, M.F. Glaesner, and H.D. Hedberg, pp. 213–234. Studies in Geology, no. 6. Tulsa: Am. Assoc. Pet. Geol.

HICKEY, L.J. 1980. Paleocene stratigraphy and flora of the Clark's Fork Basin. In *Early Cenozoic Paleontology and Stratigraphy of the Bighorn Basin, Wyoming, 1880–1980*, pp. 33–49. *Univ. Mich., Pap. Paleontol.* 24.

HOLTZMAN, R.C. 1978. Late Paleocene mammals of the Tongue River Formation, western North Dakota. *N.D. Geol. Surv. Rep. Investig.* 65:1–88.

HOLTZMAN, R.C., and WOLBERG, D.I. 1977. The Microcosmodontinae and *Microcosmodon woodi*, new Multituberculata taxa (Mammalia) from the late Paleocene of North America. *Minn. Sci. Mus., Sci. Publ., n.s.*, 4:1–13.

HUANG, X. 1978. Paleocene Pantodonta of Anhui. *Vertebr. PalAsiatica* 16:267–274.

IDEKER, J., and YAN, D. 1980. *Lestes* (Mammalia), a junior homonym of *Lestes* (Zygoptera). *Vertebr. PalAsiatica* 18:138–141.

JELETSKY, J.A. 1962. The allegedly Danian dinosaur-bearing rocks of the globe and the problem of the Mesozoic-Cenozoic boundary. *J. Paleontol.* 36:1005–1018.

JEPSEN, G.L. 1930. Stratigraphy and paleontology of the Paleocene of northeastern Park County, Wyoming. *Proc. Am. Philos. Soc.* 69:463–528.

JEPSEN, G.L. 1940. Paleocene faunas of the Polecat Bench Formation, Park County, Wyoming. *Proc. Am. Philos. Soc.* 83:217–340.

JEPSEN, G.L., and WOODBURNE, M.O. 1969. Paleocene hyracothere from Polecat Bench Formation, Wyoming. *Science* 164:543–547.

KEDVES, M. 1968. Etude palynologique des couches du Tertiaire inférieur de la region parisienne. 2. Tableau de quelques espèces et types de sporomorphs. *Pollen Spores* 10:117–128.

KEYES, C.R. 1906. Geological section of New Mexico. *Science* 23:921.

KIELAN-JAWOROWSKA, Z. 1969. Archaeolambdidae Flerov (Pantodonta) from the Paleocene of the Nemegt Basin, Gobi Desert. *Pol. Akad. Nauk., Inst. Geogr.,* 19:133–152.

KIELAN-JAWOROWSKA, Z., and DOVCHIN, N. 1969. Narrative of the Polish-Mongolian paleontological expeditions, 1963–1965. *Pol. Akad. Nauk., Inst. Geogr.,* 19:7–30.

KIELAN-JAWOROWSKA, Z., and SLOAN, R.E. 1979. *Catopsalis* (Multituberculata) from Asia and North America and the problem of taeniolabidid dispersal in the Late Cretaceous. *Acta Palaeontol. Pol.* 24:187–197.

KING, C. 1981. The stratigraphy of the London Clay and associated deposits. *Tert. Res. Spec. Pap.* 6:1–158.

KNOWLTON, F.H. 1914. Cretaceous-Tertiary boundary in the Rocky Mountain region. *Bull. Geol. Soc. Am.* 25:325–340.

KNOWLTON, F.H. 1922. The Laramie flora of the Denver Basin with a review of the Laramie problem. *U.S. Geol. Surv. Prof. Pap.* 130:1–175.

KRAUSE, D.W. 1977. Paleocene multituberculates (Mammalia) of the Roche Percée local fauna, Ravenscrag Formation, Saskatchewan, Canada. *Palaeontogr.,* Sect. A, 159:1–36.

KRAUSE, D.W. 1980. Multituberculates from the Clarkforkian Land-Mammal Age, late Paleocene–early Eocene, of western North America. *J. Paleontol.* 54:1163–1183.

KRISHTALKA, L. 1976. Early Tertiary Adapisoricidae and Erinaceidae (Mammalia, Insectivora) of North America. *Bull. Carnegie Mus. Nat. Hist.* 1:4–40.

KÜHN, O. 1940. Crocodilier- und Squamatenreste aus dem obern Paleocän von Walbeck. *Zentralb. Geol. Palaeontol.,* Sect. B, 1940:21–25.

LABRECQUE, J.L., KENT, D.V. and CANDE, S.C. 1977. Revised magnetic polarity time scale for Late Cretaceous and Cenozoic time. *Geology* 5:330–335.

LEMOINE, V. 1880. Notice géologique sur les environs de Reims. *Bull. Soc. Géol. Fr.* (3) 9:344–345.

LEMOINE, V. 1889. Etude d'ensemble sur les dents des Mammifères fossiles des environs de Reims. *Bull. Soc. Géol. Fr.* (3) 19:263–290.

LOEBLICH, A.R., JR., and TAPPAN, H. 1957. Correlation of the Gulf and Atlantic coastal plain Paleocene and Lower Eocene formations by means of planktonic Foraminifera. *J. Paleontol.* 31:1109–1137.

LOWRIE, W., and ALVAREZ, W. 1981. One hundred million years of geomagnetic polarity history. *Geology* 9:392–397.

LUCAS, S., RIGBY, K., JR., and KUES, B., eds. 1981. *Advances in San Juan Basin Paleontology.* Albuquerque: Univ. of New Mexico Press.

LYELL, C. 1833. *Principles of Geology,* Vol. 3.

MACINTYRE, G.T. 1962. *Simpsonictis,* a new genus of viverravine miacid (Mammalia, Carnivora). *Am. Mus. Novit.* 2118:1–7.

MACINTYRE, G.T. 1966. The Miacidae (Mammalia, Carnivora). Part 1. The systematics of *Ictidopappus* and *Protictis. Bull. Am. Mus. Nat. Hist.* 131:115–210.

MARIE, P. 1964. Les facies du Montien (France, Belgique, Hollande). Colloque sur le Paléogène. Bordeaux, 1962. *Fr., Bur. Rech Géol. Minières Mém.* 28:1077–1097.

MARLIÈRE, R. 1964. Le Montien de Mons; état de la question. Colloque sur le Paléogène. Bordeaux, 1962. *Fr., Bur. Rech. Géol. Minières Mém.* 28:875–883.

MATTHEW, W.D., 1937. Paleocene faunas of the San Juan Basin, New Mexico. *Trans. Am. Philos. Soc.,* n.s., 30:1–510.

MATTHEW, W.D. and GRANGER, W. 1925. Fauna and correlation of the Gashato Formation of Mongolia. *Am. Mus. Novit.* 189:1–12.

MATTHEW, W.D., GRANGER, W., and SIMPSON, G.G. 1929. Additions to the fauna of the Gashato Formation of Mongolia. *Am. Mus. Novit.* 376:1–12.

McGREW, P.O. 1963. Environmental significance of sharks in the Shotgun fauna, Paleocene of Wyoming. *Univ. Wyo., Geol. Contrib.* 2:39–41.

McKENNA, M.C. 1975. Fossil mammals and early Eocene North Atlantic land continuity. *Ann. Mo. Bot. Gard.* 62:335–353.

McKENNA, M.C. 1980A. Early history and biogeography of South America's extinct land mammals. In *Evolutionary Biology of the New World Monkeys and Continental Drift,* ed. R.L. Ciochon and A.B. Chiarelli, pp. 43–78. New York: Plenum Press.

McKENNA, M.C. 1980B. Late Cretaceous and early Tertiary vertebrate paleontological reconnaissance, Togwotee Pass area, northwestern Wyoming. In *Aspects of Vertebrate History: Essays in Honor of Edwin Harris Colbert,* ed. L.L. Jacobs, Flagstaff Mus., pp. 321–343. Flagstaff: N. Ariz. Press.

McKENNA, M.C. 1981. Early history and biogeography of South American extinct land mammals. In *Evolutionary Biology of the New World Monkeys and Continental Drift,* ed. R.L. Ciochon and A.B. Chiarelli, pp. 43–78. New York: Plenum Press.

MÉGNIEN, C., ed. 1980. Synthèse géologique du Bassin de Paris, stratigraphie et paléontologie, vol. 1. *Fr., Bur. Rech. Géol. Minières, Mém.* 101:466 pp.

NORTON, N.J. and HALL, J.W. 1969. Palynology of the upper Cretaceous and lower Tertiary in the type locality of the Hell Creek Formation, Montana, USA. *Palaeontogr., Sect B,* 125:1–64.

ORBIGNY, A. D'. 1852. *Cours élémentaire de paléontologie et de géologie stratigraphique.* Vol. 2, no. 2. Paris: V. Masson. 847 pp.

ORIEL, S.S. 1962. Main body of Wasatch Formation near La Barge, Wyoming. *Bull. Am. Assoc. Pet. Geol.* 46:2161–2173.

PABST, M. 1968. The flora of the Chuckanut Formation of northwestern Washington. *Univ. Calif., Publ. Geol. Sci.* 76:1–60.

PASCUAL, R., VUCETICH, M.G., FERNANDEZ, J. 1978. Los primeros mamiferos (Notoungulata, Henricosborniidae) de la Formacion Mealla (Grupo Salta, Subgrupo Santa Barbara). Sus implicancias filogeneticas, taxonomicas y cronologicas. *Ameghinicana.* 15:366–390.

PATTERSON, B. 1939. New Pantodonta and Dinocerata from the upper Paleocene of western Colorado. *Fieldiana, Geol.* 6:351–384.

PATTERSON, B., and McGREW, P.O. 1962. A new arctocyonid from the Paleocene of Wyoming. *Breviora* 174:1–10.

PAULA COUTO, C., DE. 1950. Novas elementos no fauna de Saõ Jose de Itaborai. *Mus. Nac. (Rio de Janeiro), Geol.* 12:1–6.

PAULA COUTO, C. DE. 1952A. Fossil mammals from the beginning of the Cenozoic in Brazil. *Bull. Am. Mus. Nat. Hist.* 99:359–394.

PAULA COUTO, C. DE. 1952B. Fossil mammals from the beginning of the Cenozoic in Brazil: Polydolopidae and Borhyaenidae. *Am. Mus. Novit.* 1559:1–27.

PAULA COUTO, C. DE. 1952C. Fossil mammals from the beginning of the Cenozoic in Brazil: Didelphidae. *Am. Mus. Novit.* 1567:1–26.

PAULA COUTO, C. DE. 1952D. Fossil mammals from the beginning of the Cenozoic in Brazil: Notoungulata. *Am. Mus. Novit.* 1568:1–16.

PAULA COUTO, C. DE. 1954. On a notostylopid from the Paleocene of Itaborai, Brazil. *Am. Mus. Novit.* 1693:1–5.

PITON, L.-E. 1940. Paléontologie du gisement éocène de Menat (Puy-de-Dôme) (flore et fauna). *Soc. Hist. Nat. Auvergne Mém.* 1:1 103.

POMEROL, C. 1969. Rapport sur les discussions au sujet de la limite Paléocène-Eocène. Colloque sur l'Eocène. *Fr., Bur. Rech. Géol. Minières Mém.* 69:447–450.

POMEROL, C. 1977. La limite Paléocène-Eocène en Europe occidentale. *C. R. Soc. Géol. France* 4:199–202.

POMEROL, C. 1978. Critical review of isotopic dates in relation to Paleogene stratotypes. In *Contribution to the Geologic Time Scale,* ed. G.V. Cohee, M.F. Glaesner, and H.D. Hedberg, pp. 235–245. Studies in Geology, no. 6. Tulsa: Am. Assoc. Pet. Geol.

POSARYSKA, K. 1965. Foraminifera and biostratigraphy of the Danian and Montian in Poland. *Pol. Akad. Nauk., Inst. Geogr.* 14:1–156.

QIU, Z., LI, C., TANG, Y., XU, Q., YAN, D., and ZHANG, H. 1977. Continental Paleocene stratigraphy of Qianshan and Xuancheng Basins, Anhui. *Vertebr. PalAsiatica* 15:85–93.

RIGBY, J.K. 1980. Swain Quarry of the Fort Union Formation, Middle Paleocene (Torrejonian), Carbon County, Wyoming: Geologic setting and mammalian fauna. *Evol. Monogr.* 3:1–178.

RIGBY, K., JR. 1981. A skeleton of *Gillisonchus gillianus* (Mammalia: Condylarthra) from the early Paleocene (Puercan) Ojo Alamo Sandstone, San Juan Basin, New Mexico, with comments on the local stratigraphy of Betonnie Tsosie. In *Advances in San Juan Basin Paleontology,* ed. S. Lucas, K. Rigby, Jr., and B. Kues, pp. 89–126. Albuquerque: Univ. New Mexico Press.

ROSE, K.D. 1975A. *Elpidophorus,* the earliest dermopteran (Dermoptera, Plagiomenidae). *J. Mammal.* 56:676–679.

ROSE, K.D. 1975B. The Carpolestidae, early Tertiary primates from North America. *Bull. Mus. Comp. Zool.* (Harv. Univ.) 147:1–156.

ROSE, K.D. 1979. A new Paleocene palaeanodont and the origin of the Metacheiromyidae (Mammalia). *Breviora* 455:1–14.

ROSE, K.D. 1980. Clarkforkian land-mammal age: Revised definition, zonation, and tentative intercontinental correlations. *Science* 208:744–746.

ROSE, K.D. 1981. The Clarkforkian land-mammal age and mammalian faunal composition across the Paleocene-Eocene boundary. *Univ. Mich., Pap. Paleontol.* 26:1–196.

ROSENKRANTZ, A. 1940. Faunaen i Cerithium Kalken og det haerdnede Skrivekridt i Stevns Klint. *Dansk. Geol. Foren., Meddel.* 9:509–514.

ROUVILLOIS, A. 1960. Le Thanétien du bassin de Paris. *Mus. Natl. Hist. Nat. Mém.*, n.s., sér. C, 8:1–151.

RUSSELL, D.E. 1962. Essai de reconstitution de la vie Paléocène au Mont de Berru. *Bull. Mus. Natl. Hist. Nat.* (2) 34:101–106.

RUSSELL, D.E. 1964. Les Mammifères paléocènes d'Europe. *Mus. Natl. Hist. Nat. Mém.*, n.s., sér. C, 13:1–324.

RUSSELL, D.E. 1967A. Le Paléocène continental d'Amérique du Nord. *Mus. Natl. Hist. Nat. Mém.*, n.s., sér. C, Part 2, 16:1–99.

RUSSELL, D.E. 1967B. Sur *Menatotherium* et l'âge paléocène du gisement de Menat (Puy-de-Dôme). Colloq. Int., *Fr., Cent. Natl. Rech. Sci.* 163:483–493.

RUSSELL, D.E. 1968. Succession, en Europe, des faunes mammaliennes au début du Tertiaire. *Fr., Bur. Rech. Géol. Minières Mém.* 58:291–296.

RUSSEL, D.E., BONDE, N., BONÉ, E., DE BROIN, F., BRUNET, M., BUFFETAUT, E., CORDY, J.M., CROCHET, J.-Y., DINEUR, H., ESTES, R., GINSBURG, L., GODINOT, M., GROESSENS, M.C., GIGASE, P., HARRISON, C.J.O., HARTENBERGER, J.-L., HOCH, E., HOOKER, J.J., INSOLE, A.N., LANGE-BADRÉ, B., LOUIS, P., MOODY, R., RAGE, J.-C., REMY, J., ROTHAUSEN, K., SIGÉ, B., SIGOGNEAU-RUSSELL, D., SPRINGHORN, R., SUDRE, J., TOBIEN, H., VIANEY-LIAUD, M., and WALKER, C.A. 1982. Tetrapods of the Tertiary Basin of Northwest Europe. Geol. Jb., Abt. A, 60:5–77.

RUSSELL, D.E., LOUIS, P., and POIRIER, M. 1966. Gisements nouveaux de la faune cernaysienne (Mammifères paléocènes de France). *Bull. Soc. Géol. Fr.* (7) 8:845–856.

RUSSELL, L.S. 1929. Paleocene vertebrates from Alberta. *Am. J. Sci.* 17:162–178.

RUSSELL, L.S. 1967. Paleontology of the Swan Hills area, north-central Alberta. *R. Ont. Mus., Life Sci., Contrib.* 71:1–31.

SAPORTA, G. DE. 1868. Prodrome d'une flore fossile des travertins anciens de Sézanne. *Soc. Géol. Fr. Mém.* (2) 8:289–436.

SAVAGE, D.E. 1971. The Sparnacian-Wasatchian mammalian fauna, Early Eocene of Europe and North America. *Hess. Landesamt. Bodenforsch., Abh.* 60:154–158.

SAVAGE, D.E., and RUSSELL, D.E. 1977. Comments on mammalian paleontologic stratigraphy and geochronology; Eocene stages and mammal ages of Europe and North America. *Geobios, Mém. Spéc.* 1:47–57.

SCHIEBOUT, J.A. 1974. Vertebrate paleontology and paleoecology of Paleocene Black Peaks Formation, Big Bend National Park, Texas. *Bull. Tex. Mem. Mus.* 24:1–88.

SCHIMPER, W.P. 1874. *Traité de Paléontologie Végétale*, vol. 3. Paris: J.B. Baillière. 896 pp.

SCHORN, H. 1971. What is type Paleocene? *Am. J. Sci.* 271:402–409.

SHOEMAKER, R.E. 1966. Fossil leaves of the Hell Creek and Tullock Formations of eastern Montana. *Palaeontogr.*, Sect. B, 119:54–75.

SIGOGNEAU-RUSSELL, D. 1975. Sur la distinction des genres *Champsosaurus* et *Simoedosaurus* (Reptilia, Choristodera) et leur presence simultanée dans le Paléocène français. *C. R. Acad. Sci.*, Paris, *Sér. D*, 281:1219–1221.

SIMONS, E.L. 1960. The Paleocene Pantodonta. *Trans. Am. Philos. Soc.*, pt. 6, 50:3–98.

SIMPSON, G.G. 1929. A new Paleocene uintathere and molar evolution in the Amblypoda. *Am. Mus. Novit.* 387:1–9.

SIMPSON, G.G. 1932. A new Paleocene mammal from a deep well in Louisiana. *Proc. U.S. Natl. Mus.* 83:1–4.

SIMPSON, G.G. 1937. The Fort Union of the Crazy Mountain Field, Montana and its mammalian faunas. *Bull. U.S. Natl. Mus.* 169:1–287.

SIMPSON, G.G. 1940. Review of the mammal-bearing Tertiary of South America. *Proc. Am. Philos. Soc.* 83:649–709.

SIMPSON, G.G. 1948A. The beginning of the age of mammals in South America. *Bull. Am. Mus. Nat. Hist.* 91:1–232.

SIMPSON, G.G. 1948B. The Eocene of the San Juan Basin, New Mexico. Parts 1 and 2. *Am. J. Sci.* 246:257–282, 373–385.

SIMPSON, G.G. 1950. History of the fauna of Latin America. *Am. Sci.* 38:361–389.

SIMPSON, G.G. 1955. The Phenacolemuridae, new family of early primates. *Bull. Am. Mus. Nat. Hist.* 105:411–442.

SIMPSON, G.G. 1959. Fossil mammals from the type area of the Puerco and Nacimiento strata, Paleocene of New Mexico. *Am. Mus. Novit.* 1957:1–22.

SIMPSON, G.G. 1967. The beginning of the age of mammals in South America. Part 2. *Bull. Am. Mus. Nat. Hist.* 137:1–259.

SINCLAIR, W.J., and GRANGER, W. 1912. Notes on the Tertiary deposits of the Bighorn Basin. *Bull. Am. Mus. Nat. Hist.* 31:57–67.

SLOAN, R.E. 1970. Cretaceous and Paleocene terrestrial communities of western North America. *Proc. North Am. Paleontol. Conv.* 1969, E, pp. 427–453.

SLOAN, R.E. 1981. Systematics of Paleocene multituberculates from the San Juan Basin, New Mexico. In *Advances in San Juan Basin Paleontology*, ed. S. Lucas, K. Rigby, Jr., and B. Kues, pp. 127–160. Albuquerque: Univ. New Mexico Press.

SULIMSKI, A. 1969. Paleocene genus *Pseudictops* Matthew, Granger and Simpson, 1929 (Mammalia) and its revision. *Pol. Akad. Nauk., Inst. Geogr.* 19:101–132.

SZALAY, F.S. 1969. Mixodectidae, Microsyopidae, and the insectivore-primate transition. *Bull. Am. Mus. Nat. Hist.* 140:193–330.

SZALAY, F.S. 1973. New Paleocene primates and a diagnosis of the new suborder Paromomyiïormes. *Folia. Primatol.* 19:73–87.

SZALAY, F.S. 1974. A new species and genus of early Eocene primate from North America. *Folia. Primatol.* 22:243–250.

SZALAY, F.S., and McKENNA, M.C. 1971. Beginning of the age of mammals in Asia: The late Paleocene Gashato fauna, Mongolia. *Bull. Am. Mus. Nat. Hist.* 144:271–317.

TAYLOR, L. 1981. The Kutz Canyon local fauna, Torrejonian

(middle Paleocene) of the San Juan Basin, New Mexico. In *Advances in San Juan Basin Paleontology,* ed. S. Lucas, K. Rigby, Jr., and B. Kues, pp. 242–263. Albuquerque: Univ. New Mexico Press.

THALER, L. 1964. Sur l'utilisation des Mammifères dans la zonation de Paléogène de France. Colloque sur le Paléogène. *Fr., Bur. Rech. Géol. Minières Mém.* 28:985–989.

THALER, L. 1977. Etat des recherches sur la faune de Mammifères du Montien de Hainin (Belgique). *Géobios., Mém. Spéc.,* 1:57–58.

TOMIDA, Y., and BUTLER, R.F. 1980. Dragonian mammals and Paleocene magnetic polarity stratigraphy, North Horn Formation, central Utah. *Am. J. Sci.* 280:787–811.

TONG, Y. 1978. The late Paleocene Taizicun fauna from the Turfan Basin. Acad. Sinica, I.V.P.P., *Mem.* 13:82–101.

TONG, Y. 1979. A late Paleocene primate from South China. *Vertebr. PalAsiatica* 17:65–70.

TONG, Y. 1979B. The new materials of archaeolambdids from South Jiangxi. In *Mesozoic and Cenozoic Red Beds of South China,* pp. 377–381. Acad. Sinica, I.V.P.P. and Nanjing Inst. Geol. and Paleont. Beijing, China. Science Press.

TONG, Y., and WANG, J. 1980. Subdivision of the Upper Cretaceous and Lower Tertiary of the Tantou Basin, the Lushi Basin and the Lingbao Basin of western Henan. *Vertebr. PalAsiatica* 18:21–27.

TSCHUDY, R.H. 1969. Palynological correlation and provincialism in the Late Cretaceous and early Tertiary of the Raton Basin. *J. Paleontol.* 43:899–900.

TSENTAS, C. 1981. Mammalian biostratigraphy of the middle Paleocene (Torrejonian) strata of the San Juan Basin: Notes on Torreon Wash and the stratotype of the *Pantolambda* and *Deltatherium* faunal "zones." In *Advances in San Juan Basin Paleontology,* ed. S. Lucas, K. Rigby, Jr., and B. Kues, pp. 264–292. Albuquerque: Univ. New Mexico Press.

VAN HOUTEN, F.B. 1944. Stratigraphy of the Willwood and Tatman Formations in northwestern Wyoming. *Bull. Geol. Soc. Am.* 55:165–210.

VAN VALEN, L. 1960. Deltatheridia, a new order of mammals. *Bull. Am. Mus. Nat. Hist.* 132:1–126.

VAN VALEN, L. 1967. New Paleocene insectivores and insectivore classification. *Bull. Am. Mus. Nat. Hist.* 135:221–284.

VAN VALEN, L. 1978A. The beginning of the age of mammals. *Evol. Theory* 4:45–80.

VAN VALEN, L., and SLOAN, R.E. 1965. The earliest primates. *Science* 150:743–745.

VIANEY-LIAUD, M. 1979. Les Mammifères montiens de Hainin (Paléocéne moyen de Belgique). Part 1. Multitubercules. *Palaeovertebr.* 4:117–139.

WATELET, A. 1866. *Description des Plantes Fossiles du Bassin de Paris.* Paris: J.B. Baillière. 264 pp.

WEAVER, D.W. 1969. The limits of Lyellian series and epochs. Colloque sur l'Eocéne. *F., Bur. Rech. Géol. Minières Mém.* 69:283–286.

WEIGELT, J. 1939. Die Aufdeckung der bisher ältesten tertiären Säugetierfauna Deutschlands. *Nova Acta Leopold.* (2) 7:515–529.

WEIGELT, J. 1940. Die erste paläozäne Säugetierfauna Deutschlands. *Forsch. Fortschr.* 16:8–10.

WEIGELT, J. 1942. Paläontologie als Heuristik. Methodisches zur Entdeckung der ersten paläocänen Säugetierfauna. *Jenaische Zeit. Naturw.* 75:243–278.

WEIGELT, J. 1947. Biometrische Studien an paläocänen Säugetieren und ihre Bedeutung fur die Beurteilung des Evolutionsgeschehens. *Forsch. Fortschr.* 21–23:88–89.

WILMARTH, G. 1925. Geologic time classification of United States Geological Survey compared with other classifications. *Bull. U.S. Geol. Surv.* 769:1–138.

WILSON, R.W. 1956A. The condylarth genus *Ellipsodon. Univ. Kans., Mus. Nat. Hist. Publ.* 9:105–116.

WILSON, R.W. 1956B. A new multituberculate from the Paleocene Torrejon fauna of New Mexico. *Trans. Kans. Acad. Sci.* 59:76–84.

WOLFE, J.S., and HOPKINS, D.E. 1967. Climatic changes recorded by Tertiary land floras in northwestern North America. In *Tertiary Correlations and Climatic Changes in the Pacific,* pp. 67–76. Pac. Sci. Congr., 11th.

WOOD, H.E., CHANEY, R.W., CLARK, J., COLBERT, E.H., JEPSEN, G.L., REESIDE, J.B. JR., and STOCK, C. 1941. Nomenclature and correlation of the North American continental Tertiary. *Bull. Geol. Soc. Am.* 52:1–48.

WOOD, R.C. 1967. A review of the Clark Fork vertebrate fauna. *Breviora* 257:1–30.

XU, Q. 1976A. New materials of Anagalidae from the Paleocene of Anhui (A). *Vertebr. PalAsiatica* 14:174–184.

XU, Q. 1976B. New materials of Anagalidae from the Paleocene of Anhui (B). *Vertebr. PalAsiatica* 14:242–251.

YAN D., and TANG, Y. 1976. Mesonychids from the Paleocene of Anhui. *Vertebr. PalAsiatica* 14:252–258.

ZHANG, Y. 1978. Two new genera of condylarthran phenacolophids from the Paleocene of Nanxiong Basin, Guangdong. *Vertebr. PalAsiatica* 16:267–274.

ZHANG, Y. 1979. A new genus of phenacolophids. In *Mesozoic and Cenozoic Red Beds of South China,* pp. 373–376. Acad. Sinica, I.V.P.P. and Nanjing Inst. Geol. and Paleont. Beijing, China. Science Press.

ZHANG, Y. 1980. A new tillodont-like mammal from the Paleocene of Nanxiong Basin, Guangdong. *Vertebr. PalAsiatica* 18:126–130.

ZHANG, Y., ZHENG, J., and DING, S. 1979. An account of Condylarthra from the Paleocene at Jiangxi Province. In *Mesozoic and Cenozoic Red Beds of South China,* pp. 382–386. Acad. Sinica, I.V.P.P. and Nanjing Inst. Geol. and Paleont. Beijing, China. Science Press.

CHAPTER FOUR

Eocene Mammalian Faunas

(*Mammals About 54 to 35 Million Years Ago*)

CONTENT OF THE EOCENE; PARIS AND LONDON BASINS

The Eocene, one of the most thoroughly studied series-epochs, is represented by thick successions of fossiliferous marine and nonmarine strata on all continents and on many continent-type islands. Radiometric dates now available indicate that Eocene was an interval of Earth's history extending about 54–35 million years before the present.

Sir Charles Lyell, in his *Principles of Geology*, Vol. 3, 1833, subdivided the Tertiary "Epoch" into three "periods": Eocene, Miocene, and Pliocene, with Pliocene further subdivided into *Older* and *Newer*. These subdivisions were based on percentage of Linnaean species of living "testacea" in the fossil assemblages that had been collected from marine sedimentary rocks by various workers in Western Europe. Lyell commented on this procedure as follows (p. 50):

It has been more usual for geologists to give tables of characteristic shells; that is to say, of those found in the strata of one period and not common to any other. These typical species are certainly of the first importance, and some of them will be seen figured in the plates, illustrative of the different tertiary eras; but we are anxious, in this work, to place in a clear light, a point of the greatest theoretical interest, which has been often overlooked or controverted, viz., the identity of many living and fossil species, as also the connexion of the zoological remains of deposits formed at successive periods.

In several other places in his publications, Lyell indicated the great significance he attached to the geochronologic continuity of species, for it was by using these, in his opinion, that a comparison between successive intervals was afforded.

The molluscan fossils from the Paris and London basins were the means for defining and recognizing Lyell's Eocene (p. 55):

Eocene period.—The period next antecedent we shall call Eocene, from ἠώς, aurora, and χαινος, recens, because the extremely small proportion of living species contained in these strata, indicates what may be considered the first commencement, or *dawn* of the existing state of animate creation. To this era the formations first called tertiary, of the Paris and London basins, are referrible. . . .

The total number of fossil shells of this period already known, is one thousand two hundred and thirty-eight, of which number forty-two only are living species, being nearly in the proportion of three and a half in one hundred. . . . In the Paris basin alone, 1122 species have been found fossil, of which thirty-eight only are still living.

In the appendix of his third volume (1833), Lyell published Deshayes's list of "testacean" species, composed principally of pelecypods and gastropods but with a few other molluscs and some foraminifers also that had been found as fossils in the Tertiary of Europe. Unfortunately, this list gives only the species that were known from more than one series-epoch and the species that were known to carry on to the present. Only about 210 of the total of 1238 species were given for the Eocene, and the 1000-plus species that Lyell and Deshayes thought to be confined to Eocene can only be surmised from later publications such as Deshayes (1860, 1864, 1866), Cossman (1886–1913). A survey of these later publications indicates that the species Lyell and Deshayes listed, as well as the other species that Deshayes probably recognized in 1833, had been obtained from scattered collecting sites representing what is now known to be the Thanetian (Upper Paleocene), Sparnacian and Cuisian (Lower Eocene), Lutetian (Middle Eocene), and Bartonian in its widest sense (Upper Eocene) stages. Our colleagues who have specialized in studies of the marine invertebrate fossils and stratigraphy of the Paris Basin inform us that it is impossible to ascertain locality information for the species that Deshayes may have identified, and the original Deshayes collection seems to have been scattered into many museum collections.

Lyell never swerved from his principle and method of subdividing the Tertiary by means of percentage of living species of marine molluscs in the fossil aggregates. As late as 1863, in his *Antiquity of Man* (p. 5), he wrote: "But the reader must bear in mind that the terms Eocene, Miocene and Pliocene were originally invented with reference purely to conchological data, and in that sense have always been and are still used by me." Thus the Lyellian subdivisions of Tertiary should be recognized and their names applied only on the basis of percentage of living species of marine molluscs in a bulk aggregate (that is, about 1000 species) of fossils from a bulk section of strata if the resultant geochronology is to be labeled purely Lyellian in its construct. Delineation of a boundary for Eocene, Miocene, Pliocene, and so on is not obtainable by Lyellian method.

Mammalian paleontologists are fortunate that Eocene was typified in the Paris and London basins, for especially in the Paris Basin, the nonmarine sedimentary beds, bearing fossils of mammals, alternate and interfinger with marine strata that produced the "testacean" standard for Eocene. Thus, no tenuous correlation is involved in recognizing *type-Eocene land mammals*. Stratigraphers since the time of Lyell have recognized lower, middle, and upper Eocene molluscan and mammalian assemblages, all from strata in the environs of Paris.

A survey of the many Eocene mammalian faunas of the world is best prefaced by a résumé of the stratigraphy and paleontology of the type Eocene. Figure 4-1 shows the

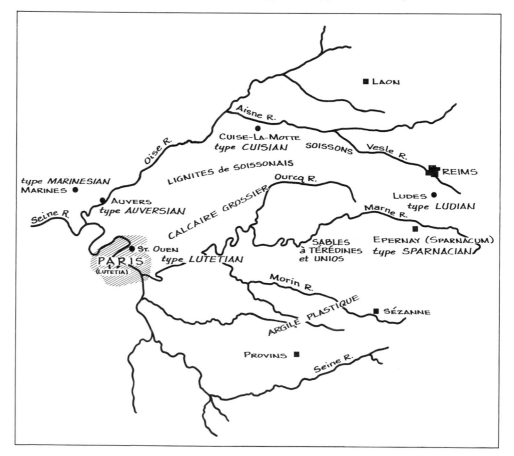

FIGURE 4-1 *Type Localities, Districts, Formations, and Stages in the Eocene of the Paris Basin, France.*

important districts, localities, formations, and stages, with special emphasis on the types. The area shown is the approximate extent of the Ile de France and is also the area of bedrock exposure of Tertiary strata. We follow, with slight modification, the terminology of Blondeau, Feugueur, and Pomerol (1965) for the stages and formations of the Paris Basin.

Eocene Succession of the Paris Basin

Stages	Substages	Representative Formations
Bartonian	Ludian	Marls of Ludes with *Pholadomya ludensis,* Paris Gypsum (*) and associated marls
	Marinesian	Marines and St. Ouen (*) Formations
	Auversian	Auvers and Beauchamps Sands (*)
Lutetian		Calcaire Grossier complex (*)
Ypresian	Cuisian	Cuise-Pierrefonds Sands, Belleu Sandstone, Aizy Sands, Sands with Teredines and Unios (*), Laon Clays
	Sparnacian	Sinceny Sands (*), Auteuil Sands, Lignites du Soissonais (*), Argile Plastique (*)

(*) Formations from which fossils of land mammals have been obtained.

Paris Basin

Note that land mammals have been found in all the recognized substages. So close is this stratigraphic tie-in that there is no need for a separate mammal-age chronology in the Paris Basin. It is only when correlations are made with other European districts, especially the nonmarine districts, that the mammal ages are useful.

The Paris Basin was the scene of relatively continuous deposition throughout the Eocene (except for the deeper-water marine lithotope), but during certain intervals there was drastic change in sedimentary environments from district to district; lithotopes of neritic-littoral, lagoonal, estuarine, paludal, lacustrine, and fluvial environments frequently grade into each other within the Basin. When the physical stratigraphy is decipherable on a regional scale, as within the Lutetian Stage, there is excellent opportunity to compare the geochronologies based on divers classes of organisms—especially charophytes, ostracods, benthonic and planktonic foraminifers, molluscs, and mammals.

The lowest Eocene strata are assigned to the Sparnacian Stage of Dollfus (1890; see also Dollfus, 1905), which we use, following Blondeau et al. (1965). Sparnacian was based on lignitic strata—the Lignites du Soissonais—exposed below the yellow sands of the Cuisian Stage of Dollfus. The Roman name for Epernay, France, was Sparnacum, and the Sparnacian is typified in exposures of the Lignites du Soissonais at Mt. Bernon, a hill on the southeast edge of Epernay. To the southwest of Epernay, the Sparnacian is represented by the Argile Plastique, claystone beds that are quarried for ceramics. The Lignites du Soissonais and the Argile Plastique have produced fossils of land vertebrates. In the district of Epernay these Sparnacian land vertebrates, comprising fishes, amphibians, reptiles, birds, and mammals, have been found in relative abundance in lenticular bodies of shelly argillaceous sand that may be termed *faluns*. Each falun may attain a thickness of several meters. Each is characterized by a mixture of complete and pulverized brackish-water molluscs and commingled thick-walled calcareous tubes formed by *Teredina* (Pelecypoda) and small calcareous concretions. All these components are held together in a matrix of clay and sand. We believe that these lenses were deposited in estuaries. Mutigny, Pourcy, and Avenay (see Figure 4-7) are good examples of falun. The Cernay "conglomérat" (late Paleocene), previously described, is also a sandy to argillaceous falun.

During the Cuisian subage, a great blanket of sand, locally a sandy lime mud, was deposited in neritic to brackish-water zones. The type of the Cuisian Stage, the sands of Cuise-la-Motte, is in the richly fossiliferous shallow-water marine facies of the Cuisian. Recently, however, a land mammal has been reported from this facies (Crochet, 1978). The renowned Sables à Térédines et Unios of the Epernay district, overlying type Sparnacian and representing a brackish-water facies, have yielded most of the land vertebrates typifying the Cuisian. The Sables à Térédines et Unios Formation resembles the Cernaysian and Sparnacian faluns in its abundant brackish-water molluscs and mollusc fragments associated with teredine tubes in a sandy matrix, but the numerous complete unionid shells and lesser lenticularity of the formation indicate a genesis different from that of the faluns. We suggest that the Sables were deposited in a well-agitated and well-aereated lagoon or in a restricted shallow-water embayment of the sea and that such areas were subjected to frequent flooding by fresh water. The abundance of herbivorous, carnivorous, and insectivorous mammals in these faluns and sand blankets of the early Eocene tells us that forested lands and savannas were nearby.

The Middle Eocene, Lutetian land-vertebrate localities of the Paris Basin are usually the product of infrequent burial of a carcass or the deposit of part of a carcass in a sandy to limy mud in the neritic zone. However, concentration of fossil bones and teeth is known from this segment of the Paris Basin succession at La Défense, a suburb on the west side of Paris. Another has been discovered (still unpublished) in the Department of Aisne. The freshwater limestone facies of the southeastern part of the Paris Basin merits further intensive prospecting for Lutetian faunas.

The lower part of the Upper Eocene (Early Bartonian) in the Paris Basin has yielded some excellent but generally isolated land-mammal specimens, for example, the type of *Anchilophus desmaresti* from the lacustrine St. Ouen Limestone. Arcis-le-Ponsart, at the base of the limestone, and Grisolles, intercalated medially, have produced small but important assemblages of land mammals, although the best vertebrate faunas of this age are in other areas.

The uppermost Eocene (Ludian = Priabonian), in the Paris Gypsum and associated marls, represents a singular environment of accumulation of fossil mammals and birds. The milieu of these articulated skeletons from the ''Gypse de Montmartre,'' made known to the world by the baron Cuvier and his followers, has no counterpart in the remainder of the record of fossil higher vertebrates. Evidently complete bloated? carcasses floated into these highly mineralized waters and remained intact during burial, through an episode of low-energy deposition of fine clastic sediment and chemical precipitates.

London Basin

The lowest part of the Eocene succession is very similar in both the London area and the Paris Basin. Dollfus noted this in 1880. As you can see in Figure 4-2, the Reading beds of the western areas are essentially a mass of clay not unlike the Argile Plastique. Eastward, the Reading grades into the more heterogeneous Woolwich, which contains marine as well as nonmarine beds and fossils. Except for lack of abundant lignite beds, the Woolwich is the counterpart of the Lignites du Soissonais. Localized lenses of coarser clastic sediments in the base of the Woolwich and in the base of the overlying London Clay suggest estuarine and lagoonal sites of deposition, also comparable to the strata of the Paris Basin. The Tertiary stratigraphy of the British Islands has been described by Curry et al. (1978). And yet more recently, King (1981) has synthesized the stratigraphy and correlations of the London Clay and associated strata. His conclusions represent the state of knowledge and the consensus (evidently) regarding aging and correlation of late Paleocene and early Eocene units in the London Basin, as based on the contained planktonic paleoorganisms. The main result of the approach of the planktonic (deeper-water marine stratigraphy) workers is to place the Paleocene-Eocene boundary of Western Europe above the Sparnacian Stage-Age and its approximate correlatives—in brief, about one stage-age higher and later than we have it.

Davis and Elliot (1957) gave us a succinct account of the fossil content of the London Clay: small wood fragments and large amounts of fruits and seeds . . . *Teredo*-bored wood, gastropods, pelecypods, *Nautilus*, . . . some brachiopods, crabs, and rare corals and echinoderms, . . . abundant fish, crocodile scutes, turtles, some *Coryphodon*, and birds. Only scattered fossils of land mammals have been

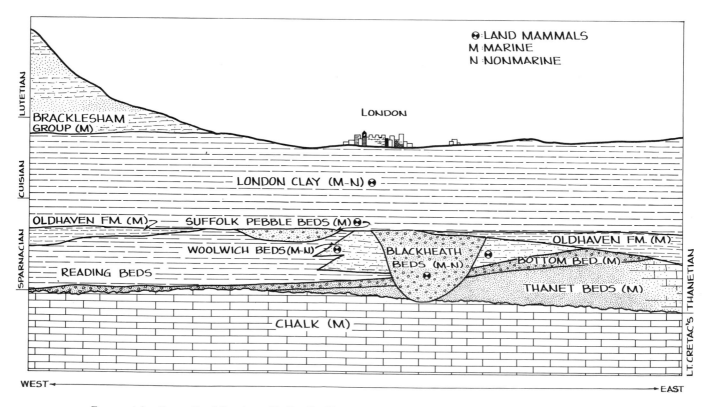

FIGURE 4-2 *Generalized Stratigraphic Profile Showing Paleocene and Eocene Units of the London Basin.*

found: the types of the genotypic species of *Hyracotherium* Owen, *Coryphodon* Owen, and *Platychoerops* Charlesworth.

SPARNACIAN (=Neustrium of Fahlbusch, 1976)

As mentioned previously, we use Dollfus's stage-age name Sparnacian because of the direct tie-in between land-mammal faunas and the type of his stage at Epernay. Stratigraphers who work with planktonic microfossils in the deeper-water marine facies do not like Sparnacian as a stage name because it was based primarily on lithologic attributes (not unlike many other stage names that we have inherited) and because the type strata cannot be characterized "biostratigraphically" by planktonic fossils. Also, we remind the reader that this latter group of workers consistently places Sparnacian and its equivalents in the Paleocene. See King (1981).

A characterizing aggregate of mammalian genera for the Sparnacian of France, Belgium, and England is *Parectypodus, Peradectes, Amphiperatherium, Apatemys, Palaeosinopa, Archaeonycteris, Platychoerops, Teilhardina, Pelycodus, Phenacolemur, Esthonyx, Paramys, Pachyaena, Oxyaena, Palaeonictis,* cf. *Viverravus, Phenacodus, Paschatherium, Coryphodon, Hyracotherium, Lophiaspis,* cf. *Bunophorus* or *Protodichobune.* Readers specializing in studies of North American Early Eocene mammals will realize immediately that this array of genera is almost identical with a contemporaneous array in North America. (We shall examine this phenomenon later in the chapter.)

The most complete local fauna representing the Sparnacian, which fortunately is in the Lignites du Soissonais close to Epernay, is adjacent to the village of Mutigny. The mammalian component of the Mutigny local fauna comprises the following taxa. (The asterisk indicates genera that are also found in early Eocene of North America.)

Mutigny Local Fauna

MULTITUBERCULATA
  Neoplagiaulacidae
    Parectypodus *
    Ectypodus *
MARSUPIALIA
  Didelphidae
    Peradectes louisi Crochet, 1979*
    P. mutignienensis Crochet, 1979
    P. russelli Crochet, 1979
    Amphiperatherium dormaalense (Quinet, 1964)
    Peratherium constans Teilhard, 1927A

(Note that asterisk indicates that the genus is also found in North America.)

CIMOLESTA
  Palaeoryctidae
    cf. *Didelphodus* *
  Inc. sed.
    cf. *Hyracolestes*
PANTOLESTA
  Pantolestidae
    Palaeosinopa *
APATOTHERIA
  Apatemyidae
    Apatemys mutiniacus Russell, Godinot, Louis, and Savage, 1979*
    A. sigogneaui Russell, Godinot, Louis, and Savage, 1979
    Heterohyus sp.   —Russell, Godinot, Louis, and Savage (1979)
CREODONTA
  Hyaenodontidae
    cf. *Prototomus* *
CARNIVORA
  Miacidae
    cf. *Miacis* *
    cf. *Viverravus* *
INSECTIVORA
  Erinaceoidea
    Macrocranion *
  Soricoidea
    cf. *Leptacodon* *
    cf. *Nyctitherium* *
DERMOPTERA
  Placentidentidae
    Placentidens lotus Russell, Louis, and Savage, 1973
CHIROPTERA
  Icaronycterididae
    Icaronycteris? menui Russell, Louis, and Savage, 1973
  Palaeochiropterygidae
    Archaeonycteris brailloni Russell, Louis, and Savage, 1973
  Fam. indet.
    Ageina tobieni Russell, Louis, and Savage, 1973
PRIMATES
  Plesiadapidae
    Platychoerops daubrei (Lemoine, 1880)
  Paromomyidae
    Phenacolemur fuscus Russell, Louis, and Savage, 1967
  Adapidae
    Pelycodus cf. *eppsi* (Cooper, 1932)[1]
PRIMATES?
  Uintasoriciidae?
    Berruvius cf. *lasseroni* Russell, 1964
ARCTOCYONIA
  Arctocyonidae, gen. indet.
  Paroxyclaenidae
    cf. *Paroxyclaenus*
TILLODONTIA
  Esthonychidae
    Esthonyx
ARTIODACTYLA
  Dichobunidae
    Protodichobune?

CONDYLARTHRA
  Hyopsodontidae
    Paschatherium
    *Hyopsodus**
  Phenacodontidae
    Phenacodus cf. *teilhardi* Simpson, 1929*
PERISSODACTYLA
  Equidae
    *Hyracotherium**
  Chalicotheriidae
    Lophiaspis cf. *maurettei* Depéret, 1907
RODENTIA
  Ischyromyoidea
    Paramys ageiensis Michaux, 1964B*
    P. woodi Michaux, 1964B
    Pseudoparamys teilhardi Michaux, 1964B
    Meldimys louisi (Michaux, 1964B)   —Michaux
      (1968)
    Microparamys nanus (Teilhard, 1927B)   —Michaux
      (1968)*

Note that 21 of the 36 identified genera in this local fauna (58 percent) are also found in the early Eocene (Wasatchian) fauna of North America. Before considering this phenomenon, however, let us review the composite Sparnacian mammalian fauna of western Europe.

Composite Sparnacian Mammalian Fauna

MULTITUBERCULATA
  Neoplagiaulacidae
    *Parcetypodus**
    *Ectypodus**
    E. childei (Kuhne, 1969A)
    E. aff. *childei*   —Godinot (1981)
MARSUPIALIA
  Didelphidae
    Peratherium constans Teilhard, 1927A
    P. matronense Crochet, 1979
    Peradectes louisi Crochet, 1979*
    P. russelli Crochet, 1979
    P. mutigniensis Crochet, 1979
    Amphiperatherium brabantense Crochet, 1979
    A. aff. *brabantense*   —Godinot (1981)
    A. maximum Crochet, 1979
    A. bourdellense Crochet, 1979
    A. goethei Crochet, 1979
LEPTICTIDA
  Leptictidae
    Adunator   —Godinot (1978)
CIMOLESTA
  Palaeoryctidae
    Didelphodus cf. *absarokae* (Cope, 1881)   —Godinot
      (1981)*
    cf. *D.*
  Inc. sed.
    cf. *Hyracolestes*

PANTODONTA
  Coryphodontidae
    Coryphodon eocaenus Owen, 1846*
    C. oweni Hebert, 1856A
PANTOLESTA
  Pantolestidae
    *Palaeosinopa**
    P. osborni (Lemoine, 1891)   (Cuisian?)
APATOTHERIA
  Apatemyidae
    Eochiromys landenensis Teilhard, 1927
    Apatemys mutiniacus D. Russell, Godinot, Louis, and Savage, 1979*
    A. sigogneaui D. Russell, Godinot, Louis, and Savage, 1979
    A. teilhardi D. Russell, Godinot, Louis, and Savage, 1979
    Heterohyus sp.
CREODONTA
  Oxyaenidae
    *Oxyaena**
    Palaeonictis gigantea Blainville, 1842*
  Hyaenodontidae
    Prototomus cf. *mordax* (Matthew, 1915)*   —Godinot (1981)
    P. cf. *palaeonictides* (Lemoine, 1880)
    Proviverra eisenmanni Godinot, 1981
CARNIVORA
  Miacidae
    Miacis latouri Quinet, 1966*
    *Didymictis**
    cf. *Vulpavus**
    cf. *Uintacyon**
    cf. *Viverravus**
INSECTIVORA
  Erinaceoidea
    cf. *Macrocranion nitens* (Matthew, 1918)*
    Dormaalius vandebroeki Quinet, 1966
    Neomatronella luciannae D. Russell, Louis, and Savage, 1975*
    "*Adapisorex*" *anglicus* Cooper, 1932
  Soricoidea
    cf. *Leptacodon**
    Nycticonodon casieri Quinet, 1964
    N. caparti Quinet, 1964
    "*Gypsonictops*" *dormaalensis* (Quinet, 1969)
DERMOPTERA
  Placentidentidae
    Placentidens lotus D. Russell, Louis, and Savage, 1973
CHIROPTERA
  Icaronycteridae
    Icaronycteris? menui D. Russell, Louis, and Savage, 1973*
  Palaeochiropterygidae
    Archaeonycteris brailloni D. Russell, Louis, and Savage, 1973
  Fam. indet.
    Ageina tobieni D. Russell, Louis, and Savage, 1973
PRIMATES
  Plesiadapidae
    Plesiadapis russelli Gingerich, 1976
    P. aff. *remensis* Lemoine, 1887
    Platychoerops daubrei (Lemoine, 1880)
  Paromomyidae
    Phenacolemur fuscus D. Russell, Louis, and Savage, 1967*

P. lapparenti D. Russell, Louis, and Savage, 1967

Adapidae

  Pelycodus eppsi (Cooper, 1932)* [1]

  P. savagei Gingerich, 1977 [1]

  cf. *Protoadapis* sp.

  Donrussellia (Omomyidae?)   (See species listed below.)

Omomyidae

  Teilhardina belgica (Teilhard, 1927A)

  Donrussellia gallica (D. Russell, Louis, and Savage, 1967;
    Adapidae?)

  D. louisi (Gingerich, 1977) Adapidae?

  D. russelli (Gingerich, 1977) Adapidae?

  D. provincialis Godinot, 1978; Adapidae?

PRIMATES?

  Uintasoricidae?

  Berruvius cf. *lasseroni* D. Russell, 1964

ARCTOCYONIA

  Arctocyonidae

  Landenodon woutersi Quinet, 1966

TILLODONTIA

  Esthonychidae

  Esthonyx sp. indet.

ARTIODACTYLA

  Dichobunidae

  Diacodexis gazini Godinot, 1978A*

  Protodichobune

ACREODI

  Mesonychidae

  Dissacus filholi (Lemoine, 1880)* (Cuisian?)

  Pachyaena gigantea Osborn and Wortman, 1892*

CONDYLARTHRA

  Hyopsodontidae

  Lessnessina packmani Hooker, 1979 (Periptychidae?)

  Paschatherium dolloi (Teilhard, 1927A)

  P. russelli Godinot, 1978A

  Hyopsodus itinerans Godinot, 1978A*

  H. wardi Hooker, 1979

  Microhyus musculus Teilhard, 1927A

  Phenacodontidae

  Phenacodus teilhardi Simpson, 1929G*

PERISSODACTYLA

  Equidae

  Hyracotherium cuniculus Owen, 1842*

  H. vulpiceps (Owen, 1858 —acc. Hooker, personal com-
    mun., 1981)

  Propachynolophus?

  Chalicotheriidae

  Lophiaspis maurettei Depéret, 1907

RODENTIA

  Ischyromyoidea

  Paramys ageiensis Michaux, 1964B*

  P. woodi Michaux, 1964B

  P. pourcyensis Michaux, 1964B

  Pseudoparamys teilhardi (Wood, 1962)

  Meldimys louisi (Michaux, 1964B)   —Michaux
    (1968)

        —D. Russell et al. (1982)

  Microparamys nanus (Teilhard, 1927A)*

  M. russelli Michaux, 1964B

  M. chandoni Hartenberger, 1971A

  Decticadapis sciuroides Lemoine, 1891 (Cuisian?)

Of the 61 genera of mammals here recognized in the composite Sparnacian fauna, only 6 (about 10 percent) are found also in the preceding Cernaysian fauna. Perhaps another 3 or 4 percent of the Sparnacian genera could have been derived from Cernaysian stocks; the remainder appear to be immigrants. Twenty-two of the Sparnacian genera (about 36 percent) do not appear in later faunas in Europe, but 39 (64 percent) carry into the succeeding Cuisian fauna. All known Sparnacian mammalian families carry into later faunas.

Affinities and Correlations of Sparnacian Mammals

The principal Sparnacian mammal localities now known were coastal and brackish-water habitats (Paris Basin, London Basin, Belgium) or lowland lake-border habitats (Provence) in which peculiar faunal facies might be expected. Although intensive underwater sieving in six of these sites has produced a good small-vertebrate sample, the large mammals are poorly represented. Surprisingly therefore, of the approximate 61 genera of mammals now identified from the Sparnacian, an astounding number of these are known also in the larger, intermontane Wasatchian fauna of North America. Present calculation indicates that about *50 percent* of the Sparnacian mammalian genera are found also in the Wasatchian. This is a significantly higher generic identity between the two continents than what has existed at any other time in the history of mammals.

A comparison with generic identity of mammals during two other episodes of high intercontinental resem-

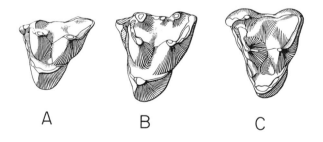

FIGURE 4-3 *Sparnacian Mammals.* <u>Peradectes louisi</u> *Crochet, 1979. Upper molars.* ×12 *approximately. Figures 7, 8, and 9 in Crochet (1980) (Published with permission of the University of Montpellier.)*

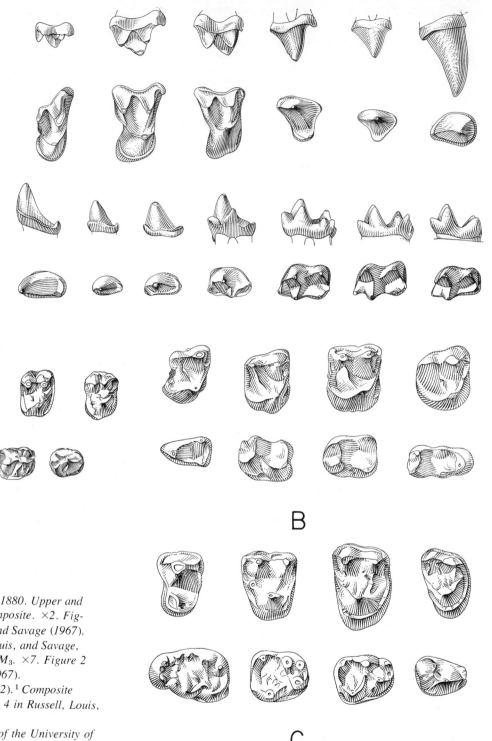

FIGURE 4-4 *Sparnacian Mammals. Icaronycteris? menui Russell, Louis, and Savage, 1973. Upper and lower dentition, composite. ×10 approximately. Figure 1 in Russell, Louis, and Savage (1973). (Published with permission of the University of California Press.)*

FIGURE 4-5 *Sparnacian Mammals.*

A *Platychoerops daubrei Lemoine, 1880. Upper and lower molars and premolars, composite. ×2. Figures 7 and 8 in Russell, Louis, and Savage (1967).*

B *Phenacolemur fuscus Russell, Louis, and Savage, 1967. Composite* P^4-M^3 *and* P_4-M_3. *×7. Figure 2 in Russell, Louis, and Savage (1967).*

C *Pelycodus cf. eppsi (Cooper, 1932).[1] Composite* P^4-M^3 *and* P_4-M_3. *×3⅓. Figure 4 in Russell, Louis, and Savage (1967).*

(All figures published with permission of the University of California Press.)

blance, the Recent and the Pleistocene, shows that about 36 percent of the 72 genera of nonvolant nonmarine mammals of greater Europe, including the Arctic and also the southern extremes of Europe, are known also in Nearctica. Only 30 percent of the 143 genera of nonvolant nonmarine mammals of Nearctica during the Pleistocene are found also in the Pleistocene of greater Europe. It is more difficult to get

comparisons with the Asian part of Palaearctica, but approximately 30 percent of the genera of living nonvolant land mammals of North America are found also in Asia, and the figure for the Pleistocene is about the same. The Sparnacian-Wasatchian generic identity is rivalled perhaps only by the similarity seen between boreal Nearctica and boreal Palaearctica during the Pleistocene. The percentage

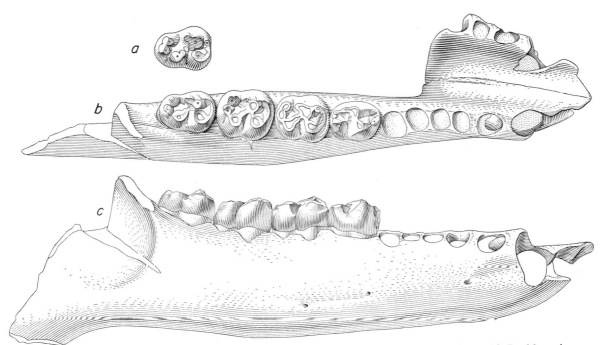

FIGURE 4-6 *Sparnacian Mammals. Phenacodus cf. teilhardi Simpson, 1929G. Lower jaw with P_4–M_3 and an isolated M_3. ×1⅓. Figure 14 in T. Rich (1971).*
(Published with permission of the University of California Press.)

FIGURE 4-7 *Representative Sparnacian Localities of the Paris Basin.*

1	Amy
2	Arey
3	Avenay
4	Boulincourt
5	Bretigny
1	Bussy
2	Canly
1	Canny-sur-Metz
6	Chailvet
7	Condé-en-Brie
1	Cuvilly
8	Fismes (Cuisian?)
9	Guny
10	Marfaux
11	Meudon
12	Mont de Berru, upper
11	Moulineux
1	Muirancourt
3	Mutigny
1	Orvillers
10	Pourcy
2	Remy
7	Saint-Agnan
13	Saint-Sauveur
14	Saron
1	Sermaize
15	Sinceny
7	Try
11	Vaugirard
16	Vauxbuin
2	Villers-sur-Coudun

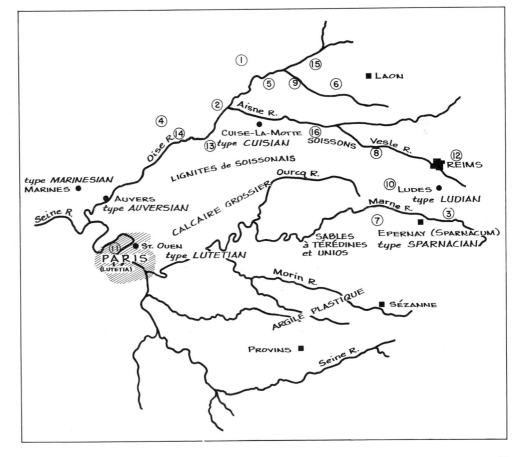

ciphers above cited for Recent and Pleistocene might vary from 5 more to 5 less than our figures when calculated by other workers, but they would be strikingly lower than the Sparnacian-Wasatchian percentage nonetheless.

We believe that perhaps as much as 85 percent of the Sparnacian land-mammal fauna was constituted from a sudden flood of immigrants from outside the European area. This horde of allochthons was a balanced fauna, occupying many niches in the economy of land life. There were tiny, medium-sized, and large forms—herbivores, omnivores, carnivores, insectivores, frugivores—arboreal dwellers as well as ground dwellers, ambulatory, cursorial, and tree-climbing locomotors, and possibly also flyers. Only a few of the recognized land-mammal adaptive types were not represented in the suite of genera that traveled into Europe about 55 million years ago. Saltatorial animals and the habitual swimmers seem not to have been present. Actually, this immigration probably predated mammalian adaptive radiation into such niches. We conclude that there was a broad terrestrial connection between Eurasia and North America and that it had to support a complete spectrum of plant communities and microclimates for this fauna to disperse.

How much of this immigrant fauna came directly across the area that is now North Atlantic and Arctic Basins from Nearctica? How much came from resident stocks in Asian Palaearctica? How much came from North America, using Asia as the dispersal highway?

Kurtén (1966) thought that the dispersal wave came directly across the region of the present North Atlantic via Greenland, Iceland, and England. McKenna (1975A and other papers) reviewed the extensive literature on the geophysics, geology and oceanography of the Northern Hemisphere and concluded that "in addition to early Eocene land continuity in the Greenland–Barents Shelf area, a subaerial route crossing the volcanic Wyville Thompson Ridge from southeastern Greenland to the Faeroes and then to Great Britain and Ireland may also have been possible for a time in the early Tertiary." There may have been some dispersal of land animals from North America to eastern Asia by way of a land connection in the area of the present Bering Strait; however, the presence of some sort of seaway (the Paleogene Sea of Ob), extending from the Arctic to the Tethyan Sea via the Turgai Straits of Kazakhstan has been recorded by geologists of the USSR (see citations by McKenna, 1975A). And this sea, represented by scattered patches of marine Paleogene strata on the geologic map of the USSR, has been visualized by McKenna and others as a barrier to early Eocene land-vertebrate dispersal between eastern Asia and Europe. Kurtén (1966, p. 4) believed that the Turgai Straits may have been intermittent.

The early Eocene land-mammal fauna of eastern Asia is poorly known but appears to combine endemic forms (Eurymylidae, Pseudictopodidae, Didymoconidae) with

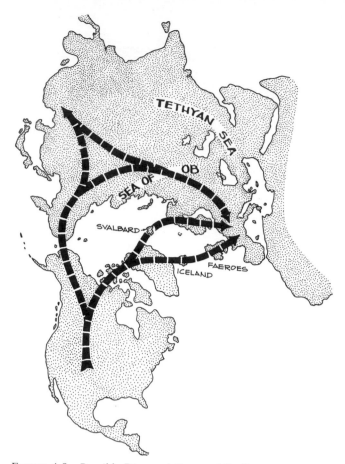

FIGURE 4-8 *Possible Dispersal Routes of Earliest Eocene Land Mammals in the Northern Hemisphere.*

families or genera of possible North American and European origin: Barylambdidae, *Coryphodon*, Omomyidae, Uintatheriidae, *Dissacus*, *Pachyaena*, *Hyopsodus*, *Propachynolophus*, *Homogalax*, "*Microparamys*." Since the genera of the above groups, except *Homogalax*, are also in the European Sparnacian or Cuisian, we think it premature to conclude that there was no possible dispersal between Asia and Europe, and our Figure 4-8 indicates this as one of the possible routes. We remind you that this figure shows the continents and islands in their *present* position as a compromise for quick recognition of geography. Modern geologic evidence indicates that Greenland–North America was appressed to Europe during Sparnacian time. (See our sketch of Eocene biogeography later in the chapter.)

CUISIAN

Cuisian mammals are known from many high-level infrequently used agricultural quarries providing sand, lignite, and fill for the vineyards that are installed on the steep slopes lower on the walls of the Marne River Valley and its tributaries. The Grauves, Mancy, Monthelon, Cuis, and Chavot

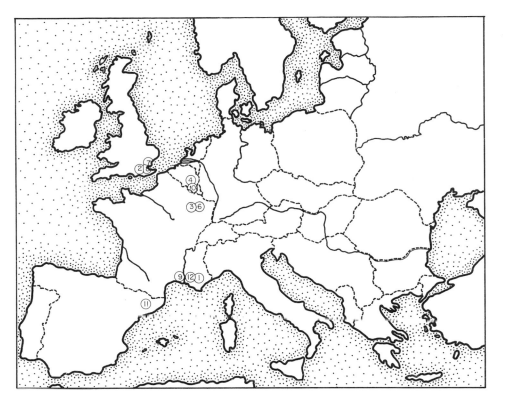

FIGURE 4-9 *Sparnacian Localities of Europe, Exclusive of Paris Basin, and Cuisian Localities of Europe.*

1   Bauduen—Sparnacian
2   Croydon—Dulwich district—Sparnacian
3   Cuise-la-Motte (type Cuisian)—Cuisian
4   Dormaal—Sparnacian
5   Eski Celtek (age?) (Turkey)
6   Grauves-Epernay district: Cuis, Chavot, Monthe-
    lon, Mont Bernon, Mancy, Grauves—Cuisian
7   Harwich—Sparnacian
8   London district: Herne Bay, Abbey Wood, Shep-
    pey, Kyson—Sparnacian and Cuisian

9   Mas de Gimel—Cuisian
10 Northeast France and Belgian localities: Laval
    (Trieu de Leval), Vertain, Vinalmont, Orp le
    Grand—Sparnacian
11 Northeast Spain localities: Montllobar, Barran de
    Forols and Castilgaleu or Noguera Pallaresa in
    Tremp Basin, Les Saleres in Ager Basin, Guëll, La
    Roca, Les Badies and Monteroda in Isabena Basin,
    Escarla in Noguera-Ribagoranza Basin—Cuisian
12 Palette—Rians district—Sparnacian

quarries are concentrated in a district on the south side of the Marne Valley and a few kilometers south of the city of Epernay. Scattered bones, teeth, and jaws of fossil vertebrates have been taken from the yellowish sands of these quarries, the Sables à Térédines et Unios, since the late 1800s. At these sites, the Sables à Térédines et Unios lie on the Lignites du Soissonais (type Sparnacian).

The recent discovery of the Sables à Térédines et Unios fauna near the village of Venteuil, on the north bank of the Marne, demonstrates its intercalation with type-Cuisian Sables de Cuise. The taxa in the Venteuil assemblage, which follow, were identified by Monsieur P. Louis (Cormicy, France) on the basis of his collection.

Venteuil Local Fauna

MULTITUBERCULATA, indet.
MARSUPIALIA
  Didelphidae
    Peratherium

CARNIVORA
  Miacidae, indet.
INSECTIVORA
  Macrocranion cf. *nitens?*
PRIMATES
  Paromomyidae
    Phenacolemur lapparenti Russell, Louis, and Savage, 1967
  Adapidae
    Protoadapis curvicuspidens Lemoine, 1878
ARTIODACTYLA
  Dichobunidae
    Protodichobune oweni Lemoine, 1878
    P. (or *Diacodexis?*)
PERISSODACTYLA
  Equidae
    Propachynolophus gaudryi (Lemoine, 1891)
    P. maldani (Lemoine, 1878)
  Lophiodontidae
    Lophiodon tapirotherium Desmarest, 1822
RODENTIA
  Ischyromyoidea

Ailuravus michauxi Hartenberger, 1975
A. remensis Hartenberger, 1975
cf. *Plesiarctomys?*
Microparamys cf. *mattaueri* Hartenberger, 1971
M. sp.
Meldimys louisi (Michaux, 1964B)
Theridomyidae
"Protadelomys"

A characterizing aggregate of mammalian genera for the Cuisian is *Peradectes, Apatemys, Palaeosinopa, Protoadapis, Prototomus, Teilhardina* (*Donrussellia*), *Propachynolophus, Lophiodon, Hyrachyus, Paramys,* and *Microparamys.* Cuisian shows new and diversified perissodactyls. It is the beginning of the great European Eocene radiation and abundance of *Lophiodon.* The diversity of the Cuisian fauna is poorly known, compared with other mammalian ages of Europe, but with continued intensive collecting, this situation should improve in the next few years.

We list the Cuisian fauna below from the cluster of localities south-adjacent to Epernay.

Composite Cuisian Mammalian Fauna

MULTITUBERCULATA
  Neoplagiaulacidae
    Parectypodus
    Ectypodus
MARSUPIALIA
  Didelphidae
    Peratherium matronense Crochet, 1979
    Peradectes louisi Crochet, 1979
    P. russelli Crochet, 1979
    P. mutigniensis Crochet, 1979
    Amphiperatherium goethei Crochet, 1979
    A. maximum Crochet, 1979
    A. bourdellense Crochet, 1979
CIMOLESTA
  Palaeoryctidae
    cf. *Didelphodus*
PANTOLESTA
  Pantolestidae
    Palaeosinopa osborni (Lemoine, 1891) (Sparnacian?)
APATOTHERIA
  Apatemyidae
    Apatemys mutiniacus Russell, Godinot, Louis, and Savage, 1979
    Heterohyus   —Russell, Godinot, Louis, and Savage (1979)
CREODONTA
  Oxyaenidae
    Oxyaena menui Rich, 1971
  Hyaenodontidae
    Prototomus palaeonictides (Lemoine, 1880A)
    Francotherium lindgreni Rich, 1971
    cf. *Tritemnodon*

CARNIVORA
  Miacidae
    Miacis
    cf. *Vulpavus*
    cf. *Uintacyon*
    cf. *Viverravus*
INSECTIVORA
  Erinaceoidea
    cf. *Macrocranion nitens* Matthew, 1918
  Soricoidea
    cf. *Leptacodon*
DERMOPTERA
  Placentidentidae
    Placentidens?
  Mixodectidae, indet.
CHIROPTERA
  Icaronycterididae
    Icaronycteris? menui Russell, Louis, and Savage, 1973
  Palaeochiropterygidae
    Palaeochiropteryx cf. *tupaiodon* Revilliod, 1917
    Archaeonycteris?
  Fam. indet.
    Ageina
PRIMATES
  Plesiadapidae
    Platychoerops richardsoni Charlesworth, 1854
    P. daubrei (Lemoine, 1880)
  Paromomyidae
    Phenacolemur lapparenti Russell, Louis, and Savage, 1967
  Adapidae
    Pelycodus savagei Gingerich, 1977[1]
    Protoadapis curvicuspidens Lemoine, 1878
    P. recticuspidens Lemoine, 1878
    Periconodon lemoinei Gingerich, 1977
    Donrussellia gallica (Russell, Louis, and Savage, 1967) (Omomyidae?)
    Agerinia roselli (Crusafont, 1967)
  Omomyidae
    Donrussellia gallica (Russell, Louis, and Savage, 1967) (Adapidae?)
ARCTOCYONIA
  Arctocyonidae, indet.
  Paroxyclaenidae
    Spaniella carezi Crusafont and Russell, 1967
TILLODONTIA
  Esthonychidae
    Esthonyx munieri (Lemoine, 1889)
ARTIODACTYLA
  Dichobunidae
    Protodichobune oweni Lemoine, 1878
    Diacodexis
  Cebochoeridae or Dacrytheriidae
    "Protodichobune" lydekkeri (Lemoine, 1891)
  Cebochoeridae or Mixtotheriidae
    Mixtotherium cf. *infans* Stehlin, 1908   —Crusafont (1973)
  Amphimerycidae
    Pseudamphimeryx renevieri Pictet and Humbert —Crusafont (1973)
CONDYLARTHRA
  Hyopsodontidae

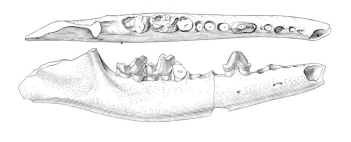

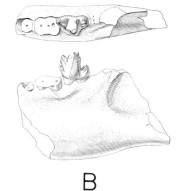

A

B

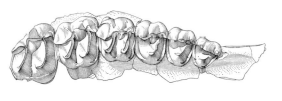

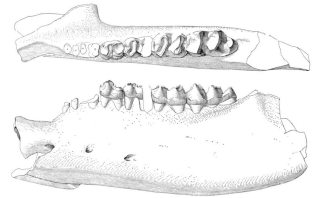

C

D

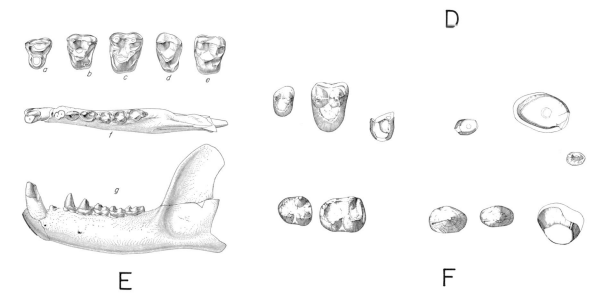

E

F

FIGURE 4-10   *Cuisian Mammals.*

   A   *Prototomus* cf. *palaeonictides* (Lemoine, 1880A). *Lower jaw.* ×1½. *Figure 4 in T. Rich (1971).*

   B   *Oxyaena menui T. Rich, 1971. Lower jaw fragment.* ×3/4 approx. *Figure 9 in T. Rich (1971).*

   C   *Lophiodon tapirotherium Desmarest, 1822. P²–M³.* ×1/2 approx. *Figure 14 in Savage, Russell, and Louis*
(1966).

   D   *Lophiodon tapirotherium Desmarest, 1822. Lower jaw.* ×1/3 approx. *Figure 19 in Savage, Russell, and Louis*
        (1966).

   E   *Protoadapis curvicuspidens Lemoine, 1878. Lower jaw and miscellaneous P⁴ and upper molars.* ×1 approx.
        Figure 10 in Russell, Louis, and Savage (1967).

   F   *Spaniella carezi Crusafont and Russell, 1967. Composite upper and lower C, premolars and molars.* ×2½
        approx. *Figures 1 and 2 in Crusafont and Russell (1967).*

(*Figures A–E inclusive published with permission of the University of California Press. Figure F published with
permission of the Muséum National d'Histoire Naturelle, Paris.*)

Paschatherium
Phenacodontidae
   Phenacodus villaltae Crusafont, 1956B
   P. cf. *teilhardi* Simpson, 1929G
PERISSODACTYLA
 Equidae
   Hyracotherium leporinum Owen, 1841
   H. vulpiceps (Owen, 1858)
   Propachynolophus gaudryi (Lemoine, 1891)
   P. maldani (Lemoine, 1878)
 Lophiodontidae
   Lophiodon tapirotherium Desmarest, 1822
 Helaletidae or Rhinocerotoidea
   Hyrachyus stehlini (Depéret, 1904)
 Chalicotheriidae
   cf. *Lophiaspis*
RODENTIA
 Ischyromyoidea
   Paramys savagei Michaux, 1964B
   P. woodi Michaux, 1964B
   P. ageiensis Michaux, 1964B
   Microparamys russeli Michaux, 1964B
   M. chandoni Hartenberger, 1971
   M. cf. *mattaueri* Hartenberger, 1971
   Pseudoparamys teilhardi (A. Wood, 1962)
   Decticadapis sciuroides Lemoine, 1891   (Sparnacian?)
   Meldimys louisi (Michaux, 1964B)
   Ailuravus michauxi Hartenberger, 1975
   A. remensis Hartenberger, 1975
 Theridomyidae
   Protadelomys
 Gliridae
   Eogliravus

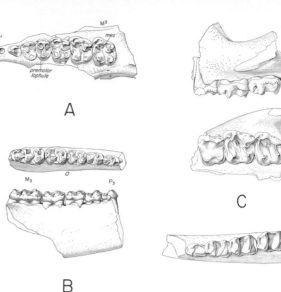

FIGURE 4-11 *Cuisian Mammals.*
  A Propachynolophus gaudryi (Lemoine, 1893). P²–<u>M³</u>. ×1/2. Figure 7 in Savage, Russell, and Louis (1965).
  B Propachynolophus gaudryi (Lemoine, 1893). Lower jaw with P₃–M₃. ×1/2. Figure 11 in Savage, Russell, and Louis (1965).
  C Hyrachyus stehlini (Depéret, 1904). M¹⁻³ and P₄–<u>M₃</u>. ×1/2. Figures 1 and 7 in Savage, Russell, and Louis (1966).
(All figures published with permission of the University of California Press.)

Of the 58 known Cuisian mammalian genera, 39 (67 percent) are shared with the preceding Sparnacian and 34 (59 percent) are not known from later faunas. Twenty-four Cuisian genera (41 percent) carry into Lutetian. Ten Cuisian families are not known in the post-Cuisian faunas of Europe: Neoplagiaulacidae, Palaeoryctidae, Oxyaenidae, Placentidentidae, Mixodectidae?, Icaronycteridae, Plesiadapidae, Esthonychidae, Arctocyonidae, and Hyopsodontidae. This is a startling number of family extinctions about the end of Cuisian time, ca. 49 million years ago. Later in this chapter, we note a similarly high family extinction at about the same time in North America. Before you conclude that this represents an especially significant ''mass extinction'' and faunal revolution caused by a special worldwide physical-chemical-environmental event, we must point out that each of these families in the European Cuisian is represented by only one species, and each of these species is represented by less than a handful of specimens! Thus, the decimation and final extermination of populations of animals assigned to these families was evidently a multi-million-year episode.

EARLY EOCENE OF ASIA

Chow and Tung (1962, 1965), Chow and Li (1965), Zheng, Tung, and Qi (1975), Zhai (1978, 1980), Zhai, Zheng, and Tong (1978) and Li, Chiu, Yan, and Hsieh (1979) cite a variety of taxa that make up an intriguing mammalian aggregate. A significant number of these Asian genera correlate with European or North American genera and thus appear to indicate an Early Eocene age for the Chinese formations and faunas. The regions yielding this aggregate extend from Xinjiang to Shandong and south through Henan and Anhui to Jiangxi and Hunan.

Composite Asian Early Eocene Mammalian Fauna

MULTITUBERCULATA
 Taeniolabididae
   Prionessus lucifer Matthew and Granger, 1925
   —Zhai (1980)

ANAGALIDA
 Eurymylidae
  Mimotona borealis Chow and Qi, 1978
  Rhombomylus turpanensis Zhai, 1978A
  R. laianensis Zhai, 1976
  Eurymylus laticeps Matthew and Granger, 1923 (late Paleocene?)
  Matutinia nitidulus Li, Chiu, Yan, and Hsieh, 1979
 Pseudictopidae
  Pseudictops lophiodon Matthew, Granger, and Simpson, 1929   —Zhai (1980)
CIMOLESTA?
  Sarcodon   —Dashzeveg and McKenna (1977)
PANTODONTA
 Pantolambdodontidae
  Archaeolambda planicanina Flerov, 1952   (late Paleocene?)
 Coryphodontidae
  Asiocoryphodon conicus Y. Xu, 1976.
  A. lophodontus Y. Xu, 1976.
  Coryphodon flerowi Chow, 1957
  C. tsaganensis Reshetov, 1976   (Late Paleocene?)
  C. ninchiashanensis Chow and Tung, 1965
  C. dabuensis Zhai, 1979
 Pastoralodontidae
  Pastoralodon lacustris Chow and Qi, 1978   —Zhai (1980)
CREODONTA
 Hyaenodontidae
  "*Sinopa*-like"   —Dashzeveg and McKenna (1977)

CARNIVORA
 Miacidae
  Xinyuictis tenuis Zheng, Tung, Yan, and Qi, 1975
DINOCERATA
 Gobiatheriidae
  Gobiatherium?
 Uintatheriidae
  Mongolotherium plantigradum Flerov, 1952
  M. efremovi Flerov, 1957   —Zhai (1980)
  Probathyopsis? sinyuensis Chow and Tung, 1962
  Pyrodon xinjiangensis Zhai, 1979
  Phenaceras lacustris Tong, 1979
  Ganatherium australis Tong, 1979
EMBRITHOPODA
  Palaeomasia kansui Ozansoy, 1966A (from Turkey)
CONDYLARTHRA
 Hyopsodontidae
  Hyopsodus orientalis Dashzeveg, 1977
ARTIODACTYLA
 Fam. indet.
  Aksyiria oligostus Gabuniya, 1973
PERISSODACTYLA
 Equidae
  Propachynolophus hengyangensis (Young, 1944)   —Li, Chiu, Yan, and Hsieh (1979)
 Brontotheriidae
  Palaeosyops
  "*Lambdotherium*"?   Zhai (1980)
 Isectolophidae
  Homogalax wutuensis Chow and Li, 1965
 Helaletidae
  Heptodon tienshanensis Zhai, 1978
  H. niushanensis Chow and Li, 1965
  Helaletes

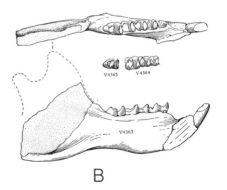

FIGURE 4-12   *Asian Early Eocene Mammals.*
 A   Anatolostylops dubius Zhai, 1978A. *P⁴–M¹.* ×*1½.* *Figure 2 in Zhai (1978A).*
 B   Rhombomylus turpanensis Zhai, 1975. *Lower jaw.* ×*1½. Figure 4 in Zhai (1978A).*
(Figures published with permission of the Institute of Vertebrate Paleontology and Paleoanthropology, Beijing.)

FIGURE 4-13   *Asian Early Eocene Mammals.*
 A   Homogalax wutuensis Chow and Li, 1965. *P³–M¹.* ×*1. Figure 1 in Chow and Li (1965).*
 B   Heptodon niushanensis Chow and Li, 1965. *P²–M³.* ×*3/4. Figure 2 in Chow and Li (1965).*
(Figures published with permission of the Institute of Vertebrate Paleontology and Paleoanthropology, Beijing.)

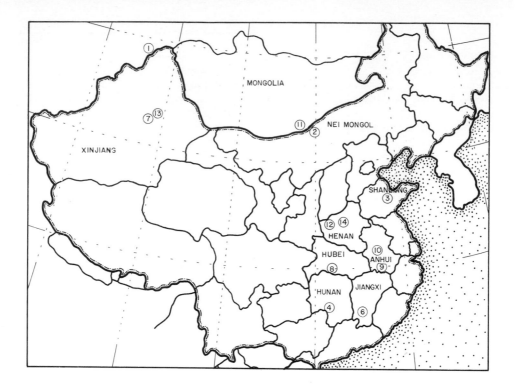

FIGURE 4-14 *Representative Early Eocene Localities and Lithostratigraphic Units of Asia.*

 1 Aksyir—Obaila Formation—Kazakhstan
 2 Bayan or Bayn Ulan Formation Sites—Nei Mongol Autonomous Region
 3 Changlo (Wutu)—Wutu and Niushan Formations—Shandong
 4 Changpiliang (Changpeiling)—Hengyang Basin—Limuping Formation—Hunan
 5 Chaybulak—Obaila Formation—Kazakhstan
 6 Chijiang Basin—Pinghu Formation—Jiangxi
 3 Chupitian Valley (Niushan, Linchu)—Wutu and Niushan Formations—Shandong
 7 Dabu Formation sites, Turfan Basin—Xinjiang
 4 Hengyang Basin—Limuping Formation—Hunan
 2 Huhebolhe Cliff (lower)—Bayn Ulan Formation—Nei Mongol
 8 Ichang (Yangchi)—Donghu Formation—Hubei
 9 Laian—Zhangshanji Formation—Anhui
 3 Linchu district (Changlo, Wutu)—Wutu and Niushan Formations—Shandong
 4 Lingcha or Ling-che (Changpiliang or Tsenpiling)—Limuping Formation—Hunan
 10 Nanjing—Zhangshanji Formation—Anhui
 11 Naran Bulak—Upper White Beds of Naran Bulak Formation—Mongolia

 6 Ninchiashan or Ninjiashan (Xinyu)—Xinyu Formation—Jiangxi
 3 Niushan (Linchu)—Wutu and Niushan Formations—Shandong
 2 Nomogen Commune—Nomogen Formation—Nei Mongol
 7 Shisanjianfang Formation Sites, Turpan Basin—Xinjiang
 12 Sichuan Basin—Yuhuanding Formation—Henan
 13 Taizicun, Turpan Basin—Taizicun Formation—Xinjiang
 14 Tantou Basin—Tantou Formation—Henan
 11 Tsagan Khushu—White Beds of Naran Bulak Formation—Mongolia
 13 Turfan Basin—Taizicun Formation—Xinjiang
 3 Wutu Basin (Changlo)—Wutu and Niushan Formations—Shandong
 10 Xuancheng Basin—Xuangtasi Formation—Anhui
 6 Yuanshui Basin—Xinyu Formation, Linjianshan Member—Jiangxi
 1 Zaisan Basin—Obaila Formation—Kazakhstan

Lophialetidae
Rhodopagus
Hyracodontidae
Triplopus
Prohyracodon
NOTOUNGULATA
Arctostylopidae
Palaeostylops iturus Matthew and Granger, 1925
—Zhai (1980)
P. macrodon Matthew, Granger, and Simpson, 1929
—Zhai (1980)
Anatolostylops dubius Zhai, 1978A
ORDER?
Didymoconidae
Hunanictis inexpectatus Li, Chiu, Yan, and Hsieh, 1979
RODENTIA
Ischyromyoidea
"Microparamys" lingchaensis (Li, Chiu, Yan, and Hsieh, 1979)
Petrokozlovia notos Shevyreva, 1972
Ctenodactylidae?
Tamquammys tantillus Shevyreva, 1971
Saykanomys chalchis Shevyreva, 1971

Dashzeveg and McKenna (1977) give a composite list of mammalian taxa from the White Beds of the Naran Bulak Formation, Nemegt Basin, and from Tsagan Khushu and Ulan Bulak in Mongolia. Most of these have been grouped with the Gashato assemblage to which they are largely equivalent. Those of the uppermost level of the Naran Bulak, however, display an "early Eocene character" and are listed here.

Mammalian Fauna from Uppermost Naran Bulak Formation

PRIMATES
Omomyidae
Altanius orlovi Dashzeveg and McKenna, 1977
CONDYLARTHRA
Hyopsodontidae
Hyopsodus orientalis Dashzeveg, 1977
PERISSODACTYLA
Equidae
Hyracotherium gabuniai Dashzeveg, 1979
Isectolophidae
cf. *Homogalax*

The array of Early Eocene mammalian genera from Asia suggests many paleozoogeographic complexities. Strong affinity with contemporary North American and European fauna is indicated by *Coryphodon, Hyopsodus*, and *"Microparamys."* Exclusively North American affinity is suggested by the uintatheres, and by *Homogalax* and *Heptodon. Propachynolophus* is known elsewhere only in the European Lower Eocene. The other genera of the Asian list are evidently endemic "holdovers" (derivatives of the Asian Paleocene mammals). We speculate that there was noteworthy dispersal of land mammals into eastern Asia from North America about the beginning of the Eocene, ca. 55–

54 million years ago, and that there was some dispersal between Europe and Asia about the same time or slightly later. (See previous discussion of the origin and affinities of the European Sparnacian fauna.)

WASATCHIAN

The Wasatchian mammalian age equals the entire Early Eocene as applied in the Nearctic nonmarine realm. Wasatchian is based on the aggregate of fossils of mammals from the upper part of the Wasatch Group of Hayden (1869), U.S. Geol. Surv. Territories, 3rd Ann. Rept., p. 191 of the 1873 ed.: "Immediately west of Fort Bridger commences one of the most remarkable and extensive groups of tertiary beds seen in the West. They are wonderfully variegated, some shade of red predominating. This group, to which I have given the name of Wasatch group, is composed of variegated sands and clays. Very little calcareous matter is found in these beds." Hayden's Wasatch Group is exposed in southwest Wyoming and adjacent Utah, and component or correlative formations and members are found in the Wind River and Bighorn basins to the north, the Washakie basin to the east, and in Colorado, North Dakota, Montana, New Mexico, and Texas. Throughout this Rocky Mountain province, the formations carrying Wasatchian fossils are characteristically sandstone-mudstone sequences banded with red mudstones, but red banding is not peculiar to Wasatchian strata.

Wasatchian-correlative floras in northwestern Nearctica indicate climates generally warmer than the climates of the Paleocene (Wolfe and Hopkins, 1967). During this time, warm-temperate to "subtropical" forests thrived from Alaska to the southeastern United States. Wasatchian mammals were diversified browsing herbivores as well as a great assortment of arboreal and ground-dwelling insectivores, frugivores, and carnivores. The forests, forest glades, and humid savannas abounded with large to very small land mammals. Insectivorous flying mammals (Chiroptera) are known, but they are not so abundant in the Wasatchian sample as from correlative localities in France. However, the beautifully preserved skeleton of *Icaronycteris index* Jepsen (1970), microchiropteran bat from the Green River Formation (lake beds) of southwestern Wyoming, is the best record known of any Paleogene mammal.

Since the time of Walter Granger's pioneer work on the "Faunal Horizons" in the nonmarine Eocene of the Rocky Mountain Region, especially Granger (1910), mammalian paleontologists have recognized three distinct subdivisions of the Wasatchian fauna and of Wasatchian time. About 1954 these subdivisions were dubbed *Graybullian, Lysitean*, and *Lostcabinian*. These mammalian subages are now used extensively in publications. They are useful and will be employed and further characterized below.

Graybullian Mammalian Subage

The "Gray Bull" fauna has, for many years, been the aggregate of mammalian species found in the North American range-zone of the earliest tapiroid, *Homogalax*. Later specialists have recognized lower, middle, and upper Gray Bull beds and faunas in the Bighorn Basin; all of these paleontologic and stratigraphic units are in the lower part of the Willwood Formation of Van Houten (1944, 1945). *Homogalax*, the classic guide fossil, is known now from the Lostcabinian (Gazin, 1962; Radinsky, 1963); and it may exist in a Lysitean equivalent in New Mexico (Froelicher Reser, 1981).

Graybullian fauna ranges through as much as 1200 feet of strata in at least two lithologic facies in the lower part of the Willwood. This is its greatest known stratigraphic range. Principal correlatives are found in the Wind River, Hoback, Powder River, Cooper Creek, Red Desert, and Washakie basins in Wyoming and in the Four Mile district of northwestern Colorado.

Composite Graybullian Fauna

MULTITUBERCULATA
  Neoplagiaulacidae
    Ectypodus simpsoni (Jepsen, 1930B)
    E. tardus (Jepsen, 1930B)
    Parectypodus   —Bown (1979)
  Eucosmodontidae
    Neoliotomus conventus Jepsen, 1940
    N. ultimus (Granger and Simpson, 1928)
MARSUPIALIA
  Didelphidae
    Peradectes chesteri (Gazin, 1952)   —Bown (1979)
    Mimoperadectes labrus Bown and Rose, 1979
      —Crochet (1979)
    Herpetotherium? mcgrewi (Bown, 1979)
LEPTICTIDA
  Leptictidae
    Prodiacodon tauri-cinerei (Jepsen, 1930B)
    P. alticuspis (Cope, 1875C)
    Palaeictops bicuspis (Cope, 1880N)
CIMOLESTA
  Palaeoryctidae
    Palaeoryctes punctatus Van Valen, 1966A
    Didelphodus absarokae (Cope, 1881S)
    D. ventanus McKenna, 1960H
PANTODONTA
  Coryphodontidae
    Coryphodon spp. (about 11 species proposed)
PANTOLESTA
  Pantolestidae
    Palaeosinopa lutreola Matthew, 1918H
    P. veterrima Matthew, 1901A
    P. didelphoides (Cope, 1881D)
    P. incerta Bown and Schankler, 1982
    Amaramnis gregoryi Gazin, 1962

APATOTHERIA
  Apatemyidae
    Apatemys bellus Marsh, 1872I, or *A. chardini* (Jepsen, 1930B)   —West (1973B)
    A. kayi (Simpson, 1929B)
TAENIODONTA
  Stylinodontidae
    Ectoganus gliriformis Cope, 1874O
    E. simplex Guthrie, 1967B
CREODONTA
  Oxyaenidae
    Oxyaena gulo Matthew, 1915D
    O. intermedia Denison, 1938
    O. forcipata Cope, 1874O
    O. aequidens Denison, 1938
    Palaeonictis occidentalis Denison, 1938
    Dipsalidictides amplus (Jepsen, 1930C)   —Denison (1938)
  Hyaenodontidae
    Prototomus mordax (Matthew, 1915D)
    P. viverrinus (Cope, 1874O)
    P. multicuspis Cope, 1875C
    Arfia opisthotoma (Matthew, 1901A)   —Van Valen (1965G)
    A. shoshonensis (Matthew, 1915D)
    Tritemnodon strenua (Cope, 1875C)
    Prolimnocyon robustus Matthew, 1915D
    P. atavus Matthew, 1915D
CARNIVORA
  Miacidae
    Miacis exiguus Matthew, 1915D
    M. latidens Matthew, 1915D
    Didymictis protenus (Cope, 1874O)
    Vassacyon promicrodon (Wortman and Matthew, 1899A)
    Uintacyon cf. *vorax* Leidy, 1872J
    U. massetericus (Cope, 1882E)
    Vulpavus australis Matthew, 1915D
    V. canavus (Cope, 1881D)
    Viverravus acutus Matthew, 1915D
    V. politus Matthew, 1915D
    Oödectes cf. *herpestoides* Wortman, 1901B
      —Bown, 1979
INSECTIVORA
  Erinaceoidea
    Talpavus nitidus Marsh, 1872G
    Talpavoides dartoni Bown and Schankler, 1982
    Macrocranion nitens (Matthew, 1918H)
    cf. *Neomatronella*
    Scenopagus hewettensis Bown and Schankler, 1982
      Schankler (1982)
  Inc. sed.
    Creotarsus lepidus Matthew, 1918H
  Soricoidea
    Leptacodon minutus (Marsh, 1872I)
    Plagioctenodon krausae Bown, 1979
    P. savagei Bown and Schankler, 1982
    Plagioctenoides microlestes Bown, 1979
    Centetodon patratus Bown and Schankler, 1982
    C. neashami Bown and Schnkler, 1982
    Parapternodus antiquus, Bown and Schankler, 1982
DERMOPTERA
  Plagiomenidae

Plagiomene multicuspis Matthew, 1918H
Worlandia inusitata Bown and Rose, 1979
PRIMATES?
  Uintasoricidae
    Niptomomys doreenae McKenna, 1960H
  Uintasoriciidae?
    Tinimomys graybulliensis Bown and Rose, 1976
    Micromomys willwoodensis Rose and Bown, 1982
  Microsyopidae
    Microsyops (Cynodontomys) alfi McKenna, 1960H
    M. angustidens Matthew, 1915F
    M. wilsoni Szalay, 1969C
PRIMATES
  Carpolestidae
    Carpolestes cf. *dubius* Jepsen, 1930C
  Paromomyidae
    Phenacolemur citatus Matthew, 1915F
    P. simonsi Bown and Rose, 1976
    P. praecox Matthew, 1915F
    P. (Ignacius) graybullianus (Bown and Rose, 1976)
  Adapidae
    Pelycodus ralstoni Matthew, 1915F [1]
    P. trigonodus Matthew, 1915F [1]
    P. mckennai Gingerich and Simons, 1977 [1]
    Copelemur praetutus (Gazin, 1962) Gingerich and Simons (1977)
  Omomyidae
    Teilhardina? americana Bown, 1976
    Tetonoides tenuiculus (Jepsen, 1930B)
    T. pearcei Gazin, 1962
    Tetonius homunculus (Cope, 1882E)
    Pseudotetonius ambiguus (Matthew, 1915F)   —Bown (1974)
    P. despairensis (Szalay, 1976)
    Anemorhysis? musculus (Matthew, 1915F) —Szalay (1976)
    Uintanius? vespertinus (Matthew, 1915F) —Szalay (1976)
ARCTOCYONIA
  Arctocyonidae
    Thryptacodon antiquus Matthew, 1915D
    T. olseni Matthew, 1915D
    Anacodon ursidens Cope, 1882E
    Chriacus gallinae Matthew, 1915D
TILLODONTIA
  Esthonychidae
    Esthonyx grangeri Simpson, 1937D
    E. spatularis Cope, 1880   —[synonym of *E. bisulcatus* Cope, 1874O acc. Bown (1979)]
DINOCERATA
  Uintatheriidae
    Probathyopsis praecursor Simpson, 1929F
    P. successor Jepsen, 1930B
ARTIODACTYLA
  Dichobunidae
    Diacodexis metsiacus (Cope, 1882E)
    D. robustus Sinclair, 1914A
    Bunophorus etsagicus (Cope, 1882E)
ACREODI
  Mesonychidae
    Dissacus navajovius Cope, 1881H
    Pachyaena gigantea Osborn and Wortman, 1892A

    P. gracilis Matthew, 1915D
    P. ossifraga Cope, 1874O
    Hapalodectes leptognathus Osborn and Wortman, 1892A
    Wyolestes apheles Gingerich, 1981   —(Family Didymoconidae, acc. Gingerich)
CONDYLARTHRA
  Hyopsodontidae
    Haplomylus speirianus (Cope, 1880E)
    Hyopsodus miticulus Cope, 1874O
    H. simplex Loomis, 1905
    H. loomisi McKenna, 1960
  Hyopsodontidae?, Mioclaenidae?
    Apheliscus sabulosus (Jepsen, 1930C)
    A. nitidus Simpson, 1937D
    A. insidiosus Cope, 1874O
  Meniscotheriidae
    Meniscotherium tapiacitis Gazin, 1965B
    Ectocion osbornianum (Cope, 1882E)
    E. parvus Granger, 1915A
  Phenacodontidae
    Phenacodus primaevus Cope, 1873D
    P. brachypternus Cope, 1882E
    P. vortmani (Cope, 1880N)
PERISSODACTYLA
  Equidae
    Hyracotherium angustidens (Cope, 1875E)
    H. spp.?
  Isectolophidae
    Homogalax protapirinus (Wortman, 1896A)
  Chalicotheriidae
    Paleomoropus jepseni Radinsky, 1964A
PHOLIDOTA?
  Metacheiromyidae
    Palaeanodon ignavus Matthew, 1918H
    P. parvulus Matthew, 1918H
  Epoicotheriidae
    Alocodontulum atopum (Rose, Bown, and Simons, 1977)
RODENTIA
  Ischyromyoidea
    Franimys amherstensis A.E. Wood, 1962B
    Paramys excavatus Loomis, 1907A
    P. copei Loomis, 1907A
    Lophiparamys murinus Matthew and Granger, 1918H
    Reithroparamys atwateri (Loomis, 1907A)
    Microparamys sp.
    Pseudotomus coloradensis A.E. Wood, 1962B

Three meticulous studies on Graybullian faunas and strata merit special note: the study of the Four Mile fauna of Colorado by McKenna (1960), the geology and mammalian paleontology of the Sand Creek lithofacies of the Willwood Formation in southeastern Bighorn Basin of Wyoming by Bown (1979), and a stratigraphic study of the Willwood Formation in the central Bighorn Basin by Schankler (1980).

McKenna prospected for fossils of small vertebrates in anthill accumulations. When the grit-sized particles of an anthill proved to be partly fossils, he quarried in nearby, poorly exposed mudstone beds, carried the quarried blocks to the closest water body and sieved the mudstone under water to concentrate the grit. This labor produced thousands of small teeth, jaw fragments, and postcranial elements from about 35 localities. The taxa, the number of specimens, and the minimum number of individuals represented by these specimens were given for each of eight principal localities. The relative stratigraphic position of each locality was not determinable, although vertical separation was probably not great. A detailed analysis of each locality sample showed McKenna that there were significant differences in the abundance of a species from site to site; thus, the groundwork was established for determining the Graybullian ecology of northwestern Colorado. The most striking demonstration from McKenna's work was the incredible abundance of fossils of small and medium-sized vertebrates in outcrops of strata that appeared superficially to be unfossiliferous.

Bown (1979) has made detailed chemical analyses supporting his conclusions regarding the environments and modes of sedimentation of the strata of the Willwood Formation, especially in the southeastern part of the Bighorn Basin. His Elk Creek facies of the Willwood (as much as 365 m thick) in the central Bighorn is characterized by bright red and orange mudstones, calcium carbonate cemented sandstones, and calcium carbonate nodules. Vertebrate fossils in this facies are found mostly in thin gray mudstone beds between thicker orange mudstones. About 145 m of Bown's Sand Creek facies of the lower Willwood are preserved in his chief area of study, the southeastern Bighorn. The Sand Creek is here characterized by relatively thin, pale, purple and gray mudstones, drab friable sandstones with little or no calcium carbonate cement, and ferric oxyhydrate nodules and concretions. Vertebrate fossils here are found in Bown's *Class A*, thin, bluish gray mudstones, which overlie mottled purple and orange mudstones. And according to Bown, these accumulated during intervals of slowed sedimentation on floodbasins, as soils were forming. Bown also concluded that contrary to conclusions of previous workers, the drab (gray) mudstones do not yield a predominantly arboreal fauna and the reddish mudstones do not yield a predominantly ungulate (less forested terrain) fauna. His *No Water* (Graybullian) fauna from the Sand Creek lithofacies appears to show ecologic difference from the contemporary fauna to the north in the Bighorn, for (to date) it lacks *Plagiomene, Miacis, Dissacus,* and *Homogalax,* and *Hyopsodus* and *Phenacodus* are scarce. Bown and Krause (1981) have substantiated and embellished Bown's 1979 conclusions.

Schankler (1980) has proposed a new zonation for the strata of Wasatchian age in the central Bighorn Basin. (See discussion of his proposals at the end of our discussion of the Wasatchian.)

Lysitean

The Lysitean subage, originally the Lysite faunal zone of Sinclair and Granger (1911), was based on the mammalian fauna from what is now recognized as the Lysite Member of the Wind River Formation (Tourtelot, 1946), north and adjacent to the villages of Lysite and Lost Cabin in the Wind River Basin, Fremont County, Wyoming. The base of the "Lysite faunal zone" was placed at the lowest occurrence of the tapiroid *Heptodon,* which presumably was just above the highest occurrence of *Homogalax.* The bottom of the overlying "Lost Cabin faunal zone" was placed on the lowest level at which the brontothere *Lambdotherium* could be found. According to Guthrie (1967B), the Lysite "zone," corresponding to the Lysite Member, which is composed of alternating beds whose surface color is brick red and white, has an exposed thickness of only 200 ft in the type area; however, the strata of Lysitean age to the north in the central Bighorn Basin are said to be about 350 ft thick. Guthrie (1967B) made surface collections in the type area of the Lysite and compiled a faunal list. His list for the Lysitean is expanded by later work in southwestern Wyoming by Gazin (1952–1967) and by continuing studies in the Bighorn Basin and Washakie Basin.

Composite Lysitean Mammalian Fauna

MULTITUBERCULATA
  Neoplagiaulacidae
    Ectypodus
    Parectypodus
MARSUPIALIA
  Didelphidae
    Peradectes
    Herpetotherium
LEPTICTIDA
  Leptictidae
    Prodiacodon
    Palaeictops cf. *pineyensis* Gazin, 1962
CIMOLESTA
  Palaeoryctidae
    Didelphodus cf. *absarokae* (Cope, 1881S)
PANTODONTA
  Coryphodontidae
    Coryphodon
PANTOLESTA
  Pantolestidae
    Palaeosinopa
APATOTHERIA
  Apatemyidae
    Apatemys bellus Marsh, 1872I, or *A. whitakeri* Guthrie, 1967B
TAENIODONTA
  Stylinodontidae
    Ectoganus cf. *simplex* Guthrie (1967B)
CREODONTA
  Oxyaenidae

Oxyaena forcipata Cope, 1874O
Hyaenodontidae
   Prototomus multicuspis Cope, 1875C
   Tritemnodon strenua (Cope, 1875C)
   Prolimnocyon elisabethae Gazin, 1952
CARNIVORA
  Miacidae
   Miacis jepseni Guthrie, 1967B
   M. latidens Matthew, 1915D
   M. exiguus Matthew, 1915D
   Didymictis protenus (Cope, 1874O)
   Uintacyon massetericus (Cope, 1882E)
   Vulpavus cf. *canavus* (Cope, 1881D)
   Oödectes iudei Guthrie, 1967B
   Viverravus gracilis Marsh, 1872G
   V. lutosus Gazin, 1952
INSECTIVORA
  Erinaceoidea
   Talpavus
   Entomolestes
   Macrocranion nitens (Matthew, 1918H)
  Inc. sed.
    cf. *Creotarsus*
  Soricoidea
    cf. *Nyctitherium*
SCANDENTIA?
  Tupaiidae?
   n. gen. (Washakie Basin)
CHIROPTERA
  Icaronycteriidae
   Icaronycteris index Jepsen, 1970 (age?)
   n. gen. (Washakie Basin)
PRIMATES?
  Uintasoricidae
   Uintasorex, sp. unident.
   Niptomomys thelmae Gunnell and Gingerich, 1981
  Microsyopidae
   Microsyops latidens (Cope, 1882E)
PRIMATES
  Paromomyidae
   Phenacolemur cf. *citatus* Matthew, 1915F
  Adapidae
   Pelycodus abditus Gingerich and Simons, 1977[1]
   Copelemur fereututus Gingerich and Simons, 1977
  Omomyidae
   Omomys minutus (Loomis, 1906A)
   cf. *Loveina*
   Anemorhysis? musculus (Matthew, 1915F)
   Absarokius abbotti (Loomis, 1906A)
   Arapahovius gazini Savage and Waters, 1978
   Tetonius cf. *homunculus* (Cope, 1882E)
   Tetonoides cf. *pearcei* Gazin, 1962
ARCTOCYONIA
  Arctocyonidae
   Thryptacodon loisi Guthrie, 1967B
   Anacodon ursidens Cope, 1882E
   Chriacus?
TILLODONTIA
  Esthonychidae

Esthonyx bisulcatus Cope, 1874O
DINOCERATA
  Uintatheriidae
   Probathyopsis
   P. lysitensis Guthrie, 1967B
ARTIODACTYLA
  Dichobunidae
   Diacodexis metsiacus (Cope, 1882A)
   Bunophorus lysitensis (Guthrie, 1967B)
   B. sinclairi Guthrie, 1966
   B. macropternus (Cope, 1882E)
   Homacodontinae, gen. indet.
ACREODI
  Mesonychidae
   Hapalodectes leptognathus Osborn and Wortman, 1892A
CONDYLARTHRA
  Hyopsodontidae
   Hyopsodus powellianus Cope, 1884O
   H. miticulus Cope, 1874O
   H. wortmani Osborn, 1902C
   H. minor Loomis, 1905
   H. latidens Denison, 1937
  Meniscotheriidae
   Meniscotherium tapiacitis Gazin, 1965B
   M. cf. *robustum* Thorpe, 1934A
   M. chamense Cope, 1874O
  Phenacodontidae
   Phenacodus vortmani (Cope, 1880N)
   P. primaevus Cope, 1873D
   P. brachypternus? Cope, 1882E
PERISSODACTYLA
  Equidae
   Hyracotherium angustidens? (Cope, 1875C)
   H. vasacciense (Cope, 1872O)
   H. tapirinum (Cope, 1875C)
   H. index (Cope, 1873D)
   H. (Xenicohippus) grangeri (Bown and Kihm, 1981)
  Isectolophidae?
  Helaletidae
   Heptodon calciculus (Cope, 1880N)
PHOLIDOTA?
  Metacheiromyidae
   Palaeanodon woodi Guthrie, 1967B
RODENTIA
  Ischyromyoidea
   Paramys copei Loomis, 1907A
   P. excavatus Loomis, 1907A
   P. francesi A. Wood, 1962B
   Franimys buccatus (Cope, 1877K)
   Lophiparamys debequensis A. Wood, 1962B
   Reithroparamys atwateri (Loomis, 1907A)
   R. debequensis A. Wood, 1962B
   Microparamys lysitensis A. Wood, 1962B
   Pseudotomus coloradensis A. Wood, 1962B
   Leptotomus loomisi A. Wood, 1962B
   cf. *Knightomys*
   cf. *Dawsonomys*

Lostcabinian

The Lostcabinian subage, equivalent to the Lost Cabin faunal zone of Granger (1910) and Sinclair and Granger (1911), was based on fossil mammals from what is now recognized as the Lost Cabin Member of the Wind River Formation in the tributaries of Badwater Creek, vicinity of the village of Lost Cabin, Wind River Basin, Fremont and Natrona Counties, Wyoming. The Lost Cabin Member differs from the underlying Lysite Member by being a purple and gray mudstone-sandstone succession, in contrast to the more brilliant red and white Lysite beds (Guthrie, 1967; Tourtelot, 1946). Lostcabinian is essentially the time interval represented by the total stratigraphic range of the genus *Lambdotherium* but further characterized by the association of *Shoshonius*, *Eotitanops*, *Hyrachyus*, *Bathyopsis*, *Stylinodon*, *Antiacodon*, *Thisbemys*, and *Leptotomus*, together with particular species of *Microsyops*, *Pelycodus*, *Copelemur*, ischyromyoid rodents, *Hyopsodus*, and *Diacodexis*.

Composite Lostcabinian Mammalian Fauna

MULTITUBERCULATA
  Neoplagiaulacidae
    Parectypodus
MARSUPIALIA
  Didelphidae
    Herpetotherium comstocki Cope, 1884O
    H. edwardi (Gazin, 1952)
    H. chesteri (Gazin, 1952)
LEPTICTIDA
  Leptictidae
    Prodiacodon tauricinerei (Jepsen, 1930B)   —Novacek (1977)
    Palaeictops bicuspis (Cope, 1880N)
    P. (= "*Parictops*") *multicuspis* (Granger, 1910A)
    P. matthewi Novacek, 1977
CIMOLESTA
  Palaeoryctidae
    Didelphodus absarokae (Cope, 1881S)
    D. altidens McKenna, Robinson, and Taylor, 1962
PANTODONTA
  Coryphodontidae
    Coryphodon singularis Osborn, 1898H
    C. ventanus Osborn, 1898H
    C. wortmani Osborn, 1898H
PANTOLESTA
  Pantolestidae
    Palaeosinopa didelphoides (Cope, 1881D)
    P. lutreola Matthew, 1918H
    P. cf. *viterrima* Matthew, 1901A
APATOTHERIA
  Apatemyidae
    Apatemys bellus Marsh, 1892I, or *A. whitakeri* Guthrie, 1967B, or *A. hurzeleri* Gazin, 1962, or all three.
TAENIODONTA

  Stylinodontidae
    Stylinodon mirus Marsh, 1874C
CREODONTA
  Oxyaenidae
    Oxyaena
    Ambloctonus cf. *major* Denison, 1938
    Patriofelis tigrina (Cope, 1880N)
    P. ulta Leidy, 1870G
  Hyaenodontidae
    Prototomus vulpecula (Matthew, 1915D)
    P. cf. *viverrinus* (Cope, 1874O)
    P. multicuspis Cope, 1875C
    Prolimnocyon elisabethae Gazin, 1952
    P. cf. *antiquus* Matthew, 1915D
    Tritemnodon strenua (Cope, 1875E)
CARNIVORA
  Miacidae
    Miacis latidens Matthew, 1915D
    M. exiguus Matthew, 1915D
    M. jepseni Guthrie, 1967B
    M. parvivorus Cope, 1872PP
    Didymictis altidens Cope, 1880N
    D. vancleveae Robinson, 1966A
    Oödectes amissadomus (Guthrie, 1967A)
    O. herpestoides Wortman, 1901B
    Vulpavus canavus (Cope, 1881D)
    V. australis Matthew, 1915D
    V. asius Gazin, 1952
    Uintacyon ascodes Gazin, 1952
    Vassacyon cf. *promicrodon* (Wortman and Matthew, 1899A)
    Viverravus acutus Matthew, 1915G
    V. gracilis Marsh, 1872G
    V. lutosus Gazin, 1952
    V. sicarius Matthew, 1909D
INSECTIVORA
  Erinaceoidea
    Talpavus sullivani Guthrie, 1971B
    Macrocranion nitens Matthew, 1918H
    Scenopagus? priscus (Marsh, 1872G)
    Scenopagus edenensis? McKenna, Robinson, and Taylor, 1962
    Centetodon pulcher Marsh, 1872   —Krishtalka and West (1979)
  Soricoidea
    Nyctitherium cf. *velox* Marsh, 1872G
PRIMATES?
  Microsyopidae
    Microsyops knightensis Gazin, 1952
    M. scottianus Cope, 1881D
    M. lundeliusi White, 1952
PRIMATES
  Paromomyidae
    Phenacolemur citatus Matthew, 1915F
    P. jepseni Simpson, 1955
  Adapidae
    Pelycodus jarrovii (Cope, 1874O)
    P. venticolis (Osborn, 1902) [1]
    Copelemur tutus (Cope, 1877K)   —Gingerich and Simons (1977)
    C. consortutus Gingerich and Simons, 1977
  Omomyidae

Shoshonius cooperi Granger, 1910A
Loveina zephyri Simpson, 1940C
Absarokius noctivagus Matthew, 1915F
Anemorhysis sublettensis (Gazin, 1952)   —Gazin, 1958B
Chlororhysis knightensis Gazin, 1958B
Omomys sheai Gazin, 1962
Huerfanius rutherfurdi Robinson, 1966A
ARCTOCYONIA
  Arctocyonidae
    Thryptacodon cf. *antiquus* Matthew, 1915D
TILLODONTIA
  Esthonychidae
    Esthonyx bisulcatus Cope, 1874O
    E. acutidens Cope, 1881D
    Megalesthonyx hopsoni Rose, 1972A
    Trogosus grangeri Gazin, 1953
DINOCERATA
  Uintatheriidae
    cf. *Bathyopsis fissidens* Cope, 1881D
    Bathyopsis
ARTIODACTYLA
  Dichobunidae
    Diacodexis cf. *secans* (Cope, 1881D)
    D. cf. *metsiacus* (Cope, 1882E)
    D. cf. *robustus* Sinclair, 1914A
    Bunophorus cf. *macropternus* (Cope, 1882E)
    B. etsagicus (Cope, 1882E)
    B. gazini Guthrie, 1971B
    Antiacodon vanvaleni Guthrie, 1971B
    A. pygmaeus (Gazin, 1952)
    Hexacodus pelodes Gazin, 1952
    H. uintensis Gazin, 1952
ACREODI
  Mesonychidae
    Pachyaena
    Hapalodectes compressus Matthew, 1909D
    H. leptognathus Osborn and Wortman, 1892A
    Mesonyx obtusidens Cope, 1872NN
CONDYLARTHRA
  Hyopsodontidae
    Hyopsodus wortmani Osborn, 1902C
    H. mentalis (Cope, 1875C)
    H. cf. *walcottianus* Matthew, 1915E
  Meniscotheriidae
    Meniscotherium chamense Cope, 1874O
    M. robustum Thorpe, 1934A
  Phenacodontidae
    Phenacodus primaevus Cope, 1873D
    P. vortmani (Cope, 1880N)
    P. brachypternus? Cope, 1882E
PERISSODACTYLA
  Equidae
    Hyracotherium index (Cope, 1873D)
    H. vasacciense (Cope, 1872O)
    H. tapirinum (Cope, 1875C)
    H. (*Xenicohippus*) *grangeri* (Bown and Kihm, 1981)
    H. (*X.*) *osborni* (Bown and Kihm, 1981)
  Brontotheriidae
    Lambdotherium popoagicum Cope, 1880N
    Eotitanops borealis (Cope, 1880N)

    E. minimus Osborn, 1919B
  Isectolophidae
    Homogalax cf. *protapirinus* (Wortman, 1896A)
  Helaletidae
    Heptodon posticus (Cope, 1882E)
    H. calciculus? (Cope, 1880N)
    H. ventorum (Cope, 1880N)
    Selenaletes scopaeus Radinsky, 1966D
  Hyrachyidae or Helaletidae
    Hyrachyus modestus (Leidy, 1870O)
PHOLIDOTA?
  Metacheiromyidae
    Palaeanodon ignavus Matthew, 1918H
  Epoicotheriidae?
    Pentapassalus pearcei Gazin, 1952
    Tubulodon taylori Jepsen, 1932
RODENTIA
  Ischyromyoidea
    Paramys wortmani A. Wood, 1962B
    P. copei Loomis, 1907A
    P. excavatus Loomis, 1907A
    P. francesi A. Wood, 1962B
    P. huerfanensis A. Wood, 1962B
    Thisbemys perditus A. Wood, 1962B
    T. nini A. Wood, 1962B
    Reithroparamys pattersoni A. Wood, 1962B
    Pseudotomus coloradensis A. Wood, 1962B
    Dawsonomys woodi Gazin, 1961B
    Knightomys senior Gazin, 1961B
    Sciuravus wilsoni Gazin, 1961B
    Lophiparamys debequensis A. Wood, 1962B
    L. woodi Guthrie, 1971B
    Microparamys
    Leptotomus huerfanensis A. Wood, 1962B
    L. grandis A. Wood, 1962B
    L. costilloi A. Wood, 1962B

Wasatchian Succession in the Bighorn Basin, Wyoming

Schankler (1980; see especially Fig. 1, p. 102), working in the central part of the Bighorn Basin, has proposed and used a *Haplomylus-Ectocion* range-zone for the stratal interval corresponding to the Lower and Middle subdivisions of the Gray Bull Faunal Zone of Van Houten (1945) and corresponding to Bown and Schankler's (1980, in press, as cited by Schankler, 1980) Lower and Middle subdivisions of their Gray Bull Biostratigraphic Zone. Schankler's overlying *Bunophorus* interval-zone (based essentially on absence of taxa characterizing the underlying or the overlying stratigraphic interval) corresponds to Van Houten's Upper part of Gray Bull Faunal Zone and to Bown and Schankler's Upper part of Gray Bull Biostratigraphic Zone. Thus, the Gray Bull member in the Bighorn Basin, as used by H. Wood and others (1941), and the Gray Bull Faunal Zone of

A

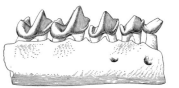

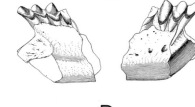

B

C

D

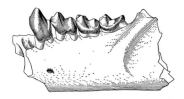

E

F

FIGURE 4-15   *Wasatchian Mammals.*

A   *Ectypodus tardus (Jepsen, 1930B).* P₄ *and* M¹. ×5. *Figure 12 in McKenna (1960H).*

B   *Apatemys. Lower jaw fragment.* ×3½. *Figure 22 in McKenna (1960H).*

C   *Leptacodon jepseni McKenna, 1960H. Lower jaw with* P₂–M₂. ×8. *Figure 24 in McKenna (1960H).*

D   *Pseudotetonius despairensis (Szalay, 1976). Lower jaw fragment with premolariform teeth.* ×3. *Figure 34 in McKenna (1960H).*

E   *Microsyops alfi (McKenna, 1960H).* M¹⁻³, P⁴–M¹, *and* M₂₋₃. ×3. *Figure 40 in Mc-Kenna (1960H).*

F   *Tetonius homunculus (Cope, 1882E). Lower jaw with* P₄–M₃. ×3. *Figure 38 in McKenna (1960H).*

G   *Apheliscus nitidus Simpson, 1937D.* P⁴–M². ×3. *Figure 60 in McKenna (1960H).*

H   *Hyracotherium angustidens (Cope, 1875E).* P²⁻⁴ *and* P⁴–M³. ×1. *Figure 62 in McKenna (1960H).*

(All figures published with permission of the University of California Press.)

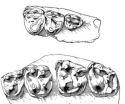

G

H

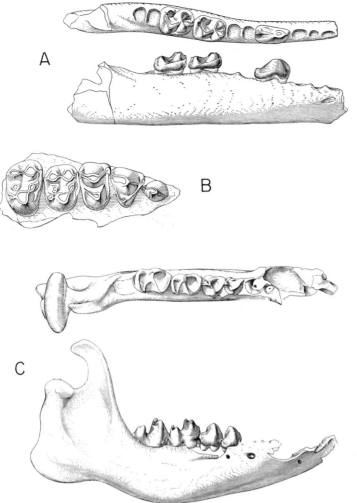

FIGURE 4-16 *Wasatchian Mammals.*

A   Haplomylus speirianus (Cope, 1880E). Lower jaw with P_3, M_{1-2}. ×3. Figure 57 in McKenna (1960H).

B   Hyopsodus loomisi McKenna, 1960H. P^2–M^2. ×3. Figure 58 in McKenna (1960H).

C   Coryphodon. Lower jaw with P_3–M_2. ×1/4. Figure 61 in McKenna (1960H).

(All figures published with permission of the University of California Press.)

FIGURE 4-17 *Wasatchian Localities and Stratigraphic Units, Exclusive of Wyoming.*

1   Cerrillos Northeast—Galisteo Formation—New Mexico—Wasatchian

2   Debeque Formation, Rifle Member Sites—Colorado—Lysitean and Lostcabinian

3   Dickinson—Golden Valley Formation—North Dakota—Graybullian

4   Four Mile—Wasatch Formation—Colorado—Graybullian

5   Huerfano Formation Sites—Colorado—("Huerfano A and B")—Lostcabinian

6   Punta Prieta—Baja California—Clarkforkian? Wasatchian?

7   San Jose Formation Sites—New Mexico—Lysitean and Lostcabinian

8   Vermilion Creek—type Vermilion Creek and Hiawatha Members of Wasatch Formation—Colorado-Wyoming—Lysitean and Lostcabinian

9   Hannold Hill Formation Sites—Texas—Wasatchian

10   Bearpaw Mountains—Montana—Graybullian

11   Powder Springs Northeast—Utah—Lostcabinian

12   Raven Ridge North—Utah—Lostcabinian

13   South Maple Canyon—Utah—Wasatchian

14   Strathcone Fiord and Bay Fiord—Ellesmere Island—Wasatchian

Van Houten (1945) are resubstantiated paleontologically and stratigraphically.

With reference to the Bighorn Basin, however, Schankler does not recognize an overlying Lysite Boistratigraphic Zone, overlaid by a Lost Cabin Zone, as he and Bown do (Bown and Schankler, 1980) but uses a *Heptodon* range-zone with Lower, Middle, and Upper divisions for the combined Lysite–Lost Cabin interval. Lysitean and Lostcabinian—bulk chapters in the history of North American land mammals, as previously characterized in this text—can't be precisely delimited in the stratigraphic succession of the Bighorn Basin as they can in other basins. And this reemphasizes that land-mammal ages are not conceived and are not applicable as paleontostratigraphic (= ''biostratigraphic'') zonation systems.

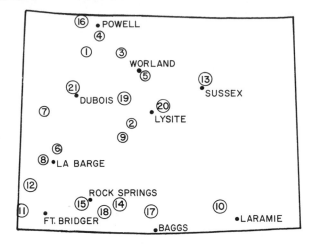

FIGURE 4-18 *Wasatchian Localities (Districts) and Stratigraphic Units in Wyoming.*

1 *Absaroka (Western Bighorn Basin)—Willwood Formation*
2 *Beaver Divide District—Wind River Formation*
3 *Bighorn Basin, Central—Willwood Formation*
4 *Bighorn Basin, Northern—Willwood Formation*
5 *Bighorn Basin, Southeastern—Willwood Formation*
6 *Big Piney (New Fork District)—Chappo, La Barge and New Fork Members of Wasatch Formation*
7 *Bondurant District—Pass Peak Formation*
8 *Chappo Oil Well—Chappo Member of Wasatch Formation*
9 *Continental Divide Basin—''Wasatch'' Formation*
10 *Cooper Creek District—Wasatch Formation*
11 *Evanston Area—Wasatch Formation*
12 *Fossil Basin—Wasatch Formation*
13 *Powder River Basin—''Wasatch'' Formation*
14 *Red Desert—Wasatch Formation*
15 *Rock Springs Uplift District—Main Body of the Wasatch Formation*
16 *Sand Coulee Area—Willwood Formation*
17 *Washakie Basin, Eastern—Niland Member of Wasatch Formation*
18 *Washakie Basin, Northern—Main Body of Wasatch, Luman Member of Green River, Niland Member of Wasatch, Wilkins Peak Member of Green River Formation*
19 *Wind River Basin, Central—Wind River Formation*
20 *Wind River Basin, Eastern (Lysite-Lost Cabin)—Wind River Formation*
21 *Wind River Basin, Western—Indian Meadows and Wind River Formations*

(Areas such as the Bighorn Basin, Wind River Basin, and Washakie Basin include hundreds of productive collecting sites.)

Gingerich (1980) proposed five divisions (''biochrons'') of the Wasatchian in the Bighorn Basin, based on successive species of the adapid primate *Pelycodus;* in chronologic order from earliest to latest, they are *P. ralstoni, P. mckennai, P. trigonodus, P. abditus,* and *P. jarrovii.* His *P. mckennai* through *P. jarrovii* ''biochrons'' correspond respectively to Lower and Upper Gray Bull, Lysite, and Lost Cabin, as he and Simons (1977) had recognized previously. But Schankler (1980) believes that the transitions between successive species of *Pelycodus* are usually too arbitrary for the species to have precise paleontostratigraphic utility.

Gingerich, Bown, Schankler, Rose, Winge, and others continue to work cooperatively on the strata and paleontology of the Bighorn Basin, and a consensus, maximally useful zonation will be developed.

Of the 113 recognized genera in the Wasatchian ($\pm$4 million years), about 36 (32 percent) are holdovers from the preceding Tiffanian-Clarkforkian (about 6 million years), and 75 (about 66 percent) do not carry into later North American faunas. Six Wasatchian families have no record in North America in post-Wasatchian time: Eucosmodontidae, Coryphodontidae, Palaeochiropterygidae, Carpolestidae, and Arctocyonidae. These doomed families, except for the coryphodontids, thus resemble the 10 families that appear to become extinct at the end of the penecontemporaneous Cuisian of Europe, in having a meager late record.

CASAMAYORAN

In southern Argentina (Patagonia) an endemic neotropical land-mammal fauna has been collected from the Casamayor Formation (''*Notostylops* beds'') and the lower part of the Sarmiento Formation of Harrington (1956). This fauna represents the Casamayoran mammal age of South America (Simpson, 1940G, 1948E, 1967A). The list of taxa for the composite fauna presented here is taken from Simpson (1967A, pp. 248–250).

Casamayoran Mammalian Fauna

MARSUPIALIA
  Didelphidae
    Coona pattersoni Simpson, 1938C
    C.? gaudryi Simpson, 1964C
  Caroloameghiniidae
    Caroloameghinia mater Ameghino, 1901
    C. tenuis Ameghino, 1901
  Bonapartheriidae   —Pascual, 1981
    Bonapartherium hinakusijum Pascual, 1980
  Borhyaenidae
    Nemolestes?
    Arminiheringia auceta Ameghino, 1902

A. cultrata Ameghino, 1902
Patene coluapiensis Simpson, 1935B
Angelocabrerus daptes Simpson, 1970D
Procladosictis anomala Ameghino, 1902
Argyrolestes peralestinus Ameghino, 1902
Polydolopidae
   Polydolops thomasi Ameghino, 1897
   P. serra Ameghino, 1902
   P. clavulus Ameghino, 1902
   P. princeps (Ameghino, 1902)
   P. bocurhor Simpson, 1948E
   Amphidolops serrula Ameghino, 1902
   Eudolops tetragonus Ameghino, 1897
   E. acuminatus (Ameghino, 1902)
   E. caroliameghinoi (Ameghino, 1903)
   Prepidolops didelphoides Pascual, 1980
   P. molinai Pascual, 1981
EDENTATA
  Dasypodidae
   Meteutatus? percarinatus Ameghino, 1902
   Utaetus buccatus Ameghino, 1902
   U. lenis (Ameghino, 1902)
   U. deustus Ameghino, 1902
   U.? laevus (Ameghino, 1902)
   "Pseudostegotherium" chubutanum Ameghino, 1902
   Coelutaetus cribellatus Ameghino, 1902
   Astegotherium dichotomum Ameghino, 1902
   Prostegotherium notostylopianum Ameghino, 1902
  Glyptodontidae
   "Glyptataelus" fractus (Ameghino, 1897)
   "Palaeopeltis" tesseratus (Ameghino, 1902)
CONDYLARTHRA
  Didolodontidae
   Didolodus multicuspis Ameghino, 1897
   D. latigonus (Ameghino, 1902)
   D. minor Simpson, 1948E
   Argyrolambda conidens Ameghino, 1904
   Paulogervaisia inusta Ameghino, 1901
   P. porca (Ameghino, 1901)
   Proectocion argentinus Ameghino, 1904
   P. precisus Ameghino, 1904
   Enneoconus parvidens Ameghino, 1901
   Asmithwoodwardia subtrigona Ameghino, 1901
   Ernestokokenia nitida Ameghino, 1901
   E. patagonica (Ameghino, 1901)
   Oxybunotherium praecursor Pascual, 1965
LITOPTERNA
  Proterotheriidae
   Josephleidya adunca Ameghino, 1901
   Ricardolydekkeria praerupta Ameghino, 1901
   Guilielmofloweria plicata Ameghino, 1901
   Anisolambda fissidens Ameghino, 1901
   A. amel Simpson, 1948E
  Macraucheniidae
   Victorlemoinea labyrinthica Ameghino, 1901
   V. emarginata Ameghino, 1901
   Ernestohaeckelia aculeata Ameghino, 1901
   E. acutidens Ameghino, 1901
NOTOUNGULATA
  Henricosborniidae
   Henricosbornia lophodonta Ameghino, 1901

H. ampla (Ameghino, 1904)
Othnielmarshia lacunifera Ameghino, 1901   —(a trigonostylopoid acc. to McKenna, 1980)
Peripantostylops minutus (Ameghino, 1901)
Notostylopidae
   Notostylops murinus Ameghino, 1897
   N. pendens (Ameghino, 1901)
   N. appressus (Ameghino, 1902)
   N. pigafettai Simpson, 1948E
   Homalostylops parvus (Ameghino, 1897)
   Edvardostroussartia sola Ameghino, 1901
Oldfieldthomasiidae
   Maxschlosseria praeterita Ameghino, 1901
   M. minima (Ameghino, 1897)
   M. rusticula (Ameghino, 1901)
   M. consumata (Ameghino, 1901)
   Oldfieldthomasia debilitata (Ameghino, 1901)
   O. parvidens Ameghino, 1901
   Ultrapithecus rutilans Ameghino, 1901
   Acoelodus oppositus Ameghino, 1901
   "Acoelodus" proclivus (Ameghino, 1902)
   Paginula parca Ameghino, 1901
Archaeopithecidae
   Archaeopithecus rogeri Ameghino, 1897
   Acropithecus rigidus (Ameghino, 1901)
Interatheriidae
   Notopithecus adapinus Ameghino, 1897
   N.? amplidens (Ameghino, 1901)
   Antepithecus brachystephanus Ameghino, 1901
   Transpithecus obtentus Ameghino, 1901
Archaeohyracidae
   Eohyrax rusticus Ameghino, 1901
   E. praerusticus Ameghino, 1902
Isotemnidae
   Pleurostylodon modicus Ameghino, 1897
   P. similis Ameghino, 1901
   P. complanatus Ameghino, 1902
   P.? recticrista (Ameghino, 1904)   —Simpson (1967A)
   Anisotemnus distentus (Ameghino, 1901)
   Acoelohyrax coronatus Ameghino, 1902
   A. complicatissimus (Ameghino, 1904)
   Isotemnus primitivus Ameghino, 1897
   I. latidens (Ameghino, 1901)
   Thomashuxleyia rostrata Ameghino, 1901
   T. externa Ameghino, 1901
Isotemnidae?
   Coelostylodon florentinoameghinoi Simpson, 1970D
ASTRAPOTHERIA
  Astrapotheriidae
   Griphodon   —Patterson, 1977
   Scaglia kraglievichorum Simpson, 1957B
TRIGONOSTYLOPOIDEA
  Trigonostylopidae
   Tetragonostylops
   Trigonostylops wormani Ameghino, 1897
   Albertogaudryia unica Ameghino, 1901   —(No-toungulate, acc. to McKenna (1980))

A. carahuasensis Vucetich and Pascual, 1977 —
　　(Notoungulate, acc. to McKenna (1980))
PYROTHERIA
　Pyrotheriidae? *inc. sed.*
　　Carolozittelia tapiroides Ameghino, 1901
　Colombitheriidae
　　Proticia venezuelensis Patterson, 1977　(age?)
INC SED.
　　Florentinoameghinia mystica Simpson, 1932L

　　This fauna demonstrates a continued and endemic evolutionary radiation from diversified stocks that are recorded in the earlier, Riochican fauna. There is no evidence for anything but complete isolation of the South American continent through Riochican and Casamayoran time so far as the record of fossil mammals is concerned. See Patterson and Pascul (1968, 1972) and Simpson (1969) for recent expanded accounts of this phenomenon.

MIDDLE EOCENE OF EUROPE— LUTETIAN

Under the subepochal designation Middle Eocene, we are placing only the Lutetian Stage-Age of the Paris Basin, as recognized by Blondeau et al. (1965), and the correlative stages, ages, faunas, and floras around the world. Middle Eocene localities produce the earliest known good record of mammals adapted to living in marine waters; these are known from Europe, Africa, Asia and North America. The faunas herein termed Early or Late Eocene in South America may be partly correlative with European Middle Eocene also, as the correlation with South American faunas is not accurate.

　　The Lutetian Stage-Age (A. de Lapparent, 1883. *Traité de Géologie.* 2nd ed., p. 989) was based on the *Calcaire Grossier* (coarse-grained limestone) and its fossils near Paris. The Roman name for Paris was *Lutetia*. This formation was made classic by Cuvier and Brongniart in their description of the geology of the Paris Basin early in the 1800s. The Calcaire Grossier is a succession of highly fossiliferous marine limestone beds and other limey rocks, in some places with a basal glauconitic sand. At Fosse, near Survilliers, glauconite from this sand was dated (K-Ar) at about 48 million years (Evernden et al., 1961) with corrected constants (Steiger and Jäger, 1977), and it was given an age of 48.7 by Bonhomme, Odin, and Pomerol (1968) by the rubidium-strontium method.

　　Many of the ancient and renowned fossil-mollusc collecting localities are in the Calcaire Grossier and the type-Lutetian Stage. This stage also produces land mammals. Some examples are a glauconitic sand and gravel at the base near Hermonville and Pévy, a marine bed with *Scutellaria* and *Palaeocarpilus* in the zone of *Orbitolites camplanula-*

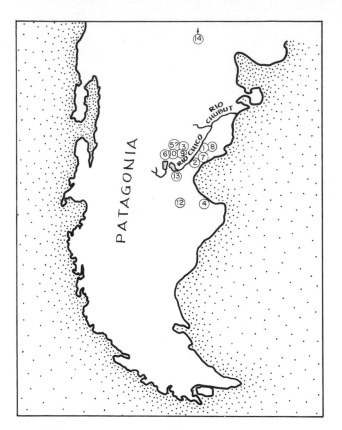

FIGURE 4-19　*Casamayoran Localities*
　1　*Cabeza Blanca = Rió Chico, primer yacimiento de Pyrotherium*
　2　*Cañadón Hondo (also Riochican)*
　3　*Cañadón Vaca*
　4　*Casamayor (=Cañadón Lobo or Cañadón Tournouër)*
　5　*Cerro Talquino (Casamayoran? and Mustersan?)*
　6　*Colhuapi (=Colhué Huapí) Norte (=Cerro del Humo and Sierra del Toro?) also includes Mustersan*
　7　*Este de Rió Chico (=probably, Pampa de Castillo East)*
　8　*Malaspina (=Rió Chico frente a Malaspina)*
　9　*Oeste de Rió Chico (includes Cañadón Vaca?)*
　10　*Pajarito (=Colhuapi Norte, in part?)—also includes Mustersan and ="Lago Musters" of some workers*
　11　*Pico Salamanca*
　12　*Pico Truncado*
　13　*Great Barranca (=Barranca)—includes Casamayoran, Mustersan, Deseadan and Colhuehuapian (=Colhué Huapí Sud and Lago Colhué-Huapí district)*
　14　*Quebrada Agua Viva—Venezuela*
(*Data from Simpson (1948C and 1967A) and from Patterson (1977)*).

tus, in the upper part of the formation at Chateau Thierry, the *Cerithium* limestone at Paris-Gentilly, miscellaneous "Calcaire grossier superieur" sites in greater Paris (e.g., Passy, type locality of *Pachynolophus*), Nanterre, and Vaugirard. At La Défense in Puteaux (a western suburb of Paris) a rich concentration of fossil land mammals has been found (Ginsburg et al., 1977). It contributes importantly to the correlative value of the type-Lutetian land-mammal assemblage. Again, there is no problem connected with using either of the terms *type-Eocene land mammals* or *type-Lutetian land mammals*.

In addition to the land mammals already mentioned, which are found in the Calcaire Grossier in the Paris Basin, the Bouxwiller (Buchsweiler) site in the Alsace Province of eastern France has yielded an important local fauna that is correlated with the mammals of the upper part of the Calcaire Grossier. The quarry at Bouxwiller exposes a lower succession of beds of greenish marly claystones and an upper succession of gray marls—both producing fossil mammals. Jaeger (1971) believed that there was no significant age difference between the two successions; Hartenberger (1969), studying the rodents, thought there was a significant difference. We follow Jaeger and include both assemblages in our composite list for Lutetian mammals. Among other local faunas included in the Lutetian, most noteworthy are the Geiseltal braunkohle assemblage from East Germany and the Messel oil shales near Darmstadt, West Germany. Beautifully preserved bat skeletons, mammal hair, bird feathers, leaves, beetles with original coloring, and the only articulated skeletons from Europe of hyracotheriine horses (*Propalaeotherium*), insectivorans, primates, and rodents have been taken from these two localities.

Composite Lutetian Mammal Fauna

MARSUPIALIA
  Didelphidae
    Amphiperatherium bastbergense Crochet, 1979
    A. giselense (Heller, 1936)
    A. goethei Crochet, 1979
EDENTATA?
    Eurotamandua joresi Storch, 1980
LEPTICTIDA?
    Leptictidium auderiense Tobien, 1962B
PANTOLESTA
  Pantolestidae
    Buxolestes hammeli Jaeger, 1970
    B. piscator W. von Koenigswald, 1980B
    Pantolestidae, gen. indet.
APATOTHERIA
  Apatemyidae
    Heterohyus armatus Gervais, 1848–1852
    H. gracilis (Stehlin, 1916)
    H. heufelderi Heller, 1930C
CREODONTA
  Hyaenodontidae
    Prodissopsalis eocaenicus Matthes, 1952
    P. ginsburgi Calas, 1969
    P. voigti Matthes, 1967
    Allopterodon theriodis (Van Valen, 1965G)
    Proviverra typica Rütimeyer, 1862C
    P. gracilis (Matthes, 1952)
    Alienetherium buxwilleri Lange-Badré, 1980

    Praecodens acutus Lange-Badré, 1980
    Oxyaenoides bicuspidens Matthes, 1967B
CARNIVORA
  Miacidae
    Quercygale macintyri (Van Valen, 1965G)
    Q. helvetica Rütimeyer, 1862C
    Paroodectes feisti Springhorn, 1980
INSECTIVORA
  Erinaceoidea
    Macrocranion tupaiodon Weitzel, 1949
    M. tenerum (Tobien, 1962B)
  Soricoidea
    Saturninia
CHIROPTERA
  Palaeochiropterygidae
    Palaeochiropteryx tupaiodon Revilliod, 1917
    P. spiegeli Revilliod, 1917
    Archaeonycteris trigonodon Revilliod, 1917
    A. revilliodi D. Russell and Sigé, 1970
    Cecilionycteris prisca Heller, 1935B
    Matthesia germanica Sigé and D. Russell, 1980
    M.? insolita Sigé and D. Russell, 1980
PRIMATES?
  Amphilemuridae
    Amphilemur leemani (Hürzeler, 1948A)
    A. eocaenicus Heller, 1935A
PRIMATES
  Paromomyidae
    Phenacolemur
  Adapidae
    Protoadapis klatti Weigelt, 1933
    P. weigelti Gingerich, 1977
    Periconodon rosselli (Crusafont, 1967) (age?)
    P. huerzeleri Gingerich, 1977
    Adapidae, gen. indet.
  Omomyidae
    Nannopithex raabi (Heller, 1930B)
    N. filholi (Chantre and Gaillard, 1897)
ARCTOCYONIA
  Paroxyclaenidae
    Kolpidodon macrognathus (Wittich, 1902)
    Pugiodens mirus Matthes, 1952
ARTIODACTYLA
  Dichobunidae
    Buxobune daubreei Sudre, 1978
    Messelobunodon messelense Franzen, 1980
    Aumelasia gabineaudi Sudre, 1980
    Hyperdichobune hammeli Sudre, 1978
    Dichobune robertiana Gervais, 1848–1852
    Meniscodon europeum Rütimeyer, 1888
  Cebochoeridae
    Cebochoerus suillus Gervais, 1848–1852
    C. ruetimeyeri Stehlin, 1908
    C. jaegeri Sudre, 1978
    C. dawsoni Sudre, 1978
  Mixtotheriidae
    Mixtotherium cf. *priscum* Stehlin, 1908
  Anoplotheriidae

Catodontherium cf. *fallax* Stehlin, 1910

C. buxgovianum Stehlin, 1910

Haplobunodon mulleri (Rütimeyer, 1862)

H. solodurense Stehlin, 1908

Dacrytheriidae

Dacrytherium cf. *elegans* (Filhol, 1884)

Tapirulus cf. *majori* Stehlin, 1910

Leptotheridium cf. *traguloides* Stehlin, 1910

Haplobunodontidae

Anthracobunodon weigelti Heller, 1929E

Rhagatherium cf. *kowalevskyi* Stehlin, 1908

Massilabune martini Tobien, 1980

ACREODI

Mesonychidae

Dissacus progressus Crusafont and Golpe, 1968

D. blayaci Stehlin, 1926

CONDYLARTHRA

Phenacodontidae

Almogaver condali Crusafont and Villalta, 1954

Phenacodus aff. *teilhardi* Simpson, 1929G

PERISSODACTYLA

Equidae

Pachynolophus boixedatensis Crusafont and Remy, 1970

P. duvali Pomel, 1847

Propalaeotherium parvulum (Laurillard, 1849)

P. isselanum Blainville, 1846

P. hassiacum Haupt, 1925

P. messelense (Haupt, 1925)

P. argentonicum Gervais, 1848–1852

P. rollinati Stehlin, 1905   ?

Lophiotherium pygmaeum Depéret, 1901

L. magnum Matthes, 1977

L. geiseltalensis Matthes, 1977

L. voigti Matthes, 1977

Anchilophus cf. *depereti* Stehlin, 1905

A.? simpsoni

Palaeotheriidae

Plagiolophus cartieri Stehlin, 1904

Paraplagiolophus codiciensis (Gaudry, 1865)

Palaeotherium eocaenum Gervais, 1875

Lophiodontidae

Lophiodon tapirotherium Desmarest, 1822

L. tapiroides Cuvier, 1812

L. cuvieri Watelet, 1864

L. medium Fischer, 1829

L. parisiense Gervais, 1848–1852

L. sardus Bosco, 1923

Isectolophidae

Paralophiodon leptorhynchus (Filhol, 1888)

P. buchsowillanum (Desmarest, 1822)

P. isselense Cuvier, 1821

Helaletidae

Chasmotherium cartieri Rütimeyer, 1862

Hyrachyus minimus (Fischer, 1829)

Atalonodon monterini Dal Piaz, 1929

Chalicotheriidae

Lophiaspis occitanicus (Desmarest, 1822)

PHOLIDOTA

Manidae

Eomanis waldi Storch, 1978

RODENTIA

Ischyromyoidea

Paramys

Ailuravus picteti Rütimeyer, 1891

A. macrurus Weitzel, 1949

Microparamys parvus (Tobien, 1954A)

Plesiarctomys hartenbergeri A. Wood, 1970

Masillamys krugi Tobien, 1954A

M. beegeri Tobien, 1954A

Theridomyidae

Protadelomys alsaticus Hartenberger, 1969

Gliridae

Eogliravus hammeli (Thaler, 1966)

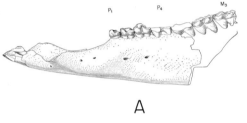

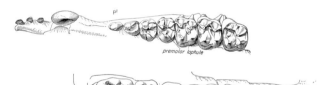

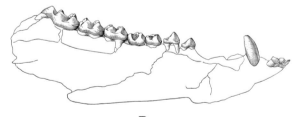

FIGURE 4-20   *Lutetian Mammals.*

A   *Pachynolophus duvali* Pomel, 1847. *P²–M³ and lower jaw dentition. ×3/4. Figures 16 and 17 in Savage, Russell, and Louis (1965).*

B   *Propalaeotherium hassiacum* Haupt, 1925. *Upper and lower dentition. ×1/2. Figures 28 and 29 in Savage, Russell, and Louis (1965).*

(*Published with permission of the University of California Press.*)

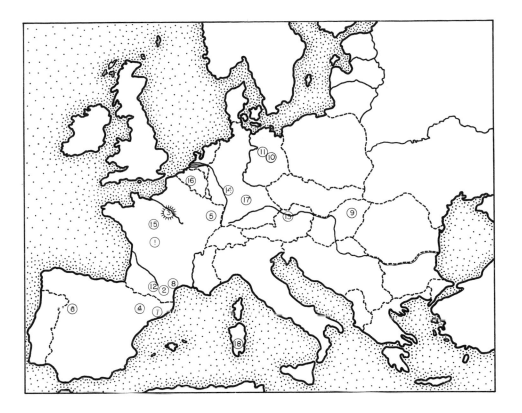

FIGURE 4-21 *Representative Lutetian Localities.*

1	Argenton-sur-Creuse	*8*	Eygalayes	*3*	Montchenot
2	Aumelas 1	*9*	Felsögalla	*4*	Montllobar 2 (Tremp Basin)
3	Bagneux (Banc vert)	*8*	Fonfroide	*3*	Nanterre
2	Bagnoles	*10*	Geiseltal braunkohle (Cecilie	*3*	Oulchy-le-Chateau
4	Benabarre		Mine)	*3*	Provins
5	Bouxwiller (Buchsweiler)	*3*	Gentilly	*15*	Saffré
4	Can Camperol	*8*	Guérin Farm	*16*	St. Gilles
	Cecilie Mine (see Geiseltal)	*11*	Helmstedt	*4*	Sant Pere Màrtir
3	Château-Thierry	*3*	Hermonville	*17*	Stolzenbach
3	Châtillon	*12*	Issel	*18*	Terras de Collu
3	Chérence	*3*	Jouy	*4*	Torrelabad
6	Corrales	*3*	Jumencourt	*4*	Tremp Basin (Montllobar)
7	Corsà	*13*	Kressenberg	*16*	Uccle
3	Coucy	*4*	La Boixedat	*3*	Vaugirard (in part)
3	Coulondres	*3*	La Défense à Puteaux	*3*	Verzenay
3	Damery 2	*4*	La Roca		
3	Dampleux	*14*	Messel		

The Lutetian mammalian fauna is characterized by a wealth of equids, palaeotheriids, *Lophiodon*, hyaenodontids, and artiodactyls. It exemplifies endemism, although some groups, for example, the palaeotheres, were apparently immigrants at about the beginning of the age.

Twenty-four of its genera (35 percent) are shared with the preceding Cuisian and 43 (62 percent) carry into the Robiacian. Three families do not appear in later faunas in Europe: Paromomyidae, Mesonychidae, and Phenacodontidae.

MIDDLE EOCENE OF ASIA

Only a few localities, some of highly questionable subepoch assignment, intimate the nature of the Middle Eocene mammalian fauna of greater Asia. In southern Asia there seems to be a mixture of taxa reflecting endemism, affinity with, and possibly dispersal from or to Europe, Africa and Nearctica. One of the most intriguing of these local faunas is from Ganda Kas (Kala Chitta Hills) in the Kuldana Formation, northwestern Pakistan—Dehm and Oettingen-Spielberg

(1958), Hussain, de Bruijn and Leinders (1978), West and Lukacs (1979) and West (1980). Here the Kuldana Formation is a succession of alternating marine and continental rocks according to West and Lukacs (1979). The Ganda Kas comprises tillodonts, creodonts, mesonychids, perissodactyls, artiodactyls, cetaceans, proboscideans and rodents. The tillodonts are of Asian, North American and European affinity. The cetaceans and proboscideans relate Pakistan to North Africa and other Tethyan-border districts of Eocene time. There is a strong challenge to future workers in this area to find skeletal materials of these early "tethytheres" to determine if they were completely adapted to locomotion in water or if they were ambulatory-amphibious, like hippopotamuses (thereby offering support to the proposed mesonychid ancestry of the cetaceans). The other taxa of Ganda Kas are either endemic or have more uncertain geographic affinities.

Ganda Kas Mammalian Fauna

CREODONTA
  Hyaenodontidae
    Paratritemnodon indicus Rango Rao, 1973
INSECTIVORA
  Erinaceoidea
    Seia shahi Russell and Gingerich, 1981
INC. SED.
    Pakilestes lathrius Russell and Gingerich, 1981
PRIMATES
  Omomyidae
    Kohatius coppens: Russell and Gingerich, 1980
TILLODONTIA
    Basalina basalensis Dehm and Oettingen-Spielberg, 1958
ARTIODACTYLA
  Fam. indet.
    Dulcidon gandaensis (Dehm and Oettingen-Spielberg, 1958)
  Helohyidae
    Khirtharia dayi Pilgrim, 1940
    Haqueina haquei Dehm and Oettingen-Spielberg, 1958
    Gobiohyus cf. *orientalis* Matthew and Granger, 1925D
  Anthracotheriidae
    Anthracokeryx
    indet. (*A.? daviesi* —West, 1980)
ACREODI
  Mesonychidae
    cf. *Honanodon* —West (1980)
CETACEA   —West (1980)
  Protocetidae
    Gandakasia potens Dehm and Oettingen-Spielberg, 1958
    Ichthyolestes pinfoldi Dehm and Oettingen-Spielberg, 1958
    Protocetus attocki West, 1980
    Basilosauridae, gen. indet.   —West (1980)
    Pakicetus inachus Gingerich and Russell, 1981
PERISSODACTYLA
  Brontotheriidae

    Eotitanops dayi Dehm and Oettingen-Spielberg, 1958
    Pakotitanops latidentatus West, 1980
  Deperetellidae
    Teleolophus? daviesi Dehm and Oettingen-Spielberg, 1958
PROBOSCIDEA   —West (1980)
  Moeritheriidae   —West (1980)
    Anthracobune pinfoldi Dehm and Oettingen-Spielberg, 1958
    Pilgrimella pilgrimi Dehm and Oettingen-Spielberg, 1958
    Lammidhania wardi (Pilgrim, 1940)
RODENTIA
  Chapattimyidae (referable to Ctenodactylidae?)
    Chapattimys wilsoni Hussain, de Bruijn, and Leinders, 1978
    C. ibrahimshahi Hussain, de Bruijn, and Leinders, 1978
    Saykanomys ijlsti Hussain, de Bruijn, and Leinders, 1978
    S. vandermeuleni Hussain, de Bruijn, and Leinders, 1978
    S. sondaari Hussain, de Bruijn, and Leinders, 1978
    S. lavocati Hussain, de Bruijn, and Leinders, 1978
    S. chalchis Shevyreva, 1972

The cetaceans in the Ganda Kas fauna seem to relate it ecologically and zoogeographically to the marine fauna of Kutch, India, to the southwest. Ganda Kas was evidently a near-shore district of an embayment of the Tethyan Sea.

R. Rao (1971A, 1972, 1973), Sahni and Khare (1971, 1972), and Sahni and Srivastava (1976, 1977) have indicated the following mammalian fauna in the Kalakot-Rajauri district of Jammu-Kashmir (Beragoa Coal Mine, Jigni Coal Field, Metka Coal Field).

Kalakot-Rajauri Mammalian Fauna

ARTIODACTYLA
  Helohyidae
    Khirtharia dayi Pilgrim, 1940
    Haqueina haquei Dehm and Oettingen-Spielberg, 1958
    Kunmunella rajaurensis Sahni and Khare, 1971
    Raoella degrai Sahni and Khare, 1971
    Bunodentus inflatus Rao, 1972
    Indohyus indirae Rao, 1971
    Indohyus kalakotensis Rao, 1971
ACREODI
  Mesonychidae
    Honanodon
PERISSODACTYLA
  Isectolophidae
    Sastrilophus dehmi Sahni and Khare, 1971
  Helaletidae
    Chasmotherium mckennai Sahni and Khare, 1972
  Helaletidae or Rhinocerotoidea
    Hyrachyus asiaticus Rao and Obergfell, 1973
  Lophialetidae
    Schlosseria radinskyi Sahni and Khare, 1972
    Kalakotia simplicidenta Rao, 1972
RODENTIA
  Ischyromyoidea?
    Metkamys blacki Sahni and Srivastava, 1976
    near *Microparamys*   —Sahni and Srivastava (1977)
  Chapattimyidae (Ctenodactylidae?)
    Saykanomys woodi (Sahni and Khare, 1972)

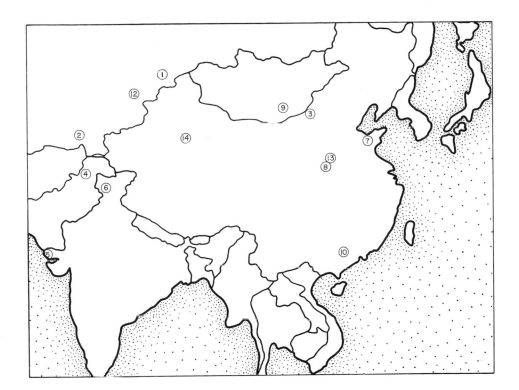

FIGURE 4-22 *Representative Middle Eocene Localities and Lithostratigraphic Units of Asia.*

*1 Aksyir II, Zaisan Basin,—Sargamys Formation—
    Kazakhstan*
2 Andarak I, II, Fergana Basin—Alay Formation—Kirgizia
3 Arshanto Formation sites—Nei Mongol Autonomous Region
*4 Attock (Ganda Kas, Lammidhan)—Kuldana Formation—
    Pakistan*
5 Babia Hill (marine)—Berwali "series"—Kutch, India
4 Basal (Lammindhan)—Kuldana Formation—Pakistan
*3 Bayn Ulan (Huhebolhe Cliff)—Arshanto Formation—Nei
    Mongol*
*6 Beragoa Coal Mine—Kalakot, etc.—Subathu Formation—
    India*
4 Chharat Formation sites—Pakistan
4 Chorlakki—Kuldana Formation—Pakistan
*7 Cicishou (Sisichou)—Xintai (Sintai)—Guanzhuang
    Formation—Shandong*
4 Domanda Formation site (marine)—Pakistan
2 Fergana Basin—Alay Formation—Kirgizia
*4 Ganda Kas (Kala Chitta Hills)—Kuldana Formation—
    Pakistan*
7 Guanzhuang—Guanzhuang Formation—Shandong
8 Haitaoyuantze (Hetaoyuan)—Sichuan—Henan (age?)
5 Harudi (marine)—Berwali "series"—Kutch, India
7 Hsi-kou—Guanzhuang Formation—Shandong
7 Hsintai (Xintai, Sintai)—Guanzhuang Formation—Shandong
3 Huhebolhe Cliff (middle)—Arshanto Formation—Nei Mongol
4 Jatta—Kuldana Formation—Pakistan
4 Jhalar—Kuldana Formation—Pakistan
6 Jigni Coal Field—Subathu Formation—India
*4 Kala Chitta Hills—Ganda Kas, etc.—Kuldana Formation—
    Pakistan*

6 Kalakot—Rajauri, etc.—Subathu Formation—India
1 Kalmakpay I, Zaisan Basin—Obaila Formation—Kazakhstan
9 Khaichin Ula II, III—Mongolia
4 Kuldana Formation sites—Pakistan
5 Kutch sites (marine)—India
4 Lammidhan (Basal)—Kuldana Formation—Pakistan
8 Linpao—Henan
10 Liuniu (fauna)—Guangxi   (Early to Middle Eocene?)
7 Mengyin Hsien—Guanzhuang Formation—Shandong
6 Metka Coal Field (Rajauri)—Subathu Formation—India
1 Obaila (Obayla), upper—Obaila Formation—Kazakhstan
5 Panandhrow—Kutch (marine)—India
6 Rajauri (Kalakot)—Subathu Formation—India
*4 Safed Tobah (Dera Ghazi Khan) (marine)—Domanda
    Formation—Pakistan*
11 Sanggau—Borneo   (Late Eocene?)
12 Shinzhaly—Kolpak Formation—Kazakhstan
3 Siaiwang County—Arshanto Formation—Nei Mongol
13 Sichuan Basin 2—Dacangfang Formation—Henan
7 Sintai   (See Hsintai.)
*7 T'an's Lokal (Tan's Locality, Meng-Yin-Hsien)—
    Guanzhuang Formation—Shandong*
3 Tukhum Formation sites (Ula Usu)—Nei Mongol
14 Turfan Basin (Cretaceous, Paleocene and Eocene)—Xinjiang
3 Ula Usu—Tukhum Formation—Nei Mongol
1 Ulan Ulasty, Zaisan Basin—Obaila Formation—Kazakhstan
7 Xintai   (See Hsintai.)
1 Zaisan Basin–Obaila and Sargamys Formations—Kazakhstan

93

This Kalakot-Rajauri fauna is believed to be about the same age as the Ganda Kas of Pakistan and is apparently earlier than the renowned fauna from the Pondaung Formation of Upper Burma.

Other Asian mid-Eocene land-mammal sites—in China, Mongolia, Kirghizia and Kazakhstan—give us the following list of mammals from northern Asia.

Mid-Eocene Mammalian Fauna from China, Mongolia, Kirghizia, and Kazakhstan

PANTODONTA
  Pantolambdodontidae
    Pantolambdodon
  Coryphodontidae
    Coryphodon flerowi Chow, 1957
    Eudinoceras obailiensis Gabunia, 1961
TILLODONTIA
  Esthonychidae
    Kuanchuanius shantunensis Chow, 1963
CREODONTA
  Hyaenodontidae
    Thinocyon? sichowensis Chow, 1975
DINOCERATA
  Uintatheriidae
    Uintatherium?
  Gobiatheriidae
    Gobiatherium mirificum Osborn and Granger, 1932
ARTIODACTYLA
  Family?
    Paraphenacodus solivagus Gabunia, 1971
ACREODI
  Mesonychidae
    Mongolonyx
    Mesonyx?
PERISSODACTYLA
  Equidae
    Propalaeotherium sinense Zdansky, 1930
  Brontotheriidae
    Microtitan?
    Desmatotitan
    Eotitanops
    cf. *Palaeosyops*
  Isectolophidae
    Isectolophus bogdulensis Reshetov, 1979
  Helaletidae
    Colodon cf. *inceptus* Matthew and Granger, 1925
    Helaletes cf. *mongoliensis* (Osborn, 1923)
    Veragromovia desmatotheroides Gabunia, 1961
  Helaletidae or Rhinocerotoidea
    Hyrachyus
  Deperetellidae
    Deperetella ferganica Belyayeva, 1962
    Teleolophus cf. *medius* Matthew and Granger, 1925
    T. beliajevi Biryukov, 1974
  Lophialetidae
    Rhodopagus minutissimus Reshetov, 1979

    Rhodopagus?
    Eoletes gracilis Biryukov, 1974
    Lophialetes expeditus Matthew and Granger, 1925
    Schlosseria magister Matthew and Granger, 1926
    Breviodon? minutus (Matthew and Granger, 1925)
    Pataecops microdon Reshetov, 1979
  Amynodontidae
    Lushiamynodon? kirghisiensis Belyayeva, 1971
    Teilhardia pretiosa Matthew and Granger, 1926
    Euryodon minimus Xu, Yan, Zhou, Han, and Zhang, 1979
    Caenolophus
  Hyracodontidae
    Triplopus cf. *implicatus* Cope, 1873
    Urtinotherium?
    Forstercooperia
  Chalicotheriidae
    Grangeria canina Zdansky, 1930
ORDER?
    Heptaconodon dubium Zdansky, 1930
RODENTIA
  Ctenodactylidae?
    Tamquammys tantillus Shevyreva, 1971
    Petrokozlovia notos Shevyreva, 1971
  Chapattimyidae
    Saykanomys chalchis Shevyreva, 1971

The similarity of this Far East composite fauna to the contemporary composite fauna of India and Pakistan supports Sahni and Khare (1972) and Sahni and Kumar (1974), who believe that the continental Indian lithospheric plate joined the Sino-Siberian plate by mid-Eocene time and that a land dispersal from north to south or south to north Asia was possible about 49 million years ago.

MIDDLE EOCENE OF AFRICA

Gouiret-el-Azib (Middle or latest Early Eocene) in Algeria has produced a small sample of land mammals that is currently being studied by J. Sudre in Montpellier, France. Dating of this locality was based on charophytes. This sample offers us the only knowledge available about the earlier Paleogene land-mammal radiation in Africa. Evidently much evolutionary diversification was under way on that great continent, diversification mostly conjectural at present. Sudre (1975) described *Azibius trerki* from the Gouiret-el-Azib, concluding that it was a prosimian primate. Gingerich (1977) thought that *Azibius* was a specialized adapid primate. Other workers believe it to be a macroscelidid. Sudre (1979) also has described several hyracoids and *Heliosius,* Order *inc. sed.,* from Gouiret-el-Azib; thus this locality is now the earliest record of hyracoids.

Sites in Egypt, Somalia, Libya, Nigeria, and Mali give us a further hint about the character of the terrestrial and aquatic vertebrate fauna of the African Middle Eocene.

Unfortunately, only Cetacea, Sirenia, Proboscidea, giant ground birds, and fishes are represented. At Gebel-el-Mokattam, near Cairo, a white limestone in the lower part of the Mokattam Formation yielded *Protocetus* (Fraas, 1904); a higher level in the Mokattam produced *Eocetus*, *Eotheroides*, and *Tomistoma* (Kellogg, 1936). In the Fayum area, southwest of Cairo, the Ravine beds with *Prozeuglodon isis* and the underlying Wadi Rayan beds with *Prozeuglodon*, *Eocetus*, *Protocetus*, *Eotheroides*, and *Protosiren* have been referred to the Middle Eocene primarily because of their associated invertebrate marine fauna. At Carcar and at Dalan in Somalia, mid-Eocene sirenians have been collected (R. Savage, 1969). In Libya, R. Savage (1969) has found a mid-Eocene sirenian at Bu-el-Haderait. R. Savage (1969) also notes the mid-Eocene *Moeritherium* from M'Bodione Dadera in Senegal, and Sudre (1979) reports condylarths from there. Andrews (1919) described *Pappocetus*, *Gigantornis*, and *Cosmochelys* and fishes from Ameki, in the Omobialla district of Nigeria. At Gao, Mali, mid-Eocene *Moeritherium* is reported by R. Savage (1969). All the above localities signify a dominantly marine near-shore paleotropical fauna thriving in the southern border of the Tethyan seaway. This seaway evidently extended continuously into southeastern Asia through mid-Eocene time.

BRIDGERIAN

The Bridgerian Mammal Age is based on the aggregate of mammalian fossils from the Bridger Formation of Hayden, southwestern Wyoming. The Bridger is a succession of lacustrine and fluvial somber gray concretionary and sandy mudstone beds intercalated with rusty brown and gray sandstone beds, some of which are stream-channel deposits. The Bridger Formation attains 675 m of thickness locally (West, 1979).

Matthew (1909), following his work with Granger in the Bridger Basin, subdivided the Bridger Formation into fossil-mammal "horizons." Matthew's horizons were essentially members of the formation, which some workers have tried to characterize further as teilzones (local stratigraphic ranges of certain taxa). Matthew's subdivisions of the Bridger have become classic in the literature. They are displayed in the table on the next page.

Without denying that significant mammalian evolution took place during the time represented by Bridger *A* through *E*, we shall treat the interval tokened by the fossil mammals of this succession as a single mammal age, the

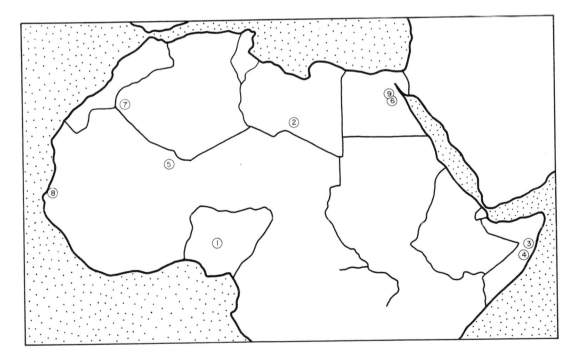

FIGURE 4-23 *Middle Eocene Localities in Africa* (*mostly marine with a land mammal or two*).

1 *Ameki, Nigeria*
2 *Bu-el-Haderait*
3 *Carcar, Somalia*
4 *Dalan, Somalia*
5 *Gao, Mali*

6 *Gebel-el-Mokattam, Egypt*
7 *Gouiret-el-Azib, Algeria*
8 *M'Bodione Dadera, Senegal*
9 *Ravine beds sites, Egypt*

~150 m	Bridger E	(with very few fossils)	
~100 m	Bridger D	upper white layer	Twin Buttes Mbr.
		Lone Tree white layer	
~90 m	Bridger C	Burntfork white layer	
		Sage Creek white layer	
~125 m	Bridger B	Cottonwood white layer	Black's Fork Mbr.
		G bed of McGrew and Sullivan (1970)	
~65 m	Bridger A		

Bridgerian. Bridger *A* through *E* probably represents less than 2 million years, according to correlated radiometric dating.

Wolfe and Hopkins (1967) and MacGinitie (1969) have characterized the flora and climate of Western Nearctica during approximate Bridgerian time as generally warm and humid, ranging from what is usually termed warm-temperate to subtropical. There were widespread woodland districts with abundant arboreal vegetation composed of "tropical" families of plants. The southern and southeastern part of Nearctica evidently thrived under comparable, and perhaps slightly warmer, climate. Axelrod and Bailey (1968, 1969) and Bailey (1964, 1966), looking at present-day altitudinal vectors and 20-year climate records around the world, concluded that such floras as these may have been in generally cooler climates than what is usually interpreted. These climates, they believe, had a high degree of *equability;* that is, they were frost-free, with an average temperature approaching 60 °F and relatively small deviation from that temperature. Climates of this sort are found today in many medium-to-high-elevation equatorial districts (e.g., Bogotá, Colombia) and in certain coastal districts of higher latitude (e.g., San Diego, California, and parts of Patagonia). Thus "tropicality" in fossil floras may reflect more of an equable-humid climate than a hot-humid climate. The wealth of Bridgerian mammals that can be interpreted as brachyodont herbivores, ungulates, and carnivorous forms with nonspecialized feet, and as arboreal animals supports the concept of widespread, lush woodland areas and equable climate.

The mammalian fauna of Bridger *A* and *D* has been carefully summarized by Gazin (1976). Our composite Bridgerian mammalian fauna listing, which follows, adds taxa from other known Bridgerian districts to Gazin's list.

Composite Bridgerian Mammalian Fauna

(From Bridger Formation *A–E* and Fowkes and Lower Washakie Formations.)

(MULTITUBERCULATA—None yet recorded, but they are in later and earlier faunas.)

MARSUPIALIA
  Didelphidae
    Herpetotherium marsupium Troxell, 1923F
    H. innominatum Simpson, 1928G   —Crochet (1979)
    H. knighti McGrew, 1959
LEPTICTIDA
  Leptictidae
    Palaeictops bridgeri (Simpson, 1959C)
    Hypictops syntaphus Gazin, 1949B (*nomen dubium*)

CIMOLESTA
  Palaeoryctidae
    Didelphodus altidens (Marsh, 1872I)   —McKenna,
      Robinson, and Taylor (1962)
PANTOLESTA
  Pantolestidae
    Pantolestes phocipes Matthew, 1909D
    P. elegans (Marsh, 1872I)
    P. longicaudus Cope, 1872N
    P. natans Matthew, 1909D
    P. intermedius Matthew, 1909D
APATOTHERIA
  Apatemyidae
    Apatemys bellus Marsh, 1872I
    A. bellulus Marsh, 1872I
    A. rodens Troxell, 1923E
TAENIODONTA
  Stylinodontidae
    Stylinodon mirus Marsh, 1874C
    S. inexplicatus Schoch and Lucas, 1981
CREODONTA
  Oxyaenidae
    Patriofelis ulta Leidy, 1870G
    P. ferox (Marsh, 1872I)
  Hyaenodontidae
    Proviverroides piercei Bown, 1982
    Limnocyon verus Marsh, 1872G
    Thinocyon velox Marsh, 1872I
    T. medius (Wortman, 1902A)
    T. comptus (Marsh, 1872I?)
    T. mustelinus Matthew, 1909D
    T. minimus Matthew, 1909D
    Proviverra rapax (Leidy, 1872B)
    P. grangeri (Matthew, 1906C)
    P. minor (Wortman, 1902A)
    P. major (Wortman, 1902A)
    Tritemnodon agilis (Marsh, 1872I)
    Machaeroides eothen Matthew, 1909D
CARNIVORA
  Miacidae
    Miacis parvivorus Cope, 1872PP
    M. latidens Matthew, 1915D
    M. sylvestris (Marsh, 1872I)
    M. hargeri (Wortman, 1901B)
    Vulpavus palustris Marsh, 1871E
    V. profectus Matthew, 1909D
    V. ovatus Matthew, 1909D
    Palaearctonyx meadi Matthew, 1909D
    Oödectes herpestoides Wortman, 1901B
    O. proximus Matthew, 1909D
    Uintacyon edax Leidy, 1872J
    U. jugulans Matthew, 1909D

U. major Matthew, 1909D
U. vorax Leidy, 1872J
U. bathygnathus (Scott, 1888B)
Viverravus gracilis Marsh, 1872G
V. sicarius Matthew, 1909D
V. minutus Wortman, 1901B

INSECTIVORA
  Erinaceoidea
    Entomolestes grangeri Matthew, 1909D
    Talpavus nitidus Marsh, 1872G
    Scenopagus edenensis (McGrew, 1959)
    S. priscus (Marsh, 1872G?)
    Centetodon pulcher Marsh, 1872I   —Krishtalka and West (1979); Lillegraven, McKenna, and Krishtalka (1981)
    C. bembicophaga Lillegraven, McKenna, and Krishtalka, 1981
  Soricoidea
    Nyctitherium serotinum (Marsh, 1872I)
    N. dasypelix (Matthew, 1909D)
    N. velox Marsh, 1872G
    Pontifactor bestiola West, 1974

CHIROPTERA
  MICROCHIROPTERA, Fam. indet.   —McKenna, Robinson, and Taylor, 1962

PRIMATES?
  Uintasoricidae
    Uintasorex parvulus Matthew, 1909D
    Alveojunctus minutus Bown, 1982
  Microsyopidae
    Microsyops elegans (Marsh, 1871D)
    M. scottianus Cope, 1881D
    M. annectens (Marsh, 1872I)
    M. schlosseri Wortman, 1903A

PRIMATES
  Paromomyidae
    Elwynella oreas Rose and Bown, 1982
  Adapidae
    Pelycodus? sp.
    Notharctus robustior (Leidy, 1872B)
    N. tenebrosus Leidy, 1870Q
    Smilodectes gracilis (Marsh, 1871D)
  Omomyidae
    Gazinius amplus Bown, 1979
    Omomys carteri Leidy, 1869B (first fossil-primate taxon described from North America)
    Hemiacodon gracilis Marsh, 1872I
    Uintanius ameghini (Wortman, 1904A)
    Washakius insignis Leidy, 1873B
    Trogolemur myodes Matthew, 1909D
    Shoshonius? laurae McGrew, 1959
    Anaptomorphus aemulus Cope, 1872TT
    A. wortmani Gazin, 1958B
    Strigorhysis rugosus Bown, 1979
    S. bridgerensis Bown, 1979
    Aycrossia lovei Bown, 1979

TILLODONTIA
  Esthonychidae
    Esthonyx acutidens Cope, 1881D
    Trogosus hillsi Gazin, 1953
    T. grangeri Gazin, 1953
    T.? latidens (Marsh, 1874C)
    T. castoridens Leidy, 1871F

    T. hyracoides (Marsh, 1873H)
    Tillodon fodiens (Marsh, 1875B)
DINOCERATA
  Uintatheriidae
    Bathyopsis middleswarti Wheeler, 1961
    Uintatherium anceps (Marsh, 1871D)
ARTIODACTYLA
  Dichobunidae
    Microsus cuspidatus Leidy, 1870Q
    Homacodon vagans Marsh, 1872G
    Lophiohyus alticeps Sinclair, 1914A
    Antiacodon pygmaeus Cope, 1872M, 1873E
    A. venustus Marsh, 1872I
    A. furcatus (Cope, 1873E)
    Neodiacodexis emryi (West and) Atkins, 1970
    Helohyus validus Marsh, 1872I
    H. plicodon Marsh, 1872I
    H. lentus (Marsh, 1871D)
    H. milleri Sinclair, 1914A
ACREODI
  Mesonychidae
    Mesonyx obtusidens Cope, 1872M
    Synoplotherium vorax (Marsh, 1876H)
    Harpagolestes macrocephalus Wortman, 1901B
CONDYLARTHRA
  Hyopsodontidae
    Hyopsodus paulus Leidy, 1870O
    H. minusculus Leidy, 1873B
    H. marshi Osborn, 1902C
    H. despiciens Matthew, 1909D
    H. lepidus Matthew, 1909D
  Phenacodontidae
    Phenacodus primaevus Cope, 1873D   —West and Atkins (1970)
PERISSODACTYLA
  Equidae
    Orohippus pumilus Marsh, 1872I
    O. agilis Marsh, 1873H
    O. sylvaticus (Leidy, 1870T)
    O. major Marsh, 1874B
    O. progressus Granger, 1908A
  Brontotheriidae
    Palaeosyops fontinalis Cope, 1873EE
    P. paludosus Leidy, 1870Q
    P. major Leidy, 1871B
    P. leidyi Osborn, 1908A
    P. grangeri Osborn, 1908A
    P. copei Osborn, 1908A
    Mesatirhinus petersoni Osborn, 1908A
    M. megarhinus (Earle, 1891A)
    Telmatherium cultridens (Osborn, Scott, and Speir, 1878A)
    T. validum Marsh, 1872G
    Manteoceras manteoceras (Osborn, 1899A)
    Limnohyops priscus Osborn, 1908A
    L. matthewi Osborn, 1908A
    L. monoconus Osborn, 1908A
  Isectolophidae
    Isectolophus latidens Osborn, Scott, and Speir, 1878A
  Helaletidae

Helaletes nanus Marsh, 1871D

H. intermedius (Osborn, Scott, and Speir, 1878A)

Dilophodon minusculus Scott, 1883A

Helaletidae or Hyrachyidae

Hyrachyus eximius Leidy, 1871K

H. princeps Marsh, 1872G

H. affinis (Marsh, 1871D)

PHOLIDOTA?

Metacheiromyidae

Metacheiromys marshi Wortman, 1903A

M. dasypus Osborn, 1904A

M. tatusia Osborn, 1904A

M. osborni Simpson, 1931A

Epoicotheriidae

Tetrapassalus mckennai Simpson, 1959A

T. proius West, 1973A

INC. SED.

Ithygrammodon cameloides Osborn, Scott, and Speir, 1878A

RODENTIA

Ischyromyoidea

Paramys delicatus Leidy, 1871L

P. delicatior Leidy, 1871L

P. cf. *excavatus* Loomis, 1907A

Reithroparamys delicatissimus (Leidy, 1871L)

R. matthewi A. Wood, 1962B

Pseudotomus robustus (Marsh, 1872I)

P. hians Cope, 1872N

Sciuravus nitidus Marsh, 1871E

S. bridgeri R. Wilson, 1938B

S. eucristidens Burke, 1937

S.? rarus R. Wilson, 1938B

Pauromys perditus Troxell, 1923B

P. schaubi A. Wood, 1959A

Microparamys minutus (R. Wilson, 1937C)

M. wyomingensis A. Wood, 1959C

M.? wilsoni A. Wood, 1962B

Tillomys senex Marsh, 1872I

T.? parvidens (Marsh, 1872I)

Leptotomus parvus A. Wood, 1959C

L. bridgerensis A. Wood, 1962B

L. grandis A. Wood, 1962B

Taxymys lucaris Marsh, 1872I

T. cuspidatus Bown, 1982

T.? progressus R. Wilson, 1938B

Mysops minimus Leidy, 1871L

M. parvus (Marsh, 1872I)

M. fraternus Leidy, 1873B

Thisbemys corrugatus A. Wood, 1962B

T. plicatus A. Wood, 1962B

Ischyrotomus horribilis A. Wood, 1962B

I. oweni A. Wood, 1962B

I. superbus (Osborn, Scott, and Speir, 1878A)

Family, *Inc. Sed.*

Prolapsus junctionis A. Wood, 1973

Bridgerian is currently identified on the basis that some of the following common genera of land mammals existed jointly: *Herpetotherium* (= "Peratherium marsupium" and other species), *Centetodon, Pantolestes, Patri-*

ofelis, Thinocyon, Proviverra (or "*Sinopa*"), *Miacis, Uintacyon, Entomolestes, Nyctitherium, Uintasorex, Microsyops, Notharctus, Omomys, Washakius, Trogosus, Uintatherium, Homacodon, Helohyus, Mesonyx, Hyopsodus, Orohippus, Palaeosyops, Mesatirhinus, Helaletes, Hyrachyus, Paramys, Sciuravus, Microparamys,* and *Ischyrotomus.* Only a few of these genera, plus some less common ones not listed above, are known from Bridgerian alone, and some of the genera may be recorded from earlier *and* later strata. Most of the species assigned to Bridgerian genera are confined to Bridgerian.

Of the 77 identified Bridgerian genera, 38 (49 percent) are shared with the preceding Wasatchian and 37 (48 percent) carry into the succeeding Uintan. Four Bridgerian families (Oxyaenidae, Esthonychidae, Phenacodontidae, and Metacheiromyidae), all sparsely represented, have no later record.

The Bridgerian fauna is noted for its numerous large perissodactyls, *Hyopsodus,* and ischyromyoid rodents. Artiodactyls appear to be much fewer compared with perissodactyls, and show only adumbration of the great morphologic-taxonomic and adaptive radiation that is witnessed in the next, Uintan mammal age. Pantodonts, so abundant as fossils in the preceding Wasatchian, were evidently exterminated in Nearctica by Bridgerian time. The exact reason for their abrupt demise ca. 50 million years ago is a mystery, but perhaps it was the result of competition from the suddenly diversified and abundant large perissodactyls—especially the brontotheriids. The uintatheres become surprisingly common in the rocks of the upper part of the Bridger Formation. Abundance and diversity of the carnivorous hyaenodontids and the miacids may be correlated with their abundant probable prey: *Hyopsodus* and the ischyromyoid rodents. Multituberculates and *Phenacolemur,* relatively abundant in Wasatchian, seem to have been occupying habitats not currently sampled in beds of Bridgerian age, for these two taxa are unknown in Bridgerian rocks but known from later strata in Wyoming and Montana (Robinson, Black, and Dawson, 1964; Black, 1967). Krishtalka and

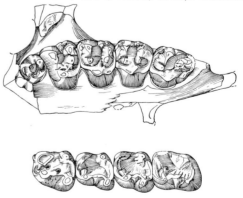

FIGURE 4-24 *Bridgerian Mammal. Leptotomus bridgerensis A. Wood, 1962B. Upper and lower dentition. ×2½. Figure 33 in A. Wood (1962B). (Published with permission of the Philadelphia Academy of Sciences.)*

FIGURE 4-25 *Bridgerian Localities.*

1 *Aldrich Creek—Ishawooa Creek—Ishawooa Mesa*
2 *Bridger Basin localities—Bridger Formation: Beaver Creek, Behunin, Birch Creek, Black's Fork, Blue Rim—Kistner 16, Burnt Fork, Burnt Fork Post Office, Buttes of Dry Creek, Church Buttes, Cottonwood Creek, Dry Creek, East Hill, Fort Bridger, George's Gorge, Granger South 1 and 2, Granger Station, Grizzly Buttes, Grizzly Buttes West, Ham's Fork, Henry's Fork, Henry's Fork Divide, Henry's Fork Hill, Lane Meadows, Little America, Lone Tree, Lone Tree North, LSV, McCann Ranch, Millersville, Notharctus Knob, Opal, Reservoir, Spanish John's Meadow, Stuck Truck, Summers Dry Creek, Trap Hutch Quarry, Twin Buttes, Twin Buttes North, Wight, 28FW, and others.*
3 *Big Sandy 1, 2, 3, 4*
4 *Black Spring Reservoir 1—Promontory Bluff, Sage Creek (Wyoming)—Sage Creek Main Quarry (Wyoming), Steamboat Springs (Wyoming)*
5 *Canoe Formation sites*

6 *Cathedral Bluff Member sites*
7 *Cedar Mountain North—York*
8 *Farson, Green, Hyopsodus Hill, Tabernacle Butte, Hyopsodus Hill East, Misery Quarry, Morrow Creek Member sites*
9 *Flattop Quarry, North Mesa 1 and 2, Vass Quarry*
10 *Green Cove 1*
11 *Haystack Mountain, lower*
12 *Nut Bed—Clarno Formation*
13 *Powder Wash—Powder Springs*
14 *Quitman South*
15 *Roslyn Formation sites (no mammals)*
17 *Sand Wash Basin*
18 *Shark River Marl, New Jersey*
19 *Soldier Summit*
21 *State Line Quarry—Thomas Canyon*
22 *Sublette County 1*
23 *W.R. Wilson Coal Mine*
24 *U.S. National Museum site (unpublished, 1981)*
25 *Base of Pruett Tuff (age?)*

Setoguchi (1977) believe that the Bridgerian sample accumulated in lowland areas, compared with the more "upland" accumulations representing Uintan local faunas such as the Badwater Creek district, but the geologic evidence for their interpretation is nebulous.

LATER EOCENE OF EUROPE

In the subepoch of the Late Eocene dating 44–35 million years ago, we place the Bartonian Stage-Age, as it is elaborated by Blondeau et al. (1965), and the correlated chronostratigraphic and geochronologic units around the world. Some of these units are termed Middle Eocene by certain workers. The Auversian (oldest), Marinesian, and Ludian (youngest) have been included in the Bartonian as sub-stages-subages, although Priabonian (from Italy) is preferred to Ludian by many workers.

A small but significant land-mammal assemblage has been obtained from the marine Auversian sands at Le Guépelle, which along with the collections from Arcis-le-Ponsart and Latilly has permitted the large collections from the Egerkingen fissure fills in Switzerland to be correlated with Auversian. Another Paris Basin local fauna, Grisolles, found intercalated in the Marinesian Calcaire de St.-Ouen, correlates well with the extensive fauna from Robiac in southern France. The type-Barton beds in southern England have produced only *Cebochoerus* cf. *helveticus*, *Plagiolophus* sp., *Palaeotherium* sp., and *Lophiodon* cf. *lautricense*. Because of this paucity, we have chosen the best-documented and best-known of the Bartonian-correlative local faunas, *Robiac*, for typifying the land-mammal age (Robiacian).

ROBIACIAN

Robiacian (=late Rhenanium of Fahlbusch, 1976) is named after the local fauna from two principal quarries in the lower part of the Calcaire de Fons southeast of the village of Robiac and about 2 km west of St. Mammert, Department of Gard, France. We are fortunate that this important locality has been reexcavated by J. Sudre and his colleagues from the University of Montpellier. Sudre used underwater sieving on about 4 tons of fossiliferous matrix and got representation of twice as many mammalian species as were known previously from this locality. His list of taxa (Sudre, 1969) now includes 46 mammalian species and is a good representative for all Robiacian local faunas. The complete mammalian assemblage, ranging through a stratal thickness of 12 m, is here listed.

Local Fauna from Robiac

MARSUPIALIA
  Didelphidae
    Amphiperatherium bourdellense Crochet, 1979
    A. fontense Crochet, 1979
    Peratherium sudrei Crochet, 1979
APATOTHERIA
  Apatemyidae
    Heterohyus sudrei Sigé, 1973
    H. nanus Teilhard, 1921
CARNIVORA
  Miacidae
    Quercygale angustidens (Filhol, 1872)
    Simamphicyon helveticus (Pictet and Humbert, 1869)

INSECTIVORA
  Soricoidea
    Saturninia mamertensis Sigé, 1976
    S. grisollensis Sigé, 1976   (Robiac?)
    S. hartenbergeri Sigé, 1976
CHIROPTERA
  Palaeochiropterygidae
    "Archaeonycteris"
    Palaeochiropteryx
  Hipposideridae?
  Family indet.
PRIMATES
  Adapidae
    Adapis sudrei Gingerich, 1979
    A. aff. *parisiensis* Blainville, 1849
  Omomyidae or Tarsiidae
    Microchoerus aff. *erinaceous* (Wood, 1846)
    Pseudoloris?
ARTIODACTYLA
  Dichobunidae
    Dichobune?
  Choeropotamidae
    Choeropotamus lautricense Noulet, 1870
  Cebochoeridae
    Cebochoerus campichii (Pictet, 1855–1857)
  Dacrytheriidae
    Tapirulus aff. *schlosseri* Stehlin, 1910
    Robiacina minuta Sudre, 1969
  Anoplotheriidae
    Catodontherium robiacense Depéret, 1906
  Xiphodontidae
    Dichodon aff. *cervinum* Owen, 1841
    Haplomeryx aff. *picteti* Stehlin, 1910
    Xiphodon castrense Kowalevsky, 1873
  Amphimerycidae
    Pseudamphimeryx aff. *renevieri* Pictet and Humbert, 1869
PERISSODACTYLA
  Equidae
    Pachynolophus sp.
    Lophiotherium robiacense Depéret, 1917
    Anchilophus cf. *desmaresti* Gervais, 1852
    A. cf. *gaudini* Pictet and Humbert, 1869
  Palaeotheriidae
    Leptolophus stehlini Remy, 1965
    Palaeotherium castrense Noulet, 1863
    P. duvali Pomel, 1853
  Helaletidae
    Chasmotherium cartieri Rütimeyer, 1862
  Lophiodontidae
    Lophiodon lautricense Noulet, 1857
RODENTIA
  Ischyromyoidea
    Plesiarctomys
    Ailuravinae, gen. indet.
  Theridomyidae
    Paradelomys tobieni (Thaler, 1966)
    Suevosciurus 1 sp.
    Suevosciurus 2 sp.
    Adelomyinae 1 sp.
    Adelomyinae 2 sp.
    Remys

Pseudoltinomys?
Gliridae
Gliravus robiacensis Hartenberger, 1965

Composite Robiacian Mammalian Fauna

MARSUPIALIA
Didelphidae
Amphiperatherium bourdellense Crochet, 1979
A. bastbergense Crochet, 1979
A. giselense (Heller, 1936)
A. minutum (Aymard, 1846)
A. lamandini (Filhol, 1876)
Peratherium sudrei Crochet, 1979
P. lavergnense Crochet, 1979
P. bretouense Crochet, 1979
APATOTHERIA
Apatemyidae
Heterohyus sudrei Sigé, 1973
H. europaeus (Rütimeyer, 1890)
H. gracilis (Stehlin, 1916)
CREODONTA
Hyaenodontidae
Hyaenodon requieni Gervais, 1848–1852
H. gervaisi Martin, 1906
H. brachyrhynchus Blainville, 1841
H. minor Gervais, 1848–1852
H. heberti Filhol, 1876
Allopterodon therioides (Van Valen, 1965G)
A. phonax (Van Valen, 1965G)
Proviverra typica Rütimeyer, 1862
Cynohyaenodon trux Van Valen, 1965G
G. ruetimeyeri Depéret, 1917
C. cayluxi Filhol, 1873
C. minor Filhol, 1877
Prototomus torvidus Van Valen, 1965G
Paroxyaena galliae (Filhol, 1881)
Oxyaenoides schlosseri (Rütimeyer, 1891)
Pterodon dasyuroides Blainville, 1839
CARNIVORA
Miacidae
Miacis
Quercygale angustidens (Filhol, 1872)
Q. helvetica (Rütimeyer, 1862)
Simamphicyon helveticus (Pictet and Humbert, 1869)
Miacidae indet.
INSECTIVORA
Soricoidea
Saturninia grandis Sigé, 1976
S. intermedia Sigé, 1976
S. mammertensis Sigé, 1976
S. grisollensis Sigé, 1976
S. hartenbergeri Sigé, 1976
Scraeva
CHIROPTERA
Palaeochiropterygidae
Palaeochiropteryx
Archaeonycteris
Hipposideridae
Hipposideros egerkingensis Revilliod, 1922
Rhinolophidae
Rhinolophus priscus Revilliod, 1920

Rhinolophoid gen. indet.
Vespertilionidae
Stehlinia ruetimeyeri (Revilliod, 1922)
S. pusilla (Revilliod, 1922)
Emballonuridae
Vespertiliavus gracilis Revilliod, 1920
PRIMATES
Adapidae
Adapis sudrei Gingerich, 1977
A. ruetimeyeri (Stehlin, 1912)
A. priscus Stehlin, 1912
A. sciureus (Stehlin, 1916)
A. aff. *magnus* Filhol, 1874
A. laharpei (Pictet and Humbert, 1869)
A. capellae (Crusafont, 1968)
Cercamonius brachyrhynchus (Stehlin, 1912) (age?)
Anchomomys gaillardi (Stehlin, 1916)
A. stehlini Gingerich, 1977
Caenopithecus lemuroides (Rütimeyer, 1862)
Protoadapis
Periconodon pygmaeus (Rütimeyer, 1890)
Omomyidae or Tarsiidae
Microchoerus erinaceus Wood, 1846
Pseudoloris isabenae (Crusafont, 1968)
P. crusafonti Louis and Sudre, 1975
P. parvulus (Filhol, 1889–1890)
Nannopithex filholi (Chantre and Gaillard, 1897)
Necrolemur zitteli Schlosser, 1888
PRIMATES?
Amphilemuridae
Amphilemur
"*Anchomomys*" cf. *grisollensis* Louis and Sudre, 1975
Gesneropithex
ARCTOCYONIA
Paroxyclaenidae
Pugiodens
ARTIODACTYLA
Dichobunidae
Dichobune robertiana Gervais, 1848–1852
Hyperdichobune langi (Rütimeyer, 1891)
H. nobilis Stehlin, 1906
Meniscodon europaeum Rütimeyer, 1888
Mouillacitherium cartieri (Rütimeyer, 1891)
M. elegans Filhol, 1882
Lophiobunodon rhodanicum Depéret, 1908
L. minervoisensis Depéret, 1908
Choeropotamidae
Choeropotamus lautricense Noulet, 1870
Cebochoeridae
Cebochoerus campichii (Pictet, 1855–1857)
C. jurensis (Stehlin, 1908)
C. suillus (Gervais, 1848–1852)
C. helveticus (Pictet and Humbert, 1869)
Tapirulus depereti Stehlin, 1910
T. majori Stehlin, 1910
T. schlosseri Stehlin, 1910
Leptotheridium traguloides Stehlin, 1910
Mixtotheriidae

Mixtotherium priscum Stehlin, 1908
M. gresslyi Rütimeyer, 1891
M. infans Stehlin, 1910
Dacrytheriidae
Catodontherium robiacense Depéret, 1906
C. buxgovianum Stehlin, 1910
C.? paquieri Stehlin, 1910
Dacrytherium priscum Stehlin, 1910
D. cf. *elegans* (Filhol, 1884)
Robiacina minuta Sudre, 1969
Haplobunodontidae
Haplobunodon mulleri (Rütimeyer, 1862)
H. solodurense Stehlin, 1908
Anthracobunodon louisi Sudre, 1978
Rhagatherium kowalevskyi Stehlin, 1908
Xiphodontidae
Paraxiphodon cournovense Sudre, 1978
Dichodon lugdunensis Sudre, 1972
D. cuspidatum Owen, 1848
D. ruetimeyeri Stehlin, 1910
Haplomeryx picteti Stehlin, 1910
H. egerkingensis Stehlin, 1910
Xiphodon castrense Kowalevsky, 1873
Amphimerycidae
Pseudamphimeryx schlosseri (Rütimeyer, 1891)
P. renevieri Pictet and Humbert, 1869

Equidae
Pachynolophus duvali Pomel, 1847
P. cesserasicus Gervais, 1848–1852
Propalaeotherium helveticum Savage, Russell, and Louis, 1965
P. isselanum Blainville, 1846
P. parvulum (Laurillard, 1849)
Anchilophus dumasi (Gervais, 1848–1852)
A. desmaresti Gervais, 1848–1852
A. depereti Stehlin, 1905
A. gaudini Pictet and Humbert, 1869
Lophiotherium robiacense Depéret, 1917
L. pygmaeum Depéret, 1901
Palaeotheriidae
Plagiolophus annectens (Owen, 1847)
P. cartieri Stehlin, 1904
Paraplagiolophus sp.
Palaeotherium castrense Noulet, 1863
P. eocaenum Gervais, 1875
P. magnum Cuvier, 1804
P. ruetimeyeri Stehlin, 1904
P. siderolithicum (Pictet and Humbert, 1869)
P. pomeli Franzen, 1968
Leptolophus stehlini Remy, 1965
Brontotheriidae
Brachydiastematherium transylvanicum Böckh and Maty, 1875
Isectolophidae

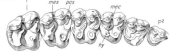

A

B

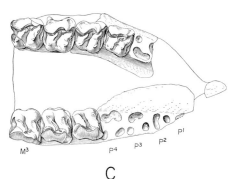

C

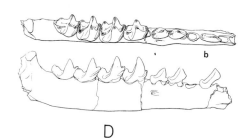

D

FIGURE 4-26   *Robiacian Mammals.*

A   Peratherium sudrei Crochet, 1979. Upper molars. ×5. Figures 170 and 171 in Crochet (1980).

B   Lophiotherium robiacense Depéret, 1917. Upper and lower dentition. ×1/2. Figures 39 and 40 in Savage, Russell, and Louis (1965).

C   Anchilophus desmaresti Gervais, 1848–1852. P⁴–M³, M¹⁻³, and lower P. ×3/8. Figure 41 in Savage, Russell, and Louis (1965).

D   Saturninia hartenbergeri Sigé, 1976. Lower jaw with C, P₂₋₄, M₁₋₃. ×2½. Figure 42 in Sigé (1976).

(Figure A published with permission of the Université des Sciences et Téchniques du Languedoc. Figures B and C published with permission of the University of California Press. Figure D published with permission of the Muséum National d'Histoire Naturelle, Paris.)

Paralophiodon leptorhynchus (Filhol, 1888)
P. isselense (Fischer, 1829)
Lophiodontidae
  Lophiodon lautricense Noulet, 1851
  L. rhinocerodes Rütimeyer, 1862
  L. cf. *cuvieri* Watelet, 1864
  L. thomasi Depéret, 1906
  L. glandicus Astre, 1960
  L. tapirotherium Desmarest, 1822
Helaletidae
  Chasmotherium cartieri Rütimeyer, 1862
Helaletidae or Hyrachyidae
  Hyrachyus minimus (Fischer, 1829)
Hyracodontidae
  Prohyracodon orientalis Koch, 1897
Chalicotheriidae
  Lophiaspis occitanicus Cuvier, 1812

RODENTIA
Ischyromyoidea
  Plesiarctomys spectabilis (Major, 1873)
  P. hurzeleri Wood, 1970
  Ailuravus picteti Rütimeyer, 1891
  A. stehlinschaubi Wood, 1976
Theridomyidae
  Paradelomys crusafonti (Thaler, 1966)
  P. sp.
  Theridomys varleti Hartenberger and Louis, 1976
  Protadelomys cartieri (Stehlin and Schaub, 1951)
  P. lugdunensis Hartenberger, 1969
  Pseudoltinomys
  Treposciurus

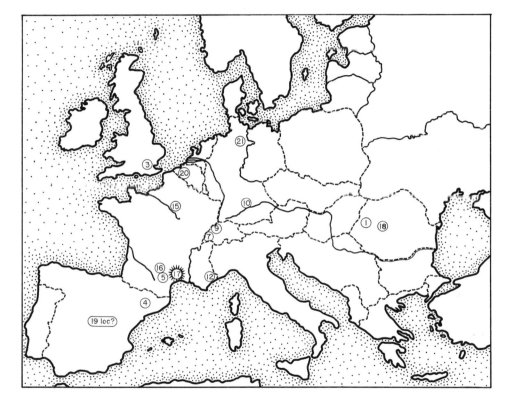

FIGURE 4-27 *Representative Robiacian Localities.*

1   Andrashaza   (age?)
2   Aumelas
3   Bracklesham Beds sites
4   Capella-Benabarre
*5   Castrais Sites: Alzieux, Bouffard, Braconnac,
    Campans, Castayral, Castres, Castres-Roc-de-Lu-
    nel, La Badayre, La Bartie, La Ferrière, La Fosse,
    La Guittardie-Montdragon, La Marcelle, La Mas-
    sale, La Maurianne, La Milette, Lautrec general,
    Lautrec propriété Causse, Le Bretou, Les Bessous,
    Molinière, Montespieu, Peyregoux, Realmont,
    Roc-de-Lunel, Saix, Sicardens, Vielmur-sur-l'A-
    gout, Viviers-la-Montagne, etc.*
5   Cesseras—La Livinière
7   Chamblon-Eclepens (Mormont)
8   Echelles

9   Egerkingen
10 Heidenheim   (age?)
11 Issel
12 La Croix
13 Lissieu (in part)
14 Moutiers
*15 Paris Basin localities: Arcis-le-Ponsart, Latilly,
    Batignolles, Berville, Gare du Nord, Le Guépelle,
    Le Ruel, Nogent l'Artaud, Parc Monceau, "Sables
    moyens," St. Denis, St. Ouen, Sergy, Grisolles*
16 Quercy site—le Bretou
17 Robiac
18 Sacel   (age?)
19 Soterranya   (age?)
20 Forest
21 Egeln   (age?)

Elfomys tobieni (Thaler, 1966)
E. cf. *parvulus*
Sciuroides
Suevosciurus romani Hartenberger, 1973
    S. *russelli* Hartenberger and Louis, 1976
Gliridae
    Gliravus robiacensis Hartenberger, 1965
Family inc. sed.
    Remys minimus Hartenberger, 1973

A characterizing assemblage of genera for the Robiacian is *Plesiarctomys, Anchilophus, Lophiotherium, Choeropotamus, Cebochoerus, Xiphodon,* and most of the genera listed for the Lutetian. Differences from the Lutetian are largely at species level, as are the differences from the succeeding Headonian mammal age. However, the giant lophiodons of the Robiacian represent the final stand of this European perissodactyl family.

We are able to distinguish only approximate correlatives of Robiacian among the mammalian faunas of other parts of the world. Robiacian may be contemporaneous, at least partly, to the "Middle" Eocene faunas of Asia and Africa, for example, Ganda Kas and Kalakot-Rajauri. And Robiacian time may be about the same interval in earth's history as the early Uintan mammal age in North America, but there is no "tight" faunal correlation to substantiate such a fine correlation.

Robiacian, like the Lutetian, was a time of predominantly evolution-in-situ for European land mammals. It was the heyday and final appearance of the giant lophiodontids. Practically all the recognized species and genera of the Robiacian appear to have likely ancestors in the Lutetian faunas. There was continued but seemingly stabilized abundance and no further significant taxonomic radiation of the established stocks of European perissodactyls (equoids and tapiroids), accompanied by evident ascendancy in numbers and diversity of the artiodactyls and theridomyoid rodents. Many families of Robiacian artiodactyls were essentially endemics: cebochoerids, anoplotheriids, and xiphodontids. The tremendous biomass of herbivores in the samples relative to carnivores and insectivores must token a taphonomic bias favoring preservation and collecting of these probable herd ungulates and the rodents.

Brachydiastematherium, brontotheriid from the Siebenbürgen–Transylvanian Alps area of northern Romania, was the only member of this family—so abundant and diversified in Asia—appearing in the Eocene of Europe, according to present records.

Of the 84 known Robiacian genera, about 43 (51 percent) are holdovers from the Lutetian fauna, and about 53 (63 percent) carry into Headonian faunas. Thirty-one of the Robiacian genera (37 percent) and 4 families (Paroxy-

claenidae, Lophiodontidae, Isectolophidae, and Hyracodontidae) do not appear in the later faunas of Europe.

LATER EOCENE—LUDIAN AND HEADONIAN

The Ludian Substage-Subage was proposed as a stage by A. de Lapparent (1893, *Traité de Géologie,* 3d ed., p. 1219). Later (5th ed.) de Lapparent tried to suppress the name. This unit was based on the gypsiferous clay and marl beds bearing *Pholadomya ludensis* near Ludes, northeastern Paris Basin. The general equivalents of the Ludian are in the upper part of the succession typifying the Bartonian Stage in southern England (Headon, Osborne, and Bembridge Beds), in at least the upper part of the Priabonian Stage of northern Italy, in the lower part of the Tongrian Stage of Belgium, in the lower part(?) of the Lattorfian Stage in Germany, and in the Almaian Stage in the USSR. The Ludian and its equivalents are controversial with respect to recognition of an Eocene-Oligocene boundary also. This continued controversy is the aftermath of nondiscriminate typification of the Oligocene Series-Epoch by Beyrich (1854), compounded by uncertainties in correlation from one depositional basin to another across Europe and to other parts of the world. For example, Denizot (1968) assigned the renowned fossiliferous section at Montmartre in Paris (the "horizon with *Lucina inornata,* the Middle Gypsum and the Montmartre Gypsum") to the lower part of the Tongrian Stage (typified in Belgium), above the Ludian as characterized above, and to the lowest Oligocene. Thaler (1965) also placed the Montmartre, containing his biochronologic zone of *Isoptychus aubreyi* (a theridomyid rodent) in the Lower Oligocene. And Schmidt-Kittler (1971) puts the proposed succession of La Débruge-to-Montmartre-to-Frohnstetten (rodent faunas) into the Lower Oligocene. We have chosen, however, to follow Blondeau et al. (1965), Cavelier (1979), and Brunet (1979) by assigning the Montmartre, together with its general correlatives in Germany (Frohnstetten), in the south of France (Euzet and La Débruge), and in the south of England (Headon Hill–Headon Cliffs) to the Upper Eocene Subseries.

Floras from the Bartonian Stage of England have been studied by Chandler (1925, 1960, 1961) and Chateauneuf (1980) and indicate a constant equable climate for coastal western Europe. Rich woodlands and woodland-savanna must have provided the dominant scene of the European lowlands through the Late Eocene and into the Oligocene. Abundance of brachyodont perissodactyls and artiodactyls supports this reconstruction of the Late Eocene landscape.

Bosma (1974) proposed Headonian as a stage on the basis of her studies, with A.N. Insole, of the Eocene strata on the Isle of Wight, England. The 1975 International Symposium on mammalian stratigraphy of the European area (Fahlbusch, 1976) adopted Headonian for use as a mammal age also. It is named from the fossiliferous Headon Beds on the Isle of Wight and on the north-adjacent mainland of southern England. The collection of fossils from these beds is the most diversified sample of land mammals from the late Eocene of western Europe. Generally correlative and famous sites such as Montmartre (thought to be a little later than the Headon), Euzet-les-Bains, and La Débruge in France have produced an enormous amount of palaeotheriid, artiodactyl, and rodent specimens. Some of these specimens have been dispersed on exchange to many of the paleontological study collections in various parts of Europe and North America.

The list of taxa from the Headon Beds, according to Crochet (1980), Bosma (1974), and Bosma and Insole (1972, 1976), arranged by our system, comprises the following.

Headonian Mammalian Fauna

MARSUPIALIA
    Didelphidae
        Amphiperatherium spp.
PANTOLESTA
    Pantolestidae
        Cryptopithecus major (Lydekker, 1887)
        Dysperna hopwoodi Cray, 1973
CREODONTA
    Hyaenodontidae
        Hyaenodon minor Gervais, 1848–1852
CARNIVORA
    Miacidae
        Quercygale hastingsiae (Davies, 1884)
INSECTIVORA
    Soricoidea
        Scraeva woodi Cray, 1973
        S. hatherwoodensis Cray, 1973
        Eotalpa anglica Sigé, Crochet, and Insole, 1977
PRIMATES
    Adapidae
        Adapis stintoni Gingerich, 1977
        A. magnus Filhol, 1874
    Tarsiidae or Omomyidae
        Microchoerus erinaceus Wood, 1846
        Pseudoloris parvulus (Filhol, 1889–1890)
PRIMATES?
    Amphilemuridae
        ''*Anchomomys*'' cf. *grisollense* Louis and Sudre, 1975
ARTIODACTYLA
    Choeropotamidae
        Choeropotamus depereti Stehlin, 1908
    Cebocheoridae
        Cebochoerus
    Dacrytheriidae
        Dacrytherium ovinum (Owen, 1857)
        Catodontherium
    Haplobunodontidae
        Haplobunodon lydekkeri Stehlin, 1908

Anthracotheriidae
        Diplopus? aymardi Kowalevsky, 1873
    Xiphodontidae
        Dichodon cuspidatum Owen, 1848
        D. cervinum Owen, 1841
    Amphimerycidae
        Pseudamphimeryx hantonensis Cooper, 1928
PERISSODACTYLA
    Equidae
        Anchilophus
    Palaeotheriidae
        Palaeotherium magnum Cuvier, 1804
        P. muehlbergi Stehlin, 1904
        P. duvali Pomel, 1853
        P. curtum Cuvier, 1812
        Plagiolophus annectens (Owen, 1847)
        P. minor (Cuvier, 1804)
RODENTIA
    Ischyromyidae
        Plesiarctomys
    Theridomyidae
        Theridomys pseudosiderolithicus (de Bonis, 1964)
        Sciuroides ehrensteinensis Schmidt-Kittler, 1971
        Treposciurus intermedius (Schlosser, 1884)
        T. mutabilis Schmidt-Kittler, 1971
        Suevosciurus palustris (Misonne, 1957)
        Paradelomys quercyi (Schlosser, 1884)
        Pseudoltinomys sp.
        Thalerimys fordi (Bosma and Insole, 1972)
    Gliridae
        Gliravus fordi Bosma and de Bruijn, 1979
        G. priscus Stehlin and Schaub, 1951
    Inc. sed.
        Ectropomys exiguus Bosma and Schmidt-Kittler, 1972

Composite Headonian Fauna

MARSUPIALIA
    Didelphidae
        Amphiperatherium minutum (Aymard, 1846)
        A. bourdellense Crochet, 1979
        A. lamandini (Filhol, 1876)
        A. exile (Gervais, 1848–1852)
        A. fontense Crochet, 1979
        A. ambiguum (Filhol, 1877)
        Peratherium perrierense Crochet, 1979
        P. cayluxi Filhol, 1877
        P. lavergnense Crochet, 1979
        P. cuvieri (Fischer, 1829)
        P. elegans (Aymard, 1846)
LEPTICTIDA
    Leptictidae
        Pseudorhynchocyon cayluxi Filhol, 1892
        Leptictis
PANTOLESTA
    Pantolestidae
        Cryptopithecus major (Lydekker, 1887)

105

C. siderolithicus Schlosser, 1890
C. alcimonensis Heissig, 1977
Dyspterna hopwoodi Cray, 1973
APATOTHERIA
  Apatemyidae
    Heterohyus sudrei Sigé, 1973
    H. nanus Teilhard, 1921
CREODONTA
  Hyaenodontidae
    Parapterodon lostangensis Lange-Badré, 1979
    Paroxyaena galliae (Filhol, 1881)
    Pterodon dasyuroides Blainville, 1839
    Prototomus minor (Filhol, 1877)
    Quercytherium tenebrosum Filhol, 1880
    Hyaenodon minor Gervais, 1848–1852
    H. brachyrhynchus Blainville, 1841
    H. requieni Gervais, 1848–1852
    H. heberti Filhol, 1876
    Cynohyaenodon cayluxi Filhol, 1873
    C. leenhardti Martin, 1906
    Hyaenodontipus praedator Ellenberger, 1980   (ichnites)
CREODONTA?
    Galethylax blainvillei Gervais, 1859
CARNIVORA
  Miacidae
    Quercygale angustidens (Filhol, 1872)
    Q. hastingsiae (Davies, 1884)
    Miacis exilis Filhol, 1876
    Simamphicyon helveticus (Pictet and Humbert, 1869)
  Amphicyonidae
    Cynodictis lacustris Gervais, 1848–1852
    C. parisiensis (Blainville, 1841)
  Ursidae
    Pseudocyonopsis antiquus Ginsburg, 1966
  Viverridae
    cf. *Palaeoprionodon*
INSECTIVORA
  Erinaceoidea
    Amphidozotherium cayluxi Filhol, 1877
    Paradoxonycteris soricodon Revilliod, 1922
  Soricoidea
    Saturninia gracilis Stehlin, 1940
    S. beata (Crochet, 1974)
    S. tobieni Sigé, 1976
    Scraeva woodi Cray, 1973
    S. hatherwoodensis Cray, 1973
    Eotalpa anglica Sigé, Crochet, and Insole, 1977
CHIROPTERA
  Emballonuridae
    Vespertiliavus
  Rhinolophidae
    Rhinolophus cf. *priscus* Revilliod, 1920
  Hipposideridae
    Palaeophyllophora quercyi Revilliod, 1917
    P. oltina (Delfortrie, 1872)
    Hipposideros cf. *morloti* (Pictet, 1855)
    H. schlosseri Revilliod, 1917
  Vespertilionidae

Stehlinia gracilis Revilliod, 1919
S. minor (Revilliod, 1922)
cf. *Tadarida* sp.
"Vespertilio" serotinoides (Blainville, 1839)
PRIMATES
  Adapidae
    Adapis magnus (Filhol, 1874)
    A. parisiensis Blainville, 1849
    A. stintoni Gingerich, 1977
    Cercamonius brachyrhynchus (Stehlin, 1912)   (age?)
    Anchomomys quercyi (Stehlin, 1916)   (age?)
    Pronycticebus gaudryi Grandidier, 1904   (age?)
    Protoadapis filholi Gingerich, 1977
    "P." ulmensis (Schmidt-Kittler, 1971)
  Tarsiidae or Omomyidae
    Necrolemur antiquus Filhol, 1873
    Microchoerus erinaceus (Wood, 1845)
    M. ornatus Stehlin, 1916
    Pseudoloris parvulus (Filhol, 1890)
PRIMATES?
  Amphilemuridae
    "Anchomomys" latidens (Teilhard de Chardin, 1921)
ARTIODACTYLA
  Dichobunidae
    Dichobune leporina Cuvier, 1822
    Moiachoerus simpsoni Golpe-Posse, 1971   —(cf.
      Acotherulum, acc. to Sudre, pers. commun.,
      1981)
    Mouillacitherium schlosseri Sudre, 1978A
    Hyperdichobune spinifera (Stehlin, 1908)
    H. spectabilis Stehlin, 1910
  Choeropotamidae
    Choeropotamus depereti Stehlin, 1908
    C. parisiensis Cuvier, 1822
    C. affinis Gervais, 1859
    C. lautricensis Noulet, 1870
    C. sudrei Casanovas-Cladellas, 1975
  Cebochoeridae
    Cebochoerus cf. *lacustris* Gervais, 1850
    C. minor Gervais, 1876
    C. fontensis Sudre, 1978
    Acotherulum quercyi (Stehlin, 1908)
    A. saturninum Gervais, 1859
    A. pumilum (Stehlin, 1908)
    A. cadurcensis (Filhol, 1877)
    A. simpsoni (Golpe-Posse, 1971)
  Mixtotheriidae
    Mixtotherium cuspidatum Filhol, 1880
    M. depressum (Filhol, 1884)   (age?)
    M. leenhardti Stehlin, 1908   (age?)
    M. quercyi (Filhol, 1888)   (age?)
  Dacrytheriidae
    Dacrytherium ovinum (Owen, 1857)
    D. saturnini Stehlin, 1910
    Robiacina quercyi Sudre, 1977
    R. lavergnensis Sudre, 1977
    Leptotheridium lugeoni Stehlin, 1910
    Tapirulus hyracinus Gervais, 1850
    T. perrierensis Sudre, 1978B
  Anoplotheriidae
    Anoplotherium commune Cuvier, 1822

A. latipes (Gervais, 1849)

A. laurillardi Pomel, 1851

Diplobune secundaria (Cuvier, 1822)

D. primaevus (Filhol, 1873)

Anoplotheriipus lavocati Ellenberger, 1980
  (ichnites)

A. similicommunis Ellenberger, 1980   (ichnites)

A. compactus Ellenberger, 1980   (ichnites)

Anoplotheriidae?

Diploartiopus longipes Ellenberger, 1980   (ichnites)

Haplobunodontidae

Haplobunodon lydekkeri Stehlin, 1908

Amphirhagatherium frohnstettense (Kowalevsky, 1873)

Rhagatherium valdense Pictet, 1855

Anthracotheriidae

Elomeryx crispus (Gervais, 1848)

Diplopus? aymardi Kowalevsky, 1873

Thaumastognathus quercyi Filhol, 1890

Prominatherium dalmatinum (Meyer, 1854)   (age?)

Xiphodontidae

Dichodon cuspidatum Owen, 1848

D. frohnstettense Meyer, 1852

D. cervinum (Owen, 1846)

D. stehlini Sudre, 1973

Xiphodon intermedium Stehlin, 1910

X. gracile Cuvier, 1822

Paraxiphodon teulonensis Sudre, 1978

Haplomeryx zitteli Schlosser, 1886

H. picteti Stehlin, 1910

H. euzetensis Depéret, 1917

H. obliquus (Cuvier, 1822)

Amphimerycidae

Pseudamphimeryx hantonensis Cooper, 1928

P. pavloviae Stehlin, 1910

P. renevieri (Pictet and Humbert, 1869)

P. valdensis Stehlin, 1910   (incl. *"P."* *decedens* Stehlin, 1910?)

Amphimeryx murinus (Cuvier, 1822)

Gelocidae

Paragelocus suevicus Schlosser, 1902

Gelocus minor Pavlov, 1900

Cainotheriidae

Oxacron courtoisi (Gervais, 1848–1852)

Paroxacron valdense (Stehlin, 1910)

PERISSODACTYLA

Equidae

Lophiotherium cervulum Gervais, 1849

L. robiacense Depéret, 1917

Pachynolophus garimondi Remy, 1967

P. cayluxi (Filhol, 1888)

P. lavocati Remy, 1972

Propalaeotherium sp.

Anchilophus dumasi (Gervais, 1849)

A. gaudini (Pictet and Humbert, 1869)

A. radegondense (Gervais, 1848–1852)

Palaeotheriidae or Equidae

Lophiopus rapidus Ellenberger, 1980   (ichnites)

L. latus Ellenberger, 1980   (ichnites)

Palaeotheriidae

Palaeotherium renevieri Stehlin, 1904

P. siderolithicum (Pictet and Humbert, 1869)

P. crassum Cuvier, 1805

P. magnum Cuvier, 1804

P. medium Cuvier, 1804

P. curtum Cuvier, 1812

P. muehlbergi Stehlin, 1904

P. duvali Pomel, 1853

P. franzeni Casanova, and Santafé, 1980

Plagiolophus annectens (Owen, 1847)

P. minor (Cuvier, 1804)

P. fraasi v. Meyer, 1852

Palaeotheriipus similimedius Ellenberger, 1980
  (ichnites)

RODENTIA

Ischyromyoidea

Plesiarctomys spectabilis (Major, 1873)

P. gervaisi Bravard in Gervais, 1850

Theridomyidae

Paradelomys crusafonti Thaler, 1966

P. spelaeus Hartenberger, 1973A

Treposciurus intermedius (Schlosser, 1884)

T. mutabilis (Schmidt-Kittler, 1971)

Microsuevosciurus minimus (Major, 1873)

Suevosciurus fraasi (Major, 1873)

S. ehingensis Dehm, 1937A

Estellomys cansouni Hartenberger, 1971

E. ibericus (Thaler, 1966)

Sciuroides intermedius Schlosser, 1884

S. quercyi Schlosser, 1884

S. ehrensteinensis Schmidt-Kittler, 1971

S. siderolithicus (Pictet and Humbert, 1869)

Theridomys (Theridomys) perrealensis Vianey-Liaud,
  1977

T. (T.) euzetensis (Depéret, 1917)

T. (T.) pseudosiderolithicus de Bonis, 1964

T. (T.) golpei Hartenberger, 1973A

Archaeomys (Blainvillimys) rotundidens (Schlosser,
  1884)

Patriotheridomys altus Vianey-Liaud, 1974

Pseudosciurus praecedens Schmidt-Kittler, 1971

P. suevicus Hensel, 1856

Thalerimys fordi (Bosma and Insole, 1972)

T. headonensis (Bosma, 1974)

Elfomys parvulus Hartenberger, 1973A

Pseudoltinomys mamertensis Hartenberger, 1973A

P. cuvieri (Pomel, 1852)

P. gaillardi Lavocat, 1951C

P. phosphoricus Hartenberger, 1973A

P. gousnatensis Vianey-Liaud, 1976

Oltinomys platiceps (Filhol, 1877)

Gliridae

Gliravus priscus Stehlin and Schaub, 1951

G. devoogdi Bosma and de Bruijn, 1979

G. robiacensis Hartenberger, 1965

G. meridionalis Hartenberger, 1971

Inc. sed.

Remys garimondi Thaler, 1966

Pairomys crusafonti Thaler, 1966

Ectropomys exiguus Bosma and Schmidt-Kittler, 1972

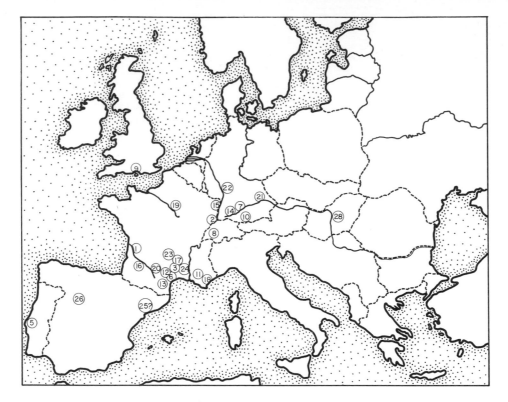

FIGURE 4-28   *Representative Headonian Localities.*

1	*Baby (Moulin Baby)*	10	*Messkirch*	3	*Quissac*
2	*Brunnstatt*	15	*Minfeld Tunnel Mine 14*	21	*Raitenbuch*
3	*Celas*	13	*Mireval-Lauvagais*	2	*Rixheim*
4	*Ehrenstein 1a, 2, 3*	16	*Monlis*	22	*Rot-Malsch*
3	*Euzet-les-Bains*	17	*Mont Anis*	23	*St.-Aubin-d'Eymet*
5	*Feliqueira Grande*	11	*Mormoiron*	1	*St.-Bonnet-de-Rochefort, etc.*
6	*Fons, in part*	8	*Mormont-Entreroches*	1	*Ste.-Sabine, etc.*
7	*Frohnstetten*	1	*Moulin Baby (Baby)*	6	*St.-Gely du Fesc*
8	*Gösgen Canal*	14	*Neuhausen (in part)*	24	*St.-Jean-de-Maruejols*
9	*Headon Hill—Headon Cliffs*	18	*Nice*	25	*San Cugat de Garadons*
10	*Herrlingen 2, 3—Ulm*	1	*Payrade*	26	*San Morales*
11	*La Débruge*	14	*Pfaffenweiler*	27	*Selva*
12	*La Grave*	13	*Pont d'Assou*	24	*Tavel*
13	*Les Ondes*	19	*Paris Basin Localities*	28	*Üröm (age?)*
14	*Mähringen*	20	*Quercy district localities*		

Of the 97 Headonian genera, 53 (55 percent) were found in the Robiacian, preceding, and 58 (60 percent) are not found in the later faunas of Europe. Nine families of the Headonian apparently do not survive into the later faunas of Europe. This represents a big turnover in the artiodactyl and perissodactyl families primarily (for detail see chapter on Oligocene).

LATER EOCENE OF ASIA

The fossil mammal localities of Asia, here grouped under the designation *Later Eocene,* are found in Mongolia, China, Korea, Japan, Burma, Kazakhstan, and the Island of Timor.

These faunas probably token a geochronologic interval corresponding to the Robiacian and Headonian in Europe, but there are so few genera in common between the two areas that direct faunal correlation is impossible. A better correlation is afforded with the Uintan and Duchesnean of North America (described next in this chapter).

Fossiliferous districts in Mongolia and in the adjacent Chinese autonomous region of Nei Mongol (Inner Mongolia) have made important contributions to knowledge of fossil mammals. See especially Matthew and Granger (1924B, 1925F), Osborn (1924E, 1925E), Osborn and Granger (1932A), Radinsky (1964, 1965A, 1967A), Dawson (1964), Mellett and Szalay (1968), and Szalay (1969B). The samples from the Irdin Manha and Shara Murun districts and formations are known best. Radinsky (1964) has summarized the problems regarding geographic designations, assignments to formations, and relative ages of the many localities in this area. Later work in other provinces of northern China (Shanxi and especially Henan) has added greatly to the taxonomic list of the Later Eocene of Asia. To the aggregate from China and Mongolia we shall add the mammals found in Later Eocene of Kazakhstan (Zaisan Basin), Kirghizia (Toruaïgyr), and Far East Province (Artëm).

Later Eocene Mammalian Fauna in Northern Asia

LEPTICTIDA
    Ictopidium lechei Zdansky, 1930
ANAGALIDA
    Eurymylidae
        Hypsimylus beijingensis Zhai, 1977
LAGOMORPHA
    Leporidae
        Gobiolagus tolmachovi Burke, 1941
        Shamolagus medius Burke, 1941
        S. grangeri Burke, 1941
        Desmatolagus
        Lushilagus lahoensis Li, 1965
PANTODONTA
    Pantolambdodontidae
        Pantolambdodon inermis Granger and Gregory, 1934
        P. fortis Granger and Gregory, 1934
    Coryphodontidae
        Eudinoceras mongoliensis Osborn, 1924
TILLODONTIA
    Esthonychidae
        Adapidium huanghoense Young, 1937
PANTOLESTA
    Pantolestidae, gen. indet.
CREODONTA
    Oxyaenidae
        Sarkastodon mongoliensis Granger, 1928
    Hyaenodontidae
        Hyaenodon yuanchuensis Young, 1937
        Propterodon morrisi (Matthew and Granger, 1924)
        Pterodon cf. *dahkoensis* Chow, 1975
        "*Pterodon*" *hyaenoides* (Matthew and Granger, 1925)
        Prolaena parva Xu, Yan, Zhou, Han, and Zhang, 1979
CARNIVORA
    Miacidae
        Miacis lushiensis Chow, 1975
        M. invictus Matthew and Granger, 1925
    Amphicyonidae
        Cynodictis
    Ursidae
        Pachycynodon?   —Tang and Qiu (1979)
        Cephalogale   —Tang and Qiu (1979)
    Canidae?
        Chailicyon crassidens Chow, 1975
    Felidae
        cf. *Eusmilus*
PRIMATES
    Families?
        Lushius qinlinensis Chow, 1961
        Hoanghonius stehlini Zdansky, 1931

ARCTOCYONIA
    Arctocyonidae
        Paratriisodon henanensis Chow, 1959
        P. gigas Chow, Li, and Chang, 1973
        Andrewsarchus mongoliensis Osborn, 1924
    Paroxyclaenidae
        Kiinkerishella zaisanica Gabunia and Birjukov, 1978
DINOCERATA
    Gobiatheriidae
        Gobiatherium
ACREODI
    Mesonychidae
        Honanodon hebetis Chow, 1965
        H. macrodontus Chow, 1965
        Lophoodon lushiensis (Chow, 1965)
        Hapalodectes serus Matthew and Granger, 1925
        Harpagolestes? orientalis Szalay and Gould, 1966
        Harpagolestes koreanicus Shikama, 1943
        cf. *Mesonyx*
        Metahapalodectes makhchinus Dashzeveg, 1976
        Mongolonyx robustus Dashzeveg, 1976
        cf. *Pachyaena*
TAENIODONTA
    Stylinodontidae
        Stylinodon?
        Chungchienia sichuanica Chow, 1963
ARTIODACTYLA
    Dichobunidae
        Dichobune?
    Family?
        Lantianus xiehuensis Chow, 1964
    Entelodontidae
        Eoentelodon likiangensis Zhang, You, Ji, and Ding, 1978
    Anthracotheriidae
        Anthracotherium
        Ulausuodon parvus Hu, 1963
        Anthracokeryx sinensis (Zdansky, 1930)
        Anthracothema minima Y. Xu, 1962
        Anthracosenex ambiguus Zdansky, 1930
        Bothriodon
    Helohyidae
        Gobiohyus orientalis Matthew and Granger, 1925D
        G. pressidens Matthew and Granger, 1925D
        G. robustus Matthew and Granger, 1925D
        Indohyus? yuanchuensis (Young, 1937)   —Coombs and Coombs (1977)
    Leptomerycidae   —Webb and Taylor (1980)
        Archaeomeryx optatus Matthew and Granger, 1925D
        Xinjiangmeryx parvus Zheng, 1978
PERISSODACTYLA
    Equidae
        Gobihippus menneri Dashzeveg, 1979
    Brontotheriidae
        Microtitan? mongoliensis (Osborn, 1925)
        Protitan grangeri (Osborn, 1925)
        P. bellus Granger and Gregory, 1943
        P. robustus Granger and Gregory, 1943
        P. obliquidens Granger and Gregory, 1943

P.? koreanicum (Takai, 1939)

Desmatotitan tukhumensis Granger and Gregory, 1943

Epimanteoceras formosus Granger and Gregory, 1943

E. praecursor Janovskaya, 1953

Dolichorhinoides angustidens Granger and Gregory, 1943

Metatitan primus Granger and Gregory, 1943

Telmatherium?

Manteoceras?

Rhinotitan mongoliensis (Osborn, 1923)

R. kaiseni (Osborn, 1925)

R. andrewsi (Osborn, 1925)

R. orientalis Janovskaya, 1957

Pachytitan ajax Granger and Gregory, 1943

Parabrontops gobiensis Granger and Gregory, 1943

Titanodectes minor Granger and Gregory, 1943

T. ingens Granger and Gregory, 1943

Protembolotherium efremovi Tang, You, Xu, Qiu, and Hu, 1974

Arctotitan honghoensis Chow, 1978

Metatelmatherium cristatum Granger and Gregory, 1943

M. parvum Granger and Gregory, 1943

Gnathotitan berkeyi Osborn, 1925

Helaletidae

  Colodon hodosimai Takai, 1939

  C.? grangeri (Tokunaga, 1933)

  Helaletes mongoliense (Osborn, 1923)

  H. fissus (Matthew and Granger, 1925)

Helaletidae or Rhinocerotoidea

  cf. *Hyrachyus*

Deperetellidae

  Teleolophus sichuanensis Xu, Yan, Zhou, Han, and Zhang, 1979

  T. medius Matthew and Granger, 1925

  T. liankanensis Zheng, 1978

  T. magnus Radinsky, 1965

  Deperetella cristata Matthew and Granger, 1925

  D. depereti (Zdansky, 1930)

  D. kungeica Tarasov, 1968

  Diplolophodon similis Zdansky, 1930

Lophialetidae

  Breviodon minutus (Matthew and Granger, 1925)

  Lophialetes expeditus Matthew and Granger, 1925

  Simplaletes ulanshirensis Qi, 1980

  S. xianensis Zhang and Qi (in press)

  S. sujiensis Qi, 1980

  Parabreviodon dubius Reshetov, 1975

  Schlosseria cf. *magister* Matthew and Granger, 1926

Lophialetidae?

  Rhodopagus pygmaeus Radinsky, 1965

  R.? minimus (Matthew and Granger, 1925)

Hyracodontidae

  Pappaceras

  Triplopus progressus (Matthew and Granger, 1925)

  T. zhukovi (Tarasov, 1968)   —Radinsky (1967)

  T.? proficiens (Matthew and Granger, 1925)

  Prohyracodon meridionale Chow and Y. Xu, 1961

  Imequincisoria mazhuangensis Wang, 1976

  I. macracis Wang, 1976

Forstercooperia grandis (Peterson, 1915)

F. cf. *confluens* (H. Wood, 1963)

F. totadentata H. Wood, 1938

Juxia borissiaki (Belyayeva, 1957)

Guixia simplex You, 1977

Amynodontidae

  Caenolophus promissus Matthew and Granger, 1925

  C. obliquus Matthew and Granger, 1925

  C. makii (Tokunaga, 1926)   —Takai (1939)

  Lushiamynodon menchiapuensis Chow and Y. Xu, 1965

  L. wuchengensis Wang, 1978

  L. sharamurunensis Y. Xu, 1966

  Sianodon honanensis Chow and Xu, 1965

  S. ulausuensis Y. Xu, 1966

  S. sinensis (Zdansky, 1930)

  S. bahoensis Y. Xu, 1965

  S. mienchiensis Chow and Y. Xu, 1965

  Sharamynodon mongoliensis (Osborn, 1936)

  Gigantamynodon promissus Y. Xu, 1966

  cf. *Cadurcodon*

  Zaisanamynodon borisovi Belyayeva

  Procadurcodon orientalis Gromova, 1960

  Amynodon watanabei Tokunaga, 1929

Chalicotheriidae

  Litolophus major (Zdansky, 1930)

  Schizotherium   —Belyayeva et al. (1974)

  Eomoropus quadridentatus Zdansky, 1930

RODENTIA

Ctenodactylidae

  Tsinlingomys youngi Li, 1963

  Yuomys elegans Wang, 1978

  Y. cavioides Li, 1975

  Advenimus bohlini Dawson, 1964

  A. burkei Dawson, 1964

  Saykanomys chalchis Shevyreva, 1972

  Petrokozlovia notos Shevyreva, 1972

  P. glambus (Shevyreva, 1972)

Cricetidae

  "*Cricetodon*" *schaubi* (Zdansky, 1930)

Zapodidae

  Plesiosminthus

ORDER?

Didymoconidae

  Kennatherium shirensis Mellett and Szalay, 1968

  Mongoloryctes acutus (Matthew and Granger, 1925)

Southeastern Asian Later Eocene faunas are fewer than those to the north, but interesting aggregates are known from Gaungxi and Yunnan (Ding, Zheng, Zhang, and Tong, 1977; Tang, 1978; Zhang, You, Ji, and Ding, 1978; Zheng and Chi, 1978; and Tang and Qiu, 1979).

Composite Southeastern Asian Later Eocene Mammalian Fauna

PANTODONTA

Coryphodontidae

  Eudinoceras crassum Tong and Wang, 1977

TILLODONTIA

Esthonychidae, gen. indet.

CREODONTA
  Hyaenodontidae
    Pterodon dahkoensis Chow, 1975
    Propterodon?
CARNIVORA
  Ursidae
    Pachycynodon?
    Cephalogale
  Canidae?
    Chailicyon crassidens Chow, 1975
  Felidae
    Eusmilus?
ARCTOCYONIA
  Arctocyonidae
    Andrewsarchus crassum Ding, Zheng, Zhang, and Tong, 1977
ACREODI
  Mesonychidae
    Guilestes acares Zheng and Chi, 1978
    cf. *Harpagolestes*
    Honanodon hebetis Chow, 1965
ARTIODACTYLA
  Helohyidae
    Gobiohyus
  Choeropotamidae, gen. indet.
  Tayassuidae, gen. indet.
  Entelodontidae
    Eoentelodon yunnanense Chow, 1958
    E. likiangensis Zheng and Chi, 1978
  Anthracotheriidae
    Probrachyodus panchuoensis Y. Xu and Chiu, 1962
    Anthracothema rubricae Pilgrim and Cotter, 1916
    Anthracokeryx birmanicus Pilgrim and Cotter, 1916
    A. moriturus Pilgrim, 1928

    Bothriodon chyelingensis Y. Xu, 1977
    Huananothema chengbiensis Tang, 1978
    Heothema angusticalxia Tang, 1978
    H. imparilica Tang, 1973
    H. media Tang, 1978
    H. bellia Tang, 1978
  Tragulidae or Hypertragulidae
    Indomeryx cotteri Pilgrim, 1928   —Tang et al. (1974)
    I. youjiangensis Qiu, 1978
    Notomeryx besensis Qiu, 1978
CONDYLARTHRA
  Phenacodontidae
    Eodesmatodon spanios Zheng and Chi, 1978
PERISSODACTYLA
  Brontotheriidae
    Dianotitan lunanensis (Chow and Hu, 1959)
    Protitan cf. *robustus* Granger and Gregory, 1943
    Rhinotitan quadridens Y. Xu and Chiu, 1962
    Metatelmatherium cf. *browni* Colbert, 1938
  Deperetellidae
    Diplolophodon similis Zdansky, 1930
    D. birmanicum (Pilgrim, 1925)
    Deperetella dienensis Chow, Chang, and Ting, 1974
    Teleolophus cf. *medius* Matthew and Granger, 1925
  Helaletidae
    Helaletes mongoliensis (Osborn, 1923)   —Radinsky
    (1965)
  Lophialetidae
    Lophialetes expeditus Matthew and Granger, 1925
    Breviodon sahoensis Chow, Chang, and Ting, 1974

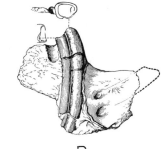

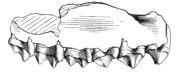
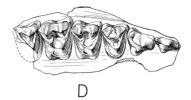
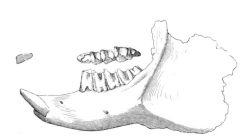

FIGURE 4-29 *Later Eocene Mammals of Asia.*

  A Shamolagus medius Burke, 1941. Upper and lower jaws and dentition. ×1. Figures 3 and 4 in Li (1965).

  B Chungchienia sichuanica Chow, 1963. Jaw with premolariform tooth. ×1/5 approximately. Figure 1 in Chow (1963).

  C Honanodon hebetis Chow, 1965. Upper and lower teeth. ×1/4. Figure 1 in Chow (1965).

  D Lantianus xiehuensis Chow, 1964. P²–M³. ×1½. Figure 1 in Chow (1964).

(Figures published with permission of the Institute of Vertebrate Paleontology and Paleoanthropology, Beijing.)

A

C

B

D

Schlosseria
Amynodontidae
  Caenolophus medius Chow, 1957
  cf. *Paramynodon birmanicus* (Pilgrim and Cotter, 1916)
  cf. *Gigantamynodon*
  Amynodon lunanensis Chow, Xu, and Zhen, 1964
  A. altidens Y. Xu and Chiu, 1962
  Teilhardia pretiosa? (Matthew and Granger, 1926)
  Paracadurcodon?
  Huananodon hui You, 1977
Hyracodontidae
  Prohyracodon meridionale Chow and Y. Xu, 1961
  P. cf. *orientale* Koch, 1897
  P. progressa Chow and Y. Xu, 1961
  Forstercooperia shiwopuensis Chow, Chang, and Ting, 1974
  Ilianodon lunanensis Chow and Y. Xu, 1961
  Indricotherium parvum Chow, 1958
  Juxia sp. unident.
  Guixia simplex You, 1977
Chalicotheriidae
  Eomoropus quadridentatus Zdansky, 1930
  Litolophus ulterior (Chow, 1962)
  Lunania youngi Chow, 1957
RODENTIA, fam. unident.

Small samples of perissodactyls from Korea (Hosan Coal Field) and from the Japanese Archipelago (Numata-Uryu in Hokkaido, Ube Coal Field in Honshu) indicate that these eastern areas were a part of the Later Eocene Palaearctic region (Takai, 1950).

Later Eocene of Southern Asia

Among the later Eocene mammalian faunas of Asia, the assemblage from the Pondaung Formation, southwest of Mandalay in Burma, has special significance because of its singular geographic location in the Tethyan-border paleotropical region. Moreover, it is the earliest record currently known for anthropoid(?) primates. See Colbert (1938), Ba Maw, Ciochon, and Savage (1979). The Pondaung fauna comprises gastropods, fishes, trionychid and carettochelyoid turtles, lizards, *Pristichampsus*-like and other crocodilians, and the following mammals.

Pondaung Mammalian Fauna

CARNIVORA
  Viverridae?
PRIMATES
  ANTHROPOIDEA?
    Pondaungia cotteri Pilgrim, 1927
    Amphipithecus mogaungensis Colbert, 1937
ARTIODACTYLA
  Anthracotheriidae
    Anthracohyus choeroides Pilgrim and Cotter, 1916

Anthracothema pangan (Pilgrim and Cotter, 1916)
  A. palustre (Pilgrim and Cotter, 1916) (synonym of *A pangan?*)
  A. rubricae (Pilgrim and Cotter, 1916)
  A. crassum (Pilgrim and Cotter, 1916)
  Anthracokeryx birmanicus Pilgrim and Cotter, 1916
  A. tenuis Pilgrim and Cotter, 1916
  A. ulnifer Pilgrim, 1928
  A. moriturus Pilgrim, 1928
  A.? lahiri Pilgrim, 1928
Traguloidea
  Indomeryx cotteri Pilgrim, 1928
  I. arenae Pilgrim, 1928
PERISSODACTYLA
  Brontotheriidae
    Sivatitanops cotteri Pilgrim, 1925
    S. birmanicum (Pilgrim and Cotter, 1916)
    S.? rugosidens Pilgrim, 1925
    Metatelmatherium? browni Colbert, 1938
    M.? lahirii (Pilgrim, 1925)
  Fam. inc. sed.
    Indolophus guptai Pilgrim, 1925
  Deperetellidae
    Diplolophodon birmanicum (Pilgrim, 1925)
  Amynodontidae
    Paramynodon birmanicus (Pilgrim and Cotter, 1916)
    P. cotteri (Pilgrim, 1925)
RODENTIA
  Fam. indet.

The abundant and varied anthracotheres of the Pondaung indicate faunal and zoogeographic affinity with North Asian faunas of the same age and with the early Oligocene fauna of North Africa. The abundant brontotheres and amynodont rhinos also show affinities with Asian Palaearctica and also with Nearctica. The two anthropoids(?) suggest that Burma may have been a principal theater of origin and early diversification of the general ancestral stocks for modern monkeys, apes, and people. It will be exciting to learn in future years whether the theater of anthropoid origins centered in the Paleo-Tropics of Asia on the northern shores of the Tethyan seaway or was also represented in Africa on the southern shores of this seaway.

Although there was evidently a great evolutionary radiation and a great abundance of brontotheriid and tapiroid perissodactyls in the later Eocene of central and northern Asia (probably all Asia), equids (horses) are not represented. The later Eocene distribution of equids is mystifying. They were present in the very similar inland faunas of North America at this time but were not represented in the near-shore Californian faunas. A different suite of equid genera had abundant representation in near-shore, contemporary, palaeothere-artiodactyl-theridomyid faunas of western Europe. Environmental causes such as differences in climate and vegetation seem an improbable explanation for this odd geographic discontinuity. We can only speculate that special horse competitors among the tapiroids dominated in Asia and in California.

FIGURE 4-30 *Representative Late Eocene Localities of Asia.*

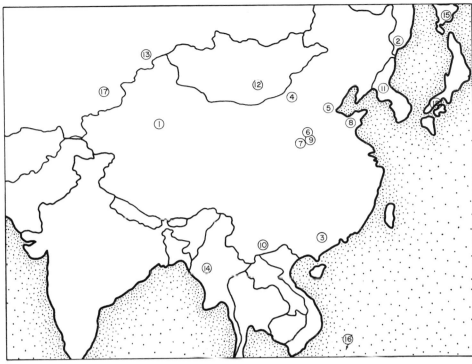

1   Anjihai South—Xinjiang
2   Artëm—Far East USSR
3   Baise (Bose) Basin—Guangxi
4   Baron Sog Mesa—Nei Mongol
5   Beijing—Beijing
4   Camp Margetts—Nei Mongol (Huhebolhe Cliff)
6   Chaili-Yuanchu—Shanxi
5   Changxindian (Gaodian)—Beijing
7   Chenjiawucun (Kangwan-gou)—Shaanxi
4   Chimney Butte Quarry—Buck-shot's Quarry—Nei Mongol
9   Chiyuan (Jiyuan)—Henan
9   Chuankou Formation site—Henan
8   Cicishou (Xintai)—Shandong
10 Dahimapan—Iliang—Yunnan
3   Dongjun (fauna)—Baise Basin—Guangxi
1   Dzungaria Basin—Xinjiang
5   Gaodian (Changxindian)—Beijing
7   Gaopocun (Shilihegou)—Shaanxi
3   Gungkang Formation sites—Guangxi
6   Hoangho (North Bank)—Shanxi
11 Hosan Coal Field—Kokaido—Korea
6   Hoti or Heti (fauna)—Shanxi
10 Hsiaoshaho—Iliang—Yunnan
6   Hsinan—Shanxi, Henan
4   Huhebolhe Cliff (upper)—Nei Mongol
9   Hunshuihe Formation site—Henan
4   Hutogein Valley—Nei Mongol
10 Iliang District—Yunnan
4   Irdin Manha Formation sites—Nei Mongol
9   Jentsen (Menchiapu)—Henan
6–9 Jiyuan Basin—Henan, Shanxi
7   Kangwangou (Chenjiwacun)—Xiehu Commune—Shaanxi
10 Kaofanpei-Iliang—Yunnan
12 Kholboldshi—Mongolia
13 Kiinkerish 1 (Zaisan Basin)—Kazakhstan
11 Kokaido—Korea
8   Kuanchuang (Guanzhuang) Formation sites—Shandong
7   Lantian (Lantien) district—Shaanxi
1   Liankan (Turfan Basin)—Xinjiang

1   Lianmuqin Pass—Xinjiang
10 Lunan Basin—Yunnan
9   Lushih (Lushi) Basin—Henan
7   Maoxicum (Sian or Xian)—Shaanxi
9   Menchiapu—Jentsun—Henan
14 Mogaung Northwest—Burma
13 Mogila Sultan (Zaisan Basin)—Kazakhstan
14 Myaing district—Burma
4   North Mesa—Nei Mongol
15 Numata-muru—Hokkaido, Japan
6   Ostpinghi—Henan and Shanxi
10 Panchiao—Yunnan
14 "Pangan" = Phan Kan—Burma
14 Phan Kan—Burma
2   Primorsky Kray—Artëm, USSR
6–9 River Section—Henan-Shanxi
16 Sanggau—Borneo
7   Sangyancun (Linlong County)—Shaanxi
7   Shahezigou or Shahegou—Lantian—Shaanxi
4   Shara Murun Formation sites—Nei Mongol
7   Shiliheyou (Gaopocun)—Shaanxi
10 Shiwopu—Daliko—Yunnan
7   Sian (Xian)—Shaanxi

9   Sichuan Basin—Xingdayuan—Henan
9   Tantout Formation sites—Henan
4   Telegraph Line Camp—Nei Mongol
3   Tientong region—Guangxi
9   Tongbo region—Henan
17 Toruaïgyr—Kirghizia
13 Tschaibulak—Kazakhstan
1   Tsungaria (Dsungaria) Basin—Xinjiang
4   Tukhum Lamasery and Tukhum North—Nei Mongol
9   Tungchang (Jiyuan)—Henan
1   Turfan Basin—Xinjiang
18 Ube Coal Field (Okinogama Colliery)—Honshu, Japan
4   Ulan Shireh Formation sites—Nei Mongol
4   Ula Usu—Nei Mongol
6   Weihoebene—Shanxi and Henan
14 Wetkya (="Wetcha")—Burma
9   Wucheng Basin—Henan
7   Siehu Commune—Shaanxi
7   Yinpocun (Xian)—Shaanxi
3   Yongle (Yungle) Basin—Gaungxi
6–9 Yuanchu or Yuanqu Basin—Shanxi-Henan
13 Zaisan Basin—Kazakhstan

Northern Asia may well have been the theater of origin for several modern families of land mammals during late Eocene time, that is, sabre-cats, rhinocerotids, and cricetids. The latest Eocene faunas of northern Asia are most impressive for their great numbers and great diversity of ungulates, especially among the Perissodactyla. This may be a reflection of the more open savanna habitats that are being sampled in the fossil record of the late Eocene of that region.

LATER EOCENE OF AFRICA

Scattered marine and near-shore lagoonal and estuarine sedimentary rocks in North Africa have produced elasmobranchs, bony fishes, birds, reptiles, archaeocete whales, sirenians, and proboscideans, giving us a restricted picture of the vertebrates that lived on that continent during Late Eocene. The best sample is from the lower part of the marine-nonmarine succession in the Fayum district and northeastward to the environs of Cairo in Egypt. Here the rocks—called Ravine, Birket-el-Qurun, Qasr-el-Sagha, and Maadi (= Upper Mokattam) formations—have yielded *Prozeuglodon* and *"Dorudon"* (archaeocetes), *Eotheroides* (sirenian), *Moeritherium* and *Barytherium* (proboscideans), associated in places with an abundance of siluroids ("catfish") and pristids ("saw-snout sharks"), plus dermochelyid, chelonid, and pelomedusid turtles, pythons and other snakes, and crocodiles. In these strata, though not directly associated in all occurrences, is found a neritic fauna of molluscs and echinoids.

The Late Eocene vertebrates from Senegal and Mali in northwest Africa and from Libya (R. Savage, 1969) indicate a westward extension of this archaeocete-sirenian-proboscidean fauna. The further westward extension of at least the marine part of this fauna across the Atlantic is indicated by the Late Eocene records of very similar archaeocetes and sirenians in the southeastern United States and in the Greater Antilles. The nonmarine components of this fauna appear to represent a lowland assemblage, thriving in relative isolation on the southern border of the Tethyan seaway.

UINTAN

The Uintan Mammal Age is based on the mammalian fauna from the Uinta Formation of the Uinta Basin, northeastern Utah, comprising Uinta "A," "B," and "C" of Peterson (in Osborn, 1895A) or the Wagonhound Member (= "A" plus "B") and the Myton Member (= "C") of H. Wood (1934). We include in this age the time represented by the fauna from the Brennan Basin Member and the lower part of the Dry Gulch Creek Member of the Duchesne River Formation, these members conformably overlying the Uinta Formation in the type area. Brennan Basin and Dry Gulch Creek—Andersen and Picard (1972)—are approximately equal in total vertical range to the poorly defined Randlett and Halfway members of the Duchesne River in earlier usage.

This thick stratal type section and its contained local faunas represent an interval much longer than any other North American Mammal Age (ca. 9 million years). The Uintan is probably subdivisible into useful subages, and the Uinta–Duchesne River succession probably can be zoned into useful chronozones when the stratigraphic paleontology becomes better known. Although the specialists working with Uintan faunas use such designations as *Earlier* and *Later,* we shall discuss it as a very long mammal age of generally unified mammalian-faunal character.

Composite Uintan Mammalian Fauna

(Includes taxa that have been designated Early and Late Uintan.)[2]
MULTITUBERCULATA
  Neoplagiaulacidae
    Parectypodus lovei Sloan, 1966
    Ectypodus —Krishtalka and Black, 1975
MARSUPIALIA
  Didelphidae
    Herpetotherium cf. *knighti* (McGrew, 1959)
    H. cf. *marsupium* Troxell, 1923F
    Nanodelphys californicus (Stock, 1936A)
    N. cf. *minutus* McGrew, 1937A
LEPTICTIDA
  Leptictidae
    Palaeictops, n. sp. Novacek, ms. (Tepee Trail)
LAGOMORPHA
  Leporidae
    Mytonolagus petersoni Burke, 1934B
CIMOLESTA
  Palaeoryctidae
    Batodonoides powayensis Novacek, 1973
PANTOLESTA
  Pantolestidae
    Pantolestidae, genus unident. (Hendry Ranch)
  Pantolestoidea, *inc. sed.*
    Simidectes medius (Peterson, 1919A)
    S. merriami Stock, 1933E
APATOTHERIA
  Apatemyidae
    Apatemys bellus Marsh, 1872I
    Stehlinella uintensis (Matthew, 1921A) —Matthew (1929E)
TAENIODONTA
  Stylinodontidae
    Stylinodon mirus Marsh, 1879C —Schoch and Lucas (1981)
CREODONTA
  Hyaenodontidae

Limnocyon douglassi Peterson, 1919A
L. potens Matthew, 1909D
Oxyaenodon dysclerus Hay, 1902A
Apataelurus kayi Denison, 1938
CARNIVORA
  Miacidae
    Miacis gracilis Clark, 1939D
    M. longipes (Peterson, 1919A)
    Uintacyon robustus (Peterson, 1919A)
    Prodaphaenus scotti Wortman and Matthew, 1899A
    Tapocyon occidentalis Stock, 1934C
    "*Viverravus*" (*Plesiomiacis*) *progressus* (Stock, 1934C)
    Procynodictis vulpiceps Wortman and Matthew, 1899A
    Eosictis avinoffi Scott, 1945A
INSECTIVORA
  Erinaceoidea
    Macrocranion robinsoni Krishtalka and Setoguchi, 1977
    Scenopagus?
    Entomolestes?
    Centetodon hendryi Lillegraven, McKenna, and Krishtalka, 1981
    C. aztecus Lillegraven, 1979–1042
    Ankylodon   —Krishtalka, 1976A
    Sespedectes singularis Stock, 1935D
    Proterixoides davisi Stock, 1935D
    Crypholestes vaughni (Novacek, 1973)
    Talpavus duplus Krishtalka, 1976A
  Soricoidea
    Nyctitherium christopheri Krishtalka and Setoguchi, 1977
    Micropternodus
    Domnina cf. *gradata* Patterson and McGrew   —Krishtalka and Setoguchi (1977)
  Talpidae?
    Apternodus cf. *illifensis* Galbreath, 1953   —Krishtalka and Setoguchi (1977)
    Oligoryctes   —Krishtalka and Setoguchi (1977)
  Inc. Sed.
    Aethomylus simplicidens Novacek, 1973
DERMOPTERA
  Plagiomenidae? (Hendry Ranch —Black and Dawson, 1966)
DERMOPTERA?
  Thylacaelurus cf. *montanus* L. Russell, 1954A
PRIMATES?
  Uintasoricidae
    Uintasorex montezumicus Lillegraven, 1976
    U. cf. *parvulus* Matthew, 1909D   —Krishtalka (1978)
  Microsyopidae
    Microsyops kratos Stock, 1938C
    M. cf. *annectens* (Marsh, 1872I)   —Lillegraven (1980)
    Craseops sylvestris Stock, 1934B
PRIMATES
  Paromomyidae
    Phenacolemur mcgrewi Robinson, 1968
    P. shifrae Krishtalka, 1978
  Adapidae
    Pelycodus   —Lillegraven, 1980
    Notharctus cf. *robustior* (Leidy, 1872B)   —Golz–Lillegraven (1977)
  Omomyidae
    Trogolemur   —Krishtalka (1978)

Ourayia uintensis (Osborn, 1895A)
Mytonius hopsoni Robinson, 1968B
Chumashius balchi Stock, 1933C
Omomys carteri Leidy, 1869B
Stockia powayensis (Gazin, 1958B)   —(Lillegraven (1980) = *Omomys*)
Macrotarsius siegerti Robinson, 1968B
M. jepseni (Robinson, 1968B)   —Krishtalka (1978)
Hemiacodon, sp. near *gracilis* Marsh, 1872I   —Lillegraven (1980)
Washakius woodringi (Stock, 1938C)
Dyseolemur pacificus Stock, 1934A
Uintanius cf. *ameghini* (Wortman, 1904A)
DINOCERATA
  Uintatheriidae
    Uintatherium cf. *anceps* (Marsh, 1871D)
    Eobasileus cornutus (Cope, 1872W)
    Tetheopsis speirianus (Osborn, 1881A)
    T. ingens (Marsh, 1884F)
ARTIODACTYLA
  Dichobunidae
    Pentacemylus leotensis Gazin, 1955
    P. progressus Peterson, 1932
    Mytonomeryx scotti Gazin, 1955
    Apriculus praetaritus Gazin, 1955
    Hylomeryx quadricuspis (Peterson, 1919A)
    H. annectens Peterson, 1919A
    Auxontodon pattersoni Gazin, 1958A
    Bunomeryx elegans Wortman, 1898A
    B. montanus Wortman, 1898A
    Mesomeryx grangeri Peterson, 1919A
    Tapochoerus egressus (Stock, 1934B)   —McKenna (1959B)
    Achaenodon insolens Cope, 1873D
    A. robustus Osborn, 1883A
    A. uintense (Osborn, 1895A)
    Parahyus vagus Marsh, 1876H
  Leptochoeridae
    Leptochoerid, genus indet.   —Black (1978)
  Agriochoeridae
    Protoreodon pumilus (Marsh, 1875B)
    P. parvus Scott and Osborn, 1887A
    P. paradoxicus (Scott, 1898C)
    P. minor Scott, 1899A
    P. petersoni Gazin, 1955
    P. pacificus Golz, 1976
    Diplobunops matthewi Peterson, 1919A
    D. crassus Scott, 1945A
    D. vanhouteni Gazin, 1955
  Protoceratidae
    Leptotragulus proavus Scott and Osborn, 1887A
    L. medius Peterson, 1919A
    L. clarki Gazin, 1955
    Leptoreodon marshi Wortman, 1898A
    L. major Golz, 1976
    L. pusillus Golz, 1976
    L. leptolophus Golz, 1976
    L. (*Hesperomeryx*) *edwardsi* Stock, 1936B

115

Poabromylus golzi Black, 1978

Toromeryx marginensis J. Wilson, 1974

Camelidae   —J. Wilson (1974)

Oromeryx plicatus Marsh, 1894L

Poëbrodon kayi Gazin, 1955

P. californicus Golz, 1976

Protylopus petersoni Wortman, 1898A

P. pearsonensis Golz, 1976

P. stocki Golz, 1976

P.? robustus Golz, 1976

P.? annectens Peterson, 1919A

Malaquiferus tourteloti Gazin, 1955

Camelodon arapahovius Granger, 1910A

Merycobunodon littoralis Golz, 1976

Hypertragulidae

Simimeryx hudsoni Stock, 1934E

Leptomerycidae

Hendryomeryx wilsoni Black, 1976

ACREODI

Mesonychidae

Mesonyx

Harpagolestes breviceps Thorpe, 1923B

H. uintensis (Scott, 1888B)

H. leotensis Peterson, 1931A

H. immanis Matthew, 1909D

CONDYLARTHRA

Hyopsodontidae

Hyopsodus uintensis Osborn, 1902E   (incl. *H. fastigatus* L. Russell and Wickenden, 1979   —Krishtalka, 1979)

PERISSODACTYLA

Equidae

Epihippus gracilis Marsh, 1871D

E. parvus Granger, 1908A

E. uintensis Marsh, 1875B

E. intermedius Peterson, 1931

Brontotheriidae

Metarhinus earlei Osborn, 1903A

M. riparius Riggs, 1912A

M.? pater Stock, 1937A

Rhadinorhinus abbotti Riggs, 1912A

R. diploconus (Osborn, 1895A)

Sphenocoelus uintensis Osborn, 1895A

Dolichorhinus longiceps Douglass, 1910A

D. intermedius Osborn, 1908A

D. heterodon Douglass, 1910A

Sthenodectes incisivus (Douglass, 1910A)

S. priscus Peterson, 1934A

S. australis J. Wilson, 1977

Manteoceras uintensis Douglass, 1910A

Protitanotherium emarginatum Hatcher, 1895A

Diplacodon progressum Peterson, 1934A

D. elatum Marsh, 1875B

Eotitanotherium osborni (Peterson, 1914A)

Telmatherium carnutum

T. altidens Osborn, 1908A

Heterotitanops parvus Peterson, 1914A

Notiotitanops mississippiensis Gazin and Sullivan, 1942.

Isectolophidae

Isectolophus annectens Scott and Osborn, 1887A

I. cuspidens (Peterson, 1919A)

Helaletidae

Dilophodon leotanus (Peterson, 1932)

D. minusculus Scott, 1883A

Colodon woodi (Stock, 1936C)

C. kayi (Hough, 1955)

Amynodontidae

Amynodon reedi Stock, 1939A

A. advenum Marsh, 1875B

A. intermedius Osborn, 1890D

Metamynodon mckinneyi J. Wilson and Schiebout, 1981

Ceratomorpha, family *inc. sed.*

Schizotheroides parvus Hough, 1955

Hyracodontidae

Triplopus cubitalis Cope, 1880I

T. implicatus Cope, 1873U

T.? douglassi (H. Wood, 1934)

T. rhinocerinus H. Wood, 1927B

T. obliquidens Scott and Osborn, 1887A

Forstercooperia grandis (Peterson, 1919A)

Epitriplopus uintensis (Peterson, 1919A)   —Radinsky (1967B)

Chalicotheriidae

Eomoropus amarorum (Cope, 1881G)

E. annectens Peterson, 1919A

RODENTIA

Ischyromyoidea

Ischyromys?   —Black, 1971A

Ischyrotomus petersoni (Matthew, 1910B)

I. compressidens (Peterson, 1919A)

I. eugenei Burke, 1935B

I. littoralis R. Wilson, 1949A

Leptotomus leptodus (Cope, 1873U)

L. mytonensis A. Wood, 1962B

L. sciuroides (Scott and Osborn, 1887A)

L. caryophilus R. Wilson, 1940A

L. (="*Tapomys*" *tapensis* (R. Wilson, 1940A)

L. guildayi Black, 1971A

Rapamys fricki R. Wilson, 1940A

R. wilsoni Black, 1971A

Microparamys dubius (A. Wood, 1949B)

M. minutus (R. Wilson, 1937C)

M. tricus (R. Wilson, 1940A)

Reithroparamys (="*Uriscus*") *californicus* (A. Wood, 1962B)

R. gidleyi (Peterson, 1919A)

Spurimus scotti Black, 1971A

S. selbyi (Peterson, 1919A)

Mytonomys robustus (Peterson, 1919A)

M. burkei (R. Wilson, 1940A)

Manitsha johanniculi A. Wood, 1974

Thisbemys uintensis (Osborn, 1895A)

T. medius (Peterson, 1919A)

T. plicatus A. Wood, 1962B

Lophiparamys sp. indet.   —A. Wood, 1973

Janimus rhinophilus Dawson, 1966

Sciuravus altidens Peterson, 1919A

S. popi Dawson, 1966

S. powayensis R. Wilson, 1940B

Cylindrodontidae

Pareumys grangeri Burke, 1935B
P. milleri Peterson, 1919A
P. lewisi Black, 1974
P. cf. *guensburgi* Black, 1970A
P.? troxelli Burke, 1935B
Pseudocylindrodon tobeyi Black, 1970B
Mysops boskeyi A. Wood, 1973
Aplodontidae?
Eohaplomys tradux Stock, 1935A
E. matutinus Stock, 1935A
E. serus Stock, 1935A
Protoptychidae
Protoptychus hatcheri Scott, 1895E
Eomyidae
Namatomys fantasma Lindsay, 1968A
Geomyidae?
Griphomys alecer R. Wilson, 1940B
G. toltecus Lillegraven, ms., 1980
Zapodidae
Simimys simplex R, Wilson, 1935B
Family, *inc. sed.*
Prolapsus sibilatoris A. Wood, 1973

Of the 132 genera here recognized in the giant, 9-million-year-long Uintan fauna, about 37 (28 percent) are holdovers from the Bridgerian. Only 26 Uintan genera (about 20 percent) are known in later faunas of North America. Ten families (Palaeoryctidae, Stylinodontidae, Plagiomenidae, Uintasoricidae, Microsyopidae, Paromomyidae, Adapidae, Uintatheriidae, Hyopsodontidae, and Isectolophidae) are not known in post-Uintan faunas of North America; however, most of these families may appear in the succeeding Duchesnean (latest pre-Oligocene) some day. Duchesnean is the smallest sample tokening a mammal age in the North American Cenozoic.

A Few Speculations on the Uintan Fauna

1. *Mytonolagus,* earliest record of rabbitlike mammals in North America, immigrated into North America from eastern Asia in Uintan time.
2. The Uintan creodonts, miacids, insectivorans, omomyids, uintatheriids, artiodactyls, brontotheriids, helaletids and rodents—also recorded in the later Eocene of eastern Asia—were endemics.
3. *Harpagolestes* (mesonychid), entelodonts?, *Helaletes* (tapiroid), *Triplopus* (hyracodontid), *Forstercooperia* (hyracodontid), *Amynodon* (rhinocerotoid) and *Eomoropus* (chalicothere) dispersed from Asia to North America or from North America to Asia (or both ways) during Uintan time.
4. There was a filtered dispersal between northwestern North America and northeastern Asia during the Uintan; perissodactyls were the most successful of the filterers.
5. Multituberculates, marsupials, leptictidans, cimolestans, pantolestans?, apatotheres, taeniodonts, dermopterans, primates, leptochoerids, agriochoerids, protocer-

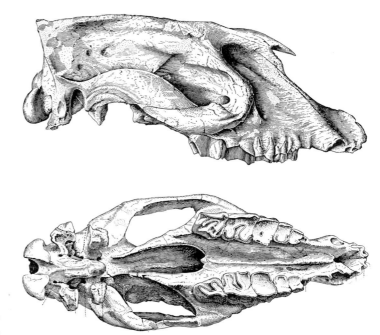

FIGURE 4-31 *Amynodontopsis bodei Stock, 1933A. Skull. × 1/5 approx. Figure 1 in Stock (1933A). (Published with permission of the National Academy of Sciences, USA.)*

atids, camelids, leptomerycids, hyopsodontids, equids, isectolophids, ischyromyoids, cylindrodontids, aplodontids, protoptychids, geomyids(?), and zapodids of the Uintan can be derived by evolution in situ from preexisting North American stocks.

General Remarks on the Uintan Fauna

The composite, 9-million-year-old Uintan paleofauna shows provincial variation from the intermontane lacustrine and fluvial basins of the northern Rockies to the intermontane volcaniclastic depositional areas of southwest Texas and on to the near-sea shoreline lithotopes of southwestern California. Nevertheless, there is a strong, overall faunal-taxonomic similarity through these districts, and there is no reason to suspect anything but relatively unrestricted land-vertebrate dispersal across North America during this interval. Lillegraven (1979) believes, however, that "decrease in tropicality, dilation of temperate realms, and increase in aridity, temporally associated with epeirogenic uplift, increased mountain building, and volcanism in the late Eocene greatly restricted land vertebrate intracontinental dispersal," and he concluded that late Uintan local faunas show much greater endemism than the earlier local faunas. Lillegraven (1979) also reviewed the geological literature on rocks of Uintan age in the southwestern United States and in northwestern Mexico. He reminds us that because of the

FIGURE 4-32   Uintan Localities.

1   Badwater Creek—Hendry Ranch district (in part)
2   Beaver Divide district: Burley Anticline, Green Cove, Oil Mountain, Wagon Bed Spring
3   Camp San Onofre
4   Candelaria
5   Carlsbad—Oceanside district: Chestnut Avenue, Laguna Riviera
6   Chisos Mountains—Mule Ear Peaks
7   Castle Rock—Gazin Locality—Needle Creek Divide
8   Guanajuato—Marfil
9   Haystack Mountain, upper
10   Magdalena Southwest
11   Sage Creek—Cook Ranch

12   San Diego district: Ardath Shale, Dog Spring, Fletcher Parkway, Friars Formation, Lake Murray, Mission San Diego, Mission Valley, Poway, Scripps, Tecolote Canyon
13   Sand Wash Basin (in part)
14   Simi Sespe localities: Brea Canyon, Tapo Ranch
15   Swift Current Creek
16   Tepee Trail Formation, Unit 24
17   Uinta Basin localities: Uinta and lower Duchesne River Formations sites, Kennedy's Hole, Leland Bench Draw, Leota Ranch, Myton Pocket, Ouray Agency, Randlett Point, White River Pocket, etc.
18   Whistler Squat
19   Laredo (Lake Casa Blanca)

post-Eocene strike-slip lithospheric plate movements, the present-day southern California Uintan mammalian fossils were animals that lived at least 360 km to the southeast. This geographic shift through time adds great complexity to our interpretation of Uintan dispersals.

The Uintan fauna is now characterized by a wealth and a great variety of insectivorans, primates, artiodactyls, perissodactyls, and rodents. "Explosive" taxonomic radiation of artiodactyls and rodents was well under way by Uintan time, just as it was at the same time in Eurasia. We visualize that savannas with forested stream valleys were the predominant habitat for Uintan mammals, but the abun-

dant micromammals of probable arboreal adaptation testify to the presence of extensive woodlands also.

Black and Dawson (1966, p. 321) noted:

Vertebrate paleontologists have long recognized that North American mammalian development during the Cenozoic can be divided into two broad phases. The first occurred during the early Cenozoic—the Paleocene and Eocene—and involved the dominance of a variety of archaic mammals living together with other groups that led into the more advanced and persistent families of the second phase. This second phase began sometime during the late Eocene and resulted in the extinction and replacement of the earlier assemblages by families that have either survived into the Recent or were of a generally more modern aspect than the archaic forms. In the study of group after group of mammals, attention is shifted to the late Eocene–early Oligocene, an interval spanning ten to twelve million years, for evidence on extinction of some groups and origin of others. . . . When the total number of fossil and Recent mammalian families, 273 recognized at present, is compared to the number of families appearing in the late Eocene and early Oligocene, 71, one finds that about 26 percent of the total number of mammalian families are first recorded from this interval.

Lillegraven (1972), considering these phenomena in relation to the records of fossil vegetation and paleoclimates, concluded (p. 261):

High numbers of first appearances (a kind of index to rates of evolutionary diversification) occurred in both mammals and flowering plants in the late Eocene to early Oligocene and in the earlier half of the Miocene. . . . The dramatic taxonomic changes of the earlier parts of the Oligocene (*i.e.*, high rates of appearance of new mammalian and plant taxa plus high rate of extinction of archaic mammals) correlate well with a general deterioration of the world's climate (lowering of mean annual temperature and decrease of equability) at that time. Contemporaneous adjustments can be observed between climatic change, rates of evolutionary diversification in plants and mammals, and rates of extinction in mammals.

Among the "modern" families of mammals—those represented in the Quaternary—the Didelphidae, some of the families of the Chiroptera, the Erinaceidae, Gliridae, Tarsiidae?, and Equidae have pre-Uintan (and correlatives) record. The Leporidae, Ursidae, Felidae, Camelidae, Rhinocerotidae, Cricetidae, Geomyidae?, and Zapodidae? seem to appear first in Uintan or correlative time. Other "modern" families of mammals seem to have post-Uintan first appearance, many appearing about the beginning of Chadronian (= Eo-Oligocene). In view of the known differences in morphologic scope of the recognized families of mammals, however, these documentations may not be as significant in the master evolutionary tree of the Mammalia as the array of family names might lead us to believe.

MUSTERSAN

The Mustersan mammal age was named from the small paleofauna obtained from localized deposits, called Musters Formation, frequently channeled into the underlying Casamayor Formation around Lago Musters and Lago Colhue-Huapi in the southern part of the Chubut Province of Argentina (Simpson, 1933Q, 1940G, 1948E, 1967A). The composite Mustersan mammalian fauna, listed herewith, is taken from Simpson (1967A).

Mustersan Mammalian Fauna

MARSUPIALIA
  Borhyaenidae
    Pharsophorus? cretaceus (Roth, 1903)
EDENTATA
  Megalonychidae?
    Proplatyarthrus longipes Ameghino, 1905
  Dasypodidae
    Machlydotherium asperum Ameghino, 1902
    M. ater Ameghino, 1902
    Meteutatus? attonsus (Ameghino, 1902)
    Pseudeutatus clypeus Ameghino, 1902
    P. depictus (Ameghino, 1902)
    P. circundatus (Ameghino, 1902)
    P. cuneiformis (Ameghino, 1902)
LITOPTERNA
  Proterotheriidae
    Polymorphis lechei Roth, 1899
    P. alius (Ameghino, 1901)
    Polyacrodon ligatus Roth, 1899
    Xesmodon langi (Roth, 1899)
    X.? prolixus (Roth, 1899)
    Heteroglyphis dewoletsky Roth, 1899
NOTOUNGULATA
  Notostylopidae
    Otronia muhlbergi Roth, 1901
  Oldfieldthomasiidae
    Tsmanichoria cabrerai Simpson, 1936
  Interatheriidae
    Notopithecus? sp.
    Guilielmoscottia plicifera Ameghino, 1901
  Archaeohyracidae
    Eohyrax? platyodus (Ameghino, 1904)
    Pseudhyrax eutrachytheroides Ameghino, 1901
    P. strangulatus (Ameghino, 1901)
    Bryanpattersonia nesodontoides (Ameghino, 1901)
      —Simpson (1967A)
    B. sulcidens (Ameghino, 1902)
  Archaeohyracidae?
    Eohegetotherium priscum Ameghino, 1901
  Isotemnidae
    Acoelohyrax? coalitus (Ameghino, 1901)
    A.? coarctatus (Ameghino, 1901)
    Periphragnis harmeri Roth, 1899
    P.? circunflexus (Ameghino, 1901)
    Rhyphodon lankesteri Roth, 1899
    Distylophorus alouatinus (Roth, 1902)
  Notohippidae
    Eomorphippus obscurus Ameghino, 1901
    Interhippus deflexus Ameghino, 1902   —*nomen dubium*, acc. Simpson (1967A)
ASTRAPOTHERIA
    Astraponotus assymmetrus Ameghino, 1901
TRIGONOSTYLOPOIDEA
    Trigonostylops gegenbauri (Roth, 1899)   —*nomen dubium*, acc. Simpson (1967A)
PYROTHERIA

Propyrotherium saxeum Ameghino, 1901
Griphodon peruvianus Anthony, 1924   —Patterson
     (1977), mid?-Eoc., Peru

Little can be said about the Mustersan fauna, except that it was endemic and closely related (and derived from) the preceding Casamayoran fauna. It may have existed during the interval 48–40 million years ago, but this is only a crude estimate.

Colombitherium Hoffstetter (1970) from the Gualanday Formation of Colombia, a pyrothere, may represent either Mustersan or Divisaderan in northern South America.

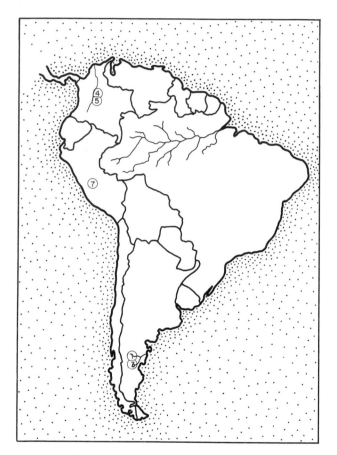

FIGURE 4-33 *Mustersan and Divisaderan Localities.*
1   Cerro Talquino
1   Colhuapi (Colhué Huapí) Norte
3   Great Barranca
1   Pajarito
4   Tama   (age?)
5   Gualanday Northeast   (age?)
7   Chicoca   (mid? Eocene)

DIVISADERAN

The small mammalian fauna from the Divisadero Largo Formation, Mendoza Province, Argentina, is currently the basis for tentative recognition of a latest Eocene land-mammal age in South America (Marshall, Pascual, Curtis, and Drake, 1977). Simpson, Minoprio, and Patterson (1962) and Patterson and Pascual (1968) were more reluctant to recognize a discrete Divisaderan, although Pascual was the first to propose it. Combined evidence from the paleontology and from general radiometric dating associated with the succeeding Deseadan land-mammal age shows, however, that Divisaderan is a post-Mustersan interval earlier than 36 million years before present (constants from Steiger and Jager, 1977); hence, we recognize it as a latest Eocene interval.

Mammalian Fauna from the Divisadero Largo Formation

(From Simpson, Minoprio, and Patterson, 1962.)
MARSUPIALIA
   Groeberidae
      Groeberia minoprioi Patterson, 1952
LITOPTERNA
   Adianthidae
      Adiantoides leali Simpson and Minoprio, 1949
   Proterotheriidae?
      Phoradiadius divortiensis Simpson, Minoprio, and Patterson, 1962
NOTOUNGULATA
   Oldfieldthomasiidae?
      Brachystephanus postremus Simpson, Minoprio, and Patterson, 1962
      Xenostephanus chiottii Simpson, Minoprio, and Patterson, 1962
      Allalmeia atalaensis Rusconi, 1946D and 1946E
   Mesotheriidae
      Trachytherus? mendocensis (Simpson and Minoprio, 1949)
   Hegetotheriidae
      Ethegotherium carrettei (Minoprio, 1947)
Order?
   Family?
      Acamana ambiguus Simpson, Minoprio, and Patterson, 1962

Although the Divisaderan mammal fauna differs from the preceding Mustersan and the succeeding Deseadan by its peculiar genera, it shows no indication of immigration of allochthonous mammals and appears to be only an evolutionary modification of the residual South American Eocene fauna.

DUCHESNEAN

Duchesnean is probably the most hotly debated and certainly the most weakly substantiated of the Nearctic land-mammal ages we adopt. Admittedly, the faunas herein assigned to Duchesnean could be termed later or latest Uintan or "eo-Chadronian," and such designation would be as useful as the system we are using. J. Wilson (1977), for example, could find no basis for recognizing a Duchesnean age in the excellent succession of Late Eocene into Early

Oligocene faunas of Trans-Pecos Texas. It is indeed mysterious to us why there is so little taxonomic-evolutionary change among land mammals during the 10 million years here labeled Uintan and Duchesnean, for this was an interval of notable climatic and vegetative change in various provinces of Nearctica (Wolfe, 1971, 1978; Wolfe and Hopkins, 1967; Lillegraven, 1972; and Black and Dawson, 1966).

Duchesnean was proposed as a North American provincial time term by H. Wood et al. (1941), based on the Duchesne River Formation of northeastern Utah. L. Kay (1934) had previously identified, from lowest to highest, the Randlett, Halfway, and Lapoint horizons as informal subdivisions of the Duchesne River Formation and had listed fossil-mammal taxa from each "horizon." Andersen and Picard (1972) properly defined, typified, and described useful members of the formation. The *Brennan Basin Member* is the basal unit; it is equivalent stratigraphically to the Randlett and lower Halfway of Kay (1934) and has produced a mammalian fauna practically indistinguishable from the underlying and interdigitating upper part of the Uinta Formation not far away. The *Dry Gulch Creek Member* of Andersen and Picard equals the upper two-thirds of Kay's Halfway and has produced *Epihippus intermedius* Peterson, *Eosictis avinoffi* Scott, *Protitanotherium,* and *Protoreodon pumilus,* according to various authors. Kay's Lapoint horizon was validated as *Lapoint Member* by Andersen and Picard (the member is slightly more restricted stratigraphically) and has produced the fauna from Twelvemile Wash that West et al. (1983?) and others select as typical Duchesnean. This fauna, according to Andersen and Picard, modified slightly and arranged in our system, includes the following.

Duchesnean Fauna from Twelvemile Wash

LEPTICTIDA?
   Protictops alticuspidens Peterson, 1934B
CREODONTA
   Hyaenodontidae
      Hyaenodon cf. *vetus* Stock, 1933B,   —Mellett (1977)
      Hessolestes ultimus Peterson, 1931A
ARTIODACTYLA
   Entelodontidae
      Brachyhyops wyomingensis Colbert, 1937C   —J. Wilson (1971)
   Protoceratidae
      Poabromylus kayi Peterson, 1932
   Agriochoeridae
      "*Agriochoerus*" *maximus* Emry, 1981
   Hypertragulidae
      Simimeryx minutus Peterson, 1931A
PERISSODACTYLA
   Brontotheriidae
      Duchesneodus uintensis (Peterson, 1931B)
         Lucas and Schoch (1982)

   Hyracodontidae
      Epitriplopus medius Peterson, 1934B
      Hyracodon primus Peterson, 1934B
RODENTIA
   Cylindrodontidae
      Pareumys guensburgi Black, 1970A
   Eomyidae
      Protadjidaumo typus Burke, 1934A

This Twelvemile Wash assemblage in the Dry Gulch Creek Member leaves much to be desired for faunal-taxonomic typification of Duchesnean. Although the brontothere *Teleodus* was used by some as the index taxon for the age, it is known now only in the Chadronian (succeeding) faunas (J. Wilson, 1977). The overlying *Starr Flat Member* of the Duchesne River Formation is unfossiliferous. It could be Duchesnean or post-Duchesnean.

California Institute of Technology locality 150 (= Pearson Ranch; several quarries), in the lower part of the Sespe Formation of Ventura County, southern California, was listed by H. Wood et al. (1941) as the correlative of the Duchesne River fauna. The Pearson Ranch sample is now in the collections of the Natural History Museum of Los Angeles County. Mammals from Pearson Ranch, following Golz-Lillegraven (1977), in our system of classification, are as follows.

Duchesnean Fauna from Pearson Ranch

MARSUPIALIA
   Didelphidae?
LEPTICTIDA
   Lepticitidae
PANTOLESTA
   Simidectes merriami Stock, 1933B
CREODONTA
   Hyaenodontidae
      Hyaenodon vetus Stock, 1933A
      H. venturae Mellett, 1977
INSECTIVORA
   Erinaceoidea
      Sespedectes singularis Stock, 1935C
      Proterixoides davisi Stock, 1935C
      Insectivora, unident.
PRIMATES
   Omomyidae
      Chumashius balchi Stock, 1933D
ARTIODACTYLA
   Agriochoeridae
      Protoreodon pacificus Golz, 1976
   Camelidae   —J. Wilson (1974)
      Protylopus pearsonensis Golz, 1976
      Eotylopus
      oromerycine?
   Protoceratidae

121

Leptoreodon?
Hypertragulidae
  Simimeryx hudsoni Stock, 1934D
  S. sp.
  artiodactyl, unident.
PERISSODACTYLA
  Brontotheriidae
    Duchesneodus californicus (Stock, 1935D)
  Amynodontidae
    Amynodontopsis bodei Stock, 1933C
  Hyracodontidae?
    triplopine?
RODENTIA
  Ischyromyoidea
    Microparamys tricus (R. Wilson, 1940A)
    "paramyine, gen. and sp. indet."
    "ischyromyid"

Cylindrodontidae
  Pareumys near *milleri* Peterson, 1919A
  Presbymys lophatus R. Wilson, 1949
  "cylindrodontid"
Geomyidae?
  Griphomys near *alecer* R. Wilson (1940B)
Zapodidae?
  Simimys simplex (R. Wilson, 1935A)
  Rodentia, unident.

Krishtalka (1979) has described *Hyopsodus sholemi* from a "Duchesnean" assemblage in Wyoming.

Of the genera represented by more than a very few fossils and known from the type Duchesnean and from Pearson Ranch, *Hyaenodon, Eotylopus, Duchesneodus, Amynodontopsis, Hyracodon, Presbymys,* and *Protadjidaumo* are not known from the preceding Uintan faunas. *Hyaenodon, Eotylopus, Duchesneodus, Hyracodon,* and *Protadjidaumo* are known from the succeeding Chardonian faunas.

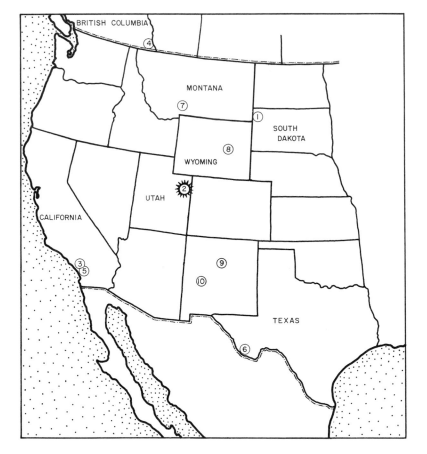

FIGURE 4-34 *Duchesnean Localities.*

1 Antelope Creek—Slim Buttes
2 Duchesne River Formation (in part)—Uinta Basin
3 Hartman Ranch
4 Kishenehn Formation site   (age?)
5 Pearson Ranch
3 Pine Mountain sites
1 Reva Gap—Slim Buttes
    Sespe Formation (lower) sites
    (See Hartman and Pearson Ranch.)

2 Teleodus Quarry—Duchesne River
2 Tridell East—Uinta Basin
2 Vernal West
6 Skyline (Mahgarita) (age?)
7 Shoddy Springs—Three Forks
8 Badwater 20

EOCENE MARINE MAMMALS

The earliest record of marine mammals, based on currently accepted datings and correlations, comprises a few teeth named *Ishatherium subathuensis* by Sahni and Kumar (1980). These teeth are from a shelly limestone in the Subathu Formation, Simla Hills, in the Simla Himalayas area of India, and the limestone has been correlated with the Ypresian of western Europe on the basis of its foraminifers. Sahni and Kumar (1980) refer *Ishatherium* to the dugongid sirenians.

By middle Eocene time (Lutetian and correlatives), the Sirenia and Cetacea appear in the record, fully adapted to continuous life in marine waters. All agree that these marine adapters were derived from the earlier land mammals, but the definitive story of this great morphologic-physiologic evolutionary transition is yet to be unveiled. It has been speculated that sirenians and cetaceans arose in the shallow coastal waters of the ancient Tethyan seaway, including its westernmost extension, signified by the "Tethyan" rocks and marine faunas in present-day southeastern North America. We know of no compelling evidence opposing this interpretation.

Consensus holds that cetaceans had ancestry in the mesonychids (or perhaps we should say more properly that cetaceans and mesonychids are "sister groups," according to McKenna, 1975B). The fresh-to-brackish-to-marine middle Eocene strata of Pakistan and India are producing documentation supporting the theory that cetaceans and mesonychids are close together phylogenetically (West, 1980). McKenna (1975B) and others agree that sirenians stem from an ancestry that might be called early tethytherian—formerly called phenacodontoid by some workers—and that proboscideans and desmostylians are the most closely related groups phyletically. Early representatives of these groups, exclusive of the desmostylians, inhabited the Tethyan area.

Middle Eocene Marine Mammals of the World

CETACEA
  ARCHAEOCETI
    Protocetidae
      Protocetus Fraas, 1904
      Pappocetus Andrews, 1920
      Eocetus Fraas, 1904
      Prozeuglodon Andrews, 1906
      Pakicetus Gingerich and Russell, 1981
SIRENIA
  Prorastomidae
    Prorastomus Owen, 1855   (age?)
  Protosirenidae
    Protosiren Abel, 1904
  Dugongidae
    Protumolherium veronense Zigno, 1875
    Sirenavus hungaricus Kretzoi, 1941D   (age?)
    Anisosiren pannonica Kordos, 1979   (age?)
    Eotheroides Palmer, 1899

The slightly later, early Bartonian marine mammal fauna includes:

CETACEA
  ARCHAEOCETI
    Zygorhiza wanklyni (Seeley, 1876)
    Basilosaurus Harlan, 1834 (incl. *Zeuglodon* Owen, 1839?)
    Pachycetus robustus Van Beneden, 1883
    P. humilis Van Beneden, 1883

Later Eocene in North America

The Castle Hayne Formation of North Carolina, Santee Limestone of South Carolina, Ocala Limestone of Georgia and Florida, Jackson Formation or Group of Louisiana, Alabama, and Mississippi, and Yegua Formation of Texas—all marine and all later Eocene—have yielded fossils referable to the following archaeocete whales (Kellogg, 1936):

- *Zygorhiza kochii*
- *Basilosaurus cetoides*
- *Pontogeneus brachyspondylus*
- *Dorudon serratus*

Because of the close affinities of these genera and species with forms from the North African marine Later Eocene, various workers have proposed a late Eocene shallow-water connection between the Mediterranean-Tethyan province and the southeastern coastal United States.

A vertebra identified as archaeocete from Vancouver Island, British Columbia (Kellogg, 1936; Barnes and Mitchell, 1978) seems to be the only possible late-Eocene record of whales from the Pacific Basin.

STRATIGRAPHIC DEMONSTRATIONS

Europe   It has been difficult to find direct superposition of one mammal-bearing stratum on an older mammal-bearing stratum in France's Eocene because extensive exposures of bedrock in this forested and cultivated terrain are lacking. This predicament remains today even though the vertical succession of formations, substages, and stages is well known and well mapped, especially in the Paris Basin. In the environs of Epernay, the Cuisian sites in the Sables à Térédines et Unios lie directly on brackish-water claystone and siltstone beds of the type Sparnacian, but the Sparnacian at these sites has not yet produced fossils of mammals. The producing Sparnacian sites near Epernay (Mutigny, Avenay) underlie soil or Lutetian? carbonaceous claystone and lacustrine limestone.

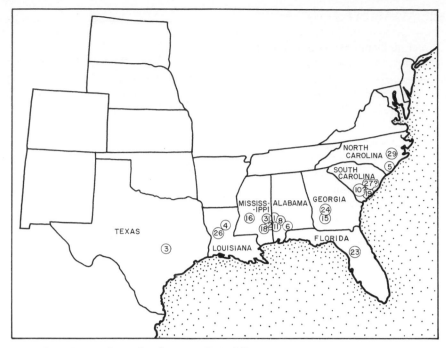

FIGURE 4-35 *Marine Late Eocene Sites, Southeastern United States.*

1 Ben Turner Place	*13 Dead Level—Clarksville*	*23 Oakhurst Lime Company*
2 Buckatunna River West	*14 Fail—Ben Turner Place*	*Quarry*
3 Burleson County	*15 Flint River*	*24 Ouachita River—Caldwell*
4 Caldwell Parish	*16 Hinds County*	*Parish*
5 Castle Hayne Station	*17 Isney West—Ben Turner Place*	*25 Perry South*
6 Cedar Reach Rock	*18 Jasper County*	*26 Red River*
7 Choctaw County—Ben Turner	*19 Masyk Plantation—Fair*	*27 Santee Chalk sites*
Place	*Spring*	*28 Suggsville-Clarksville*
8 Clarksville Southwest	*20 Melvin East—Buckatunna*	*29 Wadsworth Pit*
9 Cocoa—Ben Turner Place	*River*	*30 Washington County—Court*
10 Cooper Marls site	*21 Melvin Southeast—Bucka-*	*House*
11 Court House (Old)	*tunna River*	*31 Quitman South (brontothere)*
12 Creagh Plantation—Clarks-	*22 Millsaps College East—Hinds*	
ville	*County*	

The Cuisian, in turn, is demonstrably underlies the Calcaire Grossier in many places in the eastern Paris Basin.

Sudre (1969) reviewed the stratal succession near St. Mamert, southern France, in which the type-Robiac fauna in the lower part of the Calcaire de Fons overlies a formation assigned to the Lutetian and underlies Fons 4 and other Headonian sites in the upper part of the Calcaire de Fons.

The lower Hamstead (Hempstead) beds—early Oligocene—overlie strata carrying mammals of Headonian age in the south of England.

Asia. Qi (1980) divides the Tertiary strata of Huhebolhe Cliff ("Camp Margetts"), Nei Mongol Autonomous Region, China, into the following succession:

- Upper, Irdin Manha (Late Eocene), about 5 m thick

- Middle, Arshanto (Middle Eocene with 35 mammalian species), about 35 m
- Lower, Bayan Ulan (Early Eocene with *Mongolotherium* etc.), about 36 m

Radinsky (1964) had previously noted from the American Museum Expeditions field notes of the 1920s and from publications such as Matthew and Granger (1926) and Berkey and Morris (1927) that Irdin Manha Formation and fossils overlie Arshanto Formation and fossils. Granger and Berkey (1922) and Radinsky (1964) also pointed out that Irdin Manha underlies in places Houldjin Gravels, the latter bearing Oligocene mammals. And in the Shara Murun district, the authors cited above noted a succession comprising Tukhum Formation with Middle? Eocene mammals underlying Shara Murun Beds with Late Eocene mammals, underlying Ulan Gochu beds with Early Oligocene mammals.

124

Africa. Late Eocene marine with *Barytherium* and *Moeritherium* (Proboscidea)—the Qasr-el-Sagha Formation—are subjacent to the Jebel-el-Qatrani Formation and its renowned Oligocene land-mammal fauna.

North America. Several depositional basins of the Rocky Mountain region show beautiful stratal succession representing some or all recognized mammal ages of the Eocene:

1. Bighorn Basin has a complete Puercan through Lostcabinian sequence in the Polecat Bench and Willwood Formations, representing a 15-million-year interval.
2. Wind River Basin has Graybullian (early Wasatchian) through Lostcabinian in the Indian Meadows and Wind River formations.
3. Uinta Basin has Uintan, Duchesnean, and Chadronian in stratigraphic succession.

Probably the greatest relatively uninterrupted homoclinal nonmarine stratal succession of the entire world, however, is the one recorded across the northern segment of the Washakie Basin in southwestern Wyoming. Its makeup is shown in the accompanying chart.

UINTAN	Adobe Town Member of Washakie Formation
BRIDGERIAN	Kinney Rim Member of Washakie Formation
	Laney Member of the Green River Formation
	Cathedral Bluffs Member of the Wasatch Formation
LOSTCABINIAN	Wilkins Peak Member of the Green River Formation
	Tipton Member of the Green River Formation (no mammals)
	Niland Member of the Wasatch Formation
	Luman Tongue of the Green River Formation
LYSITEAN	Main Body of the Wasatch Formation
GRAYBULLIAN	
CLARKFORKIAN	
TIFFANIAN	"Fort Union Formation"
TORREJONIAN (not represented)	with fossil plants
PUERCAN (not represented)	
LANCIAN	"Lance" Formation

The Washakie succession, calibrated by mammal ages as shown, represents an interval of about 30 million years, 70–40 million years ago in the history of geologic and biologic phenomena in North America. (See Figure 4-36.)

South America Strata of Mustersan age lie in channels on the Casamayor Formation (Casamayoran) and underlie the Deseado Formation (Deseadan-Oligocene) in the Great Barranca of Patagonia.

DATING AND CORRELATION

Although scattered potassium-argon radiometric dates from glauconites and from volcanic rocks in the Eocene of Europe and North America often lack direct stratigraphic tie-in with mammal-bearing strata, they generally support the following year-age assignments.

In Europe

- Headonian began about 39 million years before present and lasted until about 35 m.y.b.p.
- Robiacian extends from about 44 to about 39 m.y.b.p.
- Lutetian extends from about 49 to about 44 m.y.b.p.
- Cuisian extends from about 51 to about 49 m.y.b.p.
- Sparnacian extends from about 54 to about 51 m.y.b.p.

In North America

- Chadronian extends from about 38 to about 32 m.y.b.p.
- Duchesnean extends from about 40 to about 38 m.y.b.p.
- Uintan extends from about 48 to about 40 m.y.b.p.
- Bridgerian extends from about 50 to about 48 m.y.b.p.
- Wasatchian extends from about 54 to about 50 m.y.b.p.

See Berggren, McKenna, Hardenbol, and Obradovich (1978) for a slightly different Eocene chronology.

Correlation within the Eocene and correlation among Eocene subdivisions of Europe and North America are shown in Figures 4-37, 4-38, and 4-39.

BIOGEOGRAPHY AND THE EOCENE MILIEU

Our diagram Figure 4-40 presents a greatly generalized plan for the distribution of continents and seas about 55 million years ago. For reasons detailed in the preceding discussion of the Sparnacian of Europe, we believe that one or more land corridors extended through the present-day North At-

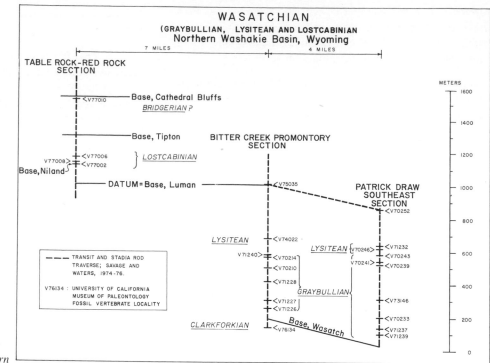

FIGURE 4-36 *Stratigraphic Sections in the Bitter Creek Area, Southwestern Wyoming.*

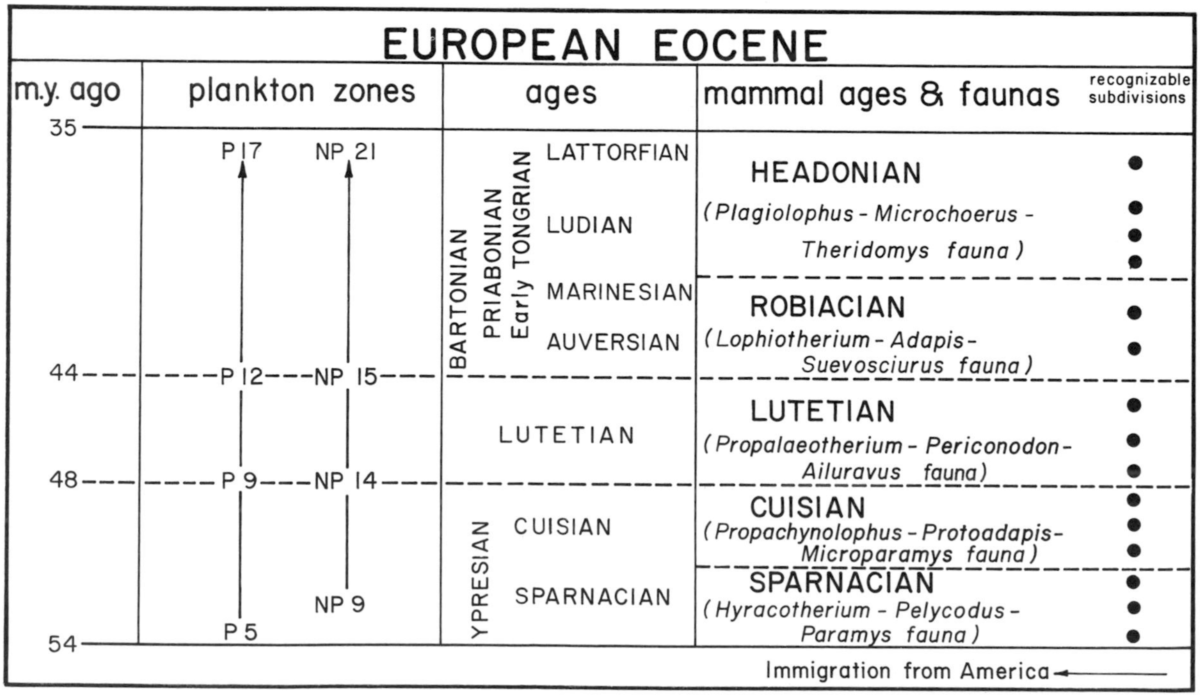

EUROPEAN EOCENE

m.y. ago	plankton zones		ages		mammal ages & faunas	recognizable subdivisions
35 —	P 17	NP 21	BARTONIAN PRIABONIAN Early TONGRIAN	LATTORFIAN	**HEADONIAN** *(Plagiolophus - Microchoerus - Theridomys fauna)*	● ● ● ●
				LUDIAN		
				MARINESIAN	**ROBIACIAN** *(Lophiotherium - Adapis - Suevosciurus fauna)*	● ●
44 — — —	P 12 — — NP 15			AUVERSIAN		
				LUTETIAN	**LUTETIAN** *(Propalaeotherium - Periconodon - Ailuravus fauna)*	● ● ●
48 — — —	P 9 — — NP 14		YPRESIAN	CUISIAN	**CUISIAN** *(Propachynolophus - Protoadapis - Microparamys fauna)*	● ● ●
	NP 9			SPARNACIAN	**SPARNACIAN** *(Hyracotherium - Pelycodus - Paramys fauna)*	● ● ●
54 —	P 5				Immigration from America ◄───	

FIGURE 4-37 *Correlations Within the Eocene of Europe.*

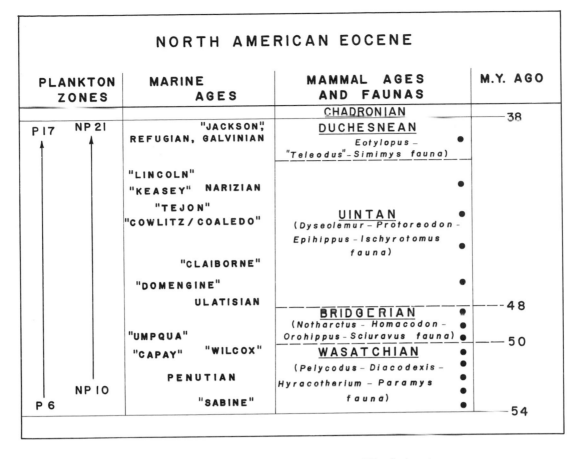

FIGURE 4-38 *Correlations Within the Eocene of North America.*

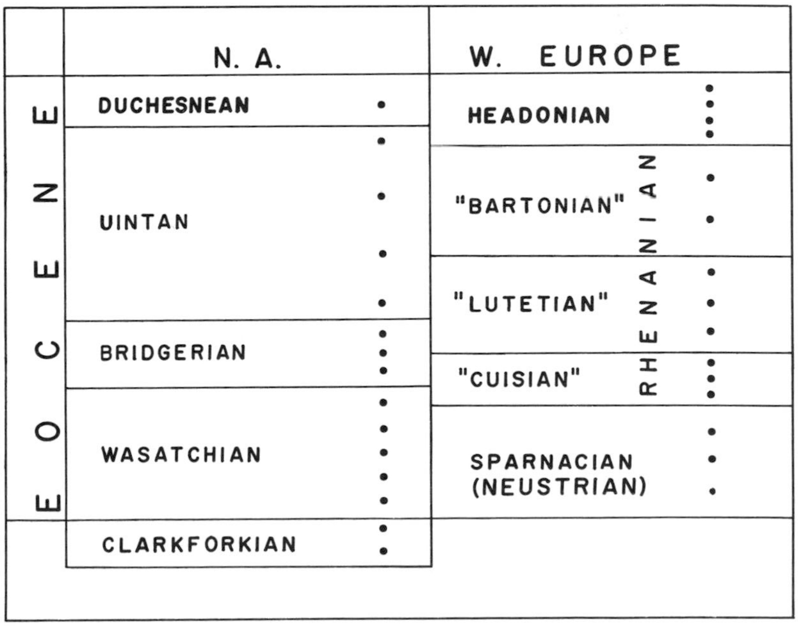

	N. A.		W. EUROPE	
E	DUCHESNEAN	•	HEADONIAN	• • •
N		•		
	UINTAN	•	"BARTONIAN"	•
E		•		•
O		•	"LUTETIAN"	• •
C	BRIDGERIAN	• •	"CUISIAN"	• •
E	WASATCHIAN	• • • •	SPARNACIAN (NEUSTRIAN)	• •
E	CLARKFORKIAN	• •		

(W. Europe column labelled vertically: RHENANIAN)

FIGURE 4-39 *Eocene Correlations Between Europe and North America.*

lantic–Arctic area connecting North America to northwestern Europe. Seaways or other barriers to land-vertebrate dispersal lay between Europe and Africa and between Europe and Asia. North America may have been connected through the Bering district to Asia. The Indian subcontinent had not yet effectively joined the main Asian continent. South America and Antarctica-Australia remained isolated from all other continents.

Soon after this time, by about 48 million years ago, great changes in land and sea had evolved. Most significant of these changes was the completion of the North Atlantic Ocean and its uninterrupted connection into the Arctic Sea, which stopped a flow of land mammals between the Western and Eastern hemispheres. And in the brief interval of only 3–6 million years, European mammals became remarkably dissimilar in ensemble of genera to their North American counterparts.

By about 41 million years ago—late Uintan in North America, Headonian in Europe, Late Eocene in Asia and Africa—the distribution of land and sea was approximately

as diagrammed in Figure 4-41. Eurasia, now including the Indian plate as part of its land continuum, was isolated by seaways and oceans, except for a possible restricted connection to the Alaska region of North America. Forty-one million years ago was a time of isolation for most continents. Africa, Antarctica, Australia, and South America seem to have been surrounded by sea barriers to land-vertebrate dispersal. On these isolated continents, the overprint of developing seasonality of climates caused a veritable revolution in the composition of land-mammal faunas. This revolution took place from about 48 million years ago to about 35 million years ago, mid-Eocene into Oligocene.

The principal floras from the Early Eocene of Europe have been recovered from the Lignites du Soissonais (Watelet, 1866; Fritel, 1910), the Woolwich, Reading, Oldhaven, and Blackheath Beds (Chandler, 1961–1964), the Belleu Sandstone (Watelet, 1866), the sands of Cuise-la-Motte (Koeniguer, 1968) and especially the London Clay (Reid and Chandler, 1933; Edwards, 1936; Chandler, 1961–1964). Earliest Eocene floras indicate a continuation of the warm, moist, near-coast climate of the preceding Paleocene, with a trend toward the probably warmer and more moist climate recorded in the London Clay. Stratigraphers

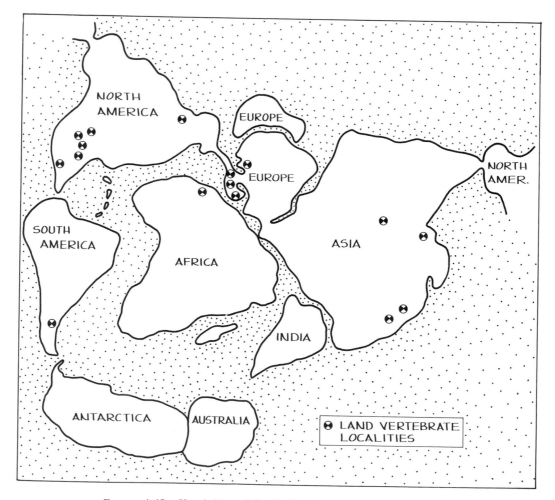

FIGURE 4-40 *Sketch-Map of the World, 55 Million Years Ago.*

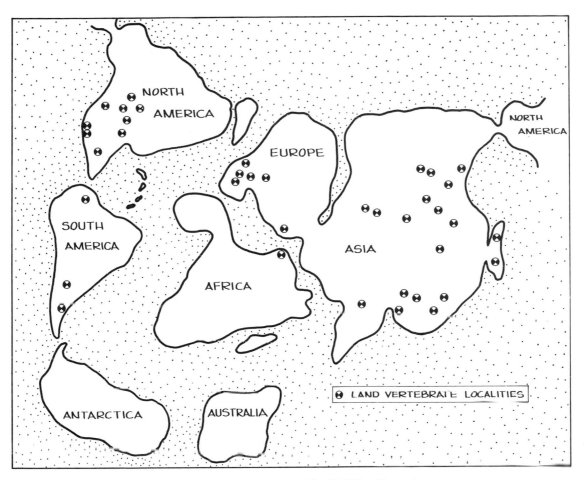

FIGURE 4-41 *Sketch-Map of the World, 41 Million Years Ago.*

and paleobotanists have related the trend toward "tropicality" to a readvance of the Tethyan seaway; Chandler (1964) has pointed out the great complexities, however, and hesitates to make this simple correlation. The well-studied London Clay nut-and-fruit flora is of tropical rain forest aspect, with abundant lianas. It is said to have a strong Indo-Malayan affinity. *Nipa*, the "strand line" palm of the paleotropics, also known in the Eocene of the Western Hemisphere, is abundant, in contrast to its previous absence. Chandler (1964) noted that 70 percent of the 500 species of the London Clay flora are extinct but 73 percent of the mostly modern genera live today in the Indo-Malayan area. She referred the reader to Moseley (1892) for his description of the present-day coast of New Guinea depicting parallel climatic conditions.

The more temperate floras across northern Palearctica were termed *Greenland* (earlier), evolving into the *Turgaian* (Eocene) by Krystofovich (1935, 1955). The more tropical floras to the south he named *Gelinden* (earlier), evolving into *Poltavian* (Eocene). Rasky (1956) and Kovacs (1957) have described lower Eocene "tropical" rain forest in the Hungarian region. Axelrod (1967) has summarized the interpreted climates of Early Eocene and earlier Tertiary from rocks in presently arid regions. These floras

from "southern California, southern Arizona, southern Nevada, coastal Peru-Chile, southern Australia, north Africa, southern Russia and western India, all reveal forests that suggest precipitation near 50 inches at minimum, and probably a dry season of moderate intensity." The floras of middle Nearctica (to be discussed later in this chapter) complement this picture. It appears, therefore, that through great latitudinal range during the Early Eocene, the world's life thrived in moist, warm, nonextreme climates.

Middle Eocene floras of Europe are best represented by samples from the Bournemouth Beds and Bracklesham Beds in England (Chandler, 1963), and especially the famous braunkohle of Geiseltal in East Germany (Beyn, 1940; Gallwitz and Krutsch, 1953; Krumbiegel and Schmidt, 1968). Abundance of ferns and certain families of plants now found in more tropical latitudes in these fossil floras suggests that the Middle Eocene in Europe was a time of continued warm and wet climate without great extremes in temperature. The Middle Eocene may have been the time of greatest "tropicality" during the Cenozoic of Europe, and it was the time of greatest *equability* in the sense of Axelrod and Bailey (1968, 1969).

Wolfe (1971, p. 54) stated: "Almost all paleobotanists who have been concerned with Tertiary climates have

129

agreed that the Eocene was the warmest epoch of the Tertiary. . . . More recent work . . . in Alaska, Washington, and Oregon indicates that there probably were major climatic fluctuations in the Eocene.'' MacGinitie (1969, 1974) has concluded that the Middle Eocene floras of southwestern Wyoming and adjacent Utah and Colorado, generally correlative with Bridgerian, indicate a more mesic climate (drier compared with previous Tertiary climates). The diversified river floodplain and channel deposits of the slightly later Uinta–Duchesne River succession indicate that in the interval 48–35 million years ago the climate of the northern Rocky Mountain region of the United States was drier and had more seasonality, that is, alternate wetting and drying, than previously in the Eocene.

Wolfe and Hopkins (1967) and Wolfe (1971, 1978) concluded that a great climatic deterioration (cooling) occurred in an interval of 1–2 million years approximately 35–34 million years ago. Wolfe (1971) first termed this cooling an *Oligocene deterioration;* later (1978), believing that the Eocene-Oligocene boundary should be placed at 33 m.y.b.p. because of conclusions by fossil-plankton workers, he called this purportedly short cooling interval the *terminal Eocene event.* (Wolfe's terminal Eocene is in the beginning of the Oligocene in the chronology we use in this book.)

We conclude, following data compiled by Axelrod, Wolfe, MacGinitie, and the authors cited by these writers, that during the interval 48–34 m.y.b.p. (Uintan into Chadronian mammal ages in North America, Robiacian into early Oligocene mammal ages in Europe), which involves probably the later half of the Eocene plus earliest Oligocene, climates of the continental areas were becoming more provincial, more seasonal, more diversified geographically, and drier in many if not most inland districts of the world. Correlatively, this ''revolution'' in climate profoundly modified the vegetation and the fauna in various provinces. Within this changing milieu, there were many significant extinctions and originations of mammalian groups (Lillegraven, 1972). Many modern families of land mammals first appeared during the later part of this interval, ca. 40–34 million years ago.

LITHOSTRATIGRAPHIC UNITS

Adobe Town Member, Washakie Formation—Bridgerian and Uintan—Wyoming

Aksyir Formation—Late Eocene (and Early Oligocene?)—Kazakhstan

Alamo Creek Basalt, Chisos Formation—Uintan?—Texas

Alay Formation—Middle Eocene—Kirghizia

Ali Usu Formation—middle? Eocene—Nei Mongol

Almy Formation—Tiffanian into Wasatchian—Wyoming

Apt, couches d'—Headonian—France

''Arencun'' Formation (See Lumeiyi)—Late Eocene—Yunnan

Argile Plastique—Sparnacian—Paris Basin

Arshanto Formation—Middle Eocene—Nei Mongol

Aryan, lignite d'—Headonian—France

Aycross Formation—Bridgerian—Wyoming

Babia Stage, shell limestone—Middle Eocene—Kutch, India

Baca Formation—Bridgerian into Chadronian—New Mexico

Bailuyuan Formation—Late Eocene and Early Oligocene—Shaanxi

Banc Vert, Calcaire Grossier—Lutetian—Paris Basin

Barton Clay and Sand—Bartonian (Robiacian)—England

Bayan or Bayn Ulan (Bayanwulan) Formation—Early Eocene—Nei Mongol

Beauce, Calcaire de—Robiacian—Paris Basin

Beauchamp, Sables (or Grès) de—Robiacian—Paris Basin

Bembridge Limestone or Chalk or Marls or Beds—Headonian—England

Berru, lignite du Mont de—Sparnacian—Paris Basin

Blackheath Beds—Sparnacian—England

Black's Fork Member, Bridger Formation—Bridgerian—Wyoming

Boyabat lignites—Late Eocene—Turkey

Brackelsham Beds or Group—Cuisian into Bartonian—England

Brasles, sables de—Sparnacian—Paris Basin

Braunkohle (Geiseltal)—Lutetian—Germany

Brennan Basin Member, Duchesne River Formation—Uintan—Utah

Bridger Formation—Bridgerian—Wyoming

Brockenhurst Beds—Headonian—England

Bruxelles, Sables de—Lutetian—Belgium

Buck Hill Group—Late Bridgerian?, Uintan into Chadronian—Texas

Bulldog Hollow Member, Fowkes Formation—Bridgerian—Wyoming

Bumbin-Nuru Beds—Early Eocene—Mongolia

Calcaire de Beauce—Robiacian—Paris Basin

Calcaire de Ducy—Robiacian—Paris Basin

Calcaire de St. Ouen—Robiacian—Paris Basin

Calcaire d'Issigeac—Headonian—France

Calcaire Grossier—Lutetian—Paris Basin

Canoe Formation—Bridgerian—Texas

Casamayor Formation—Casamayoran—Argentina

Castellnou, lignite de—Headonian?—Spain

Castle Hayne Formation (marine)—Late Eocene—North Carolina

Castrais, sables de—Robiacian—France

Cathedral Bluffs Tongue or Member, Wasatch Formation—Lostcabinian? and Bridgerian—Wyoming

Célas Sandstone—Headonian—France

Chaili Member, Heti Formation—Late Eocene—Shanxi-Henan

Chharat Formation (''series'')—Early and Middle Eocene—Pakistan

Chisos Formation—Uintan—Texas

Chuankou Formation—Middle and Late Eocene—Henan

Chugouyu Formation—Late Eocene—Henan

Clarno Formation—Bridgerian and Duchesnean or Early Chadronian—Oregon

Climbing Arrow Formation—Duchesnean—Montana

Colmena Tuff—Uintan—Texas

Conglomérat de Meudon—Sparnacian—Paris Basin

Creechbarrow Limestone—Robiacian—England

Croydon Beds—Sparnacian—England

''Cuchara'' Formation—Lostcabinian—Colorado

Dabu Formation—Early Eocene—Xinjiang

Dacangfang Formation—Middle Eocene—Henan

Debeque Formation—Tiffanian into Wasatchian—Colorado

Devils Graveyard Formation, Buck Hill Group—Late Bridgerian into Chadronian—Texas

Divisadero Largo Formation—Divisaderan—Mendoza, Argentina

Domanda Formation—Middle Eocene—Pakistan

Donghu Formation—Early Eocene—Hubei

Dongjun (Dong Jun) Formation—Late Eocene—Guangxi

Douglass Creek Member, Green River Formation—Wasatchian—Utah

Dry Gulch Creek Member, Duchesne River Formation—Uintan—Utah

Duchesne River Formation—Uintan and Duchesnean—Utah

Ducy, calcaire de—Robiacian—Paris Basin

Dulwich plastic clay—Sparnacian—England

Earnley Formation, Bracklesham Group—Lutetian—England

Elk Creek facies, Willwood Formation—Wasatchian—Wyoming

Erquelinnes sands—Late Landenian and Sparnacian—Belgium

Eureka Sound Formation—Wasatchian—Ellesmere Island

Fanchuang "series"—Late Eocene—China

Gypse de Paris—Late Eocene—Paris Basin

Flagstaff Limestone, upper—Wasatchian—Utah

Fowkes Formation—Bridgerian—Wyoming

Fresville sables de—Lutetian—France

Friars Formation—Uintan—California

Fronsadais, molasse de—Headonian—France

Galisteo Formation—Wasatchian and Duchesnean—New Mexico

Gashato Formation—Late Paleocene and Early Eocene—Mongolia

Gehlbergschichten, Grünsande de—Robiacian?—Germany

Geiseltal Braunkohle—Lutetian—Germany

Golden Valley Formation—Graybullian—North Dakota

Gour Lazib—Middle Eocene—Algeria

Gray Bull (Greybull) beds (Lower Willwood Formation)—Graybullian—Wyoming

Green Cove ash—Late Wasatchian or Early Bridgerian—Wyoming

Green River Formation—Wasatchian into Duchesnean—Wyoming, Colorado, Utah

Gualanday Formation—Mustersan?, Divisaderan?—Colombia

Guanajuato red beds—Uintan?—Mexico

Guanzhuang Formation—Middle Eocene—Shandong

Haitaoyahtze Formation—Late Eocene—Likwanchiao Basin, Henan

"Halfway" Member, Duchesne River Formation—Uintan?—Utah

Headon Beds—Headonian—Isle of Wight, England

Hendry Ranch Member, Tepee Trail Formation—Uintan—Wyoming

Hengstbury Beds—Robiacian—England

Hetaoyuan Formation—Late Eocene—Henan

Heti Formation—Late Eocene—Henan

Hiawatha Member, Knight or Wasatch Formation—Wasatchian—Wyoming

Honghe Formation—Late Eocene—Shaanxi

Honglishan (or Wulungu?) Formation—Late Eocene—Xinjiang

Hosan Formation—Late Eocene—Korea

Huerfano Formation—Lostcabinian—Colorado

Hunshuihe Formation—Late Eocene—Henan

Huntingbridge Formation—Robiacian—England

Indian Meadows Formation—Wasatchian—Wyoming

Irdin Manha Formation—Late Eocene—Nei Mongol

Issigeac, calcaire de—Headonian—France

Jiguan Formation—Late Eocene—Henan

Khirthar "series"—Middle Eocene—India

Kingsbury Conglomerate—Wasatchian—Wyoming

Kinney Rim Member, Washakie Formation—Bridgerian—Wyoming

Kishenehn Formation—Duchesnean—British Columbia

Knight (Wasatch) Formation—Wasatchian—Wyoming

Kolpak Formation—Middle Eocene—Kazakhstan

Kuanchuang (Guanzhuang) Formation—Middle Eocene—Shandong

Kuldana Formation—Middle Eocene—Pakistan

Laney Member, Green River Formation—Bridgerian—Wyoming

Lapoint Member, Duchesne River Formation—Duchesnean—Utah

Lede, sables de—Lutetian—Belgium

Liankan (Lian-Kan) Formation—Late Eocene—Turpan Basin, Xinjiang

Lignite d'Aryan—Headonian—France

Lignites de Vaugirard—Sparnacian—Paris Basin

Lignites et Grès du Soissonais—Sparnacian—Paris Basin

Limuping Formation—Early Eocene—Hunan

Lincha Formation—Early Eocene—Hunan

Linjianshan Member, Xinyu Formation—Early Eocene—Jiangxi

Lishigou Formation—Late Eocene—Henan

Liuniu Formation—Early–Middle Eocene?—Guangxi

London Clay—Cuisian—England

Lookout Mountain Conglomerate Member, Wasatch Formation—Wasatchian—Wyoming

Lost Cabin Member, Wind River Formation—Lostcabinian—Wyoming

Lower red beds, Lower Member, Naran Bulak Formation—Late Paleocene or Early Eocene—Mongolia

Lumbrera Formation—Casamayoran—Argentina

Lumeiyi Formation—Late Eocene—Yunnan

Luse Formation—Late Eocene—Xinjiang

Lushi Formation—Late Eocene—Henan

Lysite Member, Wind River Formation—Lysitean—Wyoming

Maadi Formation (marine)—Late Eocene—Egypt

Main Body of Wasatch Formation—Graybullian and Lysitean—Wyoming

Mam Creek Member, Debeque Formation—Wasatchian—Colorado

Manti facies, Green River Formation—Eocene—Utah

Maojiado or Maojiapo Formation—Middle? Eocene—Henan

Marsh Farm Formation—Lutetian—England

Melanienkalk at Mühlhausen—Headonian—Germany

Melawi Formation—Middle or Late Eocene—Borneo

Meudon, Conglomérat de—Sparnacian—Paris Basin

Mission Valley Formation—Uintan—California

Molasse du Fronsadais—Headonian—France

Montchenot, lignites de—Lutetian—Paris Basin

Morrow Creek Member, Green River Formation—Bridgerian—Wyoming

Mugrosa Formation—Mustersan—Colombia

Musters Formation—Mustersan—Argentina

Myton Member, Uinta Formation—Uintan—Utah

Naduo (Nadu, Nado) Formation—Late Eocene—Baise district, Guangxi

Naran Bulak Formation—Late Paleocene and Early Eocene—Mongolia

New Fork Member, Wasatch Formation—Lostcabinian—Wyoming

Niland Tongue or Member, Wasatch Formation—Lostcabinian—Wyoming

Ningjiashan Member—Early Eocene—Jiangxi

Niushan Formation—Early Eocene—Shandong

Nomogen Formation—Late Paleocene or Early Eocene—Nei Mongol

Norwood Tuff—Duchesnean?—Utah

''Notostylops'' zone—Casamayoran—Argentina

Obaila (Obayla) Formation—Early and Middle Eocene—Kazakhstan

Osborne Beds—Headonian—Isle of Wight, England

Ostricourt, sables d'—Cernaysian and Sparnacian—France

''Panchiao Stage''   (See Lumeiyi)—Late Eocene?—Yunnan

Parachute Creek Member, Green River Formation—Bridgerian—Utah

Pass Peak Formation—Wasatchian—Wyoming

Pechelbronnschichten—Headonian and? Early Oligocene—Alsace and adjacent Germany

Pinghu Formation—Early Eocene—Jiangxi

Pondaung Formation—Late Eocene—Burma

Poway Group—Uintan—California

Pruett Formation—Uintan—Texas

''Randlett Member,'' Duchesne River Formation—Uintan—Utah

Ravine beds (marine with a land mammal)—Middle and Late Eocene—Egypt

Red Beds ''series''—Cretaceous into Miocene with Mid?-Eocene *Griphodon*—Peru   —Patterson (1977)

Red Desert Member, Wasatch Formation—Wasatchian—Wyoming

Rencun Member, Heti Formation—Late Eocene—Henan, Shanxi

''Rose Canyon Shale'' or Formation—Uintan—California

Roslyn Formation—Bridgerian?—Washington (turtle and fish)

Sables à Térédines et Unios—Cuisian—Paris Basin

Sage Creek Formation—Uintan—Montana

St. Ouen, calcaire de—Robiacian—Paris Basin

Sand Butte Bed, Laney Member, Green River Formation—Bridgerian—Wyoming

Sand Creek facies, Willwood Formation—Graybullian—Wyoming

San Jose Formation—Tiffanian and Wasatchian—New Mexico and Colorado

Santiago Formation—Uintan and? Duchesnean (in part)—California

Sargamys Formation—Middle Eocene—Kazakhstan

Selsey Formation, Bracklesham Group—Lutetian—England

Sespe Formation, lower—Uintan and Duchesnean—California

Shara Murun Formation—Late Eocene—Nei Mongol

Shark River Marl—marine and a land mammal, Bridgerian—New Jersey

Shi-san-jian-fang (Shih-san-Chien-Fang) Formation—Early Eocene—Turpan Basin, Xinjiang

Shuantasi (Xuangtasi) Formation—Early Eocene—Anhui

Sinyu (Xinyu) Formation, middle—Eocene—Jiangxi

Slim Buttes Formation—Duchesnean—South Dakota

Soissonais, lignites et grès du—Sparnacian—Paris Basin

Stadium Conglomerate—Uintan—California

Subathu Formation—Middle and Late Eocene—Jammu and Kashmir, India

Suffolk pebble beds—Sparnacian—England

Swift Current Creek beds—Uintan—Saskatchewan

Taizicun Formation—Early Eocene—Xinjiang

Tandou or Tantou Formation—Early Eocene—Henan

Ta-Tsang-Fang (Da-can-fang or Dacangfang) Formation—Middle Eocene—Henan

Tepee Trail Formation—Uintan—Wyoming

Terras de Collu—Lutetian—Sardinia

Tipton Tongue or Member, Green River Formation—Lostcabinian—Wyoming

Trujillo Formation—Early? Eocene—Venezuela

Tukhum Formation or red clay—Middle Eocene—Nei Mongol

Twin Buttes Member, Bridger Formation—Bridgerian—Wyoming

Uglov Formation—Late Eocene—Artëm

Uinta Formation—Uintan—Utah

Ulan Gochu beds—Late Eocene—Nei Mongol

Ulan Shireh Formation—Late Eocene—Nei Mongol

Ulungu Formation—Late Eocene—Xingjiang

Vaugirard, lignites de—Sparnacian—Paris Basin

Wadi Rayan beds (marine with a land mammal)—Middle Eocene—Egypt

Wagonbed Formation—Uintan—Wyoming

Wagonhound Member, Uinta Formation—Uintan—Utah

Wasatch Formation or Group—Tiffanian into Bridgerian—Wyoming and Utah

Washakie Formation—Bridgerian and Uintan—Wyoming

Wemmel, sables de—Lutetian—Belgium

White Beds, upper member, Naran Bulak Formation—Early Eocene—Mongolia

Wiggins Formation—Uintan—Wyoming

Willwood Formation—Clarkforkian into Lostcabinian—Wyoming

Wind River Formation—Graybullian through Lostcabinian—Wyoming

Wind River Formation tuff—Lostcabinian—Wyoming

Wittering Formation—Cuisian—England

Woolwich Beds—Sparnacian—England

Woolwich Bottom Beds—Sparnacian—England

Wulidui Formation—Late Eocene—Henan

Wulungu? (or Honglishan?) Formation—Late Eocene—Xinjiang

Wutu Formation—Early Eocene—Shandong

Xiangshan Formation—Late Eocene—Yunnan

Xinyu (Sinyu) Formation—Early Eocene—Jiangxi

Xuangtasi Formation—Early Eocene—Anhui

Yegua Formation, Claiborne Group (marine)—Late Eocene—Texas

Ypres, Argiles d'—Cuisian—Belgium

Yuanchü (Yuanqu) Formation—Late Eocene—Shanxi, Henan

Yu-Huang-Ding (Yuhuangding or Yu-Huang-Ting) Formation—Early Eocene—Henan

Zhangshanji Formation—Early Eocene—Anhui

EOCENE LOCALITIES

This glossary of alphabetically arranged locality names can help the beginning researcher to locate sites that are mentioned in an article without being described geographically.

We can only claim that our list covers most of the published names. There are literally thousands of unpublished Eocene locality names in the archives of various museums and paleontologic institutes around the world. (The University of California Museum of Paleontology alone has approximately 500 unpublished Eocene mammal localities.) Also, many workers prefer not to publish information about a locality, fearing or having experienced destruction of the scientific value of the site by hobbyist or commercial collectors who may be directed to the site by published geographic data.

Abbey Wood—Sparnacian—England
Absaroka or Absaroka Mountains area—Tiffanian into Chadronian—Wyoming
Aconin—Sparnacian—France
Ager Basin   (See Les Saleres.)
Aizy—Jouy—Lutetian—Paris Basin
Aksyir, upper and lower—Late Eocene into Early Oligocene—Kazakhstan
Aldrich Creek and Ishawooa Creek—Bridgerian—Wyoming
Alheit Pocket and Alheit Pocket West—Graybullian—Colorado
Alkali Creek sites—Lostcabinian—Wyoming
Almagre "facies" and fauna—Lysitean and Lostcabinian—New Mexico
Almy   (See Red Canyon.)
Alzieux (Castrais)—Robiacian—France
Ameki—Middle Eocene—Nigeria (marine)
Amy   Sparnacian—Paris Basin
Andarak—Middle Eocene—Kazakhstan
Andrashaza—Robiacian?—Romania
Antelope Creek (Slim Buttes)—Duchesnean—South Dakota
Anthill Quarry—Graybullian—Colorado
Apt   (See La Débruge.)
Archuleta Draw—Lostcabinian—Colorado
Arcis-le-Ponsart—Robiacian—Paris Basin   (= Mont de Bery of P. Louis)
Arcueil—Early Eocene—Paris Basin
Ardath Shale sites—Uintan—California
Argenteuil—Headonian—Paris Basin
Argenton-sur-Creuse—Lutetian—France
Arminto district—Lostcabinian—Wyoming
Arroyo Blanco South I and II—Lysitean and Lostcabinian?—New Mexico
Arshanto Formation sites—Middle Eocene—Nei Mongol
Arsy—Sparnacian—Paris Basin
Artëm or Artiom Valley, near Vladivostok in Primorsky Kray Territory—Late Eocene?—"Far East USSR"
Aryan   (See St.-Jean-de-Marvejols.)
Aschersleben   (See Helmstedt.)
Attock District   (See Lammidhan, Kala Chitta Hills, etc.)
Aubrelong 2 (Quercy)—Headonian—France
Aumelas 1—Lutetian—France
Avenay—Sparnacian—Paris Basin
Avignon   (See La Débruge.)
Azillanet—Headonian—France
Babia Hill West   (See Kutch.)
Baby (= Moulin Baby?)—Headonian—France
Bach, in part? (Quercy)—Headonian—France
Badger Basin—Graybullian—Wyoming
Badland Gulch—Lostcabinian—Wyoming

Badwater Creek district—Uintan and Duchesnean—Wyoming
Baggs North, 1, 2, 3—Lostcabinian—Wyoming
Bagneux—Lutetian—Paris Basin
Bagnoles—Lutetian—France
Baise Basin—Late Eocene—Gaungxi
Banc Vert—Lutetian—Paris Basin
Banjo Anthills—Graybullian—Wyoming
Banjo CAG, Flats 1, 2, 3, Banjo Pocket or Quarry—Graybullian—Wyoming
Baranda Southeast—Middle Eocene—Kutch, India
Baron Sog Mesa—Late Eocene—Mongolia
Barranca   (See Great Barranca.)
Barran de Forels—Cuisian—Spain
Barton Cliff—Robiacian—England
Basal   (See Lammidhan)—Middle Eocene—Pakistan
Bas-Charnier—Headonian—France
Baser Bend South—Uintan—Utah
Batignolles—Robiacian—Paris Basin
Bauduen—Sparnacian—France
Bay Fiord District—Wasatchian—Ellesmere Island
Bayn Ulan Formation sites (Huhebolhe Cliff)—Early Eocene—Nei Mongol
Bearpaw Mountains—Graybullian?—Montana
Bear River   (See Evanston.)
Beauchamp—Robiacian?—Paris Basin
Beaver Creek Vicinity—Lostcabinian—Wyoming
Beaver Divide Conglomerate sites—Uintan—Wyoming
Beaver Divide District—Wasatchian into Chadronian—Wyoming
Beduer—Headonian—France
Beijing (Peking)—Late Eocene—Beijing
Bembridge Beds sites (Binstead)—Headonian—England
Benabarre—Lutetian—Spain
Beragoa Coal Mine—Middle Eocene—India
Berru, Mont de—Cernaysian and Sparnacian sites—Paris Basin
Berville—Robiacian—France
Bighorn Basin, many sites—Puercan through Wasatchian—Wyoming
Big Piney District—Lostcabinian—Wyoming
Big Sandy River 1, 2, 3, 4—Bridgerian—Wyoming
Big "W"—Graybullian—Wyoming
Biniamar—Headonian?—Mallorca
Binstead—Headonian—England
Birch Creek—Bridgerian—Wyoming
Birdseye Creek—Lostcabinian—Wyoming
Birket-el-Qurun Beds sites (marine and a land mammal)—Late Eocene—Egypt
Bison Basin 2—Wasatchian—Wyoming
Bitter Creek Station Vicinity—Wasatchian—Wyoming
Black Buttes Vicinity—Late Cretaceous, Tiffanian, Clarkforkian and Wasatchian—Wyoming
Blackheath Beds sites   (See Abbey Wood.)
Black Mountain or Promontory Bluff—Lostcabinian—Colorado
Black's Fork sites—Bridgerian—Wyoming
Black Spring Reservoir 1—Bridgerian—Wyoming
Blue Rim, Kistner 16—Bridgerian—Wyoming
Bodenheim—Headonian?—Germany
Bognor Regis (London Clay)—Cuisian—England
Bondurant District—Tiffanian into Wasatchian—Wyoming
Bone Hill (Head of Elk Creek)—Graybullian—Wyoming

Bonneville Southwest Quad. 1—Lostcabinian—Wyoming

Bouffard (Castrais)—Robiacian—France

Bosc-Nègre (Quercy)—Headonian—France

Bose Basin   (See Baise Basin.)

Boulder Vicinity—Lostcabinian—Wyoming

Boulincourt—Sparnacian—Paris Basin

Boussac de la Salle—Caylus (Quercy)—Headonian—France

Bouxwiller (Buchsweiler)—Lutetian—France

Bown Bonanza—Graybullian—Wyoming

Bozeman—Wasatchian—Wyoming

Bracklesham Group sites (Hampshire Basin)—Cuisian to Robiacian—England

Braconnac (Castrais)—Robiacian—France

Bramford—Sparnacian—England

Brasles, Sables de—Sparnacian—Paris Basin

Braunkohle sites   (See Geiseltal, Cecilie Mine.)

Brea Canyon—Uintan—California

Bretigny—Sparnacian—Paris Basin

Brunnstatt—Headonian—Alsace

Bruxelles—Lutetian?—Belgium

Bu el-Haderait (marine)—Libya

Buffalo Basin (Bighorn Basin subdivision) sites—Wasatchian—Wyoming

Burley Anticline—Uintan—Wyoming

Burnham (London Clay)—Cuisian—England

Burnt Fork—Bridgerian—Wyoming

Bussy—Sparnacian—Paris Basin

"Buttes of Dry Creek"—Bridgerian—Wyoming

Byrd Draw—Lostcabinian—Wyoming

Cabeza Blanca (= Rio Chico, primer yacimiento de *Pyrotherium*)—Casamayoran and Deseadan—Argentina

Cajarc (Quercy)—Headonian—France

Calcaire Grossier sites—Lutetian—Paris Basin

Calpet South—Wasatchian—Wyoming

Camp—Graybullian—Wyoming

Campans (Castrais)—Robiacian—France

Campbell Quarry—Graybullian—Wyoming

"Camp Margetts" area (Huhebolhe Cliff)—Early into Late Eocene—Nei Mongol

Camp San Onofre (Camp Pendleton)—Uintan—California

Cañadón Colorado—Mustersan—Argentina

Cañadón Hondo—Riochican and Casamayoran—Argentina

Cañadón Tournouër   (See Casamayor.)

Cañadón Vaca—Casamayoran—Argentina

Canal—Graybullian—Wyoming

Can Camperol—Lutetian—Spain

Candelaria—Uintan—Texas

Canly—Sparnacian—Paris Basin

Canny-sur-Metz—Sparnacian—Paris Basin

Canoe Formation sites—Bridgerian—Texas

Canquel, lower—Casamayoran—Argentina

Canyon Creek Butte (birds)—Lostcabinian?—Wyoming

Capella (Benabarre)—Robiacian—Spain

Cardalu (Quercy)—Headonian?—France

Carlsbad East—Uintan—California

Carter Mountain—Lostcabinian?—Wyoming

Casamayor (Cañadón Tournouër or Lobo)—Casamayoran—Argentina

Castalon—Uintan—Texas   (also later Arikareean)

Castayral—Robiacian?—France

Castellnou—Headonian?—Spain

Castelnau-Valence—Headonian—France

Castigaleu (Noguera Pallaresa) (turtles)—Cuisian—Spain

Castle Rock Southwest—Uintan—Wyoming

Castrais (district) sites—Robiacian—France

Castres (Castrais)—Robiacian—France

Castres-Roc-de-Lunel—Robiacian—France

Cathedral Bluffs Member sites—Lostcabinian? and Bridgerian—Wyoming

Cave Gulch Head 1 and 2—Lysitean—Wyoming

Caylus (Quercy)—Headonian—France

CCC Draw   (See Milligan's Arroyo.)

Cecilie Mine (Geiseltal)—Lutetian—Germany

Cedar Mountain North—Bridgerian—Wyoming

Célarié—Headonian—France

Célas—Headonian—France

Çeltek (Eski Çeltek)—Early Eocene—Turkey

Central ENHER (Noguera-Ribagoranzana)—Cuisian—Spain

Cerrillos Northeast—Wasatchian—New Mexico

Cerro Blanco—Casamayoran—Argentina

"Cerro del Humo"   (See Colhuapi Norte.)

Cerro Negro—Casamayoran—Argentina

Cerro Talquino—Casamayoran?, Mustersan?—Argentina

Cesseras—Robiacian—France

Chaili—Yuanchu—Late Eocene—Shanxi

Chailvet—Sparnacian—Paris Basin

Chamblon at Yverdon—Robiacian?—Switzerland

Changlo District (See Wutu.)

Changpiliang (Changpeiling)—Early Eocene—Hunan

Changxindian (Gaodian)—Late Eocene—Beijing

Chappo Member sites—Wasatchian, in part—Wyoming

Château-Thierry—Lutetian—Paris Basin

Châtillon—Lutetian—Paris Basin

Chavot—Cuisian—Paris Basin

Chelles—Headonian—Paris Basin

Chérence—Lutetian—Paris Basin

Chery-Chartreuve—Robiacian—Paris Basin

Chestnut Avenue—Uintan—California

Chharat Formation sites—Middle Eocene—Pakistan

Chicoca farmhouse—Middle? Eocene—Peru

Chijiang Basin—Late Paleocene or Early Eocene—Jiangxi

Chimney Butte Quarry (= "Buckshot's Quarry")—Late Eocene—Nei Mongol

Chinzhily—Late Eocene—Kazkhstan

Chisos Northeast—Graybullian?—Texas

Chiyuan Basin—Late Eocene—Henan

Chorlakki Village—Middle Eocene—Pakistan

Chuankou Formation site—Middle to Late Eocene—Henan

Chugouyu Formation site—Late Eocene—Henan

Chupitian Valley (Niushan—Linchü)—Early Eocene—Shandong

Church Buttes—Bridgerian—Wyoming

Cicekdag—Late? Eocene—Turkey

Cicishou—Middle or Late Eocene—Shandong

Circle Draw, Head—Lysitean?—Wyoming

Clark Fork Basin   (See Sand Coulee.)

Clignancourt—Headonian—Paris Basin

Côanac 1—Headonian—France

Côja (Coimbra)—Headonian—Portugal

Colhuapi (Colhué-Huapí) Norte—Casamayoran and Mustersan—Argentina

Colhuapi Sud (See Great Barranca.)

Collsuspina (San Cugat de Vavadens)—Headonian or early Oligocene—Spain

Concots (Quercy)—Headonian—France

Condé-en-Brie—Cuisian—Paris Basin

Continental Divide Basin—Wasatchian—Wyoming

Cook Ranch, Montana (See Sage Creek Formation sites.)

Cooper Creek—Wasatchian—Wyoming

Cormeilles-en-Parisis—Headonian—Paris Basin

Corrales (near Zamora)—Lutetian—Spain

Corsà—Cuisian and Lutetian—Spain

Costa de la Vila (Barcelona) (See Santpedor.)

Cottonwood Creek Bridger—Bridgerian—Wyoming

Coucy—Lutetian—Paris Basin

Coulondres—Lutetian—Paris Basin

Craig 6 (Four Mile)—Graybullian—Colorado

Creechbarrow Hill—Robiacian—England

Crowheart Butte South—Lostcabinian—Wyoming

Croydon—Sparnacian—England

Cuis—Cuisian—Paris Basin

Cuise-la-Motte—Cuisian—Paris Basin

Cuny—Sparnacian—Paris Basin

Currant Southwest (Grant Range)—Bridgerian?—Nevada

Cuvilly—Sparnacian Paris Basin

Dabu Formation sites—Early Eocene—Xinjiang

Dad—Lostcabinian—Wyoming

Dahimapan (Iliang district)—Late Eocene—Yunnan

Damery 1, reworked Cuisian fossils—Paris Basin Lutetian

Damery 2—Lutetian—Paris Basin

Dampleux—Lutetian—Paris Basin

Davis Draw South—Lysitean—Wyoming

"Davis Ranch" = Locy Ranch = Pine Ridge—Wasatchian—Wyoming

"Deardorff Hill"—Lysitean—Colorado

Défense à Puteaux—Lutetian—Paris Basin (La Défense à Puteaux)

Dera Ghazi Khan (marine)—Middle Eocene—Pakistan

Despair Quarry—Graybullian—Colorado

Dickinson—Graybullian—North Dakota

Dielsdorf—Headonian—Switzerland

Divisadero Largo Formation sites—Divisaderan—Argentina

Djebel Coquin (see Dor el Talha)

Dobie Butte (Univ. Wyo. 13)—Graybullian—Wyoming

Dog Spring District—Uintan—California

Dongjun fauna—Baise Basin—Late Eocene—Gaungxi

Dor el Talha (Djebel Coquin)—Late Eocene and Early Oligocene—(marine and nonmarine)—Libya

Dormaal—Sparnacian—Belgium

Dorr—Lysitean or early Lostcabinian—Wyoming

Dorsey Creek Head (Wardell's Ranch)—Graybullian—Wyoming

Dorsey Creek Southeast—Graybullian—Wyoming

Douglass Creek Member site (See Ostutah.)

Dragon—Vernal Stage Road East—Uintan—Utah

Dry Creek—Bridgerian—Wyoming

Dry Muddy Creek 1, 2, 3, 4 and North—Lostcabinian—Wyoming

Dry Well—Graybullian—Wyoming

Dubois North—Lostcabinian—Wyoming

Duchesne River Formation sites—Uintan and Duchesnean—Utah

Dug Springs Divide—Lostcabinian—Colorado

Dulwich—Sparnacian—England

Dzungaria Basin (See Ulan Bulak.)

East Fork Crossing—Graybullian—Wyoming

East Fork East—Graybullian—Wyoming

East Wittering—Cuisian—England

Echelles—Robiacian—France

Eclépens (Mormont)—Robiacian—Switzerland

Egeln—Lutetian?, Robiacian?—Germany

Egerkingen (and subsites)—Robiacian—Switzerland

Ehrenstein 1a, 2, 3—Headonian—Germany

Elderberry Canyon–Ely—Bridgerian—Nevada   —Fouch (1979)

Elk Creek (and tributaries) sites—Graybullian—Wyoming

Elk Mountain West—Graybullian—Wyoming

Ellesmere Island localities (Bay Fiord district)—Wasatchian

El Pueyo (Isabena)—Cuisian—Spain

Entreroches (Mormont)—Headonian—Switzerland

Erquelinnes 2—Sparnacian—Belgium

Escamps (Quercy)—Headonian and Early Oligocene—France

Escarla Huesca (Noguera Ribagorzana)—Cuisian—Spain

Este de Rio Chico (= Pampa de Castillo East?)—Casamayoran—Argentina

Eureka Sound Formation sites—Wasatchian—Ellesmere Island

Euzet-les-Bains (= St. Hippolyte-de-Caton)—Headonian—France

Evanston—Wasatchian—Wyoming

Eygalayes—Lutetian—France

Farson—Bridgerian—Wyoming

Feligueira Grande—Headonian—Portugal

Felsögalla (marine)—Lutetian—Hungary

Fenton—Graybullian—Wyoming

Fergana Basin—Middle Eocene—Kirghizia

Ferry Cliff—Sparnacian—England

Fifteen Mile Creek—Wasatchian—Wyoming

Filain (See Pargny-Filain.)

Fishbourne—Headonian—Isle of Wight, England

Fisherman (Fish) Creek—Graybullian—Wyoming

Fismes—Sparnacian?—Paris Basin

Five Mile Creek—Graybullian—Wyoming

Flat Top Mountain North—Lostcabinian—Wyoming

Flattop Quarry—Bridgerian—Wyoming

Fletcher Parkway district—Uintan—California

Fonfroide—Lutetian—France

Fons 1, 2, 4—Robiacian and Headonian—France

Fontenay-sous-Bois—Headonian—France

Forest—Lutetian—Belgium

Fort Bridger—Bridgerian—Wyoming

Fossil Basin (Fossil Butte, Kemmerer West, etc.)—Wasatchian—Wyoming

Fossil Creek—Lostcabinian—Colorado

Four Mile Localities (See Alheit Pocket, Anthill Quarry, Kent Quarry, Despair Quarry, Third Hill, Sand Quarry, Timberlake Quarry.)—Graybullian—Colorado

Four Points Hill—Lysitean—Colorado

Freighter Gap 1—Lostcabinian—Wyoming

Fressenneville—Lutetian—France

Friars Formation sites—Uintan—California

Frohnstetten—Headonian—Germany

Eocene Mammalian Faunas

Fürstenau—reworked Cuisian fossils—Germany

Galisteo Formation sites—Wasatchian and Duchesnean—New Mexico

Ganda Kas (Kala Chitta Hills, Attock district)—Middle Eocene—Pakistan

Gao—Late Paleocene and Middle Eocene—Mali

Gaodian (Changxindian)—Late Eocene—Beijing

Garcia Cañon—Lostcabinian—Colorado

Gardner Butte district—Lostcabinian—Colorado

Gare du Nord—Robiacian—Paris Basin

Gargas (fauna) (See La Débruge)—Headonian—France

Garrigue (La Garrigue, Quercy)—Headonian—France

Gascou (Quercy)—Headonian—France

Gazin Locality—Uintan—Wyoming

Gebel (Djebel) Zelten West (= Bu el Haderait?) (marine)—Middle Eocene—Libya

Geiseltal (Cecilie Mine, Grube Halle, etc.)—Lutetian—Germany

Gentilly—Lutetian—Paris Basin

Girardot West—Divisaderan?, Mustersan?—Colombia

Golden Valley Formation sites   (See Dickinson.)

Gonnesa—Lutetian—Sardinia

Gösgen-Canal and Pump Station—Headonian—Switzerland

Gouiret el Azib—Early or Middle Eocene—Algeria

Gousnat—Headonian—France

Grand Reng (See Erquelinnes 2.)

Granger South 1, 2—Bridgerian—Wyoming

Granger Station—Bridgerian—Wyoming

Grauves—Cuisian—Paris Basin

Gray Bull (or Greybull) beds and fauna—Graybullian—Wyoming

Great (or High) Barranca—Casamayoran through Colhuhuapian—Argentina

Great Divide Basin—Lostcabinian—Wyoming

Green—Bridgerian—Wyoming

Green Cove 1—Bridgerian—Wyoming

Green Cove Northeast—Uintan—Wyoming

Grisolles—Robiacian—Paris Basin

Grizzly Buttes district—Bridgerian—Wyoming

Grube Prinz von Hessen—Messel—Lutetian—Germany

Gualanday Northeast—Mustersan?, Divisaderan?—Colombia

Guanajuato—Uintan?—Mexico   (See Marfil.)

Guanzhuang—Middle Eocene—Shandong

Güell (Isabena Valley)—Cuisian—Spain

Guépelle (Le Guépelle)—Robiacian—Paris Basin

Guérin-Farm—Lutetian—France

Guny—Sparnacian—Paris Basin

Guriliyn-Gobi Cauldron—Middle? Eocene—Mongolia

Gurnard—Headonian—England

Hackberry Hollow—Graybullian—Wyoming

H.A. Hector Quarry—Wasatchian—Colorado

Haitaoyuantze (Sichuan Basin)—Late Eocene—Henan

Halfway ''horizon''—Uintan?—Utah

Hampshire Basin   (Isle of Wight, Bracklesham Group sites, Headon Beds)

Ham's Fork—Bridgerian—Wyoming

Hamstead (= Yarmouth)—Headonian—England

Harrison Ranch North—Lostcabinian—Wyoming

Hartman Ranch—Duchesnean—California

Harudi (marine)—Middle Eocene—Kutch, India

Harwich—Sparnacian—England

Hateg Depression—Headonian?—Romania

Hatley Ranch—Wasatchian—New Mexico

Hausero's Ranch—Lostcabinian—Colorado

Haystack Mountain, lower—Bridgerian—Wyoming

Haystack Mountain, upper—Uintan—Wyoming

Headon Beds sites (upper and lower)—Headonian—Isle of Wight, England

Headon Hill—Headon Cliff—Headonian—Isle of Wight, England

Heart Mountain East—Graybullian—Wyoming

Heidenheim—Robiacian?—Germany

Helmstedt—Robiacian?—Germany

Hendry Ranch Member sites   (See Badwater Creek.)

Hengstbury Head—Robiacian—England

Hengyang Basin (Changpiliang)—Early Eocene—Hunan

Henry's Fork District, Divide, Hill—Bridgerian—Wyoming

Hentung—Early Eocene—Hunan

Hermonville—Lutetian—Paris Basin

Herne Bay—Sparnacian—England

Herrlingen 2, 3—Headonian—Germany

Hetaoyuan (Sichuan)—Late Eocene—Henan

High Acres Ranch—Lostcabinian—Wyoming

High Schindler Hill—Lysitean or Lostcabinian—Colorado

Hilltop—Graybullian—Wyoming

Hoangho, North Bank—Late Eocene—Shanxi

Hoback Basin Rim—Lysitean—Wyoming

Hoe Ranch Quad. 1—Graybullian—Wyoming

Hordle Cliffs (Hordwell)—Headonian—England

Hôsan Coal Field   (See Kokaido.)

Hoti (fauna)—Late Eocene, Yuanchu Basin, Shanxi

Hsiaoshaho (Iliang district)—Late Eocene—Yunnan

Hsi-Kou—Middle Eocene—Shandong

Hsinan—Late Eocene—Shanxi, Henan

Hsintai   (See Sintai.)

Huerfano A sites   (Silverstine Ranch)—Lostcabinian—Colorado

Huerfano B sites   Gardner Butte district—Lostcabinian—Colorado

Huerfano Muddy Divide—Lostcabinian—Colorado

Huhebolhe Cliff—Early into Late Eocene—Nei Mongol

Huisheim—Headonian—Germany

Hunshuihe Formation site—Late Eocene—Henan

Hutogein Valley—Late Eocene—Nei Mongol

Hyopsodus Hill (Tabernacle Butte)—Bridgerian—Wyoming

Hyopsodus Hill East—Bridgerian—Wyoming

Hyracotherium Slide—Lysitean?—Colorado

Ichang—Early Eocene—Hubei

Iliang district (Ilyan)—Late Eocene—Yunnan

Indian Point—Wasatchian—Wyoming

In Tafidet (marine and nonmarine)—Middle Eocene—Mali

Irdin Manha Formation sites—Late Eocene—Nei Mongol

Isabena Basin or Valley   (See Güell, La Roca, Les Badies, Monteroda.)

Ishawooa Creek and Aldrich Creek—Bridgerian—Wyoming

Ishawooa Mesa—Bridgerian—Wyoming

Isle of Sheppey—Sparnacian—England

Isle of Wight (many formations and localities)—Middle and Late Eocene—England

Issel—Lutetian—France

Issigeac—Headonian—France

Jabron—Headonian—France

James Park—Sparnacian or Cuisian—England

Jatta—Middle Eocene—Pakistan

Jaulgonne—Lutetian—Paris Basin

Jenkins Draw district—Graybullian—Wyoming

Jentsen or Jentsun (See Mienchi)—Late Eocene—Henan

Jeumont (See Erquelinnes 2.)

Jhalar—Middle Eocene—Pakistan

Jigni Coal Field—Middle Eocene—India

Jim Creek—Graybullian and Lostcabinian—Wyoming

Jiyuan Basin—Late Eocene—Henan

Jouy (Aizy-Jouy)—Lutetian—Paris Basin

Jumencourt (Coucy-le-Chateau)—Lutetian—Paris Basin

Kala Chitta Hills or Range (See Kuldana Formation sites,
  Ganda Kas.)

Kalakot or Kalakkot (also, Rajauri)—Middle Eocene—India

Kalmakpay—Middle Eocene and Oligocene—Kazakhstan

Kaofenpei (Iliang district)—Late Eocene—Yunnan

Kemmerer West—Wasatchian—Wyoming

Kennedy's Hole South—Uintan—Utah

Kent Quarry (Four Mile)—Graybullian—Colorado

Khaichin II, III (= Khaychin Ula II, III)—Middle Eocene—Mongolia

Kholboldzhi (= Kholobolchi-Nor)—Late Eocene—Mongolia

Kiinkerish—Late Eocene or Early Oligocene—Kazakhstan

Kirby Draw East—Wasatchian—Wyoming

Kirby Draw West—Wasatchian—Wyoming

Kishenehn Formation site—Duchesnean?—British Columbia

Kistner 16 (See Blue Rim.)

Knight Station—Lysitean—Wyoming

Kohat region—Middle Eocene—Pakistan

Kokaido (Hôsan Coal Field)—Late Eocene—Korea

Kressenberg—Middle Eocene—Austria

Kuanchuang ''series'' sites (See Kuan Chiang, Tan's Locality,
  Meng-Yin-Hsien) Late Eocene—Shandong

Kuldana Formation sites—Middle Eocene—Pakistan (Kala
  Chitta, etc.)

Kutch (Gujarat, Harudi, Babia Hill)—marine Middle Eocene—
  India

Kyson—Sparnacian—England

La Ametlla (Ager)—Cuisian—Spain

La Badayre (La Badaïre) (Castrais)—Robiacian—France

La Barge (Big Piney)—Lostcabinian—Wyoming

La Bartie (Labarthie)—Robiacian—France

La Boixedat—Lutetian—Spain

La Bouffie (Quercy)—Headonian—France

Lacey's Farm Quarry—Headonian—England

La Croix—Robiacian—France

La Débruge—Headonian—France

La Défense à Puteaux (Paris)—Lutetian—Paris Basin

La Ferrière (Castrais)—Robiacian—France

La Fosse (Castrais)—Tobiacian—France

Lago Colhué-Huapí (See Great Barranca.)

Lago Musters (See Pajarito or Cerro del Humo.)

La Grave—Headonian—France

Laguerres—Robiacian—Spain

La Guittardie à Montdragon—Robiacian—France

Laguna de la Bombilla (Paso de los Indios)—Casamayoran—Argentina

Laguna del Mate—Mustersan—Argentina

Laguna Riviera—Uintan—California

Lai'an—Early Eocene?—Anhui

Lake Murray district—Uintan—California

La Livinière—Robiacian—France

Lamandine-Basse (Quercy)—Headonian—France

La Marcelle—Robiacian—France

La Massale—Robiacian—France

La Maurianne (Castrais)—Robiacian—France

La Milette (Castrais)—Robiacian—France

Lammidhan—Middle Eocene—Pakistan

La Montague-du-Charbon—Headonian—France

La Muette—Robiacian—Paris Basin

Lane Meadows—Bridgerian—Wyoming

Langles—Headonian—France

Lantern Hill—Graybullian—Wyoming

Lantian (Lantien-Xian)—Late Eocene—Shaanxi

Laon district—Lutetian—France

Lapoint ''horizon''—Duchesnean—Utah

Laredo—Uintan—Texas

Largo facies sites—Wasatchian—New Mexico

Larnagol (Quercy)—Headonian—France

La Roca—Cuisian—Spain

Lascours (Gard)—Headonian—France

La Soucarede—Middle? Eocene—France

Las Violetas—Riochican—Argentina

Latilly—Robiacian—France

Laure—Robiacian—France

Lautrec district—Robiacian—France

Laval (Trieu de Leval)—Sparnacian—Belgium

Lavergne (Quercy)—Headonian—France

La Veta East—Wasatchian—Colorado

Lebratieres 1—Headonian—France

Le Bretou (Quercy)—Robiacian—France

Lee-on-Solent—Lutetian—England

Le Guépelle—Robiacian—Paris Basin

Leland Bench Draw—Uintan—Utah

Le Marronnier—Headonian—France

Leota Ranch—Uintan—Utah

Le Ruel—Robiacian—France

Le Saillant (Saillans)—Headonian—France

Les Badies (Isabena Basin)—Cuisian—Spain

Les Bessous (Castrais)—Robiacian—France

Les Clapies—Headonian—France

Les Echelles—Robiacian—France

Les Lornbarts (See Sabarat East.)

Les Matelles—Robiacian—France

Les Ondes—Headonian—France

Les Pradigues—Headonian—France

Les Salères (Lerida)—Cuisian—Spain

Les Sorcières—Headonian—France

Likiang—Late Eocene—Yunnan

Likwanchiao Basin (Sichuan)—Late and Middle? Eocene—
  Henan

Linchu district (Changlo-Wutu)—Shandong—Early Eocene

Ling-che (= Changpiliang, etc.)—Early Eocene—Hunan

Linpao—Middle? Eocene—Henan

Lissieu—Robiacian—France

Little America—Bridgerian—Wyoming

Little Dry Creek—Bridgerian—Wyoming

Little Sand Coulee 1, 2—Graybullian—Wyoming

Liuniu fauna—Early-Middle Eocene—Guangxi

Livry—Headonian—Paris Basin

Llamaquique (Oviedo)—Headonian—Spain

Locy Ranch—Graybullian—Wyoming

London Clay sites—Cuisian—England

Lone Tree district—Bridgerian—Wyoming

Lookout Mountain Conglomerate Member site—Wasatchian—Wyoming

Lost Cabin district (Member and fauna)—Lostcabinian—Wyoming

Lost Yale—Graybullian—Wyoming

Loupoigne—Lutetian—Belgium

Lovell South   (See Foster Gulch.)

Lower Sand Creek Divide—Graybullian—Wyoming

Lulyan (Lu-liang) region—Late Eocene—Xinjiang

Lunan Basin—Late Eocene—Yunnan

Lushih (Lushi)—Late Eocene—Henan

Lybyer Ranch—Lysitean—Wyoming

Lysite district—Lysitean—Wyoming

Magdalena Southwest—Uintan—New Mexico

Mähringen-Ulm—Headonian—Germany

Maïlibay (Buran)—Middle and Late Eocene—Kazakhstan

Malaspina (Rio Chico frente a Malaspina)—Casamayoran—Argentina

Malpérié (Quercy)—Headonian—France

Manacor (Mallorca)—Headonian—Mallorca

Mancy—Cuisian—Paris Basin

Maoxicum Sian—Late Eocene—Shaanxi

Maransart—Lutetian—Belgium

Marconi Road—Graybullian—Wyoming

Marfaux—Sparnacian—Paris Basin

Marfil (Guanajuato)—Uintan?—Mexico

Mary's Hill—Graybullian—Wyoming

Mas de Gimel—Cuisian—France

Mas de Piquet—Lutetian—France

Mas Saintes Puelles (= Villeneuve la Comptal?)—Headonian—France

Mauremont   (See Mormont.)

Mazou—Eocene—France

M'Bodione Dadere—Middle or Late Eocene (marine)—Senegal

McCann Ranch Northwest—Bridgerian—Wyoming

McCulloch Peaks—Wasatchian—Wyoming

McDaniel Ranch East—Wasatchian—New Mexico

McGrew Locality 19   (See Opal.)

Melsbroek—Lutetian—Belgium

Memerlein (Quercy)—Headonian and Oligocene—France

Menchiapu—Late Eocene—Henan

Meng-Yin-Hsien (Meng Yin Valley)—Middle? Eocene—Shandong

Meniscotherium Hollow—Lostcabinian?—Colorado

Mercer Ranch (''*Meniscotherium* quarry'')—Wasatchian—New Mexico

Messel—Lutetian—Germany

Messkirch (Mösskirch)—Headonian—Germany

Methamis—Headonian—France

Metka Coal Field—Middle Eocene—India

Meudon—Sparnacian—Paris Basin

Mienchih (Jentsen)—Late Eocene—Shanxi, Henan

Millersville—Bridgerian—Wyoming

Milleson Draw South—Lostcabinian—Wyoming

Milligan's Arroyo—Lostcabinian—Colorado

Minfeld Tunnel Mine 14—Headonian—Germany

Mireval-Lauvagais—Headonian—France

Misery Quarry—Bridgerian—Wyoming

Mission San Diego—Uintan—California

Mogaung Northwest—Late Eocene—Burma

Möhren 6, 10—Headonian—Germany

Molinière (Castrais)—Robiacian—France

Moneta North—Lostcabinian—Wyoming

Monlis—Headonian—France

Mont Anis—Headonian—France

Mont Bernon—Sparnacian (no mammals) and Cuisian—Paris Basin

Montchenot—Lutetian—Paris Basin

Mont de Berru, upper—Sparnacian—Paris Basin

Monte Duello (Zuelo)—marine Middle Eocene—Italy

Monte Promina—Headonian?—Yugoslavia

Monteroda (Isabena Basin)—Cuisian—Spain

Montespieu (Castrais)—Robiacian—France

Monthelon—Cuisian—Paris Basin

Montllobar—Cuisian—Spain

Montllobar 2—Lutetian—Spain

Montmagny—Headonian—Paris Basin

Montmartre—Headonian—Paris Basin

Monument—Wasatchian—Wyoming

Mormoiron—Headonian—France

Mormont–Eclépens-Entreroches—Headonian and Robiacian?—Switzerland

Morrow Creek Member sites—Bridgerian—Wyoming

Mouillac (Quercy)—Headonian and Oligocene—France

Moulin Baby—Headonian—France

Moulineux—Sparnacian—Paris Basin

Moutiers—Headonian—Switzerland

Muddy Creek—Lostcabinian—Wyoming

Muddy Creek Bridge Southwest—Lostcabinian—Wyoming

Muddy Ridge Northwest 1, 2—Lostcabinian—Wyoming

Mühlhausen   (See Rixheim.)

Muirancourt—Sparnacian—Paris Basin

Mule Ear Peaks—Uintan—Texas

Musters Formation sites—Mustersan—Argentina

Mutigny—Sparnacian—Paris Basin

Myaing district—Late Eocene—Burma

Myton Pocket—Uintan—Utah

Nace—Lostcabinian—Wyoming

Naduo (Nado) Formation sites—Late Eocene—Baise Basin, Guangxi

Nanjing—Early Eocene—Anhui

Nanterre—Lutetian—Paris Basin

Naran-Bulak, Upper White Beds—Early Eocene—Mongolia

Nederockerzeel—Lutetian—Belgium

Needle Creek Divide—Uintan—Wyoming

Neuhausen at Tuttlingen—Headonian (in part)—Germany

Neuilly-Vexin—Headonian—Paris Basin

Neustadt—Headonian (and early Oligocene?)—Germany

New *Coryphodon*—Graybullian—Colorado

New Fork district—Lostcabinian—Wyoming

Nice—Headonian—France

Ninchiashan or Ninjiashan (Xinyu)—Early Eocene—Jiangxi

Niushan-Linchü district (Chupitian Valley)—Early Eocene—Shandong

Nipple Hollow—Lostcabinian?—Colorado

Nipple Hollow Southwest—Lostcabinian—Colorado

Nivelles—Lutetian—Belgium

Nogent l'Artaud—Robiacian—Paris Basin

Noguera Pallaresa (See Castigaleu, Sosis.)

Nomogen Commune—Late Paleocene and Early Eocene—Nei Mongol

Nordhausen—Headonian (or Early Oligocene)—Germany

North Mesa—Late Eocene—Nei Mongol

Numata-muru—Late Eocene—Hokkaido, Japan

Nut Bed—Bridgerian—Oregon

Oak Creek (Castillo Pocket)—Lostcabinian—Colorado

Obaila (Obayla)—Middle and Late Eocene—Kazakhstan

Obergösgen—Headonian—Switzerland

Oeste de Rio Chico—Casamayoran—Argentina

Oil Mountain South—Lostcabinian?—Wyoming

Oil Mountain Southeast—Uintan—Wyoming

Oilspud—Graybullian—Wyoming

Oil Well—Graybullian—Wyoming

Okie Trail—Lostcabinian—Wyoming

Omobialla district   (See Ameki.)

Ongar—Cuisian—England

Opal (McGrew 19)—Bridgerian—Wyoming

Opphem—Lutetian—Belgium

Oregon Buttes Northwest—Wasatchian—Wyoming

Orp-le-Grand—Sparnacian—Belgium

Orsmael   (See Dormaal.)

Orvillers—Sparnacian—Paris Basin

Osborne Beds sites (Hampshire Basin, Isle of Wight)—Headonian—England

Ostpinghi—Late? Eocene—Henan, Shanxi

Ostrand—Headonian—Germany

Ostutah—Wasatchian—Utah

Otto Basin Crossing, East-Southeast—Graybullian—Wyoming

Otto North—Graybullian—Wyoming

Otto South—Graybullian—Wyoming

Oulchy-le-Chateau—Lutetian—Paris Basin

Ouray Agency—Uintan—Utah

Owl Creek, South Fork—Uintan—Wyoming

Pajarito (Colhuapi Norte, in part?)—Casamayoran and Mustersan—Argentina

Palette—Sparnacian—France

Pampa de Castillo East   (See Este de Rio Chico.)

Panandhro West   (See Kutch.)

Panchiao (Pankiao)—Late Eocene—Yunnan

Pangan   (See Phan Kan.)

Pantin—Headonian—Paris Basin

Parc Monceau—Robiacian—Paris Basin

Pargny-Filain—Lutetian—Paris Basin

Paris   (See Montmartre.)

Paso de los Indios (Laguna de la Bombilla)—Casamayoran—Argentina

Passel—Sparnacian—Paris Basin

Pass Peak Formation sites—Wasatchian—Wyoming

Passy—Sparnacian—Paris Basin

Pat O'Hara Creek—Graybullian—Wyoming

Patrick Draw Northeast sites—Wasatchian—Wyoming

Payrade—Headonian—France

Pazigyi Village Southwest—Late Eocene—Burma

Pearson Ranch—Duchesnean—California

Peel Common—Cuisian—England

Peking   (See Beijing.)

Peréal (Perréal) (= La Débruge)

Perrière (Quercy)—Headonian—France

Pevy—Lutetian—Paris Basin

Peyregoux (Castrais)—Robiacian—France

Pfaffenweiler—Headonian—Germany

Pfister Ranch—Wasatchian—Wyoming

Phan Kan—Late Eocene—Burma

Piceance Creek Basin—Wasatchian—Colorado

Pico Salamanca—Casamayoran—Argentina

Pico Truncado—Casamayoran—Argentina

Pine Mountain sites (Sespe)—Duchesnean?—California

Poble vell de l'Amettla—Lutetian—Spain

Pondaung Formation sites—Late Eocene—Burma

Pont d'Assou—Headonian—France

Pourcy—Sparnacian—Paris Basin

Poway Conglomerate or Group sites—Uintan—California

Powder River Basin sites—Graybullian—Wyoming

Powder Springs Northeast—Lostcabinian—Utah

Powder Wash—Bridgerian—Utah

Pradigues (Quercy) (Les Pradigues)—Headonian—France

Prajous (Quercy)—Headonian into Oligocene—France

Preator's Ranch South (Burlington South)—Wasatchian—Wyoming

Primorsky Kray (Artëm)—Late Eocene—Far East USSR

Primrose Hill—Cuisian—England

Princeton Bone Hill—Graybullian—Wyoming

Privrasky—Lostcabinian?—Wyoming

Promontory Bluff   (See Black Mountain.)

Prospect Quarry—Headonian—Isle of Wight, England

Provins—Lutetian—Paris Basin

Pumpkin Buttes   (See Pfister Ranch.)

Purple Valley 1, 2, 3, 4—Graybullian—Wyoming

Qasr-el-Sagha district (marine with land mammal)—Late Eocene—Egypt

Quebrada Agua Viva—Early? Eocene—Venezuela

Quercy district (100+ localities)—Robiacian into Oligocene—France

Quissac—Headonian—France

Quitman South—Late Eocene (marine with *Notiotitanops*)—Mississippi

Raitenbuch—Headonian?—Germany

Rajauri (Kalakot)—Middle Eocene—India

Ralston Northwest—Graybullian—Wyoming

Rand—Wasatchian—Utah

Randlett "horizon"—Uintan—Utah

Randlett Point Northeast—Uintan—Utah

Rattlesnake 1, 2—Graybullian—Wyoming

Raven Ridge North—Wasatchian—Utah

Réalmont (Castrais)—Robiacian—France

Reculusa Ranch Northwest (Reculusa Blowout)—Graybullian—Wyoming

Red Bluff Wash—Uintan—Utah

Red Canyon district—Wasatchian—Wyoming

Red Creek—Clarkforkian? and Wasatchian—Wyoming

Red Desert 1—Graybullian—Wyoming

139

Red Desert district—Wasatchian—Wyoming
Remy—Sparnacian—Paris Basin
Renaud-Croute—Headonian—France
Repeu del Guaita (Siena de Montllobar)—Lutetian—Spain
Reva Gap (Slim Buttes)—Duchesnean—South Dakota
Rians—Sparnacian—France
Riedisheim—Headonian—Germany
Rifle Member sites—Wasatchian—Colorado
Rifle South-Southeast—Wasatchian—Colorado
Rixheim—Headonian—Germany
Robiac Nord et Sud (= St. Mammert)—Robiacian—France
Roc de Lunel (Castrais)—Robiacian—France
Roc des Carteirades (Souvignargues)—Headonian—France
Rocking Chair Ranch—Lostcabinian—Wyoming
Rock Springs district sites—Wasatchian—Wyoming
Roda district (Isabena Basin)—Lutetian—Spain
Romainville—Headonian—Paris Basin
Rosières 2, 4 (Quercy)—Headonian—France
Roslyn Formation sites (fish and turtles)—Bridgerian?—Washington
Rot-Malsch—Headonian—Germany
Roydon (Hampshire Basin, marine)—Headonian—England
Sabarat East—Cuisian?, Lutetian?—France
Sacel—Robiacian?—Rumania
Safed Tobah (Dera Ghazi Khan district) (marine?)—Middle Eocene—Pakistan
Saffre (Bois-Couët)—Lutetian—Paris Basin
Sage Creek—Bridgerian—Wyoming
Sage Creek, Main Quarry—Bridgerian—Wyoming
Sage Creek Formation sites—Uintan—Montana
Saikan Range (Zaisan Depression)—Middle Eocene—Kazakhstan
St.-Agnan—Sparnacian and Cuisian—Paris Basin
St.-Antonin (Quercy)—Headonian—France
St.-Aubin d'Eymet and St. Aubin-de-Cadaleich—Headonian—France
St.-Bonnet-de-Rochefort—Headonian—France
St.-Capraise d'Eymet—Headonian—France
St.-Ciers—Headonian—France
St.-Clément-la-Riviere—Cuisian—France
St.-Croix-de-Beaumont—Headonian—France
St. Denis—Robiacian—Paris Basin
Ste.-Néboule (Quercy)—Headonian—France
Ste-Radegonde   (See La Débruge.)
Ste.-Sabine—Headonian—France
St. Gely du Fesc—Headonian—France
St. Gilles—Lutetian—Belgium
"St. Hippolyte de Caton"   (See Euzet-les-Bains.)
St.-Jean-de-Marvejols—Headonian—France
St. Joe East—Graybullian—Wyoming
St.-Josse-Ten-Noode—Lutetian—Belgium
St.-Leu-Taverny—Headonian—Paris Basin
St. Mamert   (See Robiac.)
St. Martin-de-Villeréal—Headonian—France
St. Ouen—Robiacian—Paris Basin
St. Papoul—Headonian?—France
St. Remi-Geest—Lutetian—Belgium
St. Saturnin   (See La Débruge.)

St.-Sauveur—Sparnacian—Paris Basin
St.-Verena—Headonian?—Switzerland
Saissan or Zaisan Basin or Depression—Early and Middle and Late Eocene—Kazakhstan
Saïx (Castrais)—Robiacian—France
Salesmes—Headonian—France
Sallent—Headonian—Spain
Salta Group sites—Casamayoran, in part—Argentina
San Cugat de Gavadons—Headonian—Spain
Sand Butte (= Bitter Creek Station Southwest?)—Wasatchian—Wyoming
Sand Coulee Basin (Big)—Tiffanian into Graybullian—Wyoming
Sand Coulee Basin 1—Wasatchian—Wyoming
Sand Creek Divide—Graybullian—Wyoming
Sand Draw near Barrel Spring   (See Oil Mt. South.)
San Diego Mission   (See Mission San Diego.)
Sand Quarry (Four Mile)—Graybullian—Colorado
Sand Wash Basin—Bridgerian and Uintan—Colorado
Sänggau (Seboemban Oeloe)—Late? Eocene—Borneo
San Jose Formation sites—Tiffanian and Wasatchian—New Mexico and Colorado
San Morales—Lutetian into? Headonian—Spain
Sannois—Headonian (also Early Oligocene)—France
Santes Creus—Lutetian—Spain
Sant Miquel—Lutetian—Spain
Santpedor (Barcelona)—Headonian—Spain
Sant Père Martir—Lutetian—Spain
Sanzoles—Lutetian—Spain
Sargomysse—Middle to Late Eocene—Kazakhstan
Saron—Sparnacian—Paris Basin
Saugron—Headonian—France
Saussenac—Headonian—France
Schaerbeek (Kattepoel)—Middle to Late Eocene—Belgium
Sconce (Isle of Wight)—Headonian—England
Scripps Formation sites—Uintan—California
Seafield (Isle of Wight)—Headonian—England
Section 31 (Univ. Mich. loc.)—Lostcabinian—Colorado
Selva—Headonian—Mallorca
Sergy—Robiacian—Paris Basin
Sermaize—Sparnacian—Paris Basin
Sespe Formation sites (lower Sespe)—Uintan and Duchesnean—California
Sète—Headonian? (also Villafranchian)—France
Sézanne-Broyes—Cuisian—Paris Basin
Shahezigou (Lantian district)—Late Eocene—Shaanxi
Shara Murun district—Late Eocene—Nei Mongol
Shark River Marl site—marine Middle Eocene and Bridgerian mammal—New Jersey
Sheppey (Isle of Sheppey)—Cuisian—England
Shisanjianfang Formation site—Early Eocene—Xinjiang
Shiwopu, Dahko (Lunan Basin)—Late Eocene—Yunnan
Shoddy Springs—Uintan—Montana
Shooty Gulch—Lostcabinian—Colorado
Shoshone River Canyon—Graybullian—Wyoming
Shotgun Butte Northeast—Graybullian—Wyoming
Shteir Quarry (Four Mile)—Graybullian—Colorado
Sian (or Xian, Lantien district)—Late Eocene—Shaanxi
Sicardens (Castrais)—Robiacian—France
Sichuan (Sichwan) Basin—Middle? and Late Eocene—Henan
Sierra del Toro   (See Colhuapi Norte.)
Silveirinha—Sparnacian—Portugal

Silverstine Ranch South—Lostcabinian—Colorado
Sinceny—Sparnacian—Paris Basin
Sindou D (Quercy)—Headonian—France
Sintai (Hsintai)—Sisichou—Middle Eocene—Shandong
Sisichou—Middle (and Early?) Eocene—Shandong
Sisters Hill East—Wasatchian—Wyoming
Siziwang County—Middle Eocene—Nei Mongol
Slick Creek Quarry—Graybullian—Wyoming
Slim Buttes district and Formation (also Antelope Creek)—Duchesnean—South Dakota
Soissons district—Sparnacian—Paris Basin
Soldiers Summit—Bridgerian—Utah
Sommières   (See Souvignargues.)
Sosis (Huesca)—Headonian—Spain
Soterranya—Middle? to Late? Eocene—Spain
South Maple Canyon—Wasatchian—Utah
Souvignargues (Sommières)—Roc des Carteirades—Headonian—France
Spanish John Meadow—Bridgerian—Wyoming
Spring City Southwest—Duchesnean—Utah
Squaw Buttes district—Wasatchian—Wyoming
State Line Quarry—Bridgerian—Wyoming
Steamboat Mountain East—Wasatchian—Wyoming
Stetten—Lutetian—Germany
Stolzenbach—Lutetian—Germany
Stonehenge Sandstone Quarry—Graybullian—Wyoming
Strathcona Fiord district—Wasatchian—Ellesmere Island
Subathu Formation sites—Middle Eocene—Jammu and Kashmir, India
Sublette County 1—Bridgerian—Wyoming
Suffolk Pebble Beds site   (See Kyson.)
Summers Dry Creek—Bridgerian—Wyoming
Supersite Quarry—Graybullian—Wyoming
Sussex district—Graybullian—Wyoming
Swift Current Creek—Uintan—Saskatchewan
Sydenham (Woolwich Beds)—Sparnacian—England
Tabernacle Butte district—Bridgerian—Wyoming
Table Rock—Lostcabinian—Wyoming
Taizicun Formation sites—Late Paleocene and Early Eocene—Xinjiang
Tama—Mustersan?—Colombia
Tan's Locality   (See Men-Yin-Hsien.)
Tantou Basin and Formation sites—Late Paleocene or Early Eocene or both—Henan
Tapo Ranch and Tapo Canyon—Uintan—California
Tappan Creek—Lostcabinian—Wyoming
Tatman Mountain—Wasatchian—Wyoming
Tavel—Headonian—France
Tecolote Canyon—Uintan—California
Telegraph Line Camp (Irdin Manha)—Late Eocene—Nei Mongol
Teleodus Quarry—Duchesnean—Utah
Ten-Mile Creek, Head—Graybullian—Wyoming
Tepee Trail Formation, Unit 24—Uintan—Wyoming
Terras de Collu—Lutetian—Sardinia
Third Hill (Four Mile)—Graybullian—Colorado
Thomas Canyon   (See State Line Quarry.)
Three Forks   (See Shoddy Springs.)
Thryptacodon—Graybullian—Wyoming
Timberlake Quarry (Four Mile)—Graybullian—Colorado
Timor Island—Late Eocene
Tinimomys Hills—Graybullian—Wyoming

Tipton Butte—Lysitean?—Wyoming
Tongbo—Late Eocene—Henan
Tonque—Duchesnean—New Mexico
Torrelabad (Isabena Basin)—Lutetian—Spain
Toruaïgyr—Late Eocene—Kirghizia
Totland Bay—Headonian—England
Trelew-Gaiman District—Riochican and Colhuehuapian—Argentina
Tremp Basin (See Montllobar)—Cuisian and Lutetian—Spain
Tridell East—Duchesnean—Utah
Trieu-de-Leval   (See Laval.)
Try—Sparnacian (in part?)—Paris Basin
Tsagan Khushu—Cretaceous into Late Eocene—Mongolia
Tschaikbulak (Chaybulak)—Late Eocene—Kazakhstan
Tschi-Lih (Beijing)—Late Eocene—Beijing
Tsungaria (Dsungaria) Basin—Middle and Late Eocene—Xinjiang
Tufal—Headonian—France
Tukhum North—Late Eocene—Nei Mongol
Tungchang (Chiyuan)—Late Eocene—Henan
Turpan Basin—Paleocene, Eocene and Oligocene—Xinjiang
Twelvemile Wash—Duchesnean—Utah
Twin Buttes district—Bridgerian—Wyoming
Twisty Turn Hollow—Graybullian—Wyoming
Two Head Hill Quarry—Graybullian—Wyoming
Ube Coal Field (Okinoyama Colliery)—Late Eocene—Japan
Uccle—Lutetian—Belgium
Ulan Bulak—Late Eocene—Mongolia
Ulan Shireh—Late Eocene—Nei Mongol
Ula Usu (Shara Murun district)—Late Eocene—Nei Mongol
Ulken-Ulasty—Middle Eocene—Kazakhstan
UM SC-69—Wasatchian—Wyoming
UM Sub-Wy 2, 4, 16, 23, 27, 28, 29—Wasatchian—Wyoming
Unit 24   (See Tepee Trail Formation, Unit 24.)
Urcel—Sparnacian?—France
Urtyn Obo—Middle? Eocene—Nei Mongol
USGS D-1033   (See Flattop Quarry.)
USGS D-1034   (See Vass Quarry.)
Valle Hermoso—Casamayoran—Argentina
Vass Quarry—Bridgerian—Wyoming
Vaugirard—Sparnacian and Lutetian—Paris Basin
Vauxbuin—Sparnacian—Paris Basin
Vermilion Creek—Wasatchian—Colorado and Wyoming
Vernal West—Duchesnean—Utah
Verrière de Roches—Headonian?—Switzerland
Vertain—Cernaysian and Sparnacian—Paris Basin
Verzenay—Lutetian—Paris Basin
Vielmur-sur-l'Agout (Castrais)—Robiacian—France
Villeneuve-la-Comptal   (See Mas Saintes Puelles.)
Villers-sur-Coudun—Sparnacian—Paris Basin
Villiers-Adam—Headonian—Paris Basin
Vinalmont—Sparnacian—Belgium
Vitry-sur-Seine—Headonian—Paris Basin
Viviers-la-Montagne (Castrais)—Robiacian—France
Vladivostok—Artëm (Primorsky Kray)—Late Eocene—Far East USSR
Vlierzele—Lutetian—Belgium
Wadi Kraus Quarry—Graybullian—Wyoming

Wagon Bed Spring Northeast 1—Uintan—Wyoming
Wagon Bed Spring South—Uintan—Wyoming
Waltman Draw West—Lostcabinian—Wyoming
Walton—Sparnacian?—England
Washakie Basin sites—Tiffanian through Uintan—Wyoming
Weihoebene—Late Eocene—Shanxi, Henan
Wetkya (Wetcha)—Late Eocene—Burma
Whistle Creek—Graybullian—Wyoming
Whistle Creek, Head—Lostcabinian—Wyoming
Whistler Squat—Uintan—Texas
White Cliff Bay 2A, 2B—Headonian—England
White Hill—Lostcabinian—Wyoming
White River Pocket—Uintan—Utah
Wiggins Formation sites—Uintan—Wyoming
Wilderness Quad 1—Lostcabinian—Wyoming
Williams Creek—Lostcabinian—Colorado
Wind River Basin sites—Wasatchian—Wyoming
Wolwe—St. Etienne—Lutetian—Belgium
Wood locality—Duchesnean—Wyoming
Woolwich Beds sites—Sparnacian—England
Worland District (also Ten-Mile Creek)—Graybullian—Wyoming
W.R. Wilson Coal Mine (Princeton Basin)—Bridgerian—British Columbia
Wucheng Basin—Late Eocene—Henan
Wutu (Udu)—Early Eocene—Shandong
Wyo. Univ. 1646 (Red Desert)—Lostcabinian—Wyoming
Xuancheng Basin—Late Paleocene or Early Eocene—Anhui
Yale 31, 40, 185—Lysitean—Wyoming
Yale 97b, 290, 348, 358—Graybullian—Wyoming
Yangchi (Ichang)—Late Eocene—Hebei
Yarmouth (Bouldnor Cliff, Hamstead)—Headonian—England
Yateley—Lutetian—England
Yongle Basin—Late Eocene—Gaungxi
York—Bridgerian—Wyoming
Yuanchü or Yuanqu Basin—Late Eocene—Shanxi, Henan
Yuanshui Basin—Early Eocene—Jiangxi
Zaventhen—Lutetian—Belgium
Zaisan Depression or Basin—Early to Late Eocene—Kazakhstan

ENDNOTES

1. This species has been referred to *Cantius* Simons by Gingerich and Haskin, 1981. *Univ. Mich., Mus. Paleontol., Contrib.,* 25:327–337.
2. We thank Mr. Mark Mason, Department of Paleontology, University of California, for his corrections of our Uintan faunal list.

BIBLIOGRAPHY

ANDERSEN, D.W., and PICARD, M.D. 1972. Stratigraphy of the Duchesne River Formation (Eocene-Oligocene?), Northern Uinta Basin, Northeastern Utah. *Bull. Utah Geol. Miner. Surv.* 97:1–29.

ANDREWS, C.W. 1919. A description of a new species of zeuglo-dont and of leathery turtle from the Eocene of Southern Nigeria. *Proc. Zool. Soc. Lond.* 1919:309–319.

ATKINS, E.G. 1970. See West, R.M., and Atkins, E.G.

AXELROD, D.I., and BAILEY, H.P. 1968. Cretaceous dinosaur extinction. *Evolution* 22:595–611.

AXELROD, D.I., and BAILEY, H.P. 1969. Paleotemperature analysis of Tertiary floras. *Palaeogr., Palaeoclimatol., Palaeoecol.* 6:163–195.

BAILEY, H.P. 1964. Toward a unified concept of the temperate climate. *Geogr. Rev.* 54:516–545.

BAILEY, H.P. 1966. The mean annual range and standard deviation as measures of dispersion of temperature around the annual mean. *Geogr. Ann., Ser. A.,* 48A:183–194.

BA MAW, CIOCHON, R.L., and SAVAGE, D.E. 1979. Late Eocene of Burma yields earliest anthropoid primate, *Pondaungia cotteri. Nature* 282:65–67.

BARNES, L.G., and MITCHELL, E. 1978. Cetacea. In *Evolution of African Mammals,* ed. J.J. Maglio and H.B.S. Cooke, chap. 29, pp. 582–602. Cambridge: Harvard Univ. Press.

BELYAYEVA, E.I., RESHETOV, V.I., and TROFIMOV, B.A. 1973. Basic stages in the late Mesozoic-Paleogene of central Asia. Moskovsk. O-vo. Ispyt. Prir., *Byull. Otdil. Geol.* 1 for 1973.

BERGGREN, W.A., MCKENNA, M.C., HARDENBOL, J., and OBRADOVICH, J. 1978. Revised Paleogene polarity time scale. *J. Geol.* 86:67–81.

BEYN, W. 1940. Die Einschaltung geformter Pflanzenreste in das Braunkohlenprofil des Geiseltales. *Nova Acta Leopold.* 58:377–438.

BEYRICH, E. 1854. Uber die Stellung der hessischen Tertiär-bildungen. *Monatsb. Akad. Wiss. Berlin.* 1854:664–666, Oligocän.

BLACK, C.C. 1967. Middle and Late Eocene mammal communities: A major discrepancy. *Science* 156:62–64.

BLACK, C.C. 1970. Paleontology and geology of the Badwater Creek area, central Wyoming. Part 5. The cylindrodont rodents. *Ann. Carnegie Mus.* 41:201–214.

BLACK, C.C. 1971. Paleontology and geology of the Badwater Creek area, central Wyoming. Part 7. Rodents of the family Ischyromyidae. *Ann. Carnegie Mus.* 43:179–217.

BLACK, C.C. 1978. Paleontology and geology of the Badwater Creek area, central Wyoming. Part 14. The artiodactyls. *Ann. Carnegie Mus.* 47:223–259.

BLACK, C.C., and DAWSON, M.R. 1966A. A review of the Late Eocene mammalian faunas from North America. *Am. J. Sci.* 264:321–349.

BLACK, C.C., and DAWSON, M.R. 1966B. Paleontology and geology of the Badwater Creek area, central Wyoming. Part 1. History of field work and geological setting. *Ann. Carnegie Mus.* 38:297–307.

BLONDEAU, A., CAVELIER, C., FEUGUEUR, L., and POMEROL, C. 1965. Stratigraphie du Paléogène du bassin de Paris en relation avec les bassins avoisinants. *Bull. Soc. Géol. Fr.* (7) 7:200–221.

BLONDEAU, A., GRUAS-CAVAGNETTO, C., LE CALVEZ, Y., and LEZAUD, L. 1976. Etude paléontologique du sondage de Cuise (Oise). *Bull. Inform. Géologues Bassin Paris* 13 (2):3–32.

BONHOMME, M., ODIN, G.S., and POMEROL, C. 1968. Age des formations glauconieuses de l'Albien et de l'Eocène du Bassin de Paris. Colloque sur l'Eocène. *Fr., Bur. Rech. Géol. Minières Mém.* 58:339–346.

BOSMA, A.A. 1974. Rodent biostratigraphy of the Eocene-Oligocene transitional strata of the Isle of Wight. *Utr. Micropaleontol. Bull. Spec. Publ.* 1:1–126.

BOSMA, A.A., and DE BRUIJN, H. 1979. Eocene and Oligocene Gliridae (Rodentia, Mammalia) from the Isle of Wight, England. Part 1. The *Gliravus priscus–Gliravus fordi* lineage. *Proc. Ned. Akad. Wet., Koninkl., Ser. B.*, 82:367–384.

BOSMA, A.A., and INSOLE, A.N. 1972. Theridomyidae (Rodentia, Mammalia) from the Osborne Beds (Late Eocene), Isle of Wight, England. *Ned. Akad. Wet., Ser. B*, 75:133–144.

BOSMA, A.A., and INSOLE, A.N. 1976. Pseudosciuridae (Rodentia, Mammalia) from the Osborne Beds (Headonian), Isle of Wight, England. *Ned. Akad. Wet., Ser. B.*, 79:1–8.

BOWN, T.M. 1974. Notes on some Early Eocene anaptomorphine primates. *Univ. Wyo., Geol. Contrib.* 13:19–26.

BOWN, T.M. 1976. Affinities of *Teilhardina* (Primates, Omomyidae) with description of a new species from North America. *Folia. Primatol.* 25:62–72.

BOWN, T.M. 1979. Geology and mammalian paleontology of the Sand Creek facies, lower Willwood Formation, (Lower Eocene), Washakie County, Wyoming. *Wyo. Geol. Surv. Mem.* 2:1–151.

BOWN, T.M., and KRAUSE, M.J. 1981. Vertebrate fossil-bearing paleosol units (Willwood Formation, lower Eocene, northwestern Wyoming, U.S.A.): Implications for taphonomy, biostratigraphy, and assemblage analysis. *Palaeogeogr., Palaeoclim., Palaeoecol.* 34:31–56.

BOWN, T.M., and ROSE, K.D. 1976. New Early Tertiary primates and a reappraisal of some Plesiadapiformes. *Folia. Primatol.* 26:109–138.

BOWN, T.M., and SCHANKLER, D.M. 1982. A review of the Proteutheria and Insectivora of the Willwood Formation (lower Eocene), Bighorn Basin, Wyoming. *U.S. Geol. Surv., Bull.* 1523:1–79.

BRADLEY, W.H. 1959. Revision of stratigraphic nomenclature of Green River Formation of Wyoming. *Bull. Am. Assoc. Pet. Geol.* 43:1072–1075.

BRADLEY, W.H. 1964. Geology of Green River Formation and associated Eocene rocks in southwestern Wyoming and adjacent parts of Colorado and Utah. *U.S. Geol. Surv. Prof. Pap.* 496-A:1–86.

BROWN, R., and PECORA, W.T. 1949. Paleocene and Eocene strata in the Bearpaw Mountains, Montana. *Science* 109:487–489.

BRUNET, M. 1979. Les grandes Mammifères chefs de file de l'immigration oligocène et le problème de la limite Eocène-Oligocène en Europe. Paris: Edit. Fondation Singer-Polignac. 325 pp.

BURKE, J.J. 1935. Preliminary report on fossil mammals from the Green River formation in Utah. *Ann. Carnegie Mus.* 25:13–14.

BUTLER, P.M. 1947. An arctocyonid from the English Ludian. *Nat. Hist. Ann. Mag.* 13:691–701.

CAPPETTA, H., HARTENBERGER, J.-L., SIGÉ, B., and SUDRE, J. 1968. Une faune de vertébrés de la zone de Cuis dans l'Eocène continental du Bas-Languedoc (gisement du Mas de Gimel, Grabels, Hérault). *Bull. Bur. Rech. Géol. Minières* (2) 3:46–48.

CARBAJAL, E., PASCUAL, R., PINEDO, R., SALFITY, J.A., and VUCETICH, M.G. 1977. Un nuevo mamífero de la formación Lumbrera (Grupo Salta) de La Comarca de Carahuasi (Salta, Argentina). Edad y correlaciones. *Publ. Mus. Municip. Cien. Nat., Mar del Plata "Lorenzo Scaglia,"* 2:148–163.

CAVELIER, C. 1979. La limite Eocène-Oligocène en Europe occidentale. *Univ. L. Pasteur, Strasbourg Mém.* 54:1–280.

CHANDLER, M.E.J. 1961. The lower Tertiary floras of southern England. Pts. 1 and 2. Br. Mus. (Nat. ist.), Publ.

CHANDLER, M.E.J. 1963. The lower Tertiary floras of southern England. Pt. 3. Br. Mus. (Nat. Hist.), Publ.

CHATEAUNEUF, J.J. 1980. Palynostratigraphie et paléoclimatologie de l'Eocène supérieur et de l'Oligocène du Bassin de Paris (France). *Fr., Bur. Rech. Géol. Minières Mém.* 116:1–360.

CHOW, M. 1961B. A new tarsioid primate from the Lushi Eocene, Honan. *Vertebr. PalAsiatica* 3:3–4.

CHOW, M. 1963B. Tillodont materials from Eocene of Shantung and Honan. *Vertebr. PalAsiatica* 7:97–104.

CHOW, M. 1965B. Mesonychids from the Eocene of Honan. *Vertebr. PalAsiatica* 9:290–291.

CHOW, M. 1975. Some carnivores from the Eocene of China. *Vertebr. PalAsiatica* 13:168–176.

CHOW, M., CHANG, Y., and TING, S. 1974. Some Early Tertiary Perissodactyla from Lunan basin, East Yunnan. *Vertebr. PalAsiatica* 10:262–273

CHOW, M., and CHIU, C. 1964. An Eocene giant rhinoceros. *Vertebr. PalAsiatica* 3:266–267.

CHOW, M., and LI, C. 1965. *Homogalax* and *Heptodon* of Shantung. *Vertebr. PalAsiatica* 9:19–21.

CHOW, M., LI, C., and CHANG, Y. 1973. Late Eocene mammalian faunas of Honan and Shansi with notes on some vertebrate fossils collected therefrom. *Vertebr. PalAsiatica* 11:179–190.

CHOW, M., and TUNG, Y. 1962. Notes on some new uintathere materials of China. *Vertebr. PalAsiatica* 6:371–374.

CHOW, M., and TUNG, Y. 1965. A new coryphodont from the Eocene of Sinyu, Kiangsi. *Vertebr. PalAsiatica* 9:114–121.

CHOW, M., and XU, Y. 1965. Amynodonts from the Upper Eocene of Honan and Shansi. *Vertebr. PalAsiatica* 9:199–203.

CHOW, M., and XU, Y. 1967. New primitive true rhinoceroses from the Eocene of Iliang, Yunnan. *Vertebr. PalAsiatica* 5:291–305.

CHOW (OR ZHOU) M., ZHANG, Y., WANG, B., and DING, S. 1977. Mammalian fauna from the Paleocene of Nanxiong Basin, Guangdong. *Palaeont. Sinica., Ser. C2*, 153:1–100.

CLARK, J., BEERBOWER, J.R., and KIETSKE, K.K. 1967. Oligocene sedimentation, stratigraphy and paleoclimatology in the Big Badlands of South Dakota. *Fieldiana: Geol. Mem.*, 5:1–158.

COLBERT, E.H. 1938. Fossil mammals from Burma in the American Museum of Natural History. *Bull. Am. Mus. Nat. Hist.* 74:255–436.

COOMBS, M.C. 1971. Status of *Simidectes* of the Late Eocene of North America. *Am. Mus. Novit.* 2455:1–41.

COOMBS, M.C., and COOMBS, W.P., JR. 1977. Dentition of *Gobiohyus* and a reevaluation of the Helohyidae (Artiodactyla). *J. Mammal.* 58:291–308.

COOMBS, W.P., JR., and COOMBS, M.C. 1979. *Pilgrimella*, a primitive Asiatic perissodactyl. *J. Linn. Soc. Lond.* 65:185–192.

COOPER, C.F. 1932. The genus *Hyracotherium*. A revision and description of new specimens found in England. *Philos. Trans., R. Soc. Lond., Ser. B*, 221:431–448.

CROCHET, J.-Y. 1977. Les didelphidés paléogènes holarctiques: historique et tendances évolutives. *Géobios, Mém. Spéc.*, 1:127–134.

143

CROCHET, J.-Y. 1979. Diversité systematique des Didelphidae (Marsupialia) européens tertiaires. *Géobios* 11:365–378.

CROCHET, J.-Y. 1980. Les Marsupiaux du Tertiaire d'Europe. Paris: Edit. Fondation Singer-Polignac. Pp. 1–279.

CRUSAFONT, M., DE RENZI, M., and CLAVELL, E. 1968. Les grands traits d'une coupure Cretacé-Paléocène-Eocène en sud des Pyrenées (Isabena). *Fr., Bur. Rech. Géol. Minières Mém.* 58:591–596.

CURRY, D., ADAMS, C.G., BOULTER, M.C., DILLEY, F.C., EAMES, F.E., FUNNELL, B.M. and WELLS, H.K. 1978. A correlation of Tertiary rocks in the British Isles. *Geol. Soc. Lond., Spec. Report* 12:1–72.

DASHZEVEG, D. 1977. On the first occurrence of *Hyopsodus* Leidy, 1870, (Mammalia, Condylarthra) in the Mongolian Peoples' Republic. Mesozoic and Cenozoic Faunas, Floras and Biostratigraphy of Mongolia. *Trans. Joint Sov.-Mongol. Exped.* 4:7–13. (In Russian.)

DASHZEVEG, D. 1979A. The find of *Homogalax* (Perissodactyla, Tapiroidea) in Mongolia and its stratigraphical significance. *Bull. Moscow Soc. Naturalists, Geol. Sect.*, 54:105–111. (In Russian.)

DASHZEVEG, D. 1979-0413. Discovery of a hyracothere in Mongolia. *Paleont. J.* 3:108–113. (In Russian.)

DASHZEVEG, D. 1979-0412. On an archaic representative of the equoids (Mammalia, Perissodactyla) from the Eocene of central Asia. Mesozoic and Cenozoic Faunas of Mongolia. *Trans. Joint. Sov.-Mongol. Paleont. Exped.* 8:10–22. (In Russian.)

DASHZEVEG, D. 1980. New pantodonts from the Eocene of Mongolia. *Paleont. J.* 2:108–115. (In Russian.)

DASHZEVEG, D., and MCKENNA, M.C. 1977. Tarsioid primate from the early Tertiary of the Mongolian Peoples' Republic. *Acta Palaeontol. Pol.* 22:119–137.

DAVIS, A.G., and ELLIOT, G.F. 1957. The palaeogeography of the London Clay sea. *Proc. Geol. Assoc. (Lond.)* 68:255–277.

DAWSON, M.R. 1964. Late Eocene rodents (Mammalia) from Inner Mongolia. *Am. Mus. Novit.* 2191:1–15.

DAWSON, M.R. 1966. Additional Late Eocene rodents (Mammalia) from the Uinta Basin, Utah. *Ann. Carnegie Mus.* 38:97–114.

DAWSON, M.R. 1967. Middle Eocene rodents (Mammalia) from northeastern Utah. *Ann. Carnegie Mus.* 39:327–370.

DAWSON, M.R. 1977. Late Eocene rodent radiation: North America, Europe, Asia. *Géobios* 1:195–210.

DEHM, R., and OETTINGEN-SPIELBERG, T. 1958. Paläontologische und geologische Untersuchengen im Tertiär von Pakistan. 2. Die mitteleocänen Säugetiere von Ganda Kas bei Basal in Nordwest-Pakistan. *Bayer. Akad. Wiss., Math. Naturwiss. Kl., Abh.*, 91:1–54.

DENIZOT, G. 1968. Bartonien, Ludien et Tongrien. *Bur. Rech. Géol. Minières Mém.* 58:533–552.

DING, S., ZHENG, J., ZHANG, Y., TONG, Y. 1977. The age and characteristic of the Liuniu and the Dongjun faunas, Bose Basin of Guangxi. *Vertebr. PalAsiatica* 15:35–45.

EDWARDS, W.N. 1936. The Flora of the London Clay. *Proc. Geol. Assoc. (Lond.)* 47:22–31.

ELLENBERGER, P. 1980. Sur les empreintes de pas des gros Mammifères de l'Eocène supérieur de Garrigues–Ste.-Eulalie (Gard). *Palaeovertébr., Mém. Jubil. R. Lavocat*, pp. 37–78.

FAHLBUSCH, V. 1976. Report on the International Symposium on mammalian stratigraphy of the European Tertiary. *Newsl. Stratigr.* 5:160–167.

FEUGUEUR, L. 1963. L'Yprésien de Bassin de Paris. Essai de *monographie stratigraphique*. Carte Geol. Detail France, Mém. 568 pp.

FRAAS, E. 1904. Neue Zeuglodonten aus dem unteren Mitteleocänen von Mokattam bei Cairo. *Geol. Palaont. Abh.* 10:1–24.

FRANZEN, J.L. 1968. Revision der Gattung *Palaeotherium* Cuvier, 1804. Freiburg: Albert-Ludwigs Univ. *Naturwiss. Math. Fak.* Pp. 1–186.

FRANZEN, J.L. 1981. Das erste skelett einer Dichobunide (Mammalia, Artiodactyla), geborgen aus mitteleozänen Olschiefern der ''Grube Messel'' bei Darmstadt (Deutschland), S-Hessen. *Senckenberg. Lethaea* 61:299–353.

GABUNIA, L.K. 1978. Concerning the remains of Eocene mammals from the Obayla Formation of the Zaysan Depression. *Tr. Akad. Nauk. Gruz. SSR, Inst. Paleobiol.*, 7:15–28. (Translated by G. Shkurkin, 1979.)

GALLWITZ, H., and KRUTZSCH, W. 1953. Material zur biostratonomie der Geiseltal. *Nova Acta Leopold.* 16:71–126.

GARIMOND, S., REMY, J.A., and SUDRE, J. 1975. Nouvelles données sur le renouvellement des faunes de mammifères à l'Eocène supérieur, d'après les gisements de Fons (Gard). *Fr., Cent. Natl. Rech. Sci., Colloq. Int.*, 218:611–625.

GAZIN, C.L. 1961. New sciuravid rodents from the lower Eocene Knight Formation of western Wyoming. *Proc. Biol. Soc. Wash.* 74:193–194.

GAZIN, C.L. 1962. A further study of the Lower Eocene mammalian faunas of southwestern Wyoming. *Smithson. Misc. Collect.* 144:1–98.

GAZIN, C.L. 1965B. A study of the Early Tertiary condylarthran mammal, *Meniscotherium*. *Smithson. Misc. Collect.* 149:1–98.

GAZIN, C.L. 1965C. An endocranial cast of the Bridger Middle Eocene primate, *Smilodectes gracilis*. *Smithson. Misc. Collect.* 149:1–14.

GAZIN, C.L. 1976. Mammalian faunal zones of the Bridger Middle Eocene. *Smithson. Misc. Collect.* 153:1–90.

GINGERICH, P.D. 1976. Cranial anatomy and evolution of early Tertiary Plesiadapidae (Mammalia, Primates). *Univ. Mich., Mus. Paleontol. Pap. Paleontol.* 15:1–140.

GINGERICH, P.D. 1977. New species of Eocene primates and the phylogeny of European Adapidae. *Folia. Primatol.* 28:60–80.

GINGERICH, P.D. 1978. New Condylarthra (Mammalia) from the Paleocene and early Eocene of North America. *Univ. Mich., Mus. Paleontol. Contrib.* 25:1–9.

GINGERICH, P.D., ed. 1980. Early Cenozoic paleontology and stratigraphy of the Bighorn Basin, Wyoming. *Univ. Mich., Mus. Paleontol. Pap. Paleontol.* 24:1–146.

GINGERICH, P.D. 1981. Radiation of early Cenozoic Didymoconidae (Condylarthra, Mesonychia) in Asia, with a new genus from the early Eocene of western North America. *J. Mammal.* 62:526–538.

GINGERICH, P.D., and GUNNELL, G.F. 1979. Systematics and evolution of the genus *Esthonyx* (Mammalia, Tillodontia) in the Early Eocene of North America. *Univ. Mich., Mus. Paleontol. Contrib.* 25:125–153.

GINGERICH, P.D., and HASKINS, R.A. 1981. Dentition of early Eocene *Pelycodus jarrovii* (Mammalia, Primates) and the generic attribution of species formerly referred to *Pelycodus*. *Univ. Mich., Mus. Paleontol. Contrib.* 25:327–337.

GINGERICH, P.D., and RUSSELL, D.E. 1981. *Pakicetus inachus*, a new archaeocete (Mammalia, Cetacea) from the early-middle Eocene Kuldana Formation of Kohat (Pakistan). *Univ. Mich., Mus. Paleontol. Contrib.* 25:235–246.

GINGERICH, P.D., and SIMONS, E.L. 1977. Systematics, phylogeny, and evolution of Early Eocene Adapidae (Mammalia, Primates) in North America. *Univ. Mich., Mus. Paleontol. Contrib.* 24:245–279.

GINSBURG, L., MENNESSIER, G., and RUSSELL, D.E. 1967. Sur l'âge éocène inférieur des sables bleutés du Haut-Var et sur ses conséquences. *C. R. Soc. Géol. Fr.* 1967. Pp. 272–274.

GINSBURG, L., MONTENAT, C., and POMEROL, C. 1965. Découverte d'une faune de Mammifères terrestres dans les couches marines de l'Auversien (Bartonien inférieur) du Guépelle (Val d'Oise). *C. R. Acad. Sci.*, Paris, *Sér. D*, 260:3445–3446.

GINSBURG, L., ARQUES, J., DE BROIN, F., LE CALVEZ, Y., MOUTON, J., OBERT, D., PRIVE-GILL, C., and ROUCAN, J.-P. 1977. Découverte d'une faune de Mammifères dans le Lutétien supérieur de La Défense à Puteaux près de Paris (Hte-de-Seine). *C. R. Soc. Géol. Fr.* 6:311–313.

GODINOT, M. 1978A. Diagnoses de trois nouvelles espèces de mammifères du Sparnacien de Provence. *C. R. Soc. Géol. Fr.* 6:286–288.

GODINOT, M. 1978B. Un nouvel Adapidé (Primate) de l'Eocène inférieur de Provence. *C. R. Acad. Sci.*, Paris, 286:1869–1872.

GODINOT, M. 1981. Les Mammifères de Rians (Eocène inférieur, Provence). *Palaeovertébr.* 10:43–126.

GODINOT, M., DE BRUIJN, F., BUFFETAUT, E., RAGE, J.-C., and RUSSELL, D.E. 1978. Dormaal: une des plus anciennes faunes éocènes d'Europe. *C. R. Acad. Sci.*, Paris, 287:1273–1276.

GOLZ, D.J. 1976. Eocene Artiodactyla of Southern California. *Bull. Los Ang. Cty. Mus. Nat. Hist., Sci.*, 26:1–85.

GOLZ, D.J., LILLEGRAVEN, J.A. 1977. Summary of known occurrences of terrestrial vertebrates from Eocene strata of Southern California. *Univ. Wyo., Contrib. Geol.* 15:43–65.

GRANGER, W. 1910. Tertiary faunal horizons in the Wind River Basin, Wyoming, with descriptions of new Eocene mammals. *Bull. Am. Mus. Nat. Hist.* 28:235–251.

GRANGER, W. 1914. On the names of lower Eocene faunal horizons of Wyoming and New Mexico. *Bull. Am. Mus. Nat. Hist.* 33:201–207.

GRANGER, W., GREGORY, W.K., and COLBERT, E.H. 1937. In Matthew, W.D., 1937. Paleocene faunas of the San Juan Basin, New Mexico, pp. 361–372. *Trans. Am. Phil. Soc.* 30:1–510.

GURR, P.R. 1962. A new fish fauna from the Woolwich Bottom Bed (Sparnacian) of Herne Bay, Kent. *Proc. Geol. Assoc. (Lond.)* 73:419–447.

GUSTAFSON, E.P. 1979. Early Tertiary vertebrate faunas, Big Bend area, Trans-Pecos Texas. *Simidectes* (Mammalia, Insectivora). *Texas Mem. Mus., Pearce-Sellards Ser.*, 31:1–9.

GUTHRIE, D.A. 1966. A new species of dichobunid artiodactyl from the Early Eocene of Wyoming. *J. Mammal.* 47:487–490.

GUTHRIE, D.A. 1967A. *Paeneprolimnocyon*, a new genus of Early Eocene limnocyonid (Mammalia, Creodonta). *J. Paleontol.* 41:1285–1287.

GUTHRIE, D.A. 1967B. The mammalian fauna of the Lysite Member, Wind River Formation (Early Eocene) of Wyoming. *South. Calif. Acad. Sci. Mem.* 5:1–53.

GUTHRIE, D.A. 1971. The mammalian fauna of the Lost Cabin Member, Wind River Formation (Lower Eocene) of Wyoming. *Ann. Carnegie Mus.* 43:47–113.

HARDENBOL, J., and BERGGREN, W.A. 1978. A new Paleogene time scale. In *Contribution to the Geological Time Scale*, ed. G. V. Cohee, M. F. Glaesner, and H.O. Hedberg, pp. 213–234. Studies in Geology, No. 6. Tulsa: Am. Assoc. Pet. Geol.

HARRINGTON, H.S. 1956. Argentina. Handbook of South American Geology. *Geol. Soc. Am. Mem.* 65:131–165.

HARTENBERGER, J.-L. 1968. Les Pseudosciuridae (Rodentia) de l'Eocène moyen et le genre *Masillamys* Tobien. *C. R. Acad. Sci.*, Paris, 267:1817–1820.

HARTENBERGER, J.-L. 1969. Les Pseudosciuridae (Mammalia, Rodentia) de l'Eocène moyen de Bouxwiller, Egerkingen et Lissieu. *Palaeovertébr.* 3:28–61.

HARTENBERGER, J.-L. 1971. Contribution à l'étude des genres *Gliravus* et *Microparamys* (Rodentia) de l'Eocène d'Europe. *Palaeovertébr.* 4:97–135.

HARTENBERGER, J.-L. 1973A. Etude systématique des Theridomyoidea (Rodentia) de l'Eocène supérieur. *Soc. Géol. Fr. Mém.* 117:1–72.

HARTENBERGER, J.-L. 1973B. Les rongeurs de l'Eocène d'Europe. Leur évolution dans leur cadre biogéographique. *Bull. Mus. Natl. Hist. Nat.* 132:49–70.

HARTENBERGER, J.-L., SIGÉ, B., and SUDRE, J. 1974. La plus ancienne faune de mammifères du Quercy: Le Bretou. *Palaeovertébr.* 6:177–196.

HEISSIG, K. 1979. Die hypothetische Rolle Südosteuropas bei den Säugetierwanderungen im Eozän und Oligozän. *N. Jahrb. Geol. Palaont* 2:83–96.

HOFFSTETTER, R. 1970. *Colombitherium tolimense*, pyrothérien nouveau de la formation Gualanday (Colombie). *Ann. Paléont., Vertébrés*, 56:1–23.

HOOKER, J.J. 1977. The Creechbarrow Limestone—its biota and correlation. *Tert. Res.* 1:139–145.

HOOKER, J.J. 1979. Two new condylarths (Mammalia) from the early Eocene of southern England. *Bull. Br. Mus. (Nat. Hist.) Geol.* 32:43–56.

HOOKER, J.J. 1980. The succession of *Hyracotherium* (Perissodactyla, Mammalia) in the English early Eocene. *Bull. Br. Mus. (Nat. Hist.) Geol.* 33:101–114.

HUSSAIN, S.T., DE BRUIJN, H., and LEINDERS, J.M. 1978. Middle Eocene rodents from the Kala Chitta Range (Punjab, Pakistan). *Proc. Ned. Akad. Wet., Ser. B*, 81:74–112.

JAEGER, J.-J. 1970. Pantolestidae nouveaux (Mammalia, Insectivora) de l'Eocène moyen de Bouxwiller (Alsace). *Palaeovertébr.* 3:63–82.

JAEGER, J.-J. 1971. La faune de mammifères du Lutétien de Bouxwiller (Bas-Rhin) et sa contribution à l'élaboration de l'échelle des zones biochronologiques de l'Eocène européen. *Alsace-Lorraine, Bull. Serv. Carte Géol.* 24:2–3, 93–105.

JEHENNE, Y. 1969. Etude du gisement de Saint-Capraise d'Eymet en Dordogne. *Univ. Poitiers, Bull. Sci. Terre* 10:1–42.

JEPSEN, G.L. 1930B. New vertebrate fossils from the Lower Eocene of the Bighorn Basin, Wyoming. *Proc. Am. Philos. Soc.* 69:117–131.

JEPSEN, G.L. 1930C. Stratigraphy and paleontology of the Paleocene of Northeastern Park County, Wyoming. *Proc. Am. Philos. Soc.* 69:463–528.

JEPSEN, G.L. 1963A. Eocene vertebrates, coprolites, and plants in the Golden Valley Formation of Western North Dakota. *Bull. Geol. Soc. Am.* 74:673–684.

JEPSEN, G.L. 1970. Bat origins and evolution. In *Biology of Bats*, ed. W. Wimsatt, Vol. 1, pp. 1–64. New York and London: Academic Press.

KAY, J.L. 1957. *The Eocene vertebrates of the Uinta Basin, Utah.* Guidebook, Intermountain Assoc. Pet. Geol., 8th Ann. Field Conf. pp. 110–112.

KEEFER, W.R. 1965. Stratigraphy and geologic history of the uppermost Cretaceous, Paleocene, and Lower Eocene rocks in the Wind River Basin, Wyoming. *U.S. Geol. Surv. Prof. Pap.* 495A:1–77.

KELLOGG, R. 1936. A review of the Archaeoceti. *Carnegie Inst. Wash. Publ.* 482:1–366.

KING, C. 1981. The stratigraphy of the London Clay and associated deposits. *Tert. Res., Spec. Pap.,* 6:1–158.

KITTS, D.B. 1956. American *Hyracotherium* (Perissodactyla, Equidae). *Bull. Am. Mus. Nat. Hist.* 110:1–60.

KOENIG, K.J. 1960. *Bridger Formation in the Bridger Basin, Wyoming.* Guidebook, Wyo. Geol. Assoc., 15th Ann. Field Conf. Pp. 163–168.

KOENIGSWALD, G.H.R. VON. 1967. An Upper Eocene mammal of the family Anthracotheriidae from the Isle of Timor, Indonesia. *Proc. Ned. Akad. Wet., Ser. B,* 70:529–533.

KOENIGSWALD, W. VON. 1979. Ein Lemurenrest aus dem eozänen Ölschiefer der Grube Messel bei Darmstadt. *Palaeont. Z.* 53:63–76.

KOENIGSWALD, W. VON. 1980A. Die Fossillagerstätte Messel. Literaturübersicht der Forschungsergebnisse aus den Jahren 1969–1979. *Geol. Jb. Hessen* 108:23–38.

KOENIGSWALD W. VON. 1980B. Das Skelett eines Pantolestiden (Proteutheria, Mammalia) aus dem mitteleren Eozän von Messel bei Darmstadt. *Palaeont. Z.* 54:267–287.

KOVACS E. 1957. Tropischer Farn aus dem Eozän in Ungarn. *Budapest, Univ. Sci., Ann. Rolando Eotvos* 1:185–187.

KRISHTALKA, L. 1976. Early Tertiary Adapisoricidae and Erinaceidae (Mammalia, Insectivora) of North America. *Ann. Carnegie Mus.* 47:4–40.

KRISHTALKA, L. 1978. Paleontology and geology of the Badwater Creek area, Central Wyoming. Part 15. Review of the late Eocene primates from Wyoming and Utah, and the Plesitarsiiformes. *Ann. Carnegie Mus.* 47:335–360.

KRISHTALKA, L. 1979. Paleontology and geology of the Badwater Creek area, Central Wyoming. Part 18. Revision of Late Eocene *Hyopsodus. Ann. Carnegie Mus.* 48:377–389.

KRISHTALKA, L., and BLACK, C.C. 1975. Paleontology and geology of the Badwater Creek area, Central Wyoming. Part 12. Description and review of Late Eocene Multituberculata from Wyoming and Montana. *Ann. Carnegie Mus.* 45:287–297.

KRISHTALKA, L., and SETOGUCHI, T. 1977. Paleontology and geology of the Badwater Creek area, Central Wyoming. Part 13. The late Eocene Insectivora and Dermoptera. *Ann. Carnegie Mus.* 46:71–99.

KRISHTALKA, L., and WEST, R.M. 1979. Paleontology and geology of the Bridger Formation, southern Green River Basin, southwestern Wyoming. Part 4. The Geolabididae (Mammalia, Insectivora). *Milw. Public Mus., Contrib. Biol. Geol.* 27:1–10.

KRYSTOFOVICH, A.N. 1935. A final link between the Tertiary floras of Asia and Europe. *New Phytol.* 34:339–344.

KÜHNE, W.G. 1969. A multituberculate from the Eocene of the London Basin. *Proc. Geol. Soc. Lond.* 1658:199–202.

KURTÉN, B. 1966. Holarctic land connexions in the early Tertiary. *Biol. Soc. Sci. Fenn., Comment.,* 29:1–5.

LANGE-BADRÉ, B. 1979. Les Créodontes (Mammalia) d'Europe occidentale de l'Eocène supérieur à l'Oligocène moyen. *Bull. Mus. Natl. Hist. Nat.* (Paris), *Sér. C,* 42:1–249.

LANGE-BADRÉ, B. 1980. Les Créodontes (Mammalia) de Bouxwiller (Bas-Rhin). *Ann. Paléontol.* 67:21–35.

LAPPARENT, A. DE. 1893. *Traité de Géologie,* 3d ed.

LI, C. 1963A. Paramyid and sciuravids from N. China. *Vertebr. PalAsiatica* 6:72–77.

LI, C. 1975. A new ischyromyoid rodent genus from the Upper Eocene of North China. *Vertebr. PalAsiatica* 13:58–70.

LI, C., CHIU, C., YAN, D., and HSIEH, S. 1979. Notes on some early Eocene mammalian fossils of Hengtung, Hunan. *Vertebr. PalAsiatica* 17:71–82.

LILLEGRAVEN, J.A. 1972. Ordinal and familial diversity of Cenozoic mammals. *Taxon* 21:261–274.

LILLEGRAVEN, J.A. 1979. A biogeographical problem involving comparisons of late Eocene terrestrial vertebrate faunas of western North America. In *Historical Biogeography, Plate Tectonics, and the Changing Environment,* ed. J. Gray and A.J. Boucot, pp. 333–347. Corvallis: Oregon St. Univ. Press.

LILLEGRAVEN, J.A. 1980. Primates from later Eocene rocks of Southern California. *J. Mammal,* 61:181–204.

LILLEGRAVEN, J.A., McKENNA, M.C., and KRISHTALKA, L. 1981. Evolutionary relationships of middle Eocene and younger species of *Centetodon* (Mammalia, Insectivora, Geolabididae) with a description of the dentition of *Ankylodon* (Adapisoricidae). *Univ. Wyo., Publ.* 45:1–115.

LILLEGRAVEN, J.A., and WILSON, R.W. 1975. Analysis of *Simimys simplex,* an Eocene rodent (Zapodidae?). *J. Paleontol.* 49:856–874.

LOUIS, P. 1964. Gisements nouveaux de Mammifères d'âge Eocène inférieur dans les environs d'Epernay. *Rev. Féd. Franc. Soc. Sci. Nat.* (Soc. Etude Sci. Nat., Reims), (3) 3:87–94.

LOUIS, P. 1966. Note sur un nouveau gisement situé à Condé-en-Brie (Aisne) et renfermant des restes de Mammifères de l'Eocène inférieur. *Univ. Reims, Ann. A.R.E.R.S.,* 4:108–118.

LOUIS, P. 1970. Note préliminaire sur un gisement de mammifères de l'Eocène inférieur situé route de Broyes à Sézanne (Marne). *Univ. Reims, Ann. A.R.E.R.S.,* 8:48–62.

LOUIS, P. 1976. Gisements de Mammifères bartoniens du Tardenois. Remarques sur la paléogéographie de l'Auversien du sud de Fismes. *Bull. Inform. Géologues, Bassin de Paris,* 13:41–58.

LOUIS, P. and MICHAUX, J. 1962. Présence de Mammifères sparnaciens dans les sablières de Pourcy (Marne). *C.R. Soc. Géol. Fr.* 1962:170–171.

LOUIS, P., and SUDRE, J. 1975. Nouvelles données sur les primates de l'Eocène supérieur européen. *Cent. Natl. Rech. Sci., Colloq. Internat.,* 218:805–828.

LYELL, C. 1833. Principles of Geology, Vol. 3.

LYELL, C. 1863. Antiquity of Man.

MACGINITIE, H.D. 1969. The Eocene Green River flora of northwestern Colorado and northeastern Utah. *Univ. Calif. Publ. Geol. Sci.* 83:1–103.

MARSHALL, L.G., PASCUAL, R., CURTIS, G.H., and DRAKE, R.E. 1977. South American geochronology: Radiometric time scale for middle to late Tertiary mammal-bearing horizons in Patagonia. *Science* 195:1325–1328.

MATTHES, H.W., and THALER, B. 1977. Eozän Wirbeltiere des

Geiseltales. *Martin-Luther Univ., Wissensch. Beit.*, 1977/2 (P5).

MATTHEW, W.D. 1909. Carnivora and Insectivora of the Bridger Basin. *Am. Mus. Nat. Hist. Mem.* 9:291–567.

MATTHEW, W.D., and GRANGER, W. 1924B. New Carnivora from the Tertiary of Mongolia. *Am. Mus. Novit.* 105:1–9.

MATTHEW, W.D., and GRANGER, W. 1925F. The smaller perissodactyls of the Irdin Manha formation, Eocene of Mongolia. *Am. Mus. Novit.* 199:1–9.

McGREW, P.O. 1959. The geology and paleontology of the Elk Mountain and Tabernacle Butte area, Wyoming. *Bull. Am. Mus. Nat. Hist.* 117:117–176.

McGREW, P.O., and Sullivan, R. 1970. The stratigraphy and paleontology of Bridger A. *Univ. Wyo., Contrib. Geol.*, 9:66–85.

McKENNA, M.C. 1960. Fossil Mammalia from the early Wasatchian Four Mile fauna, Eocene of northwest Colorado. *Univ. Calif. Publ. Geol. Sci.* 37:1–130.

McKENNA, M.C. 1975A. Fossil mammals and early Eocene North Atlantic land continuity. *Ann. Mo. Bot. Gard.* 62:335–353.

McKENNA, M.C. 1975B. Toward a phylogenetic classification of the Mammalia. In W.P. Luckett and F.S. Szalay, *Phylogeny of the Primates*, pp. 21–46. New York and London: Plenum Press.

McKENNA, M.C., ROBINSON, P., and TAYLOR, D.W. 1962. Notes on Eocene Mammalia and Mollusca from Tabernacle Butte, Wyoming. *Am. Mus. Novit.* 2102:1–33.

McKENNA, M.C., and SIMPSON, G.G. 1959. A new insectivore from the Middle Eocene of Tabernacle Butte, Wyoming. *Am. Mus. Novit.* 1952:1–12.

MELLETT, J.S., and SZALAY, F.S. 1968. *Kennatherium shirensis* (Mammalia, Palaeoryctoidea), a new didymoconid from the Eocene of Asia. *Am. Mus. Novit.* 2442:1–7.

MICHAUX, J. 1964. Diagnoses de quelques Paramyidés de l'Eocène inférieur de France. *C.R. Soc. Géol. Fr.* 4:153–154.

MICHAUX, J. 1968. Les Paramyidae (Rodentia) de l'Eocène inférieur du Bassin de Paris. *Palaeovertébr.* 1:136–193.

MORRIS, W.J. 1954. An Eocene fauna from the Cathedral Bluffs tongue of the Washakie Basin, Wyoming. *J. Paleontol.* 28:195–203.

NELSON, M.E., MADSEN, J.H., JR., and STOKES, W.L. 1980. A titanothere from the Green River Formation, central Utah: *Teleodus uintensis. Univ. Wyo. Contrib. Geol.* 18:127–134.

NOVACEK, M.J. 1976. Insectivora and Proteutheria of the Later Eocene (Uintan) of San Diego County, California. *Los Ang. Cty. Mus. Nat. Hist., Contrib.* 283:1–52.

NOVACEK, M.J. 1977. A review of Paleocene and Eocene Leptictidae (Eutheria: Mammalia) from North America. *PaleoBios* 24:1–42.

NOVODVORSKAJA, I.M., and JANOVSKAJA, N.M. 1977. Intercontinental distribution of mammals and their importance for paleobiogeography of Asia Eocene. *J. Palaeontol. Soc. India* 10:69–76.

OSBORN, H.F. 1895A. Fossil mammals of the Uinta Basin. *Bull. Am. Mus. Nat. Hist.* 7:71–105.

OSBORN, H.F. 1924E. *Andrewsarchus,* giant mesonychid of Mongolia. *Am. Mus. Novit.* 146:1–5.

OSBORN, H.F. 1925E. Upper Eocene and Lower Oligocene titanotheres from Mongolia. *Am. Mus. Novit.* 202:1–12.

OSBORN, H.F., and GRANGER, W. 1932A. Coryphodonts and uintatheres from the Mongolian expedition of 1930. *Am. Mus. Novit.* 552:1–16.

PASCUAL, R. 1965. Un nuevo Condylarthra (Mammalia) de edad Casamayorense de Paso de Los Indios (Chubut, Argentina).

Breves consideraciones sobre le edad Casamayorense. *Ameghiniana* 4:57–65.

PASCUAL, R. 1981. Adiciones al conocimiento de *Bonapartherium hinakusijum* (Marsupialia, Bonapartheriidae) del Eoceno temprano del noroeste Argentina. *Anais II Congr. Latin-Amer. Paleontol., Porte Alegre.* Pp. 507–520.

PATTERSON, B. 1977. A primitive pyrothere (Mammalia, Notoungulata) from the early Tertiary of northwestern Venezuela. *Fieldiana, Geol.* 33:397–422.

PATTERSON, B., and PASCUAL, R. 1968 (and 1972). Evolution of mammals on Southern Continents. 5. The fossil mammal fauna of South America. *Rev. Biol.* 43:409–451.

PETERSON, O.A. 1934B. List of species and description of new material from the Duchesne River Oligocene, Uinta Basin, Utah. *Ann. Carnegie Mus.* 23:373–389.

PFLUG, H. 1952. Palynologie und Stratigraphie der eozänen Braunkohlen von Helmstedt. *Palaeontol. Z.* 26:112–137.

QI, T. 1980. Irdin Manha Upper Eocene and its mammalian fauna at Huhebolhe Cliff in central Inner Mongolia. *Vertebr. PalAsiatica* 18:28–32.

QUINET, G.E. 1964. Morphologie dentaire des Mammifères éocènes de Dormaal. *Bull. Group. Int. Rech. Sci. Stomatol.* 7:272–294.

QUINET, G.E. 1966A. Sur la formule dentaire de deux primates du Landenien continental belge. *Bull. Inst. R. Sci. Nat. Belg.* 42:1–6.

QUINET, G.E. 1966C. Les Mammifères du Landenien continental belge. Vol. 2. *Inst. R. Sci. Nat. Belg. Mém.* 158:1–64.

QUINET, G.E. 1969. Apport de l'étude de la faune mammalienne de Dormaal. *Inst. R. Sci. Nat. Belg. Mém.* 162:1–188.

RADINSKY, L.B. 1963. Origin and early evolution of North American Tapiroidea. *Bull. Peabody Mus. Nat. Hist.* (Yale Univ.) 17:1–106.

RADINSKY, L.B. 1964A. Notes on Eocene and Oligocene fossil localities in Inner Mongolia. *Am. Mus. Novit.* 2180:1–11.

RADINSKY, L.B. 1964B. *Paleomoropus,* a new Early Eocene chalicothere (Mammalia, Perissodactyla), and a revision of Eocene chalicotheres. *Am. Mus. Novit.* 2179:1–28.

RADINSKY, L.B. 1965A. Early Tertiary Tapiroidea of Asia. *Bull. Am. Mus. Nat. Hist.* 129:181–264.

RADINSKY, L.B. 1967A. *Hyrachyus, Chasmotherium,* and the early evolution of helaletid tapiroids. *Am. Mus. Novit.* 2213:1–23.

RADINSKY, L.B. 1967B. A review of the rhinocerotoid family Hyracodontidae (Perissodactyla). *Bull. Am. Mus. Nat. Hist.* 136:1–46.

RADULESCO, C., ILIESCO, G., and ILIESCO, M. 1976. Un Embrithopode nouveau (Mammalia) dans le Paléogène de la dépression de Hateg (Roumanie) et la gólogie de la région. *Monatsh. N. Jhb. Geol. Palaeontol.* 11:290–298.

RAO, M.R. 1971. New mammals from Murree (Kalakot zone) of the Foot Hills near Kalakot, Jammu and Kashmir State, India. *J. Geol. Soc. India* 12:125–134.

RAO, M.R. 1972. Further studies on the vertebrate fauna of Kalakot, India, new mammalian genera and species from the Kalakot zone of Himalayan foothills near Kalakot, Jammu and Kashmir State, India. *Direct. Geol. Oil Nat. Gas Comm. Spec. Pap.* 1:1–22.

RAO, M.R. 1973. Notice of two new mammals from Upper Eocene

Kalakot beds, India. *Direct. Geol. Oil Nat. Gas Comm. Spec. Pap.* 2:1–6.

RASKY K. 1956. Fossil plant remains from the Lower Eocene of Transdanubia (Hungary). *Foeldt. Koezl.* 86:291–294.

RAT, P. 1965. La succession stratigraphique des Mammifères dans l'Eocène du Bassin de Paris. *Bull. Soc. Géol. Fr.* (7) 7:248–256.

REID, E.M., and CHANDLER, M.E.J. 1933. Flora of the London Clay. Br. Mus. (Nat. Hist.), Publ. London., 561 pp.

REMY, J.-A. 1972. Etude du crâne de *Pachynolophus lavocati*, n. sp. (Perissodactyla, Palaeotheriidae) des Phosphorites du Quercy. *Palaeovertébr.* 5:1–78.

RICH, T.H.V. 1971. Deltatheridia, Carnivora, and Condylarthra (Mammalia) of the Early Eocene, Paris Basin, France. *Univ. Calif. Publ. Geol. Sci.* 88:1–72.

ROBINSON, P. 1957A. Age of the Galisteo Formation, Santa Fe County, New Mexico. *Bull. Am. Assoc. Pet. Geol.* 41:757.

ROBINSON, P. 1957B. The species of *Notharctus* from the Middle Eocene. *Postilla* 28:1–26.

ROBINSON, P. 1963. Fossil vertebrates and age of the Cuchara Formation of Colorado. *Univ. Colo. Ser. Geol. Studies* 1:1–5.

ROBINSON, P. 1966. Fossil Mammalia of the Huerfano Formation, Eocene, of Colorado. *Bull. Peabody Mus. Nat. Hist.* (Yale Univ.) 21:1–95.

ROBINSON, P. 1968A. Nyctitheriidae (Mammalia, Insectivora) from the Bridger Formation of Wyoming. *Univ. Wyo. Contrib. Geol.* 7:129–138.

ROBINSON, P. 1968B. *Talpavus* and *Entomolestes* (Insectivora, Adapisoricidae). *Am. Mus. Novit.* 2339:1–7.

ROBINSON, P., BLACK, C.C., and DAWSON, M.R. 1964. Late Eocene multituberculates and other mammals from Wyoming. *Science* 145:809–811.

ROSE, K.D. 1972. A new tillodont from the Eocene upper Willwood Formation of Wyoming. *Postilla* 155:1–13.

ROSE, K.D., BOWN, T.M., and SIMONS, E. 1977. An unusual new mammal from the Early Eocene of Wyoming. *Postilla* 172:1–10. (Generic name changed in *J. Paleontol.* 52:1162, Sept. 1978.)

ROUCH, T.D. 1979. Character and paleogeographic distribution of Upper Cretaceous(?) and Paleogene nonmarine sedimentary rocks in east-central Nevada. In *Cenozoic Paleogeography of the Western United States*. Publ. of Soc. Econ. Paleontol. and Mineral, 1979.

RUSSELL, D.E. 1968. Succession, en Europe, des faunes mammaliennes au début du Tertiaire. Colloque sur l'Eocène. *Bur. Rech. Géol. Minières* 58:291–296.

RUSSELL, D.E., BONÉ, E., DE BROIN, F., BRUNET, M., BUFFETAUT, E., CORDY, J.M., CROCHET, J.-Y., DINEUR, H., ESTES, R., GINSBURG, L., GODINOT, M., GROESSENS, M.C., GIGASE, P., HARRISON, C.J.O., HARTENBERGER, J.-L., HOCH, E., HOOKER, J.J., INSOLE, A.N., LANGE-BADRÉ, B., LOUIS, P., MOODY, R., RAGE, J.C., REMY, J., ROTHAUSEN, K., SIGÉ, B., SIGOGNEAU-RUSSELL, D., SPRINGHORN, R., SUDRE, J., TOBIEN, H., VIANEY-LIAUD, M., VINKEN, R., and WALKER, C.A. 1982. *Tetrapods of the Tertiary Basin of Northwest Europe. Geol. Jb., Abt.A.,* 60:5–77.

RUSSELL, D.E., HARTENBERGER, J.-L., POMEROL, C., SEN, S., SCHMIDT-KITTLER, N., and VIANEY-LIAUD, M. 1982 The Paleontology of Europe: Mammals and stratigraphy. *Palaeovertebrata*, 1–77.

RUSSELL, D.E., and GINGERICH, P.D. 1980. Un nouveau Primate omomyidé dans l'Eocène du Pakistan. *C.R. Acad. Sci., Paris, Sér. D.,* 291:621–624.

RUSSELL, D.E., and GINGERICH, P.D. 1981. Lipotyphla, Proteutheria(?), and Chiroptera (Mammalia) from the early-middle Eocene Kuldana Formation of Kohat (Pakistan). *Univ. Mich., Mus. Paleontol., Contrib.* 25:277–287.

RUSSELL, D.E., GODINOT, M., LOUIS, P., and SAVAGE, D.E. 1979. Apatotheria (Mammalia) de l'Eocène inférieur de France et de Belgique. *Bull. Mus. Natl. Hist. Nat.* (Paris) (4) 1:203–243.

RUSSELL, D.E., LOUIS, P., and SAVAGE, D.E. 1973. Chiroptera and Dermoptera of the French Early Eocene. *Univ. Calif. Publ. Geol. Sci.* 95:1–57.

RUSSELL, D.E., LOUIS, P., and SAVAGE, D.E. 1975. Les Adapisoricidae de l'Eocène inférieur de France. Réévaluation des formes considérées affines. *Bull. Mus. Natl. Hist. Nat.* (Paris) (3) 327:129–192.

SAHNI, A. 1979. An Eocene mammal from the Subathu-Dagshai transition zone, Dharampur, Simla Hills. *Bull. Geol. Assoc. India* 12:259–262.

SAHNI, A. 1980. Eocene vertebrate faunas from the Salt Range (Pakistan), Kashmir and Simla Himalayas and China. *11th Himal. Geol. Sem., Paleontol. and Biostratigr. of Himalayas, W.I.H.G.* (abstract). Pp. 23–24.

SAHNI, A., and KHARE, S.K. 1971. Three new Eocene mammals from Rajauri district, Jammu and Kashmir. *J. Paleontol. Soc. India* 16:41–53.

SAHNI, A., and KHARE, S.K. 1972. Additional Eocene mammals from the Subathu Formation of Jammu and Kashmir. *J. Paleontol. Soc. India* 17:31–49.

SAHNI, A., and KUMAR, V. 1974. Palaeogene palaeobiogeography of the Indian subcontinent. *Palaeogeogr., Palaeoclim., Palaeoecol.* 13:209–226.

SAHNI, A., and KUMAR, V. 1980. Lower Eocene mammal, *Ishatherium subathuensis* gen. et sp. nov. from the type area Subathu Formation, Subathu, Simla Himalayas, H.P. *J. Paleontol. Soc. India* 23–24:132–135.

SAHNI, A., and MISHRA, V.P. 1975. Lower Tertiary vertebrates from western India. *Paleontol. Soc. India Monogr.* 3:1–48.

SAHNI, A., and SRIVASTAVA, M.C. 1976. Eocene rodents and associated reptiles from the Subathu Formation of northwest India. *Proc. North Am. Paleontol. Conv.* Pp. 992–998.

SAHNI, A., and SRIVASTAVA, M.C. 1977. Eocene rodents of India: Their palaeobiogeographic significance. *Géobios, Mém. Spéc.,* 1:87–97.

SAVAGE, D.E. 1977. Aspects of vertebrate paleontological stratigraphy and geochronology. In *Concepts and Methods of Biostratigraphy,* ed. E. Kauffman and J. Hazel, pp. 427–442. Stroudsburg, Pa.: Dowden, Hutchinson & Ross.

SAVAGE, D.E., RUSSELL, D.E., and LOUIS, P. 1965. European Eocene Equidae (Perissodactyla). *Univ. Calif. Publ. Geol. Sci.* 56:1–94.

SAVAGE, D.E., RUSSELL, D.E., and LOUIS, P. 1966. Ceratomorpha and Ancylopoda from the Lower Eocene, Paris Basin, France. *Univ. Calif. Publ. Geol. Sci.* 66:1–38.

SAVAGE, D.E., and WATERS, B.T. 1978. A new omomyid primate from the Wasatch Formation of southern Wyoming. *Folia Primatol.* 30:1–29.

SAVAGE, R.J.G. 1969. Early Tertiary mammal locality in southern Libya. *Proc. Geol. Soc. Lond.* 1648:98–101.

SCHANKLER, D.M. 1980. Faunal zonation of the Willwood Formation in the Central Bighorn Basin, Wyoming. *Univ. Mich. Pap. Paleontol.* 24:99–114.

SCHMIDT-KITTLER, N. 1971B. Eine unteroligozäne Primatenfauna von Ehrenstein bei Ulm. *Mitt. Bayer. Staatssamml. Paleontol. Hist. Geol.* 11:171–204.

SCHMIDT-KITTLER, N. 1971C. Odontologische untersuchunden an Pseudosciuriden (Rodentia, Mammalia) des Alttertiärs. *Bayer. Akad. Wiss. Math.-Naturw. Kl., N.F., Abh.,* 150:1–130.

SCHOCH, R.M. and LUCAS, S.G. 1981. The systematics of *Stylinodon,* an Eocene taeniodont (Mammalia). *J. Vertebr. Paleontol.* 1:175–183.

SCOTT, W.B. 1945. The Mammalia of the Duchesne River Oligocene. *Trans. Am. Philos. Soc.* 34:209–253.

SEN, S., and HEINTZ, E. 1979. *Palaeoamasia kansui* Ozansoy, 1966, Embrithopode (Mammalia) de l'Eocène d'Anatolie. *Ann. Paléontol. (Vertébrés)* 65:73–91.

SETOGUCHI, T. 1975. Paleontology and geology of the Badwater Creek area, Central Wyoming. Part 11. Late Eocene marsupials. *Ann. Carnegie Mus.* 45:263–275.

SIGÉ, B. 1974B. *Pseudorhynchocyon cayluxi* Filhol, 1892: Insectivore géant des Phosphorites du Quercy. *Palaeovertébr.* 6:33–46.

SIGÉ, B. 1975B. Insectivores primitifs de l'Eocène supérieur et Oligocène inférieur d'Europe occidentale: Apatemyidés et Leptictidés. *Colloq. Internat., Cent. Natl. Rech. Sci.* 218:653–673.

SIGÉ, B. 1976. Insectivores primitifs de l'Eocène supérieur et Oligocène inférieur d'Europe occidentale: Nyctitheriidés. *Mus. Natl. Hist. Nat.* (Paris), *Mém., Sér. C.* 34:1–140.

SIGÉ, B. 1978. La poche à phosphate de Ste.-Néboule (Lot) et sa faune de vertébrés du Ludien supérieur. 8. Insectivores et Chiroptères. *Palaeovertébr.* 8:243–268.

SIGÉ, B., CROCHET, J.-Y., and INSOLE, A.N. 1977. Les plus vieilles taupes. *Géobios, Mém. Spéc.,* 1:141–157.

SIMPSON, G.G. 1933Q. Stratigraphic nomenclature of the Early Tertiary of central Patagonia. *Am. Mus. Novit.* 644:1–13.

SIMPSON, G.C. 1940G. Review of the mammal-bearing Tertiary of South America. *Proc. Am. Philos. Soc.* 83:649–709.

SIMPSON, G.G. 1947. Holarctic mammalian faunas and continental relationships during the Cenozoic. *Bull. Geol. Soc. Am.* 58:613–688.

SIMPSON, G.G. 1948C. The Eocene of the San Juan Basin, New Mexico. Parts 1 and 2. *Am. J. Sci.* 246:165–385.

SIMPSON, G.G. 1948E. The beginning of the age of mammals in South America. *Bull. Am. Mus. Nat. Hist.* 91:1–232.

SIMPSON, G.G. 1967A. The beginning of the age of mammals in South America. Part 2. *Bull. Am. Mus. Nat. Hist.* 137:1–260.

SIMPSON, G.G. 1969. South American mammals. Biogeography and Ecology in South America, 1969. Vol. 2. *Monographiae Biologicae.* Pp. 879–909.

SIMPSON, G.G., MINOPRIO, J.L., and PATTERSON, B. 1962. The mammalian fauna of the Divisadero Largo Formation, Mendoza, Argentina. *Bull. Mus. Comp. Zool.* (Harv. Univ.) 127:239–293.

SINCLAIR, W.J., and GRANGER, W. 1911. Eocene and Oligocene of the Wind River and Bighorn Basins. *Bull. Am. Mus. Nat. Hist.* 30:83–117.

SPRINGHORN, R. 1980. *Paroodectes feisti,* der erste Miacide (Carnivora, Mammalia) aus dem Mittel-Eozän von Messel. *Palaeontol. Z.* 54:171–191.

STORCH, G. 1978. *Eomanis waldi,* ein Schuppentier aus dem Mittel-Eozän der "Grube Messel" bei Darmstadt (Mammalia: Pholidota). *Senckenb. Lethaea* 59:503–529.

STORCH, G. 1981. *Eurotamandua jovesi,* ein Myrmecophagide aus dem Eozän de "Grube Messel" bei Darmstadt (Mammalia: Xenarthra). *Senckenb. Lethaea* 61:247–289.

STORER, J.E. 1978. Rodents of the Swift Current Creek local fauna. Eocene (Uintan) of Saskatchewan. *Can. J. Earth Sci.* 15:1673–1674.

SUDRE, J. 1969. Les gisements de Robiac (Eocène supérieur) et leurs faunes de Mammifères. *Palaeovertébr.* 2:95–156.

SUDRE, J. 1972. Revision des Artiodactyls de l'Eocène moyen de Lissieu (Rhône). *Palaeovertébr.* 5:111–156.

SUDRE, J. 1975. Un prosimien du palaéogène ancient du Sahara nord-occidental: *Azibius trerki,* n.g., n. sp. *C.R. Acad. Sci., Paris,* 280:1539–1542.

SUDRE, J. 1977. L'évolution du genre *Robiacina* Sudre, 1969, et l'origine de Cainotheriidae; implications systematique. *Géobios, Mém. Spéc.,* 1:213–231.

SUDRE, J. 1978A. La poche à phosphate de Ste.-Néboule (Lot) et sa faune de vertébrés du Ludien supérieur. 9. Primates et Artiodactyles. *Palaeovertébr.* 8:269–290.

SUDRE, J. 1978B. Les artiodactyles de l'Eocène moyen et supérieur d'Europe occidentale. Systematique et évolution. Montpellier, *Ecole Pratique Hautes Etudes, Mém.* 7:1–229.

SUDRE, J. 1979. Nouveaux Mammifères éocènes du Sahara occidental. *Palaeovertébr.* 9:83–115.

SULLIVAN, R. 1980. A stratigraphic evaluation of the Eocene rocks of southwestern Wyoming. *Wyo., Geol. Surv., Rep. Invest.* 20:1–50.

SZAFER, W. 1958. Nowa flora Eocenska. In *Tatrach. Kwart. Geol.,* Warsaw 2:173–176.

SZALAY, F.S. 1969A. Mixodectidae, Microsyopidae, and the insectivore-primate transition. *Bull. Am. Mus. Nat. Hist.* 140:193–330.

SZALAY, F.S. 1969B. The Hapalodectinae and a phylogeny of the Mesonychidae (Mammalia, Condylarthra). *Am. Mus. Novit.* 2361:1–26.

SZALAY, F.S. 1969C. Uintasoricinae, a new subfamily of Early Tertiary mammals (?Primates). *Am. Mus. Novit.* 2363:1–36.

SZALAY, F.S. 1976. Systematics of the Omomyidae (Tarsiiformes, Primates) Taxonomy, phylogeny and adaptations. *Bull. Am. Mus. Nat. Hist.* 156:159–449.

SZALAY, F.S., and McKENNA, M.C. 1971. Beginning of the age of mammals in Asia: The late Paleocene Gashato fauna, Mongolia. *Bull. Am. Mus. Nat. Hist.* 144:269–318.

TAKAI, F. 1950. *Amynodon watanabei* from the latest Eocene of Japan with a brief summary of the latest Eocene mammalian faunule in eastern Asia. *Geol. Surv. Japan Rept.* 131:1–14.

THALER, L. 1965. Une échelle de zones biochronologiques pour les Mammifères du Tertiaire d'Europe. *C.R. Soc. Géol. Fr.* 1965:118.

TOBIEN, H. 1962. Insectivoren (Mammalia) aus dem Mitteleozän (Lutetium) von Messel bei Darmstadt. *Notizbl. hess. Landessamte Bodenf.* 90:7–47.

TOBIEN, H. 1969. *Kopidodon* (Condylarthra, Mammalia) aus dem

Mitteleozän (Lutetium) von Messel bei Darmstadt (Hessen). *Notizbl. hess. Landessamte Bodenf.* 97:7–37.

TOBIEN, H. 1980A. Ein anthracotherioider Paarhufer (Artiodactyla, Mammalia) aus dem Eozän von Messel bei Darmstadt (Hessen). *Geol. Jhb. Hessen* 108:11–22.

TOBIEN, H. 1980B. Taxonomic status of some Cenozoic mammalian local faunas from the Mainz Basin. *Mainzer Geowiss. Mitt.* 9:203–235.

TONG, Y. and WANG, J. 1980. Subdivision of the Upper Cretaceous and Lower Tertiary of the Tantou Basin, the Lushi Basin, and the Lingbao Basin of Western Honan. *Vertebr. PalAsiatica* 18:21–27.

TOURTELOT, H.A. 1946. *Tertiary stratigraphy in the northeastern part of the Wind River Basin, Wyoming.* U.S. Geol. Surv., Oil Gas Invest., Chart OC-22.

TROFIMOV, B.A. 1952. Concerning the genus *Pseudictops*, a unique insectivore from the lower Tertiary deposits of Mongolia. *Tr. Akad. Nauk. SSSR, Paleontol. Inst.* 41:7–12. (Translated by G. Shkurkin, 1977.)

TROFIMOV, B.A. 1953. Ancient Tertiary mammals in the Far East of the USSR. *Priroda* (1953):12:111–112. (Translated by G. Shkurkin, 1978.)

VAN HOUTEN, F.B. 1945. Review of the latest Paleocene and early Eocene mammalian faunas. *J. Paleontol.* 19:421–461.

VAN VALEN, L. 1965A. Paroxyclaenidae, an extinct family of Eurasian mammals. *J. Mammal.* 46:388–397.

VAN VALEN, L. 1965B. Some European Proviverrini (Mammalia, Deltatheridia). *Palaeontol.* 8:638–665.

VAN VALEN, L. 1966. Deltatheridia, a new order of Mammals. *Bull. Am. Mus. Nat. Hist.* 132:1–126.

VAN VALEN, L. 1967. New Paleocene insectivores and insectivore classification. *Bull. Am. Mus. Nat. Hist.* 135:217–284.

VIANEY-LIAUD, M. 1977. Nouveaux Theridomyinae du Paléogène d'Europe occidentale. *C.R. Acad. Sci. Paris* 284:1277–1280.

WAHLERT, J.H. 1973. *Protoptychus*, a hystricomorphous rodent from the Late Eocene of North America. *Breviora* 419:1–14.

WEBB, S.D., and TAYLOR, D.E. 1980. The phylogeny of hornless ruminants and a description of the cranium of *Archaeomeryx*. *Bull. Am. Mus. Nat. Hist.* 167:121–157.

WEST, R.M. 1970. Sequence of mammalian faunas of Eocene age in the northern Green River Basin, Wyoming. *J. Paleontol.* 44:142–147.

WEST, R.M. 1973A. An early Middle Eocene epoicotheriid (Mammalia) from southwestern Wyoming. *J. Paleontol.* 47:929–931.

WEST, R.M. 1973B. Review of the North American Eocene and Oligocene Apatemyidae (Mammalia: Insectivora). *Texas Tech. Univ., Mus., Spec. Publ.*, 3:1–42.

WEST, R.M. 1974. New North American Middle Eocene nyctithere (Mammalia, Insectivora). *J. Paleontol.* 48:983–987.

WEST, R.M. 1976A. Paleontology and geology of the Bridger Formation, southern Green River Basin, southwestern Wyoming. Part 1. History of field work and geologic setting. *Milw. Public Mus., Contrib. Biol. Geol.* 7:1–12.

WEST, R.M. 1976B. The North American Phenacodontidae (Mammalia, Condylarthra). *Milw. Public Mus., Contrib. Biol. Geol.* 6:1–78.

WEST, R.M. 1979. Apparent prolonged evolutionary stasis in the Middle Eocene hoofed mammal *Hyopsodus*. *Paleobiology* (Paleontological Society) 5:252–260.

WEST, R.M. 1980. Middle Eocene large mammal assemblage with Tethyan affinities, Ganda Kas Region, Pakistan. *J. Paleontol.* 54:508–533.

WEST, R.M., and ATKINS, E.G. 1970. Additional Middle Eocene (Bridgerian) mammals from Tabernacle Butte, Sublette County, Wyoming. *Am. Mus. Novit.* 2404:1–26.

WEST, R.M., and LUKACS, J.R. 1979. Geology and vertebrate-fossil localities, Tertiary continental rocks, Kala Chitta Hills, Attock District, Pakistan. *Milw. Public Mus., Contrib. Biol. Geol.* 26:1–20.

WEST, R.M., MCKENNA, M.C., BLACK, C.C., BOWN, T.M., DAWSON, M.R., GOLZ, D.J., LILLEGRAVEN, J.A., SAVAGE, D.E., and TURNBULL, W.D. 1983? Eocene chronology of North America. In *Vertebrate Paleontology as a Discipline in Geochronology: Examples and Problems*, ed. M.O. Woodburne. Calif., Univ., Publ. Geol. Sci. (In preparation.)

WHEELER, W.H. 1961. Revision of the uintatheres. *Bull. Peabody Mus. Nat. Hist.* (Yale Univ.) 14:1–93.

WHITE, R.E. 1952. Preliminary analysis of the vertebrate fossil fauna of the Boysen Reservoir area. *Proc. U.S. Natl. Mus.* 102:185–207.

WILSON, J.A. 1974. Early Tertiary vertebrate faunas, Vieja Group and Buck Hill Group, Trans-Pecos Texas: Protoceratidae, Camelidae, Hypertragulidae. *Bull. Texas Mem. Mus.* 23:1–34.

WILSON, J.A. 1977. Stratigraphic occurrence and correlation of Early Tertiary vertebrate faunas, Trans-Pecos Texas. *Bull. Texas Mem. Mus.* 25:1–42.

WILSON, J.A., and SCHIEBOUT, J.A. 1981. Early Tertiary vertebrate faunas, Trans-Pecos Texas: Amynodontidae. *Texas Mem. Mus., Pearce-Sellards Ser.* 33:1–62.

WILSON, J.A., TWISS, P.C., DEFORD, R.K., and CLABAUGH, S.E. 1968. Stratigraphic succession, potassium-argon dates, and vertebrate faunas. Vieja Group, Rim Rock country, Trans-Pecos Texas. *Am. J. Sci.* 266:590–604.

WOLFE, J.A. 1971. Tertiary climatic fluctuations and methods of analysis of Tertiary floras. *Palaeogeogr., Palaeoclim., Palaeoecol.* 9:27–57.

WOLFE, J.A. 1978. A paleobotanical interpretation of Tertiary climates in the Northern Hemisphere. *Am. Sci.* 66:694–703.

WOLFE, J.A., and HOPKINS, D. 1967. Climatic changes recorded by Tertiary land floras in northwestern North America. In *Tertiary Correlations and Climatic Changes in the Pacific*, ed. K. Hatai, 25:67–76. Symp., 11th Pacif. Sci. Congr., Tokyo.

WOOD, A.E. 1962. The Early Tertiary rodents of the family Paramyidae. *Trans. Am. Philos. Soc.* 52:1–261.

WOOD, A.E. 1965B. Small rodents from the early Eocene Lysite Member, Wind River formation of Wyoming. *J. Paleontol.* 37:124–134.

WOOD, A.E. 1973. Eocene rodents, Pruett Formation, southwest Texas; their pertinence to the origin of the South American Caviomorpha. *Texas Mem. Mus., Pearce-Sellards Ser.* 20:1–40.

WOOD, H.E. 1934. Revision of the Hyrachyidae. *Bull. Am. Mus. Nat. Hist.* 67:181–295.

WOOD, H.E., CHANEY, R.W., CLARK, J., COLBERT, E.H., JEPSEN, G.L., REESIDE, J.B., JR., and STOCK, C. 1941. Nomenclature and correlation of the North American continental Tertiary. *Bull. Geol. Soc. Am.* 52:1–48.

WORTMAN, J.L. 1901. Studies of Eocene Mammalia in the Marsh

collection, Peabody Museum. Part 1. Carnivora. *Am. J. Sci.* (4) 11:333–348.

XU, Y. 1965. A new genus of amynodont from the Eocene of Lantian, Shensi. *Vertebr. PalAsiatica* 9:83–86.

XU, Y. 1966. Amynodonts of Inner Mongolia. *Vertebr. PalAsiatica* 10:123–190.

XU, Y., and CHIU, C. 1962. Early Tertiary mammalian fossils from Lunan, Yunnan. *Vertebr. PalAsiatica* 6:313–332.

XU, Y., YAN, D., ZHOU, S., HAN, S., and ZHANG, Y. 1979. The study of the fossil mammalian localities and the subdivision of the period of the Red Beds of the Liguanquiao Basin. In *Mesozoic and Cenozoic Red Beds of South China,* pp. 416–432. Acad. Sinica, I.V.P.P. and Nanjing Inst. Geol. and Paleont. Beijing, China: Science Press.

YANOVSKAYA, N.M. 1980. The brontotheres of Mongolia. *Trans. Joint Sov.-Mongol. Paleont. Exped.* 12:1–219. (In Russian.)

ZDANSKY, O. 1930. Die alttertiären Säugetiere Chinas nebst Stratigraphischen Bemerkungen. Acad. Sin., Inst. Vertebr. Palaeontol. Palaeoanthropol, *Paleontol. Sinica* (C) 2:1–87.

ZHAI, R. 1978A. Two new Early Eocene mammals from Sinkiang with remarks on the age of Gashato Formation. *Inst. Vert. Paleontol. Paleoanthropol.* (Acad. Sin.), *Mem.,* 13:102–106.

ZHAI, R. 1978B. Description of some Late Eocene mammals from Lian-kan Formation of Turfan Basin, Sinkiang. *Inst. Vert. Paleontol. Paleoanthropol.* (Acad. Sin.), *Mem.,* 13:116–125.

ZHAI, R., ZHENG, J., and TONG, Y. 1978. Stratigraphy of the mammal-bearing Tertiary of the Turfan Basin, Sinkiang. *Inst. Vert. Paleontol. Paleoanthropol.* (Acad. Sin.), *Mem.,* 13:68–81.

ZHANG, Y., YOU, Y., JI, H., and DING, S. 1978. Cenozoic stratigraphy of Yunnan. *Prof. Papers Strat. Paleontol.* 7:1–21. Beijing: Geol. Publ. House.

ZHENG, J., and CHI, H. 1978. Some latest Eocene Condylarthra mammals from Gaungxi, South China. *Vertebr. PalAsiatica* 16:97–102.

ZHENG, J., TUNG, V., and CHI, H. 1975. Discovery of Miacidae (Carnivora) in Yuanchui Basin, Kiangsi Province. *Vertebr. PalAsiatica* 13:96–104.

CHAPTER FIVE

Oligocene Mammalian Faunas

Mammals About 35 to 24 Million Years Ago

TYPIFICATION OF THE OLIGOCENE

In 1855, H. E. von Beyrich proposed the name Oligocene for Tertiary beds in the Mainz Basin of western Germany, in middle and northern Germany, and in corresponding deposits in Belgium. In the Mainz Basin, these beds had been assigned previously to either Upper Eocene or Lower Miocene. The Belgian units had been included in the Tongrian Stage (Dumont, 1839. *Bull. Acad. R. Belg., Cl. Sci.* 6:473–479) and in the Rupelian Stage (Dumont, 1849. *Bull. Acad. R. Belg., Cl. Sci.* 16:367.) Beyrich assigned a suite of characterizing strata and localities of varying age to his Oligocene. Arranged in the accepted sequence Beyrich proposed (as listed by Wilmarth, 1924), these were:

North Germany: Sands, in which occur the "mussel rocks" of Sternberg, ... Loamy sand rich in mussel shells known from borings near Crefeld.

Middle Germany: Upper yellow marine sands near Cassel, Güntersen, Luithorst, Alfeld, Hildeshelm, Osnabrück, Bünde.

Belgium: Upper rupelian system. . . .

North Germany: Septaria loam of Walle near Celle; . . . .

Belgium: Lower rupelian system. . . .

North Germany: Glauconitic loamy sands and sandy loams above the lignite near Westeregeln, Biere, Calbe a. S., Osterwiddinger. [These districts were later made the "stratotype" of

the Lattorfian Stage by Mayer-Eymar, 1893, and subsequent papers.]

Belgium: Upper tongrian system. Brackish-water deposits.

Middle Germany: Fresh-water deposits with lignite.

Mainz Basin: Entire succession of brackish-water and fresh-water beds (*Cyrene* marl, limestone containing land mollusks, snails, and *Cerithia*; *Littorinella* limestone, lignite beds, sandstone bearing fossil leaves, bone sand of Eppelsheim).

Belgium: Lower tongrian system.

North Germany: Lignite-bearing Tertiary beds of the lowlands of northeast Germany.

In citing this sequence from Wilmarth's translation, we have left out many of the strata and localities originally given; this was to emphasize that in addition to fossiliferous marine units, many brackish and freshwater deposits with their invertebrates, plants, and vertebrates were included in Beyrich's Oligocene. Unfortunately, his typification was too vague and too heterogeneous. Many of his units have been shown to correlate with either Eocene or Miocene in other places. A conspicuous error was the bone sand of Eppelsheim, which carries a late Miocene *Hipparion* fauna.

Dispute continues over the designation of uppermost or lowermost Oligocene in the various depositional basins of Europe.[1] These disputes lead to a series-epoch boundary controversy in other parts of the world. In each area there are multiple problems. See the colloquia on the Eocene and the Paleogene (*Fr., Bur. Rech. Geol. Minières* 28 in 1964, 59 in 1968, and 69 in 1969; Berggren, 1971).

Land mammals and other vertebrates have been found in many of the strata and districts originally assigned to the Oligocene in Germany and Belgium. For example, Misonne (1957, 1958) listed land mammals from Hoogbutsel and Hoeleden, Belgium (type area of the base of the upper Tongrian) that would now be called *Theridomys, Eucricetodon, Steneofiber, Peridyromys, Oxacron, Dichobune, Tapirulus, Acotherulum,* and *Eggysodon.* Fossil mammals and birds have been collected from the Boom Clay at Rupelboom, Belgium—part of the type Rupelian Stage: tapirs, rhinoceroses, sirenians, gulls, shore birds, ducks, and "geese." Lower Middle and Upper Oligocene mammals are known also from the Oligocene strata of the Mainz Basin (Tobien, 1956B, 1960D, and personal communication).

LA GRANDE COUPURE AND THE EOCENE-OLIGOCENE BOUNDARY

Stehlin (1909) made a strong written plea to the geologists and stratigraphers of Western Europe to accept a boundary between the Eocene and the Oligocene based on a great phenomenon of "the first order." This phenomenon was the striking and apparently sudden change in the faunas of ungulates (artiodactyls and perissodactyls) resulting from extermination of many Eocene genera and species as new families and genera arrived on the scene, evidently from the east. This event he dubbed *La Grande Coupure,* and he asserted that in the mammalian population of Europe it was the greatest and most sudden change known during all of Tertiary time.

The taxa that were immigrant arrivals at this time, according to Stehlin, were *Cadurcotherium* (amynodont rhinocerotoid), *Hyracodon?* and other rhinocerotoids, titanotheres?, chalicotheriids (perhaps arriving slightly later), *Entelodon, Ancodus* (an anthracothere now best called *Bothriodon*—Simpson, 1945), *Anthracotherium, Brachyodus* (anthracothere) and palaeochoerids (tayassuids). La Grande Coupure is recognized as a fact by mammalian paleontologists to this day, although Stehlin's data have been modified and supplemented by later information (Lopez and Thaler, 1974; Brunet, 1977); and Sigé, 1976). This later information indicates, as would be expected, that not all the Eocene ungulates became extinct at the same time and not all the immigrants arrived at the same time. The big issue that remains is usually not the reality of the Coupure but how long it endured and whether or not it should mark the base of the Oligocene (but see Sigé and Vianey-Liaud, 1979).

We choose to follow the school of workers who place the base of the Oligocene at La Grande Coupure, which is the approximate base of the Stampian Stage as French stratigraphers now recognize it. This is apparently the approximate base of the upper substage of the Tongrian Stage of Belgium. This boundary probably is later than most of the Lattorfian Stage-Age of northwest Germany (certain lithologic units of which were earlier listed in Beyrich's Oligocene) as it is identified and calibrated by Hardenbol and Berggren (1978) and as it is used by Tobien (1980). We are aware that other workers, for example Lopez and Thaler (1974), currently use the Grande Coupure as the marker between early and middle Oligocene.[1]

ZONATIONS OF NONMARINE OLIGOCENE IN EUROPE

The great abundance of fossil rodents in the Cenozoic localities of Western Europe has led many paleontologists, especially in France, Germany, and Holland, to become specialists on this order of mammals. Concomitantly, these specialists have proposed detailed zonations of their Cenozoic succession, basing their zonations mostly on the rodents. Recognition of subdivisions of this sort in Europe stem from the early, nonrodent work and conclusions of Stehlin (1909) or even earlier work; see the summary review by Hartenberger (1969).

For the Oligocene, Thaler (1965A) proposed six principal subdivisions, which he termed biochronologic zones. He concluded that these zones were discrete and sequential as listed, and he gave subzones "for the fresh-water molasse of Switzerland" in the upper three zones. Each of his zones or subzones for the Oligocene is named from a type locality but is defined and recognized by the presence of a particular species (or subspecies?, or forma?) of rodent. All but one of his species belong to the Theridomyidae in the sense of Hartenberger (1973A). Each taxon, presumably, is restricted to its zone or subzone. Thaler asserted further: "They are therefore units of the same nature as the biochron (duration of existence of a taxon), ..." Below is his zonation.

15. Coderet
    15.2 Kuttigen: *"Cricetodon" collatum* A, *Archaeomys laurillardi* B
    15.1 *Rickenbach: Archaeomys laurillardi* A
14. Cournon
    14.2 Aarwangen: *Blainvillimys geminatus* C
    14.1 Boningen: *Blainvillimys geminatus* B
13. Antoignt
    13.2 Oensingen: *Blainvillimys geminatus* A
    13.1 Mumliswyl: *Blainvillimys blainvillei*
12. La Sauvetat: *Blainvillimys gregarius*
11. Ronzon: *Isoptychus aquatilis*
10. Montmartre: *Isoptychus auberyi*

(Note that Thaler did not list the species actually found at Coderet, Cournon, and Antoignt.)

Hugueney (1969), Vianey-Liaud (1972, 1974) and others have used these "biochronologic zones" and have added embellishments. In 1972, Thaler added to his paleontologic-chronologic scheme for the European Tertiary, recognizing principal faunizones separated by breaks (coupures). Thus, his Oligocene was divided into an *Anthracotherium* faunizone, bounded below by the Great Oligocene Break (La Grande Coupure), separated from the later, *Amphilagus* faunizone by the Chattian Break. His "old scale of biochronologic zones" was modified and presented under the name of *scale of biozones* (pp. 418–419), as shown in the accompanying chart.

BIOZONES	Number of Subzones	Correlation of the type-levels (niveaux-types) of the biozones
16. Paulhiac (Aquitaine)	1	Chattian? Aquitanian?
15. Rickenbach (Switz.)	1	Chattian?
14. Boningen (Switz.)	3	Chattian?
13. Antoignt (Auvergne)	1	Chattian? Stampian?
12. Montalban (Spain)	1	Stampian?
11. Hoogbutzel (Belgium)	1	Tongrian? Stampian?
10. La Débruge (Provence)	1	Lattorfian? Tongrian?
9. Euzet (Languedoc)	1	Bartonian? Lattorfian?

Franzen (1968) and Hartenberger (1969) proposed and Jaeger and Hartenberger (1975) further explained the *niveau repère* (standard or reference or datum level). This is a "datum stage of evolution," a "zone ponctuelle" (Hartenberger, p. 57). It "est un instantané de l'histoire des faunes" (Jaeger and Hartenberger, 1975). As with Thaler's biozones-biochrons, the *niveau repère* is named after a type locality (e.g., Ronzon) where the index species (e.g., *Isoptychus aquatilis*) is found. Thus the *niveau repère* is proclaimed to be a phyletic-evolutionary pinpoint in the geochronologic scale, signified by a species name. Crochet (1979) has used the niveau repère exhaustively. We have previously criticized this concept (Savage and Russell, 1977) and have opined that the niveau repère, like the paleoclassical *index taxon* or *guide fossil,* fancies and asserts too much refinement for the paleontological discipline in stratigraphy and geochronology. Niveaux repères can be useful locally, however, as has been shown in Western Europe.

Fahlbusch (1976), representing the conclusions formulated in four days of review and discussion by an International Symposium on Mammalian Stratigraphy of the European Tertiary in Munich, April 11–14, 1975, published Mammal Ages (resp. Stages) and "Superstages" for the general Oligocene interval (accompanying chart). The Fahlbusch symposium did not give paleontological characterization for the proposed mammal ages but stated (p. 166): "In their paleontological content these higher units (mammal ages) are defined by that part of phylogenetic evolution

of mammals which took place during the units represented by reference localities belonging to it by definition." Fahlbusch has used these mammal ages extensively in his 1979 publication on the Eomyidae.

	Definition		Interpretation
Reference Localities	Mammal Ages (resp. Stages)	"Superstages"	
Coderet La Milloque Boningen Antoignt	ARVERNIUM	OCCITANIUM	OLIGOCENE
Heimersheim Montalban Villebramar Hoogbutsel	SUEVIUM		
		Grande Coupure	

EARLY OLIGOCENE OF EUROPE

In the category of Early Oligocene we here include two sets of faunas that Vianey-Liaud (1971B) recognized as the Hoogbutsel Level (earlier) and the Villebramar Level (later). This category also includes the Ronzon, or Hoogbutsel, biozone of Thaler (1965A, 1972) and the Ronzon "Zone" = lowest Middle Oligocene of Schmidt-Kittler (1971B). Equivalent terms, more or less, are *Early Suevian* and *Earliest Occitanian* (Fahlbusch, 1976). The species from Hoogbutsel, Soumailles, Ronzon, Sainte Marthe, Aubrelong 1, Ravet, Mazan, Fontaines-de-Vaucluse, Hamstead Beds, Möhren 13, Renheim 1, Villebramar, Mas d'Agenais, Roqueprune 2, Mas de Got, La Plante 2, Schelkingen 1, and Lobsann provide the list for the composite fauna of the Early Oligocene of Europe.

Composite Early Oligocene Mammalian Fauna of Europe

MARSUPIALIA
  Didelphidae
    Peratherium elegans (Aymard, 1846)
    P. lavergnense Crochet, 1979
    Amphiperatherium minutum (Aymard, 1846)
    A. ambiguum (Filhol, 1877)
    A. exile (Gervais, 1848–1852)
    A. frequens (Meyer, 1846)
    A. lamandini (Filhol, 1876)
LAGOMORPHA
    Shamolagus franconicus (Heissig and Schmidt-Kittler, 1975)
PANTOLESTA
    Cryptopithecus siderolithicus Schlosser, 1890
    C. alcimonensis Heissig, 1977
    Dysperna woodi Hopwood, 1927

CREODONTA
 Hyaenodontidae
  Hyaenodon leptorhynchus Lezair and Parieu, 1838
  H. dubius Filhol, 1873
  H. gervaisi Martin, 1906
  H. cayluxi (Filhol, 1876)   (age?)
  H. filholi Schlosser, 1887   (age?)
  Apterodon gaudryi Fischer, 1880
  Consobrinus quercyi Lange-Badré, 1979   (age?)
  Paenoxyaenoides liguritor Lange-Badré, 1979   (age?)
  Paracynohyaenodon schlosseri Martin, 1906   (age?)
CARNIVORA
 Amphicyonidae
  Cynodictis intermedius Filhol, 1876   (age?)
  C. longirostris Filhol, 1876   (age?)
  "C." palmidens (Filhol, 1882)
  Cynelos crassidens (Filhol, 1874)
  C. piveteaui Ginsburg, 1965
  Harpagophagus sanguineus Bonis, 1971
  Pseudamphicyon lupinus Schlosser, 1888
  Pseudocyonopsis antiquus Ginsburg, 1966
  P. landesquei (Helbing, 1928C)
  P. ambiguus (Filhol, 1876)
  P. quercensis Ginsburg, 1966
  Sarcocyon ferox Ginsburg, 1966
  Brachycyon gaudryi Filhol, 1872
  Haplocyon robustus Springhorn, 1977
  H. elegans Bonis, 1966B
  Amphicyon prior Springhorn, 1977   (age?)
  Amphicyanis felinoides Springhorn, 1977   (age?)
  Symplectocyon praecursor (Schlosser, 1902)
 Ursidae
  Amphicynodon cf. *speciosus* (Filhol, 1882)
  A. typicus (Schlosser, 1888)
  A. leptorhynchus (Filhol, 1877)
  A. velaunus (Aymard, 1851)
  A. gracilis (Filhol, 1877)
  Cephalogale cadurcensis (Blainville, 1839)   (age?)
  C. robusta (Filhol, 1876)   (age?)
 Inc. Sed.
  Adracon quercyi Filhol, 1884
 Procyonidae
  Plesictis crassirostris (Filhol, 1876)   (age?)
  P. major Teilhard, 1915   (age?)
  P. pygmaeus Schlosser, 1889
  P. filholi (Depéret, 1887)
  P. stenogalinus Teilhard, 1915   (age?)
  P. aff. *stenoplesictis* Teilhard, 1915   (age?)
 Mustelidae
  Mustelictis piveteaui Lange-Badré, 1969A
  Stenogale gracilis (Gervais, 1868)
  S. intermedia (Filhol, 1876)
  S. julieni (Filhol, 1879)
 Viverridae
  Stenoplesictis minor Filhol, 1882
  Palaeoprionodon lamandini (Filhol, 1880)
  P. minutus Schlosser, 1889
  P. simplex Filhol, 1883

  Anictis simplicidens (Schlosser, 1888)
 Felidae
  Eofelis edwardsi (Filhol, 1872)
  Eusmilus bidentatus (Filhol, 1873)
  Quercylurus major Ginsburg, 1979   (=*Nimravus?*)
  Nimravus intermedius (Filhol, 1872)
  N. bonali (Helbing, 1922)
  Proailurus lemanensis Filhol, 1879
  P. medius Filhol, 1879
INSECTIVORA
 Erinaceoidea
  Tetracus nanus (Aymard, 1846)
  Exodaenodus schaubi Hurzeler, 1944   (age?)
 Soricoidea
  Darbonetus aubrelongensis Crochet, 1974
  Saturninia cf. *gracilis* Stehlin, 1940
  Amphisorex primaevus Filhol, 1884   (age?)
  Myxomygale antiquua Filhol, 1890   (age?)
  Butselia biveri Quinet and Misonne, 1965
  Quercysorex primaevus (Filhol, 1833)
  Q. herrlingensis (Palmowski and Wachendorf, 1966)
CHIROPTERA
 Emballonuridae
  Vespertiliavus bourguignati (Filhol, 1876)   (age?)
  V. gracilis Revilliod, 1920
  V. schlosseri Revilliod, 1920   (age?)
  V. wingei Revilliod, 1920   (age?)
 Hipposideridae
  Palaeophyllophora oltina (Delfortrie, 1872)
  P. quercyi Revilliod, 1917
  Hipposideros dubius (Weithofer, 1887)   (age?)
  H. heliophagus (Weithofer, 1887)   (age?)
  H. morloti (Pictet, 1855)
  H. schlosseri (Revilliod, 1917)   (age?)
  H. weithoferi (Revilliod, 1917)   (age?)
  Paraphyllophora robusta Revilliod, 1922   (age?)
  "Rhinolophus" antiquus (Filhol, 1872)
 Megadermatidae
  Necromantis adichaster Weithofer, 1887   (age?)
  N. grandis Revilliod, 1920   (age?)
  N. planifrons Revilliod, 1920   (age?)
 Rhinolophidae
  Rhinolophus priscus Revilliod, 1920
  R. pumilio Revilliod, 1920   (age?)
  "R." antiquus (Filhol, 1872)   (age?)
 Vespertilionidae
  "Myotis" misonnei (Quinet, 1965)
  Stehlinia minor (Revilliod, 1922)
 Inc. Sed.
  Archaeopteropus transiens Meschinelli, 1902
PRIMATES
 Tarsiidae
  Microchoerus ornatus Stehlin, 1916
  (*Adapis, Anchomomys, Pronycticebus, Protoadapis, Necro-
   lemur,* and *Pseudoloris* are reported from various Quercy
   sites, but their age is most likely Headonian.)
ARTIODACTYLA
 Dichobunidae
  Dichobune leporina Cuvier, 1822
  D. jehennei Brunet and Sudre, 1980
  D. cf. *fraasi* Schlosser, 1902

Mouillacitherium sp.
Metriotherium paulum Stehlin, 1906 (age?)
M. minutum Brunet and Sudre, 1980
Cebochoeridae
Cebochoerus quercyi Stehlin, 1908 (age?)
Dacrytheriidae
Tapirulus hyracinus Gervais, 1850
Entelodontidae
Entelodon antiquum Répelin, 1919
E. magnus Aymard, 1846
E. deguilhemi Repelin, 1918
Tayassuidae
Palaeochoerus gergovianus (Croizet, in Blainville, 1846)
P. paronae Dal Piaz, 1930B
Anoplotheriidae
Diplobune minor (Filhol, 1877)
D. quercyi (Filhol, 1877)
D. bavarica Fraas, 1870
D. primaevus (Filhol, 1873) (age?)
Anthracotheriidae
Elomeryx woodi (Cooper, 1926)
E. porcinus (Gervais, 1859)
Anthracotherium alsaticum Cuvier, 1834
A. bumbachense Stehlin, 1910
A. monsvialense Zigno, 1887
A. calmatinum Meyer, 1854
Bothriodon velaunum (Pomel, 1852)
B. aymardi (Pomel, 1852)
B. leptorhynchus Aymard, 1846
Anthracochoerus stehlini Dal Piaz, 1931
A. fabianii Dal Piaz, 1931
Cainotheriidae
Caenomeryx procommunis (Filhol, 1877)
Plesiomeryx cadurcensis Gervais, 1873
Oxacron courtoisi (Gervais, 1848)
Leptomerycidae
Bachitherium insigne (Filhol, 1877)
B. minus Filhol, 1880
Lophiomeryx chalaniati Pomel, 1853
Prodremotherium elongatum Filhol, 1877
Amphimerycidae
Amphimeryx riparius (Aymard, 1855)
"Pseudamphimeryx" decedens Stehlin, 1910 (age?)
Gelocidae
Gelocus communis (Aymard, 1848)
Gelocus? curtus Filhol, 1876
G. villebramarensis Brunet and Jehenne, 1976
Inc. Sed.
Tragulohyus inermis Gervais, 1874
PERISSODACTYLA
Palaeotheriidae
Palaeotherium curtum Cuvier, 1812
P. medium Cuvier, 1804
P. castrense Noulet, 1863
P. muehlbergi Stehlin, 1904
Plagiolophus fraasi Meyer, 1852
P. minor (Cuvier, 1804)
Pseudopalaeotherium longirostrum Franzen, 1972
Brontotheriidae
Menodus? rumelicum Toula, 1892 (Kameno, Bulgaria)

Amynodontidae
Cadurcotherium cayluxi Gervais, 1873
C. minus Filhol, 1879–1880
Hyracodontidae
Eggysodon gaudryi Rames, 1884
E. osborni (Schlosser, 1902)
E. hoeledenensis (Misonne, 1957)
Rhinocerotidae
Ronzotherium filholi (Osborn, 1900)
R. romani Kretzoi, 1940
R. velaunum (Aymard, 1853)
Trigonias osbornii Dal Piaz, 1930C
Chalicotheriidae
Limognitherium ingens Filhol, 1880
Schizotherium modicum Gervais, 1876
PHOLIDOTA
Manidae
Necromanis quercyi Filhol, 1894 (age?)
RODENTIA
Theridomyidae
Microsuevosciurus palustris (Misonne, 1957)
Suevosciurus fraasi (Major, 1873)
S. ehingensis Dehm, 1937A
Theridomys bonduelli Lartet, 1869
T. aquatilis Aymard, 1849
Archaeomys (Blainvillimys) gregarius Schlosser, 1884
A. (B.) langei Vianey-Liaud, 1972A
A. (Taeniodus) avus Stehlin and Schaub, 1951
Elfomys medius Vianey-Liaud, 1976
Pseudoltinomys gaillardi Lavocat, 1951
P. major Vianey-Liaud, 1976
Pseudosciurus suevicus Hensel, 1856
Sciuromys cayluxi Schlosser, 1884
S. quercyi (Stehlin and Schaub, 1951)
Tarnomys quercinus Hartenberger and Schmidt-Kittler, 1976
Eomyidae
Eomys antiquus (Aymard, 1853)
Aplodontidae
Plesispermophilus angustidens Filhol, 1882
P. atavus Schmidt-Kittler and Vianey-Liaud, 1979
Sciurodon cadurcensis Schlosser, 1884
Paracitellus cingulatus Heissig, 1979
P. marmoreus Heissig, 1979
Oligopetes radialis Heissig, 1979
O. lophulus Heissig, 1979
O. obtusus Heissig, 1979
Trigonomys simplex Heissig, 1979
Sciuridae
Heteroxerus cf. *paulhiacensis* Black, 1965B
H.? lavocati Hugueney, 1969A
Sciurus cayluxi Filhol, 1884 (age?)
S. schlosseri Freudenberg, 1941 (age?)
S. dubius Schlosser, 1884 (age?)
S. aff. *giganteus* Freudenberg, 1941
Palaeosciurus goti Vianey-Liaud, 1974
P.? feignouxi (Pomel, 1853)
Castoridae
Steneofiber butselensis Misonne, 1957

Gliridae
    Gliravus aff. *priscus* Stehlin and Schaub, 1951
    G. majori Stehlin and Schaub, 1951
    G. fordi Bosma and de Bruijn, 1979
    Peridyromys micio Misonne, 1957
Cricetidae
    Eucricetodon atavus (Misonne, 1957)
    Paracricetodon dehmi Hrubesch, 1957B
    Pseudocricetodon montalbanensis Thaler, 1969A

Melissiodon aff. *schaubi* Dehm, 1935A
Inc. Sed.
    Quercymys quercinus Thaler, 1966B   (age?)

    Of the 114 early Oligocene genera in the European mammalian fauna, 39 (34 percent) are holdovers from the late Eocene Headonian fauna. Sixty-five of these early Oligocene genera (57 percent) continue into the middle Oligocene. About six families that appear in the early Oligocene of Europe appear to have no later representatives there: Pantolestidae, Tarsiidae, Cebochoeridae, Dacrytheriidae, Amphimerycidae, and Brontotheriidae.

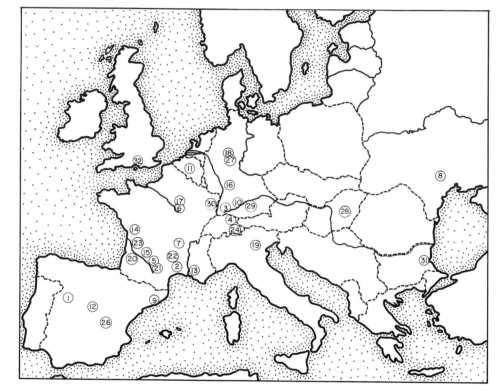

FIGURE 5-1 *Early Oligocene Localities of Europe.*

1 Aldealengua—Spain
1 Aldearrubia—Spain
2 Alès—France
3 Allstadt bei Messkirch—Germany
4 Arnegg 2—Switzerland
2 Aubrelong 1—France
2 Avejan—France
5 Belgarite IVa—France
5 Briatexte—France
6 Brie, Calcaire de, sites—France
7 Briennon, Argiles de, sites—France
8 Bugajewka (Izyum)—Ukraine, USSR
1 Cabrerizos—Spain
9 Calaf—Spain
3 Ehingen (in part)—Germany
3 Ehrenstein 1 (B)—Germany
3 Eselsberg bei Ulm—Germany
32 Hamstead—England

10 Heidenheim—Germany
11 Hoeleden—Belgium
11 Hoogbutsel—Belgium
12 Huermeces de Cerro—Spain
13 Isle-sur-Sorgues—France
14 Lagny-Thorigny—France
15 Lamandine (Quercy)—France (age?)
30 Lobsann—France
15 Malemort (in part)—France
15 Mas de Labat—France
19 Monteviale—Italy (age?)
15 Mouillac (in part)—France
16 Neustadt (Marburg)—Germany
17 Noisy-le-Sec—France (age?)
18 Nordshausen bei Kassel—Germany (age?)
20 Pinchenat—France
6 Pouquette—France
21 Puy-Laurens—France
15 Quercy district (many locali-

ties, in part)—France
15 Ravet (in part)—France
22 Ronzon—France
23 Ruch—France
26 St. Cugat de Gavadons (Cuenca de Moia)—Spain
24 St. Margrethen—Switzerland
25 St. Swithin's—England
27 Sieblos an der Wasserkuppe (Rhon)—Germany
20 Soumailles—France
28 Szendehey (marine)—Romania?
29 Treuchlingen—Germany (age?)
3 Unterer Eselsberg—Germany (age?)
17 Villejuif—France (age?)
10 Weidenstetten—Germany
10 Weissenburg 2, 3, 8, 9—Germany

Mammal faunas assigned to the Early Oligocene in Asia are found mostly in North China and in Mongolia. The composite mammalian aggregate from these regions, following Matthew and Granger (1925B), Belyayeva et al. (1974), Tang and Qiu (1979), Q. Xu (1977), Trofimov (1952B, 1956), Jiang, Wang, and Qi (1976) is:

Early Oligocene Mammalian Fauna of Asia

ANAGALIDA
Anagale?
Ardynictis furunculus Matthew and Granger, 1925B
LAGOMORPHA
Procaprolagus vetustus (Burke, 1941)　　—Gureev (1960)
Desmatolagus robustus Matthew and Granger, 1923
PANTODONTA
Coryphodontidae
Hypercoryphodon thomsoni Osborn and Granger, 1932A
CREODONTA
Hyaenodontidae
Hyaenodon emineus Matthew and Granger, 1925B
H. dubius Filhol, 1873
Pterodon mongoliensis (Dashzeveg, 1964)
CARNIVORA
Amphicyonidae
Cynodictis?
Canidae?
Ursidae
Amphicticeps
Amphicynodon (= ''*Pachycynodon*'') sp.
Felidae
Aelurogale mongoliensis Gromova, 1958
machairodontine　　—Tang and Qiu (1979)
Hoplophoneus　　—Tang and Qiu (1979)
ARTIODACTYLA
Tayassuidae, new genera　　—Tang and Qiu (1979)
Entelodontidae
Entelodon orientalis Dashzeveg, 1965
E. gobiensis (Trofimov, 1952B)
E. dirus Matthew and Granger, 1923
Eoentelodon trofimovi Dashzeveg, 1976
Archaeotherium
Choeropotamidae, n. gen.　　—Tang and Qiu (1979)
Anthracotheriidae
Probrachyodus sp.
Bothriodon tientongensis Qiu, 1977
B. chowi Y. Xu, 1961
Anthracokeryx gungkangensis Qiu, 1977
A. kwangsiensis Qiu, 1977
Brachyodus hui (Chow, 1958)
Heothema chengbiensis Tang, 1978
H. bellia Tang, 1978
Huananothema imparilica Tang, 1978
Leptomerycidae?
Miomeryx altaicus Matthew and Granger, 1925E
Hypertragulidae
Archaeomeryx
Lophiomeryx angarae Matthew and Granger, 1925E
L. gobiae Matthew and Granger, 1925E

Tragulidae?
Gobiomeryx dubius Trofimov, 1956
Bovidae?, gen. indet.　　—Jiang, Wang, and Qi (1976)
ACREODI
Mesonychidae
Mongolestes
Harpagolestes alxaensis Qi, 1975
PERISSODACTYLA
Brontotheriidae
Protitan? cingulatus Granger and Gregory, 1943
Parabrontops gobiensis Osborn, 1925E
Metatitan primus Granger and Gregory, 1943
M. progressus Granger and Gregory, 1943
M. relictus Granger and Gregory, 1943
Protembolotherium efremovi Janovskaya, 1957
Embolotherium andrewsi Osborn, 1929D
E. ultimum Granger and Gregory, 1943
E. loucksii Osborn, 1929D
E. insigne Janovskaya, 1980
E.? grangeri (Osborn, 1929D)
Hyotitan thomsoni Granger and Gregory, 1943
Titanodectes ingens Granger and Gregory, 1943
Epimanteoceras robustum (Granger and Gregory, 1943)
Menodus
Helaletidae
Colodon inceptus Matthew and Granger, 1925E
Paracolodon
Deperetellidae
Teleolophus magnus Radinsky, 1965
Amynodontidae
Amynodon giganteus Gromova, 1958
A. alxaensis Qi, 1975
Caenolophus promissus Matthew and Granger, 1925
Sharamynodon mongoliensis (Osborn, 1936)
Cadurcodon ardynensis (Osborn, 1923C)
Paracadurcodon suhaitensis Y. Xu, 1966
Gigantamynodon giganteus Y. Xu, 1961
Hypsamynodon progressus Gromova, 1954
H. tuskabakensis (Biryukov, 1963)
cf. *Metamynodon*　　—Y. Xu (1961)
Hyracodontidae
Forstercooperia ergiliinensis Gabunia and Dashzeveg, 1974
Ardynia praecox Matthew and Granger, 1923
Urtinotherium incisivum Chow and Chiu, 1963
Indricotherium intermedium Chiu, 1962
I. parvum Chow, 1958
Prohyracodon sp.
Allacerops minor Belyayeva, 1954
Symphyssorrhachis brevirostris Belyayeva, 1954 (Rhinocerotidae?)
Rhinocerotidae
Huananodon hypsodonta You, 1977
Ilianodon
Guixia youjiangensis You, 1977
Chalicotheriidae
Schizotherium avitum Matthew and Granger, 1923C
S. nabanensis Zhang, 1976
Eomoropus

RODENTIA
  Ctenodactylidae
    Hulgana ertnia Dawson, 1968A
  Cylindrodontidae
    Ardynomys olseni Matthew and Granger, 1925B
    A.? vinogradovi Shevyreva, 1972

Like the preceding Paleogene mammalian faunas of Asia, this composite Early Oligocene fauna has affinities at the level of family and genus with the roughly contempor-

aneous faunas of North America and Europe. We conclude that there was dispersal of some land mammals across Holarctica about the beginning of the Oligocene as here recognized. But Asia also maintained a strong element of endemics. The most conspicuous and inexplicable absence in the Eurasian continental Oligocene is the Equidae (horses)—so multitudinous in the Oligocene of North America, and such good travelers in earlier and later times. Recently, however, Miao (1982) has found a hyracotheriine equid—*Qianohippus*—in beds believed to be Early Oligocene in Guizhou, southern China.

The sample of small mammals other than rodents and lagomorphs from the Asian Early Oligocene is poor.

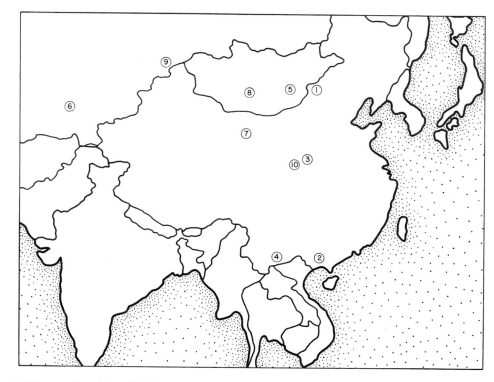

FIGURE 5-2 *Early Oligocene Localities of Asia.*

  1 *Ardyn Obo (=Erghil Obo, Erghil Dzo, Ergelyeen Dzo, Ulan Gochu Fm sites)—Nei Mongol*
  2 *Baise Basin (in part)—Guangxi*
  3 *Baishui Village—Shanxi*
  1 *Chaganbulage (Suhaitu, Haosibuerdu Basin)—Nei Mongol*
  4 *Chuching—Yunnan (Early to Middle Oligocene)*
  1 *East Mesa—Nei Mongol*
  1 *Ergelyeen Obo (Ardyn Obo)—Nei Mongol*
  2 *Gongkang Formation sites—Guangxi*
  5 *Goateg—Mongolia*

  1 *Haosibuerdu Basin (Chaganbulage)—Nei Mongol*
  1 *Houldjin Gravels sites—Nei Mongol*
  6 *Ilyiskaja Fissure—Kazakhstan*
  1 *Jhama Obo (East Mesa)—Nei Mongol*
  1 *Khul'dzhin (=Houldjin)—Nei Mongol*
  7 *Hui-Hui-pu (Shih-er-ma-cheng)—Gansu*
  9 *Kiinkerish Hill—Kazakhstan*
  8 *Koer Dzan—Mongolia*
  9 *Kusto Formation sites—Kazakhstan*
  10 *Lantian (in part)—Shaanxi*
  3 *Lok 1—Shanxi*

  4 *Loping district—Yunnan*
  4 *Lunan Basin (in part)—Yunnan*
  7 *Po-yang-ho (Shih-er-ma-cheng)—Gansu*
  4 *Qujing Basin—Yunnan (-Chuching)*
  7 *Shih-er-ma-cheng (Po-yang-ho)—Gansu*
  3 *Sian (Xian) (in part)—Shanxi*
  1 *Suhaitu (Chaganbulage, Alashan)—Nei Mongol*
  1 *Ulan Gochu (Ardyn Obo)—Nei Mongol Urtyn Obo (See Ardyn Obo.)*
  10 *Yuanqu Basin—Shanxi*
  2 *Yungle Basin—Gaungxi*
  9 *Zaysan Depression (in part)—Kazakhstan*

One may wonder whether this asymmetry in faunal composition is the result of collecting bias or whether it signifies that Asian paleohabitats were in more open savannas and were less supportive of micro-insectivores, micro-carnivores, frugivores, and forest dwellers. Absence of primates in the sample supports the second idea.

OLIGOCENE OF AFRICA

Fossil vertebrates have been collected from three principal levels in the fluvial Jebel el Qatrani (= "Fluvio-marine") Formation of the Fayum Province, southwest of Cairo, Egypt (Simons, 1968). The lowest fossiliferous layer in this stratal succession is the "Lower Fossil Wood Zone" (American Museum of Natural History and Yale University quarries A, B, C, D, E, F). About 150 ft higher in the section is Yale's quarry G level, and another 150 ft or more upsection is the "Upper Fossil Wood Zone" (Yale's quarries I, J, M, O, P, R). A basalt flow caps the Jebel el Qatrani and is about 250 ft stratigraphically above the "Upper Fossil Wood Zone." This basalt has been dated by the potassium-argon method as not less than 25 million years old (Simons, 1968), with revision of constants following Steiger and Jager (1977).

Although the "Fayum fauna" has been dated paleontologically as early Oligocene, the duration of deposition of the Jebel el Qatrani Formation may represent much of late Eocene and Oligocene time. The mammalian fauna from the "Lower Fossil Wood Zone," classification modified from Simons (1968), Simons and Gingerich (1974), and Gingerich (1978), comprises:

MACROSCELIDEA
  Macroscelididae
    Metoldobotes stromeri Schlosser, 1910
PANTOLESTA?
  Ptolemaiidae
    Ptolemaia lyonsi Osborn, 1908E
    Qarunavus meyeri Simons and Gingerich, 1974
CREODONTA
  Hyaenodontidae
    Metasinopa aethiopica (Andrews, 1906)
    Apterodon altidens Schlosser, 1910
    Pterodon africanus Andrews, 1903
    P. leptognathus Osborn, 1909F
    P. phiomensis Osborn, 1909F
    Hyaenodon brachycephalus Osborn, 1909F
CHIROPTERA
  MICROCHIROPTERA
    Vampyravus orientalis Schlosser, 1910
    gen. nov.   —Simons (1968)
PRIMATES
  Parapithecidae
    Parapithecus fraasi Schlosser, 1910
  Pliopithecidae   —Szalay and Delson (1979)
    Propliopithecus markgrafi (Schlosser, 1910) (upper level? —Gingerich, 1978)
    P. haeckeli Schlosser, 1910   (upper level?)
  Family?
    Oligopithecus savagei Simons, 1962
ARTIODACTYLA
  Mixtotheriidae
    Mixtotherium mezi Schmidt, 1913 (identification questioned later)
  Anthracotheriidae
    Rhagatherium aegyptiacum Fourtau, 1918
    Bothriogenys andrewsi Schmidt, 1913
    B. fraasi Schmidt, 1913
    B. gorringei (Andrews and Beadnell, 1902)
    B. parvus (Andrews, 1906)
    B. rugulosus Schmidt, 1913
    B. africanus (Andrews, 1899)
EMBRITHOPODA
  Arsinoitheriidae
    Arsinoitherium zitteli Beadnell, 1902
    A. andrewsi Lankester, 1903
PROBOSCIDEA
  Moeritheriidae
    Moeritherium andrewsi Schlosser, 1911
  Mammutidae
    Palaeomastodon beadnelli Andrews, 1901A
    P. barroisi Pontier, 1907
    P. intermedius Matsumoto, 1922
    P. parvus Andrews, 1905
  Gomphotheriidae
    Phiomia minor (Andrews, 1904)
    P. osborni Matsumoto, 1922
    P. wintoni (Andrews, 1905)
HYRACOIDEA
  Pliohyracidae
    Pachyhyrax championi (Arambourg, 1933)
    P. crassidentatus Schlosser, 1910
    P. pygmaeus (Matsumoto, 1922)
    Saghatherium antiquum Andrews and Beadnell, 1902
    S. sobrina Matsumoto, 1926
    Geniohyus mirus Andrews, 1904
    G. diphycus Matsumoto, 1926
    G. magnus (Andrews, 1904)
    Bunohyrax fajumensis (Andrews, 1904)
    B. major (Andrews, 1904)
    Megalohyrax eocaenus Andrews, 1903
    Titanohyrax andrewsi Matsumoto, 1921
    T. ultimus Matsumoto, 1921
    Thyrohyrax domorictus Meyer, 1973
RODENTIA
  Phiomyidae
    Phiomys andrewsi Osborn, 1908
    P. lavocati A. Wood, 1968
    Metaphiomys schaubi A. Wood, 1968
    Gaudeamus aegyptius A. Wood, 1968

The mammalian fauna from Yale's quarry G level, according to Simons (1968), Simons and Gingerich (1974), and A. E. Wood (1968), comprises:

CREODONTA
    Hyaenodontidae
        Masrasector aegypticum Simons and Gingerich, 1974
PRIMATES
    Parapithecidae
        Apidium moustafai Simons, 1962
    Pliopithecidae
        Propliopithecus haeckeli Schlosser, 1910
RODENTIA
    Phiomyidae
        Phiomys andrewsi Osborn, 1908
        P. paraphiomyoides A. Wood, 1968
        Metaphiomys schaubi A. Wood, 1968

The mammalian fauna from Yale's quarries I, J, M, O, P, and R, according to Simons (1968) and A. Wood (1968), comprises:

CREODONTA
    Hyaenodontidae
        Metasinopa fraasi Osborn, 1909F
PRIMATES
    Parapithecidae
        Simonsius grangeri (Simons, 1974)   —Gingerich
            (1978)
        Apidium phiomense Osborn, 1908
    Pliopithecidae
        Aegyptopithecus zeuxis Simons, 1965   —(this is
            Propliopithecus, acc. Szalay and Delson, 1980)
        Propliopithecus, n. sp.   —Simons (1968)
        P. chirobates (Simons, 1965)   (= "*Aeolopithecus*")
        P. haeckeli Schlosser, 1910   —(this level, acc. Gingerich, 1978)
RODENTIA
    Phiomyidae
        Phiomys aff. *paraphiomyoides* A. Wood, 1968
        Metaphiomys beadnelli Osborn, 1908
        Paraphiomys simonsi A. Wood, 1968
        Phiocricetomys minutus A. Wood, 1968

Although the Oligocene fauna of the Fayum has some elements in common with the contemporaneous and earlier fauna of Eurasia, it has many peculiar animals. It was probably isolated incompletely from most of Eurasia by the remnant Tethyan seaway, but its hyaenodontids, cebochoerids, anthracotheriids, and embrithopods indicate probable earlier dispersal of some land mammals between these two continental areas. Conspicuous by their absence in the Fa-

yum, however, are the fissiped carnivores, perissodactyls, theridomyid and cricetid rodents, and the great variety of artiodactyls of Palaearctica. The Fayum Oligocene is renowned for its many specimens of the peculiar giant herbivore *Arsinoitherium,* for its earliest-known mastodonts *Palaeomastodon* and *Phiomia,* for its diverse hyracoids, and for its many anthracotheres. It is uniquely important, thanks largely to the mid-twentieth-century work of Elwyn Simons of Yale and Duke universities, for its record of early diversification among the anthropoid primates.

Principal correlatives of the faunas from the Jebel el Qatrani Formation of Egypt are found in Libya (R. Savage, 1969A). These are Zella Oasis with *Palaeomastodon, Phiomia, Megalohyrax,* and *Bothriogenys* and the fauna from the upper marls and sandstone beds at Dor el Talha. Also, in Tunisia, Jebel (Gebel) Bou Gobrine has produced *Phiomia* and an anthracothere?; in Algeria, Isserville (questioned Oligocene) had *Phiomia pygmaeus* (Depéret). Bedeil, Somalia, produced Oligocene sirenian.

The nonmarine Oligocene in North Africa usually is represented by sandy, fluvial strata (R. Savage, 1969A).

NONMARINE OLIGOCENE OF NORTH AMERICA

The Oligocene interval in North America, as based on nonmarine phenomena, has been subdivided customarily into three mammal ages, represented by many faunas from formations and members in the well known White River Group of South Dakota and Nebraska. These ages are Chadronian, Orellan, and Whitneyan. In the last decade, radiometric datings from rocks in the overlying Arikaree Group (Gering Formation) show year ages of more than 24 million years; thus we include the earlier part of the succeeding, Arikareean land-mammal age in the Oligocene.

Exposures of the White River Group are usually scenic badlands and are one of the most productive fossil-mammal terrains in the world. The White River consists of nodular mudstones, tuffaceous siltstones, sandstones, and conglomerates. Most of the fine detritus in these sediments was derived from contemporaneous volcanic eruptions to the northwest of the depositional area. These clastic sediments were laid down on river floodplains and in channels, ponded streams, and lakes. The White River covers several thousand square miles of the northern Great Plains and adjacent Rocky Mountain provinces, and the lower part is even exposed in some of the folds of the high parts of the Rocky Mountains. This shows that the last pulse of Rocky Mountain ("Laramide") orogeny was later than early Oligocene. McKenna (1980) has documented the fact that White River was deposited over much of the Bighorn Basin of northern Wyoming and was subsequently removed by erosion.

The lower part of the Arikaree Group comprises mostly fine- to medium-grained sandstones. Hunt and Breyer

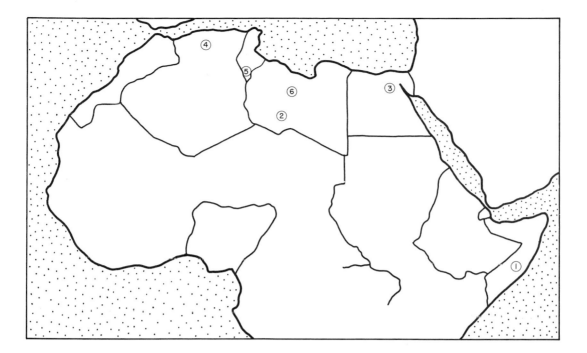

FIGURE 5-3 *Early Oligocene Localities of Africa.*
  1 Bedeil, Somalia
  2 Djebel Coquin = Dor el Talha
  2 Dor el Talha, upper—Libya
  3 Fayum local faunas—Egypt. See also Jebel el Qat-
    rani Formation sites, "Fluvio-marine" Formation
    sites

  4 Isserville—Algeria (age?)
  5 Jebel Bou-Gobrine
  6 Zella Oasis—Libya

(personal communication, 1978) are finding that these sandstones are composed of as much as 80 percent volcanic clasts, also derived from continuing volcanic activity in the Cordilleran belt to the west and northwest. Earlier Arikareean strata are found mostly in northwestern Nebraska and adjacent southwestern South Dakota (Gering, Sharps, and Monroe Creek formations and Monroe Creek equivalent in the Rosebud Formation). Probable correlatives are found in California and Oregon.

Fossil-bearing units of the White River lie in the southeastern quadrant of Wyoming (Emry, 1973) and in northeastern Colorado (Galbreath, 1953). Very similar lithology is known from south-central Montana (the Renova Formation of the Pipestone Springs district, Kuenzi and Fields, 1971). Other mammal-bearing rocks generally correlative with the lower part of the White River include fluvial sands in the Cypress Hills of Saskatchewan (L. Russell, 1978), volcaniclastic strata of the Vieja Group in southwestern Texas (J. Wilson, 1977B), red beds of the Titus Canyon Formation in southeastern California (Stock and Bode, 1935), Sespe Formation red beds in southwestern California (Stock, 1935B), and sedimentary volcanic detritus in the uppermost part of the Clarno Formation (Mellett, 1969B) and the lower part of the John Day Formation (Oligo-Miocene) of north-central Oregon (Rensberger, 1971A). All the known nonmarine rocks of the Oligocene of North America are fluvial or lacustrine (minor) in origin in inland

areas, except for late Oligocene sites in the uppermost Sespe and a probable near-coast site in northern Florida.

CHADRONIAN (LATEST EOCENE AND EARLY OLIGOCENE)

The Chadronian land-mammal age is based on the fauna from the Chadron Formation, lowest part of the White River Group, as exposed in the Big Badlands of South Dakota, the name being derived from Chadron, Nebraska. The richly fossiliferous outcrops of the Chadron in northwestern Nebraska definitely constitute part of the "type area" (H. Wood et al., 1941; Schultz and Stout, 1955; and Harksen and Macdonald, 1969B). The composite Chadronian mammalian fauna, taken from correlated localities in Saskatchewan, South Dakota, Nebraska, Colorado, Wyoming, Montana, Texas, California, and Mexico, comprises the following genera.

Composite Chadronian Mammalian Fauna

MULTITUBERCULATA
  Neoplagiaulacidae
    cf. *Parectypodus*
MARSUPIALIA
  Didelphidae
    Herpetotherium valens (Lambe, 1908A)

163

H. fugax Cope, 1873T (age?)
Nanodelphys cf. *minutus* McGrew, 1937A
N? mcgrewi L. Russell, 1972B
"*Peratherium*" *donohoei* Hough, 1961
LEPTICTIDA
  Leptictidae
    Leptictis dakotensis (Leidy, 1868J)
    L. haydeni Leidy, 1868J
    L. douglassi Novacek, 1976
    L. wilsoni Novacek, 1976
    "*L.*" *montanus* (Douglass, 1905A)
    "*L.*" *emryi*
    Prodiacodon acutidens Douglass, 1901B
    P. thompsoni (Matthew, 1903B)
LAGOMORPHA
  Leporidae
    Palaeolagus temnodon Douglass, 1901B
    Megalagus turgidus (Cope, 1873T) (age?)
    M. brachyodon (Matthew, 1903B) —(includes "*Desmatolagus dicei*" (Burke, 1936), acc. to Gawne (1978)
    Chadrolagus emryi Gawne, 1978
PANTOLESTA
  Pantolestidae
    Chadronia margaretae Cook, 1954
APATOTHERIA
  Apatemyidae
    Sinclairella dakotensis Jepsen, 1934
CREODONTA
  Hyaenodontidae
    Hyaenodon montanus Douglass, 1901B
    H. microdon Mellett, 1977
    H. vetus Stock, 1933D
    H. mustelinus Scott, 1895B
    H. crucians Leidy, 1853D
    H. horridus Leidy, 1853D
    H. megaloides Mellett, 1977
    Hemipsalodon grandis Cope, 1885H
    H. cooki Schlaikjer, 1935B
    Ischognathus savagei Stovall, 1948B
CARNIVORA
  Miacidae
    Miacis
  Amphicyonidae —Hunt, 1974
    Proamphicyon nebraskensis Hough, 1948A
    Daphoenus canadensis L. Russell, 1934A
    D. lambei L. Russell, 1934A
    D. cf. *vetus* Leidy, 1853D
    D. cf. *hartshornianus* (Cope, 1873CC)
    Daphoenocyon dodgei (Scott, 1898B) —Hough, 1948A
    Daphoenictis tedfordi Hunt, 1974
  Ursidae
    Campylocynodon personi Chaffee, 1954
    Parictis dakotensis Clark, 1936
    P. parvus Clark and Beerbower, 1967
  Procyonidae
    Plesictis priscus (Clark, 1936)
  Mustelidae

Palaeogale infelix Matthew, 1903B —(referable to *P. lagophaga,* Cope, 1873CC?)
  Canidae
    Hesperocyon gregarius (Cope, 1873T)
    H. paterculus (Matthew, 1903B)
    cf. *Mesocyon,* n. gen. —Emry (1973)
  Felidae
    Dinictis felina Leidy, 1856H
    D. fortis Adams, 1895A
    Hoplophoneus mentalis Sinclair, 1921B
    H. oharrai Jepsen, 1926A
    H. robustus Adams, 1896B
    Eusmilus
INSECTIVORA
  Soricoidea
    Centetodon magnus (Clark, 1936)
    C. chadronensis Lillegraven, McKenna, and Krishtalka, 1981
    Ankylodon —Lillegraven, McKenna, and Krishtalka (1981)
    Apternodus altitalonidus Clark, 1936
    A. mediaevus Matthew, 1903B
    A. brevirostris Schlaikjer, 1934A
    A. gregoryi Schlaikjer, 1933
    Clinopternodus gracilis (Clark, 1936) —Clark (1937)
    Micropternodus borealis Matthew, 1903B
    M (= "*Kentrogomphios*") *strophensis* T. White, 1954
    Oligoryctes cameronensis Hough, 1956
  Soricidae
    Domnina thompsoni Simpson, 1941I
  Talpidae
    Cryptoryctes kayi C. Reed, 1954A
    Oligoscalops galbreathi (C. Reed, 1956)
CHIROPTERA, gen. unident. —Emry (1973)
PRIMATES
  Omomyidae
    Macrotarsius montanus Clark, 1941B
    Rooneyia viejensis J. Wilson, 1966
ARTIODACTYLA
  Leptochoeridae
    Stibarus yoderensis Macdonald, 1955
    S. lemurinus (Cope, 1873R)
    Nanochoerus montanus (Matthew, 1903B)
  Enteledontidae
    Brachyhyops wyomingensis Scott, 1945A
    B.? viensis L. Russell, 1980
    Archaeotherium cf. *crassum* Marsh, 1873H
    A. marshi Troxell, 1920B
    A. clavum (Marsh, 1893D)
    A. mortoni Leidy, 1850A
    A. scotti Sinclair, 1921C
    A. coarctatum (Cope, 1889)
  Tayassuidae
    Thinohyus nanus Marsh, 1894L —Stirton and Woodburne (1965)
    Perchoerus minor Cook, 1922
  Anthracotheriidae
    Aepinacodon americanus (Leidy, 1856D) —Macdonald (1956C)
    A. deflectus (Marsh, 1890D)
    Heptacodon
  Agriochoeridae

Protoreodon petersoni Gazin, 1955

P. pumilus (Marsh, 1875B)

P. minimus (Douglass, 1901B)

Agriochoerus antiquus Leidy, 1850C

A. maximus Douglass, 1901B

Merycoidodontidae[2]

Merycoidodon dunagani J. Wilson, 1971A

M. lewisi Clark and Beerbower, 1967

M. forsythae Schultz and Falkenbach, 1968

M. culbertsoni (Leidy, 1848B)

Eporeodon bullatus (Leidy, 1869A)   —*Otionohyus*, Harksen and Macdonald (1969B)

Prodesmatochoerus natronensis Schultz and Falkenbach, 1954

P.? cf. meekae Schultz and Falkenbach, 1954

Bathygenys alpha Schultz and Falkenbach, 1956

B. reevsi J. Wilson, 1971A

Megabathygenys goorisi Schultz and Falkenbach, 1968

Parabathygenys paralpha Schultz and Falkenbach, 1968

Limnenetes cf. *platyceps* Douglass, 1901B

Aclistomycter middletoni J. Wilson, 1971A

Otionohyus wardi Schultz and Falkenbach, 1968

O.? vanderpooli Schultz and Falkenbach, 1968

Stenopsochoerus chadronensis Schultz and Falkenbach, 1956

S. douglasensis Schultz and Falkenbach, 1956

S. reideri Schultz and Falkenbach, 1956

Oreonetes anceps (Douglass, 1901B)

Camelidae

Poebrotherium franki J. Wilson, 1974

P. cf. *andersoni* Troxell, 1917B

Hidrosotherium transpecosensis J. Wilson, 1974

Malaquiferus   —Emry (1973)

Paratylopus primaevus Matthew, 1904B   (age?)

Hypertragulidae

Hypertragulus heikeni Ferrusquia-V., 1969

H. calcaratus Cope, 1873T

Hypisodus cf. *minimus* Cope, 1873T

Parvitragulus priscus Emry, 1978

Leptomerycidae

Leptomeryx defordi J. Wilson, 1974

L. yoderi Schlaikjer, 1935B

L. mammifer Cope, 1886U

L. esulcatus Cope, 1889I

Protoceratidae   —Patton and Taylor (1973)

Heteromeryx dispar Matthew, 1905

Poabromylus kayi Peterson, 1931

P. minor J. Wilson, 1973

"*Leptotragulus profectus*" (Matthew, 1903)

Pseudoprotoceras longinaris Cook, 1934

P. semicinctus (Cope, 1889)

P. taylori Emry and Storer, 1981

PERISSODACTYLA

Equidae

Mesohippus texanus McGrew, 1971

M. bairdii (Leidy, 1850C)

M. celer (Marsh, 1874B)

M. proteulophus Osborn, 1904B

M. grandis Clark and Beerbower, 1967

M. hypostylus Osborn, 1904B

M. latidens Douglass, 1903A

M. montanensis Osborn, 1904B

M. portentus Douglass, 1903A

M. viejensis Clark and Beerbower, 1967

Haplohippus texanus McGrew, 1953

Brontotheriidae

Teleodus avus Marsh, 1890D

Protitanops curryi Stock, 1936D

Brontops amplus (Marsh, 1890D)

B. bicornutus (Osborn, 1902B)

B. brachycephalus (Osborn, 1902A)

B. dispar Marsh, 1887B

B. robustus Marsh, 1887B

B. tyleri (Lull, 1905D)

Menodus bakeri Stovall, 1948B

M. crassicornis (Marsh, 1891E)

M. giganteus Pomel, 1849

M. marshi (Osborn, 1902B)

M. proutii (Owen, Norwood, and Evans, 1850A)

M. serotinus (Marsh, 1887B)

M. varians (Marsh, 1887B)

M. walcotti (Osborn, 1916A)

M. heloceras (Cope, 1873S)

Megacerops copei (Osborn, 1908A)

Brontotherium dolichoceras (Scott and Osborn, 1887B)

B. gigas Marsh, 1873H

B. hatcheri Osborn, 1908A

B. leidyi Osborn, 1902B

B. medium (Marsh, 1891E)

B. platyceras (Scott and Osborn, 1887B)

B. ramosum (Osborn, 1896B)

B. tichoceras (Scott and Osborn, 1887B)

Helaletidae

Colodon occidentalis (Leidy, 1863I)

C. kayi (Hough, 1950) —Harksen and Macdonald (1969B)

Hyracodontidae

Hyracodon petersoni H. Wood, 1927B

Triplopus? mortivallis (Stock, 1949)

Amynodontidae

Metamynodon chadronensis H. Wood, 1937A

Toxotherium hunteri H. Wood, 1961   —(includes *T. woodi* and *Schizotheroides jackwilsoni*, acc. Emry (1979))

Rhinocerotidae

Caenopus mitis (Cope, 1874B)

Trigonias osborni Lucas, 1900D

T. wellsi H. Wood, 1927B

T. paucidens H. Wood, 1927B

T. gregoryi H. Wood, 1927B

T. hypostylus Gregory and Cook, 1928A

Penetrigonias hudsoni Tanner and Martin, 1976

Subhyracodon copei (Osborn, 1898I)

S. trigonodus (Osborn and Wortman, 1894A)

Amphicaenopus platycephalus (Osborn and Wortman, 1894A)

PHOLIDOTA?

Epoicotheriidae

Epoicotherium unicum (Douglass, 1905A)

Manidae?   —Emry (1970)

Patriomanis americanus Emry, 1970

RODENTIA

Ischyromyoidea

Leptotomus gigans A. Wood, 1974

L. guildayi Black, 1965A
Mytonomys gaitana Ferrusquia-V. and A. Wood, 1969
Microparamys perfossus A. Wood, 1974
Ischyrotomus cf. *petersoni* (Matthew, 1910B)
Manitsha johanniculi A. Wood, 1974
M. tanka Simpson, 1941H
Ischyromys blacki A. Wood, 1974
I. douglassi Black, 1968
I. junctus L. Russell, 1972B
I. veterior (Matthew, 1903B)   —Black (1968A)
Cylindrodon fontis Douglass, 1901B
C. collinus L. Russell, 1972B
Pseudocylindrodon neglectus Burke, 1936
P. texanus A. Wood, 1974
Ardynomys occidentalis Burke, 1936
A. saskatchewanensis (Lambe, 1908A)   —Storer
   (1978)
Jaywilsonomys ojinagaensis Ferrusquia-V. and A. Wood,
   1969
J. pintoensis Ferrusquia-V. and A. Wood, 1969
Aplodontidae   —Rensberger (1975)
Prosciurus vetustus Matthew, 1903B
Pelycomys rugosus Galbreath, 1953
Pipestoneomys bisulcatus Donohoe, 1956
Cedromus jeffersoni (Douglass, 1901B) (family?)
Spurimus   (family?)
Eomyidae
Adjidaumo minutus (Cope, 1873T)
A. minimus (Matthew, 1903B)
Paradjidaumo minor (Douglass, 1903A)
P. hansonorum (L. Russell, 1972B)
Centimanomys major Galbreath, 1955A
Aulolithomys bounites Black, 1965A
Cupressimus barbarae Storer, 1978
Yoderimys lustrorum A. Wood, 1974
Y. bumpi A. Wood, 1955A
Y. burkei Black, 1965A
Y. stewarti (L. Russell, 1972B)
Viejadjidaumo magniscopuli A. Wood, 1974
Namatomys lloydi Black, 1965A
Castoridae
Agnotocastor galushai Emry, 1972
Heteromyidae
Heliscomys cf. *vetus* Cope, 1873T
Meliakrouniomys skinneri Emry, 1972A (family?)
M. wilsoni Harris and A. Wood, 1969 (family?)
Eutypomyidae
Eutypomys inexpectatus A. Wood, 1974
E. parvus Lambe, 1908A
E. cf. *thompsoni* Matthew, 1905A
Cricetidae
Eumys pristinus L. Russell, 1972B
Paracricetodon?   —Alker (1968)
Eoeumys   —Martin (1980)
Family?
Griphomys
Family?
Nonomys simplicidens (Emry and Dawson, 1972A)
   —Emry and Dawson (1973)

Subsumus candelariae A. Wood, 1974
  cf. *Simimys*
Family?
Guanajuatomys hibbardi Black and Stephens, 1973 (age?)
Floresomys guanajuatoensis Fries, Hibbard, and Dunkle, 1955

Chadronian faunas are characterized by their greatly abundant leporids (*Palaeolagus*), ischyromyoid rodents, diversified oreodonts (agriochoerids and merycoidodontids), horses (*Mesohippus*), and brontotheres. This level is the "Titanotherium beds" of early North American vertebrate paleontologists. And it was a time in North America of early appearance and abundance of many modern families of land mammals: Castoridae, Canidae, Mustelidae, Felidae, Rhinocerotidae, and Tayassuidae.

Of the 127 recognized mammalian genera of the Chadronian, only about 31 (24 percent) have been found earlier, in the Eocene. Seventy-eight (54 percent) of the Chadronian genera and about 5 families (Neoplagiaulacidae, Pantolestidae, Miacidae, Brontotheriidae, and Manidae?) are not known in the later faunas of North America.

DESEADAN

Simpson (1940G) proposed Deseadan as the stage-age name for the general faunal, chronostratigraphic, geochronologic unit that earlier workers in Argentina had termed *Pyrotherium* beds and fauna, Deseadense, Deseadiano, and so on. The Deseado Formation and its fauna (the "*Pyrotherium* beds") overlies the Musters Formation and fauna ("*Astraponotus* beds"), which in turn overlie the Casamayoran; and the Deseado underlies the Colhué-Huapí Formation and fauna ("*Colpodon* beds") in the succession of fluvial sedimentary rocks of the renowned Great (or High) Barranca, south of Lake Colhué-Huapí in Chubut Province. Thus at the Great Barranca, there is superb stratigraphic demonstration of these four stages as Carlos and Florentino Ameghino first recognized them at the end of the nineteenth century— (Loomis, 1914; Simpson, 1940G).

The fauna of the Deseado Formation includes the following taxa.

The Deseado Fauna

(After Loomis, 1914; Simpson, 1933 to 1945; Patterson,
1952A; Marshall, 1978).

MARSUPIALIA
Microbiotheriidae
Microbiotherium spp. Ameghino
Borhyaenidae
Notogale mitis (Ameghino, 1897)
Proborhyaena gigantea Ameghino, 1897
Pharsophorus lacerans Ameghino, 1897
P. tenax Ameghino, 1897
P.? antiquus (Ameghino, 1894)
Caenolestidae
Palaeothentes chubutensis Ameghino, 1897

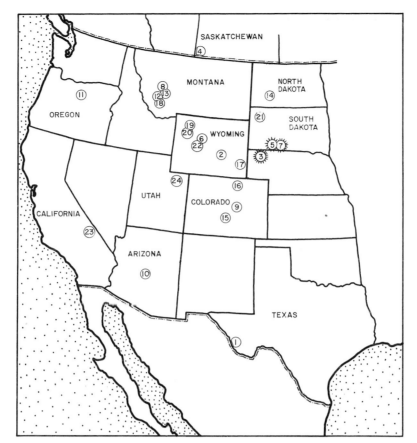

FIGURE 5-4 *Representative Chadronian Localities.*

1 *Airstrip—Texas*
2 *Alcova North, West, Southeast—Wyoming*
3 *Ant Hill–Orella (town)—Nebraska*
4 *Anxiety Butte—Saskatchewan*
1 *Ash Spring—Texas*
2 *Bates Hole–Flagstaff Rim—Wyoming*
5 *Battle Creek Breaks, Canyon, Draw—South Dakota*
6 *Beaver Divide–Cameron Spring—Wyoming*
7 *Big Badlands—South Dakota*
1 *Big Cliff—Texas*
8 *Canyon Ferry Reservoir—Montana*
9 *Castle Rock Conglomerate site—Colorado*
10 *Cave Creek —Arizona*
3 *Chadron Formation sites,* <u>Chadronia</u> *Pocket, etc.—Nebraska*
11 *Clarno Ferry–Hancock Quarry—Oregon   (earlier?)*
4 *Cypress Hills Formation sites—Saskatchewan*
12 *Diamond O—Montana*

13 *Dry Creek–Toston–Easter Lily Mine—Montana*
14 *Fitterer Ranch—North Dakota*
2 *Flagstaff Rim–Bates Hole—Wyoming*
15 *Florissant Formation sites—Colorado   (Orellan?)*
16 *Gerry's Ranch—Colorado*
17 *Goshen Hole district—Wyoming*
16 *Horsetail Creek Member sites—Colorado*
18 *Jefferson River Section—Montana*
14 *Little Badlands—North Dakota*
1 *Little Egypt—Texas*
19 *Mink Creek (Yellowstone)—Wyoming*
20 *Pilgrim Creek (Jackson Hole)—Wyoming*
12 *Pipestone Springs—Montana*
1 *Porvenir—Texas*
21 *Slim Buttes (Reva Gap)—South Dakota*
22 *South Pass City West—Wyoming*
23 *Titus Canyon—California*
24 *Vernal West—Utah*
17 *Yoder Ditch—Wyoming*

Pilchenia lucina Ameghino, 1903
Acdestis praecursor (Loomis, 1914)
Parabderites minusculus Ameghino, 1902

EDENTATA
   Palaeopeltidae   —Patterson and Pascual (1968)
      Palaeopeltis inornatus (incl. *Pseudorophodon* Kraglievich and Rivas, 1951)
   Dasypodidae
      Proeutatus lagenaformis Ameghino, 1895
      Prozaedius impressus Ameghino, 1897

P. tenuissimus Ameghino, 1902
Stenotatus ornatus Ameghino, 1897
Proeuphractus setiger Ameghino, 1886
P. laevis Ameghino, 1886
Pseudostegotherium Ameghino, 1902 (Ameghino also proposed *Archaeutatus, Amblytatus, Isutaetus, Sadypus, Hemiutatus,* and *Anutaetus*—Loomis, 1914)
Peltephilus undulatus Ameghino, 1897
P. protervus Ameghino, 1897
   Glyptodontidae

167

Glyptatelus tatusinus Ameghino, 1897

G. malaspinensis Ameghino, 1902

Mylodontidae

Orophodon hapaloides Ameghino, 1895

Octodontotherium grandis Ameghino, 1895

O. crassidens Ameghino, 1897

Megatheriidae

cf. *Proschismotherium*

Hapalops? antistis (Ameghino, 1897)

LITOPTERNA

Proterotheriidae

Deuterotherium distichum Ameghino, 1895

Eoproterotherium inaequifacies Ameghino, 1904

Macraucheniidae

Notodiaphorus crassus Loomis, 1914

Adianthidae

Proadianthus excavatus Ameghino, 1897

Tricoelodus bicuspidatus Ameghino, 1897

Coniopternium? Ameghino, 1895?

NOTOUNGULATA

Archaeohyracidae

Archaeohyrax patagonicus Ameghino, 1897

A. propheticus Ameghino, 1897

A. concentricus Ameghino, 1897

Isotemnidae

Trimerostephanos scabrus Ameghino, 1895

T. scalaris Ameghino, 1897

T. angustus Ameghino, 1897

T. biconus Ameghino, 1897

Pleurocoelodon wingei Ameghino, 1895

P. cingulatus Ameghino, 1895

Homalodotheriidae

Asmodeus osborni Ameghino, 1895

A. scotti Ameghino, 1895

Leontiniidae

Leontinia gaudryi Ameghino, 1895

L. garzoni Ameghino, 1896

Ancylocoelus frequens Ameghino, 1895

A. lentus Ameghino, 1901

A. minor Ameghino, 1901

A. armatum (Ameghino, 1895)

Scarrittia canquelensis Simpson, 1934B

Henricofilholia cingulata Ameghino, 1901

H. lustrata Ameghino, 1901

H. inaequilatera Ameghino, 1901

H. circumdata Ameghino, 1901

Notohippidae

Interhippus phorcus Ameghion, 1904

I. deflexus Ameghino, 1904

Nesohippus insulatus Ameghino, 1904

Morphippus imbricatus Ameghino, 1897

Rhynchippus equinus Ameghino, 1897

R. pumulis Ameghino, 1897

R. medianus Ameghino, 1901

Eurygenium latirostris Ameghino, 1895

E. normalis Ameghino, 1897

Toxodontidae

Proadinotherium leptognathus Ameghino, 1895

Interatheriidae

Archaeophylus patrinus Ameghino, 1897

Plagiarthrus clivus Ameghino, 1897

Cochilius fumensis Simpson, 1932G

Medistylus dorsatus (Ameghino, 1903)—Stirton, 1952C

Mesotheriidae

Trachytherus spegazzinianus Ameghino, 1889

T. conturbatus Ameghino, 1895

T. grandis (Loomis, 1914)

Proedium solitarium Ameghino, 1895

Hegetotheriidae

Prohegetotherium sculptum Ameghino, 1897

P. shumwayi Loomis, 1914

Prosotherium garzoni Ameghino, 1897

P. triangulidens Ameghino, 1897

Propachyrucos smithwoodwardi Ameghino, 1897

P. aequilatus Ameghino, 1901

ASTRAPOTHERIA

Astrapotheriidae

Parastrapotherium holmbergi Ameghino, 1895

P. ephebicum Ameghino, 1895

P. martiale Ameghino, 1901

P. insuperabile Ameghino, 1901

Synastrapotherium   —Paula Couto (1977)

PYROTHERIA

Pyrotheriidae

Pyrotherium romeri Ameghino, 1889

P. sorondoi Ameghino, 1894

Griphodon peruvianus Patterson, 1942   (age?)

CONDYLARTHRA

Didolodontidae

Protheosodon coniferus Ameghino, 1897

RODENTIA

Octodontidae

Platypittamys brachyodon A. Wood, 1949A

Echimyidae

Deseadomys arambourgi A. Wood and Patterson, 1959

D. loomisi A. Wood and Patterson, 1959

Chinchillidae

Scotamys antiquus Loomis, 1914

Dasyproctidae

Cephalomys arcidens Ameghino, 1897

C. plexus Ameghino, 1897

Litodontomys chubutensis Loomis, 1914

Eocardiidae

Asteromys punctus Ameghino, 1897

Chubutomys simpsoni A. Wood and Patterson, 1959

Erethizontidae

Protosteiromys medianus (Ameghino, 1903A)

P. asmodeophilus A. Wood and Patterson, 1959

Family?

Luribayomys

Branisamys luribayensis Hoffstetter, Lavocat, 1970A

Other faunas assigned Deseadan age include:

1. Salla-Luribay, Bolivia (Hoffstetter, 1969B, 1976; and Hoffstetter and Lavocat, 1970, 1976).

The spectacular escarpments exposing a thick section of banded reddish mudstone, limestone, sandstone, and conglomerate beds in tributary valleys to the Luribay River, southeast of La Paz, are yielding an important Deseadan fauna of didelphid? and borhyaenid marsupials, dasypodid, peltephilid, glyptodontid, mylodontid, and megalonychid edentates, proterotheriid and macraucheniid litopterns, notohippid, interatheriid, mesotheriid, archaeohyracid, and hegetotheriid notoungulates, astrapotheres, pyrotheres, echimyid, dasyproctid, and other rodents, reptiles, birds, and anurans. The most exciting discovery here was *Branisella* Hoffstetter (1969B), a ceboid and the earliest-known primate in South America. *Prozaedius, Proeutatus, Peltephilus, Glyptatelus, Cephalomys,* cf. *Deuterotherium, Proadiantus, Proadinotherium, Henricofilholia, Rhynchippus, Plagiarthrus, Trachytherus, Archaeohyrax, Prohegetotherium, Parastrapotherium,* and *Pyrotherium* relate this fauna directly to the fauna from the Deseado Formation of Argentina (Hoffstetter and Lavocat, 1976).

2. In northern South America, the Deseadan is represented by the Chaparral fauna in the Gualanday Group or Formation and the Peneyita specimen in an unnamed stratum, both in Colombia. The Chaparral includes a megalonychoid, *Protheosodon, Proadinotherium,* and an astrapothere. The Peneyita is only an isolated maxilla of *Eosteiromys,* an erethizontid rodent, (Stirton, 1953A).

Mostly, the Deseadan fauna shows continued endemic evolution among the land mammals of South America; however, two allochthonous orders appear, the Rodentia and the Primates. South American rodents and primates have provoked voluminous speculative writing, reviews, and restudies of hard-part and soft-part anatomy. There have been many appeals to the developing knowledge and speculation about the geophysical and paleogeologic history of the South Atlantic, the Caribbean, and South America—as well as just plain "arm-waving" about when and how these orders entered the Neotropical "scene." See, for example, the recent volume edited by Ciochon and Chiarelli, *Evolutionary Biology of the New World Monkeys and Continental Drift,* Plenum Press, 1980.

Wood and Patterson (1959) believed that the Deseadan rodents were caviomorph hystricomorphs and that they diversified into at least six families. This diversity indicates a noteworthy evolutionary history for the group, somewhere, in pre-Deseadan time. Was this earlier history in Africa? in South America? in Central or North America or both? elsewhere? Did these rodents have close common ancestry with Eastern Hemisphere hystricomorphs and did they float on drift-vegetation rafts across the South Atlantic from Africa? from elsewhere?

Deseadan primates (Platyrrhini, Ceboidea) are now represented by about three specimens from Salla-Luribay, Bolivia. These specimens support the taxon *Branisella* Hoffstetter. Questions put forth above for the rodents apply also to Deseadan primates. We may never know whether Deseadan ceboids were as numerous and as diversified taxonomically as the associated rodents. Fossils of primates and other forest-dwelling animals are rare. Perhaps primates entered South America slightly later than the rodents and

had not yet had time to diversify. But from the standpoint of numbers of known taxa in the two orders, primates are less diversified phyletically.

FIGURE 5-5 *Deseadan and Colhuehuapian Localities.*

1 *Acre—Brazil   (age?)*   *(Paula Couto, 1977)*
2 *Cabeza Blanca (Chubut)–Deseadan and Colhuehuapian—Argentina*
3 *Cajani (Khahajani)—Bolivia*
4 *Chaparral Deseadan?—Colombia*
4 *Coiyama–Colhuehaupian—Colombia*
5 *Chiococa–Deseadan?—Peru*
2 *Colhuapi Norte (Cerro el Humo)–Deseadan and Colhuapian—Argentina*
6 *Corocoro Basin–Colhuehuapian—Bolivia*
7 *Gaiman (Trelew district, upper)–Colhuehuapian—Argentina*
2 *Great Barranca–Deseadan, Colhuehuapian, and older—Argentina*
2 *La Flecha–Deseadan—Argentina*
8 *Peneyita–Deseadan?—Colombia*
2 *Rinconada de los Lopez (Scarritt Quarry)—Argentina*
2 *Rio Deseado district–Deseadan and Colhuehuapian—Argentina*
3 *Salla Luribay district—Bolivia*

On the paleobiochorologic questions here involved, two opposing schools of thought have arisen. Old World workers have supported Old World origin for South American rodents and primates, followed by fortuitous dispersal of these orders from Africa to South America in late Eocene or earliest Oligocene time. New World workers (especially North American mammalian paleontologists) have believed that Central American–North American origin of the caviomorphs and platyrrhines, followed by island-hopping to South America across late Eocene Caribbean seaways, was the stronger inference. The consensus that ceboids (platyrrhines) had close common ancestry with Old World catarrhines (Old World monkeys and other anthropoids) tends to strengthen the argument for Old World origin of ceboids and preceboids. And if as some writers believe, the dominant oceanic currents of late Eocene time swept from east to west across the South Atlantic and through the Caribbean to the Pacific (past the north coast of South America), another powerful argument joins the Old World school. Discussions and speculations continue. We note that the pre-Deseadan record of Central America and Africa is practically nonexistent, and we believe that the anatomic, molecular, chorologic, and so on data that do or may bear on the above arguments have been thoroughly drained of information content. Now is the time to get into the field and find proto-Deseadan rodents and primates in the critical areas.

MIDDLE OLIGOCENE OF EUROPE

We here include three sets of faunas recognized as successional within the mid-Oligocene by D. Russell et al. (1983): "Level" of Montalban (earliest), "Level" of La Ferte–Alais, and "Level" of Antoignt (latest). Included also are the Montalban and Antoignt biozones of Thaler (1972) and the Montalban, Heimersheim, and Antoignt reference localities of Fahlbusch (1976).

Composite Middle Oligocene Mammalian Fauna of Europe

MARSUPIALIA
  Didelphidae
    Peratherium elegans (Aymard, 1846)
    Amphiperatherium exile (Gervais, 1848–1852)
    A. lamandini (Filhol, 1876)
    A. ambiguum (Filhol, 1877)
    A. minutum (Aymard, 1846)
CREODONTA
  Hyaenodontidae
    Hyaenodon exiguus (Gervais, 1873)

    H. leptorhynchus Laizer and Parieu, 1838
    H. dubius Filhol, 1873
    H. brachyrhynchus Blainville, 1841
    H. gervaisi Martin, 1906
    Apterodon flonheimensis (Andreae, 1887)   —Szalay (1967)
    Thereutherium thylacodes Filhol, 1876
CARNIVORA
  Amphicyonidae
    "Cynodictis" palmidens (Filhol, 1882) (age?)
    Cynelos crassidens (Filhol, 1876)
    C. piveteaui Ginsburg, 1965
    Pseudocyonopsis ambiguus (Filhol, 1876)
    P. landesquei (Helbing, 1928C)
    Sarcocyon ferox Ginsburg, 1966
    Pseudamphicyon cayluxensis (Filhol, 1876) (age?)
    Haplocyon (age?)
    Harpagophagus sanguineus Bonis, 1971
  Ursidae
    Cephalogale
    Amphicynodon leptorhynchus (Filhol, 1876) (age?)
    A. speciosus (Filhol, 1882) (age?)
    Pachycynodon tenuis Teilhard, 1915 (age?)
    P. dubius (Filhol, 1882) (age?)
    P. boriei (Filhol, 1874) (age?)
    P. crassirostris Schlosser, 1889 (age?)
    P. filholi Schlosser, 1889 (age?)
    Adelpharctos mirus Bonis, 1971A (age?)
  Procyonidae
    Plesictis filholi Depéret, 1906
  Mustelidae
    Palaeogale felina (Filhol, 1877)
    Stenogale gracilis (Gervais, 1868)
  Viverridae
    Stenoplesictis minor Filhol, 1882
    Palaeoprionodon
  Felidae
    Nimravus intermedius (Filhol, 1872)
INSECTIVORA
  Erinaceoidea
    Tetracus nanus (Aymard, 1846)
    Neurogymnurus cayluxi Filhol, 1877
  Soricoidea
    Saturninia cf. *gracilis* Stehlin, 1940
    Srinitium marteli Hugueney, 1976
    Myxomygale cf. *antiqua* Filhol, 1890
    "Talpa"
CHIROPTERA
  Emballonuridae
    Vespertiliavus
  Hipposideridae
    Hipposideros schlosseri Revilliod, 1917
    Palaeophyllophora obtina (Delfortrie, 1872)
  Rhinolophidae
    Rhinolophus cf. *priscus* Revilliod, 1920
  Vespertilionidae
    Stehlinia minor (Revilliod, 1922)
ARTIODACTYLA
  Dichobunidae
    Metriotherium

Tayassuidae
Palaeochoerus pusillus Ginsburg, 1974
P. gergovianus (Croizet in Blainville, 1846)
Doliochoerus quercyi Filhol, 1882
Suidae?
Hemichoerus lamandini (Filhol, 1880) (age?)
Anoplotheriidae
Hyracodontherium filholi Lydekker, 1889
Anthracotheriidae
Elomeryx cluai (Depéret, 1906)
Anthracotherium bimonsvialensi-magnum Golpe-Posse, 1971
A. alsaticum Cuvier, 1822
A. bumbachense Stehlin, 1910
A. magnum Cuvier, 1822
A. seckbachense Stehlin, 1910
Cainotheriidae
Caenomeryx procommunis (Filhol, 1877) (age?)
cf. *C. elongatum* Filhol, 1877
Plesiomeryx cadurcensis Gervais, 1873
Procaenotherium
Gelocidae
Gelocus communis (Aymard, 1848)
Cryptomeryx
Leptomerycidae
Prodremotherium elongatum Filhol, 1877 (age?)
Lophiomeryx chalaniati Pomel, 1853 —(acc. to Sudre, pers. commun., 1981)
Bachitherium medium Filhol, 1880
B. curtum (Filhol, 1877)
B. minor Filhol, 1882
B. insigne (Filhol, 1877)
PERISSODACTYLA
Palaeotheriidae
Plagiolophus fraasi Meyer, 1852
P. minor Cuvier, 1804
Helaletidae
"Chasmotherium" minimum (Fischer, 1829)
Amynodontidae
Cadurcotherium cayluxi Gervais, 1873
Hyracodontidae
Eggysodon gaudryi Rames, 1884
Rhinocerotidae
Ronzotherium romani Kretzoi, 1940
R. filholi (Osborn, 1900)
Preaceratherium minus (Filhol, 1884)
Aceratherium albigense Roman, 1911
Chalicotheriidae
Schizotherium modicum Gervais, 1876
RODENTIA
Eomyidae
Eomys zitteli Schlosser, 1884
Aplodontidae
Plesispermophilus angustidens Filhol, 1882
Sciurodon cadurcensis Schlosser, 1884
Castoridae
Steneofiber
Sciuridae
"Sciurus"
Palaeosciurus goti Vianey-Liaud, 1974
Heteroxerus paulhiacensis Black, 1965B

Theridomyidae
Theridomys major Depéret, 1906
T. lembronicus Gervais, 1948–1852
Archaeomys (Archaeomys) gracilis (Schlosser, 1884)
A. (A.) gervaisi Thaler, 1966B
A. (Blain. illimys) gregarius (Schlosser, 1884)
A. (B.) heimersheimensis Bahlo, 1975
A. (B.) blainvillei (Gervais, 1848–1852)
A. (B.) helmeri Vianey-Liaud, 1972A
A. (Taeniodus) curvistriatus Pomel, 1853
A. (T.) hexalophodus Bahlo, 1972
Pseudoltinomys major Vianey-Liaud, 1975
P. gaillardi Lavocat, 1951
Elfomys nanus (Thaler, 1969A)
E. medius Vianey-Liaud, 1976
Sciuromys quercyi (Stehlin and Schaub, 1951)
Sciuromys cayluxi Schlosser, 1853
Suevosciurus ehingensis Dehm, 1937A
Pseudosciurus suevicus Hensel, 1856
Issiodoromys minor Schlosser, 1884
Gliridae
Gliravus majori Stehlin and Schaub, 1951
G. tenuis Bahlo, 1975
G. cf. *bruijni* Hugueney, 1967
Oligodyromys planus Bahlo, 1975
Glirudinus cf. *praemurinus* (Freudenberg, 1941)
Cricetidae
Eucricetodon atavus (Misonne, 1957)
E. huerzeleri Vianey-Liaud, 1972B
E. huberi (Schaub, 1925)
Pseudocricetodon montalbanensis Thaler, 1969A
P. thaleri (Hugueney, 1969A)
P. philippi Hugueney, 1971A
P. moguntiacus (Bahlo, 1975)
Paracricetodon cadurcensis (Schlosser, 1884)
P. dehmi Hrubesch, 1957B
P. confluens Schaub, 1925
P. walgeri Bahlo, 1972
P. spectabile Schlosser, 1884
Melissiodon schaubi Dehm, 1935A
Heterocricetodon helbingi Stehlin and Schaub, 1951
H. stehlini Schaub, 1925
Zapodidae
Plesiosminthus aff. *promyarion* Schaub, 1930

Of the 81 genera of land mammals recognized in the middle Oligocene fauna of Europe, 60 (74 percent) were also in the early Oligocene of the area. (Four genera—*Amphicyon, Proailurus, Protapirus,* and *Peridyromys*—are reported from early and late Oligocene and should be found in the intervening middle Oligocene. If they are counted here, the above percentage increases to 80.) Thirty-five of the middle Oligocene genera (43 percent) and 3 families of perissodactyls (Palaeotheriidae, Helaletidae, Hyracodontidae) are not known in later faunas of Europe.

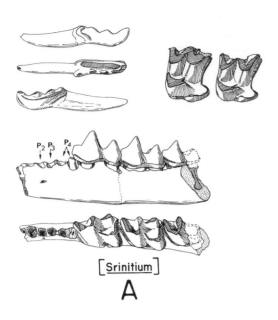

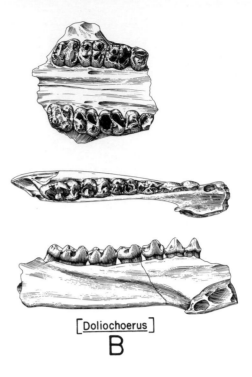

[Srinitium]

A

[Doliochoerus]

B

FIGURE 5-6   *Middle Oligocene Mammals of Europe.*

A   *Srinitium marteli* Hugueney, 1976. Isolated inci-
sors, upper molars, and a lower jaw. ×10 approximately. Figures 1–5 inclusive in Hugueney (1976).

B   *Doliochoerus quercyi* Filhol, 1882. Palatal dentition and lower dentition. ×1/2. Figures 1 and 2 in
Ginsburg (1974).

(*Figure A published with permission of the University of Montpellier. Figure B published with permission of the
Academie des Sciences, Paris.*)

Oligocene Land-Mammal Faunas on the Mediterranean Island of Mallorca

Adrover and Hugueney (1975) and Adrover, Hugueney, and
Mein (1977) have identified middle and late Oligocene land-
mammal local faunas on Mallorca (Balearic Islands) that are
singularly important to the depiction of Oligocene tectonics
and geography in the western Mediterranean Basin. According to these authors, the Eocene and early Oligocene
land fauna of Mallorca shows only European affinity. Likewise, the Sineu Mine local fauna, correlated with the later
middle Oligocene Antoignt fauna (= "level" = "biochron" = "biochronologic zone" = "mammalogic biozone"), shows only European affinity also. But the recently
described Paguera local fauna, "near the middle Oligocene–upper Oligocene passage" (Adrover and Hugueney,
1975) contains phiomyid and ctenodactylid rodents, showing African affinity. Thus, these authors conclude, although
it is known that part of Mallorca was then transgressed by
the sea, part of the island was available to an African-Bal-
earic dispersal of African rodents. They favor the interpretation that Mallorca remained emergent as an attachment to
the Betic Ranges (and structural trend) of southern Spain,
which in turn were probably continuous with the Rif-Magh-
rebin Ranges and structure of North Africa and southern

Spain. Perhaps ecological competition from autochthonous
faunas may have presented barriers to dispersal of terrestrial
vertebrates; otherwise, more African elements would have
visited Paguera.

MIDDLE OLIGOCENE OF ASIA

In the category Middle Oligocene of Asia, we include mammalian assemblages from Kazakhstan, Mongolia, and China.

Composite Middle Oligocene Asian Fauna

ORDER?
  Didymoconidae
    Didymoconus colgatei Matthew and Granger, 1924
    D. berkeyi Matthew and Granger, 1924
LAGOMORPHA
    Desmatolagus gobiensis Matthew and Granger, 1923
    "*D.*" *robustus* (Matthew and Granger, 1923)   —de
      Muizon (1977)
    Ochotonolagus argyropuloi (Gureev, 1960)
    Ordolagus teilhardi (Burke, 1941)
    Procaprolagus gobiensis Matthew and Granger, 1923
    P. radicidens (Teilhard, 1926)
    Agispelagus simplex Argyropulo, 1940

172

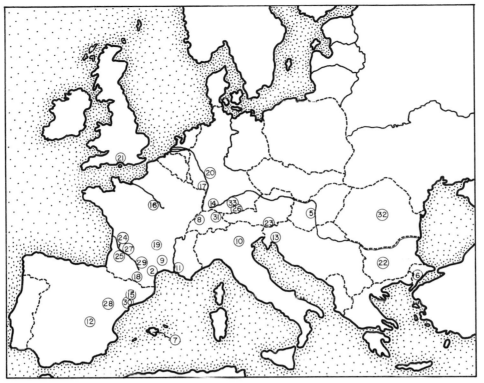

FIGURE 5-7 *Middle Oligocene Localities of Europe.*

1 Antoignt—France	*12 Cerro Arenosa—Spain*	*24 La Réole—France*
2 Assas—France	*13 Dalmatia area—Yugoslavia*	*25 Mas d'Agenais—France*
3 Aubenas-les-Alpes—France	*14 Ehingen (in part)—Germany*	*26 Miesbach—Germany (age?)*
4 Aurillac—France	*15 El Talladel (Tarrega)—Spain*	*27 Moissac—France*
5 Bakonyer Mountains, etc.—	*16 Étampes—France*	*28 Montalban—Spain*
Hungary	*16 Ferté-Alais (La Ferté-Alais)—*	*27 Quercy district—France*
6 Basyigit, etc.—Thrace	*France*	*29 Rabastens—France*
7 Biniamar, etc.—Mallorca	*17 Flonheim, etc.—Germany*	*30 Rocullaura (La Sagarra)—*
8 Blumbach (Bumbach?)-Schan-	*18 Fronsac—France*	*Spain*
gau—Switzerland (age?)	*19 Gergovie—France*	*31 Schannis—Switzerland*
9 Bournoncle-St.-Pierre—	*20 Gusterhain—Germany*	*32 Siebenburger—Ro-*
France	*21 Hamstead—England*	*mania (age?)*
10 Cadibona—Italy	*22 Kajali—Bulgaria (age?)*	*33 Solnhofen 4—Germany*
11 Céreste—France	*23 Keutchach—Austria (age?)*	*15 Tarrega (El Talladel)—Spain*

Sinolagomys cf. *major* Bohlin, 1939

Bohlinotona pusillus (Teilhard, 1926)

CREODONTA

  Hyaenodontidae

    Hyaenodon pervagus Matthew and Granger, 1924

    H. dubius Filhol, 1871

    H. ambiguus Martin, 1906   —Mellett (1968)

    H. filholi Schlosser, 1887

CARNIVORA

  Amphicyonidae

    Cynodictis? elegans Matthew and Granger, 1924

    C.? minor Janovskaya, 1970

    C.? constans (Matthew and Granger, 1924)

    Amphicyon? sp.   —Teilhard (1926)

  Ursidae

    Amphicticeps shackelfordi Matthew and Granger, 1924

    Amphicynodon teilhardi (Matthew and Granger, 1924)

Procyonidae?

  cf. *Plesictis*

Mustelidae

    Palaeogale ulysses Matthew and Granger, 1924

    P. parvula Matthew and Granger, 1924

Viverridae

    Palaeoprionodon gracilis Matthew and Granger, 1924

Felidae

    Nimravus

    Proailurus

INSECTIVORA

  Erinaceoidea

    "Ictopidium" tatagolensis (Sulimski, 1970)

    Tupaiodon morrisi Matthew and Granger, 1924

    T. minutus Matthew and Granger, 1924

    Amphechinus acridens (Matthew and Granger, 1924)

      —Sulimski (1970)

A. cf. *rectus* (Matthew and Granger, 1924)   —Mellett (1968)

A. cf. *minimus* (Bohlin, 1942)   —Mellett (1968)

Exallerix hsandagolensis McKenna and Holton, 1967

Soricoidea

   Gobisorex kingae Sulimski, 1970

ARTIODACTYLA

   Suidae or Tayassuidae

   Propalaeochoerus

   Entelodontidae

   Entelodon ordosius (Young and Chow, 1956)

   E. diconodon Trofimov, 1952

   Anthracotheriidae

   Brachyodus

   Hemimeryx turgaicus Borissiak, 1941

   Hyoboops

Leptomerycidae

   Lophiomeryx turgaicus Flerov, 1940

   Prodremotherium flerowi Trofimov, 1957

   Pseudomeryx gobiensis Trofimov, 1957

   P. hypertalonidus Trofimov, 1957

Hypertragulidae? Gelocidae?

   cf. *Miomeryx*

Cervidae

   Eumeryx culminis Matthew and Granger, 1924

Bovidae?

   Palaeohypsodontus asiaticus Trofimov, 1957

PERISSODACTYLA

   Helaletidae

   Colodon orientalis Borissiak, 1918

   Amynodontidae

   Cadurcodon zaisanensis Belyayeva, 1962

   C. kazakademius Biryukov, 1961

   Hyracodontidae

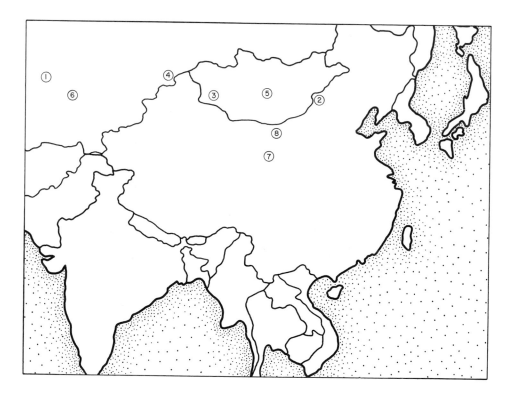

FIGURE 5-8   *Representative Middle Oligocene Localities of Asia.*

   1   Atam-Bas-Chink—Kazakhstan

   2   Baron Sog—Nei Mongol

   3   Boongin (Bungin) Gol—Mongolia

   4   Buran Formation sites—Kazakhstan

   3   Buylstyeen Khuduk—Mongolia

   1   Chelkar–Tenis Lake (Myn Say, Kur Say, Ak Say,
       etc.)—Kazakhstan

   5   Grand Canyon—Mongolia

   5   Hsanda Gol Formation sites—Mongolia

   1   Kalmakpay II–Zaisan Basin—Kazakhstan

   3   Khaitch (Khaytch) Bulak—Mongolia

   1   Kulebay (Baksy, Zhilanshik River, Shalkanura)—
       Kazakhstan

   6   Kyzyl-Kak—Kazakhstan

   7   Lingwu County—Ningxia

   4   Maylibay–Zaisan Basin—Kazakhstan

   1   Myn-Eske-Suyek—Kazakhstan

   3   Nareen Bulak—Mongolia

   4   Qingshuiyin—Ningxia

   8   St. Jacques (San-Tan-Cho)—Nei Mongol

   1   Saty (Sary Turgay River)—Kazakhstan

   5   Tatal Gol—Mongolia

   3   Ulan Ganga—Mongolia

   3   Zaram Bulak (Narin Bulak)—Mongolia

Triplopus? turgaiensis (Belyayeva, 1954)
Ardynia kazachstanensis (Belyayeva, 1952)
Eggysodon turgaicus (Borissiak, 1915)
E. minor (Belyayeva, 1954)
Indricotherium transouralicum Pavlova, 1922
I. brevicervicale (Biryukov, 1953)
Chalicotheriidae
    Schizotherium turgaicum Borissiak, 1920
    S. chucuae Gabunia, 1951
Rhinocerotidae?
    Aceratherium?   —Teilhard (1926)
RODENTIA
Aplodontidae   —Kowalski (1974); Rensberger (1974)
    Selenomys mimicus Matthew and Granger, 1923
    Plesispermophilus? lohiculus (Matthew and Granger, 1923)
    Prosciurus arboraptus Shevyreva, 1971
Castoridae
    Agnotocastor aubekerovi Lychev, 1978
    Propalaeocastor habilis Borissoglebskaya, 1967
    P. kazachstanicus Borissoglebskaya, 1967
Cylindrodontidae
    Pseudocylindrodon mongolicus Kowalski, 1974
    Tsaganomys altaicus Matthew and Granger, 1923
    Cyclomylus lohensis Matthew and Granger, 1923
    C. minutus Kowalski, 1974
    C. turgaicus (Vinogradov and Gamburian, 1952)
    Ardynomys kazachstanicus Vinogradov and Gamburian, 1952
Ctenodactylidae
    Tataromys deflexus Teilhard, 1926
    T. gobiensis Kowalski, 1974
    T. plicidens Matthew and Granger, 1923
    T. sigmodon Matthew and Granger, 1923
    T. cf. *grangeri* Bohlin, 1946
    Karakoromys decessus Matthew and Granger, 1923
    Woodomys chelkaris Shevyreva, 1971
    Yindirtemys?
Zapodidae
    Plesiosminthus asiaecentralis (Bohlin, 1946)
    P. tangingoli (Bohlin, 1946)
    P. quartus Shevyreva, 1970
Cricetidae
    Cricetops dormitor Matthew and Granger, 1923
    C. aeneus Shevyreva, 1965
    Eucricetodon asiaticus (Matthew and Granger, 1923)
      —Lindsay (1978)
Rhizomyidae
    Tachyoryctoides obrutschewi Bohlin, 1937
    T. pachygnathus Bohlin, 1937
    T. tatalgolicus Dashzeveg, 1971
Thryonomyidae
    Sayimys
Ischyromyoidea?
    Terrarboreus arcanus Shevyreva, 1971

The mid-Oligocene of Asia displays a strong holarctic element—that is, families common to Eurasia and North America—which was rejuvenated by dispersals across the Bering region at about the beginning of the Oligocene. Mixed with this are continuing Asian endemics such as the didymoconids and certain groups of rodents. Only about 21 per-cent of the 62 genera recognized in Asia are also found in North America, but about 73 percent of the Asian families are also found in North America. Early indricotheres (giant hyracodontids) are the most publicized member of the Asian mid-Oligocene fauna.

ORELLAN

The Orellan land-mammal age is based on the fauna from the Orella Member of the Brule Formation, which overlies the Chadron and constitutes the upper division of the White River Group. The type section of the Brule, like the Chadron, is in the Big Badlands of South Dakota and the renowned fossiliferous Brule of northwestern Nebraska is included as "Type area" (H. Wood et al., 1941; Schultz and Stout, 1955; and Harksen and Macdonald, 1969B). The composite Orellan mammalian fauna, taken from localities in Nebraska, South Dakota, Wyoming, and Colorado, includes the following.

Composite Orellan Mammalian Fauna

MARSUPIALIA
Didelphidae
    Herpetotherium fugax Cope, 1873T
    Nanodelphys minutus McGrew, 1937A
LEPTICTIDA
Leptictidae
    Leptictis dakotensis (Leidy, 1868J)
    L. bullatus (Matthew, 1899A)
    L. haydeni Leidy, 1868J
    L.? caniculus (Cope, 1873T)
LAGOMORPHA
Ochotonidae
    Desmatolagus gazini Burke, 1936
Leporidae
    Litolagus mollidens Dawson, 1958
    Paleolagus burkei A. Wood, 1940
    P. haydeni Leidy, 1856I
    Megalagus brachyodon (Matthew, 1903B)
    M. turgidus (Cope, 1873T)
APATOTHERIA
Apatemyidae
    Sinclairella
CREODONTA
Hyaenodontidae
    Hyaenodon crucians Leidy, 1853D
    H. cruentus Leidy, 1853D
    H. horridus Leidy, 1853D
    H. mustelinus Scott, 1894F
    H. paucidens Osborn and Wortman, 1894A
CARNIVORA
Amphicyonidae   —in the sense of Hunt (1974)
    Daphoenus hartshornianus (Cope, 1873CC)

D. vetus Leidy, 1853D
D. minimus Hough, 1948
Protemnocyon inflatus Hatcher, 1902F
Ursidae?
Brachyrhynchocyon intermedius (Loomis, 1931)
Procyonidae?
Mustelidae
Palaeogale lagophaga Simpson, 1946C
Canidae
Hesperocyon gregarius (Cope, 1873T)
Felidae
Dinictis bombifrons Adams, 1875A
D. cismontanus (Thorpe, 1920A)
D. felina Leidy, 1856H
D. squalidens (Cope, 1873T)
Hoplophoneus occidentalis (Leidy, 1896A)
H. primaevus (Leidy, 1891H)
H. robustus Adams, 1896B
Eusmilus sicarius Sinclair and Jepsen, 1927A
INSECTIVORA
Erinaceoidea
Proterix loomisi Matthew, 1903C
Geolabididae
Centetodon marginalis (Cope, 1873CC) —Lille-
graven, McKenna, and Krishtalka (1981)
Ankylodon annectens Patterson and McGrew, 1937
Soricidae
Pseudotrimylus compressus (Galbreath, 1953) —
Engesser (1979)
Domnina crassa (Scott, 1894A)
Talpidae
Proscalops tertius K. Reed, 1961
ARTIODACTYLA
Leptochoeridae
Stibarus obtusilobus Cope, 1873T
S. lemurinus (Cope, 1873T)
S. loomisi Scott, 1940B
Leptochoerus gracilis Marsh, 1894L
L. spectabilis Leidy, 1856I
Nanochoerus elegans Macdonald, 1955
N. quadricuspis (Hatcher, 1901A)
N. scotti Macdonald, 1955
Entelodontidae
Archaeotherium mortoni Leidy, 1850A
A. wanlessi Sinclair, 1921C
Entelodon? cf. *magnus* Aymardl 1846
Megachoerus? praecursor Scott, 1940B
Tayassuidae
Thinohyus nanus Marsh, 1894L —Stirton and
Woodburne (1965)
Perchoerus probus Leidy, 1856J
Anthracotheriidae
Aepinacodon rostratus (Scott, 1895E) —Macdon-
ald (1956C)
Heptacodon occidentalis Scott, 1940B
H. quadratus Scott, 1940B
Agriochoeridae
Agriochoerus antiquus Leidy, 1850C

A. latifrons Leidy, 1869A
A. major Leidy, 1856J
Merycoidodontidae[2]
Subdesmatochoerus socialis (Marsh, 1885B)
Prodesmatochoerus meekae Schultz and Falkenbach, 1954
Miniochoerus battlecreekensis Schultz and Falkenbach, 1956
M. starkensis Schultz and Falkenbach, 1956
M. affinis (Leidy, 1869A)
M. gracilis (Leidy, 1851D)
M. helprini Schultz and Falkenbach, 1956
Platyochoerus platycephalus (Thorpe, 1921F)
P. heartensis Schultz and Falkenbach, 1956
Stenopsochoerus sternbergi Schultz and Falkenbach, 1956
S. joderensis Schultz and Falkenbach, 1956
S. chadronensis Schultz and Falkenbach, 1956
S. douglasensis Schultz and Falkenbach, 1956
Parastenopsochoerus conversensis Schultz and Falkenbach,
1956
Merycoidodon culbertsonii Leidy, 1848B
M. macrorhinus (Douglass, 1903A)
M. dani Schultz and Falkenbach, 1968
Paramerycoidodon georgei Schultz and Falkenbach, 1968
P. bacai Schultz and Falkenbach, 1968
Otionohyus wardi Schultz and Falkenbach, 1968
O. bullatus (Leidy, 1869A)
O.? cedrensis (Matthew, 1901B)
Genetochoerus periculorum (Cope, 1884M)
G. norbeckensis Schultz and Falkenbach, 1968
Hadroleptauchenia eiseleyi Schultz and Falkenbach, 1968
Pseudoleptauchenia orellaensis Schultz and Falkenbach, 1968
Leptauchenia harveyi Schultz and Falkenbach, 1968 (age?)
Camelidae
Poebrotherium eximium Hay, 1902A
P. labiatum Cope, 1881I
P. wilsoni Leidy, 1847B
P. andersoni Troxell, 1917B
Paratylopus primaevus Matthew, 1904B
Hypertragulidae
Hypertragulus calcaratus Cope, 1873T
Hypisodus minimus Cope, 1873T
Leptomerycidae
Leptomeryx evansi Leidy, 1853D
PERISSODACTYLA
Equidae
Mesohippus antiquus (Schlaikjer, 1935C)
M. bairdi (Leidy, 1850C)
M. barbouri Schlaikjer, 1931
M. obliquidens Osborn, 1904B
M. trigonostylus Osborn, 1918A
Helaletidae
Colodon dakotensis Osborn and Wortman, 1895A
C.? cingulatus Douglass, 1901 —Schultz, Martin,
and Corner (1975)
Tapiridae
Protapirus simplex Wortman and Earle, 1893A
Hyracodontidae
Hyracodon nebraskensis Leidy, 1850C
Amynodontidae
Metamynodon planifrons Scott and Osborn, 1887B
Rhinocerotidae
Caenopus dakotensis Peterson, 1920A

Subhyracodon copei (Osborn, 1898I)

S. metalophus (Troxell, 1921F)

S. occidentalis (Leidy, 1851E)

S. trigonodus (Osborn and Wortman, 1891A)

Diceratherium tridactylum (Osborn, 1893C)

PHOLIDOTA?

Epoicotheriidae

Xenocranium pileorivale Colbert, 1942H

RODENTIA

Ischyromyoidea

Ischyromys typus Leidy, 1856I

I. veterior (Matthew, 1903B)   —Black (1968A)

Aplodontidae   —Rensberger (1975)

Prosciurus relictus (Cope, 1873CC)

P. aff. *saskatchewanensis* (Lambe, 1908A)

Pelycomys placidus Galbreath, 1953

Cedromus wardi R. Wilson, 1949A   (family?)

Eomyidae

Adjidaumo minutus (Cope, 1873T)

Paradjidaumo trilophus (Cope, 1873T)

Castoridae

Agnotocastor

Sciuridae, 2 genera undescribed, acc. to Harksen and Macdonald (1969B)

Geomyidae?

Diplolophus adspectans A. Wood, 1936C

Heteromyidae

Heliscomys hatcheri A. Wood, 1935C

H. senex A. Wood, 1935C

FIGURE 5-9 *Representative Orellan Localities.*

 1 *Badland Creek—Nebraska*
 2 *Bald Butte (Hat Creek Basin)—Wyoming*
 3 *Battle Creek Breaks, Canyon, Draw—South Dakota*
 3 *Big Badlands—South Dakota*
 4 *Big Hole River—Montana*
 1 *Brecht Ranch, Stock Dam—Nebraska*
 5 *Cedar Creek "facies" sites—North Dakota*
 6 *Cedar Creek Member sites—Colorado*
 7 *Chili Gulch—California   (age?)*
 8 *Cook Ranch Formation sites—Montana*
 9 *Devils Gap (Beaver Divide)—Wyoming*
 5 *Dickinson Southwest—North Dakota*

10 *Douglas Southeast (Orin Junction)—Wyoming*
 5 *Fitterer Ranch—North Dakota*
11 *Goshen Hole district—Wyoming*
11 *Harvard Fossil Reserve—Wyoming*
 2 *Hat Creek Basin—Wyoming-Nebraska*
 1 *Meng Ranch—Nebraska*
12 *Palisades Section—Montana*
13 *Point Creek Head–Slim Buttes—South Dakota*
14 *Scotts Bluff Monument district—Nebraska*
13 *Slim Buttes—South Dakota*
 1 *Toadstool Park—Nebraska*
12 *Toston (Dry Creek)—Montana*

    H. vetus Cope, 1873CC
    H. gregoryi A. Wood, 1933A
    Apletotomeus crassus Reeder, 1960
    Akmaiomys incohatus Reeder, 1960
  Eutypomyidae
    Eutypomys magnus A. Wood, 1937D
    E. thompsoni Matthew, 1905A
  Cricetidae
    Eumys elegans Leidy, 1856I
    E. obliquidens A. Wood, 1937D
    E. parvidens A. Wood, 1937D
    Eoeumys exiguus (A. Wood, 1937D)   —Martin
      (1980) (=*Paracricetodon* of Alker, 1968?)
    E. vetus (A. Wood, 1937D)   —Martin (1980)
    Coloradoeumys galbreathi Martin, 1980
    Wilsoneumys planidens (R. Wilson, 1949A)   —
      Martin (1980)

Of the 75 Orellan mammalian genera, 49 (65 percent) were in the Chadronian fauna and 25 (33 percent) do not appear later in North America. Only 1 family (Apatemyidae) does not have a post-Orellan record in North America.

LATE OLIGOCENE OF EUROPE

Three sets of faunas are described as Late Oligocene of Europe; they are regarded as successional on the basis of their contained rodent species, following D. Russell et al. (1983): "Level" of Boningen, "Level" of Rickenbach, and "Level" of Coderet. This interval is the general equivalent of the *Amphilagus* faunizone of Thaler (1972) as well as his Boningen and Rickenbach biozones, and it also equals the Boningen, La Milloque, and Coderet reference localities of Fahlbusch (1976). It also appears to be about the same interval as the Neochattian of some of the marine stratigraphers in Europe.

Composite Late Oligocene Mammalian Fauna of Europe

MARSUPIALIA
  Didelphidae
    Peratherium antiquum (Blainville, 1840)
    P. elegans (Aymard, 1846)
    Amphiperatherium exile (Gervais, 1848–1852)
LAGOMORPHA
  Ochotonidae
    Amphilagus antiquus (Pomel, 1853)
    Piezodus branssatensis Viret, 1929C
CREODONTA
  Hyaenodontidae
    Hyaenodon brachyrhynchus Blainville, 1841
    H. exiguus (Gervais, 1873)
    H. leptorhynchus Laizer and Parieu, 1838
    H. gervaisi Martin, 1906

CARNIVORA
  Amphicyonidae
    Pseudocyonopsis landesquei (Helbing, 1928C)
    Haplocyon dombrowskii Helbing, 1928C
    Haplocyonopsis crassidens de Bonis, 1973
    Ysengrinia tolosana (Noulet, 1876)
    Goupilictis minor (Dehm, 1935A)
  Ursidae
    Cephalogale brevirostris (Blainville, 1839)
    C. geoffroyi Jourdan, 1862
    C. minor Filhol, 1879
    C. bonali Helbing, 1922
    C. filholi (Munier-Chalmas, 1877)
  Procyonidae
    Plesictis branssatensis Viret, 1928
    P. milloquensis Helbing, 1928C
    P. genettoides Pomel, 1846
    P. julieni Viret, 1928
    P. dieupentalensis Ginsburg and Vidalenc, 1977
    P. cf. *stenogalinus* Teilhard, 1915
  Mustelidae
    Amphictis ambiguus Gervais, 1869
    A. nanus Teilhard, 1915
    A. barbonicus Viret, 1928
    Palaeogale sp.
    Stenogale gaillardi Viret, 1928
    Potamotherium valetoni Geoffroy, 1832
  Viverridae
    Stenoplesictis cayluxi Filhol, 1880
  Felidae
    Proailurus aff. *lemanensis* Filhol, 1879
INSECTIVORA
  Erinaceoidea
    Neurogymnurus cayluxi Filhol, 1877
    N. mediterraneus Viret, 1947
    Amphechinus arvernensis (Blainville, 1840)
  Soricoidea
    Plesiosorex soricinoides (Blainville, 1838)
    Paratalpa micheli Lavocat, 1951C
    Mygatalpa arvernensis Schreuder, 1940
    heterosoricines   —Crochet (pers. commun., 1981)
    Dinosorex huerzeleri Engesser, 1975
    Crocidosorex piveteaui Lavocat, 1951C
    Geotrypus cadurcensis (Filhol, 1884)
    "*G.*" *jungi* (Lavocat, 1951C)
CHIROPTERA
  Hipposideridae
    Hipposideros branssatensis Hugueney, 1965
    cf. *Palaeophyllophora* sp.
  Rhinolophidae
    Rhinolophus cluzeli Hugueney, 1965
    R. cf. *priscus* Revilliod, 1920
  Vespertilionidae
    Stehlinia minor (Revilliod, 1922)
    Myotis salodorensis Revilliod, 1922
    cf. *Myotis* sp.
ARTIODACTYLA
  Dichobunidae
    Metriotherium mirabile Filhol, 1882
  Anoplotheriidae
    Ephelcomenus filholi Lydekker, 1857

Tayassuidae
Palaeochoerus gergovianus (Croizet in Blainville, 1846)
P. pusillus Ginsburg, 1974
Doliochoerus quercyi Filhol, 1882
Anthracotheriidae
Anthracotherium
Elomeryx cf. *borbonicus* (Gervais, 1852)
Microbunodon minimum (Cuvier, 1822)
Cainotheriidae
Caenomeryx procommunis (Filhol, 1877)
Plesiomeryx cadurcensis Gervais, 1873
Cainotherium cf. *geoffroyi* Pomel, 1851
C. commune Bravard, 1835
Gelocidae or Leptomerycidae
Prodremotherium elongatum Filhol, 1877
Rutitherium nouleti Filhol, 1876
Leptomerycidae
Lophiomeryx chalaniati Pomel, 1853
Cervoidea
Dremotherium
Amphitragulus primaevus Schlosser, 1925
A. feningrei Schlosser, 1925
ARCTOCYONIA
Paroxyclaenidae
Kochictis cetenii Kretzoi, 1943
PERISSODACTYLA
Tapiridae
Protapirus aginense Richard, 1938
Amynodontidae
Cadurcotherium nouleti Roman
Rhinocerotidae
Ronzotherium romani Kretzoi, 1940
Aceratherium albigense Roman, 1911
A. (*Mesaceratherium*) aff. *paulhiacensis* (Richard, 1937)
Brachypotherium lemanense (Pomel, 1853)
PHOLIDOTA
cf. *Necromanis*
RODENTIA
Theridomyidae
Archaeomys (*Archaeomys*) *major* Schlosser, 1884
A. (*A.*) *intermedius* Vianey-Liaud, 1977
A. (*A.*) *laurillardi* Gervais, 1848–1852
A. (*Blainvillimys*) *geminatus* Thaler, 1966
Issiodoromys quercyi (Schlosser, 1884)
I. puffiensis Vianey-Liaud, 1976
I. pseudanaema Gervais, 1884
Columbomys lavocati Thaler, 1962
Eomyidae
Eomys zitteli Schlosser, 1884
E. major Freudenberg, 1941
Pseudotheridomys pusillus Fahlbusch, 1969B
P. aff. *parvulus* (Schlosser, 1884)
Rhodanomys schlosseri Depéret and Douxami, 1902
Aplodontidae
Plesispermophilus ernii Stehlin and Schaub, 1951
P. macrodon Schmidt-Kittler and Vianey-Liaud, 1979
P.? argoviensis Stehlin and Schaub, 1951
Sciuridae
Heteroxerus costatus (Freudenberg, 1941)
H. lavocati Hugueney, 1969A
H. paulhiacensis Black, 1965B

Palaeosciurus?
Sciurus chalaniati Pomel, 1853
"S." solitarius Hugueney, 1969A
"S."
Castoridae
Steneofiber eseri Meyer, 1846
S. dehmi Freudenberg, 1941
Gliridae
Peridyromys murinus (Pomel, 1853)
Gliravus bruijni Hugueney, 1967
Glirudinus praemurinus (Freudenberg, 1941)
G. glirulus (Dehm, 1935A)
Branssatoglis fugax (Hugueney, 1967)
B. concavidens Hugueney, 1967

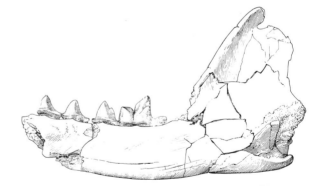

FIGURE 5-10 *Late Oligocene Mammals of Europe.*
A *Ysengrinia tolosana (Noulet, 1876). Lower jaw with*
$\overline{P_2-M_2}$. *×1/2 approx.. Figures 1 and 2 in Crouzel,
Ginsburg, and Vidalenc (1977).*
B *Plesictis genettoides Pomel, 1846. Lower jaw with*
$\overline{P_2-M_1}$. *×1½. Figures 18 and 19 in Crouzel,
Ginsburg, and Vidalenc (1977).*
*(Published with permission of the Société d'Histoire Natu-
relle, Toulouse, France.)*

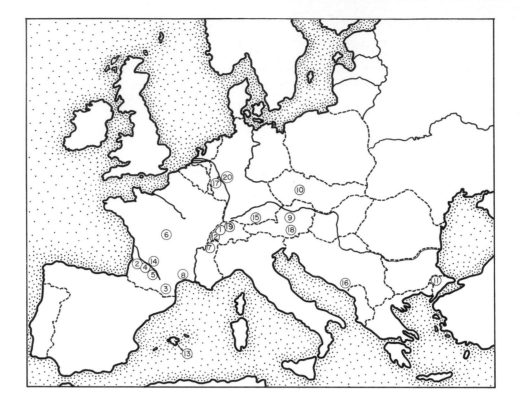

FIGURE 5-11 *Representative Late Oligocene Localities of Europe.*

1	*Aarwangen—Switzerland*	
2	*Aillas—France*	
3	*Armissan—France*	
4	*Bazin—France*	
5	*Bessens—France*	
1	*Boningen—Switzerland*	
6	*Branssat-Coderet—France*	
6	*Coderet-Branssat—France*	
5	*Dieupentale—France*	
4	*Domilhac—France*	
6	*Gannat—France*	
7	*Humilly—France*	

8	*La Colombière Fissure— France*
4	*La Milloque—France*
9	*Linz—Austria*
10	*Lukawitz-Markdorf—Czecho- slovakia*
11	*Masatly (Edirne)—Thrace*
12	*Mathod—Switzerland*
13	*Paguera I and II—Mallorca*
14	*Pech Desse (Quercy)—France*
14	*Pech du Fraysse (Quercy)— France*

15	*Peissenberg—Germany*
16	*Petnik-Ivangrad-Montene- gro—Yugoslavia*
1	*Rickenbach-Mühle—Switzer- land*
17	*Rott at Bonn—Germany*
18	*Trifail—Austria*
19	*Veltheim—Switzerland*
20	*Westerhain-Westerwald—Ger- many*

Cricetidae

Eucricetodon quercyi Vianey-Liaud, 1972B

E. huerzeleri Vianey-Liaud, 1972B

E. collatum (Schaub, 1925)

E. praecursor (Schaub, 1925)

Pseudocricetodon philippi Hugueney, 1971A

P. thaleri (Hugueney, 1969A)

Heterocricetodon gaimersheimense Freudenberg, 1941

H. schlosseri Schaub, 1925

H. cf. *helbingi* Stehlin and Schaub, 1951

Paracricetodon spectabile (Schlosser, 1884)

Melissiodon chatticum Freudenberg, 1941

M. emmerichi Schaub, 1920

M. quercyi Schaub, 1920

Adelomyarion vireti Hugueney, 1969A

Rhizospalax poirrieri Miller and Gidley, 1919

Zapodidae

Plesiosminthus schaubi Viret, 1929C

P. promyarion Schaub, 1930

Of the 79 mammalian genera known from the late Oligocene of Europe, 46 (58 percent) were in the middle Oligocene and 45 (56 percent) carry into the Agenian (early Miocene). Six families of the late Oligocene (Paroxyclaenidae, Hyaenodontidae, Dichobunidae, Anoplotheriidae, Gelocidae, and Amynodontidae) have no later record in Europe.

Certain faunas in Asia have been called Aquitanian = late Oligocene or Aquitanian = early Miocene. They are not readily distinguishable from faunas that have been called Burdigalian = early Miocene, except that possibly the mastodonts appear on the scene and the giant indricothere rhinoceroses dwindle and disappear by Burdigalian time. Data for correlating these Asian faunas with the late Oligocene or with the Agenian (= earliest Miocene) of Europe appear to be equivocal. Lacking a better name, we can only term these Asian faunas *Oligo-Miocene*, and they may be partly contemporaneous with faunas listed as early Miocene (next chapter). Georgian SSR, Kazakhstan, Mongolia, and China contribute samples to this composite fauna.

Composite Oligo-Miocene Asian Fauna

LAGOMORPHA
  Sinolagomys kansuensis Bohlin, 1937
  S. major Bohlin, 1937
  S. gracius Bohlin, 1942
  Agispelagus simplex Argyropulo, 1940
  Desmatolagus aff. *gobiensis* Matthew and Granger, 1923
CREODONTA
  Hyaenodontidae
    Hyaenodon dubius Filhol, 1873
CARNIVORA
  Amphicyonidae
    Amphicyon
  Ursidae
    Cephalogale meschethense Gabunia, 1964
  Procyonidae
    Plesictis
  Felidae
    Nimravus
INSECTIVORA
 Erinaceoidea
    Amphechinus cf. *rectus* (Matthew and Granger, 1924)
    A. acridens (Matthew and Granger, 1924)
    A. minimus (Bohlin, 1942)
ARTIODACTYLA
  Suidae
    Conohyus betpakdalensis Trofimov, 1949
  Entelodontidae
    Paraentelodon intermedium Gabunia, 1964
  Anthracotheriidae
    Anthracotherium kwablianicum Gabunia, 1964
    Telmatodon
    Hemimeryx turgaicus Borissiak, 1941
    Hyoboops
    Bothriodon
    cf. *Parabrachyodus borbonicoides* (Forster-Cooper, 1924)
  Leptomerycidae
    Prodremotherium trepidum Gabunia, 1964
    Lophiomeryx benarensis Gabunia, 1964

    L. turgaicus Flerov, 1940
  Cervidae?
    Iberomeryx parvus
ORDER?
    Didymoconus berkeyi Matthew and Granger, 1924
PERISSODACTYLA
  Helaletidae
    Colodon sp.
  Hyracodontidae
    Paraceratherium prohorovi (Borissiak, 1939)
    P. lipidus Xu and Wang, 1978
    Aprotodon borisiaki Belyayeva, 1954
    Eggysodon turgaicus (Borissiak, 1915)
    "*Prothyracodon*" *turgaiensis* (Belyayeva, 1954)
    Ardynia kazachstanensis (Belyayeva, 1954)
    A. plicidentata Gabunia, 1955
    Benaritherium callistrati Gabunia, 1955
    Dzungariotherium ordosensis Chiu, 1973
    D. turfanensis Xu and Wang, 1978
  Rhinocerotidae
    Aceratherium aralense Borissiak, 1944
    Dicerorhinus minutus (Cuvier, 1824)
    Brachypotherium
    Meschotherium mescheticum Gabunia, 1964
  Chalicotheriidae
    Schizotherium chucuae Gabunia, 1951
    Borissiakia betpakdalensis (Flerov, 1938)
RODENTIA
  Ctenodactylidae
    Tataromys sigmodon Matthew and Granger, 1923
    T. grangeri Bohlin, 1946
    T. deflexus Teilhard, 1926
    T. suni Li and Qiu, 1980
    T. cf. *plicidens* Matthew and Granger, 1923
    Leptotataromys gracilidens Bohlin, 1946
    Yindirtemys woodi Bohlin, 1946
  Cylindrodontidae
    Tsaganomys
  Castoridae
    Propalaeocastor kumbulakensis Lychev, 1970
    Steneofiber
  Cricetidae
    Eucricetodon aff. *asiaticus* (Matthew and Granger, 1923)
    Aralomys gigas Argyropulo, 1939
    Eumysodon spurius Argyropulo, 1939
    E. orlovi Argyropulo, 1939
    Argyromys aralensis (Argyropulo, 1939)
    A. woodi (Argyropulo, 1939)
    Protalactaga borissiaki Argyropulo, 1939
  Rhizomyidae
    Tachyoryctoides obrutshewi Bohlin, 1937
    T. kokonorensis Li and Qiu, 1980
  Zapodidae
    Plesiosminthus asiaecentralis (Bohlin, 1946)
    P. tangingoli (Bohlin, 1946)
    P. parvulus (Bohlin, 1946)

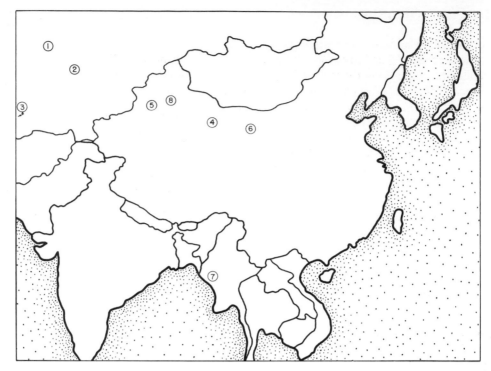

FIGURE 5-12 *Oligo-Miocene Localities of Asia.*

1	*Agispe district—Kazakhstan*
1	*Aral Sea district—Kazakhstan*
2	*Askazan-Sor (Betpak-Dala)— Kazakhstan*
3	*Benara—Georgian SSR*
2	*Betpak-Dala (Steppe)—Ka- zakhstan*
4	*Choei-Tong-Kou (Shui-Tung- Kou)—Gansu*
5	*Dzungaria Basin (in part)— Xinjiang*
5	*Feiyue (Hami)—Xinjiang*

5	*Hongshan Farm (South Ansi- hai)—Xinjiang*
6	*Kuyan (age?) and Hui Hui Pu (Oligocene?)—Ningxia*
7	*Myaing–Kyaukwet—Burma (in part?)*
7	*Pegu "series" (in part) sites— Burma*
1	*Petrovskiy Bay–Aral Sea—Ka- zakhstan*
4	*Shargaltein Gol (Shara Gol Valley)–Tsaidam Basin—Gansu*

4	*Shui-Tung-Kou (Choei-Tong- Kou)—Gansu*
4	*Taben Buluk (Yindirte)—Gansu*
4	*Tsaidam Basin (in part)— Gansu*
8	*Turfan Basin (in part)—Xin- jiang*
1	*Turgai area (in part)—Ka- zakhstan*

WHITNEYAN

The Whitneyan land-mammal age is based on the fauna from the Whitney Member of the Brule as exposed in northwestern Nebraska, with much of the typical Whitneyan fauna having been obtained from the upper part of the Brule (Poleslide Member) in the Big Badlands of South Dakota (H. Wood et al., 1941; Schultz and Stout, 1955; and Harksen and Macdonald, 1969B). The composite Whitneyan mammalian fauna, taken from sites in Nebraska and South Dakota, includes the following.

Composite Whitneyan Mammalian Fauna

MARSUPIALIA
  Didelphidae
    Herpetotherium merriami Stock and Furlong, 1922A
LEPTICTIDA
  Leptictidae
    Leptictis (Slim Buttes)

LAGOMORPHA
  Leporidae
    Palaeolagus haydeni Leidy, 1856I
    Megalagus turgidus (Cope, 1873T)
CREODONTA
  Hyaenodontidae
    Hyaenodon cf. *cruentus* Leidy, 1853D
CARNIVORA
  Amphicyonidae
    Daphoenus
  Ursidae
    Allocyon (Arikareean?)
    Parictis?
    Brachyrhynchocyon sesnoni Macdonald, 1967 (family?)
  Mustelidae
    Palaeogale? (I-75 locality)
  Canidae
    Hesperocyon gregarius (Cope, 1873T)
    H. geismarianus (Cope, 1879D)
    H. latidens (Cope, 1881C)
    H. lemur (Cope, 1879E)

Mesocyon coryphaeus (Cope, 1879F)
M. brachyops Merriam, 1906A
M. josephi (Cope, 1881C)
Pericyon socialis Thorpe, 1922C
Cynarctoides cuspidatus (Green, 1954C)
Sunkahetanka sheffleri Macdonald, 1967B
Felidae
Nimravus altidens Macdonald, 1950A
N. brachyops (Cope, 1879F)
N. bumpensis Scott and Jepsen, 1936
Dinictis bombifrons Adams, 1895A
D. felina Leidy, 1856H
Hoplophoneus
Eusmilus dakotensis Hatcher, 1895B
INSECTIVORA
Erinaceoidea
Proterix bicuspis (Macdonald, 1951E)
P. loomisi Matthew, 1903C
Soricoidea
Domnina   (Slim Buttes)
Centetodon marginalis (Cope, 1873CC)
ARTIODACTYLA
Leptochoeridae
Leptochoerus supremus Macdonald, 1955
Stibarus lemurinus (Cope, 1873R)
Nanochoerus   (Slim Buttes)
Entelodontidae
Archaeotherium altidens Sinclair, 1921C
A. latidens Troxell, 1920B
A. lemleyi Macdonald, 1951E
A. zygomaticus Troxell, 1920B
Tayassuidae
Perchoerus probus Leidy, 1856J
Chaenohyus Cope, 1879E   —Stirton and Wood-
    burne (1965)
Anthracotheriidae
Elomeryx armatus (Marsh, 1894F)
Heptacodon gibbiceps (Marsh, 1894K)
Octacodon valens Marsh, 1894F
Agriochoeridae
Agriochoerus gaudryi (Osborn and Wortman, 1893A)
Merycoidodontidae[2]
Promesoreodon scanloni Schultz and Falkenbach, 1949
Subdesmatochoerus montanus (Douglass, 1907C)
S. shannonensis Schultz and Falkenbach, 1954
Miniochoerus nicholsae Schultz and Falkenbach, 1956
M. cheyennensis Schultz and Falkenbach, 1956
M. ottensi Schultz and Falkenbach, 1956
Platyochoerus hatcreekensis Schultz and Falkenbach, 1956
Stenopsochoerus berardae Schultz and Falkenbach, 1956
Merycoidodon lambi Schultz and Falkenbach, 1968
M. galushai Schultz and Falkenbach, 1968
M. lynchi Schultz and Falkenbach, 1968
Paramerycoidodon major (Leidy, 1854A)
P. wanlessi Schultz and Falkenbach, 1968
Otionohyus hybridus (Leidy, 1869A)
O. alexi Schultz and Falkenbach, 1968
Genetochoerus geygani Schultz and Falkenbach, 1968
G. chamberlaini Schultz and Falkenbach, 1968
G. dickensonensis (Douglass, 1907C)
Pithecistes tanneri Schultz and Falkenbach, 1968

Leptauchenia harveyi Schultz and Falkenbach, 1968
L. decora Leidy, 1856I
Hadroleptauchenia primitiva Schultz and Falkenbach, 1968
Pseudocyclopidius frankforteri Schultz and Falkenbach, 1968
Camelidae
Pseudolabis dakotensis Matthew, 1904B
P. matthewi Lull, 1921C
Paratylopus primaevus Matthew, 1904B
P. campester (Matthew, 1901B)
P. matthewi (Lull, 1921C)
Protoceratidae
Protoceras celer Marsh, 1891A
Protoceratidae?
Nanotragulus   (I-75 locality)
Hypertragulidae
Hypisodus
Leptomerycidae
Leptomeryx evansi Leidy, 1853D
L. semicinctus Cope, 1889I
L. obliquidens Lull, 1922B
L. minimus Frick, 1937
PERISSODACTYLA
Equidae
Miohippus brachystylus (Osborn, 1904B)
M. crassicuspis Osborn, 1904B
M. gidleyi Osborn, 1904B
M. intermedius (Osborn and Wortman, 1895A)
M. meteulophus (Osborn, 1904B)
M. validus (Osborn, 1904B)
Helaletidae
Colodon cf. *dakotensis* Osborn and Wortman, 1895A
Tapiridae
Protapirus obliquidens Wortman and Earle, 1893A
P. validus Hatcher, 1896B
Hyracodontidae
Hyracodon nebraskensis (Leidy, 1850C)
Amynodontidae
Metamynodon planifrons Scott and Osborn, 1887B
Rhinocerotidae
Subhyracodon
Caenopus dakotensis Peterson, 1920A
Amphicaenopus platycephalus (Osborn and Wortman, 1894A)
Diceratherium tridactylum (Osborn, 1893C)
RODENTIA
Ischyromyidae
Ischyromys typus Leidy, 1956I
Aplodontidae
Pelycomys   (Slim Buttes)
Prosciurus cf. *relictus* (Cope, 1873CC)
Eomyidae
Paradjidaumo
Castoridae
Agnotocastor praetereadens Stirton, 1935
Heteromyidae
Heliscomys   (Slim Buttes)
Cricetidae
Eumys brachyodus A. Wood, 1937D
Scottimus lophatus A. Wood, 1937D

Paracricetodon exiguus (A. Wood, 1937D)   —Alker
   (1968)

Wilsoneumys planidens (R. Wilson, 1949A)   —
   Martin (1980)

Of the 71 Whitneyan mammalian genera, 50 (70 percent) were in the Orellan and 31 (44 percent) continue into early Arikareean. Three Whitneyan families (Leptictidae, Helaletidae, and Ischyromyidae), all represented by very little material, do not appear in North America later. The Eomyidae are not reported from early Arikareean but appear later.

We believe that from the beginning of the Chadronian through the early Arikareean—in other words, through the Oligocene—the North American land-mammal faunas were largely evolving in situ. Immigrant taxa are conjectural.

The Oligocene land-mammal fauna of North America is dominantly of modern aspect. All orders and 50 percent of the families have living representatives. By Whitneyan time, only 20 percent of the archaic families of the Chadronian and Orellan still remained. The herbivorous component, including seed eaters, suggests a countryside most suited to browsers. Open woodland- and brush-savanna habitats must have been conspicuous, especially in the interior of the continent. Essential absence of primates correlates with paleobotanical interpretation of drier and more temperate climate in the midcontinent. Evidently through late Eocene and Oligocene time, the formerly abundant and diversified primates retreated southward to the heavier forests or were exterminated. The highly specialized omomyid

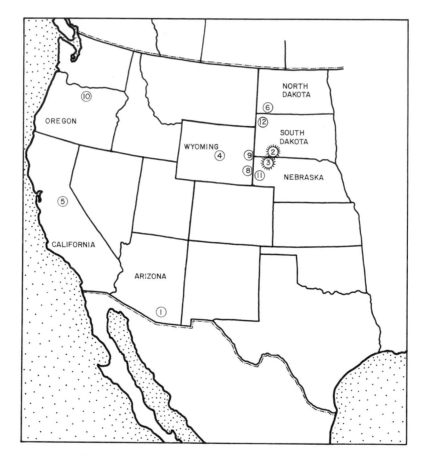

FIGURE 5-13 *Representative Whitneyan Localities.*

  1 *Atravesada—Arizona*
  2 *Big Badlands—South Dakota*
  3 *Blackburn Ranch–Broadwater area—Nebraska*
  4 *Cedar Ridge (Black-Dawson site)—Wyoming*
  5 *Chili Gulch—California (age?)*
  5 *Douglas Flat—California (age?)*
  6 *Fitterer Ranch—North Dakota*
  7 *Gainesville West-Southwest (I-75)—Florida*
  8 *Goshen Hole district—Wyoming*
  9 *Hat Creek Basin—Wyoming and Nebraska*

  7 *I-75 (See Gainesville.)*
10 *John Day Formation (lower) sites—Oregon*
  2 *Poleslide Member sites—South Dakota*
11 *Roubideaux Pass (Wildcat Ridge, Round Top)—*
    Nebraska
11 *Scotts Bluff area—Nebraska*
12 *Slim Buttes sites—South Dakota*
  3 *Toadstool Park (and Orellan)—Nebraska*
  3 *Whitney Member sites—Nebraska*

primate *Ekgmowechashala* Macdonald, now recorded from the Sharps Formation of South Dakota and from the John Day Formation of Oregon, is the sole representative of this great order in the North American Oligo-Miocene. No primates are known in North America for the last 20 million years of its history until people arrived from Asia a few thousand years before the present.

EARLY ARIKAREEAN

The Arikareean land-mammal age is based on the faunas from the Arikaree Group of northwestern Nebraska. This entire age was customarily assigned *early Miocene* age by the mammalian paleontologists of North America. Recently, however, it has been determined by radiometric dating that tuffs in the Gering Formation, lowest Arikaree, are more than 24 million years old and probably predate the Aquitanian Stage and Agenian land-mammal age of Europe. Therefore, we are referring the Gering and its faunas (and the equivalents of Gering in other districts) to latest Oligocene.

The fauna reported from the overlying Monroe Creek Formation is more or less a continuation of the fauna from the Gering. Following the recommendation of Tedford (pers. commun., 1981), we are including the Monroe Creek fauna in the Early Arikareean – latest Oligocene also. It is quite possible that the Monroe Creek may prove to be contemporaneous with part of Europe's Agenian mammal age (= Aquitanian = earliest Miocene).

Principal correlative of the Gering is the Sharps Formation (at least partly) and its mammalian fauna in adjacent South Dakota. The following composite Early Arikareean fauna is based on lists from the Gering, Monroe Creek, and Sharps formations.

Composite Early Arikareean Mammalian Fauna

(From Gering, Sharps, Monroe Creek, and equivalent formations.)

MARSUPIALIA
  Didelphidae
    Herpetotherium spindleri (Macdonald, 1963A)
LAGOMORPHA
  Desmatolagus?   —L. Macdonald (1972)
  Leporidae
    Palaeolagus hypsodus Schlaikjer, 1935C
    P. philoi Dawson, 1958
    Megalagus? cf. *primitivus* Dawson, 1958
CREODONTA
  Hyaenodontidae
    Hyaenodon brevirostris Macdonald, 1970A
CARNIVORA
  Amphicyonidae?, in preceding and later faunas
  Ursidae
    Brachyrhynchocyon douglasi Macdonald, 1970A
    Enhydrocyon crassidens Matthew, 1907A

  Canidae
    Hesperocyon leptodus (Schlaikjer, 1935C)
    Mesocyon is in preceding and later faunas
    Nothocyon roii Macdonald, 1963A
    N. geismarianus (Cope, 1879D)
    N. lemur (Cope, 1879E)
    Cynodesmus cooki Macdonald, 1963A
    Sunkahetanka geringensis (Barbour and Schultz, 1935)
    S. pahinsintewakpa Macdonald, 1963A
  Mustelidae
    Palaeogale dorothiae Macdonald, 1963A
  Felidae
    Dinictis eileenae Macdonald, 1970A
    Nimravus brachyops (Cope, 1879F)
    Eusmilus olsontau (Macdonald, 1963A)
INSECTIVORA
  Erinaceoidea
    Ocajila makpiyahe Macdonald, 1963A
    Metechinus cf. *marslandensis* Meade, 1941   —L. Macdonald (1972)
  Geolabididae
    Centetodon magnus (Clark, 1936)
  Soricidae
    Domnina greeni Macdonald, 1963A
    D. dakotensis Macdonald, 1970A
    Pseudotrimylus   —(Hutchison, 1972A)
  Talpidae
    Proscalops evelynae (Macdonald, 1963A)
    P. secundus Matthew, 1901B   —L. Macdonald (1972)
    Quadrodens wilsoni Macdonald, 1963A
PRIMATES
  Omomyidae
    Ekgmowechashala philotau Macdonald, 1963A
ARTIODACTYLA
  Leptochoeridae
    Leptochoerus
  Tayassuidae
    Chaenohyus decedens Cope, 1879E
  Entelodontidae, gen. indet.
  Anthracotheriidae
    Elomeryx garbanii Macdonald, 1970A
    Arretotherium
  Agriochoeridae
    Agriochoerus
  Merycoidodontidae [2]
    Mesoreodon cheeki (Schlaikjer, 1934)   —Schultz and Falkenbach (1949)
    M. chelonyx Scott, 1893
    Megoreodon hollandi (Douglass, 1907C)
    M. fricki Schultz and Falkenbach, 1954
    Paramerycoidodon meagherensis (Koerner, 1940)   — Schultz and Falkenbach (1968)
    Megasespia middleswarti Schultz and Falkenbach, 1968
    Pithecistes copei Schultz and Falkenbach, 1954
    P. altageringensis Schultz and Falkenbach, 1968
    P. mariae Schultz and Falkenbach, 1968
    P. brevifacies Cope, 1878C
    Leptauchenia martini Schultz and Falkenbach, 1968

L. parasimus Schultz and Falkenbach, 1968

L. margaryae Schultz and Falkenbach, 1968

Hadroleptauchenia shanafeltae Schultz and Falkenbach, 1968

H. densa (Loomis, 1925)   —Schultz and Falkenbach (1968)

H. extrema Schultz and Falkenbach, 1968

Pseudocyclopidius major (Leidy, 1856)   —Schultz and Falkenbach (1968)

P. lullianus (Thorpe, 1921)   —Schultz and Falkenbach (1968)

P. quadratus (Koerner, 1940)   —Schultz and Falkenbach (1968)

Hypsiops erythroceps (Stock, 1932)   —Schultz and Falkenbach (1949)

Merycoides cursor Douglass, 1907

M. nebraskensis Schultz and Falkenbach, 1949

Desmatochoerus anthonyi Schultz and Falkenbach (1968)

D. hatcheri (Douglass, 1907)   —Schultz and Falkenbach (1954)

D. grangeri Schultz and Falkenbach, 1954

D. wyomingensis Schultz and Falkenbach, 1954

D. sanfordi Schultz and Falkenbach, 1954

D. monroecreekensis Schultz and Falkenbach, 1954

D.? thurstoni (Stock, 1934)   —Schultz and Falkenbach (1954)

Pseudodesmatochoerus milleri Schultz and Falkenbach, 1954

P. hoffmani Schultz and Falkenbach, 1954

Cyclopidius emydinus (Cope, 1884)

C. simus Cope, 1878C

Sespia nitida (Leidy, 1869A)   —Schlutz and Falkenbach (1968)

S. heterodon (Cope, 1878C)   —Schultz and Falkenbach (1968)

S. californica (Stock, 1930)   —Schultz and Falkenbach (1968)

S. marianae Schultz and Falkenbach, 1968

S. ultima Schultz and Falkenbach, 1968

Camelidae

Oxydactylus cf. *wyomingensis* Loomis, 1936B

O.? sp.

Dyseotylopus migrans Stock, 1935

Hypertragulidae

Nanotragulus cf. *loomisi* Lull, 1922B

N. intermedius Schlaikjer, 1935C

Leptomerycidae

Leptomeryx sp.

Protoceratidae

Protoceras skinneri Patton and Taylor, 1973

P. celer Marsh, 1891A   —Patton and Taylor (1973)

PERISSODACTYLA

Equidae

Miohippus nr. *equinanus* Osborn, 1918A   —Macdonald (1970A)

M. equiceps (Cope, 1879D)

Hyracodontidae

Hyracodon apertus Sinclair, 1922C

Rhinocerotidae

Diceratherium cf. *gregorii* Peterson, 1920

D. armatum Marsh, 1875A   (age?)

RODENTIA

Eomyidae

Pseudotheridomys cf. *hesperus* R. Wilson, 1960   —L. Macdonald (1972)

Aplodontidae

Prosciurus dawsonae Macdonald, 1963A

Downsimus chadwicki Macdonald, 1970A

Allomys sharpi Macdonald, 1970A

Meniscomys hippodus Cope, 1879D

Crucimys milleri (Macdonald, 1970A)   —Rensberger (1980)

Niglarodon koerneri Black, 1961C

N. progressus Rensberger, 1981

Castoridae

Palaeocastor nebrascensis (Leidy, 1856I)

Capatanka cankpeopi Macdonald, 1963A

Capacikala gradatus (Cope, 1879D)   —Macdonald (1963A)

Mylagaulidae

Promylagaulus

Eutypomyidae

Eutypomys cf. *montanensis* A. Wood and Konizeski, 1965   —Macdonald (1970A)

Sciuridae

Tamias sp.

Protosciurus tecuyensis (Bryant, 1945)   —Black (1963B)

Geomyidae

Tenudomys? dakotensis (Macdonald, 1963A)   —Rensberger (1973A)

Entoptychus minor Cope, 1881E

Pleurolicus leptophrys Cope, 1881E

Grangerimus dakotensis Macdonald, 1963A

Sanctimus stuartae Macdonald, 1970A   —Rensberger (1973)

S. clasoni (Macdonald, 1963A)   —Macdonald (1970A); Rensberger (1973)

Florentiamys agnewi Macdonald, 1963A

Heteromyidae

Heliscomys schlaikjeri Black, 1961B

Proheteromys fedti Macdonald, 1963A

P. gremmelsi Macdonald, 1963A

P. bumpi Macdonald, 1963A

Hitonkala andersontau Macdonald, 1963A

Cricetidae

Leidymys blacki (Macdonald, 1963A)   —L. Martin (1980)

L. alicae (Black, 1961B)   —Engesser (1979)

Paciculus woodi (Macdonald, 1963A)   —L. Martin (1980)

P. insolitus Cope, 1879E

P. nebraskensis Alker, 1969

Scottimus kellamorum Black, 1961C

Geringia mcgregori (Macdonald, 1970A)   —L. Martin (1980)

G. gloveri (Macdonald, 1970A)

Zetamys   —L. Martin (1974)

Zapodidae

Plesiosminthus sp.   —L. Macdonald (1972)

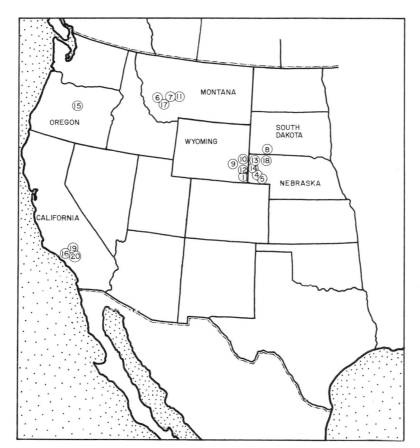

FIGURE 5-14 *Representative Early Arikareean Localities.*

1	*Albin Road—Wyoming*	*14*	*Joe Sanford Ranch (Mitchell North)—Nebraska*
2	*Ataber Ranch—Nebraska*	*15*	*John Day Formation sites (Whitneyan into Hem-*
3	*Bear Mountain—Wyoming*		*ingfordian)—Oregon*
4	*Bridgeport and Scotts Bluff area—Nebraska*	*16*	*Kew Quarry—California*
5	*Broadwater district—Nebraska*	*17*	*North Boulder Valley—Montana*
6	*Cabbage Patch local faunas (lower)—Montana*	*8*	*Rockyford district—South Dakota*
7	*Canyon Ferry (in part)—Montana*	*18*	*Rushville North—Nebraska*
8	*Cedar Pass—South Dakota*	*4*	*Scotts Bluff Monument area—Nebraska*
9	*Douglas South—Wyoming*	*15*	*Sheep Rock (in part)—Oregon*
10	*Flattop Southwest—Wyoming*	*19*	*Tecuya—California*
11	*Fort Logan (Lone Butte, White Sulphur Springs)—*	*15*	*Turtle Cove Member sites (lower)—Oregon*
	Montana	*20*	*Willard Canyon (South Mountain, in part)—Cali-*
12	*Goshen Hole district—Wyoming*		*fornia*
13	*Harrison North—Nebraska*	*8*	*Wounded Knee district (in part)—South Dakota*

Of the 82 early Arikareean mammalian genera, 40 (49 percent) were in the Whitneyan and 25 (30 percent) do not appear in later North American faunas. Three families (Hyaenodontidae, Omomyidae, and Hyracodontidae) have no later record in North America.

COLHUEHUAPIAN

At Lake Colhue-Huapi and at Gaiman, Patagonia, Argentina, in the upper part of the tuffaceous succession called the Sarmiento Group, or Formation, by some, are the *Colpodon beds* of the Ameghinos, producing the Colhue-Huapi fauna, basis for the Colhuehuapian land-mammal age (Bordas, 1936, 1939; Simpson, 1932K). The Colhue-Huapi fauna is generally endemic and comparable to the earlier faunas of the area. There are no recognized species and only a few genera in common with the preceding Deseadan, indicating a long period of evolution between the two ages. This conclusion has been substantiated by Marshall, Pascual, Curtis, and Drake (1977) whose K-Ar dates on basalts stratigraphically between Deseadan and Colhuehuapian indicate a temporal hiatus of as much as 9 million years.

187

Composite Colhuehuapian Mammalian Fauna

MARSUPIALIA
  Microbiotheriidae
    "Pachybiotherium" (Ameghino, 1902)
    Microbiotherium hernandezi Simpson, 1932K
  Borhyaenidae　　　　—after Marshall (1978)
    Sipalocyon externus (Ameghino, 1902)
    Cladosictis centralis Ameghino, 1902
    Pseudothylacynus rectus Ameghino, 1902
    Acrocyon riggsi (Sinclair, 1930)
    Borhyaena macrodonta Ameghino, 1902
    Arctodictis sinclairi Marshall, 1978
  Caenolestidae
    Halmarhippus riggsi Simpson, 1932K
    Abderites crispus Ameghino, 1902
    Micrabderites williamsi Simpson, 1932K
    Eomanodon Ameghino
    Palaeothentes Ameghino, 1887
    Parabderites Ameghino, 1902
    Pichipilus Ameghino, 1890
    Pitheculites Ameghino, 1902　　(family?)
EDENTATA
  Dasypodidae
    Prozaedius　　　　—Simpson (1945)
    Stenotatus?　　　　—Simpson (1945)
    Meteutatus?　　　　—Simpson (1945)
    Proeutatus?　　　　—Simpson (1945)
    Pseudostegotherium?　　　　—Simpson (1945)
    Stegotheriopsis Bordas, 1939
    Peltephilus　　　　—Simpson (1945)
    Parapeltocoelus Bordas, 1938
    Peltocoelus Ameghino, 1902
  Glyptodontidae
    Glyptatelus Ameghino, 1897
    Propalaeohoplophorus Ameghino, 1887
  Megalonychoidea
    Proschismotherium Ameghino, 1902
    Hapaloides Ameghino, 1902
CHIROPTERA
  Phyllostomatidae?
    Pitheculites Ameghino, 1902　　(caenolestoid?)
  Molossidae
    Tadarida faustoi Paula-Couto, 1956A　(age?)　　—
      McKenna (1980)
PRIMATES
  Cebidae
    Pitheculus　　(primate?; marsupial?)
    Homunculus
    Tremacebus
    Dolichocebus
LITOPTERNA
  Proterotheriidae
    Prolicaphrium Ameghino, 1902
    Prothoatherium Ameghino, 1902
    Licaphrops Ameghino, 1904

  Macraucheniidae
    Cramauchenia Ameghino, 1902
    Paramacrauchenia Bordas, 1939
    Theosodon Ameghino, 1887
  Adianthidae
    Adianthus Ameghino, 1891
    Proheptoconus Bordas, 1939
NOTOUNGULATA
  Leontiniidae
    Colpodon Burmeister, 1885
  Homalodotheriidae
    Homalodotherium?
  Notohippidae
    Argyrohippus fraterculus Ameghino, 1902
    Stilhippus Ameghino, 1904
    Perhippidium Ameghino, 1904
  Toxodontidae
    Proadinotherium Ameghino, 1895
  Interatheriidae
    Cochilius volvens Ameghino, 1902
    Paracochilius Bordas, 1939
    Protypotherium Ameghino, 1887
  Hegetotheriidae
    Hegetotherium Ameghino, 1887
    Pachyrukhos Ameghino, 1885
ASTRAPOTHERIA
    Parastrapotherium? Ameghino, 1895
    Astrapothericulus Ameghino, 1901
    Astrapotherium Burmeister, 1879
RODENTIA
  Erethizontidae
    Asteromys? Ameghino, 1897
    Protacaremys Ameghino, 1902
    Acaremys Ameghino, 1887
    Eosteiromys? Ameghino, 1902
    Hyposteiromys
    Parasteiromys Ameghino, 1904
    Steiromys Ameghino, 1887
  Cephalomyidae
    Cephalomys? Ameghino, 1897
    Litodontomys? Loomis, 1914
  Chinchillidae
    Perimys incavatus Ameghino, 1902
    Scotaeumys Ameghino, 1887
  Octodontidae
    Eoctodon Ameghino, 1902
  Echimyidae
    Protadelphomys Ameghino, 1902
    Paradelphomys Ameghino, 1902
    Prospaniomys Ameghino, 1902
  Dasyproctidae, gen. indet.
  Eocardidae
    Luantus Ameghino, 1899

Colhuehuapian Localities

(See map of these localities with the map of Deseadan localities.)

188

LATE? OLIGOCENE OF TASMANIA

Tedford, Banks, Kemp, McDougall, and Sutherland (1975) reported on fossil marsupials from the Geilston Travertine, Geilston Bay, near Hobart, Tasmania. Their assemblage comprises:

- Phalangeridae (likely)
- Burramyidae? (one incisor)
- Diprotodontidae
- Palorchestine, smaller than *Ngapakaldia tedfordi* Stirton

An overlying basalt yielded a K-Ar age of about 23 million years (corrected constants); thus Tedford et al. concluded that the Geilston Travertine and its fauna are Late Oligocene or older. This is now the earliest record of land mammals in the Australian region. (See location of Geilston on the map with Australian Miocene localities in Chapter 6.)

OLIGOCENE MARINE MAMMALS

Oligocene marine mammals are represented now only by relatively few fossils of the Cetacea, Sirenia, and Desmostylia, which have been found in former epicontinental, continental-shelf, or island-shelf deposits on what is now land in North America, South America, Caribbean Islands, Europe, western Asia, Australia, and New Zealand. Paleontologists agree that these mammals were "reborn" salt-water adapters, being derived phyletically from mammalian land quadrupeds. Continuing research by C. Repenning, E. Whitmore, L. Barnes, E. Mitchell, R. Fordyce, R. Savage, D. Domning, K. Rothausen, C. Ray, R. Reinhart, S. McLeod and others clearly demonstrates that the late Paleogene and Neogene record of marine mammals will be scientifically exploited, deciphered, and documented as never before in the decade to come.

The Cetacea and Sirenia have marine-adapted predecessors in strata of the upper half of the Eocene Series and indicate, therefore, general and relative diversification following the Eocene's "quantum evolutionary" transformation of total life style from nonmarine to marine milieu. The archeocete whales, structurally intermediate between mesonychids (on land) and later whales (in sea), are relatively well known from Eocene rocks in Africa, North America, Asia, Europe, and New Zealand. But they apparently dwindle during the Oligocene and become extinct by late Oligocene time. Relative to the other Oligocene cetaceans, the increase in abundance and diversity among the serrate-toothed squalodonts (Odontoceti) and the appearance of early representatives of the living suborders (Odontoceti, Mysticeti) were the most important historical events (Whitmore and Sanders, 1976; and Barnes, 1976).

R. Savage (1976) has reviewed the early sirenians. *Anomotherium* Siegfried and *Halitherium* Kaup from Europe and *Caribosiren* Reinhart from Puerto Rico are the recognized Oligocene genera of this obscure order.

Cornwallius sookensis (Cornwall) Hay (1923B) from the latest Oligocene of British Columbia is the earliest known desmostylian. The record of origin of this Pacific Basin order of mammals and the record of conversion of its members to marine existence is yet to be discovered.

The great marine radiation of pinniped carnivorans (seals, sea lions, walruses, and their extinct relatives) begins immediately after Oligocene time. Following Tedford (1976), it appears that the closest relatives of the seals (Phocidae) are the land dwelling mustelids of the Oligocene, and the closest relatives of the other pinnipeds are the land-dwelling amphicynodontine ursids ("bears") of the Oligocene.

Marine Oligocene Mammalian Genera
(After Whitmore and Sanders, 1976, and
R. Savage, 1976.)
Late Oligocene

CETACEA
  ARCHAEOCETI
    Microzeuglodon Stromer, 1903
    Patriocetus ehrlichi Rothausen, 1968
  ODONTOCETI, *Inc. Sed.*
    Agorophius Cope, 1895
    Agriocetus Abel, 1914
    Xenorophus Kellogg, 1923
  Squalodontidae
    Australosqualodon Climo and Baker, 1972
    Eosqualodon Rothausen, 1968
    Microcetus Kellogg, 1923
    Parasqualodon Hall, 1911
    Prosqualodon Lydekker, 1894
    Squalodon Grateloup, 1840
    Tangaroasaurus kakanuiensis Benham, 1935
    Metasqualodon harwoodi (Sanger, 1881)
MYSTICETI
  Cetotheriidae
    Mauicetus parki Benham, 1937A
CETACEA, *Inc. Sed.*
    Aetiocetus Emlong, 1966 (N.A.)
    Archaeodelphis Allen, 1921 (N.A.)
    Chonecetus L. Russell, 1968B (N.A.)
    Ferecetotherium Mchedlidze, 1970B (W. As.)
    Mirocetus Mchedlidze, 1970B (Azerbaijan)
MYSTICETI
  Cetotheriidae (Eu.)
SIRENIA
    Anomotherium langewieschei Siegfried, 1965 (Eu.)
    —(M. Olig., acc. Siegfried)
    Halitherium Kaup, 1838 (Eu.)
DESMOSTYLIA
    Cornwallius Hay, 1923B (N.A.)

Middle Oligocene

CETACEA
    ARCHAEOCETI
        Dorudontidae
            Kekenodon Hector, 1881   (N. Zealand)
    ODONTOCETI
        Squalodontidae
            Squalodon?
            Uncamentodon Rothausen   (*nomen nudum?*)
            Oligosqualodon Rothausen   (*nomen nudum?*)
            Oligodelphis Ashlanova and Mchedlidze, 1968   (*nomen nudum?*)
            Microcetus ambiguus (Meyer, 1840)
            Eosqualodon langewieschei Rothausen, 1968
            "*Squalodon*" *Microzeuglodon wingei* Ravn, 1926
    MYSTICETI
        Cetotheriidae
            Mauicetus Benham, 1929
            Cetotheriopsis tobieni Rothausen, 1971
SIRENIA
        Anomotherium langewieschei Siegfried, 1965
        "*Manatherium delheidi*" (Hartlaub, 1886)   —Bahlo
            and Tobien (1981)
        Halitherium schinzi Kaup, 1838
        Caribosiren turneri Reinhart, 1959

Early Oligocene

CETACEA
    ARCHAEOCETI
        Platyosphys Kellogg, 1936
        archaeocetes, indet.
SIRENIA
        Lophiodolodus chaparralensis Stirton, 1947B (sirenian?)   —Condylarthra, acc. Stirton
        cf. *Halitherium*

STRATIGRAPHIC DEMONSTRATION

The succession of mammal ages in formations within the White River Group, overlaid by early Arikareean Gering and Monroe Creek formations in northwestern Nebraska and overlaid by Sharps and Monroe Creek equivalent in the Rosebud Formation in southwestern South Dakota, is the unrivalled (if not only!) stratigraphic demonstration of land-mammal ages for the Oligocene. This sequence, well known to North American mammalian paleontologists, is shown in the accompanying chart.

South Dakota	*Nebraska*	*Land-Mammal Ages*
Rosebud Fm. (in part)	Monroe Creek Fm.	Early Arikareean
Sharps Fm.	Gering Fm.	
Brule Fm.	Brule Fm.	Whitneyan Orellan
Chadron Fm.	Chadron Fm.	Chadronian

DATING AND CORRELATION

A number of radiometric (K–Ar) dates are available for Chadronian rocks in North America:

Tuffs in the succession at Flagstaff Rim, Wyoming (Bates Hole–Lone Tree Gulch) show the following (arranged in stratigraphic order):

Uppermost ash =
    Ash J of M. Skinner ..32.4 million years before present
    Ash G of M. Skinner ....33.5 m.y.b.p.
    Ash F of M. Skinner ....34.6 and 36.6 m.y.b.p.
    Ash B of M. Skinner   34.2 and 36.1 m.y.b.p.

In the Vieja Group and Garren Group of Trans-Pecos Texas

Mitchell Mesa Fm. ....   34.8 ± 1.8 m.y.b.p.
Brite Ignimbrite .... 33.9 ± 1.1

    Airstrip local fauna in Capote Mt. Tuff
Bracks Ignimbrite ....   37.4 ± 1.2 and 37.8

    Little Egypt local fauna in Upper Chambers Tuff, Porvenir local fauna in Chambers Tuff

Buckshot Ignimbrite ....   35.6 ± 2.0 and 36.1 ± 2.3

    The Texas radiometric dates are from J. Wilson (1977B).

BIOGEOGRAPHY AND THE MILIEU

Our diagram, Figure 5–18, shows a crude plan for the distribution of continents and seas about 35 million years ago, about the beginning of Oligocene time. We are presenting only gross diagrams for the large continental areas—not accurate details, orientations, and projections for subcontinental areas. See Smith and Briden (1977) for detailed map projections proposed for different times in the last 200 million years of earth history.

By early Oligocene time the geography of the world had become similar to Recent configuration, except that epeiric seas encroached on limited districts of the Atlantic, Gulf, and Pacific coasts of North and Central America. South America, Australia, and Antarctica remained island continents; however, South America was close enough to Central America and to the Caribbean paleoislands (some say close enough to Africa) to be colonized through fortuitous dispersal (island-hopping) by rodents and by ceboid primates. Africa was almost isolated by seas, but there is suggestion from the early Oligocene land-mammal fauna of Egypt of a filtered dispersal of some groups (for example, anthracotheriids and hyaenodontids) between Eurasia and Africa at about this time. The resurgence of similarity between land mammals of Eurasia (especially striking for Europe) and North America at the beginning of the Oligocene or at the

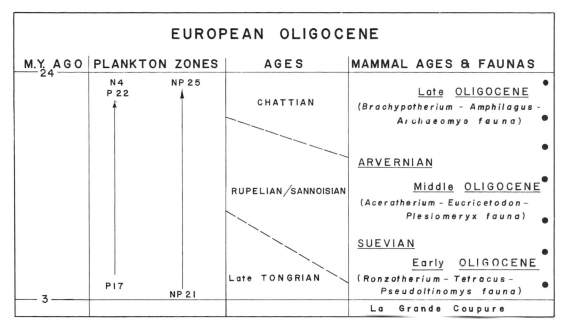

FIGURE 5-15 *Correlations Within the Oligocene of Europe.*

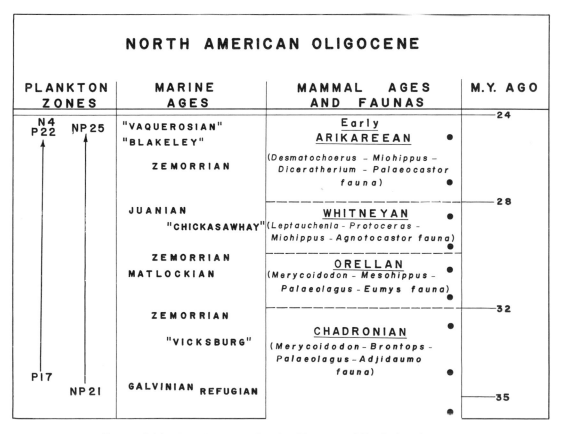

FIGURE 5-16 *Correlations Within the Oligocene of North America.*

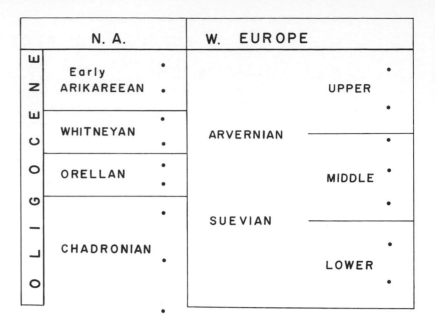

	N. A.		W. EUROPE	
O L I G O C E N E	Early ARIKAREEAN	• •	UPPER	•
			ARVERNIAN	•
	WHITNEYAN	• •		•
	ORELLAN	• •	MIDDLE	•
				•
		•	SUEVIAN	
	CHADRONIAN			•
		•	LOWER	•
		•		•

FIGURE 5-17 *Oligocene Correlations Between Europe and North America.*

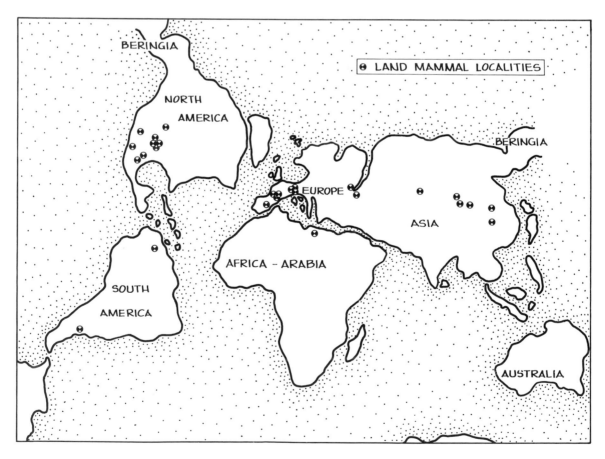

FIGURE 5-18 *Sketch-Map of the World 35 Million Years Ago.*

end of the Eocene is noteworthy. It is exemplified among the hyaenodontids, several families of fissiped carnivorans, anthracotheriids?, tapirids?, rhinocerotids, and probably several families of rodents. And this indicates that Beringia must have been an effective land bridge of land-vertebrate dispersal between North America and Eurasia.

Stratigraphers and paleogeographers have shown us that Europe at that time was a maze of inland seaways and islands. Among all the subcontinental areas, it was the most dissimilar to its present-day geographic appearance.

The early Oligocene distribution of continents and seas changed only in small details through the ensuing 15–20 million years.

Paleofloras representing the Oligocene in North America have been described from Alaska (Wolfe, 1966–1978; and Wolfe, Hopkins, and Leopold, 1966), from Washington and Oregon (Wolfe, 1978), from the Rocky Mountain Region (MacGinitie, 1953; and Becker, 1961), and from the Gulf Coast (Berry, 1916). These floras indicate coastal "tropical" conditions on the Gulf Coast (Vicksburg, Apalachicola, and Catahoula floras), drier and more diversified habitats in the Rocky Mountain province (Florissant flora), and again, more humid and more uniform climate in the low coastal areas of the West (Weaverville, Comstock, Goshen, and Willamette floras). There were more conifers and deciduous broad-leaved trees in the northernmost inland and in the later floras (Salmon, Ruby River Basin, and Bridge Creek floras). Wolfe asserted (1978, p. 700): "One of the major aspects of early Oligocene floras at middle to high latitudes [now Wolfe is designating these floras as late Eocene] is their lack of diversity, which was followed by enrichment during the remainder of the Oligocene." Earlier, Wolfe (1971) and Lillegraven (1972) had

concluded that there was a "dramatically rapid deterioration in temperature and equability in the early Oligocene, with return to a relatively mild climate by the late Oligocene and early Miocene" (Lillegraven, 1972). But the noteworthy shift toward cooler and drier and more variable or seasonal climate (as documented by the late Oligocene Bridge Creek flora, especially) must have been conducive to development of savannas in areas previously occupied by Eocene woodlands.

Leaf floras are well represented in the Oligocene of Europe and the literature of them is voluminous. Schorn (pers. commun., 1982) concludes that they show about the same trend in climate that has been demonstrated for the North American Oligocene: a dramatic change to a greater percentage of "cool tropical" and "warm temperate" plants, compared with the more tropical plants of the Eocene, followed by a warming trend in the later Oligocene.

ENDNOTES

1. Through the past 5 years one has needed a "weekly bulletin" to keep abreast the various proposals for year-age of the Eocene-Oligocene boundary! Current estimates for this date in marine strata range from 32 to 38 million years before present. Most of these varying dates have been based directly or indirectly on radiometric ages from glauconites associated with or correlated with the top of the Priabonian Stage (= "Late Eocene") or the bottom of the Rupelian Stage (= "Early Oligocene") or both in western Europe. Herewith is a summary of some of the more recent estimates for this date.

Marine Eocene-Oligocene Boundary Variance

Authors	Bases	Year-Age Estimate
Armentrout, J. 1981. *Geol. Soc. Am. Spec. Pap.* 184:137–147.	"Eocene/Oligocene boundary is placed at the Priabonian/Rupelian Stage boundary which falls within planktonic foraminiferal Zone P17 and within lower part of calcareous nannoplanktonic Zone NP21. . . ." Pomerol (1978), Odin (1978), Harris (1979), and Wolfe (1981) are cited as concluding this boundary to be more recent than 38 m.y. on "data from carefully selected glauconites and from igneous rocks. . . ." Radiometric data from Oregon and Washington are compatible with this younger dating.	32 m.y.
Alvarez, W., Alvarez, L.W., Asaro, F., and Michel, H.V. 1982. *Science* 215:886–888.	"Near the Eocene/Oligocene boundary." "Approximately synchronous with terminal Eocene extinctions. . . ."	about 34

Authors	Bases	Year-Age Estimate
Berggren, W.A., McKenna, M.C., Hardenbol, J., and Obradovich, J.D. 1978. *J. Geol.* 86:67–81.	Exact reason for this dating not specified—presumably correlated with marine planktonic foraminiferal zonation in Europe by Berggren . . . i.e., boundary between Priabonian and Rupelian.	37
Fullagar, P.D., Harris, W.B., and Winters, J. 1980. *Geol. Soc. Am., Abstr. with Prog.*, 8:430.	Rb-Sr isochron ages from glauconites in marine rocks of N. and S. Carolina indicate that Claiborne [stage]–Jackson [stage] boundary is between 35 and 37 m.y. and that Eocene-Oligocene boundary is less than 34 m.y.	Less than 34
Glass, B.P., and Crosbie, J.R. 1982. *Bull. Am. Assoc. Pet. Geol.* 66:471–476.	Microtectite layer in Caribbean has age of 34.2 ± 0.6 on fission-track dating of the microtectites and K-Ar and fission-track dating of the N.A. tectites. "Extrapolation from the microtectite layer to the overlying Eocene/Oligocene boundary indicates an age of 32.3 ± 0.9 for the E./O. boundary as defined at each site in the Initial Reports of the Deep Sea Drilling Project."	32.3 ± 0.9
Hardenbol, J., and Berggren, W.A. 1978. *Am. Assoc. Pet. Geol., Geologists Stud. in Geol.* 6:213–234.	Glauconite dates on Priabonian (Eocene) and Rupelian (Oligocene) in Europe indicate Eoc.-Olig. boundary date of about 37.	37
Harris, W.B., Fullagar, P.D., and Dischinger, J.B. 1979. *Geol. Soc. Am., Abstrs. with Prog.*, 11:439.	Glauconites from Castle Hayne Limestone, N. Carolina, give dates "in good agreement with recent placement of the boundary at 33 m.y. in Europe."	Not more than 33
La Brecque, J.L, Kent, D.V., and Cande, S.C. 1977. *Geology* 5:330–335.	Eoc.-Olig. boundary = Bartonian or Priabonian-Rupelian boundary = 38.	38
Lowrie, W., and Alvarez, W. 1981. *Geology* 9:392–397.	Eoc.-Olig. boundary = Priab.-Rupel. boundary, following La Brecque et al. and following Ness et al. (1980).	38
Lowrie, W., Alvarez, W., Napoleone, G., Perch-Nielsen, K., Premoli-Silva, I., and Toumarkine, M. 1982. *Bull. Geol. Soc. Am.* 93:414–432.	"Eocene/Oligocene boundary, defined again by planktonic foraminifera and calcareous nannofossils, lies in the negative interval between anomalies 13 and 15." They also put the Eoc.-Olig. boundary between the *Pseudohastigerina micra* Zone (above) and the *Globigerinatheka cerroazulensis* Zone (below) and between the NP 19/20 and NP21, and between CP15 and CP16.	38
Prothero, D.R., Denham, C.R., and Farmer, H.G. 1982. *Geology:* 10:650–653	"Magnetostratigraphy and magnetostratigraphic correlation with European marine microfossil zonation (=NP20/NP21 boundary and Priabonian/Rupelian boundary) 'suggests' 37 for Eoc./Olig. boundary," following Berggren et al. (1978).	37

2. Through the Oligocene and Miocene chapters we are listing all the genera and species of merycoidodont oreodonts that were recognized by Schultz and Falkenbach (1941, 1947, 1948, 1949, 1950, 1954, 1956 and 1968). We agree with most of our colleagues that Schultz and Falkenbach proposed too many taxa, but a re-study of this huge family is a lifetime's work for somebody. The Schultz and Falkenbach lists are especially valuable as indicators of the exceeding abundance and diversification of this endemic North American herbivorous group, ranging from about 40 to about 8 million years before present.

OLIGOCENE MAMMAL-BEARING LITHOSTRATIGRAPHIC UNITS

Agenais, molasse de—Oligocene—France

Agua de la Piedra Formation—Deseadan—Argentina

Ahearn Member, Chadron Formation—Chadronian—South Dakota

Aksyir Formation—Early Oligocene (and Late Eocene)—Kazakhstan

Alzeyer Meeressande—Middle Oligocene—Germany

Armissan, couches à vegétaux d'

Assas Formation—Early and Middle Oligocene—France

Astéries, calcaire à—Oligocene—France

Atala beds—Deseadan—Mendoza, Argentina

Auriferous gravels—Orellan and Whitneyan?—California

Baishuicun Formation—Early Oligocene—Shanxi

Baron Sog Formation—Middle and Late Oligocene—Mongolia

Beaver Divide Conglomerate Member, Chadron? or White River Formation—Chadronian—Wyoming

Bembridge Beds—Early Oligocene—England

Big Sand Draw lentil, White River Formation—Chadronian—Wyoming

Bone Basin Member, Renova Formation—Chadronian—Montana

Boom, Argile de—Middle Oligocene—Belgium

Boutersem, sables de—Middle Oligocene—Belgium

Breitscheid braunkohlen—Late Oligocene—Germany

Briatexte, molasse—Early Oligocene—France

Brie, calcaire de—Early Oligocene—France

Briennon, argiles de—Early Oligocene—France

Brule Formation—Orellan and Whitneyan—Nebraska and South Dakota

Buran Formation—Middle Oligocene—Kazakhstan

Caijiachong Formation—Early Oligocene—Yunnan

Capote Mountain Tuff—Chadronian—Texas

Castillon Formation—Early Oligocene—France

Castle Rock Conglomerate—Chadronian—Colorado

Catahoula Formation—Whitneyan—Texas

Cedar Creek Member of "facies"—Orellan—North Dakota

Chadron Formation—Chadronian—Nebraska, South Dakota, Colorado, Wyoming, North Dakota

Chaganbulage Formation—Early Oligocene—Nei Mongol

Chambers Tuff—Chadronian—Texas

Changi (Chanji, Changki) Formation—Late Oligocene—China

Chiococa red beds—Deseadan or earlier—Peru

Chiuppano lignites—Middle Oligocene—Italy

Clarno Formation, upper—Chadronian (or Duchesnean)—Oregon

Clermont, limagne de—Late Oligocene—France

Climbing Arrow Member, Renova Formation—Chadronian—Montana

Cook Ranch Formation—Orellan—Montana

Corbula Beds—Middle and Late Oligocene—Romania

Crazy Johnson Member, Chadron Formation—Chadronian—South Dakota

Cstakar "complex"—Oligocene—Hungary

Cypress Hills Formation—Chadronian—Saskatchewan

Cyrenenmergeln—Middle Oligocene—Germany

Deep River Formation (in part)—Early Arikareean—Montana

Deseado Formation—Deseadan—Argentina

Duchesne River Formation, uppermost—Chadronian?—Utah

Dunbar Creek Member, Renova Formation—Orellan—Montana

Duras, argiles de—Oligocene—France

Echelsbacher Flöz coal bed—Early? Oligocene—Germany

Florissant Formation—Chadronian? Orellan?—Colorado

"Fluvio-marine beds"   (See Jebel el Qatrani Formation.)

Fontainbleau, sables de—Middle Oligocene—France

Garren Group (in part)—Chadronian—Texas

Gering Formation—Early Arikareean—Nebraska and Wyoming

Gongkang Formation—Early Oligocene—Gaungxi

Gualanday Formation—Deseadan or earlier—Colombia

Hamstead (Hampstead, Hempstead) Beds—Early Oligocene—England

Horsetail Creek Member, Brule or White River Formation—Chadronian and Orellan—Colorado

Houldjin Gravel—Middle Oligocene—Nei Mongol

Hsanda Gol Formation—Middle Oligocene—Mongolia

"*Indricotherium* beds"—Middle Oligocene—Kazakhstan

Issoire, limagne de—Middle Oligocene—France

Jebel el Qatrani Formation—Early and? Middle? Oligocene—Egypt

Jirilgo Formation and fauna (lateral variant of Hsanda Gol?)—Middle Oligocene—Mongolia

John Day Formation, lower—Whitneyan and Early Arikareean—Oregon

Karakaya lignites—Early? Oligocene—Thracian Turkey

Keatley Volcanics (in part)—Chadronian—Utah

Kiinkerish Formation (in part?)—Oligocene?—Kazakhstan

Kisceller Clay—Middle Oligocene—Hungary, Romania

Kusto Formation—Early Oligocene—Kazakhstan

Lapoint Member, Duchesne River Formation—Duchesnean? and Chadronian?—Utah

Linzer Sande—Middle Oligocene to "Oligo-Miocene"—Austria

Lobsann, calcaire de—Early Oligocene—France

Luribay Conglomerates—Deseadan—Bolivia

Marbly, argiles de—Middle Oligocene—France

Marnes Blanches—Early Oligocene—Paris Basin

Marnes Supragypseuses—Early and? Middle? Oligocene—Paris Basin

Marseille, Argiles de—Middle Oligocene—France

Melanienton—Early Oligocene and? earlier?—Germany

Melker Sand, lower—Late Oligocene—Germany

"*Metamynodon* sandstone facies"—Chadronian and Orellan—Nebraska and South Dakota

Molasse Rouge (Plateau des Bornes)—Middle to Late Oligocene—France

Monument Creek Formation—Chadronian—Colorado

Naugondai Formation—Early Oligocene—Nei Mongol
(=U. Urtyn Obo fm.?)

Paudèze Braunkohlen—Late Oligocene—Switzerland

Peanut Peak Member, Chadron Formation—Chadronian—South Dakota

Pechelbronnschichten—Early Oligocene and earlier?—Alsace, France

Pegu Formation (in part)—Oligocene—Burma

Pernes, sables de—Early Oligocene—France

Piedmont Area Conglomerate—Late Oligocene—Italy

Pintura beds—Argentina—Oligocene? Miocene?

Poleslide Member, Brule Formation—Whitneyan—South Dakota

Prietos Formation—Chadronian—Chihuahua, Mexico

Qingshuiyin Formation—Middle Oligocene—Ningxia

Quercy, phosphorites de—Robiacian into mid-Oligocene—France

Ratinger Tone (clay)—Middle Oligocene—Germany and? Belgium?

Renova Formation—Chadronian and Orellan—Montana

Rosebud Formation, lowermost—Early Arikareean—South Dakota

Rupel Clay (Rupelton)—Middle Oligocene—Belgium

Salla Beds (Couches de Salla)—Deseadan—Bolivia

Santa Lucia Formation—Oligocene?—Uruguay

Scenic Member, Brule Formation—Orellan—South Dakota

Schieferton—Late Oligocene or Oligo-Miocene or both—Austria

Sespe Formation upper—Whitneyan? and Early Arikareean—California

Shanda Gol (See Hsanda Gol.)

Sharps Formation—Early Arikareean (and earlier?)—South Dakota

Szapar Formation (lignites)—Hungary—Oligocene

Taoshuyuanzi Formation—Late Oligocene—Xinjiang

Tavsanli lignites—Late? Oligocene—Turkey

Teca Formation—Oligocene?—Argentina

Tecuya Formation, lower—Early Arikareean—California

Tientong Sandstone—Oligocene—Guangxi

Titus Canyon Formation (in part?)—Chadronian—California

Toston beds—Orellan?—Montana

Tsaichiachung marls—Early Oligocene—Yunnan

Tsingshiyi Formation—Late Oligocene—Gansu

Turtle Cove Member, John Day Formation (lower part)—Early Arikareean—Oregon

"Turtle-Oreodon layer," Brule Formation—Orellan—South Dakota

Ulan Gochu Formation—Early Oligocene—Nei Mongol

Urtyn Obo Formation—Early Oligocene—Nei Mongol

Vachères, calcaire de—Middle Oligocene—France

Vaulruz Sandstone—Middle Oligocene—Switzerland

Vertessomlya lignite—Oligocene—Hungary

Vista Member, White River Formation—Whitneyan—Colorado

Westerwaldes braunkohlen—Late Oligocene—Germany

White River Group or Formation—Chadronian, Orellan, and Whitneyan—South Dakota, Nebraska, Wyoming, Colorado, North Dakota

Whitney Member, Brule Formation, White River Group—Whitneyan—Nebraska and South Dakota

Xiaotun Formation—Early Oligocene—Yunnan

Yoder Formation or Member or "facies"—Chadronian—Wyoming

OLIGOCENE LOCALITIES

Aarwangen—Late Oligocene—Switzerland

Agispe (Agyspe)—Late Oligocene or Oligo-Miocene—Kazakhstan

Aillas (Gironde)—Late Oligocene—France

Airstrip—Chadronian—Texas

Alamos Canyon—Early Arikareean—California

Albin Road—Early Arikareean—Wyoming

Alcova district—Chadronian—Wyoming

Aldealengua (Salamanca)—Early Oligocene—Spain

Aldearrubia (Salamanca) Early Oligocene—Spain

Alès (Languedoc)—Early Oligocene—France

Alharting—Late Oligocene or Oligo-Miocene—Austria

Altstadt at Messkirch—Early Oligocene—Germany

Alua—Middle Oligocene—Kazakhstan

Alzey (Mainz Basin)—Middle to Late Oligocene—Germany

Ant Hill (near Orella)—Chadronian—Nebraska

Antoignt (Limagne d'Issoire)—Middle Oligocene—France

Anvers—Early or Middle Oligocene—Belgium

Anxiety Butte—Chadronian—Saskatchewan

Aral Sea district (Agispe, etc.)—Late Oligocene and Oligo-Miocene—Kazakhstan

"Ardyn Obo" (Ergelyeen Dzo)—Early Oligocene—Mongolia

Armissan (Aude)—Late Oligocene—France (with flora)

Arnegg 2—Early Oligocene or earlier—Germany

Arner Ranch—Chadronian—Texas

Ash Spring—Chadronian—Texas

Askazan-sor (Betpak-Dala)—Late Oligocene or? Oligo-Miocene—Kazakhstan

Assas (Hérault)—Middle or Late Oligocene—France

Atabery Ranch—Early Arikareean—Nebraska

Atam-Bas-Chink (Lake Chelkar–Teniz East)—Middle Oligocene—Kazakhstan

Atravesada—Whitneyan—Arizona

Aubenas-les-Alpes (Basses-Alpes)—Middle Oligocene—France

Aubrelong 1 (Quercy)—Early Oligocene—France

Aurillac—Middle Oligocene—France

Autrac—Oligocene—France

Auzon (Gard)—Middle? Oligocene—France

Avejan—Early Oligocene—France

Baden–Unter Rauschbachgraben—Late Oligocene—Switzerland

Badland Creek—Orellan—Nebraska

Bad Tölz—Late Oligocene—Germany

Baise (Bose) Basin—Early Oligocene—Gaungxi

Baishui village (Yuanqu Basin)—Early Oligocene—Shanxi

Bakonyer Bergen (well core)—Oligocene?—Hungary

Bald Butte (Hat Creek Basin)—Orellan—Wyoming

Balm (Solothurn)—Middle Oligocene—Switzerland

Banffy-Hünyad (Komitat Kolosz)—Oligocene—Hungary-Romania

Bannwil (Bern)—Late Oligocene—Switzerland

Baron Sog (Houldjin Gravel)—Middle Oligocene—Nei Mongol

Basel-"Gellertstrasse"—Late Oligocene—Switzerland

Basel–St. Jakoh—Late Oligocene—Switzerland

Basyigit (Kesan)—Oligocene—northwest Thrace

Bates Hole   (See Flagstaff Rim, Lone Tree Gulch.)

Battle Creek Breaks or Canyon or Draw—Chadronian and Orellan—South Dakota

Bautersem   (See Hoogbutsel.)

Bazin (Lot-et-Garonne)—Late Oligocene—France

Bear Mountain—Early Arikareean—Wyoming

Beauville—Oligocene—France

Beaver Divide Conglomerate Member site—Chadronian—Wyoming

Bedeil—Early? Oligocene—Somalia

Bekas Megyer—Middle Oligocene—Hungary-Romania

Belgarric 1 (= Belgarite?) (Quercy)—Middle Oligocene—France

Belgarite I Va (Quercy)—Early? Oligocene—France

Benara—Late Oligocene—Georgian SSR

Benissons-Dieu (Loire)—Middle? Oligocene—France

Berliner Strasse—Middle or Late Oligocene—Germany

Bernloch 1(A)—Middle? Oligocene—Germany

Bernloch 1(B)—Middle Oligocene—Germany

Bessens (Tarn-et-Garonne)—Late Oligocene—France

Botpak Dala (Steppe)—Late Oligocene—Kazakhstan

Between Head of Big Corral Draw and Cottonwood Creek—Orellan—South Dakota

Big Badlands—Chadronian, Orellan and Whitneyan—South Dakota

Big Cliff—Chadronian—Texas

Big Corral Draw—Chadronian, Orellan, and Whitneyan—South Dakota

Big Hole River—Orellan—Montana

Bill Grimm Ranch—Orellan—South Dakota

Biniamai (mine)—Oligocene—Mallorca

Bird Cage Gap—Early Arikareean—Nebraska

Blackburn Ranch—Whitneyan—Nebraska

Black Hank's Canyon—Early Arikareean—Nebraska

Blauen—Oligocene—Switzerland

Blaymont (Lot-et-Garonne)—Late Oligocene—France

Blue Creek Valley—Early Arikareean—Nebraska

Blue Gulch Section—Chadronian—Wyoming

Blumbach at Schangnau (Bumbach?)—Middle Oligocene—Switzerland

Bodajk Kajmat—Middle Oligocene—Hungary

Bodenheim—"Latdorfian = late Headonian = Early Oligocene"—Germany (We regard this as late Eocene.)

Bohlen—Middle Oligocene—Belgium?

Bone Coulee—Chadronian—Saskatchewan

Boningen (Solothurn)—Late Oligocene—Switzerland

Bons—Oligocene—France

Boom (Ostflandern-Antwerp district)—Middle Oligocene—Belgium

Boongin Gol (Bungin Gol)—Oligocene—Mongolia

Boudes—Oligocene—France

Boujarc (Gard)—Middle? Oligocene—France

Bouldnor Cliff—Early Oligocene—Isle of Wight, England

Bourg de Visa—Oligocene—France

Bournoncle–St.-Pierre (Hte. Loire)—Middle? Oligocene—France

Bourret (Tarn-et-Garonne)—Late Oligocene—France

Branssat (Branssat-Coderet) (Allier)—Late Oligocene—France

Brecht Ranch (also Brecht Stock Dam and Walter Brecht Ranch)—Chadronian and Orellan—Nebraska

Breitscheid (Mainz Basin)—Late Oligocene—Germany

Briatexte (Tarn)—Early Oligocene—France

Bridge Creek—Early Arikareean—Oregon

Bridgeport district—Early Arikareean—Nebraska

Brie, calcaire de, site—Early Oligocene—France

Briennon (Loire), argiles de, site—Early Oligocene—France

Broadwater district—Orellan into Early Arikareean—Nebraska

Brons (Cantal)—Late? Oligocene—France?

Buda-Ujlak—Middle Oligocene—Hungary

Bugajewka (Izjum or Izyum district)   (marine?)—Early Oligocene—Ukraine, USSR

Bull Canyon (Harrisburg West)—Early Arikareean—Nebraska

Bumbach (Bern)—Middle Oligocene—Switzerland

Bungin (Boongin) Gol—Oligocene—Mongolia

Buran Formation sites (Zaysan Basin)—Middle Oligocene—Kazakhstan

Burcht (in type Rupelian)—Middle Oligocene—Belgium

Burgmagerbein 1—Late Oligocene—Germany

Burgmagerbein 2—Middle Oligocene—Germany

Buylstyeen Khuduk—Middle Oligocene—Mongolia

Cabeza Blanca (Chubut)—Deseadan—Argentina

Cabrerizos (Salamanca)—Early Oligocene—Spain

Cadibona—Middle Oligocene—Italy

Cain Creek, Head—Chadronian—South Dakota (71 Table South, Imlay N.)

Cajani (Khahajani)—Deseadan—Bolivia

Calaf (Barcelona)—Early Oligocene—Spain

Calamin—Late Oligocene—Switzerland

Calavon (see St. Martin-de-Castillon)

Calf Creek Valley—Chadronian—Saskatchewan

Cameron Spring—Chadronian—Wyoming

Campins (Barcelona)—Middle? Oligocene—Spain

Canquel, upper—Deseadan—Argentina

Canyon Ferry Reservoir—Chadronian?—Montana

Cap de Combe (Lot-et-Garonne)—Late Oligocene—France

Capellier—Oligocene—France

Carrascosa del Campo (Cuenca)—Late Oligocene—Spain

Castle Rock Conglomerate site—Chadronian—Colorado

Castle Rock South (Minatare East)—Early Arikareean—Nebraska

Catahoula Formation site   (See Derrick Farm.)

Cave Creek—Chadronian—Arizona

Cavuslu—Oligocene—Thracian Turkey

Cedar Bluffs East—Orellan—South Dakota

Cedar Butte–Indian Creek—Chadronian—South Dakota

Cedar Creek "facies"—Orellan—North Dakota

Cedar Creek Member sites—Orellan—Colorado

Cedar Pass district—Whitneyan and Early Arikareean—South Dakota

Cedar Ridge (Black-Dawson site)—Whitneyan—Wyoming

Cereste—Middle Oligocene—France

Cerro Arenosa (Carrascosa)—Middle Oligocene—Spain

Cerro del Humo West or upper—Deseadan—Argentina

Cestayrol—Oligocene—France

Ceva (Piedmont area)—Late Oligocene—Italy

Chadron Formation sites and fauna—Chadronian—South Dakota, Nebraska, Wyoming, Colorado, North Dakota

Chadronia Pocket—Chadronian—Nebraska

Chaganbulage (Suhaitu)—Early Oligocene—Nei Mongol and Ningxia

Chalk Butte South (Atabery Ranch)—Early Arikareean—Nebraska
Challonges
Chamberlain Pass—Orellan—South Dakota
Champeix—Oligocene—France
Champveut—Late Oligocene—Switzerland
Chantras—Oligocene—Thracian Greece
Chaparral—Deseadan—Colombia
Chapelins   (See Les Chapelins.)
Chaptuzat—Late Oligocene—France
Chauffours (Les Chaufours, Puy-de-Montdoury, Tour de Boulade)—Late Oligocene—France
Chavronay (Vaud)—Middle? Oligocene—Switzerland
Chelkar-Teniz (Myn Say, Kur-Say, Ak-Say, etc.)—Middle Oligocene—Kazakhstan
Cheyenne River district—Chadronian—South Dakota
Chiavon (Vicenza)—Middle Oligocene—Italy
Chibrac—Oligocene—France
Chili Gulch—Orellan or Whitneyan—California
Chindzhila River, Right Bank—Late? Oligocene—Kazakhstan
Chiococa (red beds)—Deseadan?—Peru
Chiuppano (Vicenza)—Early or Middle Oligocene—Italy
Choei-tong-kou   (See Shui-tung-kou.)
Chuching (Qujing)—Early? Oligocene—Yunnan
Clarno Ferry (Clarno Mammal bed, Hancock Quarry)—Chadronian or Duchesnean—Oregon
Cluzel Quarry (Coderet-Branssat)—Late Oligocene—France
Coderet-Branssat—Late Oligocene—France
Colluspina—Early? Oligocene or earlier—Spain (Collususpina)
Comberatière—Oligocene—France
Conata area—Chadronian and Orellan—South Dakota
Conglomerate Creek Valley—Chadronian—Saskatchewan
Cook Ranch Formation sites—Orellan—Montana
Cottonwood Creek—Orellan and Whitneyan—South Dakota
Cottonwood Pass—Chadronian, Orellan, and Whitneyan—South Dakota
Cournon—Middle Oligocene—France
Coyote Creek—Early Arikareean—Nebraska
Crabtree Bluff, vicinity—Early Arikareean—Montana
Crawford district—Chadronian and Orellan—Nebraska
Crow Buttes—Orellan—South Dakota
Cuny Table—Orellan—South Dakota
Cypress Hills Formation sites—Chadronian—Saskatchewan
Dallet   (See Pont-du-Chateau.)
Dalmatia (area)—Oligocene—Yugoslavia
Daudens, hameau de—Late Oligocene—France
Dengkou district—Middle Oligocene—Nei Mongol
Deep River—Early Arikareean—Montana
Derrick Farm—Whitneyan—Texas
Devils Gap (Beaver Divide)—Chadronian and Orellan—Wyoming
Dgulfa—Middle? Oligocene—Caucasus
Diamond O—Chadronian—Montana
Dickinson Southwest—Orellan—North Dakota
Dieupentale (Tarn-et-Garonne)—Late Oligocene—France
Digoin—middle? Oligocene—France
Dillon Pass—Chadronian, Orellan, and Whitneyan—South Dakota

Dinslaken—Middle Oligocene—Belgium? (marine)
Djebel Coquin   (See Dor el Talha.)
Domilhac (Lot-et-Garonne)—Late Oligocene—France
Donguz-Tau—Middle Oligocene—Kazakhstan
Dor el Talha, upper—Early Oligocene—Libya
Douglas Flat—Whitneyan?—California
Douglas South—Early Arikareean—Wyoming
Douglas Southeast (Orin Junction)—Orellan—Wyoming
Dry Creek (Toston)—Chadronian—Montana
Duffel (in type Rupelian)—Middle Oligocene—Belgium
Dulliken-Bergmatt (Solothurn)—Late? Oligocene—Switzerland
Duras—Oligocene—France
Dzheman (Gora River and Mountain)—Oligocene?—Kazakhstan
Dzungaria Basin—Late Oligocene—Xinjiang
Eagle Nest Butte—Early Arikareean—South Dakota
Easter Lily Mine—Chadronian?—Montana
East Mesa   (See Jhama Obo.)
Ebnat-Kappel (St. Gallen)—Late Oligocene—Switzerland
Echelsbacher Brücke—Middle Oligocene—Germany
Eckelsheim—Middle Oligocene—Germany
Egeres (Komitat Kolosz)—Late Oligocene—Hungary
Ehingen 1–14—Middle Oligocene and Early Oligocene—Germany
Ehrenstein 1(B)—Early Oligocene—Germany
Ehrenstein 4—Late Oligocene—Germany
Elmali—Oligocene—Thracian Turkey
El Talladell (Tarrega)—Middle? Oligocene—Spain
Ergelyeen Dzo (Erghil Obo, ''Ardyn Obo'', etc.)—Early Oligocene—Nei Mongol
Eriz (Bern)—Middle or Late Oligocene—Switzerland
Eselsberg at Ulm—Early Oligocene—Germany
Espenhain—Middle Oligocene—Belgium   (marine?)
Espinosa de Henares (Guadalajara)—Early or Middle Oligocene—Spain
Essert-Pittet (Vaud)—Middle or Late Oligocene—Switzerland
Etampes—Middle Oligocene—France
Everson Ranch—Orellan—Nebraska
Eytier—Early Oligocene—Switzerland?
Fall-Tobel—Late Oligocene—Germany
Fayum faunas—Oligocene—Egypt
Feilbingert—Middle Oligocene—Germany
Feiyue—Late Oligocene—Xinjiang
Ferté-Alais (Seine-et-Oise)—Middle Oligocene—France
Fitterer Ranch—Chadronian, Orellan, Whitneyan—North Dakota
Flagstaff Rim sites—Chadronian—Wyoming
Flat Top Butte (Harrison Ranch)—Chadronian—South Dakota
Flattop Southwest—Early Arikareean—Wyoming
Flonheim (Ufhofen, Flonheim, Alzey)—Middle Oligocene—Germany
Florissant Formation sites—Chadronian or Orellan—Colorado
Florsheim (Hochheim)—Middle Oligocene (marine)—Germany
''Fluvio-marine beds''   (See Jebel el Qatrani.)
Fontainbleau, sables de, sites—Middle Oligocene—France   (marine)
Fontaine-de-Vaucluse—Middle Oligocene—France
Forcalquier Northeast—Middle Oligocene—France
Fresnes-les-Rungis—Early Oligocene—France
Fronsac—Oligocene—France
Fumel (Lot-et-Garonne)—Late Oligocene—France
Gabsheim—Middle Oligocene—Germany
Gaimersheim at Ingolstadt (Bavaria)—Late Oligocene—Germany

Gainesville West-Southwest (I-75)—Whitneyan—Florida

Gaisbach—Late Oligocene—Austria

Ganna (Bakony Mountains)—Oligocene—Hungary

Gannat (Allier)—Late Oligocene—France

Gebel or Jebel Bou Gobrine—Early? Oligocene—Tunisia

Genebrières 1 and 2 (Tarn-et-Garonne)—Middle Oligocene—France

Gergovie (limagne de Clermont)—Middle? Oligocene—France

Gerignoz, Ravin de (Fribourg)—Middle? Oligocene—Switzerland

Gerry's Ranch—Chadronian and Orellan—Colorado

Ginestous (Haute-Garonne)—Middle? Oligocene—France

Goateg—Early Oligocene—Mongolia

Gongkang Formation sites—Early Oligocene—Guangxi

Goshen Hole district—Chadronian through Early Arikareean—Wyoming

Grafenmühle 1—Middle Oligocene—Germany

Granath Quarry, Collection 30—Chadron—South Dakota

Grand Canyon—Middle Oligocene—Mongolia

Gravelinnes—Early Oligocene—France

Green Cove Vicinity—Chadronian or Orellan—Wyoming

Grenchen I and II (Solothurn)—Middle Oligocene—Switzerland

Grepiac (Haute Garonne)—Middle Oligocene—France

Gressy (Vaud)—Late Oligocene—Switzerland

Gschwender Tobel I and II—Late Oligocene—Germany

Gua Teg—Oligocene—Mongolia

Gunzenheim (Bavaria)—Late Oligocene—Germany

Gussenstadt—Late Oligocene—Germany

Gusternhain (Hesse)—Middle? Oligocene—Germany

Hadcock Ranch (Missouri River West, Sewell Lake South)—Chadronian—Montana

Hami—Early Oligocene—Xinjiang

Hampstead (Hamstead, Hempstead)—Oligocene—England

Hanson Ranch—Chadronian—Saskatchewan

Haosibuerdu Basin (See Chaganbulage)—Early Oligocene—Ningxia

Harmanli—Oligocene—Thracian Turkey

Harney Springs East—Orellan—South Dakota

Harrisburg Southwest—Early Arikareean—Nebraska

Harrison North—Chadronian, Orellan, and Early Arikareean—Nebraska

Harrison Ranch—Chadronian—South Dakota

Hart Table—Chadronian—South Dakota

Harvard Fossil Reserve (Torrington)—Orellan—Wyoming

Hasköy—Oligocene—Thracian Turkey

Hat Creek Basin—Orellan and Whitneyan—Wyoming and Nebraska

Hat Creek Store Northeast—Orellan—Wyoming

Hautcastel (Lot-et-Garonne)—Middle? Oligocene—France

Hautevigne (Lot-et-Garonne)—Late Oligocene—France

Hay Creek—Orellan and Whitneyan—South Dakota

Heidenheim—Early Oligocene—Germany

Heimersheim at Alzey—Middle to Late Oligocene—Germany

Hemiksem (in type Rupelian)—Middle Oligocene—Belgium

Hempstead (See Hampstead.)

Henry Morgan Quarry—Chadronian—Nebraska

Hermosa—Whitneyan—South Dakota

Herrlingen 1—Middle Oligocene—Germany

Hochberg at Jungnau—Early Oligocene—Germany

Hochheim (Flörsheim)—Late Oligocene—Germany

Hodges Basin (Interior East)—Orellan—South Dakota

Hoeleden (Brabant)—Early Oligocene—Belgium

Hogback Mountain Northwest—Early Arikareean—Nebraska

Hohe Rhonen (Zug)—Late? Oligocene—Switzerland

Hongshan Farm—Late Oligocene—Xinjiang

Hoogbutsel (Brabant)—Early Oligocene—Belgium

Horn Northwest—Orellan—Nebraska

Horse Creek Basin (66 Mountain South) Whitneyan and Early Arikareean—Wyoming

Horsetail Creek Member sites—Chadronian and Orellan—Colorado

Hsanda Gol Formation sites (See also Loh, Tatal Gol)—Middle Oligocene—Mongolia

Hubbard Gap district—Early Arikareean—Nebraska

Huermeces del Cerro (Guadalajara)—Early Oligocene—Spain

Humilly (Haute Savoie)—Late Oligocene—France

Hunter 115 and 117—Chadronian—Saskatchewan

Hunter Quarry—Chadronian—Saskatchewan

Hutenmacher Table East—Chadronian—South Dakota

I-75 (Gainesville West-Southwest)—Whitneyan—Florida

Ibrice—Oligocene—Thracian Turkey

Iliff—Chadronian—Colorado

Ilyiskaya Fissure—Early Oligocene—Kazakhstan

Imlay District—Orellan—South Dakota

Inca—Oligocene—Mallorca

Indian Creek—Chadronian—South Dakota

Indian Stronghold Northeast—Orellan and Whitneyan—South Dakota

Interior District—Orellan and Whitneyan—South Dakota

Iren Dabasu Basin (Houldjin gravel and Baron Sog Formation sites)—Middle and Late Oligocene—Nei Mongol

Isle d'Albi—Oligocene—France

Isle-sur-Sorgues (Vaucluse)—Early Oligocene—France

Isserville—Early Oligocene?—Algeria

Issoire, environs (Puy-de-Dôme)—Middle Oligocene—France

Itardies (Tarn-et-Garonne)—Middle Oligocene—France

Itier (Eytier?)—Oligocene—France

Ivangrad (Petnik)—Late Oligocene—Yugoslavia

Jebel (Gebel) Bou Gobrine—Early Oligocene—Tunisia

Jebel el Qatrani Formation sites—Early and? Middle? Oligocene—Egypt

Jefferson River Section—Chadronian—Montana

Jeriah North—Early Arikareean—Wyoming

Jhama Obo (East Mesa)—Early Oligocene—Nei Mongol

Jirilgo Formation sites (Hsanda Gol)—Middle Oligocene—Mongolia

Joder West—Orellan—Nebraska

Joe Sanford Ranch (Mitchell North)—Early Arikareean—Nebraska

John Day Formation sites—Whitneyan, Early Arikareean, and later—Oregon

Jussat—Oligocene—France

Kadoka South—Early Arikareean—South Dakota

Kajali—Oligocene?—Bulgaria

Kameno-Burgas—Early Oligocene—Bulgaria

Kansu District, West—Late Oligocene—Gansu

Karabürcek—Oligocene—Thracian Turkey

Karakaya—Oligocene—Thracian Turkey

Kara-Tau—Middle Oligocene—Kazakhstan

Kara–Turgay River (Alua village)—Middle Oligocene—Kazakhstan

Kesan Northeast—Early Oligocene—Thracian Turkey

Keutchach—Oligocene—Austria

Kew Quarry—Early Arikareean?—California

Khaitch Bulak—Middle Oligocene—Mongolia

Khalyun Basin and Khalyun Somon—Middle? Oligocene—Mongolia

Khatan Khayrkhan (Ulan Sayr)—Middle? Oligocene—Mongolia

Khoer-Dzan (Dzamyn-Udo)—Early Oligocene—Mongolia

Khul'dzhin—Middle Oligocene—Mongolia (-Houldjin)

Kiinkerish Hill—Early Oligocene—Kazakhstan

Kirby Creek—Late Oligocene—Vancouver Island, British Columbia   (marine)

Kirikali—Oligocene—Thracian Turkey

Kisceller Ton sites (Obuda, Buda-Ujlak, Szepvolgy, Bekas megyer, etc.)—Middle Oligocene—Hungary-Romania

Klein-Blauen (Bern)—Middle Oligocene—Switzerland

Klemke Place (Spring Draw)—Orellan—Nebraska

Koer Dzan—Early Oligocene—Mongolia

Kontich (in type Rupelian)—Middle Oligocene—Belgium

Kube Table South—Chadronian—South Dakota

Kulebay (Baksy Zhilanshik River, Shalkanura)—Middle Oligocene—Kazakhstan

Kurbalik assemblage—Late Oligocene—Turkey

Kur-Say Gorge (Chelkar-Teniz area)—Middle Oligocene—Kazakhstan

Kustos Formation site—Early Oligocene—Kazakhstan

Kütigen at Aarau (Aargau)—Late Oligocene—Switzerland

Kuyan—Late? Oligocene—Ningxia

Kyzyl-Kak (Dzhezkazgan-Karaganda province)—Middle Oligocene—Kazakhstan

Lacam (Lot-et-Garonne)—Late Oligocene—France

Lacayani—Deseadan—Bolivia

La Colombière Fissure (Hérault)—Late Oligocene—France

La Comberatière (Lot-et-Garonne)—Late? Oligocene—France

La Conversion—Oligocene—Switzerland

Lac Pelletier—Chadronian—Saskatchewan

La Fabrique—Late Oligocene—Switzerland

La Ferté-Alais—Middle Oligocene—France

La Flecha (Santa Cruz)—Deseadan—Argentina

Lagny-Thorigny—Early Oligocene—France

Lake Alice Northwest—Early Arikareean—Nebraska

Lamandine-Quercy—Middle? Oligocene (in part?)—France

La Milloque (Lot-et-Garonne)—Late Oligocene—France

Lamontgie—Oligocene—France

Langenlonsheim—Middle Oligocene—Germany

Lantian—Early Oligocene (in part)—Shanxi

La Plante 2 (Quercy)—Middle Oligocene—France

La Réole (Garonne Basin)—Middle Oligocene—France

La Ricardie (Lot-et-Garonne)—Middle Oligocene—France

La Roche-sur-Foron—Middle Oligocene—France

La Sauvetat (mixed collection)—Middle Oligocene—France

La Tuilière (Bareme syncline)—Middle Oligocene—France

La Tuque (Lot-et-Garonne)—Middle Oligocene—France

Lausanne—Late? Oligocene—Switzerland

Lawson Ranch—Orellan—Wyoming

Le Bey—Late Oligocene—Switzerland

Le Bourgadot (Landes)—Middle Oligocene—France

Lebratières 3 (Lot)—Middle Oligocene—France

Le Cammas (Haute-Garonne)—Late? Oligocene—France

Ledingham Ranch (Tunnel Hill)—Early Arikareean—Nebraska

Lemdes—Oligocene—France

Le Moulinet—Late Oligocene—Switzerland

"Leptauchenia zone"—Whitneyan—South Dakota, Nebraska, etc.

Les Chapelins (Apt)—Middle Oligocene—France

Les Chauffours (Puy-de-Dome)—Late Oligocene—France

Les Matelles—Middle Oligocene—France

Les Milles (Aix Basin)—Middle Oligocene—France

Les Peries—Oligocene—France

Le Talent—Late Oligocene—Switzerland

Lewellen—Whitneyan—Nebraska

Lewis Creek—Orellan—Colorado

Liegenden—Middle? Oligocene—Austria?

Ligmaringen—Oligocene—Germany

Lingwu County—Middle Oligocene—Ningxia

Linz—Late Oligocene—Austria

Little Badlands—Chadronian, Orellan, and Whitneyan?—North Dakota

Little Badlands–Oelrichs—Whitneyan—South Dakota

Little Corral Draw—Orellan—South Dakota

Little Egypt—Chadronian—Texas

Little Lone Tree Gulch Section—Chadronian—Wyoming

Little Muddy Creek—Early Arikareean—Wyoming

Little Pipestone—Chadronian—Montana

Lloseta—Middle? Oligocene—Mallorca

Lobsann (Alsace, Bas Rhin)—Early Oligocene—France

Loh (Campsite)—Middle Oligocene—Mongolia

Lok 1—Early Oligocene—Shanxi

Lone Butte–White Sulphur Springs—Early Arikareean—Montana

Lone Butte (Government Slide Butte)—Chadronian?—Wyoming

Lone Tree Gulch sites—Chadronian—Wyoming   (Bates Hole)

Loping district—Early or Middle Oligocene—Yunnan

Losenegg (Bern)—Middle Oligocene—Switzerland

Lovagny (Haute-Savoie)—Middle Oligocene—France

Lower Martin Canyon—Orellan—Colorado

Lukawitz-Markdorf—Late Oligocene—Czechoslovakia

Lunan Basin—Early Oligocene—Yunnan

Lungte—Oligocene—Ningxia

Lusk Northeast—Early Arikareean—Wyoming

Lyman East—Orellan and Whitneyan—Nebraska

Malemort (Vaucluse)—Early and Middle Oligocene—France

Malhat—Middle Oligocene—France

Manacor—Middle? Oligocene—Mallorca

Manosque—Oligocene—France

Manrak Range—Oligocene—Kazakhstan

Mantle Ranch—Chadronian—Montana

Marseille   (See St. Henri.)

Masatly (Edirne)—Late Oligocene—Thrace

Mas d'Agenais (Marmande)—Middle Oligocene—France

Mas-de-Got, loc. A (Quercy)—Middle Oligocene—France

Mas de Labat (Lot)—Early Oligocene—France

Mas de Pauffié (Lot)—Middle Oligocene—France

Massongy—Late? Oligocene—France

Matelles   (See Les Matelles.)

Mathod (Vaud)—Late Oligocene—Switzerland

Maunayne (Lot)—Middle Oligocene—France
Mazan (1) (Vaucluse)—Middle Oligocene—France
McCarty's Mountain—Chadronian—Montana
Mège (Tarn-et-Garonne–Quercy)—Middle Oligocene—France
Mellinger—Orellan—Colorado
Meng Ranch—Orellan—Nebraska
Merisov-Krivadia (Komitat Hunyad)—Oligocene—Hungary
Meshtitsa—Middle? Oligocene—Bulgaria
"Metamynodon sandstone facies"—Chadronian and Orellan—
    South Dakota, Nebraska, etc.
Microfauna Loc.—Chadronian—South Dakota
Miesbach—Oligocene—Germany
Miller Basin—Cain Creek—Orellan—South Dakota
Milles   (See Les Milles.)
Milloque   (See La Milloque.)
Mills Falls–U.S. Highway 85—Orellan—Wyoming
Mine-des-Roys (Pont-du-Château)—Late Oligocene—France
Mink Creek (Yellowstone)—Chadronian—Wyoming
Mitchell North–Blackburn Ranch—Whitneyan—Nebraska
Möhren 4—Early Oligocene—Germany
Möhren 13 and 19—Middle Oligocene—Germany
Moissac (Tarn-et-Garonne)—Middle Oligocene—France
Molly and Mabel—Chadronian—Colorado
Monchsdeggingen 2—Middle Oligocene—Germany
Montagny (Vaud)—Late Oligocene—Switzerland
Montalban (Aragon)—Middle Oligocene—Spain
Montans (Tarn)—Middle Oligocene—France
Montclair-de-Quercy (Tarn-et-Garonne)—Middle Oligocene—
    France
Monteviale (Vicentino)—Early or Middle Oligocene—Italy
Montsegur (Gironde)—Oligocene—France
Monument Creek Formation site—Chadronian—Colorado
Moriken (Aargau)—Late Oligocene—Switzerland
Mottnig (Krain)—Late Oligocene—Yugoslavia
Mouillac (Tarn-et-Garonne–Quercy)—Early and Middle Oligo-
    cene—France
Muddy Creek–Spanish Mine—Early Arikareean—Wyoming
Muddy Mountain South—Chadronian—Wyoming
Mumliswyl—Middle Oligocene—Switzerland
Münchenstein—Late Oligocene—Switzerland
Murgenthal (Aargau)—Late Oligocene—Switzerland
Murs (Vaucluse)—Middle Oligocene—France
Myaing–Kyaukwet—Oligocene?—Burma
Myn-Eske-Suyek (Myneske Suïek)—Middle Oligocene—Ka-
    zakhstan
Myn-Say Gorge—Middle Oligocene—Kazakhstan
Nant d'Avril (Genève)—Late Oligocene—Switzerland
Nareen Bulak—Oligocene—Mongolia
Near Orella—Chadronian—Nebraska
Neustadt (Marburg)—Early Oligocene or earlier—Germany
Niel (type Rupelian)—Middle Oligocene—Belgium
Node North—Orellan—Wyoming
Noeveren (in type Rupelian)—Middle Oligocene—Belgium
Noisy-le-Sec—Early Oligocene or earlier—France
Nonette—Oligocene—France
Norbeck Pass Area—Orellan—South Dakota
Nordshausen at Kassel—Early Oligocene or earlier—Germany
Norman Ranch—Chadronian—Nebraska
North Boulder Valley (Cold Springs P.O.)—Early Arikareean—
    Montana
Obuda—Middle Oligocene—Hungary (Romania?)

Oelingerthal (Orlingertal)—Early Oligocene—Germany
Oelrichs–Little Badlands—Whitneyan (and Orellan?)—South
    Dakota
Oensingen-Ravellen (Bern)—Middle Oligocene—Switzerland
Offenbach (Offenbacher Hafen)—Middle Oligocene—Germany
Old Woman Creek—Orellan—Wyoming
Orbeil   (See Chauffours.)
Orella Member, fauna, etc.—Orellan—Nebraska, South Dakota,
    etc.
"Oreodon zone"—Orellan—South Dakota, Nebraska, etc.
Orin Junction (Douglas Southeast)—Orellan—Wyoming
Orsonette (Limagne d'Issoire)—Middle Oligocene—France
Oulen I, II (Vaud)—Late Oligocene—Switzerland
Paguera I and II—Late Oligocene—Mallorca
Palisades Section—Orellan—Montana
Pass Creek West—Early Arikareean—South Dakota
Pasture Creek West–Conata Southwest—Orellan—South Dakota
Paudeze—Late Oligocene—Switzerland
Pawnee Buttes area—Orellan and later—Colorado
Pech Blanc (Narbonna)—Late Oligocene—France
Pechbonnieu (Haute-Garonne)—Middle Oligocene—France
Pech Desse (Quercy)—Late Oligocene—France
Pech du Fraysse (Quercy)—Middle Oligocene—France
Pech-Grabit (Quercy) (Lot)—Middle Oligocene—France
Pegu "series" sites   (See Myaing, Maingyaung, etc.)
Peissenberg (Bavaria)—Late Oligocene—Germany
Peneyita—Deseadan?—Colombia
Pena—Chadronian?—Utah
Perignat—Oligocene—France
Perne—Oligocene—France
Petnik (Montenegro)—Late Oligocene—Yugoslavia
Petrovskiy Bay (Aral Sea, north shore)—Late Oligocene—Ka-
    zakhstan
Peublanc (Allier)—Late Oligocene—France
Pfarr-Alpe—Late Oligocene—Germany
Pilgrim Creek (Jackson Hole)—Chadronian—Wyoming
Pinchenat (Lot-et-Garonne)—Early Oligocene—France
Pine Bluff South—Orellan—Wyoming
Pine Ridge—Orellan—Nebraska
Pinnacles North—Whitneyan and Early Arikareean—South Da-
    kota
Pipestone Springs—Chadronian—Montana
Plaissan (Bas-Languedoc)—Late Oligocene—France
Planes   (See St.-Victor-la-Coste.)
Plesching at Linz—Late Oligocene—Austria
Point Creek Head (Slim Buttes)—Orellan—South Dakota
Poleslide Member sites—Whitneyan—South Dakota
Pont-du-Château (Puy-de-Dôme)
Pont-Ste.-Marie—Oligocene—France
Porcupine Creek Canyon, Mouth—Early Arikareean—South Da-
    kota
Porcupine North—Early Arikareean—South Dakota
Porvenir—Chadronian—Texas
Potato Creek district—Orellan and Whitneyan—South Dakota
Pouquette (Dordogne)—Early Oligocene—France
Pu-yang-ho (near Shih-er-ma-cheng)—Oligocene—Gansu
P.U. 1016E (Bear Creek)—Orellan—South Dakota
Pumpkin Creek Valley (Wildcat Ridge)—Whitneyan—Nebraska

Pustavum—Oligocene—Hungary

Puy-de-Montdoury (Puy-de-Dome)—Late Oligocene—France

Puy-Laurens (Tarn)—Early Oligocene—France

Quebrada Fiera (Malargüe)—Deseadan—Argentina

Quercy area, many localities—Robiacian into Middle Oligocene—France

Quinn Draw—Chadronian—South Dakota

Quinn Ranch—Chadronian—Texas

Quiver Hill–Kadota South—Orellan and Whitneyan and Early Arikareean—South Dakota

Qujing (Chuching) Basin—Early Oligocene—Yunnan

Rabastens (Tarn)—Middle Oligocene—France

Rances—Late Oligocene—Switzerland

Rancho Gaitan (Chihuahua)—Chadronian—Mexico

Ravellen (See Oensingen.)

Ravet (Quercy)—Early to Middle Oligocene—France

Ravin de Gerignoz (Fribourg)—Late Oligocene—Switzerland

Raynal (Quercy)—Middle Oligocene—France

Réal (Gironde)—Late Oligocene—France

Redington Gap area—Early Arikareean—Nebraska

Red Water Creek East—Orellan—South Dakota

Regensberg (Zurich)—Late Oligocene—Switzerland

Reva Gap district—Orellan—South Dakota

Rickenbach-Mühle (Solothurn)—Late Oligocene—Switzerland

Rigal Jouet (Quercy)—Middle Oligocene—France

Rincoñada de los Lopez (Chubut)—Deseadan—Argentina (includes Scarritt Pocket)

Rio Corcobado—Oligocene?—Argentina

Rio Deseado area—Deseadan and other ages—Argentina

Rocallaura (La Segarra–Montblanch)—Middle Oligocene—Spain

Rochette at Lausanne (Waadtland)—Late Oligocene—Switzerland

Rocks Springs South—Orellan—South Dakota

Rockyford East 1 and 2—Early Arikareean—South Dakota

Romagnat—Oligocene—France

R. O. M. localities, (Cypress Hills)—Chadronian—Saskatchewan

Ronheim 1—Middle Oligocene—Germany

Ronzon (Haute Loire)—Early Oligocene—France

Roqueprune 2 and 3 (Tarn-et-Garonne)—Middle Oligocene—France

Rossloch—Middle Oligocene—Germany

Rott at Bonn—Late Oligocene—Germany

Rottenbuch—Late Oligocene—Germany

Roubideaux Pass (Wildcat Ridge)—Whitneyan—Nebraska

Roundhouse Rock—Early Arikareean—Nebraska

Round Top (Wildcat Ridge)—Whitneyan—Nebraska

Ruch (Gironde)—Early Oligocene—France

Rufi-Schannis—Middle? Oligocene—Switzerland

Ruine Balm (Solothurn)—Middle? Oligocene—Switzerland

Rumst (in type Rupelian)—Middle Oligocene—Belgium

Rupelmonde (type Rupelian)—Middle Oligocene—Belgium

Rushville North—Early Arikareean—Nebraska

Rutba North—"Paleogene"—Irak

Sacanana (Chubut)—Colhuehuapian—Argentina

Saddle Horse Pass East—Orellan—South Dakota

Sage Creek Basin—Chadronian?, Orellan, and Whitneyan—South Dakota

Sage Creek Pass—Chadronian, Orellan, and Whitneyan—South Dakota

Sage Ridge—Chadronian—South Dakota

St.-André (Marseille) (Bouches-du-Rhone)—Middle Oligocene—France

St.-Cernin—Oligocene—France

St.-Cugat de Gavadons (Cuenca de Moia)—Early Oligocene—Spain

Ste.-Marthe d'Eymet (Dordogne)—Early Oligocene—France

Ste.-Quitterie (Lot-et-Garonne)—Late Oligocene—France

St.-Germain-Lembron—Oligocene—France

St.-Henri (Bouches-du-Rhône)—Middle Oligocene—France

St. Jacques (San-tao-cho)—Middle Oligocene—Ordos, Nei Mongol

St. Margrethen (St. Gallen)—Early Oliogocene—Switzerland

St.-Martin-de-Casselvi de Briatexte (Tarn)—Middle Oligocene—France

St. Martin-de-Castillon (loc. C, near Calavon)—Middle Oligocene—France

St.-Menoux—Oligocene—France

St. Nicolas-Waas (in type Rupelian)—Middle Oligocene—Belgium

St. Petersinsel (Bern)—Late? Oligocene—Switzerland

St.-Pourcain-sur-Bebre—Middle Oligocene—France

St.-Sulpice—Oligocene—France

St.-Swithin's (Isle of Wight)—Early Oligocene—England

St.-Victor-la-Coste (Gard)—Late Oligocene—France

St.-Vincent-de-Barbeyrargues—Middle Oligocene—France

Salla Luribay—Deseadan—Bolivia

Salvagnac—Oligocene—France

Sand Creek—Orellan, Whitneyan, and Early Arikareean—Nebraska

San-tao-cho (Ordos)—Middle Oligocene—Nei Mongol

San Marcos Pass—Early Arikareean—California

Sarèle (Alès Basin)—Middle Oligocene—France

Sary Su River—Middle Oligocene—Kazakhstan

Sauzet (Gard)—Middle Oligocene—France

Scarritt Pocket (Chubut)—Deseadan—Argentina

Scenic East–Chamberlain Pass—Orellan—South Dakota

Scenic Member sites—Orellan—South Dakota

Scenic South—Orellan—South Dakota

Schangnau–"Hombach" (Bern)—Late Oligocene—Switzerland

Schannis (St. Gallen)—Middle Oligocene—Switzerland

Schelklingen 1—Middle Oligocene—Germany

Schindelwies–Graben—Late Oligocene—Germany

Schleifmühle I and II—Late Oligocene—Germany

School of Mines Canyon North—Orellan—South Dakota

Scottsbluff City North—Whitneyan—Nebraska

Scotts Bluff Monument area—Orellan, Whitneyan, and Early Arikreean—Nebraska

Seamon Hills Southwest—Orellan—Wyoming

Seckbach (Mainz Basin)—Middle Oligocene—Germany

Section 36—Chadronian—Colorado

Selva Mine—Middle? Oligocene—Mallorca

Septarienton sites—Middle Oligocene—Belgium

Seventy-one (71) Table South—Chadronian—South Dakota

Shable Butte—Chadronian—Colorado

Shack Draw—Orellan—Wyoming

Shalkanura—Early? and Middle Oligocene—Kazakhstan

Shara-gol Valley (Shargaltein) (North Tsaidam Basin)—Late Oligocene—Gansu

Sharps Formation sites—Early Arikareean—South Dakota

Sheep Mountain Range West—Orellan—South Dakota

Sheep Mountain Table area—Orellan, Whitneyan, and Early Arikareean—South Dakota

Sheep Rock (in part)—Early Arikareean—Oregon

Shih-er-ma-cheng (Po-yang-ho) (in part?)—Early? Oligocene—Gansu

Shintusay (Shalkanura) (Shintus Ravine–Lake Shintus)—Middle Oligocene—Kazakhstan

Shoemaker Draw (Big Badlands)—Chadronian—South Dakota

Shui-tung-kou (Choei-Tong-Kou)—Late Oligocene—Gansu

Sian—Early? Oligocene—Shanxi

Siebenburger (Zitadell zu Kolzsvar)—Middle? Oligocene—Romania

Sieblos an der Wasserkuppe (Rhon)—Early Oligocene—Germany

Sigean (Aude)—Late Oligocene and Early Miocene—France

Silar Ranch (Pass Creek)—Orellan—South Dakota

Sineu Mine—Middle Oligocene—Mallorca

Sipahi—Oligocene—Thracian Turkey

Sixty-six (66) Mountain Southwest (Hagie)—Early Arikareean—Wyoming

Slim Buttes—Duchesnean, Chadronian, Orellan, and Whitneyan—South Dakota

Smith River Valley (White Sulphur Springs)—Early Arikareean—Montana

Solignat—Oligocene—France

Solnhofen 4—Middle Oligocene—Germany

Son Fe—Oligocene—Mallorca

Sorbier   (See Peublanc.)

Soulce (Bern)—Middle Oligocene—Switzerland

Soumailles (Lot-et-Garonne)—Early Oligocene—France

South Ansihai—Early Oligocene—Xinjiang

South Heart South—Orellan—North Dakota

South Mountain—Whitneyan? and Early Arikareean—California

South Pass City West—Chadronian—Wyoming

Sprendlingen—Middle Oligocene—Germany

Spring Draw—Chadronian and Orellan—South Dakota

Spring Gulch (Sage Creek West)—Chadronian—Montana

Statzendorf—Middle? Oligocene—Austria

Streamboat Rock—Early Arikareean—Nebraska

Steendorp (in type Rupelian)—Middle Oligocene—Belgium

Steine-Berg I, II, III—Late Oligocene—Germany

Stinking Water Creek—Early Arikareean—Wyoming

Stubbs Ferry (Helena Northeast)—Early Arikareean—Montana

Sugar Loaf—Chadronian—Nebraska

Suhaitu (Chaganbulage-Alashan district)—Early Oligocene—Nei Mongol

Sür—Oligocene—Hungary

Suscevaz (Vaud)—Late Oligocene—Switzerland

Syurel' River, right bank—Middle Oligocene—Kazakhstan

Szapvolgy—Middle Oligocene—Hungary-Romania?

Taben-Buluk (Yindirte)—Late Oligocene—Gansu

Tal-Holz—Late Oligocene—Germany

Tarrega (El Talladel)—Middle Oligocene—Spain

Tatal Gol—Middle Oligocene—Mongolia

Tatarcedit—Oligocene—Thracian Turkey

Tavsanli—Late Oligocene—Turkey

Tecuya—Early Arikareean—California

Terhagen (in type Rupelian)—Middle Oligocene—Belgium

Thayngen (Schaffhausen)—Late Oligocene—Switzerland

Thezels (Lot)—Late Oligocene—France

Thompson Creek—Chadronian—Montana

Tientong (= Tienyang?)—Oligocene—Guangxi

Titus Canyon—Chadronian—California

Toadstool Park—Orellan and Whitneyan—Nebraska

Tokca Assemblage—Early and Middle? Oligocene—Turkey

Toloma—Deseadan—Bolivia

Tomerdingen (in part?)—Early Oligocene—Germany

Torrington—Orellan—Wyoming

Tortmola (Dulygaly-Zhilanshik River)—Middle Oligocene—Kazakhstan

Toston (Dry Creek)—Orellan—Montana

Tremain East–Horse Creek Basin—Early Arikareean—Wyoming

Treuchlingen (Bavaria)—Early?, Middle? Oligocene—Germany

Trifail (Steiermark)—Late Oligocene—Austria

Trigonias Quarry—Chadronian—Colorado

Tsaidam Basin North—Late Oligocene—Gansu

Tsakhir—Middle Oligocene—Mongolia

Tschelkar Teniz Lake (Myn Sai)—Middle Oligocene—Kazakhstan

Tschigirin (Chigirin) (Ukraine)—Early Oligocene—USSR (marine)

Tunnel Hill (Ledingham Ranch)—Early Arikareean—Nebraska

Turpan Basin, many localities—Late Oligocene—Xinjiang

Turgai area, many localities—Middle and Late Oligocene—Kazakhstan

Turtle Cove (in part)—Early Arikareean—Oregon

Tyree Basin–Loc. 39, D. S. School Mines—Orellan—South Dakota

Uffhofen (Flonheim-Alzey-Mainz)—Middle Oligocene—Germany

Ulan Ganga—Middle? Oligocene—Mongolia

Ulan Gochu   (See Ergelyeen Dzo, etc.)—Early Oligocene—Nei Mongol

Unterer Eselsberg—Early Oligocene or earlier—Germany

Urtyn Obo—Early Oligocene—Nei Mongol

USNM DW-3—Orellan—Nebraska

USNM SX-22—Orellan—Nebraska

Varages—Oligocene—France

Vaulruz (Freiburg)—Middle? Oligocene—Switzerland

Vaumas (Sologne bourbonnaise)—Middle Oligocene—France

Vaux—Late Oligocene and Early Miocene—Switzerland

Vaylats (Quercy)—Middle Oligocene—France

Vehringen (Vehringendorf, Veringstadt)—Middle Oligocene—Germany

Veltheim (Aargau)—Late Oligocene or Early Miocene—Switzerland

Vendèze (Haute Auvergne)—Middle Oligocene—France

Verdun-sur-Garonne (Tarn-et-Garonne)—Late Oligocene—France

Vermeil—Early Oligocene—France

Vernal West—Chadronian (Duchesnean?)—Utah

Verneuil (Allier)—Late Oligocene—France

Versoix—Late Oligocene—Switzerland

Vieja local faunas—Chadronian—Texas (Porvenir, etc.)

Villebramar (Lot-et-Garonne)—Middle Oligocene—France
Villejuif—Early Oligocene or earlier—Paris Basin
Vodable—Oligocene—France
Vufflens-le-Château—Late Oligocene—Switzerland
Wagon Bed Spring district—Chadronian—Wyoming
Wagon Bed Spring North 1 and 2—Chadronian—Wyoming
Waitzen—Oligocene—Hungary
Waldenberg (Brochene Fluh)—Late Oligocene—Switzerland
Waldenberg–Humbel (Baselland)—Late Oligocene—Switzerland
Walter Brecht Ranch (See Brecht Ranch.)
Wamblee (Wanblee?) district—Orellan and Early Arikareean—
    South Dakota
Warbonnet Creek—Orellan—Nebraska
Warbonnet Ranch—Orellan—Nebraska
Warnsdorf—Late Oligocene—Czechoslovakia
Weidenstetten—Early Oligocene—Germany
Weissenburg sites—Early Oligocene or earlier—Germany
Wendelsheim—Middle Oligocene—Germany
West Branch Calf Creek—Chadronian—Saskatchewan
Westernhain (Westerwald)—Late Oligocene—Germany
West Fork (Big Corral Draw, Cottonwood Creek)—Whitneyan—
    South Dakota
White Buttes—Chadronian and Orellan—North Dakota
White Clay district—Orellan and Early Arikareean—Nebraska
White Horse Creek—Chadronian—Nebraska
White River Drainage—Orellan—Nebraska
White Sulphur Springs vicinity—Early Arikareean—Montana
Whitman P. O. (Node North, Hat Creek Basin)—Orellan—Wyo-
    ming
Wila Wila—Deseadan—Bolivia
Wildcat Ridge North (Round Top, Roubideaux Pass)—Whit-
    neyan—Nebraska
Willard Canyon (South Mountain)—Whitneyan? and Early
    Arikareean—California
Willow Creek—Early Arikareean—Wyoming
Wind Creek (Sheep Mountain East)—Orellan—South Dakota
Wolferstadt 2—Early Oligocene or earlier—Germany
Wolfwil (Solothurn)—Late Oligocene—Switzerland
Wöllstein—Middle Oligocene—Germany
Wounded Knee district—Early Arikareean and other ages—South
    Dakota
Wright Gap East (Pumpkin Creek Basin)—Early Arikareean—
    Nebraska
Wynau I–V (Bern)—Middle Oligocene—Switzerland
Yale Quarry G, I, etc.—Early to? Middle? Oligocene—Egypt
Yaylagöne—Oligocene—Thracian Turkey
Yoder (Yoder Ditch)—Chadronian—Wyoming
Yuanqu Basin—Early Oligocene—Shanxi
Yungle Basin—Early Oligocene—Guangxi
Yverdon (Vaud)—Late Oligocene—Switzerland
Zaram Bulak (Narim Bulak, Kharaat Ula Range)—Oligocene—
    Mongolia
Zaysan Depression—Early to Late Oligocene—Kazakhstan
Zella Oasis—Early Oligocene—Libya
Zitadell zu Kolzsvar (Siebenburger)—Middle to Late Oligo-
    cene—Romania
Zovencedo (Vicentino)—Middle? Oligocene—Italy

BIBLIOGRAPHY

ADROVER, R., and HUGUENEY, M. 1975. Des rongeurs (Mammalia) africains dans une faune de l'Oligocène élèvé de Majorque (Baléares, Espagne). *Mus. Hist. Nat. Lyon, Nouv. Arch.*,
13 (suppl.):11–13.

ADROVER, R., HUGUENEY, M., and MEIN, P. 1977. Fauna africana oligocena y nuevas formas endemicas entre los micromam-
iferos de Mallorca (Nota preliminar). *Soc. Hist. Nat. Baleares,
Bol.*, 22:137–149.

ALF, R. 1962. A new species of the rodent *Pipestoneomys* from
the Oligocene of Nebraska. *Breviora* 172:1–7.

ALKER, J. 1968. The occurrence of *Paracricetodon* Schaub (Cricetidae) in North America. *J. Mammal.* 49:529–530.

ARGYROPULO, A.I. 1940. (A survey of the findings of Rodentia
(Tertiary) on the territory of USSR and of the contiguous regions of Asia.) *Priroda* 1940, no. 12. Pp. 74–82. (In Russian.)

BAHLO, E. 1975. Die Nagetierfauna von Heimersheim bei Alzey
(Rheinhessen, Westdeutschland) aus dem Grenzbereich Mittel-
Oberoligozän und ihre stratigraphische Stellung. *Abh. Hess.
Landesamt. Bodenforsch.* 71:1–128.

BARNES, L.G. 1976. Outline of eastern North Pacific fossil cetacean assemblages. *Syst. Zool.* 23:321–343.

BECKER, H.F. 1961. Oligocene plants from the upper Ruby River
basin, southwestern Montana. *Geol. Soc. Am. Mem.* 82:1–127.

BELYAYEVA, Y.I., TROFIMOV, B.A., and RESHETOV, V.Y. 1974.
(Fauna and biostratigraphy of the Mesozoic and Cenozoic of
Mongolia. *Joint Soviet-Mongolian Paleontological Expedition
Works*, issue 1. Pp. 19–45.) (Translated by G. Shkurkin, 1977.)

BERGGREN, W.A. 1971. Tertiary boundaries. In *Micropaleontol-
ogy of the Oceans*, ed. B.M. Funnell and W.R. Riedel, pp.
728–738. Cambridge: Cambridge Univ. Press.

BERRY, E.W. 1916. The flora of the Catahoula sandstone. *U.S.
Geol. Surv. Prof. Pap.* 98:227–251.

BEYRICH, H.E. VON. 1855. Stellung der Hessischen Tertiärbil-
dungen. *Akad. Wiss., Berlin, Monatsb. f. 1854.* Pp. 640–666.

BLACK, C.C. 1961C. New rodents from the early Miocene deposits of Sixty-six Mountain, Wyoming. *Breviora* 146:1–7.

BLACK, C.C. 1965A. Fossil mammals from Montana. Pt. 2. Rodents from the early Oligocene Pipestone Springs local fauna.
Ann. Carnegie Mus. 38:1–48.

BLACK, C.C. 1968A. The Oligocene rodent *Ischyromys* and discussion of the family Ischyromyidae. *Ann. Carnegie Mus.*
39:273–305.

BLACK, C.C. 1974. Paleontology and geology of the Badwater
Creek area, central Wyoming. Pt. 9. Additions to the cylindrodont rodents from the late Eocene. *Ann. Carnegie Mus.* 45:151–
160.

BLACK, C.C., and STEPHENS, J.J., III. 1973. Rodents from the
Paleogene of Guanajuato, Mexico. *Occas. Pap. Mus. Texas
Tech. Univ.* 14:1–10.

BONIS, L. de. 1974. Premières données sur les Carnivores Fis-
sipedes provenant des fouilles récentes dans le Quercy.
Palaeovertébr. 6:27–32.

BONIS, L. de, CROCHET, Y., RAGE, J.-C., SIGÉ, B., SUDRE, J.,
and VIANEY-LIAUD, M. 1973. Nouvelles faunes de vertébrés
oligocènes des phosphorites du Quercy. *Bull. Mus. Natl. Hist.
Nat.* 174. *Sci. de la Terre* 28:105–113.

BORDAS, A.F. 1939A. Diagnosis sobre algunos mamiferos de las
capas con *Colpodon* del Valle del Rio Chubut (Republica Argentina). *Physis.* 14:413–434.

BOSMA, A.A., and BRUIJN, H. de. 1979. Eocene and Oligocene Gliridae (Rodentia, Mammalia) from the Isle of Wight, England. Part 1. The *Gliravus priscus–Gliravus fordi* lineage. *Proc. Ned. Akad. Weten., Koninkl., Ser. B.*, 82:367–384.

BRUNET, M. 1970. Villebramar (Lot-et-Garonne): très important gisement de vertébres Stampien inférieur du Bassin d'Aquitaine. *C. R. Acad. Sci., Paris*, 270-D:2535–3538.

BRUNET, M. 1977. Les mammifères et le problème de la limite Eocène-Oligocène en Europe. *Geobios, Mém. Spéc.*, 1:11–27.

BRUNET, M., JEHENNE, Y., and RINGEADE, M. 1977. Note préliminaire concernant la découverte d'une faune et d'une flore du niveau de Ronzon dans l'Oligocène inférieur du Bassin d'Aquitaine. *Géobios* 10:109–112.

CARBONNEL, G., CHATEAUNEUF, J.-J., FEIST-CASTEL, M., DE GRACIANSKY, P.-C., and VIANEY-LIAUD, M. 1972. Les apports de la paléontologie (Spores et pollens, charophytes, ostracodes, mammifères) à la stratigraphie et à la paléogeographie des molasses de l'Oligocène supérieur de Barrême (Alpes d'Haute Provence). *C. R. Acad. Sci., Paris*, 275-D:2599–2602.

CAVELIER, C., and POMEROL, C. 1976. Proposition d'une échelle stratigraphique standard pour le Paléogène. *Newsl. Stratigr.* 5.

CHAFFEE, R.G. 1954. *Campylocynodon personi*, a new Oligocene carnivore from the Beaver Divide, Wyoming. *J. Paleontol.* 28:43–46.

CHANDLER, M.E.J. 1964. The lower Tertiary floras of southern England. 4. A summary of findings in the light of recent botanical observations. London: Brit. Mus. (Nat. Hist.). 151 pp.

CHANEY, R.W. 1925. A comparative study of the Bridge Creek flora and the modern redwood forest. *Carnegie Inst. Wash. Publ.* 349:1–22.

CHANEY, R.W. 1927. Geology and paleontology of the Crooked River Basin, with special reference to the Bridge Creek flora. *Carnegie Inst. Wash. Publ.* 346:45–138.

CHAROLLAIS, J., GINET, C., HUGUENEY, M., and MÜLLER, J.-P. 1981. Sur la présence de dents de mammifères à la base et dans la partie supérieure de la Molasse rouge du plateau des Bornes (Hte.-Sav., France). *Helv. Eclog. Geol.* 74:37–52.

CHIU, C. 1973. A new genus of giant rhinoceros from Oligocene of Dzungaria, Sinkiang. *Vertebr. PalAsiatica* 11:182–191.

CHOW, M. 1957. On some Eocene and Oligocene mammals from Kwangsi and Yunnan. *Vertebr. PalAsiatica* 1:201–214.

CHOW, M. 1958. Some Oligocene mammals from Lunan, Yunnan. *Vertebr. PalAsiatica* 2:263–267.

CHOW, M., and CHIU, C. 1963. New genus of giant rhinoceros from Oligocene of Inner Mongolia. *Vertebr. PalAsiatica* 7:230–239.

CLARK, J. 1937. Art. 21. The stratigraphy and paleontology of the Chadron Formation in the Big Badlands of South Dakota. *Ann. Carnegie Mus.* 25:261–350.

CLARK, J., BEERBOWER, J.R., and KIETZKE, K.K. 1967. Oligocene sedimentation, paleoecology and paleoclimatology in the Big Badlands of South Dakota. *Fieldiana, Geol. Mem.*, 5:1–158.

COLOM, G. 1961. La paléoecologia des lacs du Ludien-Stampien inférieur de l'Ile de Majorque. *Rev. Micropaléontol.* 4:17–29.

CROCHET, J.-Y. 1974. Les insectivores des Phosphorites du Quercy. *Palaeovertébr.* 6:109–159.

CROCHET, J.-Y. 1980. *Les Marsupiaux du Tertiaire d'Europe*. Vols. 1 and 2. edit. Found. Singer-Polignac, Paris, 279 pp.

CROUZEL, F., GINSBURG, L., and VIDALENC, D. 1977. Les carnivores fissipèdes du Stampien terminal de Dieupentale (Tarn-et-Garonne). *Bull. Soc. Hist. Nat. Toulouse* 113:207–229.

CRUSAFONT, M., GINSBURG, L., and TRUYOLS, J. 1962. Mise en évidence du Sannoisien dans la Haute Vallée du Tage (Espagne). *C. R. Acad. Sci., Paris*, 255-D:2155–2157.

DAL PIAZ, G.B. 1932. I mammiferi dell'Oligoceno veneto. *Anthracotherium monsvialense*. Ist. Geol. Univ. Padova Mem. 10:1–63.

DASHZEVEG, D. 1974. The chalicothere *Schizotherium avitum* Matthew and Granger from Oligocene deposits in the East Gobi (locality Ergiliindzo) and a short review of the vertebrate fauna from the same locality. In *Mesozoic and Cenozoic Faunas and Biostratigraphy of Mongolia. Trans. Joint Soviet-Mongolian Expedition* 1:74–79. (In Russian.)

DAWSON, M.R. 1968A. Oligocene rodents (Mammalia) from East Mesa, Inner Mongolia. *Am. Mus. Novit.* 2324:1–12.

DEHM, R. 1950. Zur Eozän-Oligozän Grenze. *Monatsh. Neues Jahrb. Geol. Palaeontol.* 1950:193–200.

DONOHOE, J.C. 1956. New aplodontid rodent from Montana Oligocene. *J. Mammal.* 37:264–268.

DUMONT, A. 1839. Rapport sur les travaux de la carte géologique de la Belgique pour l'année 1839. *Bull. Acad. R. Belg., Cl. Sci.*, 16:367.

EMLONG, D. 1966. A new archaic cetacean from the Oligocene of northwestern Oregon. *Bull. Univ. Oregon Mus. Nat. Hist.* 3:1–51.

EMRY, R.J. 1970. A North American Oligocene pangolin and other additions to the Pholidota. *Bull. Am. Mus. Nat. Hist.* 142:455–510.

EMRY, R.J. 1972A. A new heteromyid rodent from the early Oligocene of Natrona County, Wyoming. *Proc. Biol. Soc. Wash.* 85:179–190.

EMRY, R.J. 1972B. A new species of *Agnotocastor* (Rodentia, Castoridae) from the early Oligocene of Wyoming. *Am. Mus. Novit.* 2485:1–7.

EMRY, R.J. 1973. Stratigraphy and preliminary biostratigraphy of the Flagstaff Rim area, Natrona County, Wyoming. *Smithson. Contrib. Paleobiol.* 18:1–43.

EMRY, R.J. 1975. Revised Tertiary stratigraphy and paleontology of the western Beaver Divide, Fremont County, Wyoming. *Smithson. Contrib. Paleobiol.* 25:1–20.

EMRY, R.J. 1978. A new hypertragulid (Mammalia, Ruminantia) from the early Chadronian of Wyoming and Texas. *J. Paleontol.* 52:1004–1014.

EMRY, R.J. 1979. Review of *Toxotherium* (Perissodactyla: Rhinocerotoidea) with new material from the Early Oligocene of Wyoming. *Proc. Biol. Soc. Wash.* 92:28–41.

EMRY, R.J., and DAWSON, M.R. 1972. A unique cricetid (Rodentia, Mammalia) from the early Oligocene of Natrona County, Wyoming. *Am. Mus. Novit.* 2508:1–14.

EMRY, R.J., and STORER, J.E. 1981. The hornless protoceratid *Pseudoprotoceras* (Tylopoda: Artiodactyla) in the Early Oligocene of Saskatchewan and Wyoming. *J. Vert. Paleontol.* 1:101–110.

ENGESSER, B. 1975. Revision der euopäischen Heterosoricinae (Insectivora, Mammalia). *Helv. Eclog. Geol.* 68:649–671.

FAHLBUSCH, V. 1970. Populations verschiebungen bei tertiären

Nagetieren, eine Studie an oligozänen und miozänen Eomyidae Europas. *Verlag. Bayerisch. Akad. Wiss.* 145:1–136.

FAHLBUSCH, V. 1973. Die stammesgeschichtlichen Beziehungen zwischen den Eomyiden (Mammalia, Rodentia) Nordamerikas und Europas. *Mitt. Bayer. Staatssamml. Palaontol. Hist. Geol.* 13:141–175.

FAHLBUSCH, V. 1976. Report on the International Symposium on mammalian stratigraphy of the European Tertiary in Munich, April 11–14, 1975. *Newsl. Stratigr.* 5:160–167.

FAHLBUSCH, V. 1979. Eomyidae—Geschichte einer Säugetierfamilie. *Z. Paläontol.* 53:8897.

FLEROV, K.K., and JANOVSKAJA, N.M. 1971. (The ecological complexes of the mammals of Asia and their zoogeographical characteristics. In *Recent Problems in Paleontology.*) Moscow. "Nauka". Pp. 7–31. (In Russian.)

FORSTEN, A. 1970. *Mesohippus* from the Chadron of South Dakota, and a comparison with Brulean *Mesohippus bairdii* Leidy. *Biol. Soc. Scient. Fenn. Comment,* 31:1–22.

FRANZEN, J.-L. 1968. Revision der Gattung *Palaeotherium* (Perissodactyla, Mammalia). Inaugural Dissert. Naturwiss.-math. Fak., Albert-Ludwigs-Univ. Vol. 1, pp. 1–181; Vol. 2, 35 pls., 15 tables.

FRANZEN, J.-L. 1972. *Pseudopalaeotherium longirostratum* n. g., n. sp. (Perissodactyla, Mammalia) aus dem unterstampischen Kalkmergel von Ronzon (Frankreich). *Senckenberg. Lethaea* 53:315–331.

FRIES, C., JR., HIBBARD, C.W., and DUNKLE, D.H. 1955. Early Cenozoic vertebrates in the red conglomerate at Guanajuato, Mexico. *Smithson. Misc. Coll.* 123:1–15.

GALBREATH, E.C. 1953. A contribution to the Tertiary geology and paleontology of northeastern Colorado. *Univ. Kans., Paleontol. Contrib., Vertebr.,* Art. 4, 13:1–120.

GALBREATH, E.C. 1955. A new eomyid rodent from the lower Oligocene of northeastern Colorado. *Trans. Kans. Acad. Sci.* 58:75–78.

GAWNE, C.E. 1978. Leporids (Lagomorpha, Mammalia) from the Chadronian (Oligocene) deposits of Flagstaff Rim, Wyoming. *J. Paleontol.* 52:1103–1118.

GINGERICH, P.D. 1978. The Stuttgart collection of Oligocene primates from the Fayum Province of Egypt. *Z. Palaeontol.* 52:82–92.

GINSBURG, L. 1965. Les amphicyons des Phosphorites du Quercy. *Ann. Paléontol.* 52:1–44.

GINSBURG, L. 1969. Une faune de Mammifères terrestres dans le Stampien marin d'Etampes (Essonne). *C. R. Acad. Sci., Paris,* 268-D:1266–1268.

GINSBURG, L. 1973. Les tayassuidés des Phosphorites du Quercy. *Palaeovertébr.* 6:55–85.

GINSBURG, L. 1979. Revision taxonomique des Nimravini (Carnivora, Felidae) de l'Oligocène des Phosphorites du Quercy. *Bull. Mus. Natl. Hist. Nat., Paris,* (4)1:35–49.

GLIBERT, M., and HEINZELIN, J. de. 1952. Le gîte des vertébrés Tongriens de Hoogbutsel. *Bull. Mus. Hist. Nat. Belg.* 28:1–22.

GORROÑO, R., PASCUAL, R., and POMBO, R. 1979. Hallazgo de mamiferos eogenos en el sur de Mendoza. *Actas 7. Congr. Geol. Argentino.* 2:475–487.

GRADZINSKI, R., KAZMIERCZAK, J., and LEFELD, J. 1968. Geo-graphical and geological data from the Polish-Mongolian paleontological expeditions. *Palaeontol. Pol.* 19:33–84.

GREEN, M. 1941. A study of the Oligocene Leporidae in the Kansas University Museum of Vertebrae Paleontology. *Trans. Kans. Acad. Sci.* 45:229–247.

GROMOVA, V. 1952. (On primitive carnivores of the Paleogene of Mongolia and Kazakhstan.) *Tr. Akad. Nauk SSSR, Paleontol. Inst.* 41:51–77. (In Russian.)

GROMOVA, V. 1960. (On a new family (Tshelkariidae) of primitive Carnivora (Creodonta) of the Oligocene of Asia.) *Tr. Akad. Nauk SSSR, Paleontol. Inst.* 77:41–78. (In Russian.)

GUREEV, A.A. 1960. (Oligocene Lagomorpha of Mongolia and Kazakhstan.) *Tr. Akad. Nauk SSSR, Paleontol. Inst.* 77:5–34. (In Russian.)

HARDENBOL, J., and BERGGREN, W.A. 1978. A new Paleogene time scale. In *Contribution to the Geological Time Scale,* ed. G.V. Cohee, M.F. Glaesner, and H.D. Hedberg, pp. 213–234. Studies in Geology, no 6. Tulsa: Am. Assoc. Pet. Geol.

HARKSEN, J.C., and MACDONALD, J.R. 1969B. Type sections for the Chadron and Brule Formations of the White River Oligocene in the Big Badlands, South Dakota. *S. Dak. Geol. Surv. Rept. Investig.* 99:1–23.

HARTENBERGER, J.-L. 1969. Les Pseudosciuridae (Mammalia, Rodentia) de l'Eocène moyen de Bouxwiller, Egerkingen et Lissieu. *Palaeovertébr.* 3:27–61.

HARTENBERGER, J.-L. 1971. La systématique des Theridomyidae (Rodentia). *C. R. Acad. Sci., Paris,* 273-D:1917–1920.

HARTENBERGER, J.-L. 1973A. Etude systématique de Theridomyoidea de l'Eocène supérieur. *Soc. Géol. Fr. Mém.* 117:1–76.

HARTENBERGER, J.-L., SIGÉ, B., and SUDRE, J. 1974. La plus ancienne faune de Mammifères du Quercy: le Bretou. *Palaeovertébr.* 6:177–196.

HARTENBERGER, J.-L., SIGÉ, B., SUDRE, J., and VIANEY-LIAUD, M. 1970. Nouveau gisements de vertébrés dans le bassin Tertiaire d'Alès (Gard). *Bull. Soc. Géol. Fr.* 12:879–885.

HEISSIG, K. 1977. Neues material von *Cryptopithecus* (Mammalia, Pantolestidae) aus dem Mitteloligozäne von "Möhren 12" in Mittelfranken. *Mitt. Bayer. Staatssamml. Paläontol. Hist. Geol.* 17:213–225.

HEISSIG, K. 1979. Die fruhsten Flughornchen und primitive Ailuravinae (Rodentia, Mammalia) aus dem suddeutschen Oligozän. *Mitt. Bayer. Staatssamml. Paläontol. Hist. Geol.* 19:139–169.

HELMER, D., and VIANEY-LIAUD, M. 1970. Nouveaux gisements de Rongeurs dans l'Oligocène moyen de Provence. *C. R. Soc. Géol. Fr.* 2:45–46.

HOFFSTETTER, R. 1969B. Un primate de l'Oligocène inférieur sud-americain: *Branisella boliviana,* gen. et sp. nov. *C. R. Acad. Sci., Paris, Sér. D,* 269:434–437.

HOFFSTETTER, R. 1976. I. Introduction au Déséadien de Bolivie. In Hoffstetter and Lavocat (1976).

HOFFSTETTER, R., and LAVOCAT, R. 1970. Découverte dans le Déséadien de Bolivie de genres pentalophodontes appuyant les affinités africaines des rongeurs caviomorphes. *C. R. Acad. Sci. Paris, Sér. D,* 271:172–175.

HOFFSTETTER, R., and LAVOCAT, R. 1976. Rongeurs caviomorphs de l'Oligocène de Bolivie. *Palaeovertébr.* 7:1–90.

HOUGH, J. 1961. Review of Oligocene didelphid marsupials. *J. Paleontol.* 35:218–228.

HOUGH, J. 1956. A new insectivore from the Oligocene of the

Wind River Basin, Wyoming, with notes on the taxonomy of the Oligocene Tenrecoidea. *J. Paleontol.* 30:531–541.

HOUGH, J., and ALF, R. 1956. A Chadron mammalian fauna from Nebraska. *J. Paleontol.* 30:132–140.

HU, C. 1961. The occurrence of *Parabrontops* in Hami, Sinkiang. *Vertebr. PalAsiatica* 1:41–42.

HU, C. 1962. Cenozoic mammalian fossil localities in Kansu and Ningshia. *Vertebr. PalAsiatica* 6:162–172.

HUGUENEY, M. 1968. Les Gliridae (Rodentia) de l'Oligocène supérieur de St.-Victor-la-Coste (Gard). *Palaeovertébr.* 2:1–16.

HUGUENEY, M. 1969. Les Rongeurs (Mammalia) de l'Oligocène supérieur de Coderet-Bransat (Allier). *Univ. Lyon, Fac. Sci., Thèse*, 596:1–227.

HUGUENEY, M. 1975. Les Castoridae (Mammalia, Rodentia) dans l'Oligocène d'Europe. *Fr., Cent. Natl. Rech. Sci., Colloq. Int.*, 218:791–804.

HUGUENEY, M. 1976. Un stade primitif dans l'évolution des Soricinae (Mammalia, Insectivora): *Srinitium marteli* nov. gen., nov. sp. de l'Oligocène moyen de Saint-Martin-de-Castillon (Vaucluse). *C. R. Acad. Sci., Paris, Sér. D.*, 282:981–984.

HUNT, R.M., JR. 1974. *Daphoenictis*, a cat-like carnivore (Mammalia, Amphicyonidae) from the Oligocene of North America. *J. Paleontol.* 48:1030–1047.

JAEGER, J.J., and HARTENBERGER, J.-L. 1975. Pour utilization systématique de niveaux-repères en biochronologie mammalienne. Montpellier, *3e Réunion Annuel Sci. Terre*. p. 201.

JAMBOR, A., KORPAS, L., KRETZOI, M., PALFAVY, M.I., and RAKOSI, L. 1969. A dunántúli oligocén képzödmények rétegtani problémái. *Hung., Foeldt. Intez., Evi. Jel.* (1969):141–154.

JIANG, Y., WANG, B., and QI, T. 1976. Stratigraphy of the early Oligocene Chaganbulage Formation, Haosibuerdu Basin, Ningxia. *Vertebr. PalAsiatica* 14:35–41. (In Chinese.)

KOWALSKI, K. 1974. Middle Oligocene rodents from Mongolia. *Palaeontol. Pol.* 30:147–178.

KRETZOI, K. 1940. Altertiaire Perissodactylen aus Ungarn. *Ann. Mus. Natl. Hungarici, Pars Min. Geol. Palaeontol.* 33:87–98.

KRETZOI, K. 1943. *Kochictis centennii*, n.g., n. sp., ein altertümliche creodonta aus dem oberoligozän siebenbürgens. *Foeldt. Koezl.* 73:190–195.

KREUZER, H., DANIELS, C.H.V., GRAMANN, R., HARRE, W., and MATTIAT, B.B. 1973. K-A dates of some glauconites of the N.W. German Tertiary Basin. *Fortschr. Mineral.* 50:94–95.

KUENZI, W.D., and FIELDS, R.W. 1971. Tertiary stratigraphy, structure and geologic history, Jefferson Basin, Montana. *Bull. Geol. Soc. Am.* 82:3373–3394.

KUSS, S.E. 1965. Revision der europäischen Amphicyoninae (Canidae, Carnivora, Mammalia) ausschliesslich der voroberstampischen Formen. *Abh. Sitz. Heidelb. Akad. Wiss., Mathnaturw.* Kl. 1. Pp. 1–168.

LAVOCAT, R. 1976. II. Rongeurs du Bassin déséadien de Salla-Luribay. In Hoffstetter and Lavocat (1976).

LAVROV, V.V., and BASHANOV, V.S. 1959. (Results of the geological and paleontological studies on the Tertiary beds of the Zajsan Depression.) *Vestn. Akad. Nauk Kaz. SSR* 1:55–59.

LILLEGRAVEN, J.A. 1970. Stratigraphy, structure, and vertebrate fossils of the Oligocene Brule Formation, Slim Buttes, northwestern South Dakota. *Bull. Geol. Soc. Am.* 81:831–850.

LILLEGRAVEN, J.A. 1972. Ordinal and familial diversity of Cenozoic mammals. *Taxon.* 21:261–274.

LOOMIS, F. B. 1914. *The Deseado Formation of Patagonia*. Concord, N.H.: Amherst College Publs. 232 pp.

LOPEZ, N., and THALER, L. 1974. Sur le plus ancien lagomorphe européen et la "Grande Coupure" Oligocène de Stehlin. *Palaeovertébr.* 6:243–251.

LUTTIG, G., and THENIUS, E. 1961. Uber einen Anthracotheriiden aus dem Alttertiär von Thrazien (Griechenland). *Z. Paläontol.* 35:179–186.

MACDONALD, J.R. 1951E. Additions to the Whitneyan fauna of South Dakota. *J. Paleontol.* 25:257–265.

MACDONALD, J.R. 1951C. The fossil vertebrata of South Dakota. In *Guidebook, 5th Field Conf., Soc. Vert. Paleontol., western South Dakota*, pp. 63–74. Rapid City: S. Dak. Sch. Mines.

MACDONALD, J.R. 1955. The Leptochoeridae. *J. Paleontol.* 29:439–459.

MACDONALD, J.R. 1956C. The North American Anthrocotheres. *J. Paleontol.* 30:615–645.

MACGINITIE, H.D. 1953. Fossil plants of the Florissant Beds, Colorado. *Carnegie Inst. Wash. Publ.* 599:1–198.

MARSHALL, L.G. 1978. Evolution of the Borhyaenidae, extinct South American predaceous marsupials. *Univ. Calif., Publ. Geol. Sci.* 117:1–89.

MARTIN, L.D. 1974. New rodents from the Lower Miocene Gering Formation of western Nebraska. *Univ. Kans., Mus. Nat. Hist., Occ. Pap.* 32:1–12.

MARTINI E., and RITZKOWSKI, S. 1969. Die Grenze Eozän/Oligozän in der typus-region des Unteroligozäns (Helmstedt-Egeln-Latdorf). *Colloque sur l'Eocène*, 1968, *But. Rech. Géol. Min., Mém.* 69:233–237.

MCGREW, P.O. 1953. A new and primitive early Oligocene horse from Trans-Pecos Texas. *Fieldiana: Geol.* 10:167–171.

MCGREW, P.O. 1971. Early Tertiary vertebrate faunas, Vieja Group, Trans-Pecos Texas: Equidae. Pt. 2. *Mesohippus* from the Vieja Group, Trans-Pecos Texas. *Tex. Mem. Mus., Pearce-Sellards Ser.* 18:6–11.

MCKENNA, M.C. 1980. Remaining evidence of Oligocene sedimentary rocks previously present across the Bighorn Basin, Wyoming. In *Early Cenozoic Paleontology and Stratigraphy of the Bighorn Basin, Wyoming*, pp. 139–143. *Univ. Mich., Pap. in Paleontol.* 24:143–146.

MELLETT, J.S. 1968. The Oligocene Hsanda Gol Formation, Mongolia: A revised faunal list. *Am. Mus. Novit.* 2318:1–16.

MELLETT, J.S. 1969B. A skull of *Hemipsalodon* (Mammalia, Deltatheridia) from the Clarno formation of Oregon. *Am. Mus. Novit.* 2387:1–9.

MIAO, D., 1982. Early Tertiary fossil mammals from the Shinao Basin, Panxian County, Guizhou Province. *Acta Palaeontol. Sinica* 21:526–536.

MISONNE, X. 1957. Mammifères oligocènes de Hoogbutsel et de Hoeleden. 1. Rongeurs et Ongulés. *Bull. Mus. Nat. Hist. Belg.*, 33:1–16.

MISONNE, X. 1958. Faune du Tertiaire et du Pleistocène inférieur de Belgique (Oiseaux et Mammifères). Données paléontologiques. *Bull. Mus. R. Hist. Nat. Belg.* 34:1–36.

NIKOLOV, I. 1967. Neue oberoligozäne Arten der Gattung *Elomeryx. Abh. Neues Jhb. Geol. Paläont.* 128:205–214.

OSBORN, H.F. 1929. *Embolotherium*, gen. nov., of the Ulan Go-chu, Mongolia. *Am. Mus. Novit.* 353:1–20.

OZANSOY, F. 1962. Les Anthracothériens de l'Oligocène inférieur de la Thrace orientale (Turquie). *Bull.* (Turkey) *Min. Res. Expl. Inst.* 58:85–96.

OZANSOY, F. 1964. Le niveau du Sannoisien et sa faune mammalienne de la Thrace orientale (Turquie) dans le système de l'Oligocène d'Europe. *Bur. Rech. Géol. Minières Mém.* 28:991–999.

OZANSOY, F. 1969. Yeni bir *Palaeomasia kansui* Boyabat (Sinop) Eosen fosil Memeli biozonu ve paleontolojik belegeleri. *Bell. Turk. Tarih Kurumu,* 33:581–585.

PATTERSON, B. 1952A. Un nuevo y extraordinario marsupial deseadiano. *Rev. Mus. Mun. Cien. Nat. Trad. Mar del Plata* 1:39–44.

PATTON, T.H. 1969A. An Oligocene land vertebrate fauna from Florida. *J. Paleontol.* 43:543–546.

PATTON, T.H., and TAYLOR, B.E. 1973. The Protoceratinae (Mammalia, Tylopoda, Protoceratidae) and the systematics of the Protoceratidae. *Bull. Am. Mus. Nat. Hist.* 150:347–414.

QI, T. 1975. An early Oligocene mammalian fauna of Ningxia. *Vertebr. PalAsiatica* 13:217–224.

QUINET, G.E. 1965A. Un condylarthre de Hoogbutsel. *Bull. Inst. R. Sci. Nat. Belg.* 51:1–5.

QUINET, G.E., and MISONNE, X. 1965. Les insectivores zalambdodontes de l'Oligocène inférieur Belge. *Bull. Inst. R. Sci. Nat. Belg.* 51:1–15.

REED, C.A. 1956. A new species of the fossorial mammal *Arctoryctes* from the Oligocene of Colorado. *Fieldiana: Geol.* 10:305–311.

REED, C.A., and TURNBULL, W.D. 1965. The mammalian genera *Arctoryctes* and *Cryptoryctes* from the Oligocene and Miocene of North America. *Fieldiana: Geol.* 15:99–170.

REED, K.M. 1961. The Proscalopinae, a new subfamily of talpid insectivores. *Bull. Mus. Comp. Zool.* (Harvard Univ.) 125:473–494.

REMY, J.A., and THALER, L. 1967. Une faune de vertébrés de l'Oligocène supérieur dans les phosphorites du groupe d'Uzès. *C. R. Soc. Géol. Fr.* 4:161–163.

RENSBERGER, J.M. 1971A. Entoptychine pocket gophers (Mammalia, Geomyidae) of the early Miocene John Day formation, Oregon. *Univ. Calif., Publ. Geol. Sci.* 90:1–209.

RENSBERGER, J.M. 1975. *Haplomys* and its bearing on the origin of the aplodontid rodents. *J. Mammal.* 56:1–14.

RENSBERGER, J.M. 1980. A primitive promylagauline rodent from the Sharps Formation, South Dakota. *J. Paleontol.* 54:1267–1277.

REPENNING, C.A. 1967. Subfamilies and genera of the Soricidae. *U.S. Geol. Surv., Prof. Pap.* 565:1–74.

RUSSELL, D.E., HARTENBERGER, J.-L., POMEROL, C., SEN, S., SCHMIDT-KITTLER, N., and VIANEY-LIAUD, M. 1982. The Paleogene of Europe, *Palaeovertebrata,* 1–77

RUSSELL, L.S. 1972. Tertiary mammals of Saskatchewan. Part 2: The Oligocene fauna, non-ungulate orders. *R. Ont. Mus., Life Sci. Contrib.* 84:1–97.

RUSSELL, L.S. 1978. Tertiary mammals of Saskatchewan. Part 4: The Oligocene anthracotheres. *R. Ont. Mus., Life Sci. Contrib.* 115:1–16.

SAVAGE, D.E., and RUSSELL, D.E. 1977. Comments on mammalian paleontologic stratigraphy and geochronology; Eocene stages and mammal ages of Europe and North America. *Geobios, Mém. Spéc.,* 1:47–57.

SAVAGE, R.J.G. 1967A. Early Tertiary mammal localities in southern Libya. *Proc. Geol. Soc. Lond.* 1657:167–171.

SAVAGE, R.J.G., 1976. Review of early Sirenia. *Syst. Zool.,* 25:344–351.

SCHIEBOUT, J.A. 1977. *Schizotheroides* (Mammalia: Perissodactyla) from the Oligocene of Trans-Pecos Texas. *J. Paleontol.* 51:455–458.

SCHMIDT-KITTLER, N. 1971B. Odontologische Untersuchungen an Pseudosciuriden (Rodentia Mammalia) des Alttertiärs. *Abh. Bayer. Akad. Wiss., Math.-Naturwiss., Kl.,* 150:1–131.

SCHMIDT-KITTLER, N., and VIANEY-LIAUD, M. 1975. Les relations entre les faunes de rongeurs d'Allemagne du Sud et de France pendant l'Oligocène. *C. R. Acad. Sci., Paris,* 281-D:511–514.

SCHMIDT-KITTLER, N., and VIANEY-LIAUD, M. 1979. Evolution des Aplodontidae oligocènes europeéns. *Palaeovertébr.* 9:33–82.

SCHULTZ, C.B., and FALKENBACH, C.H. 1954. Desmatochoerinae, a new subfamily of oreodonts. *Bull. Am. Mus. Nat. Hist.* 105:143–256.

SCHULTZ, C.B., and FALKENBACH, C.H. 1956. Miniochoerinae and Oreonetinae, two new subfamilies of oreodonts. *Bull. Am. Mus. Nat. Hist.* 109:373–482.

SCHULTZ, C.B., and FALKENBACH, C.H. 1968. The phylogeny of the oreodonts, Parts 1 and 2. *Bull. Am. Mus. Nat. Hist.* 139:1–498.

SCHULTZ, C.B., and STOUT, T.M. 1955. Classification of Oligocene sediments in Nebraska. A guide for the stratigraphic collecting of fossil mammals. *Bull. Nebr. State Mus.* 4:16–52.

SETOGUCHI, T. 1978. Paleontology and geology of the Badwater Creek Area, Central Wyoming. Part 16. The Cedar Ridge local fauna (Late Oligocene). *Bull. Carnegie Mus. Nat. Hist.* 9:1–61.

SHEVYREVA, N.S. 1971. (New rodents from the middle Oligocene of Kazakhstan and Mongolia) *Tr. Akad. Nauk SSSR, Paleontol. Inst.* 130:70–86. (In Russian.)

SIGÉ, B. 1975. Insectivores primitifs de l'Eocène supérieur et Oligocène inférieur d'Europe occidentale; Apatemyidés et Leptictidés. *Fr., Cent. Natl. Rech. Sci., Colloque Int.* 218:653–673.

SIGÉ, B. 1976A. Insectivores primitifs de l'Eocène supérieur et Oligocène inférieur d'Europe occidentale; Apatemyidés et Leptictidés. *Fr., Cent. Natl. Rech. Sci., Colloque Int.* 218:653–673.

SIGÉ, B. 1976A. Insectivores primitifs de l'Eócène supérieur et Oligocène inférieur d'Europe occidentale. Nyctitheriidés. *Mus. Natl. Hist. Nat. Mém., Sér. C,* 34:1–140.

SIGÉ, B. 1976B. Les insectivores et chiroptères du Paleogène moyen d'Europe dans l'histoire des faunes de Mammifères sur ce continent. *Paléobiol. Continent.* 7:1–25.

SIGÉ, B., and VIANEY-LIAUD, M. 1979. Impropriété de la Grande Coupure de Stehlin comme support d'une limite Eocène-Oligocène. *Newsl. Stratigr.* 8:79–82.

SIMONS, E.L. 1968. Part 1. African Oligocene mammals: Introduction, history of study, and faunal succession. In Simons and A. Wood (1968), pp. 1–21.

SIMONS, E.L., and GINGERICH, P.D. 1974. New carnivorous

mammals from the Oligocene of Egypt. *Ann. Geol. Surv. Egypt* 4:157–166.

SIMONS, E.L., and WOOD, A.E. 1968. Early Cenozoic mammalian faunas, Fayum Province, Egypt. *Bull. Peabody Mus. Nat. Hist.* (Yale Univ.) 28:1–105.

SIMPSON, G.G. 1932K. Some new or little-known mammals from the *Colpodon* beds of Patagonia. *Am. Mus. Novit.* 575:1–12.

SIMPSON, G.G. 1940G. Review of the mammal-bearing Tertiary of South America. *Proc. Am. Philos. Soc.* 83:649–709.

SIMPSON, G.G. 1945I. The principles of classification and a classification of mammals. *Bull. Am. Mus. Nat. Hist.* 85:1–350.

SKINNER, S.M., and GOORIS, R.J. 1966. A note on *Toxotherium* (Mammalia, Rhinocerotidae) from Natrona County, Wyoming. *Am. Mus. Novit.* 2261:1–12.

SMITH, A.G., and BRIDEN, J.C. 1977. *Mesozoic and Cenozoic paleocontinental maps.* Cambridge Earth Sci. Ser. Pp. 1–63.

SPRINGHORN, R. 1977. Revision der alttertiären europäischen Amphicyonidae (Carnivora, Mammalia). *Palaeontogr., Abt. A.* 158:26–113.

STEHLIN, H.G. 1909. Remarques sur les faunules de mammifères des couches éocènes et oligocènes du Bassin de Paris. *Bull. Soc. Géol. Fr.* 9:488–520.

STEIGER, R.H., and JAGER, E. 1977. Subcommission on geochronology: Convention on the use of decay constants in geo- and cosmochronology. *Earth Planet. Sci. Lett.* 36:359–362.

STEINENGER, R., RÖGL, F., and MARTINI, E. 1976. Current Oligocene/Miocene biostratigraphic concept of the central Paratethys. *Newsl. Stratigr.* 4:174–202.

STIRTON, R.A. 1953A. Vertebrate paleontology and continental stratigraphy in Colombia. *Bull. Geol. Soc. Am.* 64:603–622.

STIRTON, R.A., and WOODBURNE, M.O. 1965. Revision of the Oligocene and Miocene peccaries. *Geol. Soc. Am. Spec. Pap.* 82:281. (Abstract.)

STOCK, C. 1935B. Artiodactyla from the Sespe of the Las Posas Hills, California. *Carnegie Inst. Wash Publ.* 453:119–125.

STOCK, C., and BODE, F.D. 1935. Occurrence of Lower Oligocene mammal-bearing beds near Death Valley, California. *Proc. Natl. Acad. Sci.* 21:571–579.

STORER, J.E. 1978. Rodents of the Calf Creek local fauna (Cypress Hills Formation, Oligocene, Chadronian), Saskatchewan. *Saskatch. Cult. Youth Nat. Hist. Mus. Contrib.* 1:1–54.

SULIMSKI, A. 1970. On some Oligocene insectivore remains from Mongolia. *Palaeontol. Pol.* 21:53–72.

SUTTON, J.F., and BLACK, C.C. 1975. Paleontology of the earliest Oligocene deposits in Jackson Hole, Wyoming. Part 1. Rodents exclusive of the Family Eomyidae. *Ann. Carnegie Mus.* 45:299–315.

SYCH, L. 1975. Lagomorpha from the Oligocene of Mongolia. *Paleontol. Pol.* 33:183–200.

SZALAY, F.S., and DELSON, E. 1979. *Evolutionary History of the Primates.* New York: Academic Press. 580 pp.

TEDFORD R.H. 1976. Relationship of pinnipeds to other carnivores (Mammalia). *Syst. Zool.* 25:363–374.

TEDFORD, R.H., BANKS, M.R., KEMP, N.R., McDOUGALL, I., and SUTHERLAND, F.L., 1975. Recognition of the oldest known fossil marsupials from Australia. *Nature,* 255:141–142.

THALER, L. 1965A. Une échelle de zones biochronologiques pour les Mammifères du Tertiaire d'Europe. *C. R. Soc. Géol. Fr.* 4–5:118.

THALER, L. 1966. Les Rongeurs fossiles du Bas-Languedoc dans leurs rapports avec l'histoire des faunes et la stratigraphie du Tertiaire d'Europe. *Mus. Natl. Hist. Nat. Mém., n.s., Sér. C.,* 27:1–284.

THALER L. 1972. Datation, zonation et Mammifères. *Bur. Rech. Géol. Minières Mém.* 77:411–424.

TOBIEN, H. 1956B. Eine stampische Kleinsäugerfauna aus der Grenzregion Schleichsand/Cyrenenmergel von Heimersheim bei Alzey (Rheinhessen). *Z. Dtsch. Geol. Ges.* 106:565–566.

TOBIEN, H. 1960D. Säugetierpäläntologische Daten zur Alterstellung des Hessischen Melanientones. *Z. Dtsch. Geol. Ges.* 112:590.

TOBIEN, H. 1971. Mikromammalier aus dem alttertiären Melanienton von Nordhessen. Part 1: Marsupialia, Insectivora, Primates. *Sond. Notizbl. Hess. Landes. Bodenf. Wiesbaden* 99:1–29.

TOBIEN, H. 1980. Taxonomic status of some Cenozoic mammalian local faunas from the Mainz Basin. *Mainzer geowiss. Mitt.* 9:203–235.

TROFIMOV, B.A. 1952. (New entelodontids from Mongolia and Kazakhstan.) *Tr. Akad. Nauk SSR, Paleontol. Inst.* 41:144–154. (In Russian.)

VIANEY-LIAUD, M. 1972. L'évolution du genre *Theridomys* à l'Oligocène moyen. Intérêt biostratigraphique. *Bull. Mus. Natl. Hist. Nat.* (3) 98. *Sci. de la Terre* 28:295–372.

VIANEY-LIAUD, M. 1974. Les rongeurs de l'Oligocène inférieur d'Europe *Palaeovertébr.* 6:197–241.

VIANEY-LIAUD, M. 1976. Evolution des rongeurs à l'Oligocène en Europe occidentale. *C. R. Acad. Sci., Paris,* 284:1277–1280.

VOLLMAYR, T. 1966. Oberoligozäne Gliridae (Rodentia, Mammalia) aus der suddeutschen Faltenmolasse. *Mitt. Bayer. Staatsl. Paläontol. Mont. Hist. Geol.* 6:65–107.

WHITMORE, F.C., JR., and SANDERS, A.E. 1976. Review of the Oligocene Cetacea. *Syst. Zool.* 25:304–320.

WILMARTH, M.G. 1924. Geologic time classification of U.S. Geological Survey compared with other classifications. *Bull. U.S. Geol. Surv.* 769:1–138.

WILSON, J.A. 1971. Early Tertiary vertebrate faunas, Vieja Group, Trans-Pecos Texas: Agriochoeridae and Merycoidodontidae. *Bull. Texas Mem. Mus.* 18:1–83.

WILSON, J.A. 1974. Early Tertiary vertebrate faunas, Vieja Group and Buck Hill Group, Trans-Pecos Texas: Protoceratidae, Camelidae, Hypertragulidae. *Bull. Texas Mem. Mus.* 23:1–34.

WILSON, J.A. 1977A. Early Tertiary vertebrate faunas, Big Bend area, Trans-Pecos Texas: Brontotheriidae. *Texas Mem. Mus., Pearce-Sellards Ser.* 25:1–15.

WILSON, J.A. 1977B. Stratigraphic occurrence and correlation of early Tertiary vertebrate faunas, Trans-Pecos Texas. Part 1: Vieja Area. *Bull. Texas Mem. Mus.* 25:1–42.

WOLFE, J.A. 1966. Tertiary plants from the Cook Inlet region, Alaska. *U.S. Geol. Surv. Prof. Pap.* 398-B:1–32.

WOLFE, J.A. 1971. Tertiary climatic fluctuations and methods of analysis of Tertiary floras. *Palaeogeogr., Palaeoclimatol., Palaeoecol.* 9:27–57.

WOLFE, J.A. 1978. A paleobotanical interpretation of Tertiary climates in the Northern Hemisphere. *Am. Sci.* 66:694–703.

WOLFE, J.A., HOPKINS, D., and LEOPOLD, E. 1966. Tertiary stratigraphy and paleobotany of the Cook Inlet region, Alaska. *U.S. Geol. Surv. Prof. Pap.* 398-A:1–29.

WOOD, A.E. 1937. The mammalian fauna of the White River Oligocene. Part 2. Rodentia. *Trans. Am. Philos. Soc.* 28:155–269.

WOOD, A.E. 1940. The mammalian fauna of the White River Oligocene. Part 3. Lagomorpha. *Trans. Am. Philos. Soc.* 28:271–362.

WOOD, A.E. 1968. Part 2. The African Oligocene Rodentia. In Simons and A. Wood (1968), pp. 23–105.

WOOD, A.E. 1970A. The early Oligocene rodent *Ardynomys* (Fam. Cylindrodontidae) from Mongolia and Montana. *Am. Mus. Novit.* 2418:1–17.

WOOD, A.E. 1974. Early Tertiary vertebrate faunas, Vieja Group, Trans-Pecos Texas: Rodentia. *Bull. Texas Mem. Mus.* 21:1–112.

WOOD, A.E., and PATTERSON, B. 1959. The rodents of the Deseadan Oligocene of Patagonia and the beginnings of South American rodent evolution. *Bull. Mus. Comp. Zool.* (Harvard Univ.) 120:279–428.

WOOD, H.E., CHANEY, R.W., CLARK, J., COLBERT, E.H., JEPSEN, G.L., REESIDE, J.B., JR., and STOCK, C. 1941. Nomenclature and correlation of North American continental Tertiary. *Bull. Geol. Soc. Am.* 52:1–48.

XU, Y. 1966. Amynodonts of Inner Mongolia. *Vertebr. Pal-Asiatica* 10:123–190.

YALCINLAR, I. 1954. Les gisements de Mammifères et d'autres vertébrés fossiles de Turquie. *C. R. Congr. Géol. Int., Alger.,* 15:139–147. 19th Sess., Sect. 13, pt. 3.

CHAPTER SIX

Miocene Mammalian Faunas

Mammals About 24 to 5 Million Years Ago

TYPIFICATION AND CORRELATION

Sir Charles Lyell (1833) based his Miocene on the percentage of extant species of marine molluscs in the fossil assemblages from:

1. The basin of southwestern France in the vicinity of the Gironde Estuary, in the old province of the Aquitaine, an area usually called the Bordeaux-Dax district. The district includes strata referred to the Aquitanian and Burdigalian (marine) stages.
2. The Piedmont–Superga Hills area, outskirts of Turin, northwestern Italy (Aquitanian into Messinian marine strata).
3. The Vienna Basin. This district includes strata formerly termed Aquitanian, Vindobonian (Helvetian plus Tortonian), Sarmatian, Pannonian, and Pontian. These would now be assigned to the Eggenburgian, Ottnangian, Karpatian, Badenian, Sarmatian, Pannonian, and Pontian stages respectively.
4. The Faluns of Touraine, fossiliferous marine marls of the Helvetian and Tortonian stages in the valley of the Loire River in northwest-central France. The Faluns, which Deshayes collected and Lyell discussed at length, are regarded as stratotype Miocene by many French

stratigraphers, although most of the molluscs in Lyell's percentage figures were from Bordeaux-Dax (594 species; Lyell, 1833, Vol. 3, Appendix 1 by Deshayes, pp. 1–49). Lyell's Miocene was conceived on the basis of 1021 pre-Darwinian species of marine molluscs, of which about 176 were represented in the Recent, about 18 percent.

Fossil land-mammal assemblages have been obtained from stream and lake deposits that interfinger with the marine strata in the "type-Miocene" districts. See Crouzel (1957), for example. As with Paleocene, Eocene, and Oligocene, therefore, no extended correlation is necessary for recognition of type-Miocene land mammals. And the beautiful aggregates of land-mammal fossils from the cave and fissure deposits of France, Germany, and Austria—as well as from the swamp and lake deposits of eastern and southeastern Europe—are correlated by species identity with the type-Miocene assemblages.

We shall not relabor the history of controversies about limits and content of the Miocene of western Europe. Rather, we shall simply follow the current consensus among marine stratigraphers that the following stages, as they are identified in various districts and arranged in stratigraphic order, compose the Miocene in France, Spain, Italy, Switzerland, Germany, and Austria—that is, the type-Miocene area of western Europe:

Messinian	Selli (1971)
Tortonian	Cita (1971B)
Serravallian	Boni and Selli (1971)
Langhian	Cita (1971A)
Burdigalian	Vigneaux (1971)
Aquitanian	Vigneaux and Marks (1971)

The authors listed provide a summary description of the stage. The generally accepted years-before-present range for these stages-ages is about 24–5.

Bernor (1983), following Berggren and Van Couvering (1974), notes that some of the fossiliferous strata of western Europe assigned to Aquitanian are probably older than the proposed neostratotype [our term] in France; thus he uses 23.5 m.y. as the date before present for beginning Miocene. As a result, some of the faunas we include in earliest Miocene he leaves in latest Oligocene.

Bernor (1983) proposes six land-mammal faunal provinces to which the European mammal zones of Mein (1975, 1979) and the land-mammal ages of Ginsburg (1975C) and Fahlbusch (1976) apply.

1. Western and Southern Europe.
2. Eastern and Central Europe.
3. Romania and Western USSR.
4. "Sub-Paratethyan" province (includes Turkey and Saudi Arabia). Agenian not represented.
5. North Africa. Agenian not represented.
6. Siwalik area (North India and adjacent Pakistan). Agenian not represented.

We believe that in addition, the European mammal ages will be useful in Bernor's North Asia and Southeast Asia provinces when their land-mammal faunas become better known.

Diverse provincial climates and vegetative provinces must have developed across North Africa and Eurasia during the Miocene (see Bernor's review for the literature on this subject). Nevertheless, the rapidly evolving and rapidly dispersing Miocene land mammals are providing the best chronologic correlation data among the provinces of this huge supercontinent.

The mammal zones and subzones of Mein (1975, 1979) have become the standard for later work by European stratigraphers concerned with land-mammal geochronology of the Neogene of Europe. Daams and Freudenthal, however, conclude that at least part of Mein's zonation is misleading, and they propose another subdivision. We shall cite Mein's data for his zones as we summarize the early, middle, late, and latest Miocene mammal ages of Europe and adjacent areas.

SUBDIVISION OF NONMARINE MIOCENE IN EUROPE

In cooperation with the International Symposium on Mammalian Stratigraphy of the European Tertiary in Munich (April 1975), Ginsburg (1975C) and Fahlbusch (1976) published subdivisions for the European terrestrial Miocene. They correlated these subdivisions with the Neogene mammal zones of Mein (1975, 1979) as shown in the chart on the opposite page.[1]

Crusafont had previously proposed Vallesian (Crusafont, 1950) and Turolian (Crusafont, 1965, as a replacement term for Pikermian) as mammal ages representing respectively the earlier and later parts of the *Hipparion* faunas of Spain. Many European workers have demonstrated that the Miocene land-mammal ages of Europe correlate approximately with Aquitanian through Messinian in the marine sequence. Thus, we are using Agenian through Turolian as a subdivision of the Miocene nonmarine phenomena. Vallesian and Turolian and their correlatives around the world have previously been called *Early Pliocene* and *Middle Pliocene* respectively by most mammalian stratigraphers (see Chiu Chan-siang et al., 1979A, for example). The Miocene (Aquitanian through Messianian; Agenian through Turolian) is a 19-million-year interval. The Miocene and the Eocene are the longest epochs of the Cenozoic.

Daams, Freudenthal, and Van de Weerd (1977) defined and characterized the Aragonian as a new stage in continental Miocene strata of central Spain. They implied in their abstract that Aragonian is the temporal equivalent of Orleanian plus Astaracian (Mein's MN 3 through 8): "The Aragonian is stratigraphically higher than those beds that are conventionally considered as continental Aquitanian, and it is lower than the Vallesian." In their range chart for the Aragonian, however, they show correlation with MN 3 through 6 only, and not recognizing Orleanian and Astaracian, they give a threefold zonation of Aragonian based on occurrences of certain species of rodents (Cricetidae, Gliridae, Eomyidae) and of lagomorphs (upper zone only). Van de Weerd and Daams (1979) and Daams (1981) later include Mein's MN 3 through 8 in their Aragonian, not differentiating MN 7 from MN 8.

AGENIAN

The Agenian land-mammal age (Agenium of Fahlbusch, 1976) is based on the species from Paulhiac, Montaigu-le-Blin, and Laugnac in the Calcaire or Sables d'Agenais, Agen region, margins of the old Aquitaine (marine) province of southwestern France, and on correlated faunas in other parts of western Europe. Bernor (1983) leaves "earliest Agenian" in the Oligocene. Paulhiac and correlative faunas produce Mein's (1975) zone MN 1 with "formes caractéristiques de lignées évolutives" (*Rhodanomys transiens, Pseudocrice-*

Neogene Mammal Zones	Reference Localities	Mammal Ages	"Superstages"	Faunal Events	Interpretation
MN 13	Arquillo				
MN 12	Los Mansuetos	Turolium			
MN 11	Crevillente 3		Catalonium		
MN 10	Masia del Barbo 2B	Vallesium			
MN 9	Can Llobateres			Hipparion	
MN 8	Anwil				
MN 7	Steinheim	Astaracium			
MN 6	Sansan		Aragonium		MIOCENE
MN 5	Las Planas 4B				
MN 4b	Vieux Collonges				
MN 4a	La Romieu	Orleanium			
MN 3b	Artenay				
MN 3a	Wintershof-West			Anchitherium	
MN 2b	Laugnac				
MN 2a	Montaigu	Agenium			
MN 1	Paulhiac				

todon thaleri, Melissiodon emmerichi, and Propalaeochoerus elaverensis), his Titanomys-Piezodus association, and the first appearance of Titanomys, Apeomys, Heteromyoxus, Hyotherium, Diceratherium, Brachydiceratherium, Paraceratherium, and Neoentelodon. Montaigu and correlatives produce Mein's zone MN 2a with Vasseuromys priscus, Eucricetodon gerandianus, E. haslachense, Melissiodon schlosseri, and Marcuinomys roquesi and with the first appearance of Ritteneria, Vasseuromys, and Marcuinomys. His zone 2b, based on Laugnac and correlatives, has Vasseuromys rugosus, Eucricetodon aquitanicus, and Ritteneria manca. It also demonstrates the Titanomys-Prolagus association and the first appearance of Prolagus.

Our composite Agenian fauna, combines the species Mein listed for zone MN 1, 2a, and 2b with species from correlative localities in France, Spain, Switzerland, and Germany. Agenian is the approximate correlative of the Eggenburgian of Paratethys (Steininger, Rögl, and Martini, 1976).

Composite Agenian Fauna

MARSUPIALIA
  Didelphidae
    Amphiperatherium frequens (Meyer, 1846)
    A. exile (Gervais, 1848)   —[pre-Agenian according to Crochet, 1979]
LAGOMORPHA
  Ochotonidae
    Amphilagus antiquus (Pomel, 1853)
    A. ulmensis Tobien, 1974
    Piezodus branssatensis Viret, 1929C
    P. tomerdingensis Tobien, 1975
    Titanomys visenoviensis Meyer, 1846
    T. calmaensis Tobien, 1974
    T. parvulus Meyer, 1848

    Prolagus vasconiensis Viret, 1930?
    P. praevasconiensis Ringeade, 1979
    Marcuinomys roquesi Lavocat, 1951C
CARNIVORA
  Amphicyonidae
    Amphicyon cf. astrei Kuss, 1962
    Ysengrinia tolosana (Noulet, 1876)
    Cynelos rugosidens (Schlosser, 1899)
    C. lemanensis (Pomel, 1846)   —Ginsburg (1977)
    Pseudocyon sansaniensis (Meyer, 1849)   (Lartet?)
    Pseudocyonopsis landesquei (Helbing, 1928C)
    Haplocyon (Parhaplocyon) elegans de Bonis, 1973
    H. (Haplocyon) crucians (Filhol, 1879)
    Haplocyonopsis crassidens de Bonis, 1973
    Haplocyonoides mordax Hürzeler, 1940
  Ursidae
    Cephalogale ursinus de Bonis, 1973
    C. ginesticus Kuss, 1962
    Amphictis aginensis de Bonis, 1973
    Broiliana nobilis Dehm, 1950
  Procyonidae?
    Plesictis robustus Pomel, 1846
    P. palustris Pomel, 1853
    P. cultellatus de Bonis, 1973
    P. solidus de Bonis, 1973
    P. laugnacensis de Bonis, 1973
  Mustelidae
    Potamotherium valletoni Geoffroy, 1833
    Plesiogale angustifrons Pomel, 1853
    Palaeogale minuta (Gervais, 1848–1852)
    P. waterhousi (Pomel, 1853)
    cf. Martes laevidens Dehm, 1950
    Paragale hurzeleri Petter, 1967B   —de Beaumont
      (1968); Tedford (1976)
  Viverridae
    Herpestides antiquus (Blainville, 1842)
    H. collectus de Bonis, 1973
    Semigenetta

213

Felidae
Proailurus lemanensis Filhol, 1879
INSECTIVORA
Erinaceoidea
Dimylus paradoxus Meyer, 1846
Amphechinus edwardsi (Filhol, 1879)
"Erinaceus" —Tobien (1980)
Cordylodon sulcatus Stephan-Hartl, 1972
Dimylechinus bernouillii Hürzeler, 1944
Plesiodimylus
cf. *Pseudogalerix*
Scaptogale Trouessart, 1897, 2 spp. —Hutchison (1974), superfamily?
Soricoidea
Proscapanus primitivus Hutchison, 1974
Paratalpa micheli Hugueney, 1972
P.? brachychir (Meyer, 1846) —Hutchison (1974)
Talpa meyeri Schlosser, 1887
T.? minuta (Blainville, 1838)
Teutonotalpa meyeri (Schlosser, 1887) —Hutchison (1974)
Geotrypus tomerdingensis (Tobien, 1929) —Hutchison (1974)
Carposorex sylviae Crochet, 1975
Dinosorex neumayrianus (Schlosser, 1887) —Engesser (1975)
"Sorex" pusilliformis Doben-Florin, 1964 —Repenning (1967C)
Clapasorex sigei Crochet, 1975
C. bonisi Crochet, 1975
Heterosorex
Crocidosorex antiquus (Pomel, 1853) —Crochet (1975)
C. piveteaui Lavocat, 1951
C. thauensis Crochet, 1975
Soricella discrepans Doben-Florin, 1964
Plesiosorex cf. *soricinoides* Viret, 1940A —R. Wilson (1960)
CHIROPTERA
Megadermatidae
Megaderma brailloni Sigé, 1968
Hipposideridae
Hipposideros (*Brachipposideros*) cf. *collongensis* Depéret, 1892
H. (*B.*) *dechaseauxi* Sigé, 1968
H. (*Pseudorhinolophus*) *bouziguensis* Sigé, 1968
Rhinolophidae
Palaeonycteris praecox (Meyer, 1845)
P. insignis (Meyer, 1845)
P. robustus
P.? reinachi (Kindelin, 1900)
Rhinolophus lemanensis Revilliod, 1920
Molossidae
Nyctinomus stehlini Revilliod, 1920 (*Tadarida?*)
Vespertilionidae
cf. *Myotis*
ARTIODACTYLA

Tayassuidae
Palaeochoerus meissneri (Meyer, 1841)
P. gergovianus (Croizet and de Blainville)
P. typus Pomel, 1847
Suidae
Hyotherium major (Pomel, 1847)
Listriodon
Xenohyus venitor Ginsburg, 1980
Aureliachoerus aurelianensis (Stehlin, 1899)
Cainotheriidae
Cainotherium laticurvatum (Geoffrey St.-Hilaire, 1833)
Cervoidea
Amphitragulus boulangeri Pomel, 1853
Dremotherium sp.
PERISSODACTYLA
Tapiridae
Tapirus (*Palaeotapirus*) *intermedius* (Filhol, 1885)
T. moguntiacus R. von Koenigswald, 1930
T. douvillei
Rhinocerotidae
Brachypotherium aginense (Repelin, 1917)
B.? flörsheimense (Heller, 1933)
B. lemanense (Pomel, 1853)
Proaceratherium minutum (Cuvier, 1824) —Ginsburg and Hugueney (1980)
Aceratherium (*Mesaceratherium*) *paulhiacensis* (Richard, 1937)
A. cf. *platyodon* Mermier, 1895
Brachydiceratherium cf. *aurelianensis* (Nouel) (early Orleanian?) —Ginsburg and Antunes (1979)
Diceratherium pleuroceros (Duvernoy, 1852)=
Pleuroceros pleuroceros? —Ginsburg (1975C)
Prosantorhinus tagicus (Roman 1911) —Ginsburg and Antunes (1979)
Indricotherium (age?)
RODENTIA
Eomyidae
Ritteneria manca Stehlin and Schaub, 1951
Ligerimys aff. *antiquus* Fahlbusch, 1970
Rhodanomys schlosseri Depéret and Douxami, 1902
Apeomys tuerkheimae Fahlbusch, 1968A
Pseudotheridomys parvulus (Schlosser, 1884)
Sciuridae
Sciurus chalaniati Pomel, 1853 (pseudosciurine?)
S. costatus Freudenberg, 1947
Palaeosciurus feignouxi (Pomel, 1851)
Heteroxerus paulhiacensis Black, 1965
H. lavocati Hugueney, 1969
Blackia —Mein (1979)
Castoridae
Steneofiber eseri (Meyer, 1846)
S. castorinus Pomel, 1851
Theridomyidae
Issiodoromys pseudanaema Gervais, 1848
Cricetidae
Eucricetodon hochheimensis (Schaub, 1925)
E. collatum (Schaub, 1925)
E. gerandianus (Gervais, 1848)
E. aquitanicum Baudelot and de Bonis, 1968
E. cetinensis Daams, 1976

Pseudocricetodon cf. *thaleri* (Hugueney, 1969)
Mellisiodon? cf. *dominans* (Dehm, 1951)
Gliridae
  Glis truyolsi Daams, 1976
  Branssatoglis fugax (Hugueney, 1969)
  B. concavidens Hugueney, 1967
  Glirudinus aff. *bouziguensis* (Thaler, 1966B)
  G. glirulus (Dehm, 1935)
  G. cf. *modestus* (Dehm, 1950)
  Nievella mayri Daams, 1976
  Peridyromys murinus (Pomel, 1853)
  P. occitanus Baudelot and de Bonis, 1966
  P. brailloni (Thaler, 1966)
  Microdyromys monspeliensis Aguilar, 1977
  Vasseuromys rugosus Baudelot and de Bonis, 1966
  V. priscus de Bonis, 1973
  Pseudodryomys aljaphi Hugueney, Collier, and Huin, 1978
  P. ibericus de Bruijn, 1966A
  P. simplicidens de Bruijn, 1966A
  Myomimus murinus (Pomel, 1851)
  Armantomys   —Daams (1976)
Zapodidae
  Plesiosminthus myarion Schaub, 1930

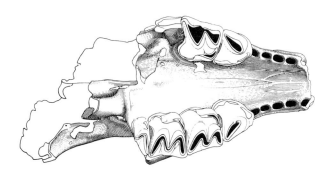

FIGURE 6-1   *Agenian mammal. Clapasorex sigei Crochet, 1975. Portion of skull. ×10. Figure 14 in Crochet (1975).*
(Published with permission of the C.N.R.S. of France.)

A

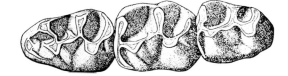

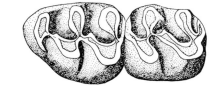

B

C

FIGURE 6-2   *Agenian rodents.*
    A   *Peridyromys murinus* (Pomel, 1853). *P⁴–M³. ×10¾ approx. Figure 7-10 in de Bonis (1973)*
    B   *Eucricetodon aquitanicam* Baudelot and de Bonis, 1968. M_{1-3} *and* M^{1-2}. *×10¾ approx. Figures 4-6 and 4-7 in de Bonis (1973).*
    C   *Rhodanomys schlosseri* Depéret and Douxmani, 1902. $P_1–M_3$, *×10¾ approx. Figure 14-6 in de Bonis (1973).*
(Published with permission of the Muséum National d'Histoire Naturelle, Paris.)

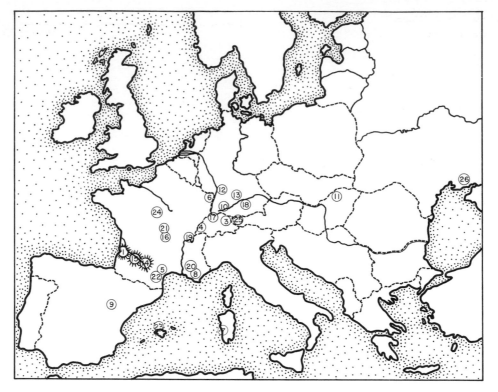

FIGURE 6-3   *Representative Agenian Localities.*

1 Aillas-Gans—France	*16 Marcoin (Puy-de-Dôme)—France*
3 Appenzell-Kaubach–Teufen—Switzerland	*17 Messen (Solothurn)—Switzerland*
4 Boudry I, II, III (Neuchâtel)—Switzerland	*18 Pappenheim (Eichstädt district)—Germany*
5 Bouzigues (Hérault)—France	*19 Pyrimont-Challonges (Ain-et-Savoie)—*
6 Budenheim–Kastrich–Mainz district—	*France*
Germany	*20 Réal–La Bastidonne (Vaucluse)—France*
7 Campidano—Sardinia	*21 St. Gérand-le-Puy (various sublocalities)—*
8 Cap Janet (Base du Rhône)—France	*France*
9 Cetina de Aragon (Zaragosa)—Spain	*22 Saint Jean–Narbonne district (Aude)—*
10 Eggingen–Ulm district—Germany	*France*
11 Felsö-Palfalva (Salgotarjan district)—	*23 Santa Margarita—Mallorca*
Hungary	*24 Selles-sur-Cher (Loire-et-Cher)—France*
12 Frankfort-Niederrad–Nordbassin—Germany	*25 Unter-Staudach (Vorarlberg at Bildstein)—*
13 Hochstadt (Hessen)—Germany	*Austria*
14 La Brète (Agenais)—France	*26 Kherson district—Ukraine, USSR*
15 Laugnac (Lot-et-Garonne)—France	

EARLY MIOCENE OF ASIA

We are currently unable to distinguish Asia mammalian faunas that may be contemporaneous with the Agenian of Europe from Asian faunas that may be contemporaneous with the later, Orleanian of Europe. Accordingly, we list the Asian faunas under the rubric *Early Miocene.*

Early Miocene Mammalian Faunas of Asia

CREODONTA
  Hyaenodontidae
    Megistotherium ingens (Forster-Cooper)
    Hyainailouros bugtiensis (Pilgrim, 1908)

CARNIVORA
  Amphicyonidae
    Amphicyon shahbazi (Pilgrim, 1910)
    A. cooperi Pilgrim, 1932
    Afrocyon   (age?)   —R. Savage (1978)
  Ursidae?
    Metarctos? bugtiensis (F.-Cooper)
ARTIODACTYLA
  Tayassuidae
    Palaeochoerus pascoei Pilgrim
    P. japonicus Takai, 1954
  Suidae
    Listriodon affinis (Pilgrim, 1908)
    Bunolistriodon lockarti (Pomel)

Xenochoerus jeffreysi Forster-Cooper

Bugtitherium grandincisivum Pilgrim, 1908 —R. Savage (1967)

Anthracotheriidae

  Anthracotherium bugtiense Pilgrim, 1907

  A. punjabiense Lydekker, 1877

  A. mus Pilgrim, 1908

  A. silistrense (Pentland, 1828)

  A. exiguum F.-Cooper, 1924

  A. adiposum F.-Cooper, 1924

  A. sminthos F.-Cooper, 1913

  Brachyodus japonicus Takai (*Parabrachyodus?*)

  Hyoboops naricus Pilgrim, 1910

  H. palaeindicus (Lydekker, 1877)

  H. longidentatus (Pilgrim, 1908)

  H. minor F.-Cooper, 1924

  Bothriodon ramsayi (Pilgrim, 1908)

  Gonotelma shahbazi (Pilgrim, 1908)

  G. major F.-Cooper, 1924

  Telmatodon hugtiensis Pilgrim, 1907

  T. orientale F.-Cooper, 1924

  Glasmodon gracilis F.-Cooper

  Hemimeryx speciosus Pilgrim, 1908

  H. lydekkeri (F.-Cooper, 1913)

  H. turgaicus Borisyak, 1941

  Parabrachyodus pilgrimi F. Cooper, 1915 (We list all of Forster-Cooper's proposed species of *Parabrachyodus*, although we doubt the validity of most of them.)

  P. gandoiensis F.-Cooper, 1915

  P. giganteus (Lydekker, 1883)

  P. hypotamoides (Lydekker, 1883)

  P. platydens (F.-Cooper)

  P. orientalis (F.-Cooper)

  P. indicus (F.-Cooper)

  P. strategus (F.-Cooper)

  P. borbonicoides (F.-Cooper)

  P. obtusus (F.-Cooper)

Tragulidae

  Tragulus sivalensis

Gelocidae

  Prodremotherium? beatrix Pilgrim, 1912

  Gelocus? gajense Pilgrim, 1912

  Gelocus indicus

Cervoidea

  Palaeomeryx minoensis (Matsumoto)

  Stephanocemas colberti Young, 1937C

  Dicrocerus cf. *elegans* Lartet

  Progiraffa exigua (family?) —Pilgrim, 1908

Bovidae

  Gobicerus mongolicus Sokolov, 1952

PERISSODACTYLA

Equidae

  Anchitherium aurelianense Cuvier, 1822

  A. hypohippoides Matsumoto

Tapiridae

  Palaeotapirus yagii Matsumoto

Amynodontidae

  Cadurcotherium indicum Pilgrim, 1910

  Metamynodon bugtiensis Forster-Cooper, 1922B

Rhinocerotidae

  Aceratherium gajense Pilgrim, 1912

  A. bugtiense (Pilgrim, 1911)

  A. abeli F.-Cooper

  A. blanfordi (Lydekker, 1884)

  A. depereti Borisyak

  Chilotherium smithwoodwardi F.-Cooper

  Diceratherium shahbazi Pilgrim, 1910

  D. naricum Pilgrim, 1910

  Indricotherium bugtiense (F.-Cooper, 1911A) (*Paraceratherium?*)

  Paraceratherium churlandense F.-Cooper

  Baluchitherium osborni (*Indricotherium?*)

  Brachypotherium fatehjangense (Pilgrim, 1910)

  B. aurelianensis (Nouel, 1866)

  B. pugnator Matsumoto

  Prosantorhinus tagicus (Roman, 1911)

  Coelodonta mongoliensis (Osborn)

Chalicotheriidae

  Schizotherium pilgrimi Cooper (*Chalicotherium?*)

  Phyllotillon naricus Pilgrim, 1908

  P. betpakdalensis (Flerov, 1938)

PROBOSCIDEA

Deinotheriidae

  Deinotherium pentapotamiae Falconer and Lydekker

Gomphotheriidae

  Gomphotherium angustidens (Cuvier, 1806)

  G. pandionis (Falconer, 1857)

  G. cooperi (Osborn, 1932)

  G. annectens (Matsumoto)

  G. subtapiroides (Schlesinger, 1917)

  G. inopinatus (Borisyak and Belyayeva, 1928)

  G. mongoliensis (Osborn, 1924)

  G. tologojensis (Belyayeva, 1952)

RODENTIA

Castoridae

  "Dipoides" anatolicus (Ozansoy and R. Savage, 1967)

From the Xining (Sining) Basin of Qinghai (Tsinghai), Li and Qiu Zhu-ding (Chiu Chu-ting) (1980) have listed the following mammals.

Early Miocene Mammals from China

LAGOMORPHA

Ochotonidae

  Sinolagomys pachygnathus Li and Qiu Zhu-ding, 1980

Leporidae, gen. indet.

CARNIVORA

Mustelidae, gen. indet.

ARTIODACTYLA

Bovidae

  Oioceros xiejiaensis Li and Qiu Zhu-ding, 1980

PERISSODACTYLA

Rhinocerotidae

  Brachypotherium

RODENTIA

Sciuridae, gen. indet.

Cricetidae

  Eucricetodon youngi Li and Qiu Zhu-ding, 1980

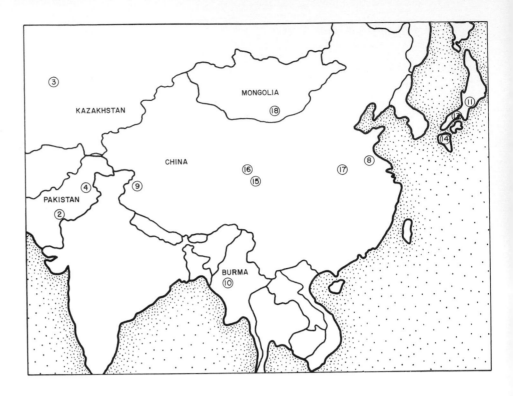

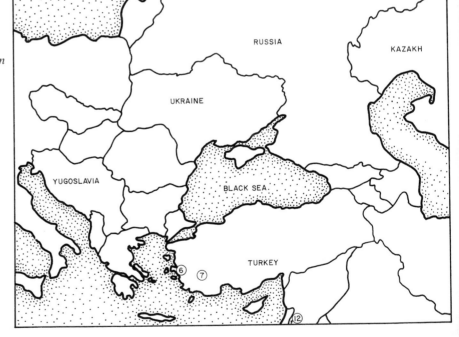

FIGURE 6-4 *Early Miocene Localities of Asia and Europe.*

2   *Bugti beds sites (Dera Bugti, Bugti Hills, Chur Lando, Kumbhi, Gandoi, Khajuri, Gaj, etc.)—Baluchistan*

3   *Djilancik (Dzhilanchik) River, upper?—Kazakhstan*

4   *Fatehjang—Pakistan*

5   *Funaoka—Japan ("Eostegodon")*

6   *Izmir—Turkey*

7   *Kale assemblage—Turkey*

8   *Jiangsu North—China   (age?)*

9   *Ladakh (Jammu-Kashmir)—India*

10   *Myaing-Mainyoung (in part)—Burma*

11   *Mino province—Japan*

12   *Negev—Israel*

13   *Okayama Prefecture—Japan*

14   *Sasebo Formation site—Kyushu, Japan*

15   *Sining Basin—Qinghai, China*

16   *Taben Buluk–Yindirte—Gansu, China*

17   *Tung-Sha-Po—Henan, China   (age?)*

18   *Ulan-Tologoj—Mongolia*

3   *Ust'-Urt, Northwest—Kazakhstan (age?)*

Zapodidae

   Plesiosminthus xiningensis Li and Qiu Zhu-ding, 1980

   P. huangshuiensis Li and Qiu Zhu-ding, 1980

   P. lajeensis Li and Qiu Zhu-ding, 1980

Ctenodactylidae

   Tataromys suni Li and Qiu Zhu-ding, 1980

Rhizomyidae

   Tachyoryctoides kokonorensis Li and Qiu Zhu-ding, 1980

EARLY RUSINGAN

A set of mammalian faunas from southwest, east, and north Africa, known incompletely since the late 1800s and early 1900s, has usually been assigned to the broad category Early Miocene. Szalay and Delson (1979) proposed the name Rusingan for these faunas in the sub-Saharan part of Africa. In this region, some of the associated rocks have been dated radiometrically with ages ranging between 23 and 17 million years before the present (Bishop, Miller, and Fitch, 1969; Van Couvering and Miller, 1969). On this basis, the Rusingan is contemporaneous with the Agenian and Orleanian of Europe, North Africa, and Asia. Faunas at Bukwa–Mt. Elgon and at Karungu in East Africa are associated with radiodates of about 23 million and are here designated Early Rusingan. And the faunas from Namibia in southwest Africa are grouped provisionally in this interval also. There seems to be little justification, however, for separating these assemblages from the slightly later and typical Rusingan faunas such as Rusinga, Koru, and Songhor of east Africa on the basis of their paleontology. See Cooke (1978, p. 29, Fig. 2.6) for the currently accepted chronological ordering of the African Miocene faunas; see also our map of African Miocene localities, Figure 6-8, for geographic position of Early Rusingan sites.

Early Rusingan Mammalian Fauna

(= Namibian of Hendey, 1974.)

LAGOMORPHA
  Ochotonidae
    Austrolagomys inexpectatus Stromer, 1924
CREODONTA
  Hyaenodontidae
    Pterodon kaiseri (Stromer, 1924)
CARNIVORA
  Felidae
    Afrosmilus africanus (Andrews, 1914A)   —R. Savage (1978)
PRIMATES
  Pliopithecidae
    Dendropithecus macinnesi (Le Gros Clark and Leakey, 1950)
  Hominidae   —Szalay and Delson (1979)
    Dryopithecus (Proconsul) nyanzae (L. G. Clark and Leakey, 1950)
    D. (Limnopithecus) legetet (Hopwood, 1933A)
ARTIODACTYLA
  Cervoidea
    Propalaeoryx austroafricanus Stromer, 1926
HYRACOIDEA
  Pliohyracidae
    Pachyhyrax championi (Arambourg, 1933A)
  Procaviidae
    Prohyrax tertiarius Stromer, 1924
  Myohyracidae
    Myohyrax oswaldi Andrews, 1914A
PERISSODACTYLA
  Rhinocerotidae
    Chilotherium
    Dicerorhinus

PROBOSCIDEA
  Deinotheriidae
    Prodeinotherium hobleyi (Andrews, 1911)
  Gomphotheriidae
    Gomphotherium?
RODENTIA   —Lavocat (1978)
  Theridomyidae
    Neosciuromys africanus Stromer, 1922C
  Diamantomyidae
    Diamantomys luederitzi Stromer, 1922C
    Pomonomys dubius Stromer, 1922C
  Pedetidae
    Parapedetes namaquensis Stromer, 1924
  Phiomyidae
    Paraphiomys pigotti Andrews, 1914
    P. stromeri
  Myophiomyidae   —Lavocat (1978)
    Phiomyoides humilis Stromer, 1924
  Bathyergoididae   —Lavocat (1978)
    Bathyergoides neotertiarius Stromer, 1924
  Bathyergidae
    Paracryptomys mackennae Lavocat, 1973

NONMARINE MIOCENE OF NORTH AMERICA

North American mammal ages that are assigned to the Miocene Epoch are Late Arikareean, Hemingfordian, Barstovian, Clarendonian, and Hemphillian. Before European stratigraphers included the Messian Age of the Mediterranean area in the Miocene, the approximate North American equivalents, Clarendonian (later part) and Hemphillian, had been termed *early* and *middle* Pliocene respectively.

Many different environments and habitats are represented by the nonmarine Miocene rocks and faunas of North America: coastal plains (Florida, Texas, California), coastal mountains (California, Mexico), streamside and lakeside communities in or adjacent to regions of active volcanism (West Coast states, Mexico), open savanna or prairie broken by streamside communities (Great Plains States), and hilly or near-mountain areas in and around woodland or chaparral or desert-border vegetation (West Coast states, Rocky Mountain states, Mexico).

In the North American region, the Miocene is the time in which hypsodont and cursorial horses, camels, and antilocaprids appeared in great numbers and became prominent elements in the faunas. This phenomenon has been correlated with the spread of grasslands and the outfall of great volumes of volcanic-silicic dust in the western half of the continent. The tough siliceous grasses and herbaceous vegetation of the savannas and prairies and the dust-covered vegetation created an abrasive environment for these grazing and browsing herbivores that must have been strongly selective for hypsodonty. Yet at the same time and in other areas, there were many and greatly varied brachyodont and hypso-brachyodont selenodont artiodactyls, especially in faunas representing the earlier half of the Miocene, and these habitats provided a continued abundance of softer, more succulent plant food.

LATE ARIKAREEAN

Arikareean was proposed by H. Wood et al. (1941) as a new "provincial time term, based on Arikaree Group of western Nebraska, Agate being the most typical locality." It was unfortunate in many ways that this mammal age was keyed to the existing stratigraphic nomenclature, for the upper limit of the Arikaree Group of some later workers does not include Agate, "the most typical locality." We follow the continuing work of R. Hunt and J. Breyer on the latest Oligocene and early Miocene of western Nebraska and include the succession of mammalian faunas from the Gering, Monroe Creek, Harrison Formations, and Upper Harrison Formation of Peterson (1906) as the basis for the Arikareean land-mammal age.

The subage Late Arikareean is based on the aggregate of mammalian taxa from the Harrison Formation and a formation superjacent to the Harrison that has been called Upper Harrison or Lower Marsland, which is now being mapped, defined, and named by R. Hunt and J. Breyer. The fauna from the Upper Harrison lithostratigraphic unit is assigned early Hemingfordian age by Woodburne et al. (1974). These sandy formations are the upper part of the Arikaree Group of Nebraska and Wyoming. Noteworthy localities in this type area include the *Stenomylus* Quarry and miscellaneous sites in the Harrison Formation of Nebraska, Harrison Formation sites in the Goshen Hole district of Wyoming, the Harper Quarry (Hunt, 1978), and the world-renowned Agate Springs–Cook Ranch quarries in Hunt and Breyer's superjacent formation (= "Upper Harrison"). Correlative faunas are known from Texas, South Dakota, Montana, Oregon, and California?.

The above formations, as well as the John Day Formation of Oregon, are composed largely of clasts of volcanic origin and are alluvial deposits. Latest Oligocene and early Miocene was an episode of tremendous volcanic eruption in western North America, and the accumulation of airborne volcanic dust in the streams and lakes was the dominant aspect of nonmarine sedimentation.

Composite Late Arikareean Mammalian Fauna

MARSUPIALIA
 Didelphidae
  Herpetotherium cf. *spindleri* (Macdonald, 1963A)
  H. youngi (McGrew, 1937A)

LAGOMORPHA
 Ochotonidae
  Gripholagomys   —Green (1972A)
  Oreolagus, sp. indet.   —Nichols (1976)
 Leporidae
  Archaeolagus acaricolus Dawson, 1958
  A. ennisianus (Cope, 1881C)   —Dawson (1958)
  A. cf. *macrocephalus* (Matthew, 1907A)
  A. buangulus Dawson, 1958   —M. Stevens (1969A)
  Megalagus primitivus (Schlaikjer, 1935)   —Dawson (1958)
  M. dawsoni Black, 1961C
CARNIVORA
 Amphicyonidae
  Cynelos sp.   —Hunt (1972A)
  Mammacyon obtusidens Loomis, 1936A   (family and age?)
  Ysengrinia?   —Hunt (1972A)
  Daphoenodon superbus Peterson, 1906C   (age?)
  D. notionastes Frailey, 1979
  D. robustum (Peterson, 1910A)   —Woodburne et al. (1974)
  Paradaphoenus Matthew, 1899   (*nomen nudum?; age?*)
  Pericyon socialis Thorpe, 1922C   (family and age?)
  Temnocyon altigenis Cope, 1879D
  T. percussor Cook, 1909D
 Ursidae
  Cephalogale   —Tedford et al. (1982?)
  Enhydrocyon stenocephalus Cope, 1879B
  E. crassidens Matthew, 1907A
  E. oregonensis Thorpe, 1922C
  Plesictis   —Macdonald (1970A)   (Procyonidae?)
  Zodiolestes daimonelixensis Riggs, 1942   —(Procyonidae by Hough (1948))
  Amphicynodon (= "*Pachycynodon*")
  Parictis primaevus Scott, 1893A
  Allocyon loganensis C. Merriam, 1930
  Brachyrhynchocyon intermedius (Loomis, 1931)
 Canidae
  Hesperocyon gregorii (Matthew, 1907A)
  Nothocyon latidens (Cope, 1881C)
  N. harlowi Loomis, 1932A
  N. annectens Peterson, 1906C   —Stevens (1977)
  N. lemur (Cope, 1879E)
  N. vulpinus Matthew, 1907A
  N. geismarianus (Cope, 1879D)
  Tomarctus?   —Macdonald (1970A)
  Cynodesmus vulpinus (Matthew, 1907A)
  Neocynodesmus delicatus (Loomis, 1932A)   —Macdonald (1963A)
  Mesocyon robustus Matthew, 1907A
  M. coryphaeus (Cope, 1879F)
  M. brachyops Merriam, 1906A
  M. josephi (Cope, 1881C)
  Philotrox condoni Merriam, 1906A
  Cynarctoides acridens (Barbour and Cook, 1914A)   —McGrew (1938A)
  Bassariscops achoros Frailey, 1979
  Aletocyon multicuspis Romer and Sutton, 1927A
  Borocyon robustus Peterson, 1910A   (family?)
 Mustelidae
  Promartes olcutti Riggs, 1942

P. lepidus (Matthew, 1907A)

P. vantasselensis (Loomis, 1932A)

P. gemmarosae (Loomis, 1932A)

P. darbyi (Thorpe, 1921C)

Paroligobunis? —Stevens (1977)

Paroligobunis petersoni Loomis, 1932A (age?)

P. simplicidens (Peterson, 1906C)

Oligobunis? *crassivultis* (Cope, 1879F)

Aelurocyon brevifacies Peterson, 1906C

Felidae

  Nimravus brachyops Cope, 1878F

  Dinaelurus crassus Eaton, 1922A (valid?)

INSECTIVORA

  Erinaceoidea

    Amphechinus horncloudi (Macdonald, 1970A) —T. Rich and Rasmussen (1973)

    Brachyerix? hibbardi Stevens, 1977

    Stenoechinus tantalus T. Rich and Rasmussen, 1973

    Parvericius montanus Koerner, 1940

  Soricoidea

    Proscalops intermedius Barnosky, 1982

    Mesoscalops?

    Domnina, sp. indet. —Nichols (1976)

    Pseudotrimylus

ARTIODACTYLA

  Tayassuidae

    Cynorca sociale (Marsh, 1875B)

    Perchoerus or *Chaenohyus* (age?)

    Perchoerus or *Thinohyus* (age?)

    Desmathyus or *Hesperhys* (age?)

  Entelodontidae

    Archaeotherium trippensis Skinner, Skinner, and Gooris, 1968

    Dinohyus hollandi Peterson, 1905C

  Anthracotheriidae

    Arretotherium acridens Douglass, 1901B

  Agriochoeriidae

    Agriochoerus ferox (Cope, 1879E)

    A. bullatus Thorpe, 1821E

    A. guyotianus (Cope, 1879D)

    A. macrocephalus (Cope, 1879E)

    A. ryderanus (Cope, 1881C)

    A. trifons Cope, 1884M

  Merycoidodontidae

    Promerycochoerus inflatus (Thorpe, 1921E) — Schultz and Falkenbach (1949)

    P. barbouri Schultz and Falkenbach, 1949

    P. carrikeri Peterson, 1906C

    P. minor Douglass, 1903A

    P. montanus (Cope, 1884M)

    P. latidens Thorpe, 1921A —Schultz and Falkenbach (1949)

    P. superbus (Leidy, 1870P)

    P. macrostegus (Cope, 1884M) —Schultz and Falkenbach (1949)

    Phenacocoelus typus Peterson, 1906A

    P. kayi Schultz and Falkenbach, 1950

    P.? leptoscelos (Stevens, 1969) —Stevens (1977)

    Oreodontoides oregonensis Thorpe, 1921D

    O. marshi (Thorpe, 1921A)

    O.? curtus (Loomis, 1924B)

O. stocki Schultz and Falkenbach, 1947

Submerycochoerus bannackensis (Douglass, 1907C) —Schultz and Falkenbach (1950)

Pseudomesoreodon rooneyi Schultz and Falkenbach, 1950

P. rolli Schultz and Falkenbach, 1950

P.? boulderensis Schultz and Falkenbach, 1950

Paramerychyus harrisonensis (Peterson, 1906) Schultz and Falkenbach (1947)

P. relictus (Peterson, 1924) —Schultz and Falkenbach (1947)

Hypsiops brachymelis (Douglass, 1907C) —Schultz and Falkenbach (1950)

H. breviceps (Douglass, 1907C) —Schultz and Falkenbach (1950)

H. luskensis Schultz and Falkenbach, 1950

H. johndayensis Schultz and Falkenbach, 1950

Merychyus calaminthus Jahns, 1940 (includes *M. crabilli* Schultz and Falkenbach) —Woodburne et al. (1974)

M. siouxensis Loomis, 1924B

Mesoreodon megalodon Peterson, 1906C

M. chelonyx Scott, 1893B

M. cheeki (Schlaikjer, 1934B)

M.? hesperus (Stock, 1930C) Schultz and Falkenbach (1949)

Megoreodon grandis (Douglass, 1909C)

M. hollandi (Douglass, 1907C) —Skinner, Skinner, and Gooris (1968)

Desmatochoerus curvidens (Thorpe, 1921A)

D. sanfordi Schultz and Falkenbach, 1954

D. leidyi (Bettany, 1876)

Pseudodesmatochoerus longiceps (Douglass, 1907A)

P. wascoensis Schultz and Falkenbach, 1954

P.? pariogonus (Cope, 1884M)

Superdesmatochoerus lulli (Thorpe, 1921A)

S. microcephalus (Thorpe, 1921A)

Merycochoerus

Merycoides giganteus Schultz and Falkenbach, 1949

Pseudogenetochoerus condoni (Thorpe, 1921C) — Schultz and Falkenbach (1968)

P. covensis Schultz and Falkenbach, 1968

Epigenetochoerus parvus (Thorpe, 1921C) —Schultz and Falkenbach (1968)

Eporeodon occidentalis Marsh, 1873H

E. davisi Schultz and Falkenbach, 1968

E. pacificus (Cope, 1884M)

E. longifrons (Cope, 1884M)

E. leptacanthus (Cope, 1884M)

Dayohyus trigonocephalus (Cope, 1884M) —Schultz and Falkenbach (1968)

D. wortmani Schultz and Falkenbach, 1968

Camelidae

  Miolabis californicus Maxson, 1930

  Nothokemas waldropi Frailey, 1978

  Oxydactylus campestris Cook, 1909B

  O. wortmani (Lull, 1921C)

  O. cameloides (Wortman, 1898A)

  Priscocamelus wilsoni Stevens, 1969A

Gentilicamelus sternbergi (Cope, 1879B)　　—McKenna (1966A)

Michenia agatensis Frick and Taylor, 1970

M. exilis (Matthew, and Macdonald, 1960)　　—[includes *M. australis* M. Stevens, 1977, acc. to Honey and Taylor (1978)]

Miotylopus gibbi (Loomis, 1911B)　　—McKenna and Love (1972A)

Aguascalientia wilsoni (Dalquest and Mooser, 1974)　　—M. Stevens (1977)

Stenomylus hitchcocki Loomis, 1910B

S. gracilis Peterson, 1906C

S. keelinensis Frick and Taylor, 1968

Delahomeryx browni M. Stevens, 1969A　　(family?)

Tanymykter brachyodontus (Peterson, 1904A)　　— Honey and Taylor (1978)

Protoceratidae

Protoceras neatodelpha Patton and Taylor, 1973

P. celer Marsh, 1891A　　—Patton and Taylor (1973)

Syndyoceras cooki Barbour, 1905B

Hypertragulidae

Nanotragulus ordinatus (Matthew, 1907A)　　—M. Stevens (1977)

N. loomisi Lull, 1922B

Hypertragulus　　(age?)

Leptomerycidae

Leptomeryx

Pronodens silberlingi Koerner, 1940

Cervoidea

Parablastomeryx olcotti (Matthew, 1908A)

P. falkenbachi Frick, 1937

P. advena (Matthew, 1907A)

Machaeromeryx tragulus Matthew, 1926B

Aletomeryx marshi (Lull, 1920A)　　(age?)

PERISSODACTYLA

Equidae

Miohippus equinanus Osborn, 1918A

M. cf. *gemmarosae* Osborn, 1918A

Parahippus pristinus Osborn, 1918A

P. nebrascensis Peterson, 1906C

Anchitherium agatense (Osborn, 1918A)

A. gracile (Marsh, 1892C?)　　(age?)

Archaeohippus equinanus (Osborn, 1918A)　　— Skinner et al. (1968)

Tapiridae

Miotapirus harrisonensis Schlaikjer, 1937A

Rhinocerotidae

Diceratherium　　—Macdonald (1970A)

Menoceras arikarense (Barbour, 1906C)

M. cooki (Peterson, 1906D)

Moschoedestes delahoensis Stevens, 1969A

Chalicotheriidae　　—Coombs (1978–0271)

Moropus elatus Marsh, 1877D

M. distans Marsh, 1877D　　(age?)

M. oregonensis (Leidy, 1873B)

M. hollandi Peterson, 1913A　　(age?)

RODENTIA

Aplodontidae

Niglarodon yeariani (Nichols, 1976)

N. petersonensis (Nichols, 1976)

N. blacki Rensberger, 1981

N. loneyi Rensberger, 1981

Sewelleladon predontia Shotwell, 1958B

Mylagaulidae

Promylagaulus riggsi McGrew, 1941B　　—Rensberger (1979)

P. lemhiensis Nichols, 1976

P. montanensis Rensberger, 1979

P. ovatus Rensberger, 1979

Sciuridae

Protosciurus condoni Black, 1963B

P. rachelae Black, 1963B

Protospermophilus vortmani (Cope, 1879)

P. angusticeps Matthew and Mook, 1933　　—Black (1961C)

Similisciurus maxwelli Stevens, 1977

Tamias　　—Black (1963)

Miosciurus ballovianus (Cope, 1881)　　—Black (1963)

Castoridae

Euhapsis platyceps Peterson, 1905A

E. gaulodon Matthew, 1907A

Palaeocastor simplicidens (Matthew, 1907A)

Capatanka brachyceps (Matthew, 1907A)　　—Macdonald (1963A)

Capacikala? sciuroides (Matthew, 1907A)　　—Macdonald (1963A)

Geomyidae

Ziamys tedfordi Gawne, 1975

Pleurolicus dakotensis A. Wood, 1936C

P. oregonensis (A. Wood, 1936C)　　—Rensberger (1973A)

P. sulcifrons Cope, 1878–1879D

Gregorymys riggsi A. Wood, 1936C

G. formosus (Matthew, 1907A)

G. kayi A. Wood, 1950　　(age?)

G. riograndensis M. Stevens, 1977

G.? curtus (Matthew, 1907A)

Dikkomys matthewi A. Wood, 1936C

Tenudomys macdonaldi Rensberger, 1973A

Schizodontomys sulcidens Rensberger, 1973A

Sanctimus tiptoni Macdonald, 1970A

Jimomys lulli (A. Wood, 1936F)　　—Wahlert (1976)

Entoptychus basilaris Rensberger, 1971A

E. wheelerensis Rensberger, 1971A

E. montanensis Hibbard and Keenman, 1950

E. germannorum A. Wood, 1936C

E. minor Cope, 1881E

E. cavifrons Cope, 1879D

E. transitorius Rensberger, 1971A

E. productidens Rensberger, 1971A

E. planifrons Cope, 1878–1879D

E. individens Rensberger, 1971A

E. fieldsi Nichols, 1976

E. sheppardi Nichols, 1976

Heteromyidae

Proheteromys ironcloudi Macdonald, 1970A

P. cejanus Gawne, 1975

FIGURE 6-5 *Representative Late Arikareean Localities and Stratigraphic Units.*

1 *Agate Springs—Stenomylus Quarry—Harper Quarry—miscellaneous Harrison and "Upper Harrison" Formation sites—Nebraska*
2 *Alamos Canyon (South Mountain)—upper Sespe Formation—California (age?)*
3 *Black Butte Mine—Hector Formation—California*
4 *Blue Lake Rhino Mold in Basalt—Columbia River Basalt Group—Washington (age?)*
5 *Brooksville—Florida*
6 *Buda—Florida*
7 *Cabbage Patch (Tavenner Ranch)—Renova Formation equivalent—Montana*
8 *Canyon Ferry (in part?)—Renova Formation equivalent—Montana*
9 *Castalon—Delahoe Formation—Texas*
10 *Cedar Run—Oakville Sandstone—Texas*
11 *El Rito—Abiquiu Formation—New Mexico*
8 *Fort Logan—"Fort Logan" Formation—Montana*
5 *Franklin Phosphate Pit—Florida*
12 *Gay Head Cliff—Massachusetts*
13 *Goshen Hole District—Harrison Formation—Wyoming*
1 *Harper Quarry—"Upper Harrison" Formation—Nebraska*
1 *Harrison vicinity—Monroe Creek and Harrison Formations—Nebraska*
14 *John Day Formation Sites (Kimberly and upper Turtle Cove Members)—Oregon*
15 *Madison Valley Formation Sites—Montana*
6 *Martin-Anthony Road Cut—Tampa Limestone or basal Hawthorne Formation—Florida*
16 *Peterson Creek (Lemhi Valley)—Renova Formation equivalent—Idaho*
17 *Pilgrim Creek—Wyoming (age?)*
18 *Redington—Arizona*
19 *Rizzi Ranch (in part)—Nevada*
20 *Rosebud Formation (in part) Sites—South Dakota*
21 *Santa Barbara Canyon—Pato Sandstone Member, Vaqueros Formation—California*
6 *SB-1A (Univ. Florida)—Florida*
2 *South Mountain District—Sespe Formation, upper—California (age?)*
22 *Squankum East—New Jersey*
11 *Standing Rock Quarry—Zia Sand—New Mexico*
23 *Tick Canyon Formation Sites—California*
24 *Tieton Reservoir—Keechelus Formation—Washington (age?)*
25 *Wellton—Arizona*
20 *Wounded Knee District (in part)—Harrison Formation equivalents in Rosebud Formation—South Dakota*

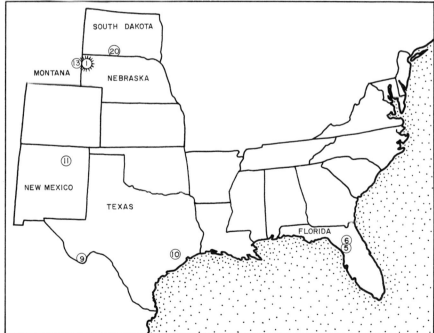

223

Heliscomys woodi McGrew, 1941C
Mookomys formicorum A. Wood, 1935C
Cricetidae
    Leidymys alicae (Black, 1961A)
    L. parvus (Sinclair, 1905A)
    L. lockingtonianus (Cope, 1881C)
    L. nematodon (Cope, 1879E)   —L. Martin (1980)
    Paciculus montanus Black, 1961C
Zapodidae
    Plesiosminthus grangeri (A. Wood, 1935A)   —Green
    (1977)

SANTACRUCIAN

The mammal-age name Santacrucian is based on the fauna from many localities in the Santa Cruz Formation of Patagonia, Argentina. We have decided not to list the many species Ameghino named from the Santa Cruz Formation, most of which were proposed before 1900. A thorough restudy of the Santacrucian species will be a multiyear project for a scholar based in Argentina. The recognized genera are in the following list.

Santacrucian Mammalian Faunas

MARSUPIALIA
  Microbiotheriidae
    Microbiotherium Ameghino, 1887
  Borhyaenidae
    Acrocyon Ameghino, 1887
    Arctodictis Mercerat, 1891
    Borhyaena Ameghino, 1887
    Cladosictis Ameghino, 1887
    Lycopsis Cabrera, 1927
    Perathereutes Ameghino, 1891
    Prothylacynus Ameghino, 1891
    Sipalocyon Ameghino, 1887
  Caenolestidae
    Abderites Ameghino, 1887
    Palaeothentes Ameghino, 1887
    Parabderites Ameghino, 1902
    Phonocdromus Ameghino, 1894
    Pichipilus Ameghino, 1890
    Stilotherium Ameghino, 1887
MARSUPIALIA?
  Necrolestes Ameghino, 1891
EDENTATA
  Megalonychidae
    Analcimorphus Ameghino, 1891
    Eucholoeops Ameghino, 1887   (family?)
    Hyperleptus Ameghino, 1891
    Megalonychotherium Scott, 1904
    Pelecyodon Ameghino, 1891
    Schismotherium Ameghino, 1887
  Megatheriidae
    Hapalops Ameghino, 1887

    Eucholoeops Ameghino, 1887   (family?)
    Planops Ameghino, 1887
    Prepotherium Ameghino, 1891
  Mylodontidae
    Analcitherium Ameghino, 1891
    Nematherium Ameghino, 1887
  Entelopsidae
    Entelops parodii Pascual, 1960
  Myrmecophagidae
    Protamandua Ameghino, 1904
  Dasypodidae
    Peltocoelus Ameghino, 1902
    Proeutatus Ameghino, 1891
    Prozaedius Ameghino, 1891
    Stegotherium Ameghino, 1887
    Stenotatus Ameghino, 1891
  Peltephilidae
    Peltephilus Ameghino, 1887
  Glyptodontidae
    Asterostemma Ameghino, 1889
    Cochlops Ameghino, 1889
    Eucinepeltus Ameghino, 1889
    Metopotoxus Ameghino, 1898
    Propalaeohoplophorus Ameghino, 1887
PRIMATES
  Cebidae
    Homunculus Ameghino, 1891
LITOPTERNA
  Proterotheriidae
    Diadiaphorus Ameghino, 1887
    Licaphrium Ameghino, 1887
    Licaphrops Ameghino, 1904
    Proterotherium Ameghino, 1883
    Thoatherium Ameghino, 1887
  Macraucheniidae
    Theosodon Ameghino, 1887
  Adianthidae
    Adianthus Ameghino, 1891
NOTOUNGULATA
  Homalodotheriidae
    Homalodotherium Flower, 1873
  Notohippidae
    Notohippus Ameghino, 1891
  Toxodontidae
    Adinotherium Ameghino, 1887
    Nesodon Owen, 1846
  Hegetotheriidae
    Hegetotherium Ameghino, 1887
    Pachyrukhos Ameghino, 1885
  Interatheriidae
    Epipatriarchus Ameghino, 1903
    Interatherium Ameghino, 1887
    Protypotherium Ameghino, 1887
ASTRAPOTHERIA
  Astrapotherium Burmeister, 1879
RODENTIA
  Erethizontidae
    Steiromys Ameghino, 1887
  Octodontidae
    Acaremys Ameghino, 1887
    Sciamys Ameghino, 1887

FIGURE 6-6 *Santacrucian Localities.*

 1 *Cabo Buen Tiempo—Patagonia*
 2 *Canadón de los Vacas—Patagonia*
 3 *Chinchinales beds sites—Argentina*
 4 *Corriguen Aike—Patagonia*
 5 *Coy Inlet—Patagonia*
 6 *Curuzcuatia (Curuzú Cuatiá)—Argentina*
 7 *Felton's Estancia—Patagonia*
 8 *Guanare West—Venezuela*
 9 *Halliday's Estancia—Patagonia*
10 *Ica River (marine)—Peru (age?)*
11 *Karaiken—Argentina ("Karaikenian" =*
    Santacrucian)
12 *La Angelina East—Patagonia*
 7 *La Costa East—Patagonia*
17 *La Cueva—Patagonia*
13 *Lago Pueyrredón—Patagonia*
14 *Laguna Blanca (in part?)—Argentina*
15 *Los Cruceros—Chile*
16 *Mauvi River—Bolivia*
17 *Monte León—Patagonia*
 2 *Monte Observación—Patagonia*
18 *Pinturas—Argentina*
19 *Quedrada Honda–Guarico—Venezuela (age?)*
20 *Rincon del Buque Patagonia*
21 *Rio Chico area (in part)—Patagonia (the*
    southern Rio Chico)
22 *Rio Colloncura—Argentina*
23 *Rio Coyle—Patagonia*
 7 *Rio Gallegos—Patagonia*
24 *Rio Negro (in part)—Argentina*
25 *Rio Santa Cruz area—Patagonia*
26 *Rio Shehuén—Patagonia*
27 *Sacaco (marine, in part)—Peru*
28 *San Francisco–Rio Güere—Venezuela*
29 *San Pedro—Venezuela*
30 *Tucupido—Venezuela*
17 *Yack-Harvey—Patagonia*
17 *Yegua Quemada—Patagonia*

*(For more detailed maps of Santacrucian localities in
Patagonia, see Marshall (1976)).*

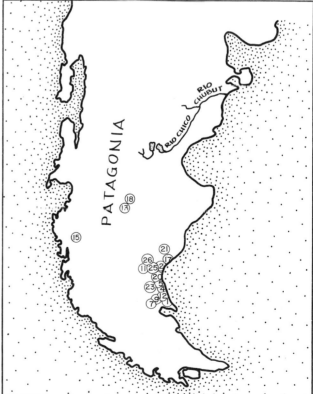

Echimyidae
  Acarechimys?
  Adelphomys Ameghino, 1887
  Spaniomys Ameghino, 1887
  Stichomys Ameghino, 1887
Chinchillidae
  Perimys Ameghino, 1887
  Prolagostomus Ameghino, 1887
Dasyproctidae
  Lomomys? Ameghino, 1891
  Neoreomys Ameghino, 1887
  Olenopsis Ameghino, 1889
  Scleromys Ameghino, 1887
Eocardidae
  Eocardia Ameghino, 1887
  Luantus Ameghino, 1899
  Phanomys Ameghino, 1887
  Schistomys Ameghino, 1887

ORLEANIAN

The Orleanian land-mammal age of Ginsburg (1975C) is based on species from Wintershof-West (subzone MN 3a of Mein, 1979), Artenay (subzone MN 3b of Mein, 1979), La Romieu (subzone MN 4a of Mein, 1979), Pontlevoy (MN 5 of Mein, 1975), and their correlatives. The characterizations of MN 3a through MN 5 (Mein, 1975, 1979) are shown in the accompanying outline.

Neogene Mammal Zone 5

Localities: Las Planas IVb, Pontlevoy, Sos, Langenmoosen, Leoben, Neudorf-Spalte, Františkovy L., Eibeswald, Paşalar, Sibnica, Mala Miliva, Lozovik.
"Formes caractéristiques de lignées évolutives": *Megacricetodon bavaricus, Democricetodon mutilus, Palaeomeryx m. pontileviensis, Pliopithecus piveteaui.*
Associations: *Dicrocerus + Bunolistriodon.*
Appearance: *Dicrocerus, Pliopithecus, Agnotherium, Heteroprox, Conohyus, Chalicotherium, Leptodontomys, Forsythia, Anchitheriomys.*

Neogene Mammal Subzone 4b

Localities: Valdemoros III, Lisboa Vb?, Baigneaux, Chevilly, Suèvres, Vieux-Collonges, Rensbach-Forsthart.
Appearance: *Pseudarctos, Ursavus, Plithocyon, Trocharion, Progenetta, Taucanamo, Cricetodon, Lartetomys, Paraglis, Anomalomys, Keramidomys, Albanensia.*
Characteristic forms of evolving lineages (given by Mein, 1975, for zone 4): *Megacricetodon collongensis, Ligerimys florancei, Eotragus artenensis.*
Associations: (given by Mein, 1975, for zone 4) *Megacricetodon + Melissiodon + Eotragus + Amphitragulus.*

Neogene Mammal Subzone 4a

Localities: Rubielos de Mora 2, Buñol Rubi, Lisboa Va, La Romieu, Dolniče, Orěchov, Čučale.
Appearance: *Schlossericyon, Bunolistriodon, Dorcatherium, Megacricetodon, Fahlbuschia, Deinotherium, Eumyarion, Neocometes, Spermophilinus.*

Neogene Mammal Subzone 3b

Localities: Moli Calopa, Artenay, Lisboa IV, Tuchořice, Nakhičevan.
Appearance: *Gomphotherium, Zygolophodon, Albanohyus, Micromeryx, Eotragus, Hyainailouros, Sansanosmilus.*
Characteristic forms of evolving lineages (given by Mein, 1975, for zone 3): *Eucricetodon infralactorensis, Ligerimys antiquus, Ligerimys lophidens, Vasseuromys rugosus, Cordylodon intercedens.*
Associations: (given by Mein, 1975, for zone 3) *Anchitherium + Brachyodus.*

Neogene Mammal Subzone 3a

Localities: Lisboa I (Univ.), Atéca 1, Chitenay, Estrepouy, Chilleurs, Bissingen, Wintershof-West, Eggenburg.
Appearance: *Anchitherium, Brachyodus, Amphimoschus, Procervulus, Lagomeryx, Palaeomeryx, Stephanocemas.*

The Orleanian is the beginning of the *Anchitherium* faunas in Europe (and North Africa, and? Asia). This beginning is evidently another of Stehlin's "coupures," based on abrupt appearance of several immigrant genera. We conclude, like previous workers, that *Hemicyon, Anchitherium, Chalicotherium, Dicrocerus, Deinotherium, Gomphotherium, Zygolophodon, Eotragus,* and other genera entered Europe at the beginning of Orleanian time or within it.

Our composite Orleanian mammalian fauna combines Mein's species for MN zones 3 through 5 with species from about 20 correlated localities in Europe. It includes the land-mammal faunas of the Touraine of France that are associated with *la mer falunienne;* Ginsburg (1970A, 1972A, and other papers) has termed these Burdigalian and Helvetian before assigning them to his Orleanian. It includes also the mammals listed by R. Savage (1967) as Burdigalian (= Early Miocene) of Europe and the assemblages assigned to the Ottnangian and Karpatian stages of the Paratethyan area (Steininger, Rögl, and Martini, 1976).

Composite Orleanian Fauna

MARSUPIALIA
  Didelphidae
    Amphiperatherium frequens (Meyer, 1846)   —W.v.
      Koenigswald (1970)
LAGOMORPHA
  Ochotonidae
    Amphilagus ulmensis Tobien, 1974
    Prolagus vasconiensis Viret, 1930
    P. schnaitheimensis Tobien, 1975
    P. oeningensis (Koenig, 1825)
    P. tobieni Lopez-M., 1977
    P. armaniacus Baudelot and Crouzel, 1976

Lagopsis verus (Hensel, 1856) —R. Savage (1967)

L. penai Royo, 1928 —Crusafont, Villalta, and Truyols (1955)

L. cadeoti Viret, 1930

Piezodus sp.

Titanomys visenoviensis Meyer, 1843? —R. Savage (1967)

CREODONTA

Hyaenodontidae

Hyainailouros sulzeri Biedermann, 1863 —Ginsburg (1980)

CARNIVORA

Amphicyonidae

Amphicyon giganteus (Schinz, 1825)

A. acutidens Dehm, 1950A

A. dietrichi Dehm, 1950A

cf. *A. steinheimensis* Fraas, 1885

A. helbingi Dehm, 1950A

A. cf. *major* Blainville, 1841

A. guttmanni Kittl, 1891

A. socialis Schlosser, 1904

A. cf. *crassidens* Pomel, 1853

A. (Ictiocyon) dehmi Crusafont and Villalta, 1955

Pseudocyon sansaniensis Lartet, 1851 —R. Savage (1967)

Cynelos bohemicus (Schlosser, 1899)

C. schlosseri (Dehm, 1950A)

Pseudarctos bavaricus Schlosser, 1899

Haplocyonoides cf. *mordax* Hürzeler, 1941A

Ysengrinia depereti (Mayet, 1908) —R. Savage (1967)

Ursidae

Thaumastocyon bourgeoisi Stehlin and Helbing, 1925C (amphicyonid?)

Hemicyon cf. *sansaniensis* Lartet, 1851

H. stehlini Hurzeler, 1944B

Herpestides aff. *antiquus* (Blainville, 1842)

Plithocyon vincenti (Mein, 1958)

P. bruneti Ginsburg, 1980

Ursavus intermedius Koenigswald, 1925A

U. elmensis Stehlin, 1917

Broiliana nobilis Dehm, 1950A

B. dehmi Beaumont and Mein, 1973

Phoberocyon aurelianensis (Frick, 1926)

P. huerzeleri Ginsburg, 1955

Alopecocyon leptorhynchus (Filhol, 1883) —[usually called *goriachensis* (Toula, 1884)]

Stromeriella franconica Dehm, 1950A

Mustelidae

Potamotherium valletoni Geoffroy, 1833

P. miocenicum (Peters, 1868)

Mionictis artenensis Ginsburg, 1968C

Palaeogale hyaenoides Dehm, 1950A

P. minuta (Gervais, 1859) —Dehm (1950A)

Palaeogale, n. sp. —Ginsburg (1970A)

Miomephitis pilgrimi Dehm, 1950A

Paralutra lorteti (Filhol, 1881) —R. Savage (1967)

Martes laevidens Dehm, 1950A

M. sainjoni (Mayet, 1908) —R. Savage (1967)

M. cf. *muncki* Roger, 1900

M. filholi Depéret

M. burdigaliensis de Beaumont, 1974

Ischyrictis zibethoides (Blainville, 1859) —R. Savage (1967)

Plesiogale angustifrons Pomel, 1846

P.? *postfelina* Dehm, 1950A

Trocharion albanense Major, 1904

Proputorius, n. sp. —Ginsburg (1970A)

Laphyctis? vorax Dehm, 1950A

L.? comitans Dehm, 1950A

Procyonidae?

Plesictis aff. *pygmaeus* Schlosser, 1887

P. aff. *sicaulensis* Viret, 1928

P. mayri Dehm, 1950A

P. vireti Dehm, 1950A

P.? *humilidens* Dehm, 1950A

Sivanasua viverroides (Schlosser, 1916)

Viverridae

Semigenetta cadeoti Roman and Viret, 1934 —R. Savage (1967)

S. mutata (Filhol, 1887)

S. repelini Helbing, 1927 —R. Savage (1967)

S. elegans Dehm, 1950A

Herpestes cf. *dissimilis* (Mayet, 1908)

Hyaenidae

Progenetta gaillardi Major, 1903

P. praecurrens Dehm, 1950A

P. cf. *crassa* (Depéret, 1892)

Plioviverrops (Protoviverrops) gervaisi Beaumont and Mein, 1972

Felidae

"Felis" vireti (Crusafont and Villalta, 1955)

Pseudaelurus turnauensis (Hoernes, 1881)

P. transitorius Depéret, 1892

P. lorteti Gaillard, 1899

P. quadridentatus Blainville, 1864

Sansanosmilus

Prosansanosmilus peregrinus Heinzmann, Ginsburg, and Bulot, 1980

INSECTIVORA

Erinaceoidea

Galerix exilis (Blainville, 1840)

Schizogalerix pasalarensis Engesser, 1980

Lanthanotherium (Rubitherium) piveteaui Crusafont and Villalta, 1955

L. lactorensis Baudelot and Crouzel, 1976

Mioechinus butleri Crusafont and Villalta, 1955

Cordylodon intercedens Müller, 1967

C. aff. *haslachense* Meyer, 1859 —R. Savage (1967)

Soricoidea

"Sorex" stehlini Doben-Florin, 1964

"S." collongensis Mein, 1958

"S." aff. *schlosseri* (Roger, 1885)

"Limnoecus" micromorphus (Doben-Florin, 1964) —Repenning (1967C)

Oligosorex dehmi (Viret and Zapfe, 1951) —Gureev (1971A)

Dinosorex zapfei Engesser, 1975

Soricella discrepans Doben-Florin, 1964

Trimylus neumayrianus (Schlosser, 1887)   —Engesser (1979)

Crocidura?   —Mein (1958)

Crocidosorex antiquus (Pomel, 1853)   —Repenning
(1967C)

Talpa minuta Blainville, 1838

Proscapanus sansaniensis (Lartet, 1851)   —R. Savage (1967)

Asthenoscapter meini Hutchinson, 1974

CHIROPTERA

Megadermatidae

Megaderma lugdunensis (Depéret, 1892)   —Sigé
(1976); Mein (1958)

Rhinolophidae

Rhinolophus aff. *lemanensis* Revilliod   —R. Savage
(1967)

PRIMATES

Pliopithecidae

Pliopithecus piveteaui Hürzeler, 1954A

ARTIODACTYLA

Tayassuidae

Taucanamo sansaniense (Lartet)

Palaeochoerus aurelianensis Stehlin   —Ginsburg
(1973, 1977B)

P. cf. *waterhousi* (Pomel, 1853)   —R. Savage (1967)

Albanohyus pigmeus (Depéret, 1892)

Suidae

Bunolistriodon lockarti (Pomel, 1848)

Hyotherium soemmeringi Meyer, 1834   —Ginsburg
(1973, 1977B)

Conohyus simorrensis (Lartet)

Aureliachoerus aurelianensis (Stehlin, 1899)

Xenohyus venitor Ginsburg, 1980

Anthracotheriidae

Brachyodus onoideus (Gervais, 1852)

Anthracotherium   —R. Savage (1967)

Cainotheriidae

Cainotherium miocaenicum Crusafont, Truyols, and Villalta,
1955

C. bavaricum Berger   —R. Savage (1967)

Hypertragulidae?

Andegameryx andegaviensis Ginsburg, 1970B

A. serum (Obergfell, 1957)   —Ginsburg (1970B)

Tragulidae

Dorcatherium guntianum Meyer, 1846

D. crassum Kaup, 1833

D. cf. *rogeri*

D. peneckei Hofmann, 1893

D. cf. *naui* Kaup, 1836

Cervoidea

Amphitragulus cf. *gracilis* Pomel, 1853   —R. Savage (1967)

A. aurelianensis Mayet, 1908

A. cf. *boulangeri* Pomel, 1853   —R. Savage (1967)

Procervulus dichotomus (Gervais, 1859)   —R. Savage (1967)

P. aurelianensis (Gervais, 1859)   —R. Savage (1967)

Dicrocerus elegans Lartet, 1851

Stephanocemas elegantulus (Roger, 1898)

S. infans Stehlin   —R. Savage (1967)

Lagomeryx pumilio Roger, 1898

L. parvulus (Roger, 1898)

L. meyeri Hofmann, 1893   —R. Savage (1967)

L. rutimeyeri Thenius

L. vallesensis Crusafont and Villalta, 1955

L. praestans Stehlin

Palaeomeryx kaupi Meyer, 1834

P. cf. *magnus* (Lartet, 1851)

Triceromeryx pachecoi Villalta, Crusafont and Lavocat,
1946B

Micromeryx

Euprox minimus (Toula, 1893)

Giraffidae

Giraffokeryx   —Ginsburg (1972A)

Bovoidea

Amphimoschus pontelevensis Bourgeois, 1873

Eotragus artenensis

E. sansaniensis (Lartet, 1851)

PERISSODACTYLA

Equidae

Anchitherium aurelianense Cuvier, 1822

Tapiridae

Tapirus   —Ginsburg (1972A)

Palaeotapirus aff. *intermedius* (Filhol, 1885)   —R.
Savage (1967)

P. helveticus Meyer, 1867   —R. Savage (1967)

Rhinocerotidae

Aceratherium platyodon Mermier, 1895

A. aff. *croizeti* (Pomel, 1853)   —R. Savage (1967)

A. cf. *tetradactylum* (Lartet, 1837)

Diceratherium douvillei Osborn

Brachypotherium (*Brachypotherium*) *brachypus* (Lartet,
1848)

B. (*B.*) *aurelianensis* (Nouel, 1866)   —Ginsburg and
Antunes (1979)

Gaindatherium   —Ginsburg and Antunes (1979)

Chilotherium ibericus Antunes, 1972

Hispanotherium matritensis (Prado, 1863)

Prosantorhinus douvillei   —Ginsburg and Antunes
(1979) (check against *Diceratherium,* above)

P. tagicus (Roman, 1911)

D. (*Lartetotherium*) *sansaniensis* (Lartet, 1851)

Dromoceratherium mirallesi Crusafont and Villalta, 1955

Indricotherium   —R. Savage (1967)   (Agenian?)

Chalicotheriidae

Phylotillon aff. *naricus* Pilgrim, 1910

Chalicotherium grande (Lartet, 1837)

PROBOSCIDEA

Deinotheriidae

Deinotherium cuvieri Kaup, 1832

Gomphotheriidae

Gomphotherium angustidens (Cuvier, 1806)

G. olisiponensis Zbyszewski, 1949

G. lusitanicus (Bergounioux, Zbyszewski, and Crouzel,
1951A)

Mammutidae

Zygolophodon pyrenaicus (Lartet, 1851)

Z. turicensis (Schinz, 1833)

FIGURE 6-7 *Representative
Orleanian Localities.*
(We are listing only a few of the
published Orleanian localities. Please
see the glossary at the end of this
chapter for a complete listing.)

1 *Alivieri (Evvoia)—Greece*
2 *Armantes I—Spain*
3 *Artenay-Auteroche (Loiret)—
France*
2 *Ateca I and III—Spain*
5 *Baigneaux-en-Beauce (Eure-et-
Loire)—France*
6 *Belluno, molasse de
(marine?)—Italy*
7 *Bissingen—Germany*
8 *Bruttelen (Bern district)—
Switzerland*
9 *Buñol Rubi—Spain*
10 *Can Mas (Papiol)—Spain*
3 *Chevilly (Loiret)*
3 *Chilleurs-au-Bois (Pithiviers
Southwest)—France*
11 *Chitenay (Blois)—France*
13 *Dolnice—Czechoslovakia*
14 *Echzell—Germany*
15 *Eggenburg—Austria*
16 *Eibiswald—Austria*
17 *Estrepouy—France*
18 *Felsö-Eszertgaly (Tarnoc West)
(marine)—Hungary*
19 *Forsthart (See Rensbach-
Forsthart)—Germany*
20 *Františkovy Lazne
(Franzenbad)—Czechoslovakia*
21 *Kirald (Matra Mountains)—
Hungary*
22 *Komotini (Thrace)—Greece*
23 *Langenmoosen (Neuberg
South)—Germany*
17 *La Romieu (Gers)—France*
23 *Las Planas IVA and B—Spain*
24 *Lectoure (Navere)—France*
25 *Leoben (Steiermark)—Austria*
26 *Lisboa or Lisbon or Lisbonne
or Lissabon (Horta de
Tripas)—Portugal*
30 *Nakhicehvan—Armeniya SSR*
24 *Navere (Lectoure)—France*
31 *Neudorf Spalte a.d. March
(Devinska Nova Ves)—
Czechoslovakia*
3 *Neuville-au-Bois (Loiret)—
France*
32 *Orechov (Brunn, Brno)—
Czechoslovakia*
3 *Pontlevoy (Thenay) (Loir-et-
Cher)—France*
19 *Rensbach (Forsthart)—
Germany*

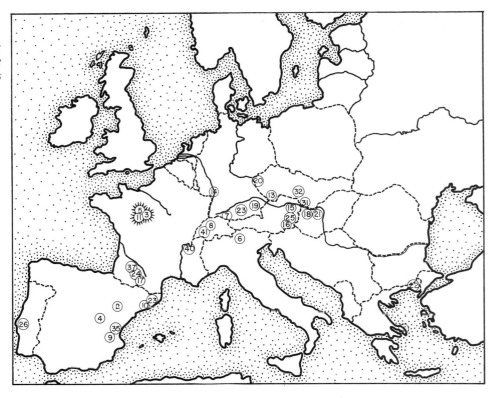

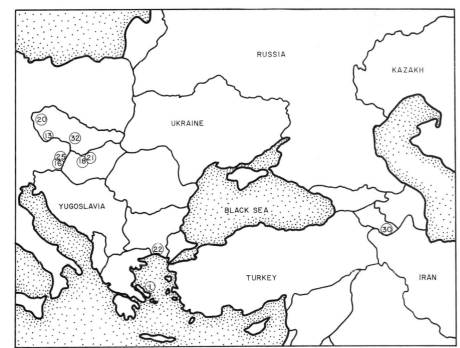

9 *Rubi (See Bunol-Rubi)—Spain*
35 *Rubielos de Mora—Spain*
21 *Salgotarjan (Matra
Mountains)—Hungary*
37 *Sos (Lot-et-Garonne)—France*
11 *Suèvres (Loir-et-Cher)—
France*
4 *Torralba de Ribota (Torralba I
and V)—Spain*

35 *Valdemoros IA, IIIA, B, D,
E—Spain*
40 *Vieux-Collonges (Rhône)—
France*
41 *Vully I, II, III (Fribourg)—
Switzerland*
7 *Wintershof-West (Eichstätt)—
Germany*

229

PHOLIDOTA
  Manidae
    Necromanis franconica (Quenstedt, 1885)   —W. v.
      Koenigswald (1969A)
    N. parva W. v. Koenigswald, 1969A
RODENTIA
  Ischyromyoidea
    Paracitellus eminens Dehm, 1950B   —Black (1966)
  Aplodontidae
    Ameniscomys selenoides Dehm, 1950B   (family?)
    Sciurodon descendens Dehm, 1950B
  Eomyidae
    Pseudotheridomys?
    Rhodanomys   —R. Savage (1967)
    Ligerimys florancei Stehlin and Schaub, 1951
    L. antiquus Fahlbusch, 1970A
    L. lophidens (Dehm, 1950B)
    L. ellipticus Daams, 1976
    Keramidomys thaleri Hugueney and Mein, 1968A
  Sciuridae
    Tamias eviensis de Bruijn, Van der Meulen, and Katsikatsos,
      1980
    Albanensia
    Blackia miocenica Mein, 1970
    Palaeosciurus fissurae (Dehm, 1950B)   —de Bruijn,
      Van der Meulen, and Katsikatsos (1980)
    Atlantoxerus blacki (de Bruijn, 1965)   —de Bruijn,
      Dawson, and Mein (1970A)
    Spermophilinus bredai (v. Meyer, 1848)
    Heteroxerus costatus (Freudenberg, 1947)
    H. vireti Black, 1965B
    H. rubricati Crusafont and Villalta and Truyols, 1955
    Ratufa obtusidens Dehm, 1950B
    Miopetaurista dehmi de Bruijn, Van der Meulen, and Katsi-
      katsos, 1980
    M. diescalidus Daams, 1977
    M. lappi (Mein, 1958)
    Aliveria brinkerinki de Bruijn, Van der Meulen, and Katsi-
      katsos, 1980
    A. luteyni de Bruijn, Van der Meulen, and Katsikatsos, 1980
  Castoridae
    Steneofiber depereti Mayet, 1908
    Anchitheriomys wiedemanni Roger
  Cricetidae
    Megacricetodon (*Megacricetodon*) *bourgeoisi* (Schaub, 1925)
    M. gregarius (Schaub, 1925)
    M. bavaricus Fahlbusch, 1964
    M. collongensis Mein, 1958
    M. (*Collongomys*) *lappi* (Mein, 1958)
    M. minor   —Daams (1976)
    Cricetodon aureus Mein and Freudenthal, 1971A
    C. meini Freudenthal, 1963
    Eucricetodon infralactorensis (Viret, 1930)
    Democricetodon mutilus Fahlbusch, 1964
    D. romieviensis Freudenthal, 1963
    D. crassus Freudenthal, 1969
    D. brevis (Schaub, 1925)
    D. aff. *hispanicus* Freudenthal, 1967

    Fahlbuschia larteti (Schaub, 1925)
    F. cf. *koenigswaldi* (Freudenthal, 1963)
    Anomalomys minor Fejfar, 1975
    Lartetomys mirabilis Mein and Freudenthal, 1971A
    L. cf. *zapfei*   —Mein and Freudenthal (1971A)
    Eumyarion helveticus (Schaub, 1925)
    E. weinfurthi (Schaub and Zapfe, 1953)
    E. valencianum Daams and Freudenthal, 1974
    Neocometes similis Fahlbusch, 1966
    N. brunonis Schaub and Zapfe, 1953
    Melissiodon dominans Dahm, 1950B
  Gliridae
    Glirudinus modestus (Dehm, 1950B)
    G. gracilis (Dehm, 1950B)
    Peridyromys gregarius (Dehm, 1950B)
    P. murinus (Pomel, 1853)
    P. cf. *occitanus* Baudelot and de Bruijn, 1966A
    Pseudodryomys simplicidens de Bruijn, 1966A
    P. ibericus de Bruijn, 1966A
    Mioglis meini (de Bruijn, 1966A)
    Glis spectabilis Dehm, 1950B
    Praearmantomys crusafonti de Bruijn, 1966A
    Microdyromys koenigswaldi de Bruijn, 1966A
    Armantomys aragonensis de Bruijn, 1966A
    Heteromyoxus schlosseri Dehm, 1950B
    Branssatoglis cadeoti Bulot, 1978
    B. astaracensis (Baudelot, 1970)   —Daams (1976)

Ginsberg and Antunes (1979) suggest that the three rhinocerotids, *Gaindatherium*, *Hispanotherium*, and *Chilotherium*, known in western Europe only on the Iberian Peninsula, could have arrived there during the Miocene. They could have come from the Anatolian region (Asia Minor) by means of a very quick dispersal via the "Alpine Arch" and Betic Massif, as the Betic Massif "was colliding with the central Iberian nucleus (Hesperid Massif)." This is an intriguing interpretation, giving birth to many unanswered questions about the paleontogeography of the Mediterranean-Parathethyan region. See, however, Dewey et al. (1973, Plate tectonics and the evolution of the Alpine System. *Bull. Geol. Soc. Am.* 84:3137–3180) to appreciate the complexity of the Mesozoic-Cenozoic geologic history of the Mediterranean area.

RUSINGAN (=ORLEANIAN) OF NORTH AFRICA

R.J.G. Savage (1967) and later workers list the following taxa from the interval in North Africa formerly designated Burdigalian.

CREODONTA
  Hyaenodontidae
    Megistotherium osteothlastes R. Savage, 1973
    Anasinopa   —R. Savage and Hamilton (1973)
    Hyainailouros fourtaui v. Koenigswald, 1948A

CARNIVORA
  Amphicyonidae, gen. indet.   —R. Savage and Hamilton (1973)
  Canidae   —R. Savage and Hamilton (1973)
    Afrocyon buroletti Arambourg, 1961B
  Felidae
    cf. *Metailurus*   —R. Savage and Hamilton (1973)
    Syrtosmilus syrtensis Ginsburg, 1978   (*Dinictis?*)
INSECTIVORA
  Erinaceoidea
    Galerix
PRIMATES
  Cercopithecidae, *inc. sed.*   —Szalay and Delson (1979)
    Prohylobates simonsi Delson, 1979–0449
    P. tandyi Fourtau   —Delson (1979)
ARTIODACTYLA
  Suidae
    Xenochoerus africanus (Stromer, 1926)   —Wilkinson (1976)
    Kubanochoerus massai (Arambourg, 1961)   —Wilkinson (1976)
    K. khinzikebirus Wilkinson, 1976
    Listriodon cf. *akatidogus* Wilkinson, 1976
    Hyotherium   —Cooke and Wilkinson (1978)
  Anthracotheriidae
    Hyoboops moneyi (Fourtau, 1918)
    H. africanus (Andrews, 1914)   —R. Savage and Hamilton (1973)
    Masritherium depereti (Fourtau, 1918)
    Gelasmodon? gracilis Forster-Cooper
  Tragulidae
    Dorcatherium libiensis Hamilton, 1973
  Cervoidea
    Canthumeryx sirtensis Hamilton, 1973
    Palaeomeryx
  Giraffidae
    Zarafa zelteni Hamilton, 1973
    Prolibytherium magnieri Arambourg, 1961A
  Bovidae
    Gazella   —R. Savage and Hamilton (1973)
    Protragocerus   —R. Savage and Hamilton (1973)
    Eotragus   —R. Savage and Hamilton (1973)
SIRENIA
  New genus   —R. Savage and Hamilton (1973)
HYRACOIDEA
  Saghatherium   —R. Savage and Hamilton (1973)
PERISSODACTYLA
  Rhinocerotidae
    Brachypotherium snowi (Fourtau)
    Aceratherium   —R. Savage and Hamilton (1973)
PROBOSCIDEA
  Deinotheriidae
    Prodeinotherium hobleyi (Andrews, 1911)
    Deinotherium cuvieri Kaup, 1832
  Gomphotheriidae
    Gomphotherium angustidens (Cuvier, 1906)
    G. spenceri (Fourtau)
    G. pygmaeus Arambourg, 1961
  Mammutidae
    Zygolophodon turicensis (Schinz, 1833)

The Rusingan fauna of North Africa appears to be essentially "African" in origin, although a few taxa are shared with Eurasia. We speculate that if there was dispersal between Eurasia and North Africa by latest Oligocene or early Miocene time, it was minimal.

RUSINGAN—Szalay and Delson (1979)

The faunas from Kenya (Rusinga, Songhor, Koru, etc.) are now renowned as a good sample of the early Miocene mammalian fauna of Africa. We have added a few species from Moghara, Egypt, to the Kenyan list.

MACROSCELIDEA
  Macroscelidae
    Rhynchocyon clarki Butler and Hopwood, 1957
    R. rusingae Butler, 1969A
    Myohyrax oswaldi Andrews, 1914
    Protypotheroides beetzi Stromer, 1922
LAGOMORPHA
  Ochotonidae
    Kenyalagomys rusingae MacInnes, 1951B
    K. minor MacInnes, 1951B
    Austrolagomys simpsoni Hopwood, 1929B
PANTOLESTA?
    Kelba quadeemae R. Savage, 1965A, 1978
CREODONTA
  Hyaenodontidae
    Teratodon spekei R. Savage, 1965A
    T. enigmae R. Savage, 1965A
    Apterodon (age?)   —R. Savage (1978)
    Pterodon africanus Andrews, 1906
    P. nyanzae R. Savage, 1965A
    Anasinopa leakeyi R. Savage, 1965A
    Hyainailouros napakensis Ginsburg, 1980
    Dissopsalis pyroclasticus R. Savage, 1965A
    Metapterodon kaiseri (Stromer)
    Leakitherium hiwegi R. Savage, 1965A
    Hyaenodon (*Isohyaenodon*) *andrewsi* R. Savage, 1965A
    H. (*I.*) *matthewi* R. Savage, 1965A
    H. (*I.*) *pilgrimi* R. Savage, 1965A
CARNIVORA
  Amphicyonidae
    Hebucides euryodon R. Savage, 1965A
    H. macrodon R. Savage, 1965A
    Afrocyon (age?)   —R. Savage (1978)
  Viverridae
    Kichechia zamanae R. Savage, 1965A
  Felidae
    Afrosmilus africanus (Andrews, 1914)   —R. Savage (1978)
INSECTIVORA
  Erinaceoidea
    Lanthanotherium   —Butler (1969A)
    Gymnurechinus leakeyi Butler, 1956A

> *G. camptolophus* Butler, 1956A
> *G. songhorensis* Butler, 1956A
> *Amphechinus rusingensis* Butler, 1956A
> *Galerix africanus* Butler, 1956A

Soricoidea
> *Crocidura*   —Butler and Hopwood (1957)

Chrysochloroidea
> *Prochrysochloris miocaenicus* Butler and Hopwood, 1957

Tenrecoidea
> *Geogale aletris* Butler and Hopwood, 1957
> *Protenrec tricuspis* Butler and Hopwood, 1957
> *Erythrozootes chamerpes* Butler and Hopwood, 1957

CHIROPTERA

MEGACHIROPTERA

Pteropodidae
> *Propotto leakeyi* Simpson, 1967B   —Walker (1969A)

MICROCHIROPTERA

Emballonuridae
> *Taphozous incognita* (Butler and Hopwood, 1957)

Hipposideridae
> *Hipposideros*   —Butler (1969A)

Megadermatidae, gen. indet.

PRIMATES   —Szalay and Delson (1979)

Lorisidae
> *Progalago dorae* MacInnes, 1943
> *P. songhorensis* Simpson, 1967B
> *Komba robustus* (Le Gros Clark and Thomas, 1952)
>   —Simpson (1967B)
> *K. minor* (Le Gros Clark and Thomas, 1952)   —
>   Simpson (1967B)
> *Mioeuoticus bishopi* Leakey, 1962D

Cercopithecidae
> *Victoriapithecus macinnesi* v. Koenigswald, 1969A
> *"V." leakeyi* (v. Koenigswald, 1969A)
> *Prohylobates tandyi* Fourtau, 1918

Pliopithecidae
> *Dendropithecus macinnesi* (Le Gros Clark and Leakey, 1950)
> *Micropithecus clarki* Fleagle and Simons, 1978   (family?)

Hominidae
> *Dryopithecus (Proconsul) africanus* (Hopwood, 1933A)
> *D. (P.) nyanzae* (Le Gros Clark and Leakey, 1950)
> *D. (P.) major* (Le Gros Clark and Leakey, 1950)
> *D. (Limnopithecus) legetet* (Hopwood, 1933A)
> *D. (Rangwapithecus) gordoni* P. Andrews, 1974
> *D. (R.) vancouveringi* P. Andrews, 1974
> *Sivapithecus africanus* (in part) Le Gros Clark and Leakey,
>   1950   —(another part is referred to *Dryopithecus*
>   *nyanzae* by Szalay and Delson, 1979)

TUBULIDENTATA

Orycteropodidae
> *Myorycteropus africanus* MacInnes, 1956

ARTIODACTYLA

Suidae   —Cooke and Wilkinson (1978)
> *Hyotherium dartevellei* (Hooijer, 1963A)
> *H. kijivium* Wilkinson, 1976
> *Kubanochoerus jeanneli* Arambourg, 1933A
> *Listriodon akatidogus* Wilkinson, 1976
> *L. akatikubas* Wilkinson, 1976

> *Mabokopithecus clarki* v. Koenigswald, 1969A
> *Lopholistriodon moruoroti* (Wilkinson, 1976)
> *Xenochoerus africanus* (Stromer, 1926)
> *Sanitherium nadirum* Wilkinson, 1976

Anthracotheriidae
> *Masritherium aequitorialis* (MacInnes, 1951)
> *Hyoboops africanus* (Andrews, 1914)

Tragulidae
> *Dorcatherium chappuisi* Arambourg, 1933A
> *D. pigotti* Whitworth, 1958
> *D. parvum* Whitworth, 1958
> *D. songhorensis* Whitworth, 1958

Gelocidae
> *Gelocus whitworthi* Hamilton, 1973

Cervoidea
> *Climacoceras africanus* MacInnes, 1936
> *Walangania africanus* (Whitworth, 1958)   —Ham-
>   ilton (1973)
> *Propalaeoryx nyanzae* Whitworth, 1958

Giraffidae
> *Palaeotragus primaevus* Churcher, 1970B   (also Ternanian)

PERISSODACTYLA

Rhinocerotidae
> *Dicerorhinus leakeyi* Hooijer, 1966B
> *Aceratherium acutirostratum* (Deraniyagala, 1951A)
>   —Hooijer (1966B)
> *Brachypotherium heinzelini* Hooijer, 1963D
> *Chilotheridium pattersoni* Hooijer, 1971

Chalicotheriidae
> *Chalicotherium rusingensis* Butler, 1965A

HYRACOIDEA   —Whitworth (1954)

Pliohyracidae
> *Pachyhyrax championi* (Arambourg, 1933A)
> *Megalohyrax?*
> *Bunohyrax*

PROBOSCIDEA

Deinotheriidae
> *Prodeinotherium hobleyi* (Andrews, 1911)

Gomphotheriidae
> *Gomphotherium* cf. *angustidens* (Cuvier, 1806)
> *Platybelodon kisumuensis* Harris and Watkins, 1974
> *Protanancus macinnesi* Arambourg, 1946A

Mammutidae
> *Zygolophodon morotoensis* Pickford and Tassy, 1980

RODENTIA   —Lavocat (1978)

Phiomyidae
> *Andrewsimys parvus* Lavocat, 1973
> *Phiomys andrewsi* Osborn, 1908

Thryonomyidae
> *Paraphiomys stromeri* (Hopwood, 1929)
> *Epiphiomys coryndoni* Lavocat, 1973

Diamantomyidae
> *Diamantomys luederitzi* Stromer, 1922

Kenyamyidae
> *Kenyamys mariae* Lavocat, 1973
> *Simonimys genovefae* Lavocat, 1973

Myophiomyidae
> *Myophiomys arambourgi* Lavocat, 1973
> *Elmerimys woodi* Lavocat, 1973

Bathyergoididae
> *Bathyergoides neotertiarius* Stromer, 1924

Bathyergidae
Proheliophobius leakeyi Lavocat, 1973
Anomaluridae
Paranomalurus bishopi Lavocat, 1973
P. soniae Lavocat, 1973
P. walkeri Lavocat, 1973
Zenkerella wintoni Lavocat, 1973
Pedetidae
Megapedetes pentadactylus MacInnes, 1957

Cricetidae
Afrocricetodon songhori Lavocat, 1973
Paratarsomys macinnesi Lavocat, 1973
Notocricetodon petteri Lavocat, 1973
Sciuridae
Vulcanisciurus africanus Lavocat, 1973
sciurid, sp.

FIGURE 6-8 *Representative Miocene Land-Mammal Localities of Africa.*

1   Adi Ugri—Rusingan—Ethiopia
2   Alengerr beds sites—Ternanian—Kenya
3   Beni Mellal—(Astaracian) Ternanian— Morocco
4   Bled ed Douarah—(Astaracian) Ternanian—Tunisia
5   Bou Hanifia Formation sites—(Vallesian) Ngororan—Algeria
6   Bukwa I and II—Rusingan—Uganda
7   Fort Ternan—Ternanian—Kenya
8   Gara Ziad—late Miocene Ngororan?— Morocco
9   Gebel Zelten—(Orleanian) Rusingan — Libya
10 Henchir Beglia (upper)—(Astaracian) Ternanian—Tunisia
10 Henchir Beglia (lower)—(Orleanian) Rusingan—Tunisia
11 Isserville-Kabylie—(Orleanian) Rusingan— Algeria
24 Kabuga beds site   (age?)—Zaire
12 Karugámania—Rusingan—Zaire
13 Karungu—Rusingan—Kenya
14 Koru—Rusingan—Kenya
14 Lothagam Group sites—Lothagamian (in part)—Kenya
2   Lukeino—Lothagamian—Kenya
15 Maboko (Island)—Ternanian?—Kenya
16 Malembe—Rusingan and marine "Burdigalian"—Cabinda
17 Melka el Ouidane (Camp Berteaux)—late Miocene, Ngororan?—Morocco
15 Mfevanganu (Island)—Rusingan—Kenya
18 Moghara (oasis or wadi)—(Orleanian) Rusingan—Egypt
19 Moroto—Rusingan—Uganda
2   Mpesida beds sites (Baringo)— Lothagamian—Kenya
20 Muruarot (Muruorot) Hill—Rusingan— Kenya
19 Napak—Rusingan—Uganda
21 Namib (Luderitz, Elisabethfelder, etc.)— Rusingan—Namibia

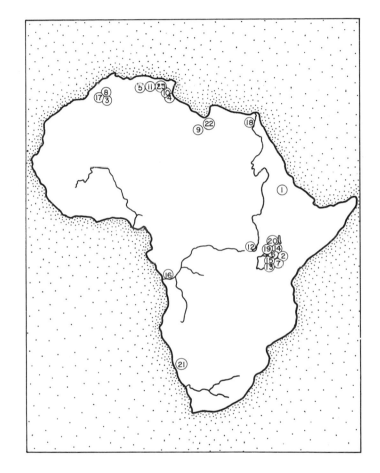

6   Ngorora—Ngororan—Kenya
15 Ombo—Rusingan—Kenya
22 Qasr es Sahabi—(Turolian) Lothagamian— Libya
14 Rusinga (Island)—Rusingan—Kenya
10 Sbeitla—(Astaracian) Ternanian—Tunisia
23 Smendou (Constantine)—(Orleanian) Rusingan—Algeria
7   Songhor—Rusingan—Kenya
18 Wadi Natrun—(Turolian) Lothagamian— Egypt

HEMINGFORDIAN

Hemingfordian was proposed by H. Wood et al. (1941) as a "new provincial time term, based on the Hemingford group, including the Marsland, and especially, the limited or lower Sheep Creek fauna (Cook and Cook, 1933, p. 38–40), and not on the formation limits as extended upward (Lugn, 1939)." Thus, Hemingfordian is based on faunas from the Marsland Formation of Schultz (1938B) but excludes the lower part of Schultz's Marsland, which equals the Upper Harrison beds of Peterson (1906), the Runningwater Formation of H. Cook (1965), the Box Butte Formation (see Galusha, 1975), and the Sheep Creek Formation.

Hemingfordian faunas record the earliest appearances of characteristically Nearctic Neogene land-mammal genera such as *Hypolagus* (hares), *Hemicyon* (and *Ursavus?*) (bears), *Merycodus* and *Ramoceros* (prongbucks), *Merychippus* and *Protohippus* (brachyhypsodont horses), *Aphelops* (long-limbed rhinos), and *Monosaulax* (beavers). These faunas perhaps are best known for their great array of genera and species of canids, soricids, merycoidodont oreodonts, camelids, cervoids, equids, sciurids, geomyids, and heteromyids. Eight of the thirty-one families of mammals here recognized in the Hemingfordian do not have living representatives.

Composite Hemingfordian Mammalian Fauna

MARSUPIALIA
 Didelphidae
  Herpetotherium   —(Green and Martin, 1976)
LAGOMORPHA
 Ochotonidae
  Oreolagus cf. *nebrascensis* McGrew, 1941B   —R. Wilson (1960)
  Cuyamalagus dawsoni Hutchison and Lindsay, 1974
  Gripholagomys lavocati Green, 1972A
 Leporidae
  Archaeolagus acaricolus Dawson, 1958
  A. primitivus (Matthew, 1907A)
  A. macrocephalus (Matthew, 1907A)
  Palaeolagus   —Wood and Wood (1937)
  Hypolagus?   —R. Wilson (1960)
  Hypolagus cf. *apachensis* Gazin, 1930   —Hutchison and Lindsay (1974)
CARNIVORA
 Amphicyonidae
  Daphaenodon cf. *superbus* (Peterson, 1906C)   —J. Wilson (1960A)
  D. iamonensis (Sellards, 1916B)
  Amphicyon idoneus Matthew, 1924C
  A. longiramus White, 1942A   —J. Wilson (1960A)
  A. frendens Matthew, 1924C
  Cynelos caroniavorus (White, 1942A)   —Tedford and Frailey (1976)
 Ursidae
  Cephalogale   —Tedford et al. (1983?)

Hemicyon johnhenryi (White, 1947)   —Tedford and Frailey (1976)
  Ursavus   —Tedford et al. (1983?)
  Absonodaphoenus bathygenus Olsen, 1958D
  Amphictis   —Tedford et al. (1983?)
 Canidae
  Nothocyon minor (Matthew, 1907A)
  Tomarctus optatus (Matthew, 1924C)   —Galusha (1975)
  T. thomsoni (Matthew, 1907A)
  T. cf. *confertus* (Matthew and Cook, 1909)   —J. Wilson (1960A)
  T. canavus (Simpson, 1932D)
  T. nobilis (Simpson, 1932D?)
  Cynarctoides acridens (Barbour and Cook, 1914A)   —Galbreath (1956); Galusha (1975)
  C. mustelinus (Matthew, 1932A)
  Euoplocyon spissidens (White, 1947)   —Tedford and Frailey (1976)
  Phlaocyon leucosteus Matthew, 1899A   (later Arikareean?)
 Mustelidae
  Craterogale   [*simus* Gazin, 1936B?]
  Promartes lepidus (Matthew, 1907A)
  Megalictis ferox Matthew, 1924C
  Brachypsalis matutinus Matthew, 1924C
  Sthenictis bellus Matthew, 1932A
  Dinogale siouxensis Cook and Macdonald, 1962
  Mionictis letifer Cook and Macdonald, 1962
  Miomustela   —Galusha (1975)
  Plionictis   —Tedford et al. (1983?)
  Mephititaxus ancipidens T. White, 1941B
  Potamotherium   —Tedford et al. (1983?)
  Leptarctus   —Tedford et al. (1983?)
 Procyonidae
  Edaphocyon pointblankensis   —J. Wilson (1960)
 Felidae
  Pseudaelurus   —Tedford et al. (1983?)
INSECTIVORA
  Inc. Sed. Arctoryctes terreneus Matthew, 1907A
 Erinaceoidea
  Brachyerix macrotis Matthew and Mook, 1933
  Amphechinus?
  Parvericius montanus Koerner, 1940
  Metechinus marslandensis Meade, 1941
 Talpidae
  Mesoscalops scopelotemus K. Reed, 1960
  M. montanensis Barnosky, 1981
  Mystipterus (= "*Mydecodon*") *martini* (R. Wilson, 1960)
  Proscalops cf. *secundus* Matthew, 1901B
  Scalopoides isodens R. Wilson, 1960
 Soricidae
  Pseudotrimylus dakotensis (Repenning, 1967C)   —Engesser (1979)
  P. roperi (R. Wilson, 1960)   —Engesser (1979)
  Domnina
  Angustidens vireti (R. Wilson, 1960)
  Antesorex compressus (R. Wilson, 1960)
  Sorex
  Wilsonosorex bateslandensis J. Martin, 1978–0815
  W. conulatus J. Martin, 1978–0815
  Limnoecus   —Hutchison and Lindsay (1974)
  Plesiosorex coloradensis R. Wilson, 1960

CHIROPTERA
  Phyllostomatidae, gen. indet.   —Hutchison and Lindsay (1974)

ARTIODACTYLA
  Tayassuidae
    Desmathyus (Hesperhys?) pinensis Matthew, 1907A
    Cynorca sociale (Marsh, 1875B)   —Patton (1969B)
    C. occidentalis Woodburne, 1969B
    C. cf. *proterva* Cope, 1867C   —Woodburne (1969B)
    Dyseohyus   —Woodburne (1969B)
  Entelodontidae
    Dinohyus hollandi Peterson, 1905C
  Anthracotheriidae
    Arretotherium fricki Macdonald and Schultz, 1956
  Merycoidodontidae
    Phenacocoelus stouti Schultz and Falkenbach, 1950
    P. cf. *leptoscelos* Stevens, 1969   Woodburne et al. (1974)
    Promerycochoerus carrikeri Peterson, 1906C
    Merycochoerus matthewi Loomis, 1924B
    M. proprius Leidy, 1858E
    Merychyus minimus Peterson, 1906C
    M. relictus Matthew and Cook, 1909
    M. arenarum Cope, 1884M
    M. verrucomalus Stevens, 1970A
    M. elegans Leidy, 1858E
    Ticholeptus tooheyi Schultz and Falkenbach, 1941
    Brachycrus laticeps (Douglass, 1900B)
    B. wilsoni Schultz and Falkenbach, 1940
    B. rusticus (Leidy, 1870O)
    B. vaughani Schultz and Falkenbach, 1941
    Mediochoerus johnsoni Schultz and Falkenbach, 1941
    Desmatochoerus newchicagoënsis Schultz and Falkenbach, 1954 (age?)
  Camelidae
    Oxydactylus exilis Matthew and Macdonald, 1960
    O. lacota Matthew and Macdonald, 1960
    O. longipes Peterson, 1904A
    O. brachyodontus Peterson, 1904A
    O. benedentatus (Hay, 1924A)   —Patton (1969B)
    O. longirostris Peterson, 1911A
    O. orarius (Patton, 1969B)   —Tedford et al. (1983?)
    Nothokemas floridanus (Simpson, 1932D)
    N. hildalgensis Patton, 1969B
    Floritragulus nanus Patton, 1969B
    F. barbouri T. White, 1947
    F. dolichanthereus T. White, 1940B
    F. texanus Patton, 1969B
    Protolabis saxeus Matthews, 1924C   —Skinner et al. (1977)
    Aepycamelus priscus (Matthew, 1924C)   —Skinner et al. (1977)
    Miolabis cf. *princetonianus* (Sinclair, 1915A)   —Galusha (1975)
    M. tenuis Matthew, 1924C   —Skinner et al. (1977)
    Michenia exilis (Matthew and Macdonald, 1960)   —Honey and Taylor (1978)
    Stenomylus gracilis Peterson, 1906C   —Frick and Taylor (1968)
    Tanymykter brachyodontus (Peterson, 1904A)   —Honey and Taylor (1978)

    Aguascalientia wilsoni (Dalquest and Mooser, 1974)   —Stevens (1977)
    Blickomylus galushai Frick and Taylor, 1968
  Protoceratidae
    Prosynthetoceras (Lambdoceras) hessei (Stirton, 1967)   —Patton and Taylor (1971)
    P. texanus (Hay, 1924A)   —Patton (1969B)
    Paratoceras, n. sp.   acc. Patton and Taylor (1973)
  Leptomerycidae
    Pseudoparablastomeryx scotti Frick, 1937
  Cervoidea
    Cranioceras dakotensis Frick, 1937
    C.? texanus (Hay, 1924A)   —Frick (1937)
    Rakomeryx yermonensis Frick, 1937
    Dromomeryx scotti Frick, 1937
    Barbouromeryx milleri Frick, 1937
    B. submilleri Frick, 1937
    B. sweeti Frick, 1937
    B. marslandensis Frick, 1937
    Aletomeryx scotti (Matthew, 1924C)
    A. lugni Frick, 1937
    A. marslandensis Frick, 1937
    A. (Sinclairomeryx) sinclairi Frick, 1937
    A. (S.) riparius (Matthews, 1924C)
    A. (S.) tedi Frick, 1937
    Blastomeryx gemmifer (Cope, 1874B)
    Problastomeryx primus (Matthew, 1908A)   —Tedford et al. (1983?)
    Pseudoblastomeryx (Parablastomeryx?) advena (Matthew, 1907A) (age?)
    P. schultzi Frick, 1937
    P. marsa Frick, 1937
    (The Blastomerycines are assigned to the Moschidae by Tedford et al. (1983?))
  Antilocapridae
    Merycodus, sp. indet.   —Galusha (1975)
    Ramoceros coronatus (Merriam, 1913C)

PERISSODACTYLA
  Equidae
    Hypohippus cf. *pertinax* Matthew, 1918A   —Galusha (1975)
    Merychippus insignis Leidy, 1856Q
    M. gunteri Simpson, 1930H   —Tedford et al. (1983?)
    "*Merychippus*" *primus* (Osborn, 1918A)   —Skinner et al. (1977)
    "*Merychippus*" *isonesus* (Cope, 1889B)   —Skinner et al. (1977)
    Anchitherium navasotae (Hay, 1924A)
    A. clarencei Simpson, 1932D
    Parahippus leonensis Sellards, 1916B
    P. cf. *tyleri* Loomis, 1908B
    Anchippus texanus Leidy, 1868I   —Tedford et al. (1983?)
    Hippodon vellicans (Hay, 1924A)
    Protohippus vetus Quinn, 1955
    Archaeohippus penultimus Matthew, 1924C
    A. blackbergi (Hay, 1924A)
  Tapiridae

Tapiravus validus (Marsh, 1871F)
Rhinocerotidae
  Diceratherium
  Aphelops meridianus (Leidy, 1865B)
  Brachypotherium americanus Yatkola and Tanner, 1979
  Menoceras falkenbachi Tanner, 1972A
  M. marslandensis Tanner, 1972A
  Floridaceras whitei H. Wood, 1964
Chalicotheriidae
  Moropus
  Tylocephalonyx skinneri Coombs, 1979
RODENTIA
  Eomyidae
    Eomys? aff. *zitteli*   —Lindsay (1974)
    Pseudotheridomys hesperus R. Wilson, 1960
    P. cuyamensis Lindsay, 1974
  Sciuridae
    Palaearctomys?
    Miospermophilus bryanti (R. Wilson, 1960)   —Black (1963B)
    M. wyomingensis Black, 1963B

Blackia Mein, sp. indet.   —Hutchison and Lindsay (1974)
Tamias sp.   —J. Martin (1976)
Protospermophilus kelloggi J. Martin (1976)
P. angusticeps (Matthew, 1933, in Matthew and Mook)   —Black (1963B)
Mylagaulidae
  Mesogaulus paniensis (Matthew, 1902D)
  M. praecursor Cook and Gregory, 1941
  M. vetus (Matthew, 1924C)
  M. novellus (Matthew, 1924C)
  Promylagaulus riggsi McGrew, 1941B
  Mylagaulodon angulatus Sinclair, 1903C
  Mylagaulus   —Quinn (1955); Galusha (1975)
Castoridae
  Monosaulax   —R. Wilson (1960); Quinn (1955)
  Anchitheriomys?   —R. Wilson (1960)

A   B

C   D

E

FIGURE 6-10   *Hemingfordian Rodents.*

A   *Miospermophilus bryanti* (R. Wilson, 1960). Lower jaw and dentition. Jaw is $\times 1^{1}/_{10}$; dentition is $\times 4^{1}/_{4}$. Figure 67 in R. Wilson (1960).

B   *Anchitheriomys*. Facial region of skull and upper dentition. Skull is $\times ^{1}/_{3}$; dentition is $\times ^{4}/_{5}$. Figure 70 in R. Wilson (1960).

C   *Proheteromys sulculcus* (R. Wilson, 1960). Lower jaw and lower dentitions. Jaw is $\times 2^{1}/_{2}$; teeth are $\times 5$. Figures 112 and 114 in R. Wilson (1960).

D   *Plesiosminthus galbreathi* (R. Wilson, 1960). Lower jaw with M_{1-3}. $\times 9$. Figure 131 in R. Wilson (1960).

E   *Pseudotheridomys hesperus* (R. Wilson, 1960). Upper teeth, lower jaw and lower teeth. $\times 5$. Jaw is $\times 1^{2}/_{3}$; teeth are $\times 5$. Figures 96–101 in R. Wilson (1960).

(*Published with permission of the University of Kansas Press.*)

A

B   C

FIGURE 6-9   *Hemingfordian Mammals.*

A   *Angustidens vireti* (R. Wilson, 1960). Lower jaw and dentition. Jaws $\times 5$, approx., teeth $\times 5^{1}/_{2}$ approx. Figure 16 in Repenning (1967).

B   *Mystipterus martini* (R. Wilson, 1960). Lower jaw. $\times 4$. Figure 32 in R. Wilson (1960).

C   *Scalopoides isodens* (R. Wilson, 1960). Humerus. $\times 2^{1}/_{5}$. Figure 39 in R. Wilson (1960).

(*Figure A published with permission of the U.S. Geological Survey. Figures B and C published with permission of the University of Kansas Press.*)

FIGURE 6-11 *Representative Hemingfordian Localities and Stratigraphic Units.*

1 Aiken (Akin) Hill (Walker County)—Texas

3 Anderson Mine—Arizona

4 Arroyo Chamisa Prospect—New Mexico—Zia Sand

5 Barger Gulch West—Colorado—Troublesome Formation

6 Barker Ranch (mostly Marine)—California—Round Mountain Silt

7 Black Bear Quarry—South Dakota—Rosebud Formation

2 Blacktail Deer Creek (Everson Creek)—Montana—Renova Formation

4 Blick Quarry—New Mexico—Zia Sand

8 Boron—California—Tropico Group

9 Box Butte Member sites—Nebraska

10 Bridgeport Quarries—Nebraska—Marsland Formaton

26 Buda—Florida

11 Caliente Mountain sites—California—Caliente Formation

12 Caper Place—Texas

4 Ceja del Rio Puerco site—New Mexico—Zia Sand

11 Cuyama Valley sites—California—Caliente Formation (lower)

13 Eastend Northwest—Saskatchewan—Cypress Hills Formation (upper)

7 Flint Hill (Black Bear Quarries)—South Dakota—Rosebud Formation

9 Foley Quarry—Nebraska—Hemingford Group

27 Franklin Phosphate Co. (Fullers Earth)—Midway—Florida

28 Gaillard Cut (Panama Canal)—Panama—Cucaracha Formation

29 Garvin Gulley ("Navasota")—Texas

14 Gay Hill—Highway 36—Texas

9 Greenside Quarry—Nebraska—Hemingford Group

9 Hilltop Quarry—Nebraska—Sheep Creek Formation, Hemingford Group

15 La Mision—Baja California, Mexico—Rosarito Beach Formation (mostly marine)

8 Logan Mine—California—Hector Formation

9 Long Quarry—Nebraska—Sheep Creek Formation

9 Marsland Northwest—Nebraska

16 Martin Canyon, Quarry A—Colorado

17 Massacre Lake—Nevada

4 Nambe Member sites—New Mexico—Santa Fe Group

28 Panama Canal—Cucaracha Formation

18 Phillips Ranch—California—Kinnick Formation

7 Porcupine South—South Dakota—Rosebud Formation

27 Quincy-Midway, etc.—Florida

9 Ravine Quarry—Nebraska—Sheep Creek Formation

19 Roll Quarry—Guernsey South—Wyoming

9 Runningwater Formation sites—Nebraska

9 Sheep Creek Formation sites—Nebraska

30 Shiloh Marl site—New Jersey

20 Split Rock—Wyoming—Split Rock Formation

9 Stonehouse Draw—Nebraska—Sheep Creek Formation

26 Thomas Farm Quarry—Florida

21 Tick Canyon Formation sites—California

22 Van Tassel South—Wyoming—Marsland Formation

11 Vedder Locality—California—Branch Canyon Formation

22 Warm Springs sites—Oregon—John Day Formation (upper)

23 White Sulphur Springs Northwest—Montana

24 Yecora—Sonora, Mexico

25 Zoyatal—Aguascalientes, Mexico

Geomyidae

Pleurolicus leptophrys Cope, 1881E

Gregorymys formosus (Matthew, 1907A)

G. curtus (Matthew, 1907A)

Grangerimus oregonensis A. Wood, 1936C

Dikkomys matthewi A. Wood, 1936C

Schizodontomys harkseni (Macdonald, 1970A) —
Rensberger (1978A); J. Martin (1976)

S. greeni Rensberger, 1973A

cf. Sanctimus   —J. Martin (1976)

Florentiamys?   —R. Wilson (1960)

Heteromyidae

Proheteromys magnus A. Wood, 1932

P. matthewi A. Wood, 1935C

P. parvus (Troxell, 1923C) (age?)

P. sulculus R. Wilson, 1960

P. floridanus A. Wood, 1932

Mookomys cf. formicorum A. Wood, 1935C

M. altifluminus A. Wood, 1931   —Lindsay (1974)

Cricetidae

Yatkolamys edwardsi L. Martin and Corner, 1980

Leidymys   —J. Martin (1976)

Paciculus cf. montanus Black, 1961C   —J. Martin (1976)

Zapodidae

Plesiosminthus (Plesiosminthus) clivosus Galbreath, 1953
—R. Wilson (1960)

P. sabrae (Black, 1958)

P. (Schaubemys) galbreathi R. Wilson, 1960 —
Green (1977)

ASTARACIAN

The Astaracian land-mammal age (Astaracium of Fahlbusch, 1976) is based on species frcm Sansan (zone MN 6 of Mein, 1975), Steinheim (MN 7 of Mein), and Anwil (MN 8). Mein's data for these zones are shown in the accompanying outline.

Neogene Mammal Zone 8

Localities: Pataniak 6, Barbera, San Quirze, Can Mata 1, lower Azambujeira, La Grive L3, St. Gaudens, Anwil, Giggenhausen, St. Stephan, Yeni Eskihisar.

Characteristic forms of evolving lineages: Deperetomys hagni, Democricetodon freisingensis, Plesiosorex schaffneri, Desmanella stehlini.

Appearance: Desmanella, Hispanomys, Palaeotragus, Protragocerus, Tetralophodon.

Neogene Mammal Zone 7

Localities: Beni Mellal, Póvoa de Santarem, Palencia, La Grive M, Simorre, Oggenhof, Steinheim, Oppole, Korèthi, Plackia, Sofca.

Characteristic forms of evolving lineages: Cricetodon albanensis, Megacricetodon gregarius, Fahlbuschia larteti, Democrice-

todon affinis, Palaeomeryx eminens, Dicerorhinus steinheimensis.

Appearance: Eurolagus, Metacordylodon, Grivasmilus, Euprox, Dinocyon, Paradiceros.

Neogene Mammal Zone 6

Localities: Madrid, Manchones, Arroyo del Val, Sansan, Sandelshausen, Göriach, Neudorf Sandberg, Byelometsketskaya, Prebeza, Çandir.

Characteristic forms of evolving lineages: Cricetodon sansaniensis, Cricetodon jotae, Megacricetodon crusafonti, Democricetodon gaillardi, Eumyarion medius, Pliopithecus antiquus, P. magnus.

Associations: Platybelodon danovi + Listriodon splendens.

Appearance: Deinotherium levius, Platybelodon, Kubanochoerus, Hypsodontus, Crouzelia, Listriodon, Conohyus, Heteroprox, Paleocricetus.

Astaracian is about the same interval as the Badenian plus Sarmatian (in part) ages of the Paratethyan area, as they are currently defined and characterized. See summary of Paratethyan ages by Steininger, Rögl, and Martini (1976). The fauna from Beni Mellal, Algeria, shares affinities with the Astaracian faunas of Europe and has been assigned to zone MN 7 by Mein. It also has affinities with the mid-Miocene fauna of sub-Saharan Africa, and we are listing it later under the rubric Ternanian.

Composite Astaracian Mammalian Fauna

LAGOMORPHA

Amphilagus ulmensis Tobien, 1974

Eurolagus fontannesi (Depéret, 1887)

Ptychoprolagus forsthartensis Tobien, 1975

Prolagus oeningensis (Koenig, 1825)

P. tobieni Lopez-Martinez, 1977

P. major Lopez-Martinez, 1977

Lagopsis verus (Hensel, 1856)

CARNIVORA

Amphicyonidae   —(included in the Ursidae by some workers)

Pseudarctos bavaricus Schlosser, 1899

Amphicyon steinheimensis Fraas, 1885

A. major Blainville, 1841

Alopecocyon leptorhynchus (Filhol, 1883) (usually called goriachensis)

Pseudocyon sansaniensis Lartet, 1851

Agnotherium grivensis (Viret, 1929)

Ursidae

Ursavus brevirhinus (Hofmann, 1887)

Hemicyon sansaniensis Lartet, 1851

Harpaleocyon sansaniensis (Frick, 1926)   —Thenius (1949H); synonym of Hemicyon, acc. Ginsburg, 1955

Hemicyon dehmi Ginsburg, 1955

H. stehlini Hürzeler, 1944B

Plithocyon armagnacensis Ginsburg, 1955

P. statzlingi (Frick, 1926)

Dinocyon   —Mitchell and Tedford (1973)

Mustelidae

Paralutra jaegeri Roman and Viret, 1934

Mionictis dubia (Blainville, 1842)
Trochotherium —Engesser (1972A)
Trochictis depereti F.-Major, 1903
Trocharion albanense F.-Major, 1903
Ischyrictis mustelinus Viret, 1933
Martes delphinensis Depéret, 1892
M. khelifensis Ginsburg, 1977
M. muncki Roger, 1900
Ailurictis jourdani Kretzoi, 1929
Presictis
Plesiomeles
Palaeomeles pachecoi Villalta and Crusafont, 1943
Taxodon sansaniensis Lartet, 1851
Potamotherium miocaenicum (Peters)
Viverridae
Herpestes filholi Gaillard, 1899
H. (Leptoplesictis) aurelianensis (Schlosser)
Semigenetta mutata (Filhol, 1887)
Hyaenidae
Progenetta gaillardi F.-Major, 1903
P. montadai (Villalta and Crusafont, 1943) —Crusafont and Petter (1969)
Plioviverrops (Mesoviverrops) gaudryi Beaumont and Mein, 1972
Hyaenictis graeca Gaudry, 1862
Felidae
Pseudaelurus tournauensis (Hoernes, 1882)
P. hyaenoides (Lartet, 1838)
P. quadridentatus Blainville, 1864 —Ginsburg and Tassy (1977); Petter (1976)
P. lorteti Gaillard, 1899
Sansanosmilus palmidens (Blainville, 1864)
INSECTIVORA
Erinaceoidea
Galerix (Parasorex) socialis (v. Meyer, 1865)
G. exilis Blainville, 1840
Lanthanotherium sansaniensis Filhol, 1888
Metacordylodon schlosseri (Andreae, 1902)
Mioechinus tobieni Engesser, 1980
M. oeningensis Butler, 1948
M. sansaniensis (Depéret) —Engesser (1980)
Soricoidea
Plesiosorex schaffneri Engesser, 1972A
P. styriacus (Hofmann, 1892)
P. germanicus
Dinosorex sansaniensis (Lartet, 1851) —Engesser (1975)
D. pachygnathus Engesser, 1972A
D. zapfei Engesser, 1975
Proscapanus sansaniensis (Lartet, 1851) —Hutchison (1974)
cf. *Scalopoides* —Hutchison (1974)
"Scaptonyx" edwardsi (Gaillard, 1899) —Hutchison (1974)
Asthenoscapter meini Hutchison, 1974
Talpa minuta Blainville, 1838
Miosorex grivensis (Depéret, 1892)
Paenelimnoecus crouzeli Baudelot, 1972
Desmanella stehlini Engesser, 1972A
D. sickenbergi Engesser, 1980
D. cingulata Engesser, 1980

Allosorex gracilidens (Viret and Zapfe, 1951)
Mygalea antiqua (Pomel, 1848)
M. jaegeri (Seemann, 1938) —Hutchison (1974)
Urotrichus? dolichochir (Gaillard, 1899) —Hutchison (1974)
Oligosorex dehmi (Viret and Zapfe, 1951) —Gureev (1971A)
Heterosorex delphinensis —Engesser (1979)
Prolimnoecus crouzeli Baudelot, 1972
CHIROPTERA
Megadermatidae
Megaderma vireti Mein, 1964
Rhinolophidae
Rhinolophus grivensis (Depéret, 1892)
R. delphinensis Gaillard, 1899
R. schlosseri Hofmann, 1893
R. ferrumequinum Schreber, 1775
Hipposideridae
Paraphyllophora? lugdunensis (Depéret, 1892)
Hipposideros collongensis (Depéret, 1892)
Asellia mariaetheresae Mein, 1958
Vespertilionidae
Scotophilus? —Engesser (1972A)
Eptesicus campanensis Baudelot, 1970
Paleptesicus priscus (Zapfe)
Miniopterus fossilis Zapfe, 1950
PRIMATES
Pliopithecidae
Pliopithecus (Pliopithecus) antiquus (Blainville, 1839)
P. (Plesiopliopithecus) lockeri Zapfe, 1961A
P. (Epipliopithecus) vindobonensis Zapfe and Hürzeler, 1957
Hominidae —Szalay and Delson (1979)
Dryopithecus fontani Lartet, 1856
Austriacopithecus weinfurteri Ehrenberg, 1937D
Crouzelia auscitanensis (Bergounioux and Crouzel, 1965) —Ginsburg (1975)
C. rhodanica Ginsburg and Mein, 1980
Sivapithecus darwini (Abel, 1902)
S.? macedoniensis (de Bonis and Melentis, 1977)
ARTIODACTYLA
Tayassuidae
Taucanamo pygmaeum (Depéret, 1892)
T. sansaniense (Lartet, 1892)
Albanohyus pigmaeus (Depéret, 1892)
Suidae
Bunolistriodon lockarti (Pomel, 1848)
Xenochoerus —Zdansky, 1909
Microstonyx choeroides (Pomel, 1848)
Korynochoerus palaeochoerus (Kaup, 1832)
Listriodon splendens Meyer, 1846
Sanitherium leobense
Conohyus simorrensis (Lartet)
Hyotherium soemmeringi Meyer, 1834
Kubanochoerus —Bernor (1983)
Tragulidae
Dorcatherium crassum (Lartet, 1851)
D. penecki Hofmann, 1892
D. naui Kaup, 1836

239

D. vindobonense Meyer, 1846
Hypertragulidae?
    Andegameryx minimus (Toula, 1884)   —Ginsburg
    (1970B)
Cervoidea
    Dicrocerus elegans Lartet, 1851
    Micromeryx flourensianus Lartet, 1851
    M. styriacus Thenius, 1950
    Euprox furcatus (Hensel, 1859)
    E. minimus (Toula, 1893)
    Lagomeryx parvulus (Roger, 1878)
    Palaeomeryx magnus Lartet, 1851
    P. bojani Meyer, 1834
    P. kaupi Meyer, 1834
    P. emenens Meyer, 1846
    Heteroprox larteti (Filhol, 1890)   —Ginsburg and
    Crouzel (1976)
    Procervulus   —Ginsburg (1975C)
    Stephanocemas elegantulus (Roger, 1878)
Giraffidae
    Palaeotragus lavocati Heintz, 1976
    P. germaini Arambourg, 1958
Bovidae
    Strogulognathus Filhol, 1890
    Miotragocerus monacensis Stromer, 1928D
    Orygotherium escheri Meyer, 1839
    Eotragus sansaniensis (Lartet, 1851)
    E. haplodon (Meyer, 1846)
    E. clavata (Schlosser, 1911)
    Gazella stehlini Thenius, 1951 —Ciric and Thenius (1959)
    Pachytragus solignaci Robinson, 1972
PERISSODACTYLA
Equidae
    Anchitherium aurelianense Cuvier, 1822
Tapiridae
    Tapirus telleri Hofmann, 1893
Rhinocerotidae
    Rhinoceros   —Engesser (1972A)
    Aceratherium tetradactylum (Lartet, 1853)
    Didermocerus or *Lartetotherium sansaniensis* (Lartet, 1848)
    Mesaceratherium simorrense (Lartet, 1851)   —
    Heissig (1976) [*Dromoceratherium* Crusafont, Villalta
    and Truyols, 1955]
    Brachypotherium brachypus (Lartet, 1848)
    Hispanotherium matritensis (Prado, 1863)   —Cru-
    safont and Villalta (1947B)
    Dicerorhinus primaevus Arambourg, 1959
    D. steinheimensis (Jäger) (emend. Roger, 1910)
    Diceros douariensis Guérin, 1966
Chalicotheriidae
    Chalicotherium grande (Lartet, 1837)
    Ancylotherium   —Rabeder (1978)
HYRACOIDEA
    Parapliohyrax mirabilis Lavocat, 1961A
PROBOSCIDEA
Deinotheriidae
    Deinotherium levius Jourdan, 1861
    D. giganteum Kaup, 1829

D. bavaricum Meyer, 1833
Gomphotheriidae
    Gomphotherium angustidens (Cuvier, 1806)
    Choerolophodon chioticus Tobien, 1980B
    Tetralophodon longirostris (Kaup, 1832)
    Rhynchotherium spenceri (Tunisia)
    Platybelodon danovi Borisyak, 1928
Mammutidae
    Zygolophodon turicensis (Schinz, 1833)   —Gins-
    burg and Tassy (1979)
RODENTIA
Eomyidae
    Keramidomys mohleri Engesser, 1972A
    K. anwilensis Engesser, 1972A
    K. carpathicus (Schaub and Zapfe, 1953)
    Eomyops catalaunicus (Hartenberger, 1966)
Sciuridae
    Atlantoxerus tadlae (Lavocat, 1961)   —Jaeger (1977)
    Albanensia albanensis (Major, 1893)
    A. sansaniensis (Lartet, 1851)
    A. grimmi (Black, 1966)   —Mein (1970A)
    Miopetaurista gibberosa (Hofmann)
    M. gaillardi Mein, 1970A
    M. neogrivensis Mein, 1970A
    Forsythia gaudryi (Gaillard, 1899)   —Mein (1970A)
    Blackia miocaenica Mein, 1970A
    Spermophilinus bredai (Meyer, 1848)
    Heteroxerus hurzeleri Stehlin and Schaub, 1951
    H. grivensis (Major, 1893)
Castoridae
    Steneofiber jaegeri Kaup, 1832
    Trogontherium minutus (Meyer, 1844)
Gliridae
    Microdyromys koenigswaldi de Bruijn, 1966A
    M. ambiguus (Lavocat, 1961)   —Jaeger (1977A)
    M. complicatus de Bruijn, 1966A (includes *M. miocaeni-
    cus* Baudelot)
    Paraglirulus werenfelsi Engesser, 1972A
    P. scalabicensis Antunes and Mein, 1977
    P. cf. lissiensis (Hugueney and Mein, 1965)
    Glirudinus gracilis (Dehm, 1950)   —de Bruijn
    (1966A)
    Eomuscardinus sansaniensis (Lartet, 1851)
    Myoglis meini (de Bruijn, 1966a)
    Muscardinus, n. sp.   —Engesser (1972A)
    M. sansaniensis (Lartet, 1851)
    M. thaleri de Bruijn, 1966A
    Branssatoglis astaracensis (Baudelot, 1970)   —
    Daams (1976)
    Pseudodryomys   —Antunes and Mein (1977)
    Armantomys aragonensis de Bruijn, 1966A
    Peridyromys hamadryas (Major, 1899)
    Tempestia
Cricetidae
    Megacricetodon schaubi Fahlbusch, 1964
    M. crusafonti (Freudenthal, 1963)
    M. cf. collongensis (Mein)
    M. similis Fahlbusch, 1964
    M. gregarius (Schaub, 1925)
    M. gersii Aguilar, 1980
    M. minor (Lartet)

Democricetodon minor (Schaub, 1925)

D. affinis (Schaub)

D. gaillardi (Schaub, 1925)

D. freisingensis Fahlbusch, 1964

D. crassus Freudenthal, 1969

D. breve (Schaub, 1925)

D. vindobonensis (Schaub and Zapfe, 1953)

Cricetodon —Engesser (1972A)

C. (Cricetodon) sansaniensis Lartet, 1851

C. (C.) albanensis Mein and Freudenthal, 1971A

Cricetodon jotae Mein and Freudenthal, 1971A

C. laskarevi Petronijevic

Fahlbuschia (Hispanomys) decedens Schaub, 1925
—Mein and Freudenthal (1971A)

F. (H.) bijugatus Mein and Freudenthal, 1971A

F. (Deperetomys) hagni (Fahlbusch, 1964)

F. (D.) rhodanicus (Depéret, 1887)

F. (Pseudoruscinomys) lavocati (Freudenthal, 1966)

Fahlbuschia larteti (Schaub, 1925)

F. darocensis (Freudenthal, 1963)

F. crusafonti Agustı, 1978

Anomalomys gaudryi Gaillard, 1900

Neocometes brunonis Schaub and Zapfe, 1953

Eumyarion helveticus (Schaub, 1925)

Cotimus latior (Schaub and Zapfe, 1953)

Middle Miocene (=Astaracian) of Anatolian Turkey

LAGOMORPHA

Ochotonidae

Prolagus cf. *oeningensis* (Koenig, 1825)

Alloptox cf. *gobiensis* Dawson, 1961A

Eurolagus fontannesi (Depéret, 1887)

CREODONTA

Hyaenodontidae

Hyainailouros

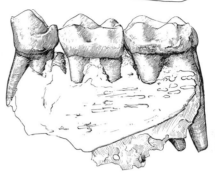

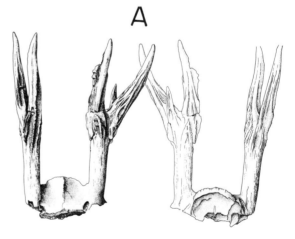

A

B

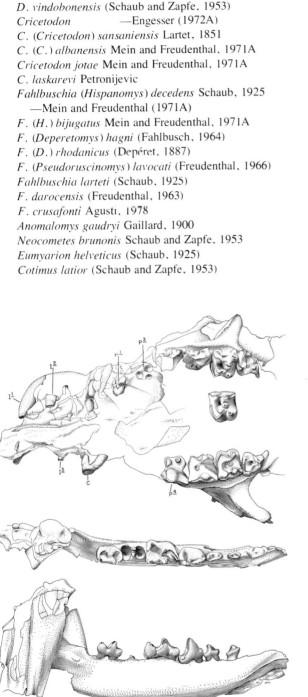

FIGURE 6-12 *Astaracian mammal. Lanthanotherium sansanlense Filhol, 1888. Facial region of skull and lower jaw.* ×2½. *Figures 16 and 19 in James (1963). (Published with permission of the University of California Press.)*

FIGURE 6-13 *Astaracian mammals.*

A <u>*Crouzelia rhodanica*</u> *Ginsburg and Mein, 1980. P₄–M₂.* ×3. *Figure 2 in Ginsburg and Mein (1980).*

B <u>*Heteroprox larteti*</u> *(Filhol, 1890). Skull cap with horn cores.* ×1/4. *Figure 4 in Ginsburg and Crouzel (1976).*

(Published with permission of the Muséum National d'Histoire Naturelle, Paris.)

FIGURE 6-14 *Representative Astaracian Localities. We are listing and plotting only a few of the published Astaracian localities. Please see the glossary at the end of this chapter for a complete listing.*

1   Antwerp or Anvers (marine)—Belgium
2   Anwil—Switzerland
3   Azambujeira (lower)—Portugal
4   Bampf-Seon—Switzerland
5   Banja Luka—Yugoslavia
6   Buchenberg (Vorarlberg)—Germany
7   Budapest—Hungary
8   Byeolometkeskaya (Caucasus)—USSR
9   Candir—Anatolian Turkey
10 Can Mata 1 (Lower Hostalets)—Spain
11 Castelnau-Barbarens (Aquitaine)—France
12 Chios (Thymiana)—Aegian Greece
13 Despotovac (Resava River)—Yugoslavia
14 Dornbach (Vienna)—Austria
15 Georgensgmünd—Germany
16 Giggenhausen—Germany
17 Goriach (Aflens)—Austria
18 Hirschtal—Switzerland
17 Kirchberg (Raab River)—Yugoslavia
19 Krivoy-Rog—USSR
20 Labastide (d'Armagnac)—France
21 La Grive–St. Alban (La Grive L3 and M)—France
22 Madrid Hidroeléctrica—Spain
23 Mallorca (Majorca)
24 Mas del Olmo (Rincon de Ademuz)—Spain
25 Mikulov (Nikolsburg)—Czechoslovakia
26 München (Flinz and Schweissanden)—Germany
27 Neudorf (Sandberg)—Czechoslovakia
28 Nikolayev—USSR
30 Nördlingen—Germany
31 Oeningen or Öhningen (Bodensee)—Germany
32 Palencia district—Spain
33 Plackia—Yugoslavia?
34 Prebeza (Nisch West)—Yugoslavia
35 Repovica (Sarajevo Southwest)—Yugoslavia
18 Rümikon (Zurich)—Switzerland

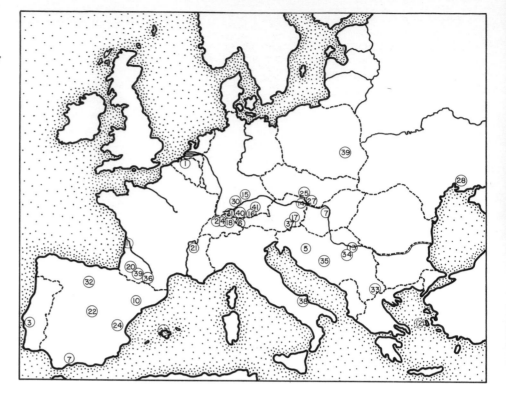

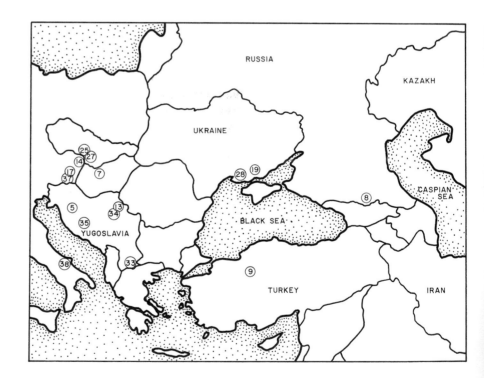

36 Saint Gaudens (Valentine)—France
37 St. Stefan (Lavant Valley)—Austria
38 San Giovannino (Fossia)—Italy
39 Sansan (Gers)—France

23 Son Fe—Mallorca
40 Steinheim (Aalbuch)—Germany
12 Thymiana (Chios Island)—Greece
41 Viehhausen (Reichenstetten)—Germany

CARNIVORA
  Amphicyonidae
    Amphicyon major Blainville, 1841
  Ursidae
    Hemicyon sansaniensis Lartet, 1851
    Ursavus primaevus (Gaillard, 1899)
    U. cf. *intermedius* Koenigswald, 1925A
    Plithocyon　　　　—Schmidt-Kittler (1976)
    agriotheriine　　　—Schmidt-Kittler (1976)
  Mustelidae
    Ischyrictis (*Hoplictis*) *anatolicus* Schmidt-Kittler, 1976
    Plesiogulo, n. sp.　　—Schmidt-Kittler (1976)
    Mionictis
    Trochictis　　　—Schmidt-Kittler (1976)
    Promephitis　　　—Schmidt-Kittler (1976)
    Proputorius　　　Schmidt-Kittler (1976)
    peruniine　　　Schmidt-Kittler (1976)
    melinine, gen. indet.　　　Schmidt-Kittler (1976)
    Anatolictis laevicaninus Schmidt-Kittler, 1976
  Procyonidae?
  Viverridae, gen. indet.
  Hyaenidae
    Protictitherium gaillardi (Forsyth-Major, 1903)　　—
      Schmidt-Kittler (1976)
    P. crassum (Depéret, 1892)　　—Schmidt-Kittler (1976)
    P. intermedium Schmidt-Kittler, 1976
    P. cingulatum Schmidt-Kittler, 1976
    Miohyaena montadai (Villalta and Crusafont, 1943)
      —Crusafont and Petter (1969A)
    Percrocuta miocenica Pavlovic and Thenius, 1965
    P. aff. *tungurensis* (Colbert, 1929A)　　—Schmidt-
      Kittler (1976)
    P. senyureki Ozansoy, 1961　　—Schmidt-Kittler
      (1976)
  Felidae
    Pseudaelurus lorteti Gaillard, 1899
    P. cf. *quadridentatus* (Blainville, 1842)　　—Schmidt-
      Kittler (1976)
    machairodontine
INSECTIVORA
  Erinaceoidea
    Schizogalerix pasalarensis Engesser, 1980
    S. anatolica Engesser, 1980
    Plesiodimylus crassidens Engesser, 1980
    Mioechinus tobieni Engesser, 1980
  Soricoidea
    soricid, gen. indet.
    Desmanella sickenbergi Engesser, 1980
    Dinosorex　　—Engesser (1980)
    Desmanodon major Engesser, 1980
    D. minor Engesser, 1980
    Prolimnoecus
CHIROPTERA
  Vespertilionidae, gen. unident.
PRIMATES
  Hominidae
    Dryopithecus sp., large
    D. sp., small
TUBULIDENTATA
  Orycteropus
PROBOSCIDEA

  Deinotheriidae
    Deinotherium
  Gomphotheriidae
    Gomphotherium angustidens (Cuvier, 1806)
    Choerolophodon pentelici (Gaudry and Lartet, 1856)
      —Tobien (1980B)
ARTIODACTYLA
  Suidae
    Taucanamo inonuensis Pickford and Erturk, 1979　(age?)
    Listriodon splendens Meyer, 1846
    Schizochoerus
    Conohyus
  Tragulidae
    Dorcatherium
  Cervoidea
    Euprox
    Palaeomeryx
    Triceromeryx
    Micromeryx or *Lagomeryx*
  Giraffidae
    Giraffokeryx
    Palaeotragus cf. *tungurensis* Colbert, 1936B
    P. cf. *primaevus*
    Samotherium
  Bovidae
    Protoryx cf. *carolinae* Major, 1891
    Prostrepsiceros cf. *rotundicornis* (Weithofer, 1888)
    Pachytragus cf. *laticeps* (Andrée, 1926)
    Oioceros
    Gazella?
    caprinae, gen. indet.
    Urmiatherium?
    ovibovine, gen. indet.
    eotragine?
    Tossunnoria
PERISSODACTYLA
  Equidae
    Anchitherium
  Tapiridae
    Tapirus
  Rhinocerotidae
    Brachypotherium brachypus (Lartet)　　—Heissig
      (1976)
    Aceratherium aff. *tetradactylum* (Lartet, 1837)
    Beliajevina tekkayai Heissig, 1974
    Hispanotherium grimmi Heissig, 1974
    Mesaceratherium simorrense (Lartet, 1848)
  Chalicotheriidae
    Chalicotherium grande (Lartet, 1837)
PROBOSCIDEA
  Gomphotheriidae
    Choerolophodon chioticus Tobien, 1980B
RODENTIA
  Sciuridae
    Atlantoxerus
    cf. *Spermophilinus bredai* (Meyer, 1848)
    sciuropterine
  Castoridae, gen. indet.

243

Hystricidae
  Hystrix
Gliridae
  Microdyromys miocaenicus (Baudelot, 1965)
  Pseudodryomys
  Myomimus dehmi (deBruijn, 1966)
Rhizomyidae
  cf. *Rhizomys*
Ctenodactylidae, gen. indet.
Dipodidae, gen. indet.
Cricetidae
  Cricetodon (Palaeocricetus)
  Megacricetodon similis Fahlbusch, 1964
  Democricetodon
  D. gaillardi (Schaub, 1925)
  Dakkamys
  Turkomys cariensis Sen and Unay, 1979
Zapodidae
  cf. *Heterosminthus*

Although we have listed the Astaracian fauna of Turkey separately, we are in accord with Mein's (1975, 1979) treatment of it as simply a geographic extension of the Astaracian fauna of Europe.

ASTARACIAN EQUIVALENT IN ASIA

(Includes species from Tung Gur, Kamlial, Chinji, Shanwang, Longshui Kou, and Jiulongthou.)

LAGOMORPHA
  Tsaganolagus wangi
  Alloptox minor
CREODONTA
  Hyaenodontidae
    Hyainailouros bugtiense (Pilgrim, 1912)
    Dissopsalis carnifex Pilgrim, 1910
    Sivapterodon lahirii (Pilgrim, 1932)
CARNIVORA
  Amphicyonidae
    Amphicyon pithecophilus Pilgrim, 1932   Chinji
    A. sindiensis Pilgrim, 1932   Kamlial
    A. palaeindicus Lydekker, 1876   Chinji
    A. confucianus
    A. tairumensis Colbert, 1939A   Tung Gur
    Vishnucyon chinjiensis Pilgrim, 1932   Chinji
  Ursidae
    Hemicyon teilhardi Colbert, 1939A   Tung Gur
  Canidae?
    Gobicyon macrognathus Colbert, 1939A   Tung Gur
  Mustelidae
    Melodon?   —Colbert, 1939A
    Eomellivora necrophila Pilgrim, 1932   Chinji
    Martes lydekkeri (Colbert, 1933A)   Chinji

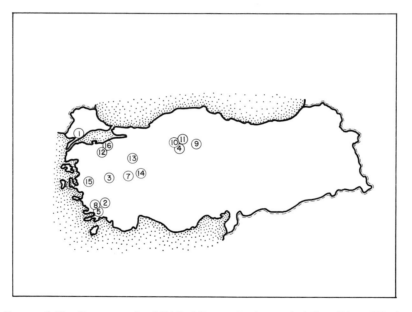

FIGURE 6-15   *Representative Middle Miocene (=Astaracian) Localities of Turkey.*

1	Arapli	*8*	Mesevle
2	Berdik	*12*	Paşalar
3	Çaliki	*8*	Sari Çay
4	Çandir	*13*	Sofça
5	Catakbağyaka	*10*	Termeyenice
6	Çifthkköy-KapBati	*4*	Tüney
7	Dumlupinar	*14*	Yaylacilar
8	Eskihisar (Yeni Eskihisar)	*8*	Yeni Eskihisar
9	Karaçay 1 and 2	*15*	Yukari Kizilca
10	Kilcak	*16*	Yürükali
11	Mahmutlar		

Sivalictis natans Pilgrim, 1932   Chinji
Vishnuonyx chinjiensis Pilgrim, 1932   Chinji
Procyonidae
  Sivanasua palaeindica Pilgrim, 1932
Viverridae
  Tungurictis spocki Colbert, 1939A   Tung Gur
  Viverra chinjiensis Pilgrim, 1932
Hyaenidae
  Percrocuta carnifex (Pilgrim, 1913)   Chinji
  P. hebeiensis Chan and Wu, 1976
  P. tungurensis (Colbert, 1939A)   Tung Gur
  Miohyaena cf. *montadai* (Villalta and Crusafont, 1943)
  Lycyaena proava (Pilgrim, 1910)   Chinji
  L. chinjiensis Pilgrim, 1932   Chinji
Felidae
  Vinayakia sarcophaga Pilgrim, 1932   Chinji
  Metailurus mongoliensis Colbert, 1939A   Tung Gur
  Sivaelurus chinjiensis (Pilgrim, 1910)   Chinji
  Vishnufelis laticeps Pilgrim, 1932
  Felidae, very small   —Pilbeam et al. (1977)
  Sansanosmilus serratus Pilgrim, 1932   Chinji
  S. rhomboidalis Pilgrim, 1932   Chinji
  S. cf. *palmidens* (Blainville, 1864)
  "Sivasmilus"   —Pilbeam et al. (1977)
  Sivasmilus copei Kretzoi, 1929   —(=*Paramachae-
    rodus*, acc. Pilbeam et al. (1977)   Chinji
PRIMATES
  Adapidae   —Gingerich and Sahni (1979)
    Sivaladapis palaeindica (Pilgrim, 1932)
  Hominidae
    Dionysopithecus shuangouensis Li Chuan-k., 1978
    "*Dryopithecus*" *keiyuanensis* Woo, 1957   —(=*Si-
      vapithecus* in part and *Ramapithecus* in part acc. Szalay
      and Delson, 1979)
    Ramapithecus punjabicus (Pilgrim, 1910)   Chinji in-
      cludes "*Bramapithecus thorpei*"
    Sivapithecus sivalensis (Lydekker, 1879)   Chinji includes
      "*Dryopithecus pilgrimi,*" "*Dryopithecus chinjiensis,*"
      "*Sivapithecus middlemissi,*" and "*Palaeosimia rugosi-
      dens*"
    Sivapithecus indicus Pilgrim, 1910   Chinji
ARTIODACTYLA
  Tayassuidae
    Palaeochoerus cf. *pascoei*
    P. perimensis (Lydekker, 1887)   —Colbert
      (1935A)
    Pecarichoerus orientalis Colbert, 1933
  Suidae
    Sanitherium schlagentweitii Meyer, 1866
    S. cingulatum Pilgrim, 1926
    Propotamochoerus salinus Pilgrim, 1926
    P. uliginosus Pilgrim, 1926
    Potamochoerus parvulus
    Hyotherium penisulus
    Listriodon pentapotamiae (Falconer, 1868)
    L. theobaldi Lydekker, 1878
    L. lishanensis
    L. guptai Pilgrim, 1926
    L. gigas
    L. lantienensis
    L. intermedius

L. mongoliensis Colbert, 1934A
L. robustus Yan, 1979
Dicoryphochoerus chisholmi Pilgrim, 1926
D. haydeni Pilgrim, 1926
D. instabilis Pilgrim, 1926
Anthracotheriidae
  Merycopotamus or *Hemimeryx pusillus* Lydekker, 1885
  Rhagatherium sindiense Lydekker, 1877
  Anthracotherium punjabiense Lydekker, 1877
  Hyoboops palaeindicus (Lydekker, 1877)
  Hemimeryx blanfordi Lydekker, 1883
  Telmatodon   —Pilgrim, 1910
Tragulidae
  Dorcabune anthracotherioides Pilgrim, 1910
  D. hyaemoschoides Pilgrim, 1915
  D. sindiense Pilgrim, 1915
  Dorcatherium majus Lydekker, 1876
  D. minus Lydekker, 1876
Cervoidea
  Lagomeryx simpsoni Jiang, 1978
  L. teilhardi Young, 1964
  L. complicidens Young, 1964
  Palaeomeryx
  Stephanocemas thomsoni Colbert, 1936A
  S. triacuminatus Colbert, 1936A
  Dicrocerus grangeri Colbert, 1936A
Giraffidae
  Palaeotragus tungurensis Colbert, 1936B
  Giraffokeryx punjabiensis Pilgrim, 1910
  Propalaeomeryx sivalensis Lydekker, 1882
  Giraffa priscilla Matthew, 1929
Bovidae
  Oioceros lishanensis
  O. grangeri Pilgrim, 1934B
  O. noverca Pilgrim, 1934B
  O.? jiulongkouensis Chan and Wu, 1976
  Protragocerus gluten (Pilgrim, 1937)
  Miotragocerus gradiens (Pilgrim, 1937)
  Kubanotragus sokolovi Gabunia, 1973
  Pseudotragus potwaricus (Pilgrim, 1939)
  Sivoreas eremita Pilgrim, 1939A
  Gazella
  Helicoportax tragelaphoides Pilgrim, 1937
  H. praecox Pilgrim, 1937
PERISSODACTYLA
  Equidae
    Cormohipparion theobaldi?   (Lydekker, 1877) Chinji?
    Anchitherium hypohippoides
    A. aurelianense Cuvier, 1822
    A. gobiense Colbert, 1939B
  Tapiridae
    Selenolophodon spectabilis
    Palaeotapirus yagii
    P. xiejiaheensis Xie Wan-ming, 1979
  Rhinocerotidae
    Aceratherium blanfordi Lydekker, 1884   —Colbert
      (1935A)
    A. perimensis Falconer and Cautley, 1847

(continued)

Plesiaceratherium gracile Young, 1937C
P. shanwangensis
Dicerorhinus cixianensis Chan and Wu, 1976
Gaindatherium browni Colbert, 1934B
Chilotherium intermedium (Lydekker, 1884)
Hispanotherium lingtungensis
Brachypotherium cf. *brachypus* (Lartet, 1848)
Tesselodon fangxianensis Yan, 1979
Chalicotheriidae
Macrotherium salinum Forster-Cooper, 1922A
M. brevirostris Colbert, 1934C
Nestoritherium? sindiense (Lydekker, 1876)   —Colbert (1935A)
PROBOSCIDEA
Gomphotheriidae
Gomphotherium gobiensis (Osborn, 1932)
G. florescens (Osborn, 1929)
Platybelodon grangeri Osborn, 1931
Mammutidae
Zygolophodon nemonguensis Chow and Chang, 1961
RODENTIA
Theridomyidae?
Diatomys shantungensis Li, 1974
Castoridae
Youngfiber sinensis
Anchitheriomys tungurensis (Stirton, 1934A)
   (= "Amblycastor")
Cricetidae
Spanocricetodon ningensis Li, 1977
Plesiocricetodon leei
Paracricetulus schaubi   (Gliridae?)

Prosiphneus lupinus A. Wood, 1936B
Copemys   —Pilbeam et al. (1977)
Megacricetodon   —Pilbeam et al. (1977)
Kanisamys indicus A. Wood, 1937B
Rhizomyidae
Rhizomyoides punjabiensis (Colbert, 1933F)
Muridae
Antemus chinjiensis Jacobs, 1978
Ctenodactylidae
Sayimys sivalensis (Hinton, 1933)   —Black (1972A)
Metasayimys intermedius Sen and Thomas, 1979
Thryonomyidae
Paraulacodus indicus Hinton, 1933
Hystricidae
Sivacanthion complicatus Colbert, 1933F
Zapodidae
Heterosminthus orientalis Schaub, 1930
Proalactaga grabaui Schaub, 1934
P. tunggurensis A. Wood, 1936B

TERNANIAN—Szalay and Delson (1979)

(Species from Fort Ternan, Loperot, Maboko, Beni Mellal, and Testour.)

Genera and species of mammals from a few localities in East and North Africa appear to be our meager representation of a fauna living in Africa about 17–12 million years ago.

246

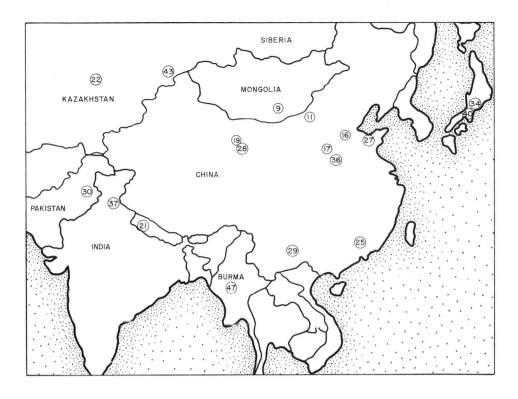

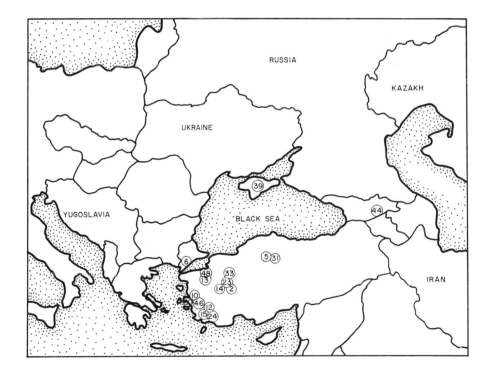

Ternanian Mammalian Fauna of Africa

LAGOMORPHA
  Ochotonidae   —Robinson and Black (1973)
CARNIVORA
  Mustelidae
    Mellidelavus   —Ginsburg (1961)
    Mellalictis mellalensis Ginsburg, 1977
  Viverridae
    Genetta   —Ginsburg (1977)
  Hyaenidae
    Ictitherium cf. *arambourgi* Ozansoy, 1965
  Felidae
    Felis   —Ginsburg (1977)
INSECTIVORA
  Erinaceoidea
    Protechinus salis Lavocat, 1961
    Galerix?   —Lavocat (1961)
    Palaeoerinaceus?   —Lavocat (1961)
  Soricoidea
    Sorex dehmi Viret and Zapfe, 1951   —Lavocat (1961)
CHIROPTERA
  Rhinolophidae
    Rhinolophus ferrumequinum Schreber, 1775   —Lavocat (1961)
  Hipposideridae
    Asellia vetus Lavocat, 1961
  Vespertilionidae   —Lavocat (1961)
  Molossidae   —Lavocat (1961)
  Family?
    Afropterus gigas Lavocat, 1961
PRIMATES
  Hominidae   —Szalay and Delson (1979)
    Dryopithecus nyanzae (Le Gros Clark and Leakey, 1950)?
    Ramapithecus wickeri (Leakey, 1962)
TUBULIDENTATA
  Orycteropus   —Bernor et al. (1979)
ARTIODACTYLA
  Suidae
    Sanitherium nadirum Wilkinson, 1976
    Lopholistriodon kidogosana Pickford and Wilkinson, 1975
  Hippopotamidae?   —Hooijer, 1970
  Cervoidea
    Climacoceras
  Giraffidae
    Palaeotragus primaevus Churcher, 1970B
    Samotherium africanum Churcher, 1970B
  Tragulidae
    Dorcatherium, n. sp.   —Gentry (1970A)
  Bovidae
    Protragocerus labidotus Gentry, 1970A
    Eotragus?
    Oioceros tanyceros Gentry, 1970A
    Pseudotragus? potwaricus (Pilgrim, 1939)   —Gentry (1970A)
    Gazella   —Gentry (1970A)
    Benicerus theobaldi Heintz, 1973

PERISSODACTYLA
  Rhinocerotidae
    Paradiceros mukiri Hooijer, 1968E
    Chilotheridium pattersoni Hooijer, 1971
HYRACOIDEA
  Procaviidae
    Parapliohyrax mirabilis Lavocat, 1961
PROBOSCIDEA
  Gomphotheriidae
    Gomphotherium angustidens (Cuvier, 1806)   —Robinson and Black (1973)
    Protanancus macinnesi Arambourg, 1946A
    Choerolophodon kisumuensis (MacInnes, 1942)
    C. ngorora (age?)
  Mammutidae
    Zygolophodon aff. *turicensis* (Schinz, 1833)
RODENTIA
  Family?
    Dubiomys mellali Lavocat, 1961
  Phiomyidae
    Paraphiomys occidentalis Lavocat, 1961
  Pedetidae
    Megapedetes   —Lavocat (1961)
  Ctenodactylidae
    Sayimys jebeli Lavocat, 1961
    Metasayimys curvidens Lavocat, 1961
    Africanomys pulcher Lavocat, 1961
    A. incertus Lavocat, 1961
    Testouromys solignaci Robinson and Black, 1973
  Cricetidae
    Leakeymys ternani Lavocat, 1964B
    Cricetodon atlasi Lavocat, 1961
    C. parvus Lavocat, 1961
    Myocricetodon cherifiensis Lavocat, 1952
  Gliridae
    Dryomys ambiguus Lavocat, 1961
  Sciuridae
    Getuloxerus tadlae Lavocat, 1961

BARSTOVIAN

The Barstovian land-mammal age was based on the aggregate of mammalian species from the Barstow Formation of the Barstow Syncline, southeastern California—"specifically on the fossiliferous tuff member in the Barstow syncline and its fauna" (H. Wood et al., 1941). Later than 1941, detailed work in this district showed that the lower part of the Barstow Formation and corresponding beds at the east end of the Calico Mountains near Yermo (15–20 miles from the Barstow) are Hemingfordian in age. This later information does not invalidate the typification of the Barstovian, however, for the "fossiliferous tuff member" overlies the Hemingfordian strata of the Barstow Formation.

    Barstovian faunas are best known as the earliest North American record of proboscideans. Gomphotheriids and mammutids arrived from Asia about 15 million years ago and remained in the Western Hemisphere until the end of Pleistocene time. Most of the Barstovian mammals are evidently descendants from Hemingfordian stock, however. This was the time of continued diversity of amphicyonids and canids. Felids, some of which may have immigrated from Asia at this time, became more conspicuous. Oreodonts and camelids may have been reduced in diversity compared with their early Miocene predecessors but individuals were exceedingly abundant in many districts. Antilocaprids and equids probably reached their acme of diversity during this time, with antilocaprids probably reaching their greatest abundance also. Among the rodents, the diversity and abundance of sciurids, heteromyids, and cricetids are noteworthy.

By Barstovian time, the land-mammal fauna is essentially modern, but 9 of the 35 families here recognized do not have living representatives; these extinct families are Amphicyonidae, Mylagaulidae, Gomphotheriidae, Mammutidae, Chalicotheriidae, Merycoidodontidae, Protoceratidae, Eomyidae, and Leptomerycidae.

Composite Barstovian Fauna

MARSUPIALIA
  Didelphidae   —Shotwell (1968)
LAGOMORPHA
  Ochotonidae
    Oreolagus
    Hesperolagomys fluviatilis Storer, 1970A
    Russellagus vonhofi Storer, 1970A
  Leporidae
    Hypolagus parviplicatus Dawson, 1958
    H. cf. *vetus* (Kellogg, 1910A)
    Panolax sanctaefidei Cope, 1874I
CARNIVORA
  Amphicyonidae
    Amphicyon longiramus White, 1942A
    A. amnicola Matthew and Cook, 1909
    A. pontoni Simpson, 1930H
    A. sinapius Matthew, 1902C
    A. ingenis Matthew, 1924C
    Pseudocyon   —Tedford et al. (1983?)
    Pliocyon medius Matthew, 1918A
    Ischyrocyon gidleyi (Matthew, 1902A)
    Euoplocyon praedator Matthew, 1924C
  Ursidae
    Hemicyon barstowensis Frick, 1926A
    H. californicus Frick, 1926A
    H. ursinus (Cope, 1875N)
    Ursavus pawniensis Frick, 1926A
  Canidae
    Aelurodon ferox Leidy, 1858E
    A. taxoides Hatcher, 1894C   —[=*Prohyaena* of Tedford and Frailey, 1976]
    A. cf. *wheelerianus* Cope, 1877K   —[=*Prohyaena* of Tedford and Frailey, 1976]
    A. francisi Hay, 1924A   —J.A. Wilson (1960)
    Epicyon saevus (Leidy, 1858E)

    E. haydeni Leidy, 1858B and 1858E
    Tomarctus brevirostris (Cope, 1873T)
    T. rurestris (Condon, 1896A)
    T. confertus (Matthew, 1918A)
    T. paulus Henshaw, 1942
    T. hippophagus (Matthew and Cook, 1909)
    T.? kelloggi (Merriam, 1911B)
    T.? cuspidatus (Thorpe, 1922E)
    "*T.*" *temerarius* (Leidy, 1858E)   —Tedford and Taylor (in prep.)
    Bassariscops willistoni (Peterson, 1924B)   —Peterson (1928)
    Cynarctus crucidens Barbour and Cook, 1914A
    C. saxatilis Matthew, 1902C
    Leptocyon vafer (Leidy, 1858E)
    Carpocyon   —Tedford and Taylor (in prep.)
  Procyonidae
    Bassariscus antiquus Matthew and Cook, 1909
    B. parvus Hall, 1927A
    B.? lycopotamicum (Cope, 1879B) (age?)   —Gregory and Downs (1951)
  Mustelidae
    Brachypsalis obliquidens Sinclair, 1915A
    B. pachycephalus (Cope, 1890I)   —Henshaw (1942)
    Martes (Plionictis) ogygia Matthew, 1901B   —Skinner et al. (1977)
    M. parviloba (Cope, in Scott and Osborn, 1890)
    M. gazini Hall, 1931
    Sthenictis dolichops Matthew, 1924C
    S. robustus (Allen, 1885A)?
    Leptarctus primus Leidy, 1856Q
    L. oregonensis Stock, 1930A
    L. wortmani Matthew, 1924C
    L. bozemanensis (Dorr, 1954A)
    Mionictis incertus Matthew, 1924C
    M. elegans Matthew, 1924C
    Miomustela madisonae (Douglass, 1929)   —Hall (1930E)
    Mustela sp.
  Felidae
    Pseudaelurus intrepidus Leidy, 1858E
    P. aeluroides Macdonald, 1954B
    P. marshi Thorpe, 1922E
    P. ("*Lynx*") *stouti* Schultz and Martin, 1972
    Sansanosmilus?
CHIROPTERA
MICROCHIROPTERA   —Lindsay (1972B)
INSECTIVORA
  Erinaceoidea
    Metechinus amplior T. Rich, 1981
    Brachyerix incertis (Matthew, 1924C)
    Parvericius montanus Koerner, 1940
    Plesiosorex latidens (Hall, 1929)   —Green (1977A)
    "*Lanthanotherium*" *sawini* (James, 1963)   —Engesser (1979)
    Amphechinus   —Storer (1975)

Plesiosorex donroosai Green, 1977A

Untermannerix copiosus T. Rich, 1981

Soricoidea

Adeloblarina berklandi Repenning, 1967C

Paradomnina relictus Hutchison, 1966A

Pseudotrimylus mawbyi (Repenning, 1967C) —Engesser (1979)

Limnoecus tricuspis Stirton, 1930

L. niobrarensis Macdonald, 1947A

Alluvisorex arcadentes Hutchison, 1966A

Mystipterus pacificus Hutchison, 1968A

Acklyoscapter longirostris Hutchison, 1968A

Scalopoides ripafodiator Hutchison, 1968A

Domninoides platybrachys (Douglass, 1903A)

D. valentiniensis Reed, 1962

Scapanoscapter simplicidens Hutchison, 1968A

ARTIODACTYLA

Tayassuidae

Dyseohyus stirtoni Woodburne, 1969B

D. fricki Stock, 1937B

Prosthennops xiphodonticus Barbour, 1925B

Cynorca proterva Cope, 1867C

C. occidentale Woodburne, 1969B

C. hesperia (Marsh, 1871D)

Desmathyus or *Hesperhys*

Merycoidodontidae

Metoreodon

Ticholeptus obliquidens (Cope, 1886H)

T. zygomaticus (Cope, 1878P)

T. rileyi Schultz and Falkenbach, 1941

T. hypsodus Loomis, 1924B

T. calimontanus (Dougherty, 1940B)

Brachycrus siouense (Sinclair, 1915A)

B. buwaldi (Merriam, 1919A)

B. laticeps (Douglass, 1907B)

Merychyus relictus Matthew and Cook, 1909

M. elegans Leidy, 1858E

Mediochoerus blicki Schultz and Falkenbach, 1941

Ustatochoerus medius (Leidy, 1858E)

U.? schrammi Schultz and Falkenbach, 1941

Camelidae

Aepycamelus procerus (Matthew and Cook, 1909)

A. leptocolon (Matthew, 1924C) —Tedford et al. (1983?)

A. giraffinus (Matthew and Cook, 1909)

Procamelus robustus (Leidy, 1858F)

P. occidentalis Leidy, 1858E

Floridatragulus hesperus Patton, 1969B

Miolabis princetonianus (Sinclair, 1915A)

M. transmontanus (Cope, 1879B)

Pliauchenia singularis Matthew, 1918A

Protolabis heterodontus (Cope, 1873R)

P. barstowensis Lewis, 1968A

Hesperocamelus alexandrae (Davidson, 1923A)

Homocamelus caninus Merrill, 1907A

Rakomylus raki Frick, 1937

Blickomylus galushai Frick and Taylor, 1968

Leptomerycidae

Pseudoparablastomeryx scotti Frick, 1937 —Taylor and Webb (1976)

P. francescita (Frick, 1937) —Taylor and Webb (1976)

Protoceratidae

Prosynthetoceras francisi Frick, 1937

P. (Lambdoceras) siouxensis Frick, 1937

P. (L.) trinitiensis Patton and Taylor, 1971

Paratoceras wardi Patton and Taylor, 1973

Cervoidea

Rakomeryx sinclairi (Matthew, 1918A)

R. raki Frick, 1937

R. jorakianus Frick, 1937

R. gazini Frick, 1937

R. kinseyi (Frick, 1937) —Tedford et al. (1983?)

Drepanomeryx falciformis Sinclair, 1915A

D. (Matthomeryx) matthewi Frick, 1937

Dromomeryx whitfordi Sinclair, 1915A

D. borealis (Cope, 1878C)

D. pawniensis Frick, 1937

D. antilopinus Scott, 1893B

D. wilsoni Frick, 1937

Cranioceras unicornis Matthew, 1918A

C. pawniensis Frick, 1937

C. granti Frick, 1937

C. cf. skinneri Frick, 1937 —Storer (1975)

C. teres (Cope, 1874I) (age?)

Barbouromeryx nebrascensis Frick, 1937

B. supernebrascensis Frick, 1937

B. pseudonebrascensis Frick, 1937

B. madisonius (Douglass, 1900A)

B. pawniensis Frick, 1937

B. trigonocorneus (Barbour and Schultz, 1934)

B.? americanus (Douglass, 1900A)

Aletomeryx gracilis Lull, 1920A

A. sinclairi Frick, 1937

Blastomeryx elegans Matthew and Cook, 1909A ⎫

B. gemmifer Cope, 1874C

B. mollis Merriam, 1911B (assigned to Blastomerycinae and

B. francesca Frick, 1937 (age?) Moschidae by Tedford

B. francescita Frick, 1937 (age?) et al. (1983?))

Parablastomeryx scotti Frick, 1937

P.? galushi Frick, 1937 ⎭

Antilocapridae

Ramoceros (Pararamoceros) howardae Frick, 1937

R. (P.) brevicornis Frick, 1937

R. osborni (Matthew, 1904A)

R. coronatus Frick, 1937

Merycodus necatus Leidy, 1854C

M. (="Cosoryx") furcatus (Leidy, 1869A)

M. alticornis Frick, 1937

M. loxocerus Furlong, 1934 —(*Paracosoryx loxocerus* (Furlong) by Tedford et al., 1983?)

M. nevadensis Merriam, 1911B

M. wilsoni Frick, 1937

M. savaronis Frick, 1937

M. sabulonis Frick, 1937

M. dawesensis Frick, 1937

Meryceros major, Frick, 1937

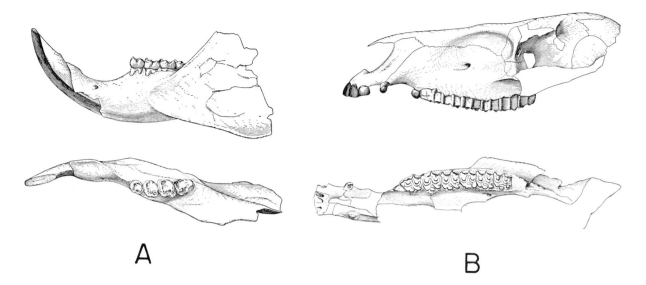

A

B

FIGURE 6-17　*Barstovian Mammals.*

　A　Protospermophilus oregonensis (Downs, 1956D). Lower jaw. ×3. Figure 4 in Downs (1956D).

　B　"Merychippus" seversus (Cope, 1879D). Skull. ×¼ approx. Figures 26 and 27 in Downs (1956D).
(Published with permission of the University of California Press.)

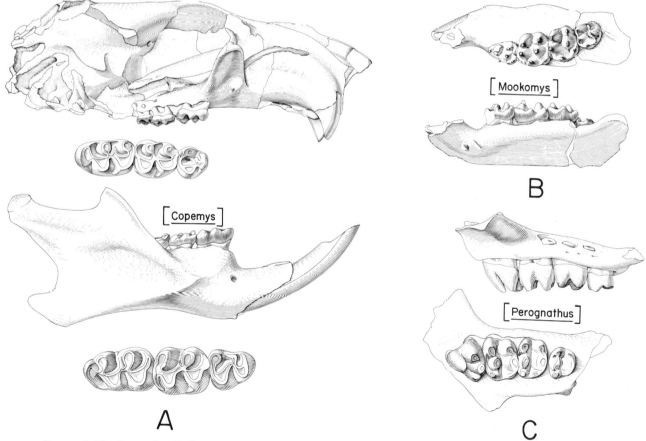

[Mookomys]

[Copemys]

[Perognathus]

A

B

C

FIGURE 6-18　*Barstovian Rodents.*

　A　Copemys longidens (Hall, 1930). Skull and lower jaw with dentitions. ×10. Figures 44 and 47 in Lindsay (1972B).

　B　Mookomys subtilis Lindsay, 1972B. Lower jaw with DP₄, M₁₋₃. ×10. Figure 21 in Lindsay (1972B)

　C　Perognathus furlongi Gazin, 1930. P⁴–M³. ×10. Figure 23 in Lindsay (1972B).
(Published with permission of the University of California Press.)

FIGURE 6-19 *Representative Barstovian Localities. (See opposite page)*

1	Alvord—California	*20*	Domengine Creek (North Coalinga)—Merychippus "zone"—California	*32*	Mollie Gulch (Lemhi Valley)—Idaho
2	Anceny—Montana			*36*	Monarch Mill Formation sites (Nevaxel)—Nevada
3	Ashville—Florida	*10*	Dome Spring (Dry Canyon district)—California		
1	Barstow District (many localities)—California	*21*	Downey—Idaho	*20*	Monocline Ridge (mostly marine)—California
4	Beatty Butte—Oregon	*2*	Dry Creek—Montana	*37*	Newport North (mostly marine)—Oregon
5	Bert Creek sites—Montana	*22*	El Gramal (Oaxaca)—Mexico	*17*	Niobrara River sites—Nebraska
6	Bijou Hills district (Springer)—South Dakota	*23*	Ellison Creek—Nevada		
7	Boyd County—Nebraska	*24*	Ely Southwest—Nevada	*17*	Norden Bridge district—Nebraska
8	Burkeville—Texas	*25*	Eubanks—Colorado	*20*	North Coalinga (Domengine Creek)—California
9	Cache Peak—California	*26*	Flint Creek sites—Montana		
10	Caliente Formation sites—California	*27*	Frank Ranch sites—Montana	*27*	Old Windmill sites—Montana
11	Calvert Formation sites (marine)—Maryland	*16*	Goodrich—Texas	*22*	Pan Am Highway II (Oaxaca)—Mexico
12	Camp Creek—Nevada	*16*	Gospel Hill—Texas	*38*	Pojoaque Member sites—New Mexico
13	Carlin District—Nevada	*11*	Greenburgh—Maryland		
14	Cedar Mountains (Stewart Valley)—Nevada	*28*	Guano Ranch—Oregon	*39*	Punchbowl Formation (in part) sites—California
2	Chalk Cliffs—Montana	*29*	Hand Hills—Alberta	*28*	Quartz Basin—Oregon
16	Cold Spring (Trinity River Pit)—Texas	*30*	Highrock Canyon—Nevada	*26*	Railroad Canyon sites—Montana
2	Confederate Gulch—Montana	*31*	Ixtapa (Chiapas)—Mexico	*40*	Red Basin I (Skull Springs)—Oregon
1	Cronise Basin—California	*17*	Joseph Jamber Farm—Nebraska	*41*	Rizzi Ranch—Nevada
17	Crookston Bridge Member sites—Nebraska	*25*	Logan County—Colorado		(also Arikareean)
18	Crooked River district—Oregon	*32*	Lemhi Valley (Mollie Gulch)—Idaho		
19	Deep River sites—Montana	*33*	Mascall Formation sites (many localities)—Oregon		
17	Devil's Gulch—Nebraska	*34*	Matatlan—Mexico		
		35	Mint Canyon Formation (lower) site—California		(continued)

M. crucensis Frick, 1937

M. warreni (Leidy, 1858E)

M. joraki Frick, 1937

M. hookwayi (Furlong, 1934)

M. crucianus Frick, 1937

M. (Submeryceros) minor Frick, 1937

M. (S.) minimus Frick, 1937

PERISSODACTYLA

  Equidae

    Hippodon Leidy, 1854C

    Anchitherium

    Hypohippus pertinax Matthew, 1918A

    H. cf. *osborni* Gidley, 1907A

    H. cf. *affinis* Leidy, 1858E

    Desmatippus integer (Matthew, 1924C)

    D. avus (Marsh, 1874B)   —Tedford et al. (1983?)

    Parahippus coloradensis Gidley, 1907A

    Merychippus perditus (Leidy, 1858E) (*Protohippus?*)

    M. sphenodus (Cope, 1889B)

    "M." isonesus (Cope, 1889B)

    "M." sejunctus (Cope, 1874P)

    "M." insignis Leidy, 1856Q   —Skinner et al. (1977)

    "M." wilsoni Quinn, 1955

    "M." eohipparion Osborn, 1918A

    "M." relictus (Cope, 1889K)

    "M." seversus (Cope, 1879D)

    "M." brevidontus (Bode, 1934B)

    "M." californicus (Bode, 1935B)

    "M." calamarius (Cope, 1875P)

    "M." stylodontus (Merriam, 1919A)

    "M." intermontanus (Merriam, 1915D)

    "Pliohippus" circulus (Quinn, 1955)

    Pliohippus campestris (Gidley, 1907A)   —Tedford et al. (1983?)

    Archaeohippus ultimus (Cope, 1886B)

    A. mourningi (Merriam, 1913G)

    A. minimus (Douglass, 1899 or 1900A?)

    "Eoequus" wilsoni (Quinn, 1955)

    "Nannippus"   —Quinn (1955)

    "Hipparion or Neohipparion"   —Quinn (1955)

    Calippus francisi (Hay, 1924A)

    C. proplacidus   —Tedford et al. (1983?)

    "Pseudhipparion"   —Quinn (1955)

    Cormohipparion goorisi MacFadden and Skinner, 1981

    Megahippus mckennai Tedford and Alf, 1962

  Tapiridae

    Tapiravus polkensis Olsen, 1960E

  Rhinocerotidae

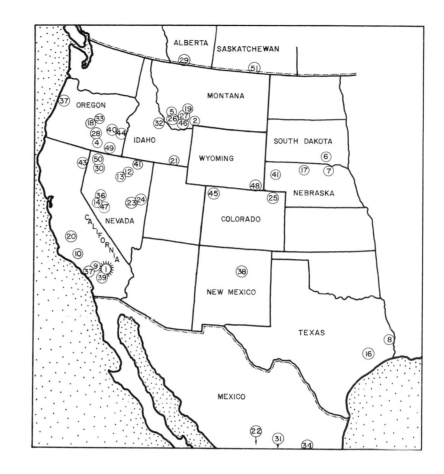

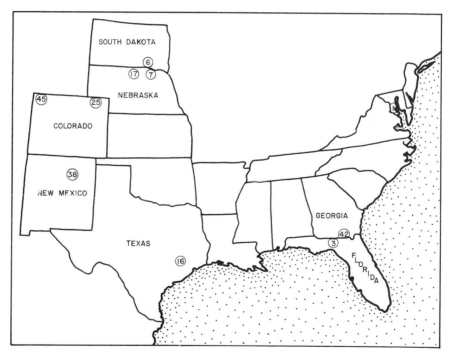

Diceratherium —Quinn (1955)

Teleoceras (Mesoceras) thomsoni Cook, 1930B

Aphelops

Peraceras

Menoceras

Chalicotheriidae

  Moropus merriami Holland and Peterson, 1914

  Tylocephalonyx —Coombs (1979)

PROBOSCIDEA

Gomphotheriidae

  Gomphotherium

Mammutidae

  Miomastodon merriami (Osborn, 1921B)

RODENTIA

Eomyidae

  Leptodontomys?

  Pseudadjidaumo stirtoni Lindsay, 1972B

  P. russelli (Storer, 1970) —Storer (1975)

  Adjidaumo quartzi Shotwell, 1967B

  Pseudotheridomys pagei Shotwell, 1967B

Aplodontidae

  Allomys stirtoni Klingener, 1968

  Liodontia alexandrae (Furlong, 1910A)

  Tardontia nevadanus Shotwell, 1970A

Sciuridae

  Tamias ateles (Hall, 1930C) —Black (1963B)

  Protospermophilus malheurensis (Gazin, 1932)

  P. angusticeps (Matthew, 1933) (age?)

  P. quatalensis (Gazin, 1930)

  P. oregonensis (Downs, 1956D)

  Spermophilus (Otospermophilus) tephrus (Gazin, 1932)

  S. (O.) primitivus (Bryant, 1945)

  Palaearctomys montanus Douglass, 1903A

  Miospermophilus —Lindsay (1972B)

  Petauristodon jamesi (Lindsay, 1972B) —Engesser (1979)

  P. minimus Lindsay, 1972B —Engesser (1979)

  P. uphami (James, 1963) —Engesser (1979)

Mylagaulidae

  Mylagaulus laevis Matthew, 1902D

Castoridae

  Monosaulax curtis (Matthew and Cook, 1909)

  M. pansus (Cope, 1874J)

  M. typicus Shotwell, 1968

  Anchitheriomys fluminus (Matthew, 1918A)

  Hystricops venustus Leidy, 1858E (age?)

Geomyidae

  Mojavemys alexandrae Lindsay, 1972B

  M. lophatus Lindsay, 1972B

  Parapliosaccomys —Lindsay (1972B)

  Jimomys labaughi Wahlert, 1976

  Prodipodomys mascallensis Downs, 1956D (heteromyid?)

  Dikkomys woodi Black, 1961C

  Lignimus montis Storer, 1970A

  Grangerimus kayi A. Wood, 1950A

Heteromyidae

Peridiomys rusticus Matthew, 1924C

P. oregonensis (Gazin, 1932)

P. borealis Storer, 1970A

Perognathus minutus James, 1963

P. furlongi Gazin, 1930

P. saskatchewanensis Storer, 1970A

P. trojectioansrum Korth, 1979–0968

Perognathoides cuyamensis A. Wood, 1936D

P. halli A. Wood, 1936D

P. eurekensis Lindsay, 1972B

P. kleinfelderi Storer, 1970A

Cupidinimus nebraskensis A. Wood, 1935C

Diprionomys agrarius A. Wood, 1935C

Mookomys altiflumenus A. Wood, 1935C

M. cf. *formicorum* A. Wood, 1935C

M. subtilis Lindsay, 1972B

Proheteromys maximus James, 1963

Cricetidae

Poamys rivicola Matthew, 1924C

Miochomys niobrarensis Hoffmeister, 1959

Copemys kelloggae (Hoffmeister, 1959)

C. pagei (Shotwell, 1976A) —Lindsay (1972B)

C. tenuis Lindsay, 1972B

C. russelli (James, 1963) —Lindsay (1972B)

C. longidens (Hall, 1930) —Lindsay (1972B)

C. barstowensis Lindsay, 1972B

C. loxodon (Cope, 1874) —Clark, Dawson, and A. Wood (1964) (age?)

Zapodidae

Plesiosminthus —Green and Holman (1977)

Macrognathomys gemmacolis Green, 1977B

Megasminthus gladiofex Green, 1977B

M. tiheni Klingener, 1966 —Green (1977B); Storer (1975)

FRIASIAN

(From manuscript by Marshall, Hoffstetter, and Pascual.)

MARSUPIALIA

Borhyaenidae

  Borhyaena?

Caenolestidae

  Abderites?

EDENTATA

Megalonychidae?

Megatheriidae

  Hapalops? Ameghino, 1887

  Eucholoeops Ameghino, 1887 (family?)

  Megathericulus Ameghino, 1904

  Promegatherium Ameghino, 1883

  Eomegatherium Kraglievich, 1926

  Pseudhapalops Pascual, 1978

  Prepotherium? Ameghino, 1891

  Diellipsodon heimi Pascual, 1978

Mylodontidae

  Neonematherium Ameghino, 1904

  Glossotheriopsis

LITOPTERNA

Proterotheriidae

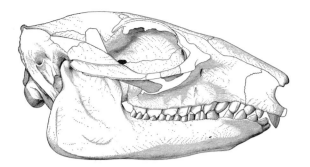
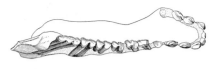
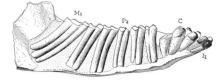

A

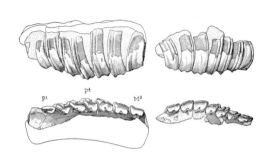
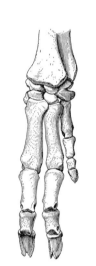

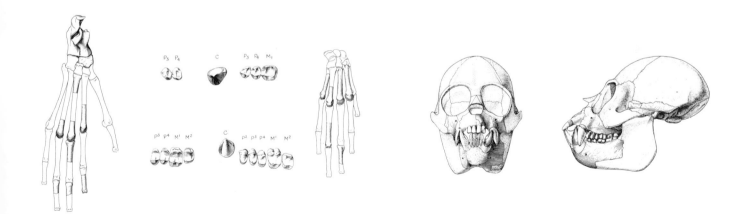

B

FIGURE 6-20 *Friasian Mammals.*

A *Miocochilius anomopodus* Stirton, 1953. *Skull, cheek teeth, feet. ×1/2. Plates 12, 13, 14, and 27 in Stirton (1953B).*

B *Cebupithecia sarmientoi* Stirton and Savage, 1951. *Reconstructed skull, teeth and feet. ×1/2. Plate 7 in Stirton (1951C).*

(Published with permission of the University of California Press.)

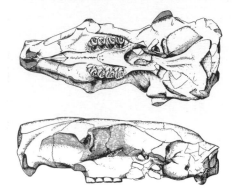

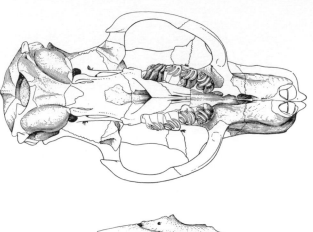

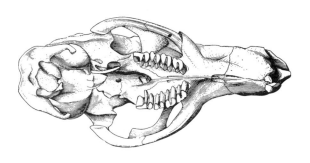

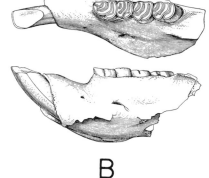

A

B

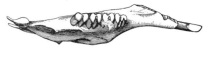

C

FIGURE 6-21   *Friasian Rodents*.

  A   <u>Scleromys schürmanni</u> Stehlin, 1940. Skull. ×3/4. Figure 3 in Fields (1957).
  B   <u>Olenopsis aequatorialis</u> (Anthony, 1922). Skull and lower jaw. ×2/3 Figures 15 and 16 in Fields (1957).
  C   <u>Prodolichotis pridiana</u> Fields, 1957. Skull and lower jaw. ×2/3. Figure 27 in Fields (1957).
(Published with permission of the University of California Press.)

Diadiaphorus Ameghino, 1887
Macraucheniidae
  Phoenixauchenia
  Theosodon Ameghino, 1887
NOTOUNGULATA
  Homalodotheriidae
    Homalodotherium Flower, 1873
  Toxodontidae
    Adinotherium Ameghino, 1887
    Hyperoxotodon?
    Nesodon Owen, 1846
    Palyeidodon obtusum Pascual, 1978
    Nesodonopsis Pascual, 1978
    Prototrigodon Kraglievich, 1930
    Stereotoxodon Ameghino, 1904
  Mesotheriidae
    Eutypotherium Roth, 1901
    Typothericulus Kraglievich, 1930

  Hegetotheriidae
    Hegetotherium Ameghino, 1887
    Pachyrukhos Ameghino, 1885
  Interatheriidae
    Interatherium Ameghino 1887
    Caenophilus Ameghino, 1903
    Epipatriarchus Ameghino, 1903
    Protypotherium Ameghino, 1887
ASTRAPOTHERIA
    Astrapotherium Burmeister, 1879
RODENTIA
  Erethizontidae
    Disteiromys
  Echimyidae
    Stichomys?
  Chinchillidae
    Prolagostomys?
  Dasyproctidae

FIGURE 6-22 *Friasian Localities.*

 1 *Carmen de Apicala—Colombia*
 2 *Cerdas (Atocha)—Bolivia*
 3 *Chubuttal—Argentina (age?)*
   Fenix—Argentina
   Laguna Blanca—Argentina
 4 *La Vela "series" site—Venezuela*
 1 *La Venta fauna (Villavieja)—Colombia*
 5 *Mauri River—Bolivia (age?)*
 6 *Mayo fauna (See Rio Mayo)—Patagonia*
   Nabón—Ecuador (age?)
 7 *Nueva Palmira—Uruguay (Friasian in part?)*
   Punta Chaparro—Uruguay (Friasian in part?)
   Punta Gorda—Uruguay (Friasian in part?)
   Quebrada Honda—Bolivia
 8 *Quehua Southeast—Bolivia*
 9 *Rio Frias—Argentina*
 6 *Rio Guenguel—Argentina*
10 *Rio Güera—Venezuela*
11 *Rio Huemules—Chile?*
 6 *Rio Mayo—Argentina*
12 *Rio Senguer—Argentina*
13 *Savonetta River—Trinidad (age?)*
14 *Suipacha—Bolivia*
13 *Trinidad (See Savonetta River.)*
 1 *Villavieja (La Venta)—Colombia*
10 *Zaraza—Venezuela*

 Neoreomys Ameghino, 1887
Dinomyidae
 Eusigmomys
 Simplimus Ameghino, 1904
Eocardiidae
 Eocardia Ameghino, 1887
 Megastus elongatus Pascual, 1978
Caviidae
 Cardiomys Ameghino, 1885

Friasian Mammalian Fauna from La Venta, Colombia

(Modified from Hirschfeld and Marshall, 1976.)
MARSUPIALIA
 Didelphidae
  Marmosa laventica Marshall, 1976
  Hondadelphys fieldsi Marshall, 1976
 Borhyaenidae
  Lycopsis longirostris Marshall, 1977
  cf. *Borhyaena*
EDENTATA
 Megalonychidae, gen. unident.
 Megatheriidae
  cf. *Hapalops*
  cf. *Prepotherium*
 Mylodontidae
  Pseudoprepotherium   —Hirschfeld and Marshall
   (1976)

  cf. *Neonematherium*
 Myrmecophagidae
  Neotamandua borealis Hirschfeld, 1976
 Dasypodidae
  Vassallia   —Robertson (1976)
 Glyptodontidae
  Asterostemma depressa Ameghino   —Krohn (1978–
   0722)
  A. cf. *venezolensis*   —Hirschfeld and Marshall (1976)
  Propalaeohoplophorus   —Hirschfeld and Marshall
   (1976)
CHIROPTERA
 Phyllostomatidae
  Notonycteris magdalenensis D. Savage, 1951
PRIMATES
 Cebidae
  Cebupithecia sarmientoi Stirton and D. Savage, 1951
  Stirtonia tatacoensis (Stirton, 1951)
  Neosaimiri fieldsi Stirton, 1951
LITOPTERNA
 Proterotheriidae, gen. indet.
 Macraucheniidae, gen. indet.
NOTOUNGULATA
 Leontiniidae, n. gen. et sp.   —J. Colwell (in prep.)
 Toxodontidae, n. gen. et sp.   —S. Johnson (in prep.)
 Hegetotheriidae?
 Isotemnidae?   —McKenna (1980)
 Interatheriidae
  Miocochilius anomopodus Stirton, 1953

257

ASTRAPOTHERIA
Xenastrapotherium kraglievichi Cabrera, 1929
Astrapotheriidae, n. gen. et sp.   —S. Johnson (in prep.)
SIRENIA
Trichechidae (Manatidae   —Hirschfeld and Marshall, 1976)
Potamosiren magdalenensis Reinhart, 1951   (syn. of *Ribodon?*)
RODENTIA
Erethizontidae
cf. *Eosteiromys*   —Fields (1957)
Caviidae
Prodolichotis pridiana Fields, 1957
Dinomyidae
Scleromys schurmanni Stehlin, 1940
S. colombianus Fields, 1957
Olenopsis aequatorialis (Anthony, 1922)   —Fields (1957)
Dasyproctidae
Neoreomys huilensis Fields, 1957
Echimyidae, gen. indet.

MIOCENE OF AUSTRALIA

Personal communications from T.H.V. and Patricia V. Rich (National Museum of Victoria and Department of Earth Sciences, Monash University) and from M.O. Woodburne (University of California at Riverside), combined with data and conclusions from Woodburne, Tedford, Plane, Turnbull, Archer, and Lundelius (1983?) have enabled us to present the following summary of nonmarine mammalian faunas representing Miocene time in Australia.

Early Miocene

Wynyardia bassiana Spencer (1900A) from the marine Fossil Bluff Sandstone of the Table Cape Group of Tasmania is the only land-mammal record referable to early Miocene. It is assigned this age on the basis of correlation of the marine strata with the marine Longfordian Stage.

Middle? Miocene

1. Lake Pinpa, Namba Formation, Ericmas fauna, South Australia (Tedford et al., 1977; T. Rich, Archer, and Tedford, 1978).
2. South Prospect B, Lake Namba, Namba Formation, Ericmas fauna, South Australia (T. and P. Rich, pers. commun., 1981).
3. Ian's Prospect, Billeroo Creek, Namba? Formation, South Australia (Archer and T. Rich, 1979).

4. Tom O's Quarry, Lake Tarkarooloo channel deposit, Tarkarooloo local fauna, South Australia (Archer and Bartholomai, 1978).
5. Lake Palankarinna district, Etadunna Formation, Ngapakaldi fauna, Tirari Desert, South Australia (Stirton, Tedford, and Woodburne, 1967B).
6. Lake Yanda, S. Australia, Namba Formation.

These localities are fossil accumulations in playa-lake and stream-channel deposits. The Ngapakaldi-Tarkarooloo-Ericmas fauna of South Australia, tentatively dated mid-Miocene, is a diversified marsupial fauna, indicating a long and unsampled prior interval of marsupial evolutionary diversification on the Australian island continent. The composite fauna is characterized by *Ngapakaldia, Pitikantia, Ektopodon,* and *Perikoala* (Woodburne et al., 1983).

Composite Middle Miocene Fauna of Australia

MONOTREMATA
Obdurodon insignis Woodburne and Tedford, 1975
MARSUPIALIA
Dasyuridae
Dasylurinja kokuminola Archer, 1982
Ankotarinja tirarensis Archer, 1976
Wakamatha tasselli Archer and T. Rich, 1979   (age?)
Peramelidae?
cf. *Perameles*   —Rich and Rich (personal commun., 1981)
Keeuna woodburnei Archer, 1976
Thylacoleonidae?   —Woodburne et al. (1983?)
Vombatidae
vombatoid (phascolarctidlike)   —Woodburne et al. (1983?)
Phascolarctidae
Perikoala palankarinica Stirton, 1957B
phascolarctid, n. gen.   —T. Rich et al. (1982)
Wynyardiidae
Namilamadeta snideri T. Rich and Archer, 1979
Ektopodontidae
Ektopodon, n. sp.   —T. and P. Rich (in prep.)
n. gen. and sp.   —T. and P. Rich (in prep.)
Phalangeridae
cf. *Trichosurus*   —T. and P. Rich (pers. commun., 1981)
cf. *Phalanger*   —T. and P. Rich (pers. commun., 1981)
form 1   —T. and P. Rich (pers. commun., 1981)
form 2   —T. and P. Rich (pers. commun., 1981)
Petauridae
petaurine   —T. and P. Rich (pers. commun., 1981)
pseudocheirines, 3 species   —Woodburne et al. (1983?)
Burramyidae   —Woodburne et al. (1983?)
cf. *Cercartetus*   —T. Rich et al. (1982)
Macropodidae
cf. *Setonix*   —T. and P. Rich (pers. commun., 1981)
macropodine, group 1   —Woodburne et al. (1983?)
macropodine, group 2   —Woodburne et al. (1983?)
cf. *Aepyrimnus rufescens*   —T. Rich et al. (1982)

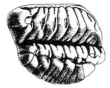

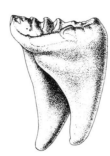

FIGURE 6-23 *Middle to Late Miocene Mammal in Australia.*
Ektopodon serratus Stirton, Tedford and Woodburne, 1967B. Lower molar. ×3⅔. Figure 7 in Stirton, Tedford, and Woodburne (1967B).
(Published with permission of the South Australian Museum, Adelaide.)

Potoroidae
cf. *Bettongia*   —T. and P. Rich (pers. commun., 1981)
Diprotodontidae
Raemeotherium yatkolai T. Rich, Archer, and Tedford, 1978
Ngapakaldia tedfordi Stirton, 1967A
N. bonythoni Stirton, 1967A
Pitikantia dailyi Stirton, 1967A
CHIROPTERA
Hipposideridae
cf. *Rhinonicteris*   —Woodburne et al. (1983?)
CETACEA
Eurhinodelphidae   —T. Rich et al. (1982)

Middle to Late Miocene

A second set of local faunas, tentatively assigned a middle-to-late Miocene age by Woodburne et al. (1983?), comprises:

1. Kutjamarpu, Wipajiri Formation, Lake Eyre Basin, South Australia (younger).
2. Kangaroo Well, Northern Territory (older).
3. Riversleigh, Carl Creek Limestone, Queensland (oldest).
4. Bullock Creek, Camfield Beds, Northern Territory (younger).
5. Batesford Quarry, Batesford Limestone, Victoria
6. Nooralccba, Queensland

The interval represented by this set of local faunas is characterized, according to Woodburne et al. (1983?), by

Neohelos, Wakaleo, Bematherium, Ektopodon, Litokoala, and *Rhizophascolonus.* The composite fauna is shown in the following list.

Composite Middle-to-Late Miocene Fauna of Australia

MARSUPIALIA
Dasyuridae
Ankotarinja   —T. Rich et al. (1982)
cf. *Antechinus*   —T. Rich et al. (1982)

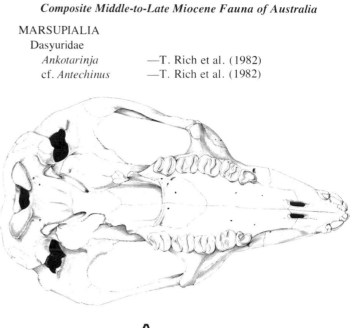

A

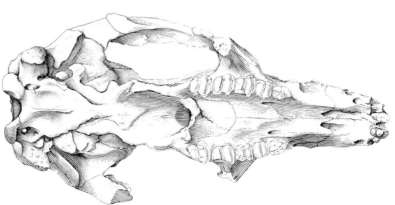

B

FIGURE 6-24 *Middle and Late Miocene Mammals of Australia.*
A Ngapakaldia tedfordi Stirton, 1967A. Skull. ×⅜. Figure 3 in Stirton (1967A).
B Kolopsis torus Woodburne, 1967A. Skull. ×¼. Figure 30 in Woodburne (1967B).
(Published with permission of the Australian Bureau of Mineral Resources.)

Peramelidae
  cf. *Peroryctes*   —T. and P. Rich (pers. commun., 1981)
  cf. *Perameles*   —T. Rich et al. (1982)
  cf. *Echymipera*   —T. Rich et al. (1982)
  cf. *Isoodon*   —T. Rich et al. (1982)
Phalangeridae   —T. Rich et al. (1982)
Thylacoleonidae
  Wakaleo oldfieldi Clemens and Plane, 1974
  W. vanderleueri Clemens and Plane, 1974
Vombatidae
  Rhizophascolonus crowcrofti Stirton, Tedford, and Woodburne, 1967B
  hypsodont wombats   —Woodburne et al. (1983?)
Phascolarctidae
  Litokoala kutjamarpensis Stirton, Tedford, and Woodburne, 1967B
Ektopodontidae
  Ektopodon serratus Stirton, Tedford, and Woodburne, 1967B
Petauridae   —Woodburne et al. (1983?)
Burramyidae   —Woodburne et al. (1983?)
Macropodidae
  Potoroinae, highly specialized   —Woodburne et al. (1983?)
  Dorcopsis-like   —Woodburne et al. (1983?)
  cf. *Dorcopsoides*   —Woodburne et al. (1983?)
  macropodinae, n. gen.   —Woodburne et al. (1983?)
Macropododoid, family?
  Wabularoo naughtoni Archer, 1979
Diprotodontidae
  Bematherium angulum Tedford, 1967A
  Neohelos tirarensis Stirton, 1967B
  N., n. sp.   —Woodburne et al. (1983?); Plane (in prep.)
  Palorchestine   —Woodburne et al. (1983?)
  Zygomaturine   —Woodburne et al. (1983?)
CHIROPTERA
  Hipposideridae   —T. Rich et al. (1982)
  Megadermatidae   —T. Rich et al. (1982)

Late Miocene

Alcoota (older?), in Northern Territory, and Beaumaris (younger?), in the Otway Basin of Victoria, are assigned as approximate Late Miocene by Woodburne et al. (1983?). Characterization is on the basis of the earliest-known *Kolopsis*, *Plaisiodon*, *Palorchestes*, and *Dorcopsoides*, coupled with absence of grazing macropodines and absence of archaic taxa (Woodburne et al., 1983?).

Composite Late Miocene Mammalian Fauna of Australia

MARSUPIALIA
Thylacinidae
  Thylacinus potens Woodburne, 1967B
Thylacoleonidae
  Wakaleo, alcootensis Archer and Rich, 1982
Vombatidae   —Woodburne et al. (1983?)

Petauridae
  Pseudocheirops n. sp.   T. Rich et al. (1982)
Macropodidae
  Hadronomas puckridgi Woodburne, 1967B
  Dorcopsoides fossilis Woodburne, 1967B
  protemnodont?   —Woodburne (1967B)
  macropodid, large   —Woodburne (1967B)
  macropodid, small   —Woodburne (1967B)
Diprotodontidae
  Pyramios alcootense Woodburne, 1967A
  Zygomaturus gilli Stirton, 1967C
  Kolopsis torus Woodburne, 1967A
  Plaisiodon centralis Woodburne, 1967A
  Palorchestes painei Woodburne, 1967B

VALLESIAN

Vallesian is based on species from Can Llobateres and correlatives (zone MN 9 of Mein, 1975) and from Masio del Barbo and correlatives (zone MN 10 of Mein). Mein's data comprise:

Neogene Mammal Zone 10

Localities: Oued Zra, Masia del Barbo, Azambujeira (upper), Villadecaballs, Soblay, Montredon, Vösendorf, Csákvár, Sebastopol, Ialovena, Varnitsa, Kastellios, Akin.

Characteristic forms of evolving lineages: *Progonomys hispanicus*, *P. lavocati*, *Pliopetaurista bressana*, *Eliomys hartenbergeri*, *Rotundomys bressanus*, *R. montisrotundi*, *Hipparion depereti*.

Associations: *Schizochoerus* + *Conohyus*; *Aceratherium simorrense* + *Pliohyrax rossignoli*.

Appearance: *Kowalskia*, *Pliopetaurista*, *Schizochoerus*, *Birgerbohlinia*, *Samotherium*, *Simocyon*, *Ictitherium*.

Neogene Mammal Zone 9

Localities: Bou Hanifia, Beglia, Can Llobateres, Pedregueras, Nombrevilla, St. Jean de Bournay, Höwenegg, Eppelsheim, Gaiselberg, Drassburg, Rudabanya, Lapouchna, Braila, Kalfa, Esme-Akçaköy.

Characteristic forms of evolving lineages: *Ruscinomys thaleri*, *Eumyarion leemani*, *Rotundomys sabadellensis*, *R. hartenbergeri*, *Chalicotherium goldfussi*.

Associations: *Hipparion* + *Eurolagus*; *Aceratherium tetradactylum* + *Dicerorhinus schleiermacheri*; *Dicerorhinus sansaniensis* + *Aceratherium incisivum*.

Appearance: *Progonomys*, *Rotundomys*, *Muscardinus*, *Machairodus*, *Indarctos*, *Hipparion*, *Microstonyx*, *Miotragoceros*, *Amphiprox*, *Diceros*.

Our composite Vallesian mammalian fauna combines Mein's species with the species from the localities Mein listed for MN9 and 10. It also includes faunas assigned to the Sarmatian Stage of Paratethys (Steininger, Rögl, and Martini, 1976).

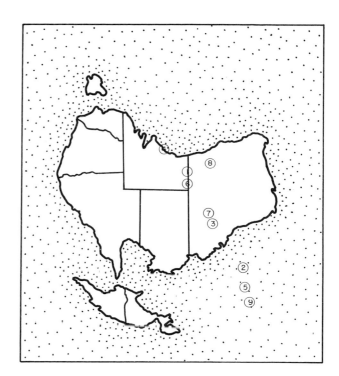

Composite Vallesian Mammalian Fauna

LAGOMORPHA
  Ochotonidae
    Eurolagus fontannesi (Depéret, 1887)   —Tobien (1974)
    Prolagus crusafonti Lopez-Martinez, 1975
    Ochotona   —Nicolas (1978)
CARNIVORA
  Amphicyonidae
    Amphicyon pyrenaicus Depéret and Rerolle, 1885, 1927
  Ursidae
    Indarctos arctoides (Depéret, 1895)   —Nicolas (1978)
    Dinocyon   —Mitchell and Tedford (1973)
    Simocyon (=''*Metarctos*'') *diaphorus* (Kaup, 1832)
    S. batalleri (Viret, 1929B)
    Ursavus depereti Schlosser, 1902   (Turolian?)
  Mustelidae
    Plesiomeles cajali Viret and Crusafont, 1955
    Eomellivora hungarica Kretzoi, 1930   —Petter (1967A)
    Limnonyx sinerizi Crusafont, 1950C
    Lutra pontica (Nordmann, 1858)   —Nicolas (1978)
    Martes melibulla Petter, 1962A
    M. aff. andersonii Schlosser, 1924   —Petter (1967A)
    M. basilii Petter, 1964
    Marcetia santigae Petter, 1967A
    Ischyrictis   —Petter (1967A)
    Sabadellictis crusafonti Petter, 1962A
    Circamustela dechaseauxi Petter, 1967A
    Taxodon cf. *sansaniensis* Lartet, 1851
    Sivaonyxhessicus (Lydekker, 1890)   —Nicolas (1978)
    Trochictis narcisoi Petter, 1976
    Trocharion albanense Major, 1903
    Hydrictis fricki Zapfe, 1949B   (*Lutra?*)
    Hadrictis fricki Pia, 1940
    Mesomephitis medius (Petter, 1963B)   —Petter (1967A)
    Promephitis maeotica Alexejew, 1916
    P. pristinidens Petter, 1962A
    Enhydriodon lluecai Villalta and Crusafont, 1946B —(includes *Sivaonyx lehmani* Crusafont and Golpe-P., 1962, acc. Repenning, 1976)
    E.? campanii Meneghini, 1863   (Turolian?)
  Viverridae
    Semigenetta ripolli Petter, 1976
  Hyaenidae
    Ictitherium prius Ozansoy, 1965
    I. cf. *orbignyi* Gaudry   —Nicolas (1978)
    Thalassictis
    Progenetta crassa (Depéret, 1892)   —Crusafont and Petter (1969)
    P. gaillardi F.-Major, 1903   —Petter (1976)
    P. montadai (Villalta and Crusafont, 1943)
    Percrocuta algeriensis (Arambourg, 1959)

261

P. senyureki Ozansoy, 1961   —Nicolas (1978)

Allohyaena kadici Kretzoi, 1938   —Ficcarelli and
   Torre (1970)

Felidae

Felis prisca Kaup, 1833   —Nicolas (1978)

F. antediluviana Kaup, 1833

Machairodus aphanistus Kaup, 1833

"*Paramachairodus*"

"*Pseudaelurus intrepidus* Leidy" (age?)   —(Pavlov, 1908); Pilgrim (1931)

INSECTIVORA

Erinaceoidea

Lanthanotherium sanmigueli Villalta and Crusafont, 1944D

Galerix   —Mein and Truc (1966)

Schizogalerix vaesendorfensis (Rabeder, 1978?)

Soricoidea

Paranourosorex excultus (Mayr and Fahlbusch, 1975)

Sorex   —Mein and Truc (1966)

Desmanella quinquecuspidata Mayr and Fahlbusch, 1975

Talpa vallesensis Villalta and Crusafont, 1944D   —
   Hutchison (1974)

CHIROPTERA

Megadermatidae

Megaderma vireti Mein, 1964

Rhinolophidae

Rhinolophus delphinensis (Gaillard, 1899)   —Mein (1964)

R. grivensis (Depéret)

Vespertilionidae

Myotis boyeri Mein, 1964

PRIMATES   —Szalay and Delson (1979)

Pliopithecidae

Pliopithecus (*Anapithecus*) *hernyaki* (Kretzoi, 1974)

Hominidae

Dryopithecus fontanni Lartet, 1856

Ramapithecus punjabicus (Pilgrim, 1910)   —(includes *R. hungaricus*, Kretzoi (1969))

Sivapithecus darwini (Abel, 1902)

S. meteai (Ozansoy, 1957)

Ouranopithecus macedoniensis   —(synonym of *S.
   meteai*, acc. to Szalay and Delson, but recognized by de
   Bonis et al. (1981))

TUBULIDENTATA

Orycteropus mauritanicus Arambourg, 1959

ARTIODACTYLA

Tayassuidae

Albanohyus pigmaeus (Depéret, 1892)

Schizochoerus vallesensis Crusafont and Villalta, 1954
   —Pickford (1978)

Suidae

Hyotherium soemmeringi Meyer, 1834   —Ginsburg (1973)

Korynochoerus palaeochoerus (Kaup, 1832)   —
   Ginsburg (1973)

Microstonyx antiquus Kaup, 1832   —Ginsburg (1973)

M. major (Pomel, 1842)   —Ginsburg (1973)

M. choeroides (Pomel, 1848)   —Hünermann (1969)

Conohyus simorrensis (Lartet)

Listriodon splendens Meyer, 1846

Nyanzachoerus devauxi (Arambourg, 1968)

Anthracotheriidae

Merycopotamus anisae Black, 1972A

Tragulidae

Dorcatherium jourdani Depéret, 1887

Cervoidea

Euprox

Micromeryx

Capreolus?

Giraffidae

Birgerbohlinia

Helladotherium

Samotherium maeoticum Korotkevich, 1978

Decennatherium   —Crusafont (1949)

Bohlinia speciosa (Roth and Wagner, 1854)   —Nicolas (1978)

Palaeotragus germaini Arambourg, 1958

P. coelophrys (Rodler and Weithofer, 1890)   —Geraads (1978)

P. rouenii Gaudry, 1867   —Geraads (1978)

Bovidae

Prostrepsiceros

Miotragocerus

Damalavus boroccoi Arambourg, 1959

Gazella pregaudryi Arambourg, 1959

G. stehlini Thenius, 1951

G. aff. *pilgrimi* Bohlin, 1935

Helicotragus rotundicornis (Weithofer, 1888)   —
   Nicolas (1978)

Tragocerus cf. *amalthea* (Roth and Wagner, 1854)
   —Nicolas (1978)

T. leskewitschi Borisyak, 1914

PERISSODACTYLA

Equidae

"*Hipparion*" *koenigswaldi* (Sondaar, 1961)   —Bernor, Woodburne, and Van Couvering (1980)

"*H.*" *africanum* (Arambourg, 1959)   —Bernor,
   Woodburne, and Van Couvering (1980)

"*H.*" *catalaunicum* (Pirlot, 1956)   —Bernor,
   Woodburne, and Van Couvering (1980)

"*H.*" (*Hipparion primigenium* Meyer, 1829, and *H. gracile*
   Kaup, 1835)   —Bernor, Woodburne, and Van
   Couvering (1980)

H. moldavicum Gromova, 1955

H. eldaricum Gabunia, 1959

Cormohipparion?

Rhinocerotidae

Aceratherium incisivum Kaup, 1832?   —Ginsburg
   (1974)

A. simorrense (Lartet, 1851)   —Santafe-Llopis (1978)

Chilotherium cf. *kowalevskii* Pavlov, 1913

Diceros neumayri Osborn, 1900

Dicerorhinus primaevus Arambourg, 1959

D. schleiermacheri (Kaup, 1832)

Chalicotheriidae

Chalicotherium goldfussi Kaup, 1833

PROBOSCIDEA

Deinotheriidae

Deinotherium giganteum Kaup, 1829

Gomphotheriidae

Gomphotherium angustidens (Cuvier, 1806)   —Tobien (1980)

Choerolophodon pentelici (Gaudry and Lartet, 1856)
   —Tobien (1980B)

Tetralophodon longirostris (Kaup, 1832)   —Tassy
  (1977); Jaeger (1977B)
Synconolophus serridentinoides Viret and Yalcinlar
  —Nicolas (1978)
Stegotetrabelodon gigantorostris (Klahn, 1922)   —
  Tobien (1980)
Mammutidae
  Zygolophodon turicensis (Schinz, 1833)   —Tobien
    (1980)
RODENTIA
  Eomyidae
    Eomyops catalaunicus (Hartenberger, 1966)   —En-
      gesser (1979)
  Sciuridae
    Albanensia aff. *grimmi* (Black, 1966)   —Mein
      (1970A)
    Pliopetaurista bressana Mein, 1970A
    Miopetaurista crusafonti (Mein, 1970A)

Heteroxerus   —Van de Weerd and Daams (1979)
Spermophilinus bredai Meyer, 1848
Atlantoxerus   —Van de Weerd and Daams (1979)
  marmotine, gen. indet.
Castoridae
  Steneofiber jaegeri Kaup, 1839
  Palaeomys castoroides Kaup, 1832
  Trogontherium minutum (Meyer, 1838)
Cricetidae
  Fahlbuschia (*Hispanomys*) *nombrevillae* Freudenthal, 1966
  F. (*H.*) *lustanicus* Schaub, 1925
  F. (*H.*) *aragonensis* Freudenthal, 1966
  F. (*Pararuscinomys*) *lavocati* Freudenthal, 1966
  F. (*Ruscinomys*) *thaleri* Hartenberger, 1966A
  Fahlbuschia crusafonti Agusti, 1978

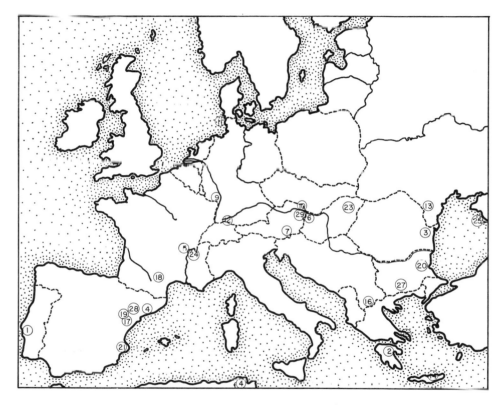

FIGURE 6-26   *Vallesian Localities. (We are listing here and plotting on the accompanying map only a few of the published Vallesian localities. See glossary at end of this chapter for a complete listing.)*

1	*Azambujeira (upper)—*	*12*	*Höwenegg—Germany*	*23*	*Rudabańya—Hungary*
	Portugal	*13*	*Kalfa—Moldavia*	*22*	*Sabadell—Spain*
2	*Biodrak (Boutia?)—Greece*	*14*	*Kastellios (Hill)—Crete*	*24*	*St.-Jean-de-Bournay—*
3	*Braila—Romania*	*15*	*Katheni (Evvoia)—Greece*		*France*
4	*Can Llobateres—Spain*	*16*	*Lefkon 1 (Macedonia)—*	*25*	*Sevastopol—Crimea, USSR*
5	*Croix-Rousse (Lyon)—*		*Yugoslavia*	*14*	*Sitia*
	France	*17*	*Masia del Barbo II—Spain*	*14*	*Sitia Southeast—Crete*
6	*Csakvar—Hungary*	*18*	*Montredon—France*	*26*	*Soblay—France*
7	*Drassburg—Austria*	*19*	*Nombrevilla—Spain*	*27*	*Varnitsa—Bulgaria*
9	*Eppelsheim—Germany*	*20*	*Nesebr—Bulgaria*	*28*	*Villadecaballs (Can*
10	*Gaiselberg (Zisterdorf)—*	*21*	*Pedregueras (Zaragosa)—*		*Bayona)—Spain*
	Austria		*Spain*	*29*	*Vösendorf (Vienna)—*
11	*Glimmerton (Sylt)—North*	*4*	*Polinya (Barcelona)—Spain*		*Austria*
	Sea				

Byzantinia nikosi de Bruijn, 1976
Megacricetodon debruijni Fahlbusch, 1964
M. ibericus (Schaub, 1944)
Cricetulodon sabadellensis Hartenberger, 1965
C. hartenbergeri (Freudenthal, 1967)
Microcricetus molassicus Mayr and Fahlbusch, 1975
Democricetodon crassus Freudenthal, 1969?
D. cf. *brevis* (Schaub, 1925)
Eumyarion leemani (Hartenberger, 1966)
Rotundomys bressanus Mein, 1975
R. freiriensis Antunes and Mein, 1979
R. mundi Calvo, Elizaga and Martinez, 1978–0197
R. montisrotundi (Schaub, 1944)
Kowalskia lavocati (Hugueney and Mein, 1965)
Myocricetodon sebouli Jaeger, 1977B
M. ouedi Jaeger, 1977B
M. trerki Jaeger, 1977B
spalacine
Gliridae
Muscardinus davidi Hugueney and Mein, 1965
M. crusafonti Hartenberger, 1966
M. hispanicus de Bruijn, 1966
M. vireti Hugueney and Mein, 1965
Glirulus lissiensis Hugueney and Mein, 1965
Myoglis aff. *meini* (de Bruijn, 1966A)
Pentaglis   —Hartenberger, 1966
Eomuscardinus vallesiensis Hartenberger, 1966
Tempestia hartenbergeri (de Bruijn)
Eliomys reductus Mayer, 1979
E. assimilis Mayr, 1979
Myomimus multicrestatus (de Bruijn, 1966)   —de
    Bruijn (1976)
Microdyromys chaabi Jaeger, 1977B
Muridae
Progonomys cathalai Schaub, 1938
P. hispanicus Michaux, 1971
P. cf. *woelferi* Bachmayer and R. Wilson, 1970
Anomalomyidae or Cricetidae
Anomalomys   —Van de Weerd and Daams (1979)

Late Miocene (= *Vallesian*) of Turkey

CARNIVORA
Ursidae, gen. indet.
Mustelidae
Promephitis
Hyaenidae
Ictitherium hipparionum (Gervais, 1859)   —Schmidt-
    Kittler (1976)
I. robustum
Percrocuta senyureki Ozansoy, 1961   —Schmidt-
    Kittler (1976)
Adcrocuta eximia (Roth and Wagner, 1848?)   —
    Kretzoi (1938)
Protictitherium crassum (Depéret, 1892)   —Schmidt-
    Kittler (1976)
Pachycrocuta cf. *salonica* (Andrews, 1918)
Felidae

Miomachairodus pseudailuroides Schmidt-Kittler, 1976
Machairodus aphanistus Kaup, 1833
Paramachairodus
Felis attica Wagner, 1857   —Schmidt-Kittler (1976)
PRIMATES
Hominidae
Sivapithecus meteai (Ozansoy, 1957)   (includes "*Ankarap-ithecus*," "*Graecopithecus freybergi*," and "*Ouranopi-thecus macedoniensis*")
ARTIODACTYLA
Suidae
Listriodon splendens Meyer, 1846
Sivachoerus giganteus (Falconer and Cautley, 1847)
Korynochoerus
cf. *Dicoryphochoerus* or *Microstonyx*
Sus erymanthius Roth and Wagner, 1854
Giraffidae
Palaeotragus cf. *decipiens* Bohlin, 1926
P. cf. *germaini* Arambourg, 1958
cf. *P. quadricornis* Bohlin, 1926
Samotherium
Decennatherium
Bovidae
cf. *Palaeoreas*
cf. *Orasius*
Gazella gaudryi Schlosser, 1904
G. deperdita? (Gervais, 1847)
Mesembriacus
Oioceros
Palaeoryx?
Urmiatherium polaki Rodler, 1889
cf. *Protragelaphus*
Helicotragus?
Prostrepsiceros
PERISSODACTYLA
Equidae
Hipparion matthewi Abel, 1926
H. mediterraneum Hensel, 1860
H. galacticum Ozansoy, 1965
H. gracile Kaup, 1833
Anchitherium
Rhinocerotidae
Diceros pachygnathus (Wagner, 1850)
D. neumayri Osborn, 1900
Chilotherium intermedium (Lydekker, 1884)
C. zernovii Borisyak, 1915
C. samium (Weber, 1905)
C. kowalevskii (Pavlov, 1913)
C. habereri (Schlosser, 1903)
Chalicotheriidae
Chalicotherium cf. *goldfussi* Kaup, 1833
HYRACOIDEA
Pliohyrax graecus (Gaudry, 1862)
TUBULIDENTATA
Orycteropus
PROBOSCIDEA
Deinotheriidae
Deinotherium giganteum Kaup, 1829
Gomphotheriidae
Choerolophodon pentelici (Gaudry and Lartet, 1856)
    —Tobien (1980B)

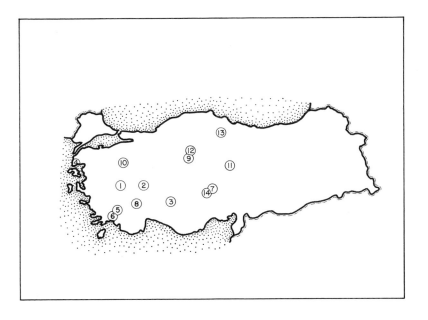

FIGURE 6-27   *Representative Late Miocene localities in Turkey.*

(=*Vallesian*)

1	Akçaköy	6	Becin	2	Kinik
2	Garkin	7	Cevril	13	Köprübaşi (age?)
3	Kayadibi	8	Denizli	8	Mahmutgazi
4	Gülpinar	9	Evcilaergillari	7	Mancusun
4	Çanakkale	9	Gökdere	14	Taşkinpaşa
		10	Harmancik	7	Yemliha
(=*Turolian*)		11	Kalinköy	9	Zivra
5	Amasya	12	Kavurca		

LATE MIOCENE (= VALLESIAN) OF EASTERN ASIA

(Ba Ho, Shihuiba, Bu Lung, Liu Ho, U.? Chinji?, Nagri, Daud Khel, Zhongxiang.)

LAGOMORPHA
Leporidae
CREODONTA
Hyaenodontidae
   cf. *Isohyaenodon*   —Pilbeam et al. (1977)
   Dissopsalis carnifex Pilgrim, 1910
   Crocuta gigantea (Schlosser, 1903)   —Colbert (1935A)
CARNIVORA
Amphicyonidae
   Amphicyon
   A. palaeindicus Lydekker, 1876   —Colbert (1935A)
   A. youngi Chen, 1981
Mustelidae
   Sivaonyx bathygnathus (Lydekker, 1884)
   Martes?   —Pilbeam et al. (1977)
   Eomellivora
Procyonidae
   Sivanasua himalayensis Pilgrim, 1932
Viverridae
   viverrine, gen. indet.
   Progenetta?   —Pilbeam et al. (1977) (Hyaenidae?)
Hyaenidae

   Percrocuta carnifex (Pilgrim, 1913)
   P. grandis Kurtén, 1957
   Ictitherium gaudryi Zdansky, 1924
Felidae
   Vinayakia nocturna Pilgrim, 1932
   Metailurus
   Epimachaerodus
   Megantereon praecox Pilgrim, 1932   —Colbert (1935A)
INSECTIVORA
Erinaceoidea
   Galerix rutlandae Munthe and West, 1980
   cf. *Echinosorex*
   Amphechinus kreuzae Munthe and West, 1980
PRIMATES
Adapidae   —Gingerich and Sahni (1979)
   Indraloris himalayensis (Pilgrim, 1932)
   Sivaladapis nagrii (Prasad, 1970)
Lorisidae? (Nagri)
Pliopithecidae
   Pliopithecus (*Krishnapithecus*) *krishnaii* Chopra and Kaul, 1979   —Ginsburg and Mein (1980)
   P. (*K.*) *posthumus* Schlosser, 1924   —Ginsburg and Mein (1980)
Hominidae   —Szalay and Delson (1979)
   Ramapithecus punjabicus (Pilgrim, 1910) (includes "*Palaeopithecus? sylvaticus*")
   Sivapithecus indicus Pilgrim, 1910

S. sivalense (Lydekker, 1879) (includes "*Dryopithecus cautleyi,*" "*Dryopithecus giganteus,*" in part, *Hylopithecus hysudricus,*" "*Sugrivapithecus salmontanus,*" and "*Ramapithecus hariensis*")

Gigantopithecus giganteus (Pilgrim, 1915)   —Szalay and Delson (1979)

TUBULIDENTATA
  Orycteropus browni Colbert, 1933B
  O. pilgrimi Colbert, 1933B

ARTIODACTYLA
  Tayassuidae
    Schizochoerus dangari (Prasad, 1970)   —Pickford (1978)
    S. gandakasensis (Pickford, 1977)
  Anthracotheriidae
    Merycopotamus or *Hemimeryx pusillus* Lydekker, 1885
    Choeromeryx Pomel, 1848   (age?)
  Tragulidae
    Dorcatherium minus Lydekker, 1876
    Dorcabune nagrii Pilgrim, 1915
  Cervoidea
    Propalaeomeryx
    Metacervulus
    Heterocemas   —Young (1937)
  Giraffidae
    Palaeotragus microdon (Koken, 1885)
    Giraffokeryx punjabiensis Pilgrim, 1910
    cf. *Sivatherium*   —Pilbeam et al. (1977)
    Samotherium   —Hamilton (1973)
    Injanatherium hazimi Heintz, Brunet and Sen, 1981
  Bovidae
    Shensispira chowi
    (antelope spp.)
    Antilospira
    Gazella
    Miotragocerus punjabicus (Pilgrim, 1910)
    Pseudotragus? sp.   —Pilbeam et al. (1977)
    boselaphine, small   —Pilbeam et al. (1977)
    Helicoportax praecox Pilgrim, 1937
    Strepsiportax glute Pilgrim, 1937   (= Astaracian?)
      U. Chinji
    Dorcadoxa porrecticornis (Lydekker, 1878)   —Pilgrim (1939A)
    cf. *Tragoreas potwaricus* Pilgrim, 1939A
    Kobikeryx atavus Pilgrim, 1939A   (age?)
    Pachyportax nagrii Pilgrim, 1939A

PERISSODACTYLA
  Equidae
    Cormohipparion theobaldi (Lydekker, 1877)   —MacFadden and Bakr (1979)
    Hipparion weihoense
    H. chiae
    H. nagriense Hussain, 1971
    Hipearian   (U.? Chinji?)   —Hussain et al. (1977)
  Tapiridae
    Tapirus
  Rhinocerotidae
    Chilotherium gracile Ringstöm, 1924

C. intermedium (Lydekker, 1884)
Dicerorhinus orientalis (Schlosser, 1921)
Aceratherium perimense Falconer and Cautley, 1847
A. blanfordi Lydekker, 1884
Gaindatherium browni Colbert, 1934B
Brachypotherium perimense (Falconer and Cautley, 1868)
  Chalicotheriidae
    Macrotherium salinum Forster-Cooper, 1922
PROBOSCIDEA
  Deinotheriidae
    Deinotherium   —Pilbeam et al. (1977)
  Gomphotheriidae
    Gomphotherium wimani (Hopwood, 1935B)   (age?)
  Stegodontidae?
RODENTIA
  Sciuridae
    Eutamias urialis Munthe, 1980
  Gliridae
    Myomimus sumbalenwalicus Munthe, 1980
  Ctenodactylidae
    Sayimys sivalensis (Hinton, 1933)
  Rhizomyidae
    Rhizomyoides pilgrimi (Hinton, 1933)   —Black (1972A)
    R. punjabiensis (Colbert, 1933F)   —Black (1972A)
    Rhizomys sivalensis Lydekker, 1878
    Kanisamys sivalensis A. Wood, 1937B
    K. indicus Wood, 1937
    K. potwarensis Flynn, 1982
    Brachyrhizomys nagrii (Hinton, 1933)
    B. micrus Flynn, 1982
    B. hehoensis Zheng, 1980
  Cricetidae
    Copemys (*Democricetodon*) sp.   —Hussain et al. (1977)
    Myocricetodon   —Hussain et al. (1977)
  Muridae
    Progonomys debruijni Jacobs, 1978
    Parapodemus   —Jacobs (1978)
    cf. "*Mastomys*" *colberti* (Lewis, 1939)   —Jacobs (1978)
    Karnimata darwini Jacobs, 1978
    Antemus cf. *chinjiensis* Jacobs, 1978
  Hystricidae
    Hystrix sivalensis Lydekker, 1878

NGOROORAN—Szalay and Delson (1979)

The Ngororan fauna is now represented only from Ngorora in Kenya, Ch'orora in Ethiopia, Upper Beglia Formation sites in Tunisia, Bou Hanifia in Algeria, and Oued Zra in Morocco.

Composite Ngororan Fauna
CARNIVORA
  Hyaenidae
    Percrocuta tobieni Crusafont and Aguirre, 1971
ARTIODACTYLA
  Cervoidea
    Climacoceras

FIGURE 6-28   (=Vallesian) of Asia.
  1 Ba Ho (Lantien)—Shaanxi
  2 Bu Lung (Biru)—Xizang (Tibet)
  3 Chinji Formation, uppermost,
     sites—Pakistan and India
     (age?)
  4 Daud Khel—Pakistan
  5 Liu Ho (Nanking)—
  3 Nagri Formation sites—
     Pakistan and India
  6 Shihuiba (Lufeng)—Yunnan

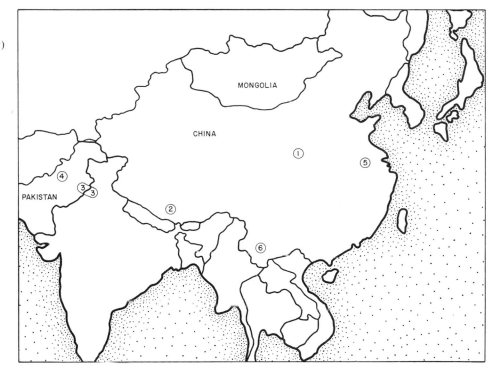

Giraffidae
  Palaeotragus
  Samotherium?
Tragulidae, 2 spp.   —Bishop and Pickford (1975)
Bovidae
  Protragocerus   —Gentry (1978)
  Pseudotragus?
  Gazella praegaudryi Arambourg, 1959
PERISSODACTYLA
  Equidae
    Hipparion
  Rhinocerotidae
    Chilotherium   —Hooijer (1971)
    Brachypotherium   —Hooijer (1971)
    Dicerorhinus   —Hooijer (1971)
RODENTIA
  Thryonomyidae
    Paraulacodus johanesi Jaeger, Michaux, and Sabatier, 1980
    Paraphiomys, 2 spp. Jaeger, Michaux, and Sabatier, (1980)
  Cricetidae
    Zramys salemi Jaeger, 1977B
    Z. haichi Jaeger and Michaux, 1973
    Z. semmensis Jaeger, 1977B
  Ctenodactylidae
    Africanomys kettarati Jaeger, 1977B

CLARENDONIAN

H.E. Wood et al. (1941) based their Clarendonian on the sample they designated Clarendon local fauna (and member?), near Clarendon, Panhandle of Texas. Since the early 1900s, various museums and institutions from the eastern and western United States have collected fossil vertebrates from this district. Workers from the Frick Laboratories of the American Museum of Natural History accumulated enormous amounts of fossil mammals from Clarendon quarries.

We follow Webb (1969) and many other workers who have considered the Clarendon fauna in recognizing that it represents the late part of the Clarendonian mammal age. And we expand the concept of Clarendonian so that it includes most of the Nebraskan faunas that Schultz, Skinner, and other workers at the University of Nebraska and at the American Museum of Natural History are assigning to the pre-type-Clarendonian and post-type-Barstovian, that is, the *Valentinian*.

Composite Clarendonian Mammalian Fauna

LAGOMORPHA
  Ochotonidae
    Hesperolagomys galbreathi Clark, Dawson, and A. Wood, 1964
  Leporidae
    Hypolagus cf. *vetus* (Kellogg, 1910A)
    H. mohavensis (Stock and Furlong, 1926)
    H. fontinalis Dawson, 1958
    H. apachensis Gazin, 1930
CARNIVORA
  Amphicyonidae
    Ischyrocyon gidleyi (Matthew, 1902A)
    I. hyaenodus Matthew, 1904C
    Hadrocyon mohavensis Stock and Furlong, 1926
      (*Ischyrocyon?*)
    Pliocyon   —Voorhies (1969B)
    Pseudocyon   —Tedford et al. (1983?) (late Barstovian?)

267

Canidae

    Strobodon stirtoni Webb, 1969C

    Cynarctus fortidens Hall and Dalquest, 1962

    cynarctus   —Voorhies (1969B)

    Epicyon littoralis (VanderHoof, 1931)   —Baskin
      (1980)

    E. saevus (Leidy, 1858E)

    E. haydeni Leidy, 1858E

    E. diabloensis (Richey, 1938)

    E. validus (Matthew and Cook, 1909A)

    E. mortifer (Cook, 1914B)

    E. aphobus (Merriam, 1919A)

    E. inflatus VanderHoof and Gregory, 1940

    Aelurodon taxoides Hatcher, 1894C   —(=*Prohy-*
      aena, Tedford et al., 1983?)

    A. marshi (Hay, 1899C)

    Carpocyon   —Tedford & Taylor (in prep.)

    Proturocyon macdonaldi Tedford and Taylor (in prep.)

    Leptocyon vafer (Leidy, 1858E)

    Vulpes   —Shotwell and D. Russell (1963)

    Tomarctus euthos (McGrew, 1935)

    T. cf. *brevirostris* Cope, 1873T   —Messenger and
      Messenger (1977)

    T. robustus Green, 1948A

Ursidae

    Hemicyon   —Webb (1969C); Voorhies (1969B)

    H. (*Plithocyon*)   —Tedford et al. (1983?)

Procyonidae

    Bassariscus parvus Hall, 1927A   —Voorhies (1969B)

    n. gen. A   —Webb, MacFadden, and Baskin (1981)

    n. gen. B   —Webb, MacFadden, and Baskin (1981)

Mustelidae

    "Lutra" pristina (Matthew, 1904C)   —Skinner et al.
      (1977)

    Martes stirtoni R.L. Wilson, 1968

    M. campestris Gregory, 1942B

    M. (*Plionictis*) *ogygia* Matthew, 1901B   —R.L.
      Wilson (1968)

    Martinogale nambiana (Cope, 1874I) (age?)   —
      Hall (1930E)

    Eomellivora   —Shotwell and D. Russell (1963)

    Sthenictis junturensis Shotwell and D. Russell (1963)

    S. lacota (Matthew, 1904C)

    Brachypsalis pristinus (Matthew and Gidley, 1906)

    Brachypsaloides modicus (Matthew, 1918A)   —
      Webb (1969C)

    Leptarctus primus Leidy, 1856Q

    Leptarctus   —Webb, MacFadden, and Baskin (1981)

    Beckia grangerensis Bryant, 1968   —[prob. *Ischyr-*
      ictis acc. to Tedford et al. (1983?)]

    Potamotherium   —Tedford et al. (1983?)

    Mionictis   —Tedford et al. (1983?)

Felidae

    Pseudaelurus cf. *marshi* (Thorpe, 1922E)   —Mes-
      senger and Messenger (1977)

    P. pedionomus Macdonald, 1948C

    Nimravides thinobates (Macdonald, 1948B)

    N. galiani Baskin, 1981

Barbourofelis whitfordi (Barbour and Cook, 1914B)

    B. morrisi Schultz, Schultz, and Martin, 1970A

    B. osborni (Merriam, 1919A)

    B. lovei Baskin, 1981

    Sansanosmilus?

CHIROPTERA

    Vespertilionidae

    Myotis   —R.L. Wilson (1968)

    Phyllostomatidae   —James (1963)

INSECTIVORA

    Erinaceoidea

    Metechinus nevadensis Matthew, 1929   —T. Rich
      (1981)

    "Lanthanotherium" sawini (James, 1963)   —En-
      gesser (1979)

    "L." dehmi (James, 1963)   —Engesser (1979)

    Untermannerix copiosus T. Rich, 1981

    Soricoidea

    Mystipterus vespertilio Hall, 1930B

    Scalopoides spp.   —Shotwell (1970A)

    Domninoides mimicus R.L. Wilson, 1968

    D. riparensis Green, 1956

    Scapanus (*Xeroscapheus*) *shultzi* Tedford, 1961

    S. (*X.*) *proceridens* Hutchison, 1968A

    Scalopus

    cf. *Gaillardia*

    Plesiosorex   —Shotwell (1970A)

    Limnoecus tricuspis Stirton, 1930C

    Alluvisorex chasseae (Tedford, 1961)

    Petenyia concisa (R.L. Wilson, 1968)   —Hibbard and
      Jammot (1971)

    Hesperosorex?   —Shotwell and D. Russell (1963)

    Tregosorex holmani Hibbard and Jammot, 1971

    Anchiblarinella wakeeneyensis Hibbard and Jammot, 1971

    Anourosorex?

ARTIODACTYLA

    Tayassuidae

    Prosthennops (=*"Hesperopithecus"*) *haroldcooki* (Osborn,
      1922A)

    P. niobrarensis Colbert, 1935H

    P. serus (Cope, 1878C)

    Dyseohyus? sp.   —Woodburne (1969B)

    Merycoidodontidae

    Ustatochoerus profectus (Matthew and Cook, 1909)

    U. skinneri Schultz and Falkenbach, 1941

    U. major (Leidy, 1858E)

    U. californicus (Merriam, 1919A)

    U. compressidens (Douglass, 1901A) (age?)

    U. medius (Leidy, 1858E)

    Camelidae

    Procamelus occidentalis Leidy, 1858E

    P. grandis Gregory, 1942

    P. robustus (Leidy, 1858E)

    P. leptognathus Cope, 1893A

    Protolabis coartatus (Stirton, 1929)   —Honey and
      Taylor (1978); includes *P. notiochorinos* Patton (1969B)

    P. heterodontus (Cope, 1873R)

    Megatylopus primaevus Patton, 1969B

    M. major (Leidy, 1886B)

    Aepycamelus bradyi Macdonald, 1956A

    Homocamelus (age?)   —Tedford (1982?)

Pliauchenia magnifontis Gregory, 1939 (ref. to *Hemiau-chenia?*)

"*Hemiauchenia*" cf. *minima* (Leidy, 1886) —Webb et al. (1981)

Nothotylopus camptognathus Patton, 1969B

Michenia yavapaiensis Honey and Taylor, 1978

Miolabis (age?)

Alforjas

Leptomerycidae

Pseudoparablastomeryx (age?)

Gelocidae —Tedford et al. (1983?)

Pseudoceras

Cervoidea

"*Yumaceras*" (*Pediomeryx*) —Webb et al. (1981)

Cranioceras clarendonensis Frick, 1937

C. skinneri Frick, 1937

C. teres Frick, 1937 (age?)

C. mefferdi Frick, 1937 (age?)

Blastomeryx elegans Matthew and Cook, 1909 — (blastomerycines assigned to Moschidae by Tedford et al. (1983?))

B. mefferdi Frick, 1937

Longirostromeryx clarendonensis Frick, 1937

L. wellsi (Matthew, 1904)

L. merriami Frick, 1937

L. serpentis Frick, 1937

L. novomexicanus Frick, 1937

L.? blicki Frick, 1937

L.? vigoratus Frick, 1937

Parablastomeryx gregorii Frick, 1937

Protoceratidae

Synthetoceras tricornatus Stirton, 1932E

Paratoceras macadamsi Frick, 1937

FIGURE 6-29 *Clarendonian Mammals.*

A *Ischyrocyon gidleyi* (*Matthew, 1902A*). *Palate.* ×3/8. *Figure 9 in Webb (1969).*

B *Ischyrocyon hyaenodus* Mat-thew, *1904C. Mandible.* ×1/6. *Figures 10 and 11 in Gregory (1942).*

C *Leptocyon vafer* (*Leidy, 1858E*). *Lower jaw.* ×1. *Figure 7 in Webb (1969).*

D *Tomarctus euthos* (*McGrew, 1935*). *Lower jaw.* ×1/2. *Figure 6 in Webb (1969).*

E *Strobodon stirtoni* Webb, 1969. *Maxillary with* P^2–M^1. ×1/2. *Figure 8 in Webb (1969).*

(*Published with permission of the University of California Press.*)

C

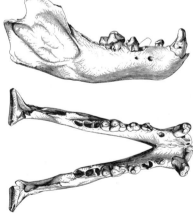

A

D

B

E

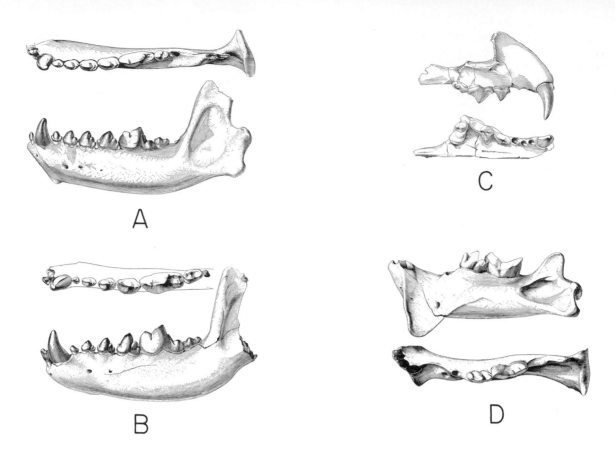

FIGURE 6-30   *Clarendonian Mammals.*

A   *Aelurodon taxoides* Hatcher, *1894C. Lower jaw.* ×*1/3. Figures 8 and 9 in Gregory (1942).*

B   *Epicyon inflatus VanderHoof and Gregory, 1940. Lower Jaw.* ×*4/10. Figure 7 in Gregory (1942).*

C   *Martes campestris Gregory, 1942B. Maxillary with C, P³–M¹.* ×*1/2. Figure 12 in Gregory (1942).*

D   *Barbourofelis whitfordi (Barbour and Cook, 1914B). Lower jaw.* ×*3/8. Figure 13 in Gregory (1942).*

(Published with permission of the University of California Press.)

Antilocapridae

Ramoceros ramosus (Cope, 1874I)   —Frick (1937)

R. hitchcockensis Frick, 1937   (age?)

R. marthae Frick, 1937

R. kansanus Frick, 1937   (age?)

R. palmatus Frick, 1937

Merycodus furcatus (Leidy, 1869A)   (*Cosoryx* is the name preferred by the Frick–American Museum group)

M. cerroensis Frick, 1937

M. ilfonsensis Frick, 1937

M. furlongi (Frick, 1937)   —(*Paracosoryx,* acc. to Tedford et al. (1983?))

Meryceros major Frick, 1937

M. crucensis Frick, 1937

M. nenzelensis Frick, 1937

M. warreni (Leidy, 1858)   (age?)

Plioceros blicki Frick, 1937   (age?)

P. floblairi Frick, 1937

P. dehlini Frick, 1937   (age?)

Proantilocapra platycornea Barbour and Schultz, 1934A   (age?)

Texoceros?

PERISSODACTYLA

Equidae

Merychippus cf. *insignis* (Leidy, 1856Q)

"M." calamarius (Cope, 1875P)   (age?)

Hypohippus affinis Leidy, 1858E

Megahippus matthewi (Barbour, 1914A)

Pliohippus supremus Leidy, 1869A

P. tantalus Merriam, 1913H

P. pernix Marsh, 1874B

Pliohippus or *Protohippus tehonensis* Merriam, 1915D

Dinohippus subvenus Quinn, 1955

D. fossulatus (Cope, 1893A)

D.? leardi (Drescher, 1941)

D.? pachyops (Cope, 1892W)

"Hippotigris" sellardsi (Quinn, 1955)

"Equus" laparensis (Quinn, 1955)

Cormohipparion occidentale (Leidy, 1856D) Skinner et al. (1977)

C. coloradense (Osborn, 1918A)

Hipparion plicatile Leidy, 1887A

H. forcei Richey, 1948

H. condoni Merriam, 1915A

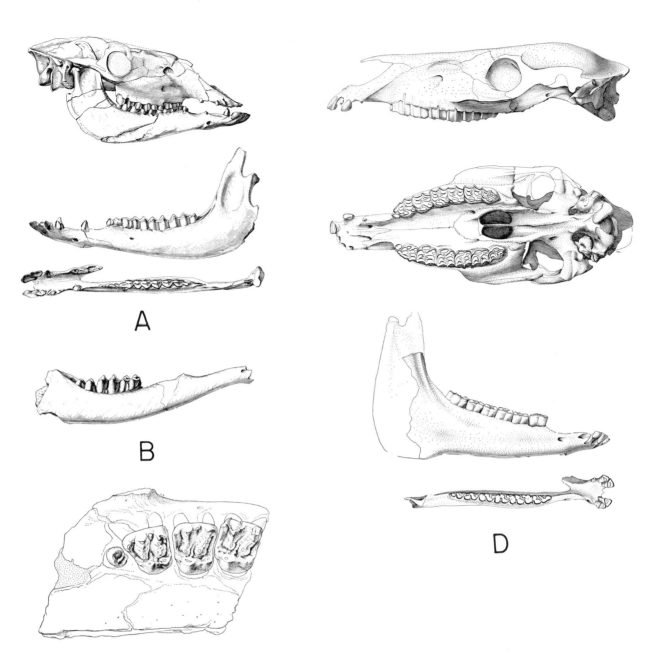

FIGURE 6-31   *Clarendonian Mammals.*

   A *Procamelus grandis* Gregory, 1942. Skull and lower jaw. Skull is ×1/8; lower jaw is ×1/6. Figures 19 and 22
      in Gregory (1942).

   B *Longirostromeryx wellsi* (Matthew, 1904). Lower jaw. ×1/2. Figure 37 in Gregory (1942).

   C *Petauristodon mathewsi* (James, 1963). P³–M². ×5½. Figures 36 and 37 in James (1963).

   D *Pseudhipparion retrusum* (Cope, 1889B). Skull and lower jaw. ×1/4. Figures 30 and 32 in Webb (1969).

(Published with permission of the University of California Press.)

H. tehonense (Merriam, 1916C)	*Calippus regulus* Johnston, 1937D
H. mohavense Merriam, 1913H	*C. anatinus* Quinn, 1955
H. trampasense Edwards, 1982	*C. optimus* Quinn, 1955
"Nannippus" ingenuum (Leidy, 1885A)	*C. placidus* (Leidy, 1869A)
Neohipparion cf. *leptode* (Merriam, 1915A) —Webb	*Astrohippus curtivallis* Quinn, 1955
et al. (1981)	*A. martini* (Hesse, 1936D)
N. cf. *minor* Sellards, 1916B —Webb et al. (1981)	*Pseudhipparion retrusum* (Cope, 1889B)

(continued)

P. gratum (Leidy, 1869A) —(referred to *Griphippus* by Tedford, pers. commun., 1982)

Tapiridae

Tapiravus polkensis Olsen, 1960E (age?)

Tapirus? johnsoni Schultz, Martin and Corner, 1975

Tapirus simpsoni Schultz, Martin and Corner, 1975 —Webb et al. (1981)

Rhinocerotidae

Teleoceras proterus (Leidy, 1885A)

T. fossiger (Cope, 1878H)

T. major Hatcher, 1894A

Aphelops malacorhinus Cope, 1878X

Peraceras superciliosus Cope, 1880K

P. troxelli Matthew, 1918A

Diceratherium jamberi Tanner, 1977

PROBOSCIDEA

Gomphotheriidae

Gomphotherium simpsoni (Stirton, 1939D)

G. nebrascensis (Osborn, 1924H)?

G. productum (Cope, 1874J)

Platybelodon cf. *barnumbrowni* Barbour, 1929

Megabelodon minor Mawby, 1968B

Eubelodon morrilli Barbour, 1914B

Tetralophodon fricki (Osborn, 1934A)

Amebelodon barbourensis (Frick, 1933) —Webb et al. (1981)

Mammutidae

Miomastodon

Pliomastodon furlongi Shotwell and D. Russell (1963)

RODENTIA

Eomyidae

Leptodontomys —Shotwell (1967B)

Aplodontidae

Tardontia occidentale (Macdonald, 1956A) —Shotwell (1970A)

Liodontia

Sciuridae

Protospermophilus quatalensis (Gazin, 1930)

Spermophilus (*Otospermophilus*) *matthewi* Black, 1963B —Green (1971)

S. junturensis (Shotwell and D. Russell, 1963)

S. (*O.*) *wilsoni* (Shotwell, 1956)

Tamias ateles (Hall, 1930C) —Black (1963B)

Eutamias —Shotwell (1970A)

Marmota vetus (age?) —Black (1963B)

Ammospermophilus fossilis James, 1963

Petauristodon jamesi (Lindsay, 1972B) —Engesser (1979)

P. mathewsi (James, 1963) —Engesser (1979)

Mylagaulidae

Mylagaulus elassos Baskin, 1980

M. monodon Cope, 1881E

M. cf. *laevis* Matthew, 1902D —G. Schultz (1977)

Epigaulus minor Hibbard and Phillis, 1945

Castoridae

Eucastor lecontei (Merriam, 1910)

E. cf. *tortus* Leidy, 1858E

E. dividerus Stirton, 1935

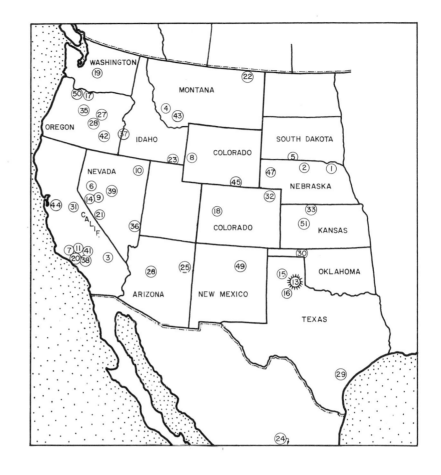

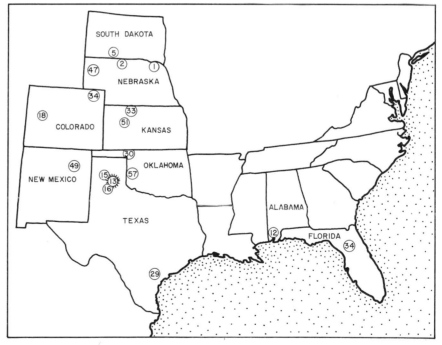

E. malheurensis Shotwell and D. Russell, 1963

E. planus Stirton, 1935

E. phillisi R.L. Wilson, 1968

Monosaulax pansus (Cope, 1874J) —Skinner et al.
(1977)

Hystricops —Shotwell (1970A)

Geomyidae

Pliosaccomys wilsoni James, 1963

Pliosaccomys Shotwell, 1967B

Lignimus hibbardi Storer, 1973

Heteromyidae

Diprionomys agrarius A. Wood, 1935C

Perognathus coquorum A. Wood, 1935C

P. furlongi Gazin, 1930

P. minutus James, 1963

Cupidinimus nebraskensis A. Wood, 1935C

Perognathoides quartus (Hall, 1930D) —A. Wood
(1935C)

Cricetidae

Tregomys shotwelli R.L. Wilson, 1968

Gnomomys saltus R.L. Wilson, 1968

Copemys pisinnus R.L. Wilson, 1968

C. russelli (James, 1963)

C. dentalis (Hall, 1930C)

C. esmeraldensis Clark, Dawson, and A. Wood, 1964

Microtoscoptes —Shotwell (1970A)

Zapodidae

Plesiosminthus —Green (1977B)

Macrognathomys nanus Hall, 1930D

CHASICOAN

Chasicoan Mammalian Genera From Argentina

(From manuscript by Marshall, Hoffstetter, and Pascual,
1980.)

MARSUPIALIA

Borhyaenidae

Chasicostylus Reig, 1957B

Pseudolycopsis Marshall, 1976

Caenolestidae

Pliolestes

EDENTATA

Megatheriidae

Hapalops Ameghino, 1887

Plesiomegatherium Roth, 1911

Mylodontidae

Octomylodon

Dasypodidae

Macroeuphractus Ameghino, 1887

Paleuphractus Kraglievich, 1934B

Chasicotatus

Peltephilidae

Epipeltephilus Ameghino, 1904

Glyptodontidae

Palaeohoplophorus Ameghino, 1883

LITOPTERNA

Proterotheriidae, gen. indet.

Macraucheniidae

Cullinia Cabrera and Kraglievich, 1931

Theosodon Ameghino, 1887

NOTOUNGULATA

Homalodotheriidae

Chasicotherium Cabrera and Kraglievich, 1931

Toxodontidae

Hemioxotodon

Ocnerotherium Pascual, 1954B

Paratrigodon Cabrera and Kraglievich, 1931

Pisanodon

Mesotheriidae

Typotheriopsis Cabrera and Kraglievich, 1931

Hegetotheriidae

Hemihegetotherium Rovereto, 1914

Paedotherium Burmeister, 1888

Pseudohegetotherium Cabrera and Kraglievich, 1931

Interatheriidae

Protypotherium Ameghino, 1887

RODENTIA

Octodontidae

Chasicomys Pascual, 1967

Echimyidae

Cercomys Cuvier, 1832

Chasichimys Pascual, 1967

Pattersomys Pascual, 1967

Chinchillidae

Lagostomopsis

Dinomyidae

Diaphoromys Kraglievich, 1931

Gyriabrus Ameghino, 1891

Potamarchus Burmeister, 1885

Tetrastylus Ameghino, 1886

Caviidae

Allocavia Pascual, 1962

Cardiomys Ameghino, 1885

Orthomyctera Ameghino, 1887

Procardiomys Pascual, 1961

Hydrochoeridae

Cardiatherium Ameghino, 1883?

Procardiatherium Ameghino, 1885

TUROLIAN

Turolian, named from the Teruel district in northern Spain, is based on species from Crevillente 3 and correlatives (zone MN 11 of Mein, 1975), Los Mansuetos and correlatives (MN 12 of Mein), and Arquillo and correlatives (MN 13). Mein's data on these zones are in the accompanying outline.

Neogene Mammal Zone 13

Localities: Khended el Ouaich, Alcoy, La Alberca, Librilla, Venta de Moro, Arquillo, Salobrena, Lissieu, Luberon, Casino, Baltavar, Polgardi, Mamay, Tudorovo, Amasya, Kinik, Samos 5.

Characteristic forms of evolving lineages: *Apodemus primaevus, Cricetus kormosi, Ruscinomys lasallei, Kowalskia lavocati, Hispanolagus crusafonti, Veterilepus hungaricus, Pannonicola brevidens, Anomalomys viretschaubi, Hippopotamus primaevus.*

Associations: *Hippopotamus + Machairodus* or *Anancus + Deinotherium.*

Appearance: *Apodemus, Pannonicoia, Amblycoptus, Veterilepus, Paraethomys, Anancus, Hippopotamus, Sus, Paracervulus, Parabos, Urmialia* (spelling?), *Iranotherium.*

Neogene Mammal Zone 12

Localities: Crevillente 5, Los Mansuetos, Concud, Crevillente 4, Ratavoux, Cobruci, Taraklia, Novo Elisavetovka, Cimislia, Grebeniki, Salonique, Samos 1 and 4, Pikermi 1, Veles, Kucukcekmece, Cobanpinar, Garkin Ilhan.

Characteristic forms of evolving lineages: *Pseudoruscinomys schaubi, Occitanomys adroveri, Valerymys turolense, Ruscinomys? hellenicus.*

Associations: *Choerolophodon + Enhydriodon.*

Appearance: *Choerolophodon, Turicius, Ancylotherium, Helladotherium, Diceros, Bohlinia, Palaeoryx, Protoryx, Helicophora, Pliocervus, Enhydriodon, Baranogale.*

Neogene Mammal Zone 11

Localities: Crevillente 3, Vivero, Los Aguanaces, Mollon, Lobrieu, Dorn-Dürkheim, Eichkogel, Kohfidisch, Grossulovo, Eldar, Berislave, Kayadibi.

Characteristic forms of evolving lineages: *Protozapus intermedius, Valerymys vireti, Parapodemus lugdunensis, Allospalax plenus, Epimeriones austriacus.*

Appearance: *Dipoides, Protozapus, Allospalax, Parapodemus, Valerymys.*

Our composite Turolian mammalian fauna combines Mein's species from MN 11, 12, and 13 with species from the localities Mein listed and other localities in Europe and North Africa. It also includes land-mammal species in assemblages assigned to the Pontian Stage of Middle and Southeast Europe, as summarized by Steininger, Rögl, and Martini (1976).

Composite Turolian Fauna

LAGOMORPHA
Ochotonidae
Prolagus michauxi Lopez-Martinez, 1975
P. crusafonti Lopez-Martinez, 1975
P. oeningensis (Koenig, 1825) —Bernor (1981)
Paludotona etruria Dawson, 1959
Leporidae
Hispanolagus crusafonti Janvier and Montenat, 1970
Veterilepus hungaricus (Kormos, 1934)
Alilepus —Solounias (1981)
CARNIVORA
Ursidae
Hemicyon —Bernor (1981)
Indarctos atticus (Weithofer, 1888)
I. ponticus (Kormos, 1913J)
Simocyon primigenius (Roth and Wagner, 1855)
S. diaphorum (Gaudry, 1860)
S. hungaricus Kretzoi —Kadic and Kretzoi (1927)

Ursavus depereti Schlosser, 1902
Helarctos bockhi (Schlosser, 1899)
Mustelidae
Mustela pentelici Gaudry, 1861
M. leporinum Khomenko, 1914
Promephitis lartetii Gaudry, 1862
P. gaudryi Schlosser, 1902 (*nomen dubium?*)
Lutra pentica (Nordmann, 1858, 1900) —Pilgrim (1931)
Sinictis? pentelici (Gaudry, 1861)
S.? jägeri (Schlosser, 1835, 1902)
Martes cf. *sansaniensis* (Lartet, 1851) —Franzen and Storch (1975)
M. muncki Roger, 1900
M. leporinum (Khomenko, 1914A) (*Mustela?*)
M. woodwardi Pilgrim, 1931
Enhydriodon campanii (Meneghini, 1863) —Repenning (1976)
E.? latipes Pilgrim, 1931 (=*E.? laticeps* of Solounias (1981)?)
Promeles palaeattica (Weithofer, 1888)
Parataxidea maraghana (Kittl, 1887)
P. polaki (Kittl, 1887) —Solounias (1981)
Baranogale adroveri Petter, 1964B
Perunium ursogulo Orlov, 1947A, 1948
Promeles palaeatticus Weithofer, 1888
Plesiogulo? —Solounias (1981)
Procyonidae?
Parailurus cf. *anglicus* (Boyd-Dawkins, 1888) (from Baroth-Kopecz)
Canidae
Canis cipio Crusafont, 1950D
Hyaenidae
Ictitherium sarmaticum Pavlov, 1908
I. viverrinum Roth and Wagner, 1857 —Solounias (1981)
I. tauricum Borisyak, 1915
I. adroveri Crusafont and Petter, 1969
Plioviverrops (*Plioviverrops*) *orbignyi* (Gaudry and Lartet, 1856)
P. (*Mesoviverrops*) *guerini* (Villalta and Crusafont, 1943) —de Beaumont and Mein (1972)
Adcrocuta Percrocuta? eximia (Roth and Wagner, 1855)
Pachycrocuta salonicae (Andrews, 1918) —Ficarelli and Torre (1970)
Hyaenictis graeca Gaudry, 1861
Thalassictis (*Lycyaena*) *chaeretis* (Gaudry, 1861) —Solounias (1981)
Thalassictis hyaenoides (Zdansky, 1920)
Thalassictis (*Lycyaena*) n. sp. —Solounias (1981)
Thalassictis wongii (Zdansky, 1924)
T. (*Lycyaena*) *parva* Khomenko, 1914A
Hyaena —Solounias (1981)
Felidae
Metailurus parvulus (Hensel, 1862) —Solounias (1981)
M. major Zdansky, 1924 —Solounias (1981)
Pseudaelurus tournauensis (Hoernes, 1881) —Franzen and Storch (1975)

Felis attica (Wagner, 1857)　　—Pilgrim (1931)

F. neas Pilgrim, 1931

F. leiodon Weithofer, 1888

Machairodus or *Epimachairodus taracliensis* Tiabinin, 1929

Machairodus aphanistus (Kaup, 1833)

M. giganteus (Wagner, 1857)

M. irtyschensis Orlov, 1936

M. schlosseri Weithofer, 1888

M.? copei Pavlov, 1914　　—Pilgrim (1931)

Paramachaerodus orientalis (Kittl, 1887)　　—Pilgrim (1931)

P. ogygia (Kaup, 1833)　　—Pilgrim (1931)

INSECTIVORA

　Erinaceoidea

　　Galerix (*Parasorex*)　　—Ballesio, Carbonnel, Mein, and Truc (1979)

　　G. exilis (Blainville)

　　G. moedlingensis Rabeder, 1973

　　G. atticus Rumke, 1976

　　Schizogalerix zapfei (Bachmayer and R. Wilson, 1970A)　　—Engesser (1980)

　　S. vosendorfensis (Rabeder, 1973)　　—Engesser (1980)

　　Erinaceus　　—Bachmayer and R. Wilson (1970A)

　　Lanthanotherium cf. *sanmigueli* Villalta and Crusafont, 1944D　　—Storch (1978)

　　Plesiodimylus cf. *chantrei* Gaillard, 1897

　Soricoidea

　　Desmana pontica Schreuder, 1940

　　Talpa gilothi Storch, 1978

　　T. vallesensis Villalta and Crusafont, 1955D

　　Desmanella crusafonti Rümke, 1974

　　D. dubia (Bachmayer and R. Wilson, 1970A)　　—Solounias (1981)

　　Petenyiella repenningi Bachmayer and R. Wilson 1970A

　　Paracryptotis?　　—Bachmayer and R. Wilson (1970A)

　　Anourosorex kormosi Bachmayer and R. Wilson, 1970A [= "*Anourosorex*," acc. to Storch (1978)]

　　Dinosorex sansaniensis (Lartet, 1851)

　　Mygalinia hungarica (Kormos, 1913)　　—Hutchison (1974)

　　Amblycoptus oligodon Kormos, 1926

　　Proscapanus sansaniensis (Lartet, 1851)　　—Bernor (1981)

CHIROPTERA

　Megadermatidae

　　Megaderma vireti Mein, 1964

　Rhinolophidae

　　Rhinolophus delphinensis Gaillard, 1899　　—Bachmayer and R. Wilson (1970A)

　Vespertilionidae

　　Samonycteris majori (Ninni, 1878)

PRIMATES

　Cercopithecidae

　　Mesopithecus pentelici Wagner, 1839

　　Colobus? flandrini (Arambourg, 1959C)　　—Delson (1980)

　Oreopithecidae

　　Oreopithecus bambolii Gervais, 1872

TUBULIDENTATA

　　Orycteropus gaudryi Forsyth-Major, 1888

ARTIODACTYLA

　Suidae

　　Korynochoerus palaeochoerus (Kaup, 1832)　　— Franzen and Storch (1976)

　　Microstonyx antiquus (Kaup, 1832)　　—Ginsburg (1973)

　　M. major (Pomel, 1847)　　—Solounias (1981); Franzen and Storch (1976)

　　Sus erymanthius Roth and Wagner, 1854

　　S. major Gervais, 1852

　　Potamochoerus hytheiordes　　—Solounias (1981)

　Anthracotheriidae

　　Merycopotamus? crusafonti (Aguirre)

　Hippopotamidae

　　Hexaprotodon? crusafonti (Aguirre)

　　H.? primaevus (Crusafont, Adrover and Golpe, 1964)

　Tragulidae

　　Dorcatherium naui Kaup, 1836

　Cervoidea

　　Cervocerus novorossiae Khomenko, 1913

　　Cervavitulus mimus　　—Franzen and Storch (1976)

　　Paracervulus cf. *australis* (de Serres, 1839)　　—Ginsburg (1975)

　　Micromeryx flourensianus Lartet, 1851

　　Palaeomeryx meyeri Hofmann, 1893　　—Bernor (1981)

　　Amphiprox anocerus (Kaup, 1833)　　—Franzen and Storch (1976)

　　Euprox　　—Bernor (1981)

　　Eucladocerus Falconer, 1868　(post-Turolian?)

　　Procapreolus loczyi (Pohlig, 1911)

　　P. concudensis (Hernandez-Pacheco, 1930)　　—Franzen and Storch (1976)

　　P. ukrainicus Korotkevitsh, 1965　(age?)

　　Pliocervus pentelici (Gaudry, 1862)　　—Solounias (1981)

　　Muntiacus pliocenicus Korotkevitsh, 1965B

　Giraffidae

　　Honanotherium speciosum (Wagner, 1861)　　—Solounias (1981)

　　H. atticum (Gaudry and Lartet, 1856)　　—Solounias (1981)

　　Helladotherium duvernoyi Gaudry and Lartet, 1856

　　Birgerbohlinia schaubi Crusafont, 1952A

　　Samotherium boissieri Forsyth-Major, 1888

　　Paleotragus rouenii Gaudry, 1861

　　P. coelophrys (Rodler and Weithofer, 1890)

　Bovidae

　　Selenoportax　　—Solounias (1981)

　　Oioceros? proaries (Schlosser, 1904)

　　O.? rodleri (Pilgrim and Hopwood, 1928)　　—Solounias (1981)

　　Oioceros rothi (Wagner, 1857)　　—Ginsburg (1975C)

　　O. atropatenes (Rodler and Weithofer, 1890)

　　O. wegneri Andree, 1926

　　Urmiatherium polaki Rodler, 1889

　　Parurmiatherium rugosifrons Forsyth-Major, 1891

　　Miotragcerus rugosifrons (Schlosser, 1904)

　　Tragoportax curvicornis Andrée, 1926

　　Tragocerus amalthea (Roth and Wagner, 1854)

　　Criotherium argalioides Forsyth-Major, 1891

Parabos cordieri (de Christol, 1832)   —Ginsburg
    (1975C); Gentry (1971A)

Palaeoryx pallasi (Wagner, 1857)

P. woodwardi Pilgrim and Hopwood, 1928

P. ingens Schlosser, 1904

P. majori Schlosser, 1904

P.? boodon (Gervais, 1853)

Protoryx crassicornis Schlosser, 1904

P. carolinae Forsyth-Major, 1891

P. laticeps Pilgrim and Hopwood, 1928

Tragoreas oryxoides Schlosser, 1904   —Pilgrim and
    Hopwood (1928)

Sporadotragus parvidens (Gaudry, 1861)

Procobus melania Khomenko, 1913

P. branneri Khomenko, 1913

Prostrepsiceros houtumschindleri (Rodler and Weithofer,
    1890)

P. woodwardi Pilgrim and Hopwood, 1928

P. rotundicornis (Weithofer, 1888) (*Helicotragus?*)
    —Pilgrim and Hopwood (1928)?

Samokeras minotaurus Solounias, 1981

Helicotragus fraasi (Andree, 1926)   —Pilgrim and
    Hopwood (1928)

Protragocerus

Tragocerus amaltheus (Roth and Wagner, 1854)   —
    Pilgrim and Hopwood (1928)

T. frolovi Pavlov, 1913

T. recticornis Andree, 1926

T. rugosifrons Schlosser, 1904

T. validus Khomenko, 1913

Hippotragus kopassii Andree, 1926

Leptotragus pseudotragoides Bohlin, 1936B

Microtragus parvidens Gaudry, 1861

M. stützeli Schlosser, 1904

Eotragus   —Bernor (1981)

Palaeoreas lindermayeri (Wagner, 1848)

Sinotragus crassicornis (Sickenberg, 1936)

Pachytragus laticeps (Andree, 1926)   (*Palaeoryx* or *Proto-*
    ryx, acc. Pilgrim and Hopwood, 1928)

Gazella schlosseri Pavlov, 1914

G. deperdita (Gervais, 1847)

G. dorcadoides Bohlin, 1935

G. capricornis Wagner, 1854

G. gaudryi Schlosser, 1904

G. mytilinii Pilgrim, 1926A

G. baltavarensis Benda, 1927

Pseudotragus capricornis Schlosser, 1904   —Pil-
    grim and Hopwood (1928)

Hemistrepsiceros zitteli (Schlosser, 1904)

Antidorcas   —Sundevall (1847)

Plesiaddax   —Schlosser (1903)

Prodamaliscus gracilidens Schlosser, 1904

Miotragocerus valenciennesi (Gaudry, 1865)   —So-
    lounias (1981)

M. monacensis Stromer, 1928

Protragelaphus theodori Bouvrain, 1978

P. skouzesi Dames, 1883

Prosinotragus kuhlmanni (Andree, 1926)   —So-
    lounias (1981)

Nisidorcas Bouvrain, 1979

PERISSODACTYLA

  Equidae

"Hipparion" spp.   —(from Mt. Luberon, Samos;
    Group 1 of Bernor, Woodburne, and Van Couvering (1980))

"H." mediterraneum (Roth and ⎫
    Wagner, 1855) ⎬ Group 2 of Bernor et al.
"H." minus (Pavlov, 1890) ⎪ (1980))
"H." proboscideum (Studer, 1911) ⎭

Hipparion prostylum Gervais, 1849 ⎫
H. dietrichi Wehrli, 1941 ⎬ Group 3 of Bernor et al.
H. turkanense Hookjer and Maglio, ⎪ (1980))
    1974 ⎭

  Tapiridae

Tapiriscus pannonicus

  Rhinocerotidae

Aceratherium incisivum Kaup, 1832   —Ginsburg
    (1975C)

A. zernovi Borisyak, 1915

Chilotherium samium (Weber, 1905)

C. schlosseri (Weber, 1905)

C. kowalevskii (Pavlov, 1913)

C. persiae (Pohlig, 1885)

Diceros pachygnathus (Wagner, 1850)

D. douariensis Guérin, 1966   (Vallesian?)

Iranotherium morgani (Mecquenem, 1908)

Dicerorhinus schleiermacheri (Kaup, 1832)

  Chalicotheriidae

Ancylotherium pentelicum Gaudry and Lartet, 1856

Chalicotherium goldfussi Kaup, 1833

HYRACOIDEA

  Procaviidae

Pliohyrax graecus (Gaudry, 1862)

P. kruppii (Fraas, 1895)

PROBOSCIDEA

  Deinotheriidae

Deinotherium giganteum Kaup, 1829

  Gomphotheriidae

Choerolophodon pentelici (Gaudry and Lartet, 1856)

Stegotetrabelodon grandincisivus   —Solounias
    (1981)

Anancus

  Mammutidae

Mammut [*"Zygolophodon"*] *borsoni* (Hays, 1834)

RODENTIA

  Eomyidae, gen. indet.   —de Bruijn (1976)

  Sciuridae

Pliopetes   —Franzen and Storch (1975)

Heteroxerus   —Van de Weerd and Daams
    (1979)

Spermophilinus turolensis de Bruijn and Mein, 1968

S. cf. *bredai* (Meyer, 1848)   —Bachmayer and R.
    Wilson (1970A)

Atlantoxerus adroveri (de Bruijn and Mein, 1968)   —
    de Bruijn, Dawson, and Mein (1970A)

Blackia   —Franzen and Storch (1975)

Pliosciuropterus? n. sp.   —Bachmayer and R. Wil-
    son (1970A)

Pliopetaurista bressana Mein, 1970   —Franzen and
    Storch (1975)

Miopetaurista   —Franzen and Storch (1975)

  Castoridae

Castor neglectus Schlosser, 1902 —Franzen and Storch (1975)

Monosaulax sansaniensis (Lartet, 1851) —Ginsburg (1975C) (Ruscinian?)

Trogontherium minutum (Meyer, 1838) —Franzen and Storch (1975)

Dipoides —Van de Weerd and Daams (1979)

Dipoides [*Steneofiber?*] *problematicus* Schlosser, 1902

Steneofiber jaegeri (Kaup, 1839) —Bachmayer and R. Wilson (1970A)

Palaeomys castoroides Kaup, 1832

P. plassi Franzen and Storch, 1975

Cricetidae

  Cricetodon (*Pseudoruscinomys*) *schaubi* (Villalta and Crusafont, 1956C)

  C. (*Ruscinomys*) *lasallei* Adrover, 1969A

  Cricetodon (*Hispanomys*) —de Bruijn (1979)

  Cricetulodon —Franzen and Storch (1975)

  Kowalskia lavocati (Hugueney and Mein, 1965) — de Bruijn (1979)

  Ruscinomys hellenicus Freudenthal, 1970

  Gerbillus? —de Bruijn (1976)

  Ischymomys ponticus Topachevskiy, Skorik, and Rekovets, 1978

  Pseudomeriones pythagorasi Black, Krishtalka, and Solounias, 1980

  Allospalax planus Kretzoi, 1970

  Prospalax petteri Bachmayer and R. Wilson, 1970A

  Pliospalax cf. *sotirisi* (de Bruijn, Dawson, and Mein, 1970)

Zapodidae

  Protozapus intermedius Bachmayer and R. Wilson, 1970A

  Eozapus —Van de Weerd and Daams (1979)

  Sminthozapus —Franzen and Storch (1975)

Muridae

  Progonomys woelferi Bachmayer and R. Wilson, 1970A

  Castillomys crusafonti Michaux, 1969 —Van de Weerd (1976)

  Parapodemus barbarae Van de Weerd, 1976

  P. gaudryi (Dames, 1883)

  P. schaubi Papp, 1947A

  P. lugdunensis Schaub, 1938 —Franzen and Storch

  Occitanomys adroveri (Thaler, 1964A) —Michaux (1969B)

  O.? neutrum de Bruijn, 1976

  O.? provocator de Bruijn, 1976

  Valerymys turoliensis Michaux, 1969B

  V. ellenbergi Michaux, 1969B (age?)

  Rhagapodemus —Van de Weerd and Daams (1979)

  Apodemus primaevus Hugueney and Mein, 1965

  Stephanomys stadii Mein and Michaux, 1979

  S. ramblensis Van de Weerd, 1976

  S. aff. *donnezani* (Depéret) —Mein et al. (1973)

  Paraethomys anomalus (deBruijn, Dawson and Mein, 1970)

  P. miocaenicus Jaeger, Michaux, and Thaler, 1975

  Gerboa —Solounias (1981)

Hystricidae

  Hystrix cf. *suevica* Stromer, 1884

  H. primigenia Wagner, 1848

Anomalomyidae, or Cricetidae, cont.

  Pterospalax

  Anomalomys gernoti Daxner-Höck, 1980

Gliridae

  Muscardinus vireti Hugueney and Mein, 1965

  M. pliocaenicus Kowalski, 1963

  Glis cf. *minor* Kowalski —Franzen and Storch (1975)

  Microdyromys

  Vasseuromys thenii Daxner-Höck and de Bruijn, 1981

  Myomimus dehmi (de Bruijn, 1966) —Daxner-Höck and de Bruijn (1981)

  Glirulus lissiensis Hugueney and Mein, 1965

  Graphiurops austriacus Bachmayer and R. Wilson, 1980

Turolian Equivalent of Turkey

LAGOMORPHA

  Ochotonidae

    Prolagus cf. *loxodus* (Gervais, 1848)

  Leporidae

    Alilepus

CARNIVORA

  Mustelidae

    Parataxidea —Schmidt-Kittler (1976)

    Eomellivora piveteaui Ozansoy, 1965

  Hyaenidae

    Ictitherium robustum Nordmann, 1858

    I. hipparionum (Gervais, 1846) —Schmidt-Kittler (1976)

    Adcrocuta eximia (Roth and Wagner, 1854) — Schmidt-Kittler (1976)

  Felidae

    Metailurus parvulus (Hensel, 1862) —Schmidt-Kittler (1976)

    Felis attica Wagner, 1857 —Schmidt-Kittler (1976)

    Paramachaerodus —Schmidt-Kittler (1976)

INSECTIVORA

  Erinaceoidea

    Schizogalerix

  Soricoidea

    Amblycoptus

    Desmanella amasyae Engesser, 1980

ARTIODACTYLA

  Suidae

    Dicoryphochoerus-Microstonyx group

  Giraffidae

    Palaeotragus or *Samotherium*

  Bovidae

    Oioceros mecquenemi Pilgrim, 1934

    Palaeoryx?

    Urmiatherium cf. *polaki* Rodler, 1889

    Gazella sp.

PERISSODACTYLA

  Equidae

    Hipparion

  Rhinocerotidae

    Chilotherium schlosseri

    Diceros neumayri Osborn, 1900

PROBOSCIDEA

  Gomphotheriidae

    Choerolophodon pentelici (Gaudry and Lartet, 1856)

    Tetralophodon cf. *grandincisivus* Viret, 1953

FIGURE 6-33 *Turolian Localities.*
(See glossary at end of this chapter
for a more complete listing of the
localities.)

1 *Alcoy (Alicante)—Spain*
2 *Ano Metochi (Macedonia)—*
   Yugoslavia
3 *Arquillo de la Fontana*
   (Aragon)—Spain
4 *Baccinello—Italy*
5 *Berislav—Ukraine*
6 *Casino—Italy*
7 *Cimislia—Romania*
3 *Concud (Tereul)—Spain*
9 *Crevillente 3–5—Spain*
10 *Cucuron—France*
11 *Dorn-Dürkheim—Germany*
12 *Gargano—Italy   (Vallesian?)*
13 *Gravitelli—Sicily*
14 *Grebeniki—Romania*
15 *Imola—Italy*
16 *Kalimantsi—Bulgaria*
17 *Kalithies—Island of Rhodes*
18 *Kohfidisch—Austria*
16 *Kromidovo—Bulgaria*
19 *Küçükçekmece (Khutchur-*
   Tchekmedje)—Thracian Turkey
20 *La Alberca (Murcia)—Spain*
20 *Librilla—Spain*
21 *Lissieu (Rhône)—France*
22 *Los Mansuetos—Spain*
10 *Luberon (Cucuron)—France*
23 *Mollon (Ain)—France*
4 *Monte Bamboli (Grosseto)—*
   Italy
24 *Novo-Elisavetovka—Ukraine*
25 *Piera—Spain*
26 *Pikermi—Greece*
27 *Polgardi—Hungary*
28 *Salobrena—Spain*
29 *Saloniki—Greece*
30 *Samos 5—Island of Samos*
31 *Sopron—Hungary*
32 *Tarakliya (Taraklia)—*
   Moldavian USSR
3 *Tereul (Ademuz)—Spain*
34 *Tiraspol—Moldavian USSR*
33 *Valdecebro I, II—Spain*
2 *Veles (Orisari, Macedonia)—*
   Yugoslavia
35 *Venta del Moro—Spain*

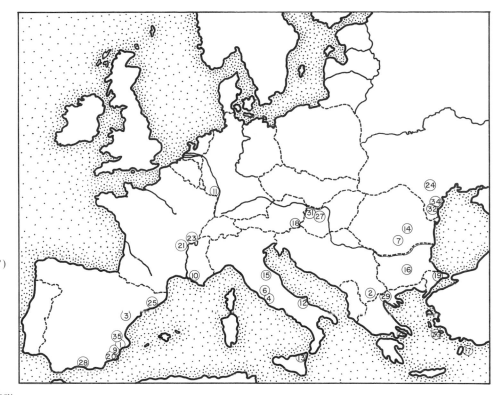

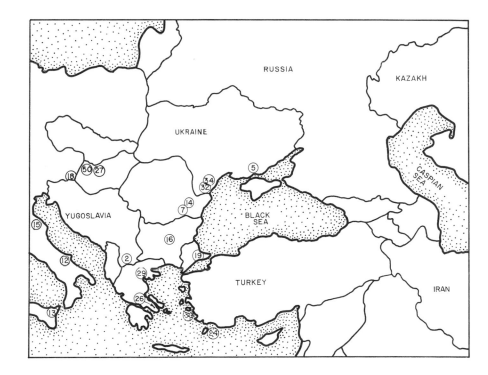

RODENTIA
Cricetidae, 2 spp.
Muridae
Apodemus cf. *primaevus* Hugueney and Mein, 1966
Paraethomys cf. *anomalus* (de Bruijn, Dawson and Mein, 1970)
Pseudomeriones

For a map of representative Turkish Turolian localities see Figure 6-27.

TUROLIAN EQUIVALENT IN ASIA

(Jiulaopo, Nantam, Paote, Oma, Dhok Pathan, Maragheh.)

LAGOMORPHA
Bellatona forsythmajori Dawson, 1967A   (age?)
Bellatona   —Chiu Chan-s. et al. (1979)
Alloptox gobiensis (Young, 1932)   —Dawson (1961A) (age?)
Ochotona cf. *lagreli* Schlosser, 1924   —Young (1935)

CARNIVORA
Amphicyonidae
Arctamphicyon lydekkeri (Pilgrim, 1910)   —Colbert (1935A)
Ursidae
Agriotherium palaeindicum (Lydekker, 1878)   —Colbert (1935A)
A.? sp. (*Hyaenarctos?* —Zdansky, 1924)   —Campbell et al. (1980)
Indarctos salmontanus Pilgrim, 1913
I. lagreli Zdansky, 1924
I. punjabensis (Lydekker, 1884)   —Colbert (1935A)
I. sinensis Zdansky, 1924
Mustelidae
Promeles palaeatticus Weithofer, 1888
Plesiogulo brachygnathus (Schlosser)
Eomellivora wimani Zdansky, 1924
Enhydriodon falconeri Pilgrim, 1931
Sinictis dolichognathus Zdansky, 1924
Promellivora punjabiensis (Lydekker, 1884)
Parataxidea polaki (Kittl, 1887)
P. sinensis Zdansky, 1924
P. maraghana (Kittl, 1887)
P. crassa Zdansky, 1924
Sivaonyx bathygnathus (Lydekker, 1884)
Martes   —Campbell et al. (1980)
Proputorius yaoguensis Chiu Zhu-ding, 1979
Mustela palaeosinensis Zdansky, 1924
Trochictis minutus Schlosser, 1924
Meles taxipater Schlosser, 1903
Melodon maraghanus (Kittl, 1887)
M. majori Zdansky, 1924
M.? incertum Zdansky, 1924
Promephitis alexejewi Schlosser, 1924
Aonyx? aonychoides (Zdansky, 1924)   —Pilgrim (1931)
"*Potamotherium*" *hasnoti* Pilgrim, 1932

Canidae
Canis antonii Zdansky, 1924
C. chihliensis Zdansky, 1924
Vulpes sinensis Schlosser, 1903
Viverridae
Vishnuictis salmontanus Pilgrim, 1932
Hyaenidae
"*Hyaena*" *variabilis* (Zdansky, 1924)
"*H.*" *honanensis* (Zdansky, 1924)
Lycyaena macrostoma (Lydekker, 1884)
L. spathulata Chiu, Huang and Guo, 1979
L.? dubia Zdansky, 1924
Ictitherium hyaenoides Zdansky, 1924
I. hipparionum (Gervais, 1846)   —Pilgrim (1931)
I. sivalense Lydekker, 1877
I. gaudryi Zdansky, 1924
I. indicum Pilgrim, 1910
I. sinense Zdansky, 1924
I. wongii Zdansky, 1924   —Chiu, Huang and Guo (1979); *Thalassictis* acc. Solounias (1981)
Adcrocuta eximia (Roth and Wagner, 1854)   —Schmidt-Kittler (1976)
A. gigantea (Schlosser, 1903)
A. mordax (Pilgrim, 1932)
Percrocuta carnifex (Pilgrim, 1913)
Hyaenictis? bosei (Matthew, 1929A)
Palinhyaena reperta Chiu, Huang, and Guo, 1979
P. imbricata Chiu, Huang, and Guo, 1979
Felidae
Felis palaeosinensis Zdansky, 1924   (*Panthera?*)
F. attica Wagner, 1857
Dinofelis abeli Zdansky, 1924
Mellivorodon palaeindicus Lydekker   (felid?)
Aeluropsis annectens Lydekker, 1884
Machairodus palanderi Zdansky, 1924
M. giganteus (Wagner, 1857)   —de Beaumont (1975)
M. tingii Zdansky, 1924
M. aphanistus (Kaup, 1833)
Paramachaerodus pilgrimi Kretzoi, 1929
P. orientalis (Kittl, 1887)   —Pilgrim (1931)
P. indicus (Kretzoi, 1929)
P. maximiliani (Zdansky, 1924)
Propontosmilus sivalensis (Lydekker, 1877)
Metailurus major Zdansky, 1924
M. minor Zdansky, 1924

INSECTIVORA
Erinaceidae
Erinaceus
Soricoidea
Anourosorex inexpectata (Schlosser, 1924)   —Repenning (1967C)
Blarinella kormosi (Schlosser, 1924)   —Repenning (1967C)
Siwalikosorex prasadi Sahni and Kumar-K., 1976

PRIMATES
Cercopithecidae
Cercopithecus hasnot (Pilgrim, 1910)   —Colbert (1935A)
Macaca sivalensis Lydekker, 1878
Mesopithecus pentelici Wagner, 1839
Hominidae   —Szalay and Delson (1979)

Sivapithecus indicus Pilgrim, 1910   (includes *"Dryopithecus? fricki"*)

S. sivalensis (Lydekker, 1879)   (includes *"Palaeopithecus sivalensis"*)

TUBULIDENTATA

Orycteropus cf. *gaudryi* Forsyth-Major, 1888   —
Campbell et al. (1980)

ARTIODACTYLA

Suidae

Microstonyx erymanthius Roth and Wagner, 1854   —
Campbell et al. (1980)

Sivachoerus prior Pilgrim, 1926

Dicoryphochoerus medius Pilgrim, 1926?

Dicoryphochoerus titan (Lydekker, 1884)

D. titanoides Pilgrim, 1926

D. vagus Pilgrim, 1926

D. vinayaki Pilgrim, 1926

Lophochoerus Pilgrim, 1926

Chleuastochoerus

Tetraconodon magnus Falconer, 1868

T. minor Pilgrim, 1926

T. mirabilis Pilgrim, 1926

Listriodon pentapotamiae (Falconer, 1868)

Propotamochoerus uliginosus Pilgrim, 1926

P. hysudricus (Stehlin, 1899)

P. ingens Pilgrim, 1926

Hyosus punjabiensis (Lydekker, 1878)

H. tenuis Pilgrim, 1926

Sivahyus hollandi Pilgrim, 1926

Sanitherium schlagentweitii Meyer, 1866

Hippohyus lydekkeri Pilgrim, 1910

H. grandis, Pilgrim

Sus comes Pilgrim, 1926

S. adolescens Pilgrim, 1926

S. praecox Pilgrim, 1926

Anthracotheriidae

Choeromeryx silistrense (Pentland, 1828)

Merycopotamus dissimilis Falconer and Cautley, 1836

Hippopotamidae

Hexaprotodon iravaticus (Falconer and Cautley, 1847)

Cervoidea

Cervavitus

Eostyloceros blainvillei Zdansky, 1924   —Chiu Zhu-d. (1979)

Dicrocerus?

Cervocerus huadeensis Chiu Zhu-d., 1979

C. novorossiae Khomenko, 1913   —Chiu Zhu-d. (1979)

Procapreolus latifrons Schlosser, 1924

Pseudaxis?   —Teilhard and Young (1931)

Metacervulus capreolus

Cervus simplicidens Lydekker, 1876

C. triplens Lydekker, 1876

Lagomeryx tsaidamensis Rosenstock and Geweih —Bohlin (1937B)

Giraffidae

Palaeotragus microdon (Koken, 1885)   —Bohlin (1927)

P. coelophrys (Rodler and Weithofer, 1896)   —
Campbell et al. (1980)

Vishnutherium iravaticum Lydekker, 1876

Helladotherium   —Gaudry (1860)

Bramatherium perimense Falconer, 1876

Giraffokeryx

Honanotherium   —Bohlin (1927)

Hydaspitherium megacephalum Lydekker, 1876

H. grande Lydekker, 1878

H. magnum Pilgrim, 1910

H. birmanicum Pilgrim, 1910

Giraffa punjabiensis Pilgrim, 1910

Samotherium cf. *neumayri* (Rodler and Weithofer, 1890) —Bohlin (1926)

S. sinense (Schlosser, 1903)   —Bohlin (1926)

S. tafeli (Killgus, 1922)   —Bohlin (1926)

Tragulidae

Dorcabune latidens Pilgrim, 1915

Dorcatherium majus Lydekker, 1876

D. minus Lydekker, 1876

Tragulus sivalensis Lydekker, 1882

Moschus   —Lydekker, 1884

M. grandaevus   —Chiu Zhu-d

Bovidae

Gazella gaudryi Schlosser, 1904

G. blacki Teilhard and Young, 1931

G. lydekkeri Pilgrim, 1937

G. altidens Schlosser, 1903

G. dorcadoides Schlosser, 1903

G. sinensis Teilhard and Piveteau, 1930

G. deperdita (Gervais, 1847)

G. paotehensis Teilhard and Young, 1931

G. rodleri

G. mongolica Dimitrieva, 1977

Kobus

Perimia falconeri (Lydekker, 1886)

Boselaphus lydekkeri Pilgrim, 1910

Tragocerus sylvaticus Schlosser, 1903

T. kokenius Schlosser, 1903

T. spectabilis Schlosser, 1903

T. gregarius Schlosser, 1903

T. browni Pilgrim, 1937

T. punjabicus Pilgrim, 1910

T. perimensis (Lydekker, 1878)   —Pilgrim (1939A)

Miotragocerus rugosifrons (Schlosser, 1904)   —
Campbell et al. (1980)

Prostrepsiceros houtumschindleri (Rodler and Weithofer, 1890)

Proleptobos birmanicus Pilgrim, 1939A

Prosinotragus tenuicornis Bohlin, 1935A

Sinotragus wimani Bohlin, 1935A

Pseudobos [*Urmiatherium?*] *gracilidens* Schlosser, 1903 —Bohlin (1935A)

Urmiatherium polaki Rodler, 1889   —Gentry (1971A)

U. intermedium Schlosser, 1903

Parurmiatherium rugosifrons Sickenberg, 1932

Plesiaddax depereti Schlosser, 1903

P.? minor Bohlin, 1935A

Tsaidamotherium hedini Bohlin, 1935F

Olonbulukia tsaidamensis Bohlin, 1937B

FIGURE 6-34 *Representative Latest Miocene (=Turolian) Localities of Asia, Exclusive of Turkey. (See opposite page)* (Nota bene: *Massive problems remain as to possible synonymies of transliterated Chinese and other Asian place names. We have found that some of these names are spelled as many as five different ways in English publications. The following list includes probably no more than a third of the published (=Turolian) localities of Asia. Probably some of the localities are slightly earlier than latest Miocene (=Turolian).*)

1	Altan Teeli—Mongolia	*10*	Ertemte (in part)—Nei Mongol	*18*	Jiulongkou (Cixian)—Hubei
2	Baode (Paote)—Shanxi			*13*	Kaiyuan (=Keiyuan? =Hsiaolungtan?) Yunnan
3	Bu Lung (Biru)—Xizang (Tibet)	*11*	Ghazgay (Khorkkabul Basin)—Afghanistan	*1*	Kalban district (Altai)— Mongolia
2	Changzhi—Shanxi	*5*	Gouochtockoula—Mongolia		
2	Changtun Basin—Shanxi	*11*	Gurgemaydan 7 and 8— Afghanistan	*15*	Kiyevskiy—Kazakhstan
2	Chinglo (in part)—Shanxi			*17*	Kou Chia Tsun (Lantien)— Shaanxi
4	Chit-Igia—Kazakhstan	*12*	Heishantou—Nei Mongol		
5	Chiton-Gol (Basin)— Mongolia	*6*	Hochêng—Gansu	*9*	Kutsay Mountain— Caucasus
		6	Hsen Shui Ho (Yungtang)— Gansu	*17*	Lantien district (in part)— Shaanxi
6	Chuan Tou Kou—Gansu	*2*	Hsihsien—Shanxi		
6	Chuhsin—Gansu	*13*	Hsiaolungtan coal field— Yunnan	*19*	Laren bed site—Jammu and Kashmir
7	Dhok Pathan (many localities)—Pakistan and India	*5*	Hung-Kureh Formation sites—Mongolia	*6*	Lihsien—Gansu
				2	Lishan—Shanxi
8	Djebel Hamrin (Bagdad)— Iraq	*14*	Irrawaddy "series" (in part) sites—Burma	*20*	Lin Ho (Nanjing)—Jiangxi
				6	Lunggia Kou—Gansu
4	Dzenema River— Kazakhstan	*15*	Irtysch—Kazakhstan	*2*	Lutzukou beds sites (Baode)—Shanxi
		16	Ischim—Siberian SSR		
9	Dzhaparidze—Caucasus USSR	*17*	Jiulaopo (Lantien)—Shaanxi	*2*	Mantankou—Shanxi

(continued)

Palaeoryx sinensis Bohlin, 1935A —(synonym of *P. pallasi* acc. Gentry, 1971A)

P. longicephalus Sokolov, 1955 —(synonym of *P. pallasi* acc. Gentry, 1971A)

Sinoryx bombifrons Teilhard and Trassaert, 1938 — (synonym of *P. pallasi* acc. Gentry, 1971A)

Protragelaphus skouzesi Dames, 1883 —Campbell et al. (1980)

Pachyportax dhokpathanensis Pilgrim, 1937

P. latidens Pilgrim, 1937

Selenoportax lydekkeri (Pilgrim, 1910) —Pilgrim (1937)

S. vexillarius Pilgrim, 1937

Tragoportax salmontanus Pilgrim, 1937

T. aiyengari Pilgrim, 1939A

T. islami Pilgrim, 1939A

Oioceros atropatus —Campbell et al. (1980)

O. rothi (Wagner, 1887) —Campbell et al. (1980)

Pachytragus crassicornis Schlosser, 1904 —Campbell et al. (1980)

Ruticeros pugio Pilgrim, 1939A

Proamphibos hasticornis Pilgrim, 1939A

P. kasmiricus Pilgrim, 1939A

P. lachrymans Pilgrim, 1939A

Kobikeryx atavus Pilgrim, 1939A

Cambayella watsoni Pilgrim, 1939A

Sivoreas erimita Pilgrim, 1939A

Lyrocerus satan Teilhard and Trassaert, 1938

Sinoreas cornucopia Teilhard and Trassaert, 1938

Paraprotoryx killgusi Bohlin, 1935A

P. minor Bohlin, 1935A

Tragoreas spp. —Dmitrieva (1977)

T.? lagreli Bohlin, 1935A

T.? palaeosinensis (Schlosser, 1903)

Antilospira licenti Teilhard and Young, 1931

Dorcadoxa porrecticornis (Lydekker, 1878) —Pilgrim (1939A)

Dorcadoxa —Chiu Zhu-d. (1979)

Antidorcas

Pachygazella grangeri Teilhard and Young, 1931

Tossunnoria pseudibex Bohlin, 1937B

Sivaceros vedicus Pilgrim, 1939A

aff. *Helicotragus vinayaki* Pilgrim, 1939A

Qurliqnoria cheni Bohlin, 1937B

Protoryx tadzhikistanica Dmitrieva, 1977

P.? planifrons Bohlin, 1935A

P.? shansiensis Bohlin, 1935A

Damalops palaeindicus (Falconer, 1859)

Torticornis ortokensis Dmitrieva, 1977

PERISSODACTYLA

Equidae

Cormohipparion theobaldi (Lydekker, 1877) —MacFadden and Bakr (1979)

Hipparion cf. *prostylum* Gervais, 1849 —Campbell et al. (1980)

H. dietrichi Wehrli, 1941

"*Hipparion*" —Group 2 of Woodburne and Bernor (1980) (*mediterraneum* and *proboscideum*)

"*H.*" —Group 3 of Woodburne and Bernor (1980)

"*H.*" —Group 4 of Woodburne and Bernor (1980) (*matthewi* Abel, 1926)

Hipparion? houfenense (Teilhard and Young, 1931)

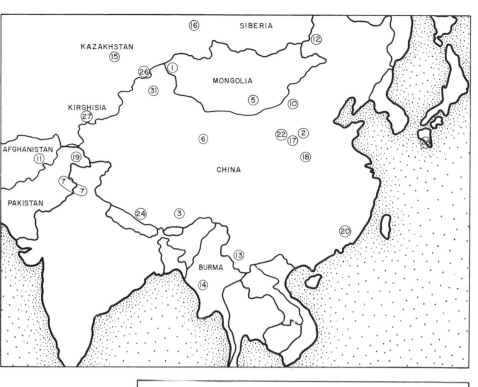

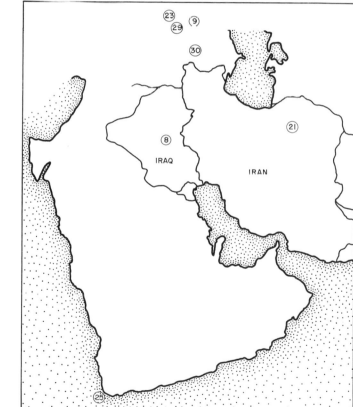

H.? plocodus Sefve, 1927
H.? punjabiense Lydekker, 1886
H.? richthofeni Koken
H.? antelopinum (Falconer and Cautley, 1849)
H.? chisholmi (Pilgrim, 1910)
H.? perimense (Pilgrim, 1940)
Anchitherium (*"Sinohippus"*) *zitteli* Schlosser, 1924

Rhinocerotidae
  Huanghotherium anlungense
  Chilotherium fenhoense
  C. anderssoni Ringström, 1924
  C. intermedium (Lydekker, 1884)
   (1980)
  C. persiae (Pohlig, 1885)

—Campbell et al.

C. gracile Ringström, 1924 —Chiu Zhu-d. (1979)
Aceratherium lydekkeri Pilgrim, 1910
A. perimense Falconer and Cautley, 1847
A. blanfordi Lydekker, 1884
A. huadoensis Chiu Zhu-d. (1979)
Diceros neumayri Osborn, 1900 —Campbell et al. (1980)
Rhinoceros planidens Lydekker, 1876
R. irvadicus
Diceratherium tsaidamense Bohlin, 1937B
D. palaeosinense Ringström, 1924 —Bohlin (1937B)
Chalicotheriidae
Ancylotherium pentelici (Gaudry, 1862)
HYRACOIDEA
Pliohyrax —Heintz, Brunet and Battail (1981)
PROBOSCIDEA
Deinotheriidae
Deinotherium indicum Falconer, 1845
D.? angustidens Koch, 1845
Gomphotheriidae
Gomphotherium hasnotensis (Osborn, 1934A)
G. connexus (Hopwood, 1935B)
Tetralophodon falconeri (Lydekker, 1880)
T. exoletus Hopwood, 1935A
T. punjabiensis (Lydekker, 1886)
T.? sinensis (Koken, 1885) (age?)
Rhynchotherium chinjiensis Osborn, 1929J
Synconolophus dhokpathanensis Osborn, 1929J
S. propathanensis Osborn, 1929J
S. corrugatus (Pilgrim, 1913)
S. ptychodus Osborn, 1929J
S. hasnoti (Pilgrim, 1913)
S.? officinalis (Hopwood, 1933?) (post-Turolian?)
Anancus perimensis (Falconer and Cautley, 1847)
Choerolophodon pentelici (Gaudry and Lartet, 1956) —Campbell et al. (1980)
Stegodontidae
Stegolophodon latidens (Clift, 1828)
S. cautleyi (Lydekker, 1886)
Stegodon? yüshiensis (Young, 1935) (post-Turolian?)
S.? adanskyi (Hopwood, 1935?) (post-Turolian?)
S. bombifrons (Falconer and Cautley, 1847)
S. clifti (Falconer and Cautley, 1847)
S. elephantoides (Clift, 1828)
Mammutidae
Zygolophodon
RODENTIA
Aplodontidae
Pseudaplodon asiatica (Schlosser, 1924)
Castoridae
Chalicomys broilii Teilhard and Young, 1931
C. anderssoni Schlosser, 1924
Dipoides majori Schlosser, 1924
Castor zdanskyi Young, 1927
Dipodidae
Paralactaga suni Teilhard and Young, 1931
Rhizomyidae
Rhizomyoides cf. *nagrii* (Hinton, 1933) —Jacobs (1978–0636)

R. saketiensis Gupta, Verma, and Tewari, 1979
Rhizomys sivalensis Lydekker, 1878
Kanisamys sivalensis A. Wood, 1937B
Pararhizomys hipparionum Teilhard and Young, 1931
Cricetidae
Ishimomys quadriradicatus Zazhigin —W. von Koenigswald, 1980
Pseudomeriones —Heintz, Brunet and Battail (1981)
Zapodidae
Heterosminthus
Prosiphneus eriksoni (Schlosser, 1924)
P. licenti Teilhard, 1926
P. sinensis Teilhard and Young, 1931
P. intermedius Teilhard and Young, 1931
Eozapus —Van de Weerd and Daams (1979)
Protozapus intermedius Bachmayer and R. Wilson, 1970A
Muridae
Parapelomys robertsi Jacobs, 1978–0636
Progonomys debruijni Jacobs, 1978–0636 (age?)
Parapodemus (age?) —Jacobs (1978–0636)
Karnimata darwini Jacobs (1978–0636) (age?)
K. huxleyi Jacobs, 1978–0636
Hystricidae
Hystrix sivalensis Lydekker, 1878

LOTHAGAMIAN

Coppens (1978) has described and defined the Lothagamian "continental" stage, based on unit 1 (members A, B, and C) of the stratigraphic section and its mammalian fauna at Lothagam, southwest of Lake Turkana, Kenya. Unit 1 underlies a basalt (identified as a sill by Behrensmeyer, 1976) that has been dated by the potassium-argon method at a little less than 4 million years and overlies volcanics dated at more than 8 million years. Smart (1976) identified the mammals from Lothagam 1, and we have added the species identified from Sahabi, Libya, to his list for a provisional Lothagamian composite fauna.

Lothagamian Mammalian Fauna

CARNIVORA
Viverridae
Civettictis?
Hyaenidae
aff. *"Euryboas"* = *Chasmaporthetes*
Felidae
feline, large
machairodontine
PRIMATES
Cercopithecidae
cf. *Parapapio*
cf. *Cercocebus*
Macaca libyca (Stromer, 1920) —Delson (1980)
Hominidae
Australopithecus?
TUBULIDENTATA
Leptorycteropus guilielmi Patterson, 1975
ARTIODACTYLA
Suidae
Nyanzachoerus syrticus (Leonardi, 1952)

N. tulotus Cooke and Ewer, 1972A
N. aff. *jaegeri* Coppens, 1971A
Anthracotheriidae
   Merycopotamus petrochii (Bonarelli, 1947B)
Hippopotamidae
   Hexaprotodon harvardi Coryndon, 1977
Giraffidae
   Palaeotragus germaini Churcher, 1979–0324
   Libytherium   —Pomel (1893)
   Giraffa
Bovidae
   cf. *Pachytragus*
   Prostrepsiceros rotundicornis Gentry, 1978
   aff. *Kobus*
   aff. *Redunca*
   aff. *Damaliscus*
   Cephalopus   —Thomas (1980)
   Miotragocerus cyrenaicus Thomas, 1979–1652
   Tragelaphus, 2 spp.
   Gazella, 2 spp.
   Antilope
   Madoqua   —Thomas (1980)
   aff. *Rhyncotragus*
   Aepyceros   —Thomas (1980)
   Leptobos syrticus Pettrochi, 1956
   Ugandax cf. *gautieri* Cooke and Coryndon, 1970   —
      Thomas (1980)
PERISSODACTYLA
  Equidae
   Hipparion turkanense Hooijer and Maglio, 1974
   H. sitifense Pomel, 1897
   H. primigenium? Meyer, 1829
  Rhinocerotidae
   Brachypotherium lewisi
   Ceratotherium praecox Hooijer and Patterson, 1972
  Chalicotheriidae
   Chemositia turgenensis Pickford, 1979
PROBOSCIDEA
  Deinotheriidae
   Deinotherium
  Gomphotheriidae
   Choerolophodon ngorora
   cf. *Tetralophodon*
   Anancus kenyensis
  Elephantidae
   Stegotetrabelodon syrticus Petrocchi, 1941
   S. orbus Maglio, 1970B
   Primelephas gomphotheroides Maglio, 1970B
   P. korotorensis (Coppens, 1965B)
RODENTIA
  Cricetidae
   Zramys hamamai Jaeger, 1977B
  Ctenodactylidae
   Irhoudia robinsoni Jaeger, 1977B

HEMPHILLIAN

H.E. Wood et al. (1941) based Hemphillian, a "new provincial time term," on the "Hemphill member of the Ogal-

lala, which includes both the Hemphill local fauna from the Coffee Ranch Quarry and the [earlier] Higgins local fauna, Hemphill County, Panhandle of Texas." Subsequently, the Hemphillian mammal age has been expanded to include faunas later than the Coffee Ranch assemblage (for example, Axtel, Yepomera, Christian Ranch, and Pinole) and faunas possibly earlier than Higgins (for example, Box T, Mulholland, Smith Valley).

The composite Hemphillian mammalian fauna includes the following.

Composite Hemphillian Mammalian Fauna

EDENTATA
  Megalonychidae
   Megalonyx curvidens Hirschfeld and Webb, 1968
   M. mathisi Hirschfeld and Webb, 1968
   Pliometanastes galushai Hirschfeld and Webb, 1968
   P. protistus Hirschfeld and Webb, 1968
  Mylodontidae
   Thinobadistes seguis Hay, 1919C
LAGOMORPHA
  Ochotonidae
   Ochotona spangclei Shotwell, 1956
  Leporidae
   Hypolagus vetus (Kellogg, 1910A)
   H. oregonensis Shotwell, 1956
   H. edensis Frick, 1921A
   H. rodmani Wagner, 1981
   Notolagus velox R. Wilson, 1937D
   cf. *Panolax sanctaefidei* Cope, 1874I   —Dawson
     (1958)
CARNIVORA
  Ursidae
   Indarctos oregonensis Merriam, Stock, and Moody, 1916A
   I. nevadensis Macdonald, 1959
   Agriotherium schneideri (Sellards, 1916B)
   A. gregorii Frick, 1921A
   Plionarctos edensis Frick, 1926A
   Simocyon marshi (Thorpe, 1922A)
  Canidae
   Tomarctus propter Cook and Macdonald, 1962
   Epicyon validus (Matthew and Cook, 1909)
   E. haydeni Leidy, 1858E
   Osteoborus cyonoides (Martin, 1928A)
   O. pugnator (Cook, 1922B)
   O. secundus (Matthew and Cook, 1909A)
   O. dudleyi (T. White, 1941A)   —Webb (1969A)
   O. direptor (Matthew, 1924C)
   O. hilli Johnston, 1939A
   O. galushai Webb, 1969A
   O. orc Webb, 1969A
   Canis davisi Merriam, 1911B
   C. condoni Shotwell, 1956
   Vulpes stenognathus D. Savage, 1941
   V. shermanensis (Hibbard, 1937E)
   Leptocyon vafer (Leidy, 1858E)
   Carpocyon   —Tedford et al. (1983?)

Procyonidae
    Bassariscus cf. *ogallalae* Hibbard, 1933A
    B. lycopotamicum (Cope, 1879B) (age?) —Gregory and Downs (1951)
    Procyon —Tedford et al. (1983?)
Mustelidae
    Enhydriodon cf. *lluecai* Villalta and Crusafont, 1945B —Repenning (1976)
    Plesiogulo marshalli (Martin, 1928A)
    P. lindsayi Harrison, 1981
    Leptarctus progressus Simpson, 1930H —(*Procyon?* —Tedford et al. (1983?))
    Leptarctus —Skinner et al. (1977)
    Taxidea —Tedford et al. (1983?)
    Pliotaxidea nevadensis (Butterworth, 1916A)
    P. garberi Wagner, 1976
    Cernictis hesperus Hall, 1935
    Eomellivora cf. *wimani* Zdansky, 1924
    Lutravus —Tedford et al. (1983?)
    Martes (*Plionictis*) *oregonensis* Shotwell, 1970A
    Pliogale furlongi (Merriam, 1911B) —Hall (1930E)
    Sthenictis
    Martinogale alveodens Hall, 1930E
Felidae
    Pratifelis martini Hibbard, 1934
    Adelphailurus kansasensis Hibbard, 1934
    Nimravides —Baskin (1981)
    Machairodus coloradensis Cook, 1922B
    Barbourofelis fricki Schultz, Schultz, and Martin, 1970A
    Pseudaelurus hibbardi Dalquest, 1969A
    Felis (*Lynx*) *proterolyncis* D. Savage, 1941
    F. longignathus Shotwell, 1956
    F. rexroadensis Stephens, 1959   U. Bone Valley Fm.
INSECTIVORA
  Soricidae
    Notiosorex —Baskin (1979)
    Limnoecus tricuspis Stirton, 1930C
    Cryptotis adamsi (Hibbard, 1953A) —Repenning (1967C)
    Paracryptotis rex Hibbard, 1950 —Repenning (1967C)
    Hesperosorex lovei Hibbard, 1957B
  Talpidae
    Gaillardia thomsoni Matthew, 1932A
    G. americana (Shotwell, 1956) —Hutchison (1968A)
    Neurotrichus? columbianus Hutchison, 1968A
    Scalopoides? —Hutchison (1968A)
    Scapanus proceridens Hutchison, 1968A
    Scalopus
Artiodactyla
  Tayassuidae
    Prosthennops serus (Cope, 1878C)
    P. kernensis Colbert, 1938A
    P. brachirostris Shotwell, 1956
    P. (*Macrogens*) *graffhami* Schultz and Martin, 1975
    Desmathyus brachydontus Dalquest and Mooser, 1980

Merycoidodontidae
    Ustatochoerus major (Leidy, 1858E) (age?) — Skinner et al. (1977)
Camelidae
    Aepycamelus —Tedford et al. (1982?)
    Procamelus robustus (Leidy, 1858F) (age?) — Skinner et al. (1977)
    Megatylopus gigas Matthew and Cook, 1909
    M. matthewi Webb, 1965C
    Hemiauchenia vera (Matthew, 1909C)
    Alforjas taylori Harrison, 1979 (incl. "*Pliauchenia hemphillensis*")
    Palaeolama guanajuatensis Dalquest and Mooser, 1980
    cf. *Camelops*, n. gen. —MacFadden, Johnson, and Opdyke (1979)
Gelocidae —G. Schultz (1977 from pers. commun. with Tedford)
    Pseudoceras
Cervoidea
    Pediomeryx hemphillensis Stirton, 1936B (= "*Yumaceras*") [*Cranioceras?*]
    P. figginsi (Frick, 1937)
    P. falkenbachi (Frick, 1937)
    Procoileus edensis Frick, 1937
Protoceratidae
    Synthetoceras tricornatus Stirton, 1932E —Patton and Taylor (1977)
Antilocapridae
    Texoceros altidens (Matthew, 1924C)
    T.? minorei —MacFadden, Johnson, and Opdyke (1979)
    Sphenophalos middleswarti Barbour and Schultz, 1941A
    S. nevadanus Merriam, 1909B
    S. (*Plioceros*) *blicki* Frick, 1937 (age?)
    Ilingoceros alexandrae Merriam, 1909B
    I. schizoceras Merriam, 1909B
    Osbornoceros osborni Frick, 1937
    Hexobelomeryx fricki Furlong, 1941A
    Hexameryx simpsoni T. White, 1941A
    H. elmorei T. White, 1942C
    Ottoceros peacevalleyensis Miller and Downs, 1974
    Antilocapra (*Subantilocapra*) *garciae* Webb, 1973
    antilocaprid, new genus —MacFadden, Johnson, and Opdyke (1979)
Bovidae
    Neotragocerus improvisus Matthew and Cook, 1909 (age?)
PERISSODACTYLA
  Equidae
    Dinohippus interpolatus (Cope, 1893A)
    D. leidyanus (Osborn, 1918A)
    D. spectans (Cope, 1880G)
    D. ocotensis (Mooser, 1957)
    D. mexicanus (Lance, 1950)
    D. edensis (Frick, 1921A)
    D. osborni (Frick, 1921A)
    D. coalingensis (Merriam, 1914C)
    D. albidens Mooser, 1968A
    Astrohippus ansae (Matthew and Stirton, 1930A)
    A. stocki Lance, 1950
    Onohippidium galushai MacFadden and Skinner, 1979
    Pliohippus?

"Pliohippus" hondurensis (Olson and McGrew, 1941)

Hippidion —MacFadden and Skinner (1979)

Hipparion lenticularis (Cope, 1893A)

H. hesperides (Mooser, 1968A)

H. aztecus (Mooser, 1968A)

H. minor (Sellards, 1916B)

"Nannippus" ingenuus (Leidy, 1885A)

Neohipparion eurystyle (Cope, 1893A)

N. leptode (Merriam, 1915A)

N. molle (Merriam, 1915A)

N. gidleyi (Merriam, 1915A)

N. phosphorum (Simpson, 1930H)

N. whitneyi Gidley, 1903B

N. floresi Stirton, 1955C

N. otomii Mooser, 1959

N. arrelanoi Mooser, 1955C

N. monias Mooser, 1964A

Calippus

Pseudhipparion or *Griphippus* —(*Griphippus*, acc. to Tedford et al. (1983?))

"Hipparion" plicatile (Leidy, 1887A) —Tedford et al. (1983?)

Cormohipparion cf. *occidentale* (Leidy, 1856D)

Protohippus —Tedford et al. (1982?)

Tapiridae

Tapirus? simpsoni Schultz, Martin, and Corner, 1975

Rhinocerotidae

Aphelops mutilus Matthew, 1924C

A. kimballensis Tanner, 1967A

A. yumensis Cook, 1930B

A. longinaris

Peraceras (age?)

Teleoceras fossiger (Cope, 1878H)

T. hicksi Cook, 1927B

T. schultzi Tanner, 1975

T. ocotensis Dalquest and Mooser, 1980

PROBOSCIDEA

Gomphotheriidae

Gomphotherium (= *"Serridentinus"*) *nebrascensis* (Osborn, 1924H)

Rhynchotherium anguirivalis (Osborn, 1926B)

Amebelodon fricki Barbour, 1927A

A. hicksi (Cook, 1922A)

Cuvieronius edensis (Frick, 1921A)

Mammutidae

Pliomastodon matthewi (Osborn, 1921B)

P. nevadanus Macdonald, 1959

P. vexillarius Matthew, 1930D —May (1982)

RODENTIA

Eomyidae

Ronquillomys wilsoni Jacobs, 1977

Leptodontomys oregonensis Shotwell, 1956

Kansasimys dubius A. Wood, 1936E

Aplodontidae

Tardontia —Shotwell (1970A)

Liodontia furlongi Gazin, 1932

Sciuridae

Marmota oregonensis Shotwell, 1956

M. vetus Black, 1963B (age?)

M. nevadensis (Kellogg, 1910A)

M. minor (Kellogg, 1910A)

Paenemarmota mexicana (R. Wilson, 1949E) — Dalquest and Mooser (1980)

Spermophilus (*Otospermophilus*) *shotwelli* (Black, 1963B)

S. (*O.*) *wilsoni* Shotwell, 1956

S. (*O.*) *gidleyi* (Merriam, Stock, and Moody, 1925A)

S. (*O.*) *argonautus* (Stirton and Goeriz, 1942)

S. (*O.*) *fricki* (Hibbard, 1942D)

S. (*O.*) *pattersoni* (R. Wilson, 1949E)

S. matachicensis (R. Wilson, 1949E)

S. (*Spermophilus*) *mckayensis* (Shotwell, 1956)

S. kimballensis Tanner, 1967B

Mylagaulidae

Mylagaulus monodon Cope, 1881E

M. kinseyi Webb, 1966A

M. sesquipedalis Cope, 1878H R. Wilson (1937B)

Castoridae

Dipoides smithi Shotwell, 1955B

D. williamsi Stirton, 1936C

D. stirtoni R. Wilson, 1934B

D. vallicula Shotwell, 1970A

Eucastor cf. *tortus* Leidy, 1858

Castor cf. *californicus* Kellogg, 1911A —Skinner et al. (1977)

Hystricops browni Shotwell, 1963B

Geomyidae

Prothomomys warrensis May, 1982

Pliosaccomys dubius R. Wilson, 1936A

P. magnus (Kellogg, 1910A)

Prodipodomys (Heteromyid?)

Parapliosaccomys oregonensis Shotwell, 1967B

Geomys

Pliogeomys buisi Hibbard, 1954A

Heteromyidae

Perognathus sargenti Shotwell, 1956

P. henryredfieldi Jacobs, 1977

P. coquorum A. Wood, 1935C

P. mclaughlini Hibbard, 1949C

Perognathus —Baskin (1979)

Perognathoides bidahochiensis Baskin, 1976

Cupidinimus magnus (Kellogg, 1910A) —A. Wood (1935C)

Diprionomys parvus Kellogg, 1910A

Dipodomys —Shotwell (1970A)

Prodipodomys kansensis (Hibbard, 1937E)

Cricetidae

Calomys (*Bensonomys*) *yazhi* Baskin, 1978

C. (*B.*) *coffeyi*

C. (*B.*) *gidleyi* Baskin, 1978

Pliotomodon primitivus Hoffmeister, 1945

Copemys —Baskin (1979)

C.? vasquezi Jacobs, 1977

Microtoscoptes disjunctus (R. Wilson, 1937A)

Paramicrotoscoptes hibbardi L. Martin, 1975

Peromyscus valensis Shotwell, 1967A

P. antiquus Kellogg, 1910A

P. pliocenicus R. Wilson, 1937A

Paronychomys alticuspis Baskin, 1979

P. lemredfieldi Jacobs, 1977

FIGURE 6-35 *Representative Hemphillian Localities.*

1 Alachua Formation sites—Florida
2 Alturas Formation sites—California
3 Amebelodon fricki Quarry—Nebraska—Kimball Formation
4 Arlington—Oregon
5 Arnett—Oklahoma
6 Ash Hollow Formation sites—Nebraska
7 Axtel—Texas
8 Bartlett Mountain—Oregon
6 Bear Tooth Slide (Frick Quarry)—Nebraska
9 Bingham Ranch—Arizona
10 Bidahochi Formation sites—Arizona
11 Black Rascal Creek—California
12 Bone Valley Formation sites—Florida
13 Borcher Gravel Pit—Kansas
5 Box T—Texas
14 Camel Canyon—Arizona
15 Chamita Formation sites—New Mexico
16 Chickasaw Creek—Alabama (Clarendonian?)
7 Christian Ranch—Texas
17 Christmas Valley—Oregon
18 Clover Creek—Idaho
19 Cobham Wharf—Virginia
5 Coffee Ranch—Texas
21 Cogswell Quarry—Kansas
7 Currie Ranch—Texas
22 Deer Lodge—Montana
23 Dry Creek (Grant County)—New Mexico
24 Dunnellon Plant No. 6—Florida
25 East Bench—Montana
26 Eastgate—Nevada
27 Edson—Kansas
28 Etchegoin Formation sites—California
29 Feltz Ranch—Nebraska
27 Found Quarry—Kansas
1 Gainesville sites—Florida
31 Ganson Farm—Kansas
32 Goleta (Michoacan)—Mexico
27 Goodland Southeast—Kansas
33 Goodnight—Texas
34 Harper—Oregon
5 Higgins—Texas
35 Hot Creek—Idaho
28 Jacalitos Formation sites—California

36 Kelly Road—Wyoming
37 Kern River Formation sites—California
3 Kimball Formation sites—Nebraska
38 Kinsey Ranch—California
39 Kissimmee River—Florida
40 Klipstein Ranch 3—California
4 Krebs Ranch—Oregon
41 Labahia Mission (Goliad State Park)—Texas
43 Lawler Ranch—California
35 Little Valley—Idaho
27 Lost Quarry—Kansas
45 Manatee County Dam Site—Florida
46 Mayflower Mine (Jefferson River)—Montana
1 McGehee Farm—Florida
47 McKay Reservoir—Oregon
11 Mehrten Formation (upper) sites—California
48 Middle Ridge (Lemhi Valley)—Idaho
49 Mixson's Bone Bed—Florida
50 Mount Eden Formation sites—California

45 Mulberry Phosphate Pit—Florida
51 Mulholland Formation sites—California
25 North Bench—Montana
11 Oakdale—California
52 Oceanside—Lawrence Canyon—California
53 Ocote (Guanajuato)—Mexico
9 Old Cabin—Redington—Arizona
54 Optima—Oklahoma
8 Otis Basin—Oregon
43 Petaluma Formation sites—California
55 Pinole Tuff sites—California
56 Rancho Lobo—Honduras
57 Rattlesnake Formation sites—Oregon
9 Redington—Arizona
58 Reynolds—Georgia
59 Rhinoceros Hill—Kansas
60 River Road—Washington
61 Rome—Oregon
62 Rooks (Plainville)—Kansas
63 Santee—Nebraska
45 Sarasota—Florida

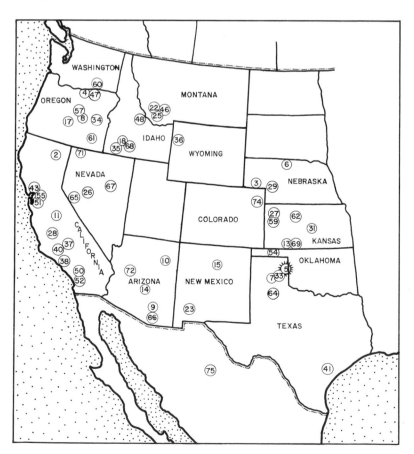

288

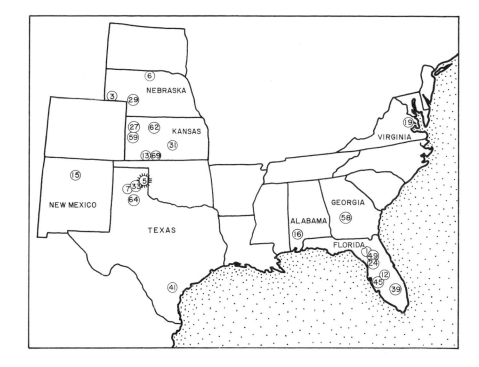

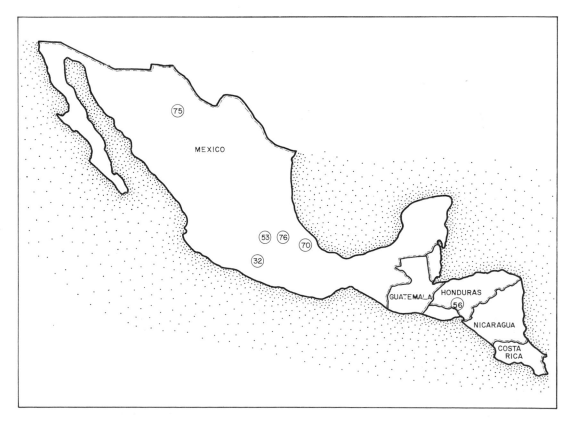

P. tuttlei Jacobs, 1977
Galushamys redingtonensis Jacobs, 1977
Oryzomys —Shotwell (1970A)
Promimomys mimus (Shotwell, 1956) —(Blancan I
of Repenning (1983?))
Propliophenacomys uptegrovensis L. Martin, 1975
P. parkeri L. Martin, 1975
Repomys gustelyi May, 1982
Zapodidae
Macrognathomys cf. *nanus* Hall, 1930D
Pliozapus solus R. Wilson, 1936A

HUAYQUERIAN

Huayquerian Mammalian Fauna (Genera Only)

(From manuscript by Marshall, Hoffstetter, and Pascual,
1980.)

MARSUPIALIA
Didelphidae
Lutreolina Thomas, 1910
Paradidelphys Ameghino, 1904
Thylatheridium Reig, 1952D
Sparassocynidae, gen. indet.
Borhyaenidae
Borhyaenidium Pascual and Bocchino, 1963
Stylocynus Mercerat, 1917
Thylacosmilidae
Achlysictis? Ameghino, 1891
Thylacosmilus Riggs, 1933B
Argyrolagidae
Microtragulus Ameghino, 1904
EDENTATA
Megatheriidae
Pronothrotherium Ameghino, 1907
Plesiomegatherium Roth, 1911
Pyramiodontherium Rovereto, 1914
Mylodontidae
Elassotherium Cabrera, 1939D
Sphenotherus Ameghino, 1891
Myrmecophagidae
Neotamandua Rovereto, 1914
Palaeomyrmedon Rovereto, 1914
CARNIVORA
Procyonidae
Cyonasua Ameghino, 1885
LITOPTERNA
Proterotheriidae
Brachytherium Ameghino, 1883
Diadiaphorus? Ameghino, 1887
Eoauchenia
Epecuenia
Macraucheniidae
Promacrauchenia Ameghino, 1904
NOTOUNGULATA
Toxodontidae
Pisanodon
Xotodon Ameghino, 1887

Mesotheriidae
Pseudotypotherium Ameghino, 1904
Typotheriopsis? Cabrera and Kraglievich, 1931
Hegetotheriidae
Hemihegetotherium Rovereto, 1914
Paedotherium Burmeister, 1888
Raulringueletia Zetti, 1972
Tremacyllus Ameghino, 1891
RODENTIA
Erethizontidae
Neosteiromys
Octodontidae
Phtoramys
Pseudoplateomys
Abrocomidae
Protabrocoma Kraglievich, 1927
Echimyidae
Cercomys Cuvier, 1832
Eumysops? Ameghino, 1888
Chinchillidae
Lagostomopsis
Dinomyidae
Diaphoromys Kraglievich, 1931
Potamarchus? Burmeister, 1885
Telicomys Kraglievich, 1921
Tetrastylopsis Kraglievich, 1931
Tetrastylus Ameghino, 1886
Caviidae
Cardiomys Ameghino, 1885
Caviodon?
Orthomyctera Ameghino, 1889
Palaeocavia Ameghino, 1889
Hydrochoeridae
Cardiatherium? Ameghino, 1883
Kiyutherium
Procardiatherium? Ameghino, 1885

MARINE MIOCENE MAMMALS

We have compiled representative but incomplete lists of the
known genera and species of marine Miocene mammals of
the world, following recent publications by Barnes, Domning, Ginsburg, Grigorescu, Kellogg, Mechedlidze, Mitchell, Ray, Reinhart, Repenning, Rothausen, and Tedford.[2]
Many of the taxa of cetaceans were proposed in Europe and
in North America before 1900. Since we lack the hundreds
of hours necessary to locate and examine the early European
literature and lack the expertise to judge the validity of early-
proposed genera and species, we have in many cases simply
listed the genera recognized in Simpson's (1945I) invaluable classification, together with the approximate ages given
therein. We itemize representative taxa under the headings
Earliest, Early, Middle, Late, and *Latest Miocene.*

Earliest Miocene Marine Mammals

(Aquitanian, Eggenburgian, ''Vaquerosian,'' late Zemorrian, Longfordian, etc., = Agenian, Late Arikareean,
Early Rusingan, approximately.)

CARNIVORA
  Enaliarctidae (North Pacific)   —Repenning, Ray, and
    Grigorescu (1979)
    Enaliarctos mealsi Mitchell and Tedford, 1973
    E. mitchelli Barnes, 1979
    Pinnarctidion bishopi Barnes, 1979
CETACEA
  ARCHAEOCETI
    Phococetus Gervais, 1876   (age?)
  ODONTOCETI
  Squalodontidae   —Rothausen (1970A) and personal
    commun. (1981)
    Prosqualodon australis Lydekker, 1893
    P. davidi Flynn, 1923
    Phoberodon arctirostris Cabrera, 1926
    Rhytisodon Paolo, 1897
    Squalodon catulli Zigno, 1876
    S. peregrinus Dal Piaz, 1977
    S. bellunense Dal Piaz, 1916
    Neosqualodon maior Rothausen, 1968A
  Platanistidae
    Allodelphis pratti L.E. Wilson, 1935
    Eoplatanides italica Dal Piaz, 1916
  Ziphiidae?
    Notocetus vanbenedeni Moreno, 1892
    Squalodelphis fabianii Dal Piaz, 1916
  Physeteridae
    Diaphorocetus Ameghino, 1892   (age?)
    Idiorophus R. Kellogg, 1925
    Scaldicetus bolzanensis Dal Piaz, 1977
  Eurhinodelphidae
    Argyrocetus Lydekker, 1893
    A. bakersfieldensis (L.E. Wilson, 1935)
    A. joaquinensis R. Kellogg, 1932
    Macrodelphinus kelloggi L.E. Wilson, 1935
    Ziphiodelphis abeli Dal Piaz, 1912
  Acrodelphidae
    Cyrtodelphis sulcatus (Gervais, 1853)
    C. gresalensis Dal Piaz, 1977
    Acrodelphis ombonii (Longhi, 1898)
  Phocaenidae
    Phocaenopsis mantelli Huxley, 1859   —Fordyce
    (1978)
  Delphinidae?
    Delphinavus newhalli Lull, 1914A
    Protodelphinus capellinii Dal Piaz, 1922
  ODONTOCETI, *Inc. Sed.*   —Barnes (1976)
    Microdelphis californicus L.E. Wilson, 1935
    gen. undet.   —Barnes (1976)
  MYSTICETI?
    Morenocetus parvus Cabrera, 1926
    gen. undet.   —Barnes (1976)
    aff. *Mauicetus*   —Fordyce (1978)
    "*Plesiocetus*" *dyticus* (Cabrera, 1926)   —Fordyce
    (1978)
    Aglaocetus moreni (Lydekker, 1893)   —Fordyce
    (1978)
SIRENIA
  Dugongidae   —Domning (1978–0365)
    Rytiodus capgrande Lartet, 1866
    Dioplotherium manigaulti Cope, 1883N   (age?)
    Metaxytherium krahuletzi Depéret, 1895

    Halitherium Kaup, 1838   (age?)
DESMOSTYLIA
    Cornwallius sookensis (Cornwall, 1922)
    Paleoparadoxia   (from California)

In earliest Miocene time, enaliarctids signified the beginning of the great radiation of otarioid carnivores in the Pacific Basin, archaeocetes were almost extinct, and squalodonts were the abundant representatives of the Odontoceti in the Atlantic. The important groups of odontocetes were extant. Mysticetes are less well known but were present, evidently in the Pacific as well as the Atlantic.

Early Miocene Marine Mammals

(Burdigalian, Langhian?, Ottnangian, Karpatian, "Temblorian" in part, Saucesian, = Orleanian or Early Aragonian, Hemingfordian, Santacrucian, and Rusingan)
CARNIVORA
  Enaliarctidae   —Repenning, Ray, and Grigorescu
    (1979)
  Canoidea   —Mitchell and Tedford (1973)
    Kolponomus clallamensis Stirton, 1960E
  Desmatophocidae (possible earliest record)   —Repenning et al. (1979)
  Odobenidae
    Gargantuodon ligerensis Ginsburg, 1969A   —(physeterid, acc. to Mein, pers. commun. (1981))
CETACEA
  ODONTOCETI
  Squalodontidae   —Rothausen (1970A)
    Squalodon grateloupi Meyer, 1843
    S. melitensis (Blainville, 1840)
    S. kelloggi Rothausen, 1968
    S. bariensis (Jourdan, 1861)
    S. zitteli Paquier, 1893
    S. dalpiazi Fabiani, 1949
    S. calvertensis Kellogg, 1923A
    S. crassus G. Allen, 1926A
    S. tiedemani J. Allen, 1887A   (age?)
    Neosqualodon assenzae Dal Piaz, 1904
    N. gastaldii Brandt, 1873
    Sachalinocetus cholmicus Dubrovo, 1971A   (age?)
  Platanistidae
    Proinia True, 1910
  Ziphiidae
    Notocetus vanbenedeni Moreno, 1892
    Cetorhynchus tetragonius (Delfortrie)
    C. macrogonius (Fischer)
  Physeteridae
    Apenophyseter patagonicus (Lydekker, 1893)
    Diaphorocetus poucheti (Moreno, 1892)
    Idiorophus Kellogg, 1925
    Scaldicetus bellunensis Dal Piaz
    Physeterula semseyi (Bockh)
    Hoplocetus curvidens Gervais, 1852   (age?)
    Physodon patagonicus Lydekker, 1893   (age?)
  Eurhinodelphidae
    Ziphiodelphis abeli Dal Piaz, 1912

Acrodelphidae
    Acrodelphis krahuletzi Abel, 1900
    Champsodelphis dationum Laurillard, 1846
    "C." valenciennesii (Brandt, 1873)
Kentriodontidae   —Barnes (1978–0087)
    Delphinodon dividum True, 1912
Delphinidae
    Iniopsis caucasia Lydekker, 1893   (age?)
    Megalodelphis magnidens R. Kellogg, 1944
    Stereodelphis brevidens Gervais and Dubreuil, 1849
MYSTICETI
Cetotheriidae
    Cophocetus oregonensis Packard and R. Kellogg, 1934
Balaenopteridae
    Plesiocetus Van Beneden, 1859
    "P." dyticus (Cabrera, 1926)
SIRENIA
Dugongidae
    Dioplotherium allisoni (Kilmer, 1965)   —Domning
       (1978–0365)
    Dusisiren reinharti Domning, 1978–0365
    Halitherium cf. *schinzi* Kaup, 1838
    Metaxytherium cordieri Christol, 1841
    M. krahuletzi Depéret, 1895
    M. lovisati Capellini, 1886
    M. studeri Meyer, 1837
    Thalattosiren petersi (Abel)
DESMOSTYLIA
    cf. *Paleoparadoxia* Reinhart, 1959
    cf. *Desmostylus* Marsh, 1888

Middle Miocene Marine Mammals

    (Langhian?, Serravallian, Early? Tortonian, Badenian, Sarmatian, Anversian, late Relizian, Luisian, late? "Temblorian," Early? Mohnian, = Astaracian or Late Aragonian, Barstovian, Friasian, Ternanian.)[3]
CARNIVORA   —Repenning and Tedford (1977)
Desmatophocidae
    Desmatophoca oregonensis Condon, 1906
    Allodesmus kernensis Kellogg, 1922
    A. kelloggi Mitchell, 1966
    A. sinanoensis (Nagao, 1941)
    A. courseni (Downs, 1956)
    A. packardi Barnes, 1972
    desmatophocids, 2 species   —Barnes (1972)
Odobenidae
    Neotherium mirum Kellogg, 1931B
Phocidae   —Ray (1976)
    Prophoca rousseaui Van Beneden
    Leptophoca lenis (True, 1906)
    L. proxima Van Beneden
    L.? vindobonensis (Toula, 1897)   —Ray (1976)
    Monotherium? wymani   —Ray (1976)
    Monotherium   —Rothausen (pers. commun., 1981)

CETACEA
ODONTOCETI
Squalodontidae   —Rothausen (1970A)
    Squalodon tiedemani J. Allen, 1887A   (age?)
    S. atlanticus (Leidy, 1856K)
    S. servatus (Meyer, 1841)
    S. antverpiensis Van Beneden, 1861
    S. pelagicus Leidy, 1869   (age?)
    Colophonodon holmesii Leidy, 1853C
    Saurocetus gibbesi Agassiz, 1848   (family?)
Platanistidae
    "Squalodon" errabundus (Kellogg, 1931B)
    Zarhachis flagellator Cope, 1868I
    Z. tysoni Cope, 1869G
    Z. velox Cope, 1869G
    Rhabdosteus latiradix Cope, 1867   (family?, age?)
Ziphiidae
    Cetorhynchus christoli Gervais, 1861
    Palaeoziphius melidensis Zbyszewski, 1954A   (age?)
    Incacetus brogii   (age?)
    Ziphioides trian Probst, 1886
    Ziphirostrum turninense Dubus, 1868   (age?)
    Anoplonassa Cope, 1869   (age?)
    Belemnoziphius compressus Huxley, 1864   (age?)
    Choneziphius Duvernoy, 1851   (age?)
    Mesoplodon Gervais, 1850   (age?)
    Proroziphius macrops Leidy, 1876   (age?)
Physeteridae
    Ontocetus emmonsi Leidy, 1859G
    O. oxmycterus Kellogg, 1925
    Aulophyseter morricei Kellogg, 1927B
    Scaldicetus caretti Dubus, 1867
    Idiophyseter merriami Kellogg, 1925B
    Orycterocetus crocodolinus Cope, 1867C
    O. quadratidens Leidy, 1853C
    Dinozipheus roemdorki Van Beneden, 1880
    Hoplocetus curvidens Gervais, 1852
Eurhinodelphidae
    Eurhinodelphis bossi Kellogg, 1925
Acrodelphidae
    Schizodelphis Gervais, 1861
    Champsodelphis acutidens (Cope, 1867C)
    C. dationum Laurillard, 1846
    C. lophogenius Vallenciennes, 1862
    Acrodelphis Abel, 1900
    Heterodelphis klinderi Brandt, 1873
Kentriodontidae Barnes, 1978–0087
    Kentriodon pernix Kellogg, 1927A
    Kampholophus serrulus Rensberger, 1969A
    Liolithax pappus (Kellogg, 1955A)
    L. kernensis Kellogg, 1931B
    Delphinodon dividum True, 1912
Delphinidae or Odontoceti, *inc. sed.*   —Barnes (1978–0087)
    Agabelus porcatus Cope, 1875H   (age?)
    Belosphys spinosus Cope, 1875H   (age?)
    Ixacanthus coelospondylus Cope, 1868G   (age?)
    Tretosphys gabbii (Cope, 1868I)
    Cetorhinops longifrons Leidy, 1877
    Pelodelphis gracilis Kellogg, 1955   (family?)
    Phocageneus venustus Leidy, 1869A

Araeodelphis natator Kellogg, 1957B
Saurodelphis? argentinus Rovereto, 1915 (age?)
Goniodelphis hudsoni Kellogg, 1944
Hadrodelphis calvertensis Kellogg, 1966 (family?)
Miotursiops mulla Deraniyagala, 1967B
Phocaenidae
Loxolithax sinuosa Kellogg, 1931B
Odontoceti, *inc. sed.* —Barnes (1976)
Oedolithax mira Kellogg, 1931B
Lamprolithax simulans Kellogg, 1931B
L. annectens Kellogg, 1931B
Nannolithax gracilis Kellogg, 1931B
Platylithax robusta Kellogg, 1931B
Grypolithax obscura Kellogg, 1931B
G. pavida Kellogg, 1931B
MYSTICETI
Cetotheriidae
Tiphyocetus temblorensis Kellogg, 1931B
Peripolocetus vexillifer Kellogg, 1931B
Parietobalaena palmeri Kellogg, 1924A
P.? securis Kellogg, 1931B
Aglaocetus patulus Kellogg, 1968
Heterocetus brevifrons (Van Beneden, 1872)
Siphonocetus clarkianus Cope, 1895A
S. expansus (Cope, 1868I)
S. priscus (Leidy, 1851F)
Diorocetus calvertensis Kellogg, 1965
D. hiatus Kellogg, 1968
Pelocetus calvertensis Kellogg, 1965
Saurocetus gibbesii Agassiz, 1848A (age?)
Cephalotropis coronatus Cope, 1896J (age?)
Cetotheriomorphis dubium Brandt, 1873
(age?)
Eucetotherium Brandt, 1872 (age?)
Herpetocetus scaldiensis Van Beneden, 1872 (age?)
Isocetus siphunculus (Cope, 1895A)
Metopocetus durinasus Cope, 1896J
Plesiocetopsis hupschii (Van Beneden) (age?)
P. megalophysum (Cope, 1895A)
P. occidentalis Kellogg, 1925B
Cetotherium rathki Brandt, 1843 (age?)
C. cephalum (Cope, 1867B)
C. leptocentrum (Cope, 1867C)
C. davidsoni (Cope, 1867Z)
C. mysticetoides (Emmons, 1858B) (age?)
C. polyporum (Cope, 1869I) (age?)
Rhegnopsis palaeatlanticus (Leidy, 1851F)
Mesocetus agrami Van Beneden, 1884
M. longirostris Van Beneden, 1880
Balaenidae
Eschrictius cephalus Cope, 1867B
E. pusillus (Cope, 1868G)
Balaenopteridae
Megaptera expansa Cope, 1868I
SIRENIA
Dugongidae
Dioplotherium allisoni (Kilmer, 1965) —Domning
(1978–0365)
Halitherium olseni Reinhart, 1976
Metaxytherium calvertense Kellogg, 1966
Trichechidae

"Potamotherium" magdalenensis (Reinhart, 1959)
(Ribodon?)
DESMOSTYLIA
Paleoparadoxia tabatai (Tokunaga, 1939) —Reinhart (1959)
Desmostylus hesperus Marsh, 1888
D. coalingensis (Reinhart, 1959) (*"Vanderhoofius"*)

Late Miocene Marine Mammals

(Late? Tortonian, Diestian, Late? Sarmatian, Pannonian, Mohnian, Early Delmontian, "Margaritan," etc. = Vallesian and Clarendonian.)

CARNIVORA
Desmatophocidae? —Repenning and Tedford (1977)
Odobenidae
Imagotaria downsi Mitchell, 1968
Otariidae
Pithanotaria starri Kellogg, 1925B
Phocidae —Ray (1976)
Phoca vindobonense Toula —Nicolas (1978)
P. pontica (Chiriac and Grigorescu, 1975)
P. maeotica Nordmann, 1858 —Nicolas (1978)
Monotherium delognii Van Beneden, 1876
M. affine Van Beneden, 1877
M. aberratum Van Beneden, 1877
"small primitive monachine" —Ray (1976)
CETACEA
ODONTOCETI
Platanistidae
"Squalodon" cf. *errabundus* (Kellogg, 1931B) —Barnes (1976)
Pachyacanthus sussi Brandt, 1873 (age?)
Hesperocetus True, 1912
Ziphiidae
Mesoplodon Gervais, 1850
Physeteridae
Physeterula dubusii VanBeneden, 1877 —Nicolas (1978)
Ontocetus oxymycterus Kellogg, 1925A
Prophyseter Abel, 1905 (age?)
Scaldicetus caretti Dubus, 1867
Thalassocetus antwerpiensis Abel, 1905 (age?)
Physeter Linnaeus, 1758
Haplocetus obesus Leidy, 1868 (age?)
Eurhinodelphidae
Eurhinodelphis longirostris Dubus, 1867 —Nicolas (1978)
Acrodelphidae
Acrodelphis fuchsii Brandt, 1900 —Nicolas (1978)
A. letochae Brandt, 1900
Heterodelphis leiodontus Papp, 1905
Cyrtodelphis (syn. of *Schizodelphis?*) *planatus* Abel, 1900
Schizodelphis sulcatus (Gervais, 1853) (age?)
Potamodelphis inaequalis Allen (age?)
Monodontidae
Delphinapterus fockii Brandt (age?)
Kentriodontidae
Liolithax kernensis Kellogg, 1931B

Liolithax —Barnes (1978–0087)
Leptodelphis stavropolitanus Kirpichnikov, 1954B
Sarmatodelphis moldavicus Kirpichnikov, 1954B
Pithanodelphis cornutus (DuBus, 1872)
Microphocaena podolica Kudrin and Tatarinov, 1965
Lophocetus repenningi Barnes, 1978–0087
L. calvertensis (Harlan, 1842)
Delphinidae
   Iniopsis caucasica Mchedlidze, 1964G
   Anacharsis orbus Bogachev, 1956 (age?)
   Macrochirifer vindobonensis Brandt, 1874
   Imerodelphis thabagarii Mchedlidze, 1959
   Delphinopsis freyeri Müller, 1853
Phocaenidae
   Loxolithax sinuosa Kellogg, 1931B
   L.? stocktoni (L. E. Wilson, 1973)
   Palaeophocaena andrussowi Abel, 1905
   Protophocaena Abel, 1905
   Phocaena euxinica (age?)
   P. communis (Pavlova) (age?)
Odontoceti, *Inc. Sed.* —Barnes (1976)
   Delphinavus newhalli Lull, 1914
   5 genera, undet.
MYSTICETI
Cetotheriidae
   Mixocetus elysius Kellogg, 1934B
   Plesiocetopsis hupschii (Van Beneden, 1880) (age)
   Cetotherium parvum Trouessart, 1898A (age?)
   C. mayeri Brandt, 1873 (age?) —Gofshtein (1965)
   C. linderi Brandt, 1871 (age?) —Gofshtein (1965)
   C. maicopicum (age?) —Mchedlidze (1964C)
   C. riabinini (age?) —Gofshtein (1965)
   Imerocetus karaganicus Mchedlidze, 1964C
   Herpetocetus scaldiensis Van Beneden, 1872 (age?)
   Cephalotropis coronatus Cope, 1896J
   Cetotheriomorphis dubium Brandt, 1873
   Eucetotherium Brandt, 1873
   Isocetus depauxi Van Beneden, 1880
   Metopocetus durinasus Cope, 1896J
   Nanisocetus eremus Kellogg, 1929B
Balaenidae
   Protobalaena palaeatlanticos Leidy, 1851
   Balaena Linnaeus, 1758
Balaenopteridae
   Megaptera miocaena Kellogg, 1922D
   Balaenopterid, gen. undet. —Barnes (1976)
   Mesoteras kerrianus Cope, 1870R
Mysticeti, *Inc. Sed.*
   "Balaenoptera" ryani (Hanna* and McLellan, 1924)
   —Barnes (1976)
   Mesocetus argillarius Roth, 1978–1036
SIRENIA
Dugongidae
   Dusisiren sp. B —Domning (1978–0365)
   D. jordani (Kellogg, 1925B) —Domning (1978–0365)
   D., sp. D —Domning (1978–0365)
   D.? ossivalense (Simpson, 1932C)

D.? floridanus (Hay, 1922B)
D.? ortegense (Kellogg, 1966)
Hesperosiren Simpson, 1932C —(synonym of *Metaxytherium calvertense?* —Domning (1978))
Indosiren javanense v. Koenigswald, 1952
"Manatus" maeoticus (Eichwald)
Trichechidae
   Ribodon limbatus Ameghino, 1883
DESMOSTYLIA
   Desmostylus —Reinhart (1976)

Latest Miocene Marine Mammals

(Messinian, Pontian, Scaldisian, "Jacalitos-Etchegoin," etc. = Turolian and Hemphillian.)

CARNIVORA
Odobenidae
   Aivukus cedrosensis Repenning and Tedford, 1977
   Dusignathus santacruzensis Kellogg, 1927C
   Pliopedia pacifica Kellogg, 1921A
   Valenictis imperialensis Mitchell, 1961
Otariidae
   Thalassoleon mexicanus Repenning and Tedford, 1977
   T. macnallyae Repenning and Tedford, 1977
   cf. *Callorhinus ursinus* (Linnaeus) —Repenning and Tedford (1977)
Phocidae —Ray (1976)
   Callophoca ambigua Van Beneden, 1876
   C. obscura Van Beneden, 1876
   Palaeophoca nystii Van Beneden, 1853
   Platyphoca vulgaris Van Beneden, 1876
   Gryphoca similis Van Beneden, 1876
   Phocanella pumila Van Beneden, 1876
   P. minor Van Beneden, 1876
   P. vitulinoides (Van Beneden, 1859)
CETACEA
ODONTOCETI
Platanistidae
   Saurodelphis argentinus Burmeister, 1891 (age?)
   Ischyrorhynchus vanbenedeni Ameghino, 1891 (age?)
   Anisodelphis Rovereto, 1915 (age?)
   Pontistes Burmeister, 1885 (age?)
   Pontivaga fischeri Ameghino, 1891 (age?)
Ziphiidae
   Anoplonassa forcipata Cope, 1869E
   Mesoplodon prorops Leidy, 1876A
   Eboroziphius
   Choneziphius liops Leidy, 1876A
   C. trachops Leidy, 1876A
   Prorozipheus macrops Leidy, 1876E
   P. chomops Leidy, 1876E
Physeteridae
   Praekogia cedrosensis Barnes, 1973B
   Scaldicetus mortezelensis Abel, 1905
   S. grandis
   Physetodon McCoy, 1879 (age?)
   Balaenodon physaloides Owen, 1846 (age?)
   Priscophyseter typus Portis, 1886 (age?)
   Physeter Linnaeus, 1758
   Kogiopsis floridana R. Kellogg, 1929C (age?)
   Dinozipheus? carolinensis (Leidy, 1877) (age?)

Acrodelphidae
  Schizodelphis sulcatus (Gervais, 1853)
Eurhinodelphidae
  Hemisyntrachelus Brandt, 1873
Monodontidae
  delphinapterine, new genus   —Barnes (1976)
Delphinidae
  Stenella or *Delphinus*   —Barnes (1976)
  aff. *Tursiops* Gervais, 1856   —Barnes (1976)
  Steno Gray, 1846
  Orcinus Fitzinger, 1860   (age?)
  Delphinus Linnaeus, 1758
Phocaenidae, 3 spp.   —Barnes (1976)
ODONTOCETI, *Inc. Sed.*   —Barnes (1976)
  Lonchodelphis occiduus (Leidy, 1868E)
  odontocetes, 3 spp.
MYSTICETI
Cetotheriidae
  Nannocetus Kellogg, 1929
  Heterocetus guiscardi Van Beneden, 1880
  Amphicetus later Van Beneden, 1880
  Cetotherium parvum Trouessart, 1898   (age?)
Balaenopteridae
  Palaeocetus sedgewicki Seeley, 1865
  Plesiocetus Gervais, 1855
  Idiocetus guicciardini Capellini, 1876
  Notiocetus romerianus Ameghino, 1891
  Balaenoptera acutorostrata Lacépède, 1804
  Burtinopsis minuta Van Beneden, 1872
  Megaptera Gray, 1846
  Megapteropsis affinis Van Beneden, 1872
Family?
  Stenodelphinae, new genus   —Barnes (1976)
Balaenidae
  Balaenula   —Barnes (1976)
  Balaenotus Van Beneden, 1872
  Balaena Linnaeus, 1758
SIRENIA
Dugongidae
  Hydrodamalis cuestae Domning, 1978–0365

STRATIGRAPHIC DEMONSTRATIONS

Many districts in western North America and on the Gulf Coastal Plain of Texas show superposition of strata bearing mammal fossils of one Miocene land-mammal age on strata with fossils of the preceding mammal age. A few districts provide stratigraphic association between continental and marine fossiliferous strata.

Thanks to the lifetime work of stratigraphic paleontologists from the University of Nebraska, the Frick Laboratories of the American Museum of Natural History, and several other organizations in the eastern United States, the area of northern and northwestern Nebraska is especially noteworthy for its stratigraphic demonstration of the entire succession of land-mammal ages for the Oligocene through Miocene of North America on an extended, regional scale.

These ages and the formations involved are shown in the accompanying chart.

Ages	Formations
HEMPHILLIAN	Kimball–upper Ash Hollow
CLARENDONIAN	Lower Ash Hollow Upper Valentine
BARSTOVIAN	Lower Valentine–Snake Creek (in part) Lower Snake Creek
HEMINGFORDIAN	Sheep Creek Box Butte Runningwater Marsland (restricted)
Late ARIKAREEAN	unnamed formation Harrison
Early ARIKAREEAN	Monroe Creek Gering
WHITNEYAN ORELLAN	Brule
CHADRONIAN	Chadron

Nebraska's faunal representation of these ages, collectively, is unrivalled.

Another demonstration that merits special comment is the Caliente Range district at the southern end of the Central Coast ranges of California. Referring to this district, Repenning and Vedder (1961) modestly claimed: "Five and possibly six North American provincial (mammalian) ages, as defined by Wood and others (1941), are represented by vertebrate faunas in upper Tertiary rocks of the eastern part of the Caliente Range. This record of superimposed mammalian assemblages is equalled at few places in the world. By tracing fossiliferous beds, faunas representing three of these ages can be correlated with assemblages of abundant marine mollusks in this same area."

We elaborate the Repenning-Vedder claim: On the northeastern side of the Caliente Range in a district that can be crossed in any direction by a two-hour hike, the interdigitation of Miocene fossiliferous marine and nonmarine strata and basalt-lava flows as shown in our Figure 6-36 can be seen. Figure 6-36 is modified from Repenning and Vedder (1961) with radiometric data (corrected constants) from Turner (1970). The array of fossils for each of the land-mammal ages in the Caliente Range is not impressive, although adequate to substantiate the datings. (See the list of identified taxa given by Repenning and Vedder, 1961.) This record of marine and continental faunas, thriving in a small, marine-shoreline district through a 20-million-year interval, is unique for the Cenozoic of the entire world so far as we

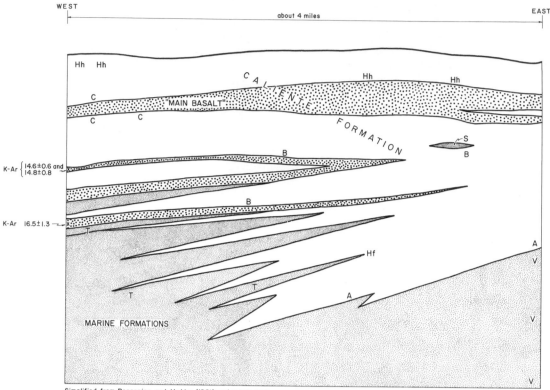

FIGURE 6-36 *Diagram of Miocene Rocks, Caliente Range, California.*
Key:
Hh, Hemphillian land-mammal age = latest Miocene.
C, Clarendonian land-mammal age = late Miocene.
S, "San Pablo" or "Margaritan" marine molluscs = late Miocene.
B, Barstovian land-mammal age = middle Miocene.
K-Ar 14.6 ±0.6, Radiometric date (corrected constants) from Turner (1970).
Hf, Hemingfordian land-mammal age = early Miocene.
T, "Temblorian" marine molluscs, early Miocene.
A, Arikareean (or early Hemingfordian) land-mammal age, earliest Miocene.
V, "Vaquerosian" marine molluscs = earliest Miocene.

can determine. And this is indeed a singular phenomenon in earth's history when viewed in its geodynamic setting; the Caliente Range district is on the western plate and is adjacent to the great San Andreas fault. It is estimated that the area of the present Caliente Range moved northwestward relative to the currently adjacent Temblor Range on the eastern plate by as much as 120 km during this 20 million years.

Another district in North America shows a beautiful exposure of diversified fossiliferous continental Miocene sedimentary rocks and associated volcanics. Here is presented an unrivalled view of the flora and fauna that existed in and around a small lake during middle to late Miocene time. This is the Stewart Valley–Cedar Mountain area of

western Nevada, and its only rival to our knowledge is the penecontemporaneous Shanwang district, Shandong, China (Hsen Hsu Hu and Chaney, 1940).

Our sketch Figure 6-37 shows the general stratigraphic profile of Miocene rocks in the present-day Stewart Valley. Water, rich in diatoms, occupied the earlier lake of this area about 16.4 million years ago. The ancient lake received and preserved the land flora of the district during a short, quiescent interval between andesitic eruptions from nearby vents. The paleoflora in the resultant diatomaceous beds (latest Hemingfordian or earliest Barstovian) includes 24 species of warm-temperate mesophytic plants (Wolfe, 1964). Live oak is the most common plant found, and the species of lobed oaks, hickory, elm, *Zelkova,* poplars, and

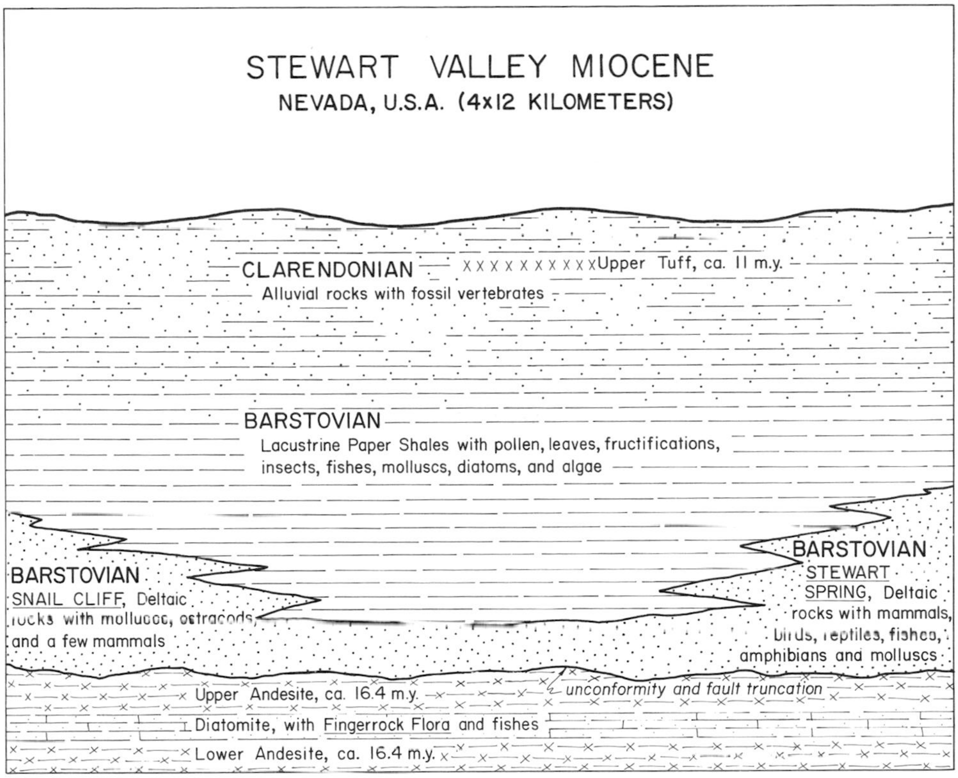

FIGURE 6-37 *Profile of the Miocene Strata in Stewart Valley, Nevada.*

maples are found also in contemporaneous floras of the Columbia Plateau to the north.

Subsequent to the last outpouring of andesitic debris, another lake formed in the district. Through an interval of 2–3 million years (during Barstovian time), it received and deposited small and periodic increments of fine dust (partly volcanic) from the air, supplemented by fine sediment discharged into the basin from several streams. Molluscs and vertebrates (Barstovian) abounded in the lake deltas and nearshore districts, while a rich insect fauna, fish fauna, and land-plant flora were being trapped and preserved at various places in the laminated strata ("paper shales") of the main lake basin. Preservation of the insects, the leaves, and the plant fructifications is outstanding. Scudder (1981) states that the quality of preservation of the insects is documented by the clear presence of (1) pigment patterns and (2) hairs, hair sockets, wing scales, and the ommatidial facets of compound eyes. This fossil insect sample (undescribed) is characterized by Scudder (1981) as shown in the accompanying chart. This is the best mid-Miocene record of insects for the entire world.

The accompanying plants, first described by Wolfe (1964), comprise 42 species dominated by live oak, spruce, and chamaecyparis cedar. Some of these plants evidently lived along the margin of the lake: water pines, poplars (4 species), willows (2 species), alders, and ironwoods. Other plants in this upper flora made up a mesophytic upland for-

Orders	Percentage of specimens in present collections
Hemiptera (true bugs)	5
Homoptera (cicadas, leaf hoppers, etc.)	10
Neuroptera (lace-winged flies, etc.)	5
Coleoptera (beetles)	Trace
Trichoptera (caddis flies)	Trace
Lepidoptera (moths and butterflies)	Trace
Diptera (gnats and true flies)	25
Hymenoptera (wasps and bees)	50

est near the lake: red firs (2 species), spruce (2 species), pines (3 species), chamacyparis cedars, oaks (3 species), Oregon grapes, currants, gooseberries, mountain mahogany, ironwoods, buffalo berries, and manzanitas.

Gradually, Stewart Valley paleo-lake was filled with sediment. By late Miocene time, stream floodplain sediments were being deposited over the older lake beds. During this final sedimentary episode, a thick layer of ash from the outburst of a distant volcano blanketed the area, and a land–vertebrate fauna (Clarendonian) lived and died on the floodplain and in the adjacent countryside.

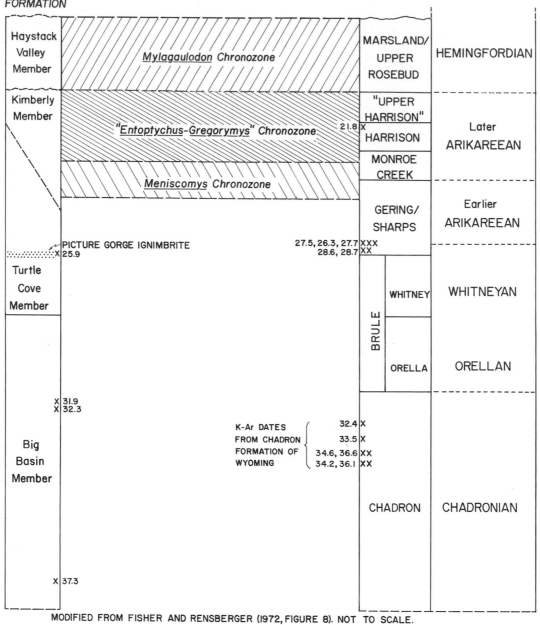

FIGURE 6-38   *Rodent Chronozones, Oregon to Nebraska. Taken from Fisher and Rensberger (1972, Figure 8).*
(Published with permission of the University of California Press.)

DATING AND CORRELATION

A general dating and correlation of Miocene fossil-mammal ages of the world is depicted in Figure 6-39. These correlations are now supported by many postassium-argon and fission-track radiometric dates from North America, Europe, Africa, and South America. You may refer to Berggren and Van Couvering (1974), Vass and Bagdasarjan (1978), Evernden et al. (1964), Turner (1970), Tedford et al. (1983?), and the references they cite for most of these radiometric dates.

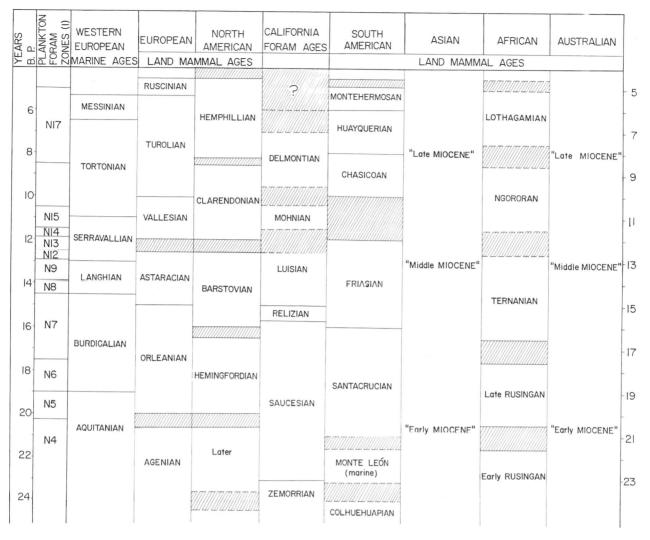

FIGURE 6-39 *General Correlations Within the Nonmarine Miocene of the World.*

BIOGEOGRAPHY AND THE MILIEU

In early Miocene time, about 20 million years ago, the western and eastern coastal belts of southern North America were largely covered with epeiric seas and there were barriers to land and sea dispersal of organisms in the Beringian and Panamanian areas. Africa maintained essentially its present-day continental outline except for slight sea encroachment in its northernmost part. Europe was still a mass of small seaways and inlets. Asia was land, as it is today, except that seas inundated parts of its east and southeast coast and adjacent archipelagos (Japan and the East Indies). South America and Australia remained island continents, with minor sea invasions in a few coastal districts.

In general, the continental plates were less emergent than in the preceding epoch. Toward the end of the Miocene, Europe witnessed recession of many of its sea inlets,

and the Mediterranean became a land-locked supersaline basin. Eurasia became emergent as Alpine and Himalayan orogenies matured. Western North America also witnessed many complex shiftings of small epeiric seaways. During the 19 million years of the Miocene, paleoorganisms of the coastal districts were constantly shifting and dispersing as a result of these geologic revolutions.

Although the marine and continental climates became slightly warmer in early-to-mid-Miocene time (Addicott, 1969; Wolfe, 1978, 1980), the general trend of continental climate was toward drier and cooler conditions. There was enormous and sporadic volcanism in the Cordillera of the Western Hemisphere, which undoubtedly had significant effect on the local climates and, in turn, on the local biota. In the 16,000-km north-south belt of rocks of the Cordillera, volcanic ash beds and lava flows are frequently intercalated with fossiliferous nonmarine and marine strata.

Thus, many opportunities arise for the radiometric chronology to be integrated with paleontologic chronologies and with paleomagnetic-reversals correlations.

The Miocene fossil land-plant record is copious, especially in western North America. It largely controls the reconstruction of Miocene habitats and climates. (See papers by R.W. Chaney, D.I. Axelrod, J.A. Wolfe, Wolfe and E. Leopold, Wolfe and Tanai, and R.S. LaMotte, and the references they cite.)

It is difficult to present a summation of Miocene climates of the entire world because the paleofloras and other data indicate a constantly changing diversity from area to area, in different latitudes, at different elevations, and at different distances from seas. Wolfe (1978) noted that many of the European and northwestern North American Neogene floras show an overall change from forests dominated by broad-leaved deciduous trees to more coniferous forests. To him this indicated a decrease in mean annual range of temperature and possibly some decline in warm-month mean temperature and mean annual temperature. But as Wolfe also noted (1978, p. 700), in some areas such as the western United States the Neogene floras are greatly complicated by altitudinal and rain-shadow factors. Habitats of the Miocene included moist lowland coastal plains, swamps, drier rocky hillsides and mountainsides, foothill basins, playa lakes and permanent lakes, higher mountain districts, river floodplains, savanna country with arid desert-border plants and arid subtropical scrub, grassland districts in some interior regions, and riparian communities.

In North America for example, Miocene mammalian faunal provinces can be identified (with vague boundaries) on the basis of peculiar aggregates of species. We are impressed, however, that mammals were (and are) uniquely independent from environmental controls compared with most other classes of organisms. The mammalian fauna of any interval during Miocene time appears to be singularly homogeneous throughout its geographic range, despite its spread through the diverse habitats listed above.

MIOCENE LITHOSTRATIGRAPHIC UNITS

Abiquiu Formation—Arikareean—New Mexico
Agenais, calcaire blanc de l', calcaire gris de l', or molasse de l'—Agenian—France
Alachua Formation—Hemphillian (and Clarendonian?)—Florida
Alcala, marnes et calcaires d'—Turolian and Ruscinian—Spain
Alengerr Beds—Ternanian—Kenya
Alturas Formation—Hemphillian—California
Alum Bluff Group—Hemingfordian—Florida and adjacent states
Anaverde Formation—Hemphillian—California
Anjou, faluns de—Orleanian—France
Anvers (Antwerp), sables d'—marine = Astaracian—Belgium
Armagnac, molasse de l'—Orleanian—France
Armagnac, sables fauves d'—Astaracian—France
Asakawa Formation—(= Astaracian)—Japan

Ash Hollow Formation—Clarendonian and Hemphillian—Nebraska
Astoria Formation—marine and Barstovian—Oregon
Avawatz Formation—Clarendonian—California
Baho Formation—Late Miocene (= Vallesian?)—Shaanxi
Balta Sands—Late Miocene—USSR
Barstow Formation—Hemingfordian and Barstovian—California
Batesland Formation—Hemingfordian—South Dakota
Battle Creek Volcanic Sandstone—Barstovian—Oregon
Bazadais, marnes à *Unio* du (= Falun de Bazas?)—Agenian—France
Beauce, calcaire de—Agenian—France
Bedrock Spring Formation—Hemphillian—California
Beglia Formation—Astaracian and Vallesian—Tunisia
Belluno, molasse de—Orleanian?—Italy
Bidahochi Formation—Clarendonian and Hemphillian—Arizona
Black Rock Sandstone Member, Sandringham Sands—Late? Miocene—Victoria, Australia
Bone Valley Formation—Clarendonian and Hemphillian—Florida
Bopesta Formation—Barstovian—California
Bou Hanifia Formation—(Vallesian) Ngororan—Algeria
Bourbonnais, sables de—Orleanian—France
Bouzigues, phosphorites de—Agenian—France
Box Butte Formation—Hemingfordian—Nebraska
Branch Canyon Formation—marine and Hemingfordian—California
Bresse, couches de la—Late Miocene—France
Bridwell Formation—Hemphillian—Texas
Briones Sandstone—marine and Barstovian—California
Browns Park Formation—Barstovian—Colorado
Bugti Member, Chitarwata Formation—Early Miocene—Baluchistan, Pakistan
Burge Sand Member or Burge Channel Sands, Valentine Formation—Clarendonian—Nebraska
Cabbage Patch beds—Late Arikareean and Early Arikareean—Montana
Cabriel, série détritique du—Turolian—Spain
Caliente Formation—Late Arikareean, Hemingfordian, Barstovian, Clarendonian, and Hemphillian—California
Calvert Formation—marine and Barstovian—Maryland
Camfield Beds—Middle to Late Miocene—Northern Territory, Australia
Camp Davis Formation—Clarendonian—Wyoming
Cañada Pilares Member, Zia Sand—Hemingfordian—New Mexico
Cap Rock Member, Ash Hollow Formation—Clarendonian—Nebraska
Carl Creek Limestone—Middle to Late Miocene—Queensland, Australia
Cedarville Formation, upper—Barstovian—California
Cerithien-schichten—marine and Agenian—Germany
Chalk Butte Formation—Hemphillian—Oregon
Chalk Hills Formation—Hemphillian—Idaho
Chama-el-rito Member, Tesuque Formation—Hemingfordian, Barstovian, and Clarendonian—New Mexico
Chamisa Mesa Member, Zia Sand—Hemingfordian—New Mexico
Chamita Formation—Hemphillian—New Mexico
Chanac Formation—Clarendonian—California
Chinchinales beds—Santacrucian—Argentina

Chinji "Formation," "zone"—(= Astaracian and? Vallesian)—Pakistan and India

Chitarwata Formation—Late? Oligocene? and Early Miocene—Baluchistan

Chokrak Formation—(= Astaracian)—Caucasus, USSR

Choptank Formation—marine and Barstovian—Maryland

Chur-Lando beds—Early Miocene—Baluchistan

Citronelle Formation—marine and Hemphillian in part—Alabama

Clarendon beds—Clarendonian—Texas

Coal Valley Formation—Clarendonian and Hemphillian—Nevada

Colloncura Formation—Santacrucian—Argentina

Contra Costa Group—Clarendonian and Hemphillian—California

Corbicula-schichten—Agenian—Germany

Couch Formation—Clarendonian—Texas

Crookston Bridge Member, Valentine Formation—Barstovian—Nebraska

Crowder Formation—Barstovian? and later—California

Cucaracha Formation—Hemingfordian—Panama

Dam Formation—(= Astaracian)—Saudi Arabia

Dawes Clay Member, Box Butte Formation—Hemingfordian—Nebraska

Deep River Formation—Arikareean—Montana

Delaho Formation—Late Arikareean—Texas

Delmore Formation—Hemphillian—Kansas

Desaguadero Formation—Santacrucian?—Bolivia

Devil's Gulch Member, Valentine Formation—Barstovian—Nebraska

Dhok Pathan beds, "zone," etc.—(= Turolian)—Pakistan and India

Diablo Formation—marine and Clarendonian—California

Digne, molasse marine de—marine and Agenian—France

Dinotheriensande—Vallesian—Germany

Djilancik (River) beds—Orleanian and Astaracian—Kazakhstan

Drewsey Formation—Hemphillian—Oregon

Edegem, sables de—marine and Astaracian—Belgium

Eggenburger Schichten—marine Eggenburgian and Orleanian—Austria

Ellensburg Formation—Clarendonian—Washington

"Emma Creek" (See Delmore Formation.)

Esmeralda Group—Barstovian and Clarendonian—Nevada

Etadunna Formation—Middle? Miocene—South Australia

Etchegoin Formation—Hemphillian—California

Flaxville Gravels—Clarendonian—Montana

Fleming Formation—Hemingfordian and Barstovian—Texas

"Fort Logan" Formation—Arikareean—Montana

Fort Randall Formation—Barstovian—South Dakota

Fossil Bluff Sandstone—Early Miocene—Tasmania

Gansu (Kansu) Formation—Late Miocene (and? into Pliocene?)—Gansu

Gers, molasse de—Orleanian?—France

Goliad Formation—Clarendonian (in part?)—Texas

Gracias Formation—Hemphillian—Honduras

Grassy Mountain Formation—Clarendonian? and Hemphillian—Oregon

Green Valley Formation—Clarendonian—California

Greit lignite—Agenian?—Switzerland

Grilos Sandstone—Astaracian—Portugal

Grimmelfinger Schichten—Astaracian—Germany

Grube lignite—Astaracian—Spain

Harrison Formation ("Lower Harrison")—Late Arikareean—Nebraska, Wyoming, and South Dakota

Haystack Valley Member, John Day Formation—Hemingfordian—Oregon

Hector Formation—Late Arikareean—California

Hemingford Group—Hemingfordian—Nebraska, Wyoming, and South Dakota

Hemphill Member or beds—Hemphillian—Texas

Herret, calcaire de—Orleanian—France

Hessenbrücker Hammer Braunkohle—Astaracian—Germany

"Hipparion red clays" or "H. richthofeni clays"—(= Turolian)—Gansu

Hiwegi Formation—Rusingan—Kenya

Hofuf Formation—(= Astaracian)—Saudi Arabia

Hollenburger Konglomerat—Astaracian—Austria

Homberger Sneckentone—Astaracian—Germany

Honda Group—Late Oligocene? and Miocene—Colombia

Horned Toad Formation—Hemphillian—California

Hsiachaohwan beds—(= Astaracian)—Kiangsu

Hsiaolungtan Formation—(= Astaracian)—Yunnan

Humboldt Group (in part)—Barstovian and Clarendonian—Nevada

Hung-Kureh Formation—Late Miocene—Mongolia

Hydrobienschichten—Agenian—Germany

Innviertler Schichten—marine Ottnangian and Orleanian—Austria

Ironside beds—Clarendonian—Oregon

Jewett Sand—marine and Late Arikareean—California

John Day Formation—Whitneyan, Arikareean, and Hemingfordian—Oregon

Johnson Member, Snake Creek Formation—Hemphillian—Nebraska

Juntura Formation—Clarendonian—Oregon

Kabuga beds—Ternanian?—Zaire

Kamlial "Formation," "zone"—(= Astaracian)—Pakistan and India

Keechelus Formation—Arikareean—Washington

Kern River Formation—Hemphillian—California

Khunuk Formation—Late Miocene—Mongolia

Kiahera Formation—Rusingan (in part?)—Kenya

Kimball Formation—Hemphillian—Nebraska

Kimberly Member, John Day Formation—Arikareean—Oregon

Kinnick Formation—Hemingfordian—California

Kirkwood Formation—marine and Late Arikareean—New Jersey

Kulu beds—Rusingan—Kenya

Lagarto Clay—Clarendonian—Texas

Lake Matthews Formation—Clarendonian—California

Landscheckenmergel—Agenian—Germany

Laren bed—Early? Miocene—Jammu-Kashmir, India

La Romieu, sables de—Orleanian—France

Laucomer Member, Snake Creek Formation—Clarendonian—Nebraska

La Vela "series"—Santacrucian—Venezuela

Laverne Formation—Clarendonian—Oklahoma

Lectoure, calcaires inférieurs de—Orleanian—France

Lengshuikou Formation—(= Astaracian)—Shaanxi

Lefkon Formation—Vallesian and Turolian—Macedonia, Greece

Litorinallenkalk—Agenian—Germany

Little White River Formation—Clarendonian—South Dakota

"Livermore Gravels"—Clarendonian into Pleistocene—California

Loh Formation—Early Miocene—Mongolia

Lothagam Group—Lothagamian and Langebaanian—Kenya

Lukeino Formation—Ngororan—Kenya

Lutzukou beds—(=Turolian)—Shanxi

Madras Gravels—Clarendonian—Oregon

Mahmoud Formation—(Orleanian) Rusingan—Tunisia

Manchar beds—Middle? Miocene—Sind, India

Maricopa Shale—marine and Clarendonian, in part—California

Marsland Formation—Hemingfordian—Nebraska

Martin Canyon beds—Hemingfordian—Colorado

Marvila, calcaire de—Astaracian—Portugal

Mascall Formation—Barstovian—Oregon

Meereswassermolasse, upper—marine Eggenburgian and Orleanian—Switzerland

Mehrten Formation—Clarendonian and Hemphillian—California

Miccosukee Formation—Barstovian—Florida

Milk Creek Formation—Clarendonian—Arizona

Mint Canyon Formation—Barstovian and Clarendonian—California

Mollon, marnes de—Vallesian—France

Monarch Mill Formation—Barstovian and Hemphillian—Nevada

Monroe Creek Formation—Arikareean—Nebraska, Wyoming, and South Dakota

Montabuzard, calcaire de—Orleanian—France

Moraga (= "Grizzly Peak") Formation—Clarendonian—California

Mt. Eden Formation—Hemphillian—California

Mpesida Beds—Ngororan—Kenya

Muddy Creek Formation—Barstovian? into Hemphillian—Nevada

Murphy Member, Snake Creek Formation—Clarendonian—Nebraska

Murree Formation—Early Miocene—Pakistan and India

Nagri "Formation," "beds," "zone"—(=Vallesian)—Pakistan and India

Nambé Member, Tesuque Formation—Hemingfordian—New Mexico

Nantan Formation—(=Turolian)—Shanxi

Ngorora Formation—Ngororan—Kenya

North Park Formation—Hemingfordian?, Barstovian?—Colorado

Nussbaum Formation—Clarendonian?—Colorado

Nyamavi beds—Rusingan—Zaire

Oakville Sandstone—Hemingfordian—Texas

Oeninger (Öhninger) Schichten—Astaracian—Germany

Ogallala Group or Formation—Clarendonian and Hemphillian—Great Plains region of United States

Ojo Caliente Sandstone Member, Tesuque Formation—Barstovian and Clarendonian—New Mexico

Olcott Formation—Barstovian—Nebraska

Orinda Formation—Clarendonian—California

Orléanais, sable de l'—Orleanian—France

Palmira Formation—Friasian?—Uruguay

Pang-Kiang Formation—(=Astaracian?)—Mongolia

Pato Red Member, Vaqueros Formation—Late Arikareean—California

Pawnee Creek beds—Hemingfordian into Clarendonian—Colorado

Payette Formation—Barstovian and Clarendonian—Oregon and Idaho

Peace Valley Formation—Hemphillian—California

Pegu Formation, upper (in part?)—Early?, Middle? Miocene—Burma

Pellecahus, sables de—Orleanian—France

Petaluma Formation—Hemphillian—California

Peterson Creek beds—Late Arikareean—Idaho

Piedra Parada Member, Zia Sand—Late Arikareean—New Mexico

Pinole Tuff—Hemphillian—California

Plush Ranch Formation—Hemingfordian (in part?)—California

Poison Creek Formation—Clarendonian—Oregon-Idaho

Pojoaque Member, Tesuque Formation—Barstovian—New Mexico

Prambachkirchen, Inviertler Serie von—marine Eggenburgian and Orleanian—Austria

Puchen beds—(=Astaracian)—Jiangsu

Puente Formation—marine and Clarendonian—California

Punchbowl Formation—Hemingfordian, Barstovian, Clarendonian, and Hemphillian—California

Pyramid Hill Sand Member, Jewett Sand—marine and Late Arikareean—California

Quibiris Formation—Hemphillian—Arizona

Raine Ranch Formation—Barstovian and Hemingfordian?—Nevada

Rattlesnake Formation—Hemphillian—Oregon

Red Valley Member, Box Butte Formation—Hemingfordian—Nebraska

Renova Formation—Oligocene and Late Arikareean—Montana

Repetto Formation—marine and Clarendonian—California

Republican River Formation—Clarendonian—Kansas

Ricardo Formation—Clarendonian—California

Rosamond Group, in part—Barstovian—California

Rosebud Formation—Arikareean and Hemingfordian—South Dakota

Round Mountain Silt Member, Temblor Formation—marine, Hemingfordian, and Barstovian—California

Runningwater Formation—Hemingfordian—Nebraska

Salt Lake Group—Hemingfordian, Barstovian, and later—Idaho-Utah

Sand Canyon Member, Sheep Creek Formation—Hemingfordian—Nebraska

Sandringham Sands, in part—Late Miocene—Victoria, Australia

San Pablo Formation or Group—marine and Clarendonian—California

Santa Cruz Formation—Santacrucian—Argentina

Santa Fe Group—Arikareean, Hemingfordian, Barstovian, Clarendonian, and Hemphillian—New Mexico

Santa Margarita Formation—marine and Clarendonian—California

Sekköy Member (=Astaracian and Vallesian)—Turkey

Serae beds—Rusingan—Ethiopia

"Sespe-Vaqueros" transition—Arikareean—California

Shanwang Formation—(=Astaracian)—Shandong

Sheep Creek Formation—Hemingfordian—Nebraska

Shiloh Marl—marine and Hemingfordian—New Jersey
Siebert Tuff, Esmeralda Group—Barstovian—Nevada
Siesta Formation—Clarendonian—California
Sinap "series"—Late Miocene and Pliocene—Turkey
Sinda beds—Rusingan—Zaire
Siwaliks "series," "Group"—Middle Miocene (= Astaracian) into Villafranchian (and later)—Pakistan and India
Six Mile Creek Formation—Barstovian, Clarendonian, and Hemphillian—Montana
Skull Ridge Member, Tesuque Formation—Barstovian—New Mexico
Skyline Tuff Member, Barstow Formation—Barstovian—California
Snake Creek Formation—Barstovian, Clarendonian, and Hemphillian—Nebraska
Sologne, sables de—Orleanian?—France
Split Rock Formation—Hemingfordian—Wyoming
Suchilquitongo Formation—Barstovian or Clarendonian—Oaxaca, Mexico
Sucker Creek Formation—Barstovian—Oregon
Süsswassermolasse, upper—marine Eggenburgian and Orleanian—Switzerland
Table Cape Group, in part—marine and early Miocene—Tasmania
Tassajara Formation—Clarendonian and Hemphillian and Blancan—California
Taylor gravel—Hemphillian—Arizona
Teewinot Formation—Hemphillian—Wyoming
Temblor Formation—marine, Hemingfordian, and Barstovian—California
Tesuque Formation—Hemingfordian, Barstovian, and Clarendonian—New Mexico
The Dalles Formation—Clarendonian—Oregon
Thin Elk Formation—Clarendonian—South Dakota
Thousand Creek Formation—Hemphillian—Nevada
Tick Canyon Formation—Late Arikareean—California
Touraine, faluns de—Orleanian, Astaracian (and marine)—France
Tropico Group, in part—Hemingfordian—California
Troublesome Formation—Hemingfordian—Colorado
Trout Creek Formation—Barstovian—Oregon
Troutdale Gravels—Clarendonian—Oregon
Truckee Group—Clarendonian, Hemphillian, and Blancan—Nevada and California
Tung Gur Formation—(= Astaracian)—Mongolia
Tungsuenkuan beds—(= Astaracian)—Jiangsu
Turgut Member—(= Astaracian)—Turkey
Turtle Butte Formation—Arikareean—South Dakota
Turtle Cove Member, John Day Formation—Arikareean—Oregon
Type Member, Snake Creek Formation—Clarendonian—Nebraska
Upper Harrison beds of Peterson—Late Arikareean—Nebraska
Valentine Formation—Barstovian and Clarendonian ("Valentinian")—Nebraska
Vetas beds—Late Oligocene or Early Miocene—Bolivia
Viehhausener Braunkohlen—Astaracian—Germany
Villavieja Formation—Friasian—Colombia
Virgin Valley Formation—Barstovian—Nevada
Waite Formation—Late? Miocene—Northern Territory, Australia

Wipajiri Formation—Middle to Late Miocene—South Australia
Wood Mountain Gravels—Barstovian—Saskatchewan
Xiejia Formation—(= Orleanian)—Qinghai
Yatağan Member—(= Vallesian and Turolian)—Turkey
Yüshe Formation—(= Turolian and later)—Shanxi?
Zia Sand—Late Arikareean and Hemingfordian—New Mexico

MIOCENE LOCALITIES

This list is not a complete recording of all published Miocene localities of the world. It will, however, give you a convenient lead to the chronological and geographical placement of many geographic names appearing without description in the current literature. It may also give you a useful appreciation of the great geographic spread and diversity of known sites.

Abeau (l'Isle d'Abeau)—Astaracian—France
Abiquiu Formation sites—Late Arikareean—New Mexico
Above Burge B—Clarendonian—Nebraska
Above Greenside Quarry—Hemingfordian—Nebraska
Above Long Quarry—Hemingfordian—Nebraska
Abstdorf-Franzensbad—Astaracian—Austria?
Acre region—Late Miocene and? later—Brazil
Adair Ranch Quarry—Hemphillian—Oklahoma
Ad Dabtiyah—(= Astaracian)—Saudi Arabia
Addicott 546—marine and Late Arikareean—California
Adelschlag—Astaracian—Germany
Adi Ugri—Rusingan—Ethiopia
Adobe Canyon—marine and Late Arikareean—California
Afyon-S.-Yaylacilar—(= Astaracian)—Turkey
Agate Northeast (Frick Quarry)—Hemingfordian?—Nebraska
Agate Spring Quarries—Late Arikareean—Nebraska
Agrico Mine—Hemphillian—Florida
Aguague Arroyo—marine and Barstovian?—California
Aiguines—Astaracian—France
Aiken (Akin) Hill—Hemingfordian—Texas
Aillas—Agenian—France
Aïn-el-hadj-Baba (Ain-el-Bey)—Late Miocene—Algeria
Ainsworth vicinity—Barstovian and? Clarendonian—Nebraska
Ajurwala—Early Miocene—Pakistan
Akcayir—(= Astaracian)—Turkey
Akhmatovo—Late? Miocene—Bulgaria
Akin (Afyon-Sandikli-A.) (= Vallesian)—Turkey
Aksakovo (= Akhmatovo?)—Vallesian?—Bulgaria
Alachua Formation sites—Hemphillian—Florida
Alais—Agenian—France
Alan—Astaracian?—France
Alcoota—Late Miocene—Northern Territory, Australia
Alcoy—Turolian—Spain
Alder Spring—Clarendonian—California
Aleksandr Dar (Krivoy Rog)—Late Miocene—USSR
Alengerr Beds sites—Ternanian—Kenya

Aletomeryx Quarry—Hemingfordian—Nebraska

Alf Quarry 5—Barstovian—California

Aliveri—Orleanian—Evia, Greece

Al Jadidah—(= Astaracian)—Saudi Arabia

Alkali Flat—Arikareean—Oregon

Alray sites—Barstovian—California

Alta-Gay   (See Kulan-Utmes.)

Altan Teli—Late Miocene—Mongolia

Althausen—Astaracian—Germany

Alturas Formation sites—Hemphillian—California

Alum Bluff—Hemingfordian—Florida

Alvord district—Barstovian—California

Amama I and II—(= Turolian) Lothagamian—Algeria

Amarillo North (= Exell?)—Clarendonian—Texas

Amasya (Aydin-Bozdogan-A.)—(= Turolian)—Turkey

Amebelodon fricki Quarry (Univ. Nebr. Ft-40)—Hemphillian—
     Nebraska

American Horse—Late Arikareean—South Dakota

AMNH 1908Q—Barstovian—Nebraska

AMNH Quarry—Clarendonian—Nebraska

Amphitheater Northwest—Clarendonian—California

Anan'yev 1—Astaracian—Ukraine, USSR

Anan'yev 2 and 3—Late? Miocene—Ukraine, USSR

Anceney—Barstovian—Montana

Anderson Mine—Hemingfordian—Arizona

Andesite Hill—Clarendonian?—Nevada

Andreyevka—Late? Miocene—Ukraine, USSR

Anglès—Orleanian—France

Anjou sites—Orleanian?—France

Ankara-Kalecik-Çandir   (See Çandir.)

Ano Metochi—Turolian—Macedonia, Greece

Antelope County North   (*Teleoceras* Catastrophe)—Clarendon-
     ian—Nebraska

Antelope Draw—Barstovian—Nebraska

Antelope Hills—Barstovian—Montana

Antelope Valley—Clarendonian?—California

Anthills of LACM—Late Arikareean—South Dakota

Antwerp (Anvers)—marine and Astaracian—Belgium

Anwil—Astaracian—Baselland, Switzerland

Apache Canyon—Clarendonian—California

Aphelops Draw—Hemphillian and older?—Nebraska

Aphelops Quarries—Hemphillian—Nebraska

Apollo Quarry (Frick)—Hemingfordian—Nebraska

Appelshofen im Ries—Astaracian—Germany

Appenzell-Kaubach—Agenian—Switzerland

Arapli (Tekirdağ-Şarköy-A.)—(= Astaracian)—Turkey

Archeman-Ali—Astaracian—Crimea, USSR

Archer—Hemphillian?—Florida

Archino—Late Miocene—Portugal

Arenas del Rey district—Late Miocene—Spain

Argenthal—Astaracian—Germany

Argithan (Konya-Argithan)—Late Miocene—Turkey

Ariere   (See Nordrand.)

Arlington—Hemphillian—Oregon

Armagnac   (See Sansan and Simorre.)

Armandville district—Barstovian?—Texas

Armantes 1 & 7—Aragonian—Spain

Armstrong Ranch—Late Arikareean—Nebraska

Arnett—Hemphillian—Oklahoma

Arquillo (de la Fontana)—Turolian—Spain   —(Rusci-
     nian = MN 15, acc. Van de Weerd and Daams, 1978)

Arrisdrift—Ternanian?—Namibia

Arrowhead Canyon—Clarendonian—Nevada

Arroyo Arenoso district—Clarendonian—New Mexico

Arroyo Chamisa Prospect—Hemingfordian—New Mexico

Arroyo del Val—Astaracian—Spain

Artenay—Orleanian—France

Artifact—Arikareean—Oregon

Artman Basin South—Late Arikareean—Oregon

Arvin South sites—Clarendonian—California

Ashbrook Pasture—Barstovian—Nebraska

Ash Hollow Formation sites—Clarendonian and Hemphillian—
     Nebraska

Ashville—Barstovian—Florida

As Sarar—(= Astaracian)—Saudi Arabia

Assif-Assermo—(= Vallesian) Ngororan—Morocco

Astoria Formation sites—Barstovian—Oregon

Ateca I and III—Orleanian—Spain

Atravesada—Miocene—Arizona

Attenfeld—Astaracian—Germany

Atzgersdorf—Astaracian—Austria

Aubignas (Aubinas)—Late Miocene—France

August Sherris Pasture—Hemingfordian or Barstovian—
     Texas

Aurignac—Astaracian—France

Auriole, Château d'—Orleanian—France

Ausson—Late Miocene?—France

Avaray—Orleanian—France

Avawatz—Clarendonian—California

Aveiras de Baixo—Astaracian—Portugal

Axelrod—Hemingfordian—California

Axtel—Hemphillian—Texas

Ayas (Ilhançayi-Ankara)—Late Miocene—Turkey

Ayvacik-Gülpinar (Çanakkale-A.-G.)—(= Vallesian)—Turkey

Azambujeira, upper—Vallesian—Portugal

Baarburg—Astaracian—Switzerland

Babai Khola—("Chinji" = Astaracian)—Nepal

Baccinello—Turolian—Italy

Bachas—Astaracian?—France

Baigneaux-en-Beauce—Orleanian—France

Bakhmutaya—Late Miocene—Moldavia, USSR

Bala-Kunduzda   (See Kiyevskiy.)

Balcic (Balchik)—Late Miocene—Bulgaria

Balçiklidere (Uşak-Esme-B.)—(= Vallesian)—Turkey

Bald Peak—Clarendonian—California

Balizac—Agenian—France

Ballard—Barstovian—California

Ballinger Canyon—Hemingfordian, Barstovian, and Clarendon-
     ian—California

Balmy–St. Clair (Lyon)—Late Miocene—France

Balta—Late Miocene—USSR

Baltavar—Turolian—Hungary

Baltringen—Astaracian—Germany

Bampf-Seon—Astaracian—Switzerland

Banja Luka—Astaracian—Yugoslavia

Bañon 5—Orleanian—Spain

Baode (Paote)—Chichiakou—(= Turolian)—Shanxi

Barger Gulch West—Hemingfordian?—Colorado

Baringo Basin—Late Miocene and Pliocene—Kenya

Barker Ranch—marine and Hemingfordian—California

Baróth-Köpecz—Turolian?—Hungary

Barrage de Bou Hanifia   (See Bou Hanifia.)

Barrieira das Pombas (Villa Nova da Rainha)—Late Miocene—Portugal

Barry Farm (Noble Farm, Navasota)—Hemingfordian—Texas

Barstow Syncline sites—Hemingfordian and Barstovian—California

Bartlett Mountain—Hemphillian—Oregon

Bartlett Reservoir—Clarendonian?—Arizona

Başbereket (Ankara-Ayas-B.)—(= Vallesian)—Turkey

Bas Neyron   (See Neyron.)

Batesland Formation sites—Hemingfordian—South Dakota

Bätterhausen—Astaracian—Switzerland

Battleship Mountain (San Ildefonso)—Clarendonian—New Mexico

Battles Well—Clarendonian?—Nevada

Baugé—Orleanian?—France

Baum—Barstovian—California

Bayles Ranch—Arikareean—Oregon

Bazas—marine and Agenian—France

Bean Creek—Clarendonian—California

Bear Creek Canyon—Barstovian—Oregon

Bear Spring—Clarendonian or Hemphillian—Arizona

Bear Tooth Slide (Frick Quarry)—Hemphillian—Nebraska

Bear Valley—Barstovian?—California

Beatty Butte—Barstovian—Oregon

Beauce, calcaire de, sites—Agenian—France

Beaugency—Orleanian—France

Beaumaris—Late Miocene—Victoria, Australia

Beaver-Laverne—Clarendonian—Oklahoma

Beaver Creek—Barstovian—Oregon

Beçin (Muğla-Milas B.)—(= Turolian)—Turkey

Bédéchan—Orleanian—France

Beecher Island—Clarendonian or Hemphillian—Colorado

Bee School—Barstovian—California

Beger-Nur (Begger-Noor)—(= Astaracian)—Mongolia

Beglia Formation sites—(= Astaracian and Vallesian) Ternanian and Ngororan—Tunisia

Beleyenice (Manisa-Saruhanli-B.)—Late Miocene—Turkey

Belgrade—Late Arikareean—Montana

Belgrade, Yugoslavia   (See Beograd.)

Bellamy—marine and Barstovian?—California

Bellegarde—Astaracian—France

Belloc–St.-Clemens—Astaracian?—France

Bellshire—Clarendonian—California

Bell Spring—Barstovian—Nevada

Belluno, molasse de, sites—marine and Orleanian?—Italy

Bellver—Vallesian—Spain

Belmont—Orleanian—France

Belmont Park—Late Arikareean—Montana

Belshaw's Ranch—Arikareean—Oregon

Belvedere—Late Miocene—Austria

Bemis—Hemphillian—Kansas

Benabato–Tung Gur—(= Astaracian)—Nei Mongol

Benghazi district—Late Miocene—Libya

Beni Mellal—(= Astaracian) Ternanian—Morocco

Benken—Astaracian—Switzerland

Benningen—Late Miocene—Germany

Benque—Astaracian?—France

Beograd vicinity—Astaracian—Yugoslavia

Bérault—Astaracian—France

Berdik (Denizli-Kale-B.)—(= Astaracian)—Turkey

Berehegur—Late Miocene—Kazakhstan, USSR

Berkeley North—Clarendonian—California

Bermersheim—Late Miocene—Germany

Bernas, bois de—Orleanian—France

Bert Creek—Clarendonian—Montana

Besyol-Manisa—Late Miocene—Turkey

Beuern—Late? Miocene—Germany

Bézian—Orleanian?—France

Bezues-Bajon—Astaracian?—France

Biberach—Astaracian—Germany

Bidahochi Formations sites—Hemphillian and Clarendonian—Arizona

Biesendorf—Astaracian—Germany

Big Beaver A—Clarendonian—Nebraska

Big Beaver C—Clarendonian—Nebraska

Big Cat Quarry—Clarendonian—California

Big Cat West—Clarendonian—California

Big Horn Canyon—Clarendonian—California

Big Pine Fault—Barstovian?—California

Big Spring Canyon—Clarendonian—South Dakota

Big Wash—Late Arikareean—Idaho

Bijou Hills   Barstovian—South Dakota

Bingham Ranch (Redington)—Hemphillian—Arizona

Biodrak   (See Biotia.)

Biotia—Vallesian—Greece

Bird Cage Gap—Late Arikareean—Nebraska

Biros—Astaracian?—France

Bissingen—Orleanian—Germany

Bjelometscheskaja   (See Byelometschetskaya.)

Black Bear Quarries—Hemingfordian—South Dakota

Blackbow Hill—Arikareean—Oregon

Black Butte I and II—Clarendonian—Oregon

Black Butte Mine—Late Arikareean—California

Black Canyon 1—Barstovian—California

Black Hank's Canyon—Late Arikareean—Nebraska

Black Hawk (or Blackhawk) Ranch—Clarendonian—California

Black Mesa East—Clarendonian—New Mexico

Black Mesa Northwest or West Side—Hemphillian—New Mexico

Black Mountain sites—Barstovian—California

Black Rascal Creek—Hemphillian—California

Black Rock sites—Arikareean—Oregon

Blacktail Deer Creek—Hemingfordian—Montana

Blagodarnoye—Late Miocene—Caucasus, USSR

Blair Junction West—Clarendonian?—Nevada

Blake—Clarendonian—California

Bled ed Douarah (Djerid)—(= Astaracian and Vallesian) Ternanian and Ngororan—Tunisia

Bleichenbach—Astaracian—Germany

Blésois—Orleanian to Astaracian—France

Blick Quarry—Hemingfordian—New Mexico

Blois—Orleanian—France

Blue Cliff—Barstovian—Nevada

Blue Jaw—Late Arikareean—Colorado

Blue Rock Spring—Barstovian and Clarendonian—California

Blue Sandstone—Clarendonian—Nevada

Blue Sandstone Flats—Clarendonian—Nevada

Bobcat Hill—Clarendonian—Nevada

Bode Ranch—Clarendonian—California

Bogadanovka—Late Miocene—Moldavia, USSR

Bogatsheny-Kishinev—Late Miocene—Moldavia, USSR

Bois de Montpellier (Castelnau–d'Anglès)—Orleanian—France

Bolero Lookout—Arikareean—California

Bolinas Creek—Clarendonian—California

Bolinas Point—marine and Barstovian—California

Bolinger Canyon sites—Clarendonian—California

Bologna Creek—Arikareean—Oregon

Bololo—Rusingan (reworked)—Zaire

Bonanza—Barstovian—California

Bone Creek—Clarendonian?—Nebraska

Bone Valley Formation sites—Hemphillian—Florida

Boneyard Canyon—Barstovian—California

Bonnefond—Astaracian—France

Bonrepos-sur-Ausonelle—Orleanian—France

Borcher Gravel Pit—Hemphillian—Kansas

Boron—Hemingfordian—California

Borschli-Balta—Late Miocene—Ukraine, USSR

Bossée—Orleanian?—France

Boucé   (See St. Gérand-le-Puy.)

Boudry—Agenian—Switzerland

Bou Hanifia (Oued el Hammam)—(= Vallesian) Ngororan—Algeria

Boulder Quarry—Barstovian—Nebraska

Boulder Valley North (Cold Springs P.O.)—Late Arikareean into Hemingfordian—Montana

Boulogne—Astaracian—France

Bouzigues, phosphorites de—Agenian—France

Box Butte Member sites—Hemingfordian—Nebraska

Box T Quarries—Hemphillian—Texas

Boyd County site (Univ. Nebr. Bd-6)—Barstovian—Nebraska

Bozeman—Barstovian?—Montana

Božurevac (Kruševac Basin)—Astaracian—Yugoslavia

Brachycrus Quarry—Barstovian—California

Brady Pocket—Clarendonian—Nevada

Braila—Vallesian

Breitenbrunn—Astaracian—Germany

Bren—Astaracian—France

Brenham district—Hemingfordian—Texas

Bresse, couches de la—Late Miocene—France

Brezhani (Struma Basin)—Late Miocene—Bulgaria

Bridge Creek sites—Arikareean—Oregon

Bridgeport Quarries—Hemingfordian—Nebraska

Bridge Ranch—Barstovian—Texas   —(Tedford et al. (1983?))

Briones Dam—Clarendonian—California

Brisbois Bluff—Arikareean—Oregon

Bristol-Alum Bluff—Late Arikareean or Hemingfordian—Florida

Brooksville—Late Arikareean—Florida

Brown Avenue—Clarendonian—California

Browns Park Formation sites—Barstovian—Colorado

Bruck-a.-Leitha—Astaracian—Austria

Brunn-Vösendorf—Late Miocene—Austria

Brüttelen—Orleanian—Switzerland

Buchberg-Eglisau—Orleanian—Switzerland

Buchenberg—Astaracian—Germany

Bückelberg—Late Miocene—Austria

Buck Quarry—Hemingfordian—Nebraska

Buda—Late Arikareean—Florida

Budapest—marine and Astaracian?—Hungary

Budenheim—Agenian—Germany

Bugti beds sites (Dera Bugti, Chur Lando)—Early Miocene—Baluchistan, Pakistan

Bukovac—Astaracian—Yugoslavia

Bukwa—Early Rusingan—Uganda

Bullock Creek—Middle? Miocene—Northern Territory, Australia

Bulumya (Konya-Hatip-B.) (= Vallesian)—Turkey

Bu Lung (Biru)—Late Miocene—Xizang (Tibet)

Buñol (Buñol Rubí)—Orleanian—Spain

Burgas—Turolian?—Bulgaria

Burgau—Astaracian—Germany

Burgdorf—Orleanian—Switzerland

Burge B Quarry, Burge Channel, Burge Quarries—Clarendonian—Nebraska

Burgschleinitz—Orleanian—Austria

Burkeville—Barstovian—Texas

Burlatskoye—Late Miocene—Caucasus, USSR

Buron (Sursee)—Astaracian—Switzerland

Burovac (Covdin)—Astaracian—Yugoslavia

Bursa–Mustafa-Kemalpaşa (See Paşalar.)

Butan 1—Clarendonian—California

Butan (Danube Basin)—Late? Miocene—Bulgaria

Butler Basin—Arikareean—Oregon

Byelometchetskaya—Astaracian?—Caucasus, USSR

Cabbage Patch—Arikareean—Montana

Cabo Buen Tiempo (= Cape Fairweather or Rudd's)—Miocene—Argentina

Cabrières d'Aigues—Late Miocene—France

Cache Peak sites—Barstovian—California

Cadeilhan—Orleanian—France

Cady Mountain fauna—Hemingfordian—California

Cajalco—Clarendonian—California

Cajon Valley district—Hemingfordian and Barstovian—California

Calatayud-Tereul area—Vallesian and Turolian—Spain

Caldecott Tunnels—Clarendonian—California

Caldes de Montbui—Vallesian—Spain

Caliente Mountain sites—Late Arikareean, Hemingfordian, Barstovian, Clarendonian, and Hemphillian—California

Çaliki (Manisa-Selendi-C.) (= Astaracian)—Turkey

Calkins—Arikareean—Oregon

Callahan Road sites—Clarendonian—California

Caloosahatchee River sites—Hemphillian?—Florida

Calvert Formation sites—marine and Barstovian—Maryland

Cambridge—Hemphillian—Nebraska

Camel Basin—Barstovian—California

Camel Canyon—Hemphillian—Arizona

Camel Jaw—Clarendonian—Nevada

Camel Pocket—Clarendonian—Nevada

Camel Quarry—Barstovian—Nebraska

Cammatta Ranch—marine and Clarendonian—California

Campbell Gravel Pit—Hemphillian—Oklahoma

Camp Berteaux   (See Melka el Ouidane.)

Camp Creek sites—Barstovian—Nevada

Camp Davis Formation site—Clarendonian—Wyoming

Campidano—Agenian?, Orleanian?—Sardinia
Cañada Pilares—Late Arikareean—New Mexico
Cañadón de los Vacas—Santacrucian—Argentina
Canakkale (Erenköy)—Late Miocene—Turkey
Can Almirall—Astaracian—Spain
Can Barba—Vallesian—Spain
Can Casablanques—Vallesian—Spain
Çandir (Ankara-Kalecik-Ç.)—(= Astaracian)—Turkey
Can Feliu—Astaracian—Spain
Can Gabarró (Polinyà)—Vallesian—Spain
Can Gonteres (Viladecaballs)—Vallesian—Spain
Can Jofresca (Terrassa)—Vallesian—Spain
Can Julià—Orleanian—Spain
Can Llobateres—Vallesian—Spain
Can Mas (Papiol)—Orleanian—Spain
Can Mata (lower Hostalets)—Astaracian—Spain
Can Perellada, N1 and N4—Vallesian—Spain
Can Ponsic (San Quirze or San Quirico)—Vallesian—Spain
Can Purull (Tarrassa)—Vallesian—Spain
Can Trullás—Vallesian—Spain
Canyada Pilares—Late Arikareean to Hemingfordian—New
   Mexico
Caper Place—Hemingfordian—Texas
Capiteux—Orleanian—France
Cap Janet—Agenian—France
Capps-Neu Pit or Quarries—Hemphillian—Oklahoma
Caravaca—Turolian—Spain
Carbona—Clarendonian—California
Cardenau-Lavardens—Orleanian—France
Carentan—Astaracian—France
Carla-Bayle—Orleanian—France
Carlin 1—Barstovian—Nevada
Carlin North—Clarendonian—Nevada
Carlin South—Barstovian—Nevada
Carmen de Apicalá—Friasian—Colombia
Carnegie—Clarendonian—California
Carnivore Canyon sites—Barstovian—California
Carry-le-Rouet—Agenian—France
Casal Vistoso Va—Astaracian?—Portugal
Casas Altas—Vallesian—Spain
Casino-Siena—Turolian—Italy
Cassagnabère—Astaracian—France
Casteani—Vallesian? and Turolian—Italy
Castelgaillard (Samaran)—Astaracian—France
Castell de Barbera—Astaracian—Spain
Castelnau-Barbarens—Astaracian—France
Castelnau–d'Angles (Bois de Montpellier)—Orleanian—France
Castelnau d'Arbieu—Astaracian—France
Castelnau-Picampeau—Astaracian—France
Castolon—Late Arikareean—Texas
Çatakbağyaka (Muğla-Yerkesik-Ç.) (= Astaracian)—Turkey
Catholic School—Clarendonian—California
Cattle Spring—Arikareean—Oregon
Caunelles—Agenian—France
Cedar Mountain sites—Clarendonian—Nevada
Cedar Run or Creek—Hemingfordian—Texas
Ceja del Rio Puerco—Hemingfordian?—New Mexico
Cendejas de la Torre—Late Miocene—Spain
Cerdagne (Das)—Late Miocene—Spain
Cerdas-Atocha—Santacrucian?—Bolivia
Cerithien-Schichten sites—Agenian—Germany

Cerro del Otero—Astaracian—Spain
Cerro del Uso—Orleanian—Spain
Cerro de San Isidro—Astaracian—Spain
Cetina de Aragón—Agenian—Spain
Çevril (Kayseri-Erkilet-Ç.) (= Vallesian and Turolian)—Turkey
Chabeuil–Les Bourbons—Vallesian—France
Chalk Cliffs—Barstovian—Montana
Chalkoutsi—Turolian?—Greece
Chalk Spring—Clarendonian—Nevada
Chama-el-rito sites—Barstovian and Clarendonian—New Mexico
Chamaret—Orleanian—France
Chamisa Mesa Member sites—Hemingfordian—New Mexico
Chamita Formation sites—Hemphillian—New Mexico
Champ de Mathieu—Orleanian—France
Chanac Formation sites—Clarendonian—California
Changpeihsien (Changpei)—(= Astaracian?)—Hubei
Changtun Basin–Wusiang—Late Miocene—Shaanxi
Changzhi—Late Miocene—Shanxi
Channay—Orleanian or Astaracian—France
Chantegré (See St. Gérand-le-Puy.)
Charmoille—Late Miocene—Switzerland
Charneca—Astaracian—Switzerland
Château de Lartigole (See Pessan.)
Chatonnay—Late Miocene—France
Chattahoochee (See Jim Woodruff Dam.)
Chavroches—Agenian—France
Chazé-Henry—Miocene—France
Chélan—Astaracian—France
Chen-Chia-Ou Village (See Lantien.)
Cherichera (Jebel Cherichera)—(= Orleanian? and Astaracian)
   Rusingan? and Ternanian—Tunisia
Chernovo (Anan'yev district)—Late Miocene—Ukraine, USSR
Cherty Bone Bed—Barstovian—California
Chevenelles—Orleanian or Astaracian—France
Chevilly—Orleanian—France
Chez-Régis à Lahas—Astaracian—France
Chia Mos-Su—Late Miocene—China
Chichiakou (See Baode.)
Chickasaw Creek—Hemphillian?—Alabama
Chicora Mine—Hemphillian—Florida
Chilleurs-au-bois—Orleanian—France
Chimishliya—Late Miocene—Moldavia, USSR
Chinchinales beds sites—Santacrucian—Argentina
Ching-an (See Quinan.)
Chinglo—(= Turolian and later)—Shanxi
Chinji "zone," "Formation," "fauna," etc.—(= Astaracian
   and? Vallesian)—Pakistan and India
Chios (See Thymiana.)
Chitenay—Orleanian—France
Chit-Irgia—Late Miocene—Kazakhstan, USSR
Chiton-Gol Basin—Late Miocene—Mongolia
Chobruchi (Tchobroutchi)—Late Miocene—Moldavia, USSR
Chões—Astaracian—Portugal
Chomateri (See Pikermi.)
Choptank Formation site (See Greenburgh.)
Chott-el-Djerid—Late Miocene—Tunisia
Christian Ranch—Hemphillian—Texas
Christmas Valley—Hemphillian—Oregon

Chuan Tou Kou—Late Miocene—Gansu
Chubuttal—Miocene—Argentina
Chuchupate—Clarendonian—California
Chunhsin—Late Miocene—Gansu
Chur Lando beds sites—Early Miocene—Baluchistan, Pakistan
Ciernat   (See St. Gérand-le-Puy.)
Çifthkkoy-KapBati (Izmir-Çeşme-Ç.)—(= Astaracian)—Turkey
Cimişlia—Turolian—Romania
Cinnabar Canyon—Barstovian?—Nevada
Cioburciu—Late Miocene—Romania
Cistern loc.—Late Arikareean—Nebraska
C.I.T. 315—Hemingfordian—California
C.I.T. 322–323—Barstovian—California
Claremont Canyon—Clarendonian—California
Claremont (Water) Tunnel—Clarendonian—California
Clarendon sites—Clarendonian—Texas
Clérieux—Astaracian—France
Cliff Gap—Hemingfordian, Barstovian, and Clarendonian—California
Cliff Southeast—Arikareean—Oregon
Climbach—Astaracian—Germany
Clover Creek—Hemphillian—Idaho
Coal Canyon—Clarendonian—Nevada
Coalinga district—marine and Barstovian, Clarendonian, and Hemphillian—California
Coal Valley Formation sites—Clarendonian and Hemphillian—Nevada
Çobanpinar—(= Turolian)—Turkey
Cobham Wharf—Clarendonian or Hemphillian—Virginia
Cobruči—(= Turolian)—Turkey?
Cocumont—Agenian—France
Coetas Creek—Clarendonian—Texas
Coffee Ranch—Hemphillian—Texas
Cogswell Quarry—Hemphillian—Kansas
Coirons—Late Miocene—France
Colclough Hill—Late Arikareean or Hemingfordian—Florida?
Cold Spring—Barstovian—Texas
Cole Highway Pit—Clarendonian—Texas
Collet-Redon—Astaracian—France
Collins Draw—Clarendonian? Hemphillian?—Kansas
Collister Road—Barstovian—California
Colloncuran = Santacrucian
Columbia—Clarendonian—California
Comanche Point sites—Clarendonian—California
Concelho   (See Monte Novo de Concelho.)
Concretion Quarry—Hemingfordian—Nebraska
Concud (Tereul)—Turolian—Spain
Conde (Smendou)—(= Orleanian?) Rusingan?—Algeria
Condom—Astaracian—France
Conference Quarry—Hemingfordian—Nebraska
Confusion Quarry—Clarendonian—Nevada
Congress Junction—Clarendonian or Hemphillian—Arizona
Conical Hill Quarry—Barstovian—New Mexico
Constantine South—Late? Miocene—Algeria
Cook's *Pliohippus* Quarry—Hemphillian—Nebraska
Coon Canyon sites—Barstovian—California
Coon Creek Head—Clarendonian—Nebraska

Coon Creek West—Clarendonian—Nebraska
Çorak Yerler (Çankiri-Ç.-Y.)—(= Vallesian)—Turkey
Corbicula-Schichten sites—Agenian—Germany
Corocoro—Santacrucian?—Bolivia
Corral Hollow—Clarendonian—California
Corriguen Aike (= Corriken Aike or Corriguen Kaik   —Marshall, 1976)—Santacrucian—Argentina
Côtière de Dombes—Vallesian—France
Cottonwood Creek—Barstovian?—Oregon
Cottonwood Creek—Arikareean—Oregon
Coueilles—Astaracian—France
Couloumé-Mondébat—Astaracian—France
Courrensan—Astaracian—France
Cowden—Clarendonian—California
Coy Inlet (= Boca del Coyle of Ameghino, Marshall, 1976)—Santacrucian—Argentina
Coyote Spring—Clarendonian—Nevada
Crazy loc.—Clarendonian—Nebraska
Creek Bottom—Clarendonian—Arizona
Creekside—Arikareean—Oregon
Crescent Valley—Clarendonian—Nevada
Crest—Astaracian—Switzerland
Crevillente—Turolian—Spain
Crocker Springs—Clarendonian—California
Croix-Rousse—Vallesian—France
Cronese—Barstovian—California
Crooked Creek—Clarendonian—Nebraska
Crooked River—Barstovian—Oregon
Crookston Bridge—Barstovian—Nebraska
Crow Canyon Creek—Clarendonian—California
Csakvar—Vallesian—Hungary
Cucuron (Mt. Léberon)—Turolian—France
Cudahy Camp—Clarendonian—California
Cuevas Basadar—Late Miocene—Spain
Cull Creek West—Clarendonian—California
Currie Ranch—Hemphillian—Texas
Curtis—Clarendonian—California
Curuzucuatia—Colhuehuapian?, Santacrucian?—Argentina
Custret (Montesquiou-sur-Losse)—Orleanian—France
Cynarctoides Quarry—Hemingfordian—New Mexico
Dadasun (Kayseri-Búnyan-D.)—(= Vallesian)—Turkey
Daggett—Hemingfordian—California
Daily—Hemingfordian and Barstovian—California
Dalles (The Dalles.)—Clarendonian—Oregon
Dalton—Hemphillian—Nebraska
Dang Valley (= Astaracian)—Nepal
Dardanelles   (See Erenköy.)
Dark Hole—Barstovian—California
Darton's Bluff—Late Arikareean—Wyoming
Das (Derdagne)—Late Miocene—Spain
Dasing—Astaracian—Germany
Datwal (Dhok Pathan)—(= Turolian)—Pakistan
Daud Khel (= Vallesian? and? Astaracian?)—Pakistan
Dayville—Arikareean—Oregon
Dayville East—Barstovian—Oregon
Dead Frog—Late Arikareean—Nebraska
Dechbetten-Regensburg—Astaracian—Germany
Dechginzhe Village—Late Miocene—Moldavia, USSR
Deep River fauna—Hemingfordian and Barstovian—Montana
Deer Butte Formation site—Barstovian—Oregon

Deer Lodge—Hemphillian—Montana
Deggenhausen—Astaracian—Germany
Delsberg—Astaracian—Switzerland
Dénezé—marine and Orleanian—France
Denizli Southwest—Turolian—Turkey
Densmore Southeast—Clarendonian—Kansas
Dera Bugti beds sites—Early Miocene—Baluchistan, Pakistan
Derby Peak—Barstovian or Clarendonian—Colorado
Dereikebir (Edirne-Uzunköprü-D.) (= Astaracian?)—Turkey
Despotovac Basin (Mala Miliva, etc.)—Orleanian—Yugoslavia
Devil's Gulch—Barstovian—Nebraska
Devil's Punchbowl—Clarendonian and Hemphillian—California
Děvínská Nová Ves (Neudorf a. d. March)—Orleanian and Astaracian—Czechoslovakia
Dhedari (Dhok Pathan)—Turolian—Pakistan
Dhok Pathan ''Stage,'' ''fauna,'' ''beds,'' ''zone''—(= Turolian)—Pakistan and India
Diablo Country Club—Clarendonian—California
Diamond Mountain-Montgomery—Barstovian?—Texas
Dicotyles—Barstovian—Oregon
Diesen—Astaracian—Germany
Dilli Place-Turkey Creek—Clarendonian—Texas
Dimitrovgrad—Late? Miocene—Bulgaria
Dinan—marine and Agenian?—France
Dinkelscherben—Astaracian—Germany
Dinohyus site—Hemingfordian—Texas
Dinotheriensande sites—Late Miocene—Germany
Ditiko I—Turolian—Macedonia, Greece
Djebel Hamrin-Bagdad—Late Miocene—Irak
Djebel m'Dilla—Late Miocene—Tunisia
Djebel Mrhile—(= Orleanian) Rusingan—Tunisia
Djebel Sehib—Late Miocene—Tunisia
Djerid   (See Draa el Djerid.)
Djilancik, middle beds sites—(= Astaracian)—Kazakhstan, USSR
Djilancik, lower beds sites—(= Orleanian)—Kazakhstan, USSR
Dobrich—Late? Miocene—Bulgaria
Doe Spring fauna—Barstovian—California
Doig Ranch West—Hemingfordian?—Colorado
Dolje—marine and Miocene—Yugoslavia
Dolnice—Orleanian—Czechoslovakia
Domengine Creek or Ranch—marine and Barstovian and Clarendonian—California
Dome Spring fauna—Barstovian—California
Donje Solnje (Skopje)—Astaracian—Yugoslavia
Dornbach—Astaracian—Austria
Dorn-Dürkheim—Turolian—Germany
Dornegg—Late Miocene—Austria
Douaria—Late Miocene—Tunisia
Doué-la-Fontaine—Vallesian—France
Douglas Quarry—Barstovian—Nebraska
Douglas South—Late Arikareean—Wyoming
Dove Springs—Clarendonian—California
Downey-Marsh Creek—Barstovian—Idaho
Dra el Djerid—Late Miocene—Tunisia
Drassburg—Pannonian and Vallesian—Austria
Drees—Arikareean—Oregon
Drinkwater—Hemphillian—Oregon
Dripping Springs (Bordon)—Barstovian?—Texas
Drizzle Basin—Arikareean—Oregon

Dry Canyon—Barstovian—California
Dry Creek—Barstovian—Oregon
Dry Creek—Hemphillian—New Mexico
Dry Prospects—Hemingfordian—Nebraska
Duboka (Zviżd Basin)—Astaracian—Yugoslavia
Dudular (Saloniki)—Turolian—Greece
Dumlupinar (Kütahya)—(= Astaracian)—Turkey
Dunham Claim—Clarendonian—Nevada
Dunnellon Plant 6—Hemphillian—Florida
Durham locality—Clarendonian—Oklahoma
Durham Quarry—Clarendonian—California
Durham Red—Late Arikareean—California
Dutch Creek (Jones Canyon)—Clarendonian—Nebraska
Dutch Fred's sites—Clarendonian—California
Dutton Ranch—Late Arikareean—Montana
Düzpinar (Manisa)—(= Vallesian)—Turkey
Dyakovo (Struma Basin)—Late Miocene—Bulgaria
Dyseohyus—Barstovian—California
Dzenema River—Late Miocene—Kazakhstan, USSR
Dzhaparidze—Late Miocene—Caucasus, USSR
Dzhenam   (See Zhenam.)
Dzhilanchik River   (See Djilancik.)
East Bench—Hemphillian—Montana
East Canyon—Arikareean—Oregon
East End Northwest—Hemingfordian—Saskatchewan
Easter (Frick)—Barstovian—California
Eastgate—Hemphillian?—Nevada
East Hilltop Quarry—Hemingfordian—Nebraska
East Jenkins Quarry—Barstovian—Nebraska
East *Pliohippus* Draw—Clarendonian—Nebraska
East Ravine—Hemingfordian—Nebraska
East Sand Quarry—Barstovian—Nebraska
East Surface Quarry—Barstovian—Nebraska
East Wall Quarry—Barstovian—Nebraska
Ebingen—Late Miocene—Germany
Ebis (Kayseri-Erkilet-E.)—(= Turolian)—Turkey
Echo Quarry—Barstovian—Nebraska
Echzell—Orleanian?—Germany
Eckingen—Agenian—Germany?
Edson—Hemphillian—Kansas
Egelhoff Quarry—Barstovian—Nebraska
Eggenburg—Eggenburgian marine and Orleanian—Austria
Eggingen—Agenian—Germany
Eglisau—Orleanian—Switzerland
Eibiswald—marine Karpatian and late Orleanian—Austria
Eichkogel—Pontian and Turolian—Austria
Eitensheim—Orleanian—Germany
El Canyet—Orleanian—Spain
El'dar—(= Turolian)—Caucasus, USSR
El Firal (Seu d'Urgell, Seo de Urgel)—late Miocene—Spain
Elgg—Astaracian—Switzerland
El Gramal—Barstovian—Oaxaca, Mexico
Elisavetovka—Vallesian—Ukraine, USSR
El Mamma du Djerid—Vallesian—Tunisia
El Hascrat—Miocene?—Libya
Elkington—Clarendonian—California
Elko Northeast—Clarendonian—Nevada

Ellensburg Formation sites—Clarendonian—Washington

Ellison Creek—Barstovian and? Clarendonian?—Nevada

Elm am Ostrand—Orleanian—Germany

El Papiol (El Canyet, etc.)—Orleanian—Spain

El Rito (Abiquiu)—Late Arikareean—New Mexico

Elsmere Canyon—marine and Clarendonian—California

El Toro—marine (= Clarendonian)—California

Ely Southeast—Barstovian—Nevada

El Ziz—Late Miocene—Tunisia

Emathla—Hemphillian?—Florida

Emmingen—Astaracian?—Germany

En Charron–Lasseran—Orleanian—France

Engelswies—Astaracian—Germany

En Pejouan—Astaracian—France

Eoulx—Agenian—France

Eppelsheim—Vallesian—Germany

Erbach—Agenian—Germany

Erd—Astaracian?—Hungary

Erdevidek—Late Miocene—Hungary

Erenköy—(= Turolian)—Turkey

E. R. Harvey 1—Barstovian—Texas

Erkertshofen—Orleanian—Germany

Erkilet—Late Miocene—Turkey

Ersandikköy (Samsun-Havza-Köprübaşi)—Late Miocene—Turkey

Ertemte—(= Turolian, in part)—Nei Mongol

Escanecrabe—Astaracian?—France

Escarpment Quarry—Clarendonian—Wyoming

Escobosa de Calatañazor (Soria)—Astaracian—Spain

Eski Bayirköy (Muğla-Yatağan-E.)—Astaracian—Turkey

Eskihisar (Muğla-Yatağan-E.)—Astaracian—Turkey

Eşme-Akçaköy (Usak)—(= Vallesian)—Turkey

Espaon—Astaracian?—France

Espinousse—Turolian—France

Espira-du-Confluent—Orleanian—France

Esselborn—Late Miocene—Germany

Estates Drive—Clarendonian—California

Estavar—Late Miocene—France

Estrepouy—Orleanian—France

Etchegoin Formation sites—Hemphillian—California

Etna–Grouse Creek—Hemingfordian?—Utah

Eubanks—Barstovian—Colorado

Euboea—Late Miocene—Greece

Eureka 5—Barstovian—California

Evci (= Evçilerağillari?)—Late Miocene—Turkey

Evçilerağillari (Ankara-Gökdere-E.)—(= Turolian)—Turkey

Exell—Clarendonian—Texas

Eylyar–Oyugi Range—Late Miocene—Caucasus, USSR

Ezerovo—Turolian?—Bulgaria

Fairfield Creek—Clarendonian—Nebraska

Fall Creek North—marine and Clarendonian—California

Fallen Derrick—marine? and Barstovian—California

Faluns de l'Anjou sites—Orleanian—France

Fangshan-Nanking—(= Astaracian)—Jiangsu

Fangxian—(= Astaracian)—Hubei

Farish Ranch—Clarendonian—Texas   —(Tedford et al. (1982?))

Far Surface Quarry—Barstovian—Nebraska

Far Well—Arikareean—Oregon

Fatehjang—(= Orleanian)—Pakistan

Fatigue locality—Clarendonian—Nebraska

Faultline—Clarendonian?—Nevada

Fault XD sites—Barstovian–California

Fay-aux-Loges—Orleanian—France

Feisternitz—Astaracian—Austria

Felsö-Tárkany—Astaracian?—Hungary

Felton—marine and Clarendonian—California

Felton's Estancia (= East Killik Aike Norte)—Santacrucian—Argentina

Feltz (= "Feldt") Ranch—Hemphillian—Nebraska

Fence Line—Clarendonian—Nebraska

Fenix—Friasian—Argentina

Fernley—Clarendonian—Nevada

Ferrocarril (San Quirico or San Quirze)—Astaracian—Spain

Feyereisen Gap (Springer)—Barstovian—South Dakota

Finca Escuela Agricultura (Caldes de Montbui)—Vallesian—Spain

Fingerrock—Barstovian—Nevada

Fingerrock Wash—Barstovian—Nevada

Firby-Holton—Barstovian—Nevada

Fisher sites—Arikareean—Oregon

Fish Lake Valley sites—Clarendonian—Nevada

Flawil—Astaracian—Switzerland

Flaxville Gravels sites—Clarendonian?—Montana

Fleming Formation sites—Barstovian—Texas

Flexure Point—Clarendonian—Nevada

Flint Creek—Barstovian—Montana

Flinz München   (See München.)

Fohnsdorf—Karpatian-Late Orleanian—Austria

Foissin (Lectoure)—Astaracian—France

Foley Quarry (Frick)—Hemingfordian—Nebraska

Fontcaude—Agenian—France

Fonte do Pincheiro—Astaracian—Portugal

Formiga (Azambuja)—Astaracian—Portugal

Forsthart—Orleanian—Germany

Fort Logan—Arikareean—Montana

Fort Logan Southeast—Hemingfordian or Early Barstovian—Montana

Fort Randall Formation site—Barstovian—South Dakota

Fort Ternan—Ternanian—Kenya

Fossil Bluff—Early Miocene—Tasmania

Foum el Guelta 1—Rusingan?—Tunisia

Foum el Guelta 2—Ternanian?—Tunisia

Found Quarry—Hemphillian—Kansas

Fox Hill—Clarendonian—Nevada

Frankfurt-Niederrad—Agenian—Germany

Frankfurt-Nordbassin—Agenian—Germany

Franklin Phosphate Company, Pit No. 2—Hemingfordian—Florida

Frank Ranch—Barstovian—Montana

Frantsfel'd—Late Miocene—Ukraine, USSR

Franzenbad—Miocene—Czechoslovakia

Freising—Astaracian—Germany

Fridensfe'd—Late Miocene—Ukraine

Frome Embayment—Middle? Miocene—South Australia

Frontenhausen—Astaracian—Germany

Frunskovka fauna—Turolian—Ukraine

Fuensaldaña—Astaracian—Spain
Fuente Podrida—Turolian—Spain
Fullers Earth sites—Barstovian—California
Funaoka—Early Miocene—Japan
Fünfkirchen—Astaracian—Germany?
Furlongs sites—Arikareean—Oregon
Gabal Boudinar   (See Henchir Beglia.)
Gabal Cherichera   (See Cherichera)
Gabaldon badlands, upper—Clarendonian into Hemphillian—New Mexico
Gabal Mrhila   (See Djebel Mrhile.)
Gabal Sehib—Late Miocene—Tunisia
Gabal Semene—Late? Miocene—Tunisia
Gaber (Sofiya)—Turolian?—Bulgaria
''Gafsa West''   (See Bled ed Douarah.)
Gaging Station—Arikareean—Oregon
Gaillard Cut (Panama Canal)—Hemingfordian—Panama
Gainesville sites—Hemphillian?—Florida
Gaiselberg (Zisterdorf)—Vallesian—Austria
Gamlitz—Astaracian—Austria
Gans—Agenian—France
Ganson Farm—Hemphillian—Kansas
Gara Ziad—Late Miocene—Morocco
Gardassa—Astaracian?—France
Garet (lac) Ichkeul—Mio-Pliocene (and? Villafranchian?)—Tunisia
Gargano—Late Miocene—Italy
Garkin (Afyon-Sanikli-G.)—(= Vallesian)—Turkey
Garvin Gulley (Navasota)—Hemingfordian—Texas
Gasr-el-Sahabi   (See Sahabi.)
Gateway—Barstovian—Oregon
Gaudernoff (Eggenburg)—Eggenburgian = Orleanian—Austria
Gaudonville (Les Bréségous)—Orleanian—France
Gay Head Cliff (Martha's Vineyard)—Late Arikareean—Massachusetts
Gay Hill (Highway 36)—Hemingfordian—Texas
Gazaupouy   (See Estrepouy.)
Gazax (Baccarisse)—Astaracian—France
Gebel Zelten—(= Orleanian) Rusingan—Libya
Gedis River—Miocene—Turkey
Geiseberg (Ulm)—Agenian?—Germany
Gélat Farm (Marignac)—Astaracian—France
Gélida (Can Julià)—Orleanian—Spain
Genkingen—Astaracian—Germany
Georgensgmünd—Astaracian—Germany
Gerlenhofen (Ulm)—Astaracian—Germany
Ghazgay (Khorkkabul Basin)—(= Turolian?)—Afghanistan
Gidley Horse Quarry—Clarendonian—Texas
Giggenhausen—Astaracian—Germany
Gihu Prefecture   (See Ôhata.)
Ginn Quarry—Hemingfordian—Nebraska
Giurcani (Falciu)—Late Miocene—Romania
Givreuil—Orleanian?—France
Gleisdorf—Late Miocene—Austria
Glenn Olson Quarry (Bijou Hills)—Barstovian—South Dakota
Glimmerton (Isle of Sylt)—Late Miocene—North Sea
Glory Gorge—Clarendonian—California
Gökdere (Ankara)—(= Turolian)—Turkey
Goldberg im Ries—Astaracian—Germany
Goldie's Honeypot Quarry—Clarendonian—Arizona

Goleta—Hemphillian—Michoacan, Mexico
Gomphothere—Clarendonian—Nevada
Goodland Southeast (Edson)—Hemphillian—Kansas
Goodnight—Hemphillian—Texas
Goodrich—Barstovian—Texas   —(Tedford et al. (1983))
Goodwin Dam—Clarendonian—California
Goose Creek—Clarendonian—Idaho
Gordon Creek Quarry—Clarendonian—Nebraska
Göriach—Astaracian = marine Badenian—Austria
Gorianwala (Dhok Pathan)—(= Turolian)—Pakistan
Gorna Susica (Bakalov-Nesseber)—Turolian—Bulgaria
Gospel Hill—Barstovian—Texas
Guochtockula—Late Miocene—Mongolia
Gracias—Clarendonian or Hemphillian—Honduras
Gramont (Bos-Barrat)—Orleanian—France
Grand-Sèrre—Late Miocene—France
Granger—Clarendonian—Washington
Grant County (Mangas Trench)—Hemphillian—New Mexico
Grassette (Gimont)—Orleanian—France
Grass Roots Quarry—Barstovian—Nebraska
Gratusous (La Romieu)—Orleanian—France
Gravitelli—Turolian—Sicily
Gray—Astaracian—France
Grayface Bluff—Arikareean—Oregon
Graywater Wash—Clarendonian—Arizona
Greaser Canyon—Barstovian—Oregon
Grebeniki—Turolian—Ukraine
Greenburgh—Barstovian—Maryland
Green Hills—Barstovian—California
Green Knoll—Clarendonian—Nevada
Greenside Quarry (Frick)—Hemingfordian—Nebraska
Greit—Agenian—Switzerland
Grilos, grès de, sites—Miocene—Portugal
Grimmelfingen—Astaracian—Germany
Grindstone—Barstovian—Oregon
Griscom Plantation (Luna Plantation)—Late Arikareean or Hemingfordian—Florida
Griswold Hills South—marine and Barstovian—California
Grizzly Canyon—Clarendonian—California
Gross-Karben—Agenian—Germany
Grosslappen—Late Miocene—Germany?
Grossulovo—Turolian—Ukraine
Grossweiffendorf—Late Miocene—Austria
Grouse Creek-Etna—Hemingfordian?—Utah
Grubbenvorst (Limburg)—Late Miocene—Belgium?
Grube lignite site   (See Mas del Olmo.)
Guamúchil (Sinaloa)—Clarendonian or Hemphillian—Mexico
Guanare West—Santacrucian—Venezuela
Guano Ranch—Barstovian—Oregon
Guernsey Southeast—Hemingfordian—Wyoming
Guiard (Tafna)—Late Miocene—Algeria
Gülpinar (Canakkale)—(= Turolian)—Turkey
Gumpersdorf (Marktl)—Vallesian—Germany
Gündlkofen—Astaracian—Germany
Guntersdorf—Astaracian—Austria
Günzburg (Neuberg)—Astaracian—Germany
Gura (Galben)—Late Miocene—Moldavia, USSR
Gurgemaydan 7 and 8—(= Turolian?)—Afghanistan

311

Gürmen—Late Miocene—Bulgaria
Guymon   (See Optima.)
Gweng (Mühldorf)—Astaracian?—Germany
Gyulu (Mendru)—Astaracian—Hungary?
Hachan—Astaracian—France
Hackleman—Arikareean—Oregon
Häder—Astaracian—Germany
Hadjidimovo—Turolian?—Bulgaria
Hahnenberg   (See Nördlingen.)
Haile—Hemphillian?—Florida
Halliday's Estancia (= East Los Pozos)—Santacrucian—Argentina
Hammam (Wadi-el-Hammam)   (See Bou Hanifia.)
Hammerschmiede—Late Miocene—Germany
Hammond Hill—marine and Barstovian—California
Hand Hills, lower—Barstovian—Alberta
Hans Johnson Quarry—Clarendonian—Nebraska
Happy Valley—Clarendonian—California
Harmancik (Kütahya-Tavsanli-H.)—(= Turolian)—Turkey
Harper—Hemphillian—Oregon
Harper Quarry—Late Arikareean—Nebraska
Harrisburg Vicinity—Clarendonian—Nebraska
Harrison Formation sites—Late Arikareean—Nebraska, Wyoming
Harris Ranch—Late Arikareean—South Dakota
Harr Obo—Late Miocene—Mongolia
Hart (Gloggnitz)—Astaracian—Austria
Hasa—Astaracian—Saudi Arabia
Haslach (Ulm)—Agenian—Germany
Hatch Gulch—Arikareean—Oregon
Hatunsaray (Konya)—(= Vallesian)—Turkey
Hatunsaray (Sekisiirti)—(= Vallesian)—Turkey
Hatvan (Matra Mountains)—Late Miocene—Hungary
Haulies—Astaracian?—France
Hausen (Sigmaringen)—Astaracian—Germany
Hauskirchen (Mistelbach)—Astaracian—Austria
Hauterives—Late? Miocene—France
Havorka (Hovorka?)—Hemingfordian—Nebraska
Hawthorne—Clarendonian—Nevada
Hay Springs Vicinity—Clarendonian and Pleistocene—Nebraska
Haystack sites—Arikareean—Oregon
Hazeltine—Barstovian—California
Hazelwood Ranch—Clarendonian—Arizona
Hazen—Clarendonian or Hemphillian—Nevada
Hedgehog Quarry—Clarendonian—California
Heggbach—Astaracian—Germany
Heilingenstadt—Astaracian—Austria
Heisenburg (Günsburg)—Astaracian—Germany
Heishatou—(= Turolian?)—Nei Mongol
Hellgate Basin sites—Barstovian—California
Hemicyon Quarry (Frick)—Barstovian—California
Hemingford Northeast—Hemingfordian—Nebraska
Hemme Hills—Hemphillian—California
Hemphill (County, fauna, sites)—Hemphillian—Texas
Henchir Beglia, upper—(= Astaracian) Ternanian—Tunisia
Henchir Beglia, lower—(= Orleanian) Rusingan—Tunisia
Henry Spring—Barstovian—Nevada
Herbasse—Astaracian—France

Hernals—Astaracian—Austria
Hernandez locality—Hemphillian—New Mexico
"Hesperopithecus" Quarry—Clarendonian—Nebraska
Hesselohe—Astaracian—Germany
Hessenbrücker Hammer—Astaracian—Germany
Hessler (Mosbach), Hessler (Wiesbaden)—Agenian—Germany
Hetch-Hetchy—Clarendonian—California
Heuberg—Late? Miocene—Germany
Hexameryx locality—Hemphillian—Florida
Heyrieux—Late Miocene—France
Hidalgo Bluff—Hemingfordian—Texas
Hidden Treasure Spring—Hemingfordian—California
Hidden Valley—Clarendonian—California
Higgins—Hemphillian—Texas
High Level Quarry—Clarendonian—California
High Point—Clarendonian—Nevada
Highrock Canyon—Barstovian—Nevada
Highway 40—Clarendonian—Nevada
Highway 93—Clarendonian?—Nevada
Hill 66—Late Arikareean—Wyoming
Hill 1284—Clarendonian—California
Hillhouse District—Clarendonian—Nevada
Hilltop Quarry—Hemingfordian—Nebraska
Hinton sites—Arikareean—Oregon
Hiramaki (Anchitherium "zone")—(= Astaracian)—Japan
Hirschtal—Astaracian—Switzerland
Hobart Mills—Clarendonian—California
Hochêng—Late Miocene—Gansu
Hochgeland—Astaracian—Germany
Hochstadt—Agenian—Germany
Hogtown Creek—Hemphillian—Florida
Hohenhöven—Astaracian—Germany
Hohenwarth (Krems)—Late Miocene—Austria
Hollabrun—Astaracian—Germany
Hollow Horn Bear Quarry—Clarendonian—South Dakota
Holton Pocket—Clarendonian—Nevada
Homberg—Astaracian—Germany
Home Station Wash—Barstovian—Nevada
Hommes—Orleanian—France
Hoover Creek sites—Arikareean—Oregon
Horse and Mastodon Quarry—Barstovian—Colorado
   (Tedford et al. (1983?))
Horse Creek Quarry—Hemingfordian—Wyoming
Horta de Tripas   (See Lisboa or Lisbon I.)
Hortu (Eskişehir-Sivrihisar-H.)—(= Vallesian)—Turkey
Hospital Creek sites—Clarendonian—California
Hostalets de Pierola, lower—Astaracian—Spain
Hostalets de Pierola, upper—Vallesian—Spain
Hot Creek—Hemphillian—Idaho
Hovorka   (See Havorka.)
Höwenegg—Vallesian—Germany
Hrabrsko—Turolian?—Bulgaria
Hsen Shui Ho (Yungtang) (= Hsienshinho?)—Late Miocene—Gansu
Hsiaolungtan coal field—Late Miocene—Yunnan
Hsiechiaho (Linchü)—(= Astaracian)—Shandong
Hsihsien—Late Miocene—Shanxi
Humbug Quarry—Hemingfordian-Barstovian—Nebraska
Hung-Kureh Formation sites—Late Miocene—Mongolia
Hunts Canyon—Barstovian—Nevada
Huntsville Northeast—Hemingfordian—Texas

Hutch Dry—Barstovian—California
Hutch Tortoise—Barstovian—California
Hydrobien Schichten sites—Agenian—Germany
Ian's Prospect—Middle? Miocene—South Australia
IBM-Bohrung—Agenian—Germany
Ilhan (Ankara)—(= Turolian)—Turkey
Ilz—Late Miocene—Austria
Imbros Island—Late Miocene—Turkey
Imola—Late Miocene—Italy
Ingolstadt—Astaracian—Germany
Ingram Creek—Clarendonian—California
Inönü (Ankara-Kazan-Sinap)—(= Vallesian)—Turkey
Inzighofen—Late Miocene—Germany?
Ione Valley sites—Clarendonian—Nevada
Ipolytarnoc—Orleanian—Hungary
Iron Canyon—Clarendonian—California
Ironside—Clarendonian—Oregon
Iron Springs—Clarendonian—Arizona
Irrawaddy "series," lower—Middle to Late Miocene (in part?)—
     Burma
Irtysch, lower—Late? Miocene—Kazakhstan
Isarbett—Late Miocene—Germany
Isayevo—Late Miocene—Ukraine
Ischim (Ishym)—(= Turolian?)—Siberian USSR
Isle-Bouzon (Landiran)—Orleanian—France
Isle-en-Dodon—Astaracian—France
Issaevo (Ananiev)—Late Miocene—Ukraine
Isserville (Kabylie) —Middle? Miocene—Algeria
Istanbul   (See Küçükçekmece.)
Ivananc—Turolian?—Bulgaria
Ixtapa—Barstovian?, Clarendonian?—Chiapas, Mexico
Izberg (Budapest)—Astaracian—Hungary
Izmir—(= Orleanian)—Turkey
Jabal Midra Ash-Shamali—(= Astaracian)—Saudi Arabia
Jacalitos Formation sites—Hemphillian—California
Jacona microfauna—Clarendonian—New Mexico   —(D.
     Chaney manuscript)
Jaligny—Agenian—France
Jamber   (See Joseph Jamber Farm.)
Jay-Em—Late Arikareean—Wyoming
JBF—Barstovian—California
JCM—Arikareean—Oregon
Jebel Cherichera   (See Cherichera.)
Jebel Semmene   (See Gabal Semene.)
Jebel Zelten   (See Gebel Zelten.)
Jeep Quarry—Hemingfordian—New Mexico
Jegun—Astaracian—France
Jenkins Quarry (Frick)—Barstovian—Nebraska
Jersey Valley—Barstovian—Nevada
Jim Swayze Quarry   (See Swayze Quarry.)
Jim Woodruff Dam—Hemingfordian—Florida
Jiulaopo (Lantien)—(= Turolian?)—Shaanxi
Jiulongkou (Cixian)—Late Miocene—Hubei
J 511 and 514—Late Miocene—Kenya
Joe's Quarry—Barstovian—Wyoming
Joe Thin Elk Gravel Pits—Clarendonian—South Dakota
John Day Formation sites—Whitneyan into Hemingfordian—Or-
     egon
John Lance sites—Clarendonian—Arizona
Johnny Kirk Spring—Arikareean—Oregon
Johnny Post—Clarendonian—Arizona

Johnson Canyon—Arikareean—Oregon
Jonas Hill Road—Clarendonian—California
Jones Canyon—Barstovian? ("Valentinian")—Nebraska
Jordansbad-(Biberach)—Astaracian—Germany
Joseph Jamber Farm—Clarendonian—Nebraska
Junction North—Arikareean—Oregon
Jungnau—Astaracian—Germany
Juniper Creek Canyon—Barstovian and Hemphillian—Oregon
Kaboor—Middle Miocene   —(R. Savage (1978))
Kabuga beds sites—Miocene?—Zaire
Kabylie (Isserville)—(= Orleanian)—Algeria
Käferberg—Astaracian—Switzerland
Kahler Basin sites—Arikareean—Oregon
Kaiser Creek—Clarendonian—California
Kaisersteinbruch—Late Miocene—Austria
Kaisheim—Agenian—Germany
Kaiyuan—Late Miocene—Yunnan
Kalban district—Late Miocene—Altai, Mongolia
Kale assemblage—(= Orleanian)—Turkey
Kalfa—Vallesian?—Moldavia, USSR
Kalimantsi (Kalimanzi)—Turolian—Bulgaria
Kalinindorf—Late Miocene—Ukraine
Kalinköy (Sivas-Derekli-K.)—(= Turolian)—Turkey
Kalithies—Turolian—Isle of Rhodes
Kalkaman Lake—Late Miocene—Kazakhstan
Kalksburg-Vienna—Astaracian—Austria
Kalmakpai—Late Miocene—Kazakhstan
Kaltenbachgraben—Orleanian—Germany
Kaltwang—Astaracian—Germany-Switzerland
Kaman—(= Turolian)—Turkey
Kamlial "Formation," "zone," etc.—(= Astaracian)—Pakistan
     and India
Kangaroo Well—Middle? Miocene—Northern Territory, Aus-
     tralia
Kani district   (See Tanohira.)
Käpffnach—Astaracian—Switzerland
Kappel—Agenian—Switzerland
Karacahasan (Ankara-K.)—(= Vallesian)—Turkey
Karaçay 1 and 2 (Çorum-Sungurlu-K.) (= Astaracian?)—Turkey
Karaceçeli (Ankara)—(= Turolian)—Turkey
Karaihen—Santacrucian—Argentina
Karaiken—Santacrucian—Argentina
Karaikenian = Santacrucian
Karain (Nevşehir-Ürgüp-K.) (= Vallesian)—Turkey
Kara–Kol River—Late Miocene—Kazakhstan
Karpewali (Dhok Pathan)—(= Turolian)—Pakistan
Karugamania—Rusingan—Zaire
Karungu—Early Rusingan—Kenya
Kastellios Hill—Vallesian—Isle of Crete
Kastrich—Agenian—Germany
Kaswanga Point (Rusinga Isle)—Rusingan—Kenya
Katheni—Vallesian—Evia, Greece
Kat Quarry—Clarendonian—Nebraska
Kaubach I—Agenian—Germany
Kavirondo Gulf region—Rusingan and later—Kenya
Kavurca (Çankiri)—(= Turolian)—Turkey
Kawsoh Mountains—Clarendonian?—Nevada
Kayadibi (Sarişik-Inleri-K.)—Late Miocene—Turkey

Kayseri district—(= Vallesian)—Turkey
Keeline North—Late Arikareean—Wyoming
Kef en Nsoura—Late Miocene—Tunisia
Keiyuan district (= Kaiyuan)   (See Hsiaolungtan.)—Late Miocene—Yunnan
Kelly Place (Mann Place)—Hemingfordian—Texas
Kelly Road—Hemphillian—Wyoming
Kendall-Mallory—Clarendonian—California
Kennesaw—Barstovian—Colorado
Kent Quarry—Barstovian—California
Kern River Formation sites—Hemphillian—California
Khanovo—Late Miocene—Bulgaria
Khenchela—(= Orleanian) and? Late Miocene?—Tunisia
Khendek el Ouaich—(= Turolian) Lothagamian—Morocco
Kherson—(= Astaracian?)—Ukraine
Kholobolchi Nor district—(= Astaracian)—Mongolia
Khorkkabul Basin (Chazgay)—(= Turolian)—Afghanistan
Khunuk Formation site—Late Miocene—Mongolia
Khutchur-Tchekmedjé West—(= Turolian)—Thracian Turkey
Kiahera Formation sites–Rusinga Island—Early? Rusingan—Kenya
Kichinev   (See Kischinev.)
Kieferstädt   (See Sosnicowice?)—Astaracian—Poland?
Kilcak (Ankara-Kalecik)—(= Astaracian?)—Turkey
Kilpatrick Quarry—Clarendonian—Nebraska
Kimball Formation sites—Hemphillian—Nebraska
King-yan-fou—Late Miocene—Gansu
Kinik (Afyon-Sandikli-K.)—(= Turolian)—Turkey
Kinsey Ranch—Hemphillian—California
Kirald—Orleanian—Hungary
Kirchberg—Astaracian?—Austria
Kirimun—Ternanian?—Africa
Kiriyet (Lunga)—Late Miocene—Moldavia, USSR
Kirkland Valley—Clarendonian or Hemphillian—Arizona
Kirkwood Formation site—Late? Arikareean—New Jersey
Kirşehir   (See Karaceçeli.)
Kischinev—marine and Late Miocene land mammals—Moldavia
Kisláng—late Miocene—Hungary
Kissimmee River—Hemphillian?—Florida
Kiva Quarry—Hemingfordian—New Mexico
Kiyevskiy—Late Miocene—Kazakhstan
Kizilhisar assemblage—Late Miocene—Turkey
Kizilirmak—Late Miocene—Turkey
Kleineisenbach—Late Miocene—Germany
Kleinfelder Farm—Barstovian?—Nebraska
Kleinhadersdorf—Badenian = Astaracian—Austria
Klipstein Ranch—Clarendonian and Hemphillian—California
Knoll Creek—Clarendonian—Nevada
Knoll Mountain 1—Clarendonian—Nevada
Koçgazi—Late Miocene—Turkey
Köflach (Voitsberger)—Karpatian = late Orleanian—Austria
Kohfidisch—Pontian = Turolian—Austria
Kolinga—Lothagamian—Lake Chad
Komotini—Orleanian—Greece
Kömürlük Dere—(= Astaracian)—Turkey
Konjic—Astaracian—Yugoslavia
Konkrovi (Dhok Pathan)—(= Turolian)—Pakistan
Konnor anticline   (See Djebel Hamrin)—Late Miocene—Irak

Köprübaşi—(= Turolian)—Turkey
Koru—Rusingan—Kenya
Köschenrüti (Seebach)—Astaracian—Switzerland
Kou Chia Tsun (Lantien)—Late Miocene—Shaanxi
Kouialnik—Late Miocene—Ukraine
Kozel'sk (Nikolayevka)—Late Miocene—USSR
Krebs Ranch—Hemphillian—Oregon
Krems—Astaracian—Austria
Krivadia—Eocene and Astaracian—Romania
Krivoy Rog (See Kozel'sk)—Late Miocene—Ukraine
Kromidovo—Turolian—Bulgaria
Kruševac Basin (Božurevac)—Astaracian—Yugoslavia
Kruševica (Kruševac?)—Astaracian—Yugoslavia
Küçükçekmece   (Kutchuk–Tchekmedje West)—Late Miocene—Thracian Turkey
Kulan-Utmes River—Late Miocene—Kazakhstan
Kul'm—Late Miocene—Ukraine
Kulu beds sites—Ternanian?—Kenya
Kundlawala, Kundvali, Kundral (Dhok Pathan)—(= Turolian)—Pakistan
Kureh Formation site   (Hung-Kureh)
Kütahya   (See Sofça.)
Kutjamarpu fauna—Middle? Miocene—South Australia
Kutsay Mountain—Late Miocene—Caucasus, USSR
Kuziriver (Kamiogawa)—marine with land mammal = Astaracian—Japan
La Alberca—Turolian—Murcia, Spain
La Angelina East—Santacrucian—Argentina
Labahia Mission—Hemphillian—Texas
Labarthe-Inard—Astaracian—France
Labastide-d'Armagnac—Astaracian—France
Labitschberg—Astaracian—Austria
Labourdette-Rozès—Orleanian—France
La Brète—Agenian—France
Lacapère à Saint-Mézard—Orleanian—France
La Chaux-de-Fonds—Astaracian—Switzerland
La Chaux–Ste. Croix—Agenian—Switzerland
La Cistérniga—Astaracian—Spain
La Costa East—Santacrucian—Argentina
La Croix Rousse   (See Croix Rousse.)
Lacualtipan—Hemphillian—Mexico
La Cueva—Santacrucian—Argentina
Ladakh—(= Orleanian?)—Jammu and Kashmir, India
La Fontana—Late Miocene—Spain
Lagarto Clay site—Late Miocene—Texas
La Gèze   (See Larroque-Magnoac.)
Lago Pueyrredón—Santacrucian—Argentina
La Grenatière—Astaracian—France
La Grive–St. Alban—Astaracian—France
La Gruas (Simorre)—Astaracian—France
Laguna Blanca—Santacrucian?—Chile
Laguna Blanca—Friasian—Argentina
Lahas—Astaracian—France
La Hidroeléctrica—Astaracian—Spain
Lahontan Reservoir—Clarendonian—Nevada
Laichingen—Astaracian—Germany
Lake Albert district—Rusingan and later—Zaire
Lake Baringo   (See Baringo)—Ngororan and Lothagamian—Kenya
Lake Bed—Barstovian—California
Lake County—Barstovian?—Oregon

Lake Palankarinna West—Middle? Miocene—South Australia
Lake Pinpa Southwest—Middle? Miocene—South Australia
Lalibetscho (Dhok Pathan)—(= Turolian)—Pakistan
Lambert North—Agenian—France
La Misión—marine and Hemingfordian—Baja California, Mexico
La Molière—Orleanian—Switzerland
Landiran (L'Isle-Bouzon)—Orleanian—France
Langau—Orleanian—Austria
Langeais—Orleanian?—France
Langenengalingen—Astaracian—Germany
Langenmoosen—Astaracian—Germany
Langenthal   (See Wischberg-L.)
Langy—Agenian—France
Lantien district—Eocene (= Astaracian) and later—Shaanxi
La Paillade—Agenian—France
Lapara Creek—Clarendonian—Texas
La Puebla de Almoradial—Late Miocene—Spain
Laren Bed site—Late Miocene—Jammu and Kashmir, India
La Roche-de-Meillard—Orleanian?—France
La Romieu—Orleanian—France
Larroque-Magnoac (La Gèze)—Astaracian—France
Lartigole   (See Pessan.)
Las Pedrizas—Turolian—Spain
Las Planas 3 & 4—Orleanian and Astaracian (Aragonian)—Spain
Lasse—Orleanian—France
Lasseran (En Charron)—Orleanian—France
Lasseube-Simorre   (See Simorre.)
Lassnitzhöhe—Pannonian = Late Miocene—Austria
Last Chance—Clarendonian—California
Last Day—Clarendonian—Nevada
Las Trampas Creek—Clarendonian—California
La Tarumba—Vallesian—Spain
Latewali Khan (Dhok Pathan)—(= Turolian)—Pakistan
La Tour-du-Pin—Late Miocene—France
Laugnac—Agenian—France
Lava Butte—Barstovian—California
Lava Mountains—Hemphillian—California
La Vardens—Astaracian—France
La Vela ''series'' site—marine and Santacrucian?—Venezuela
La Venta fauna—Friasian—Colombia
Laverne (Beaver)—Clarendonian—Oklahoma
Lawler Ranch—Hemphillian—California
Lecce—marine and Late Miocene—Italy
Lectoure (Navère)—Orleanian—France
Le Doul (Narbonne)—Agenian—France
Lefkon 1—Vallesian—Macedonia, Greece
Lehmbachmühle—Astaracian—Germany
Lehri (Dhok Pathan) (= Turolian)—Pakistan
Leiding (Pitten)—Orleanian—Austria
Le Locle—Astaracian—Switzerland
Lemhi Valley   (See Peterson Creek.)
Lengshuikou (Lintung)—(= Astaracian)—Shaanxi
Lenneberg (Mainz)—Agenian—Germany
Lenora Southwest—Clarendonian—Kansas
Leoben—Karpatian = Late Orleanian—Austria
Leptarctus Quarry—Clarendonian—Nebraska
Les Alletz—Agenian—France
Les Barres—Orleanian—France
Les Breségous   (See Gaudonville)—Orleanian—France
Les Cévennes—Agenian—France

Les Montils—Orleanian—France
Les Trois Moulins (Narbonne)—Agenian—France
Leucate Butte 1—Astaracian—France
Leuciscus turneri lake beds sites—Barstovian?—Montana
Levač Basin (Sibnica)—Astaracian—Yugoslavia
Lewis Coal Mine—Clarendonian—Nevada
Lezhanka—Late? Miocene—Siberian USSR
Liahwachen   (See Quinan)—Late Miocene—China
Lialores—Orleanian—France
Libano (Belluno)—marine and Orleanian—Italy
Librilla—Turolian—Spain
Libros—Late Miocene—Spain
Liet—Astaracian—France
Lighthill—Clarendonian—Nebraska
Lihsien—Late Miocene—Gansu
Linchü (Linqu)—(= Astaracian)—Shandong
Lindsay—Barstovian—California
Lindsay Squirrel—Clarendonian—California
Lintung   (See Lengshuikou.)
Lisboa (Lisbon) I, IV, Va, Va—Orleanian—Portugal
Lishan—Late Miocene—Shanxi
Lissieu—Lutetian and Turolian—France
Little Basin—Arikareean—Oregon
Little Beaver V, C—Clarendonian   Nebraska
Little Dike—Arikareean—Oregon
Little High Rock Canyon—Barstovian—Nevada
Little Valley—Hemphillian and later—Oregon
Little Valley—Clarendonian or Hemphillian—Idaho
Little White River—Clarendonian—South Dakota
Liu Ho (Nanjing)—Late Miocene—Jiangsu
Live Oak (SB-1A)—Late Arikareean—Florida
Lizard Quarry—Late Arikareean—Nebraska
Ljutica (Pavlovici)—Astaracian—Yugoslavia
Lobrieu—Turolian?—France
Locle (Le Locle)—Astaracian—Switzerland
Logan Butte sites—Arikareean—Oregon
Logan County—Clarendonian?—Colorado
Logan Mine sites—Hemingfordian—California
Loh Formation sites—(= Orleanian)—Mongolia
Lombez—Astaracian—France
Long Creek—Arikareean—Oregon
Long Island—Hemphillian?, Clarendonian?—Kansas
Long Pine North and East—Clarendonian—Nebraska
Long Quarry—Hemingfordian—Nebraska
Loperot—Ternanian—Kenya
Lopusna—Late Miocene—Romania
Loretto—Astaracian—Austria?
Los Aguanaces—Turolian—Spain
Los Algegares—Vallesian—Spain
Los Cruceros (Laguna de Toro)—Santacrucian?—Chile
Los Mansuetos—Turolian—Spain
Losodok—Rusingan—Kenya
Lost Quarry—Hemphillian—Kansas
Los Valles de Fuentidueña (Segovia)—Vallesian—Spain
Lothagam—Lothagamian—Kenya
Love Bone Bed and local fauna—Clarendonian—Florida
Love Mountain—Clarendonian—Nevada
Love Place or Farm—Barstovian—Texas

Lower Madison Valley—Clarendonian—Montana

Loyang   (See Tung-sha-po.)

Lozengrad (Kirklareli)—Late Miocene—Thracian Turkey

Lozère—Orleanian—France

Lozovik (Svetozarevo)—Astaracian?—Yugoslavia

Luberon   (See Mt. Luberon)—Turolian—France

Lublé—Astaracian—France

Lucenay—Late Miocene—France

Luc-sur-Orbien—Astaracian—France

Luderitz (Namib)—Early? Rusingan—Namibia

Lugnez—Late Miocene—Switzerland

Lukeino beds sites—Lothagamian—Kenya

Lull locality—Clarendonian—Nebraska

Lulu—Hemphillian?—Florida

Lumeau—Orleanian—France

Luminar   (See Nordrand-Lisboa.)

Lunggiakou—Late Miocene—Gansu

Lusk district—Late Arikareean—Wyoming

Lussan—Astaracian—France

Lustmühle (Teufen)—Orleanian—Switzerland

Lutzukou beds sites–Baode (Paote)—(= Turolian)—Shanxi

Luzinay (= Lucenay?)—Late Miocene—France

Lyden—Hemphillian—New Mexico

Lynch—Barstovian—Nebraska

Lyon (Croix Rousse)   (See Croix Rousse.)

Maboko Island—Rusingan—Kenya

MacAdams Quarry—Clarendonian—Texas

Macconens—Orleanian—Switzerland

Madaillan   (See Laugnac.)

Madison Valley Formation sites (Bozeman lake beds)—Barstovian—Montana

Madras Gravels sites—Clarendonian—Oregon

Madrid (Hidroeléctrica)   (See La Hidroeléctrica.)

Magersdorf (Hollabrunn)—Pannonian = Late Miocene—Austria

Mahmutgazi (Denizli-Çal-M.) (= Turolian)—Turkey

Mahmutlar (Ankara-Kalecik-M.)—(= Astaracian)—Turkey

Mainz (Kästrich)—Agenian—Germany

Majorca (Mallorca)—marine Middle Miocene and nonmarine Oligocene—Mallorca

Mala Miliva—Badenian = Astaracian—Yugoslavia

Malarctic-Baillasbats (Simorre)—Astaracian—France

Malembe (Cabinda)—marine and Rusingan—Cabinda, Africa

Malu ''member,'' ''shale stage''—marine Miocene—Ceylon

Mamay—Turolian—France

Mammern—Astaracian—Switzerland

Manatee County Dam Site—Hemphillian—Florida

Manchar beds of Sind—Miocene—India

Manchones—Astaracian—Spain

Mancusun (Kayseri-Bünyan-M.)—Late Miocene—Turkey

Mangas Trench (Dry Creek and Grant County, New Mexico.)

Manisa   (See Esme.)

Mannersdorf (Leithagebirge)—marine and Astaracian—Austria

Mann Place (Kelly Place)—Hemingfordian—Texas

Mantankou—(= Turolian)—Shanxi

Manthelan—Orleanian?—France

Manzanita Quarry—Clarendonian—Arizona

Maragheh (Maragha) succession and district—(= Late Vallesian? and Turolian)—Iran

Marceau—Late Miocene—Algeria

Marcel   (See Laugnac.)

Marcoin—Agenian—France

Marewa—Rusingan—Kenya

Margarethen—Astaracian—Austria

Margo—Barstovian—California

Mariatal (Hollabrunn)—Pannonian = Late Miocene—Austria

Marignac (Laspeyre)—Astaracian—France

Marigny—Orleanian—France

Marin à Lamontjoie—Orleanian—France

Marktl—Late Miocene—Germany

Markt Rettenbach—Astaracian—Germany

Marseillan—Orleanian—France

Marsland Formation sites—Hemingfordian—Nebraska

Marsland Northwest (Frick)—Hemingfordian—Nebraska

Marsland Quarry (Frick)—Hemingfordian—Nebraska

Martha's Pocket—Clarendonian—California

Martin-Anthony Roadcut—Late Arikareean—Florida

Martin Canyon Quarry A—Hemingfordian—Colorado

Martin Creek—Clarendonian—California

Martinsbrücke—Orleanian—Switzerland

Martorell—Orleanian—Spain

Marvila (Mutela)—marine and Middle to Late Miocene—Portugal

Masada del Valle—Turolian—Spain

Mascall Formation sites—Barstovian—Oregon

Mas del Olmo—Astaracian—Spain

Masia del Barbo—Vallesian—Spain

Masquefa—Astaracian—Spain

Massacre Lake—Hemingfordian—Nevada

Massenhausen (München)—Vallesian?—Germany

Mastodon Quarry (Frick)—Barstovian—Colorado

Matatlan—''Late Miocene''—Mexico

Mathew Ranch fauna—Clarendonian—California

Matillas (Cendejas de la Torre)—Late Miocene—Spain

''Matthew's Quarry 21'' = Thomson's Quarry B of 1921—Barstovian—Nebraska

Mauvi River—Santacrucian?—Bolivia

Mayall Quarries—Clarendonian—California

Mayflower Mine (Jefferson River)—Clarendonian—Montana

Mayo fauna   (See Rio Mayo)—Friasian—Argentina

Mbagathi—Rusingan—Kenya

McAllister sites—Arikareean—Oregon

McGehee Farm—Hemphillian—Florida

McKanna Spring—Barstovian—Montana   —(Tedford et al. (1983?))

McKay Reservoir—Hemphillian—Oregon

McMurray Place—Barstovian—Texas

McMurty Ranch (Saddler Creek)—Clarendonian—Texas

M'dilla—Late Miocene—Tunisia

Meadow Valley (Panaca)—Clarendonian?—Nevada

Medio Creek—Clarendonian? Hemphillian?—Texas

Mehring (Augsburg)—Astaracian—Germany

Mehrten Formation sites—Clarendonian and Hemphillian—California

Melchingen—Late Miocene—Germany

Melg or Melka el Ouidane—(= Turolian) Lothagamian—Morocco

Meluins—Clarendonian—California

Merychippus Draw—Hemingfordian and? Barstovian?—Nebraska

Merychippus "zone" (See North Coalinga)—Barstovian—California

Mesa Prospect locality—Hemingfordian—New Mexico

Mesevle (Muğla-Yatağan-M.) (= Astaracian)—Turkey

Mesoceras locality—Barstovian—Nebraska

Messen—Agenian—Switzerland

Meswa Bridge—Rusingan—Kenya

Meurnon—Late Miocene—France

Mfwanganu Island—Rusingan—Kenya

Miami Quarry (See Coffee Ranch)—Hemphillian—Texas

Michelsberg (Ulm)—Agenian—Germany

Middle Fork—Arikareean—Oregon

Middle of the Road Quarry—Hemingfordian—Nebraska

"Middle Red" sites—Barstovian and Clarendonian—New Mexico

Middle Ridge (Lemhi Valley)—Barstovian or Clarendonian—Idaho

Midway—Hemingfordian—Florida

Midway Quarry—Clarendonian?—Nebraska

Miélan—Astaracian—France

Mikhaylove—Late Miocene—Bulgaria

Milk Creek Quarry—Clarendonian—Arizona

Mill, Floor of Mill Quarry—Barstovian—Nebraska

Miller Gravel Pit—Hemphillian—Oklahoma —[Tedford, et al. (1983?)]

Minnechaduza fauna—Clarendonian—Nebraska

Mino province (See Ôhata.)

Mint Canyon Formation sites—Clarendonian and Barstovian—California

Mirabeau—Orleanian—France

Mira Loma—Clarendonian—California

Mistelbach—Pannonian = Late Miocene—Austria

Mistrals—Turolian—France

Mitridat Mountain (Kerch)—Late Miocene—Crimea, USSR

Mixson's Bone Bed—Hemphillian—Florida

Moghara Oasis or Wadi—(= Orleanian) Rusingan—Egypt

Mohari beds site—Rusingan?—Zaire

Mole Quarry—Clarendonian—California

Molayan—Late Miocene—Afghanistan

Molí Calopa (See Rubí.)—Orleanian—Spain

Mollon—Turolian—France

Mombach—Agenian—Germany

Monocline Ridge—marine and Barstovian—California

Monroe Creek Anthill—Arikareean—South Dakota?

Monroe Creek Formation sites—Arikareean—Nebraska, Wyoming, and South Dakota

Montabuzard—Orleanian—France

Montaigu-le-Blin—Agenian—France

Montastruc-sur-Baïse—Astaracian—France

Mont Chaibeut—Late Miocene—Switzerland

Mont-d'Astarac—Astaracian—France

Monteagudo—Astaracian or Late Miocene—Spain

Monte Bamboli—Turolian—Italy

Monte León—Santacrucian—Argentina

Monte Massi—Vallesian into Turolian—Italy

Monte Novo de Concelho—Late Miocene—Portugal

Monte Observación—Santacrucian—Argentina

Montesquiou-sur-Losse (Custret)—Orleanian—France

Montfaucon—Late Miocene—Switzerland

Montiron—Orleanian—France

Montjuic—Astaracian—Spain

Montmirail—Late Miocene—France

Montpellier vicinity—Agenian and Pliocene—France

Montredon—Vallesian—France

Montréjeau—Late Miocene—France

Mont Vully—Orleanian—Switzerland

Moore Reservoir—Late Arikareean—Colorado

Moranwala (Dhok Pathan)—(= Turolian)—Pakistan

Morava Pit (Plazane)—Astaracian—Yugoslavia

Morava Ranch Quarry (= Harvey Morava Quarry?)—Late Arikareean—Nebraska

Morgan sites—Arikareean—Oregon

Mörigen—Agenian—Switzerland

Moroto—Rusingan?—Uganda

Mörsingen (Upflamör)—Astaracian—Germany

Morsum Kliff (Glimmerton)—Late Miocene—Isle of Sylt, North Sea

Moscow—Barstovian—Texas

Mösskirch—Astaracian—Germany

Mouné—Astaracian—France

Mt. Abel sites—Clarendonian—California

Mt. Eden Formation sites—Hemphillian—California

Mt. Lébcron or Mt. Luberon—Turolian—France

Mt. Lewis—Clarendonian—Nevada

Mourning—Barstovian—California

Mpesida site—Ngororan to Lothagamian—Kenya

Muddy Creek East—Late Arikareean—Wyoming

Muddy Valley sites—Barstovian and Hemphillian—Nevada

Muğla district—Late Miocene—Turkey

Mulberry Phosphate Pits—Hemphillian—Florida

Mulholland Formation sites—Hemphillian—California

München district (Flinz and Schweissanden)—Astaracian—Germany

Münchsmünster—Astaracian—Germany

Muruarot (Muruorot) Hill—Rusingan—Kenya

Muruyur—Ternanian—Kenya

Mutton Mountain Southeast—Arikareean—Oregon

Myaing—Early Miocene (in part?)—Burma

Mylagaulodon—Late Arikareean and Hemingfordian?—Oregon

Nabón—Friasian?—Ecuador

Nagri "beds," "Formation," "zone," "fauna"—(= Vallesian)—Pakistan and India

Nagyvarad—Late Miocene—Hungary

Nakali—Ngororan?—Kenya

Namas River region—Late Miocene—Xinjiang

Nambe Member sites—Hemingfordian—New Mexico

Nameray—Orleanian—France

Namib—Early? Rusingan—Namibia

Nancray—Orleanian—France

Nan-ho (Paishui district)—Late Miocene—Shaanxi

Nanjing (Nanking) district (= Astaracian)—Jiangsu

Nantan (Hohsien)—(= Turolian)—Shanxi

Napak—Rusingan—Uganda

Nation Gravel Pit—Hemphillian—Oklahoma —[Tedford et al. (1983?)]

Navarette-del-Rio—Orleanian—Spain

Navasota (See Garvin Gulley)—Hemingfordian—Texas

Navère (See Lectoure.)

Naylan sites—Arikareean—Oregon

Nebelberg (Nunnigen)—Astaracian—Switzerland

Negev—Orleanian—Israel

Nemirovo (former Podol'sk Government)—Late Miocene—Crimea, USSR

Nemours—Late Miocene—Morocco

Nenzel Quarry—Clarendonian—Nebraska

Nereczi—Astaracian—Yugoslavia

Nesebur (Nesebr, Nesebar)—Late Miocene—Bulgaria

Nestelbach—Pannonian = Late Miocene—Austria

Nettle Spring (Apache Canyon)—Clarendonian—California

Nettle West Canyon—Clarendonian—California

Neuberg (See Oberstotzingen and Attenfeld)—Astaracian—Germany

Neudorf a. d. March (Děvínská Nová Ves)—Orleanian and Astaracian—Czechoslovakia

Neufeld—Pontian = Late Miocene—Austria

Neuhausen (Tuttlingen)—Headonian and Astaracian—Germany

Neuville—Orleanian—France

Nevaxel sites (Eastgate)—Hemingfordian?, Barstovian?—Nevada

Newberry—Hemingfordian and? Hemphillian?—Florida

New Chicago East—Barstovian?—Montana

Newport North—marine and Hemingfordian or Barstovian—Oregon

New Sand Quarry—Barstovian—Nebraska

New Surface Quarry—Barstovian—Nebraska

New Year Quarry (Frick)—Barstovian—California

Nexing—Sarmatian and Astaracian—Austria

Neyron—Vallesian—France

Ngapakaldi fauna—Middle? Miocene—South Australia

Ngorora beds and fauna—Ngororan—Kenya

Nicolaieff (Nikolayev)—Astaracian—USSR

Niederaichbach—Astaracian—Germany

Niederräder Schleusenkammer—Agenian—Germany

Niederstozingen—Astaracian—Germany

Nightingale Road—Clarendonian—Nevada

Nikolayev (Nicolaieff)

Nikolsburg (Mikulov)—Astaracian—Czechoslovakia

Ningting—Late? Miocene—Gansu

Ningxia province sites—Late Miocene and later—Ningxia, China

Niobrara Canyon (9 miles SW of Harrison)—Late Arikareean—Nebraska

Niobrara River fauna—Barstovian—Nebraska

Niscavit Farm—Barstovian—Texas

Noble Farm (Barry Farm)—Hemingfordian—Texas

Noevtsi—Turolian?—Bulgaria

Nombrevilla—Vallesian—Spain

Norden Bridge—Barstovian—Nebraska

Nord-Kanyô-dô—(= Astaracian)—Korea

Nördlingen—Astaracian—Germany

Nordrand-Lisboa—marine and Orleanian—Portugal

Norris Canyon—Clarendonian—California

North Bench—Hemphillian—Montana

North Coalinga—marine and Barstovian—California

North Fossil Ridge—Clarendonian—Arizona

North Rim—Clarendonian—Nebraska

North Santa Clara Canyon (Frick)—Hemphillian—New Mexico

North Silver Star Triangle—Hemphillian—Montana   —(Tedford et al. (1983?))

North Thomson Quarry—Hemingfordian—Nebraska

North Wall Quarry—Barstovian—Nebraska

North Wash—Arikareean—Oregon

Nouvelle Faculté Médecine (Plaissan)—Agenian—France

Novgorod (Seversk)—Late Miocene—Ukraine

Novocherkassk 1—Late Miocene—Caucasus, USSR

Novo-Elisavetovka (Yelizavetovka)—Turolian—Ukraine

Nowa Wieś Krôlewska (Opole) Sarmatian = Astaracian—Poland

Noyant-sous-le-Lude—Orleanian and marine—France

Nueva Palmira—Friasian or Chasicoan—Uruguay

Nussbaum Formation site—Late? Miocene—Colorado

Oak Creek—Clarendonian?—South Dakota

Oakdale—Hemphillian—California

Oak Springs—Clarendonian—California

Oakville Sandstone sites—Hemingfordian—Texas

Oberdorf—Late Miocene—Austria

Oberer Eselsberg—Agenian—Germany

Oberkirchberg—Astaracian—Germany

Oberkochen—Agenian—Germany

Oberstotzingen (Neuberg)—Astaracian—Germany

Obora   (See Tanohira.)

Observation Quarry (Frick)—Barstovian—Nebraska

Ocata (Upper Hostalets)—Vallesian—Spain

Occidental Mine—Hemphillian?—Florida

Oceanside (Lawrence Canyon)—marine and Hemphillian—California

Ocote—Hemphillian—Guanajuato, Mexico

Odessa 1—Late Miocene—USSR

Oeningen—Astaracian—Switzerland

Oettingen (Nördlingen)—Astaracian—Germany

Ogallala Formation or Group sites—Late Miocene—Great Plains region of United States

Oggenhausen (Heidenheim)—Astaracian—Germany

Oggenhof—Astaracian—Germany

Ognyanovo—Turolian?—Bulgaria

Ôhata—(= Astaracian)—Honshu, Japan

Öhningen   (See Oeningen.)

Ojo Caliente area—Barstovian?—New Mexico

Ojo Caliente Sandstone Member sites—Clarendonian—New Mexico

Olan Chorea—Late Miocene—Mongolia

Olcott Hill Quarry—Clarendonian—Nebraska

Old Cabin Quarry (Redington)—Hemphillian—Arizona

Old Kelly Behind (Behind Old Kelly Place)—Hemingfordian—Texas

Old Mill—Barstovian—Nevada

Old Stiff—Late Arikareean?—California

Old Wagon Road—Clarendonian—Montana   —(Tedford et al. (1983?))

Old Walker Place—Hemingfordian—Texas

Old Windmill—Clarendonian—Montana   —(Tedford et al. (1983?))

Olelas (Quintanelas)—Astaracian—Portugal

Oma (Chilung)—(= Turolian?)—Xizang (Tibet)

Ombo—Rusingan—Kenya

Opole (Oppeln)   (See Nowa Wieś Krôlewska.)

Optima—Hemphillian—Oklahoma

Orěchov (Brünn or Brno)—Orleanian—Czechoslovakia

Oreshets (Danube Basin)—Late Miocene—Bulgaria

Orignac—Late Miocene—France

Orinda Formation sites—Clarendonian—California

Orléanais, sable de, sites—Orleanian—France
Ornezan—Astaracian—France
Osbornoceros Quarry (Frick)—Hemphillian—New Mexico
Oschiri—Orleanian?—Sardinia
Oshkosh—Hemphillian—Nebraska
Osmaniye—Late Miocene—Turkey
Osmont sites—Barstovian—Oregon
Othmarsingen—Orleanian—Switzerland
Otis Basin—Hemphillian—Oregon
Ottenberg (Weinfelden)—Astaracian—Switzerland
Oued el Hammam   (See Bou Hanifia.)
Oued Mellegue—Late Miocene—Tunisia
Oued Zra—(= Vallesian) Ngororan—Morocco
Oum er Rbia (Rebia?)—Late Miocene—Morocco
Owl Canyon 1—Barstovian—Nevada
Owl Canyon 2—Clarendonian—Nevada
Oyster Bed—Barstovian and marine—California
Paintbrush Hill—Clarendonian—Nevada
Painted Hills—Arikareean—Oregon
Pai-Sui-Hsien   (See Nan-ho.)
Palan-Tyukan Range—Late Miocene—Caucasus, USSR
Palencia—Astaracian—Spain
Palmetto Mine—Hemphillian—Florida
Panama Canal, Station 1998/00—Hemingfordian—Panama
Pan Am Highway II—"Late Miocene"—Oaxaca, Mexico
Pang-Kiang Formation site—Miocene—Mongolia
Paote (Baode, PaoTun, PaoTe Hsien, Chichiakou)—(= Turolian)—Shanxi
Papiol   (See El Canyet and Can Mas)—Orleanian—Spain
Paplos P.O.—Arikareean—Oregon
Pappenheim (Eichstatt)—Agenian—Germany
Paridera de Vanacequia—Astaracian—Spain
Parker Pits—Hemphillian—Texas
Parlewali 1–4 (Dhok Pathan)—(= Turolian)—Pakistan
Parrot Bench Section sites—Barstovian—Montana
Parschlug—Karpatian = Orleanian—Austria
Parson's Ferry—Arikareean—Oregon
Paşalar (Bursa–Mustafa-Kemalpaşa)—(= Astaracian)—Turkey
Pataniak 6—Astaracian—Morocco
Paulhiac—Agenian—France
Paulina West—Barstovian—Oregon
Paulmy—Orleanian?—France
Pawnee Creek beds sites—Hemingfordian and Barstovian—Colorado
Payette Formation—Barstovian and Clarendonian?—Oregon Idaho
Payne Creek—Hemphillian—Florida
Pebble Creek—Clarendonian?—Nebraska
Pech Redondel (Narbonne)—Agenian—France
Pedrajas de San Esteban—Late Miocene—Spain
Pedregueras—Vallesian—Spain
Peillan—Astaracian—France
Pellecahus (La Romieu)—Orleanian—France
Pénaud   (See Saulcet.)
Penedès   (See Valles-Penedès)—Turolian—Spain
Penken (Keutschach)—Middle Miocene—Austria
Peralejos—Vallesian—Spain
Pérards   (See St.-Gérand-le-Puy.)
Perim Island—(= Turolian)—India
Pero Filho—Astaracian—Portugal
Pessan (Lartigole)—Astaracian—France

Pesth—Astaracian—Hungary
Pestzentlörinc—Late Miocene—Hungary
Petaluma Formation sites—Hemphillian—California
Peterson Creek—Late Arikareean—Idaho
Petravac—Astaracian—Yugoslavia
Petrified Stump—Barstovian—Nevada
Petropavlovsk—(= Turolian)—Siberia, USSR
Petroverovka—Late Miocene—Ukraine
Peyrecréchet (Sos)—Orleanian—France
Péyrouzet—Astaracian—France
Pfaffstetten (Hohenwarth)—Pannonian = Late Miocene—Austria
Pfullendorf (Sigmaringen)—Astaracian—Germany
Phillips Ranch—Hemingfordian—California
Pichugino 1—Late Miocene—Ukraine
Picture Gorge sites—Arikareean—Oregon
Piedra Parada Member sites—Late Arikareean—New Mexico
Pièra—Turolian—Spain
Pigeon Draw—Hemphillian—Nebraska
Pikermi (Chomateri)—Turolian—Greece
Pine Creek (Rushville South)—Clarendonian—Nebraska
Pine Valley—Barstovian and? Clarendonian?—Nevada
Ping Fan Hsien—Late Miocene—Gansu
Pinole Tuff sites—Hemphillian—California
Pinpa fauna   (See Frome Embayment)—Middle? Miocene—South Australia
Pinturas—Santacrucian—Argentina
Pirie Canyon—Barstovian—California
Plainville   (See Rooks.)
Plaissan—Agenian—France
Plantersville—Barstovian?—Texas
Pliohippus—Clarendonian—Nevada
Pliohippus Draw—Hemingfordian into Hemphillian—Nebraska
Ploski Blagoevradsko—Turolian?—Bulgaria
Plum Creek—Clarendonian—Nebraska
Plush Ranch—Hemingfordian—California
Poble Nou (St. Quirze, in part)—Astaracian—Spain
Pocket 34—Barstovian—Nebraska
Point Blank—Barstovian—Texas   —(Tedford et al. (1983?))
Poisbrunn—Astaracian—Austria
Poison Creek Formation site—Clarendonian—Oregon
Pojoaque Bluff district—Barstovian?—New Mexico
Pojoaque Member sites—Barstovian?—New Mexico
Polgardi—Turolian—Hungary
Polinyá—Vallesian—Spain
Pommiers—Late Miocene—France
Poncenat Quarry   (See St.-Gérand-le-Puy.)
Ponlat—Astaracian or Vallesian—France
Pont Boutard–Langeais—Orleanian—France
Pont de l'Herbasse—Astaracian—France
Pontigné—Orleanian—France
Pontlevoy-Thenay—Orleanian—France
Popovac-Paracín—Orleanian or Astaracian—Yugoslavia
Popovitsa—Late Miocene?—Bulgaria
Popovka—Late Miocene—Caucasus, USSR
Porcupine Creek (Porcupine P.O.)—Late Arikareean—South Dakota
Porcupine South—Hemingfordian—South Dakota

Porter One—Clarendonian—California
Port-le-Nouvelle—Astaracian—France
Port of Entry Pit—Hemphillian—Oklahoma　　　—(Tedford et al. (1983?))
Port Phillip Bay (Beaumaris)—Late? Miocene—Victoria, Australia
Pote d'Agua　(See Nordrand-Lisboa.)
Pöttmes—Astaracian—Germany
Pöttsching—Pannonian = Late Miocene—Austria
Pouylebon—Orleanian or Astaracian—France
Póvoa de Santarém—Astaracian—Portugal
Powerline Road—Clarendonian—California
Poysdorf—Astaracian—Austria
Prebeza—Astaracian? and Vallesian—Yugoslavia
Presbytère—Orleanian or Astaracian—France
Prescott Southeast—Clarendonian—Arizona
Prevelac—Late Miocene—Yugoslavia
Priay—Vallesian—France
Princeton 1000C—Barstovian—Nebraska
Promerycochoerus—Arikareean—Oregon
Prosynthetoceras Quarry (Frick)—Barstovian—Nebraska
Prottes—Pannonian = Late Miocene—Austria
Przeworno 2—Badenian = Astaracian—Poland
Pseudaelurus—Clarendonian—Nevada
Puebla de Almoradier—Turolian?—Spain
Puente de Vallecas—Astaracian—Spain
Puhsien district—Late Miocene—Shanxi
Punchbowl Formation sites—Hemingfordian, Barstovian, and Clarendonian—California
Punta Chaparro—Friasian and Chasicoan—Uruguay
Punta Gorda—Friasian and Chasicoan—Uruguay
Putnams—Arikareean—Oregon
Puy-Courny—Turolian—France
Puyméras—Late Miocene—France
Pyhra (Mistelbach)—Late Miocene—Austria
Pyramid Hill—marine and late Arikareean—California
Pyrimont-Challonges—Agenian—France
Qingyang (see Quinan)—Late Miocene—Gansu
Quarry A　(See Martin Canyon Quarry A.)
Quarry No. 2, Sinclair Draw (Frick)—Barstovian—Nebraska
Quarry Point—Clarendonian—California
Quartz Basin—Barstovian—Oregon
Quatal Canyon sites—Barstovian, Clarendonian, and Hemphillian—California
Quebrada Honda—Late Miocene?—Bolivia
Quebrada Honda—Santacrucian?—Venezuela
Quehua Southeast—Miocene—Bolivia
Quinan district—(= Astaracian or Vallesian and Turolian)—Gansu
Quincy—Hemingfordian—Florida
Quinn Canyon (Williams School)—Hemphillian?—Nebraska
Quinta da Conceicao Va₂—Orleanian?—Portugal
Quinta da Silvéria—Miocene—Portugal
Quinta das Pedreiras—Orleanian—Portugal
Quinta do Pombeiro—Orleanian—Portugal
Quiver Hill District—Late Arikareean, in part—South Dakota
Rabastens-de-Bigorre—Astaracian?—France
Racetrack—Barstovian—Nevada

Railway Quarry A—Barstovian—Nebraska　　　—(Tedford et al. (1983?))
Rainbow Basin—Barstovian—California
Rainbow Valley—Clarendonian?—Nevada
Rak Camel Quarries—Hemphillian—New Mexico
Rakomylus site—Barstovian—New Mexico
Rak Quarry—Barstovian—California
Ramiz—(= Turolian)—Turkey
Rammingen—Astaracian?—Germany
Ranch House Draw—Barstovian—Nebraska
Rancho Lobo—Hemphillian—Honduras
Rancho Ocote　(See Ocote.)
Ratavoux—Turolian—France
Rattlesnake Formation sites—Hemphillian—Oregon
Ravensberg—Astaracian—Germany
Raven's Roost—Clarendonian—California
Ravin de la Pluie—Vallesian—Greece
Ravine Quarry (Frick)—Hemingfordian—Nebraska
Ravolzhausen—Agenian—Germany
Rawhide Buttes—Late Arikareean—Wyoming
Rawhide Creek—Late Arikareean—Wyoming
Réal (La Bastidonne)—Agenian—France
Rebhubel—Astaracian—Switzerland
Rebréchien—Orleanian—France
Red Basin (Skull Springs)—Barstovian—Oregon
Red Bluff (Trinity River)—Barstovian—Texas
Red Division Quarry—Hemingfordian—California　　　—(Tedford et al. (1983?))
Red Hill—Hemingfordian?—Texas
Redington local fauna—Hemphillian—Arizona
Redington sites—Hemphillian—New Mexico
Redrock Canyon—Clarendonian—California
Reeds Ranch—Arikareean—Oregon
Reef Bed—marine and Barstovian—California
Reese River—Clarendonian—Nevada
Reichenau or Reischenau—Astaracian—Germany
Reichenstetten (Regensburg)—Astaracian or Late Miocene—Germany
Reichertshofen—Astaracian—Germany
Reinprechtspölla—Eggenburgian = Orleanian—Austria
Reisenburg—Astaracian—Germany
Relais des Cathares (Narbonne)—Agenian—France
Relea—Late Miocene—Spain
Rembach—Astaracian—Germany
Repenning's—Late Arikareean—California
Repovica—Astaracian—Yugoslavia
Rep's *Merychippus*—Barstovian—California
Rep's *Pliohippus* sites—Clarendonian—California
Reviers—Orleanian—Hungary
Reynolds—Hemphillian—Georgia
Reynolds Creek—Clarendonian?—Idaho
Rhinoceros Hill—Hemphillian—Kansas
Rhino Quarry—Hemingfordian—Nebraska
Rhodos Island　(See Kalithies)—Late Miocene
Ricardo Formation sites—Clarendonian—California
Richard Incline—Clarendonian—California
Rickstrew Ranch—Hemingfordian?—Colorado
Riedern—Astaracian—Germany
Ries—Astaracian—Germany
Rimbez—Orleanian or Astaracian—France
Rincón de Ademuz　(See Mas del Olmo.)

Rincón del Buque (Media Luna)—Santacrucian—Argentina
Rincón Quarry—Barstovian—New Mexico
Rio Chico area—Santacrucian—Argentina
Rio Colloncura—Santacrucian—Argentina
Rio Coyle (See Santa Cruz, Argentina.)
Rio del Oso (Abiquiu sites)—Barstovian—New Mexico
Rio Frias—Friasian—Argentina
Rio Gallego (See Santa Cruz, Argentina.)
Rio Guenguel—Friasian—Argentina
Rio Güere—Santacrucian or Friasian—Venezuela
Rio Huemules—Friasian—Argentina
Rio Mayo—Friasian—Argentina
Rio Negro (in part)—Santacrucian—Argentina
Rio Santa Cruz (See Santa Cruz, Argentina.)
Rio Senguer—Friasian—Argentina
Rio Shehuén—Santacrucian—Argentina
River Road—Hemphillian—Washington
Riverside West—Clarendonian—Oregon
Riversleigh fauna—Middle to Late Miocene—Queensland, Australia
Rizzi Ranch—Arikareean—Nevada
Rizzi Ranch 2—Barstovian—Nevada
Roadcut Canyon—Clarendonian—California
Road Gap Canyon—Hemingfordian—California
Roc de Couissy—Late Miocene—France
Rocky Ridge sites—Clarendonian—California
Rodent Hill (Barstow)—Barstovian—California
Roggendorf—Eggenburgian = Orleanian—Austria
Roggenstein—Astaracian—Germany
Roll Quarry—Hemingfordian—Wyoming
Romans—Astaracian—France
Rome—Hemphillian—Oregon
Romieu (See La Romieu.)
Romorantin—Orleanian?—France
Ronville—Orleanian—France
Rooks (Plainville)—Hemphillian?—Kansas
Rosebud Formation sites—Late Arikareean and Hemingfordian—South Dakota
Rosegghof (Wolfliswil)—Astaracian—Switzerland
Rosshaupten—Astaracian—Germany
Rougier—Agenian—France
Round Mountain Quarry—Clarendonian—New Mexico
Roundtop Northwest—Clarendonian—California
Rowe Creek—Arikareean—Oregon
Rozsaszentmarton—Late Miocene—Hungary
Ruan—Orleanian—France
Rubí (Molí Calopa)—Orleanian—Spain
Rubielos de Mora—Orleanian—Spain
Ruckles Creek—Barstovian?, Clarendonian?—Oregon
Rudabánya—Vallesian—Hungary
Rudio (Creek)—Arikareean—Oregon
Rudnev Quarry—Late Miocene—Ukraine, USSR
Rümikon—Orleanian or Astaracian—Switzerland
Runningwater Formation sites—Hemingfordian—Nebraska
Rusce (Vranje Basin)—Astaracian?—Yugoslavia
Rusinga Island—Rusingan—Kenya
Sabadell—Vallesian—Spain
Sacaco—marine and latest Miocene—Peru
Sadournin—Astaracian—France
Sagebrush Creek—Barstovian—Oregon
Sahabi—(= Turolian) Lothagamian—Libya

St.-Antoine—Late Miocene—France
St. Cristan—Astaracian—France
St.-Christaud—Orleanian?—France
St. Cugat del Vallés—Orleanian—Spain
St.-Fons—Late Miocene—France
St.-Frajou—Late Miocene—France
St.-Gaudens-Valentine—Astaracian—France
St. Georgensgmünd (See Georgensgmünd.)
St.-Gérand-le-Puy district—Agenian—France
St.-Jean-Narbonne—Agenian—France
St.-Jean-de-Bournay—Vallesian—France
St.-Jean-le-Vieux—Astaracian—France
St.-Just (Bari)—marine Burdigalian and Orleanian—France
St.-Laurent-de-Neste—Astaracian or Vallesian—France
St.-Lyé—Orleanian—France
St.-Maur—Orleanian or Astaracian—France
St.-Médard—Astaracian?—France
St.-Michel-en-Chasine—Miocene—France
St. Miguel de Taudell—Turolian?—Spain
St.-Nazaire-en-Royans—Orleanian—France
St. Oswald-Gratwein—Badenian = Astaracian—Austria
St.-Pé-Fabas—Astaracian?—France
St. Stefan—Sarmatian = Astaracian—Austria
St.-Suaire—Agenian—France
St.-Vincent-de-Montjoie—Orleanian—France
Saissansenke (Zaysan Depression)—Eocene and Late Miocene—Kazakhstan
Sakkul—Late Miocene—Kirghis, USSR
Saldaña—Astaracian—Spain
Salgotarjan—Orleanian—Hungary
Salkutsa—Late Miocene—Moldavia, USSR
Salmendingen—Late Miocene—Germany
Salobrena—Turolian—Spain
Saloniki A and B (Dudular)—Turolian—Greece
Samaran—Astaracian—France
Sam Houston Forest—Barstovian—Texas
Samos—Vallesian? and Turolian—Samos Island
San Clemente Island—marine and Barstovian—California
Sandberg (Neudorf-Sandberg)—marine and Astaracian—Czechoslovakia
Sand Canyon—Barstovian—Nebraska
Sand Canyon fauna—Barstovian—Colorado
Sand Canyon Quarry—Hemingfordian—Nebraska
Sandelzhausen—Astaracian—Germany
Sanders—Clarendonian—Arizona
Sand Hill—Clarendonian—Nevada
Sands—Clarendonian?—Idaho
San Francisco (Rio Güere)—Santacrucian?—Venezuela
San Giovannino—Late Miocene—Italy
San Isidro-Madrid—Orleanian?—Spain
San Juan (Rak Camel Quarries)—Hemphillian—New Mexico
Sankt Georgen—Astaracian—Germany
San Pedro (Falcón)—Santacrucian?—Venezuela
San Pedro Valley—Hemphillian and later—Arizona
San Quirico (San Quirze)—Astaracian and Vallesian—Spain
Sansan—Astaracian—France
Santa Barbara Canyon (Pato)—Arikareean—California

321

Santa Clara Canyon—Clarendonian or Hemphillian—New Mexico

Santa Cruz fauna, Formation, and area—Santacrucian—Argentina

Santa Cruz sites—Barstovian and Clarendonian—New Mexico

Santa Eugenia—Vallesian—Spain

Santa Fe—Hemphillian?—Florida

Santa Margarida N 4–5—Vallesian—Spain

Santee—Hemphillian—Nebraska

Santiga—Vallesian—Spain

Sant Miquel del Taudell—Vallesian—Spain

Sant Pere de Ribes—Astaracian—Spain

Sant Ranch—Barstovian—Montana

Sarasota—Hemphillian?—Florida

Sardara—Orleanian—Sardinia

Sari Bulak (Zaysan)—(=Astaracian?)—Kazakhstan

Sari Çay (Muğla-Milas-S.)—(=Astaracian)—Turkey

Saridjaz Basin—Late Miocene—Kirghiz, USSR

Sarilar—Vallesian?—Turkey

Şarkişla (Sivas)—(=Turolian)—Turkey

Sarremezan—Astaracian—France

Sauerbrunn—Sarmatian = Astaracian—Austria

Saulcet—Agenian—France

Sauveterre—Astaracian?—France

Savage Canyon—Barstovian—Nevada

Saverdun—Astaracian—France

Savigné-sur-Lathan—Orleanian—France

Savolyano (Struma Basin)—Late Miocene—Bulgaria

Savonetta River—Late Miocene?—Trinidad Island

SB-1A (Live Oak SB-1A)—Late Arikareean—Florida

Sbeitla—Late? Miocene—Tunisia

Sceaux—Orleanian?—France

Schaffhausen—Agenian—Germany

Schauerleiten—Orleanian—Austria

Schernfeld—Miocene—Germany

Schlagel Creek—Clarendonian—Nebraska

Schlatt—Astaracian—Switzerland

Schlechtwegen—Astaracian—Germany

Schlieren—Astaracian—Switzerland

Schlüchtern—Orleanian—Germany

Schnaitheim—Orleanian—Germany

Schonenberg—Astaracian?—Germany

Schwamendingen—Astaracian—Switzerland

Scotts Valley—marine and Clarendonian—California

SDSM V6215 (=LACM 2008)—Late Arikareean—South Dakota

SDSM V6229 (=LACM 1871)—Late Arikareean—South Dakota

Seaboard Airline R.R.—Hemingfordian—Florida

Seanwala (Dhok Pathan)—(=Turolian)—Pakistan

Sebastopol (Sevastopol)—Vallesian—Ukraine

Selah—Clarendonian—Washington

Selb—Clarendonian—California

Selbe Ranch—Clarendonian—Kansas

Selçik (Afyon-Sandikli)—(=Vallesian)—Turkey

Selim (Dzhevar)—Late Miocene—Kazakhstan

Selles-sur-Cher—Agenian—France

Selmatten—Astaracian—Switzerland

Sendai—Late Miocene—Japan

Seo (Seu) de Urgel (El Firal)—Astaracian?—Spain

Seoner Wald—Astaracian—Switzerland

Sepmes, à la Jaltière—Orleanian—France

Sépulture Chrétien (Mirabeau)—Orleanian—France

Sequence Canyon—Clarendonian and Hemphillian—California

Serafimovka—Late Miocene—Kirghiz, USSR

Serbia, miscellaneous—Miocene

Sermenaz (Bas Neyron)—Vallesian—France

Serre de Vergès—Orleanian—France

Serullah 9—(=Vallesian or Turolian)—Afghanistan

Servilly—Agenian—France

Şeylek—(=Turolian)—Turkey

Seymour Island—Early Miocene marine—West Antarctica

Shanks sites—Arikareean—Oregon

Shanwang "series" sites—(=Astaracian)—Shandong

Sharktooth Hill—marine and Barstovian—California

Sheep Creek Formation sites—Hemingfordian—Nebraska

Sheep Mountain South—Arikareean—Oregon

Sheep Rock—Arikareean—Oregon

Shehuén   (See Rio Shehuén.)

Shell Cliff—Barstovian—Nevada

Shell Ridge—marine and Clarendonian—California

Shenmu—(=Turolian)—Shaanxi

Sherman Ranch—Clarendonian—Nebraska

Sherullah   (See Serullah.)

Shiashiang (Shouyang)—(=Turolian)—Shanxi

Shields Ranch Quarry—Clarendonian—Arizona

Shihhung—(=Astaracian or Late Miocene)—Jiangsu

Shihuiba (Lufeng)—(=Vallesian?)—Yunnan

Shiloh Marl site—Hemingfordian?—New Jersey

Shimek Quarry—Hemingfordian—Nebraska

Shiogama City—Middle to Late Miocene—Japan

Shirak Steppe—Late Miocene—Caucasus, USSR

Short Ravine—Barstovian—California

Shrocks—Arikareean—Oregon

Shui-Chuan-Kou (Chingning district)—Late Miocene—Gansu

Sibnica—Badenian = Astaracian—Yugoslavia

Sidi Salem—(=Vallesian) Ngororan—Algeria

Siegfrieds—Arikareean—Oregon

Siessen—Astaracian?—Germany

Siesta Formation sites—Clarendonian—California

Sigmaringen—Astaracian—Germany

Sili—Late Miocene—Gansu

Simferopol—Late Miocene—Crimea, USSR

Simorre (many sublocalities)—Astaracian—France

Sinap "series" sites—Late Miocene and Pliocene—Turkey

Sinclair Draw East and West—Hemingfordian and Barstovian—Nebraska

Sinclair Quarry—Barstovian—Nebraska

Sinda—Rusingan—Kenya

Sining Basin—Early Miocene—Qinghai (Tsinghai)

Sining Fu—(=Astaracian)—Gansu

Sipovljani—Astaracian—Yugoslavia

Sirietz—Late Miocene—Romania

Sitia Southeast—Vallesian—Crete Island

Siwa—Rusingan or Ternanian—Kenya?

Siwaliks "series" localities—(=Astaracian into Pleistocene)—Pakistan, India, and Nepal

Skull Ridge Member sites—Barstovian—New Mexico

Skull Springs (Red Basin)—Barstovian—Oregon

Skyline Quarries (Frick)—Barstovian—California
Skyritz—Agenian—Czechoslovakia
Slug Bed—Barstovian—California
Smart Ranch—Hemphillian—Texas
Smendou (Conde-Smendou)—(= Orleanian?) Rusingan?—Algeria
Smiley's 2—Clarendonian—Washington
Smith Flat—Barstovian—Nevada
Smith Place   (See Windham Place.)
Smiths—Arikareean—Oregon
Smith Valley—Hemphillian—Nevada
Snail Cliff (= Shell Cliff)—Barstovian—Nevada
Snake Creek Formation sites—Barstovian into Hemphillian—Nebraska
Snake Quarry—Barstovian—Nebraska
Snake River East—Clarendonian?—Nebraska
Sneider Ranch—Barstovian—Oregon
Snowball Valley—Clarendonian—Nevada
Snyder Creek—Barstovian—Oregon
Soblay—Vallesian—France
Sofça (Kütahya-Sabuncu-S.) (= Astaracian)—Turkey
Soğucak-Yassiören (Ankara-Kazan-Sinap)—(= Vallesian)—Turkey
Soldier Meadow—Barstovian—Nevada
Solenhofen (Solnhofen) Fissure—Orleanian (also Oligocene)—Germany
Solera—Astaracian—Spain
Sommerein (Mannersdorf)—Astaracian—Austria
Songhor—Astaracian—Kenya
Sonnay—Vallesian—France
Sonndorf—Eggenburgian = Orleanian—Austria
Sonoita—Hemphillian?—Arizona
Sophades—Late Miocene—Macedonia, Greece
Sopron—Late Miocene—Hungary
Sorgues—Astaracian—France
Sos—Orleanian?—France
Sosnicowice—Astaracian—Poland?
Soulsbyville—Clarendonian—California
South Big Beaver—Clarendonian—Nebraska
South Canyon—Arikareean—Oregon
South Humbug Quarry—Barstovian—Nebraska
South Pine Valley—Nevada—Hemphillian
South Prospect B—Middle? Miocene—South Australia
South Reese River—Clarendonian—Nevada
Spade Flats (Lewis Ranch)—Clarendonian—Texas
Spanish Gulch—Arikareean—Oregon
Spencer Creek (Astoria)—marine and Barstovian—Oregon
Spitzberg im Ries (Nördlingen)—Astaracian—Germany
Split Rock—Hemingfordian—Wyoming
Spornegg (Baldingen)—Astaracian—Switzerland
Spray—Arikareean—Oregon
Spring Canyon—Clarendonian—Nebraska
Spring Cliff—Arikareean—Oregon
Spring Creek—Late Arikareean—Montana
Springer (Feyereisen Gap)—Barstovian—South Dakota
Spruce Mountain—Barstovian—Nevada
Squankum (Kirkwood)—Late Arikareean—New Jersey
Squaw Valley—Clarendonian—Nevada
Ssoskut—Astaracian—Hungary
Stallhofen (Voitsberg)—Astaracian—Germany
Standing Rock Quarry—Late Arikareean—New Mexico

Starlight (Frick)—Barstovian—California
Stätzling (Augsburg)—Astaracian—Germany
Stavropol (Stauropol)—Late Miocene—Caucasus, USSR
Steamboat Canyon (Cedarville)—Barstovian—California
Steckborn—Astaracian—Switzerland
Steepside Quarry (Frick)—Barstovian—California
Stegersbach—Pannonian = Late Miocene—Austria
Stein a. Rhein—Astaracian—Switzerland
Stein Gallery—Clarendonian—Arizona
Steinheim—Astaracian—Germany
Stephens Creek Road—Barstovian—Texas
Stewart Spring—Barstovian—Nevada
Stewart Valley district—Barstovian and Clarendonian—Nevada
Stirton Wash—Barstovian—California
Stonehouse Draw—Hemingfordian—Nebraska
Stormy's Camp sites—Clarendonian—California
Straight Cliff Prospect—Hemingfordian—New Mexico
Stratiyevka—Late Miocene—Ukraine
Stratzing (Krems) Pannonian = Late Miocene—Austria
Strawberry Canyon—Clarendonian—California
Stroud Claim—Hemphillian—Idaho
Struma Basin (Kalimantsi)—Turolian—Bulgaria
Stubblefield—Arikareean—Oregon
Subirats—Orleanian—Spain
Sucker Creek (Succor Creek)—Barstovian—Oregon
Suevres—Orleanian—France
Suipacha (Potosi)—Friasian?—Bolivia
Süleymanli (Çankiri)—(= Vallesian)—Turkey
Sunbeam—Barstovian—Colorado
Sunrise—Clarendonian—Nebraska
Survey Quarry (Frick)—Barstovian—Nebraska
Süssen—Astaracian—Germany
Sutton Mountain—Arikareean—Oregon
Sutton Ranch—Arikareean—Oregon
Sveti-Wrač (Strumatal)—Late Miocene—Bulgaria
Swayze Quarry—Hemphillian—Kansas
Swift Mine—Hemphillian—Florida
"Switchyard B" (Seaboard Airline R.R.)—Late Arikareean?—Florida
Sycamore Creek—Clarendonian—California
Sylt   (See Morsumkliff.)
Sylvania Mountain—Clarendonian—Nevada
Taben-Buluk—Early? Miocene—Gansu
Tablat "Rotmouten"—Astaracian—Switzerland
Tadla Beni Amir—marine and Late? Miocene—Morocco
Taghar—(= Vallesian?)—Afghanistan
Tairum Nor (Tung Gur)—(= Astaracian)—Nei Mongol
Tallahassee—Hemingfordian—Florida
Tambach—Rusingan—Kenya
Tampa Bay—Hemphillian—Florida
Tannenberg (St. Gallen)—Astaracian—Switzerland
Tanohira—Middle to Late Miocene—Japan
Tanovka—Late Miocene—Ukraine, USSR
Tapasuna (Mangual)   Hemphillian—Honduras
Tappianwali (Dhok Pathan)—(= Turolian)—Pakistan
Tarakliya (Taraklia)—Turolian—Moldavia, USSR
Target Quarry—Hemingfordian—Nebraska

Miocene Mammalian Faunas

Tarkarooloo (Lake Tarkarooloo)—Middle? Miocene—South Australia

Tarrasa—Vallesian—Spain

Taşkinpaşa (Nevşehir-Ürgüp)—(= Turolian)—Turkey

Tassajara Creek—Clarendonian—California

Tataros—Late Miocene—Hungary

Tatuno Kuti—Late Miocene—Japan

Tauq-Kirkuk (Djebel Hamrin)—Late Miocene—Irak

Tavenner Ranch 2—Late Arikareean—Montana

Tavers—Orleanian—France

Taylor Creek Drainage—Late Arikareean—New Mexico

Taylor Gravel site—Hemphillian—Arizona

Tchernigow—Late Miocene—USSR

Tebessa—(= Orleanian) Rusingan—Algeria

Tedford Pocket (Stewart Spring)—Barstovian—Nevada

Teewinot Formation site—Hemphillian—Wyoming

Tehuichila—Hemphillian—Mexico

Tejon Hills sites—Clarendonian—California

Temblor Formation sites—marine and Barstovian—California

Tenevo—Late Miocene—Bulgaria

Teplitsky—Late Miocene—Ukraine

Tereul (Ademuz)—Turolian—Spain

Tereul (Calatayud)—Turolian—Spain

Termeyenice (Ankara-Hasayaz-T.)—(= Astaracian)—Turkey

Terral d'en Matiès—Turolian—Spain

Terrassa—Vallesian—Spain

Tesla district—Clarendonian—California

Testour—(= Astaracian)—Ternanian—Tunisia

Teufen—Agenian—Switzerland

Thenay (Pontlevoy)—marine and Orleanian—France

The Pits—Hemphillian—Nebraska

Thessaloniki—Late Miocene—Greece

Thil—Orleanian—France

Thin Elk Formation sites—Clarendonian—South Dakota

Thistle Quarry—Hemingfordian—Nebraska

Thomas Farm—Hemingfordian—Florida

Thomson Quarry—Hemingfordian—Nebraska

Thomson's Quarry B of 1921—Barstovian—Nebraska

Thousand Creek Formation sites—Hemphillian—Nevada

Thousand Springs—Clarendonian?—Nevada

Thymiana—Astaracian—Chios Island, Greece

Tick Canyon Formation sites—Late Arikareean—California

Tiefernitzgraben—Late Miocene?—Austria

Tieton Reservoir—Arikareean—Washington

Timanovka—Late Miocene—Ukraine

Timok Basin (Vrbica, Zvezdan)—Astaracian—Yugoslavia

Tiraspol—Late Miocene—Moldavia, USSR

Tire (Ismir-Torbali)—Astaracian—Turkey

Tjisand—Late Miocene—Java

Tokoum (Sari)—Late Miocene—Iran

Tolosa—Turolian—Spain

Toluca—Hemphillian?—Mexico

Tomerdingen—Agenian—Germany

Tom O's Quarry—Middle? Miocene—South Australia

Tonagra—Late Miocene—Greece

Tong Xin—Late Miocene—Ningxia

Tonopah—Barstovian—Nevada

Tonto Basin—Clarendonian or Hemphillian—Arizona

Toril—Astaracian—Spain

Torralba de Ribota I and V—Orleanian—Spain

Torrent de Febulines N10, N11, N15—Vallesian—Spain

Tortajada—Turolian—Spain

Tosun-nor district—Late Miocene—Gansu

Touraine, faluns de, sites—Orleanian—France

Tour la Reine—Late Miocene—Greece

Tournan (Simorre)—Astaracian—France

Tournus—Astaracian—France

Toussieu—Late Miocene—France

Town Bluff—Barstovian—Texas

Tozeur, Oasis de—Late Miocene—Tunisia

Tracy district—Clarendonian—California

Trail Creek—Clarendonian—Wyoming

Trail Prospect—Clarendonian—Arizona

Trauenzinen—Astaracian—Austria?

Tréteau—Agenian—France

Trieblitz—Miocene—Czechoslovakia

Trimmelkam—Astaracian—Austria

Trinay—Orleanian—France

Trinchera Norte Autopista, N11, N12—Vallesian—Spain

Trinchera sur Autopista—Vallesian—Spain

Trinity River Pit—Barstovian—Texas

Trinxera (See San Quirico)—Astaracian—Spain

Triolite—Clarendonian—Nevada

Tripol'ye—Late Miocene—Ukraine

Trojan Quarry—Barstovian—Nebraska

Troublesome Formation sites—Hemingfordian—Colorado

Trout Creek—Barstovian—Oregon

Troutdale Gravels site—Clarendonian?—Oregon

Truchtelfingen—Late Miocene—Germany

Truckee Canal West—Clarendonian—Nevada

Tsaidam Basin—Late Miocene—Gansu

Tschäppel—Astaracian—Switzerland

Tschobroutschi—Late Miocene—Moldavia, USSR

Tsilin River—Late Miocene—Mongolia

Tub Creek—Arikareean—Oregon

Tuchetse—(= Turolian)—Nei Mongol

Tuchorice—Orleanian—Czechoslovakia

Tuco (Simorre)—Astaracian—France

Tucupido—Santacrucian?—Venezuela

Tudorovo—Turolian—Moldavia, USSR

Tul'chino—Late Miocene—Ukraine

Tüney (Ankara-Kalecik) (= Astaracian)—Turkey

Tung Gur—(= Astaracian)—Mongolia

Tung-sha-po (Loyang)—(= Astaracian)—Henan

Tungsing—Middle to Late Miocene—Ningxia

Tungsuenkuan beds sites—(= Astaracian)—Jiangsu

Tunnel Rock East—Clarendonian—Nebraska

Turkana district—Rusingan—Kenya

Turlock Lake—Hemphillian—California

Turtle Butte Formation sites—Late Arikareean—South Dakota

Turtle Buttes—Clarendonian—South Dakota

Turtle Canyon—Clarendonian—Nebraska

Turtle Cove—Arikareean—Oregon

Turtle Pocket—Hemphillian—Arizona

Tuttlingen—Late Miocene—Germany

Tutzing—Astaracian—Germany

Two Mile Bar—Clarendonian—California

Tyrrell sites—Clarendonian—Washington

Tyshkan River—Late Miocene—Kazakhstan

Tyul'-Kul'-Say—Left Miocene—Kazakhstan
Tzehsien—Late Miocene—Hubei
Uchidheri (Dhok Pathan)—(= Turolian)—Pakistan
Udabno—Late Miocene—Georgia, USSR
Ueberlingen—Late Miocene—Germany
Ufhusen—Astaracian—Switzerland
Ulan-Tologoj—(= Orleanian?)—Mongolia
Ulaş (Muğla-Milas) (= Astaracian)—Turkey
Ulengo River South—Late Miocene—Xinjiang
Ulm (Michelsberg)—Astaracian—Germany
Undingen—Late Miocene?—Germany
Undorf (Regensburg)—Astaracian—Germany
Unity—Clarendonian—Oregon
Universidade Católica—Orleanian—Portugal
UNSM BX-59A—Late Arikareean—Nebraska
Unter-Staudach—Agenian—Switzerland?
Upper County Road—Clarendonian—Nebraska
Upper Dry Canyon—Barstovian—California
Upper Parrot Bench Section—Hemphillian—Montana
Uptegrove Quarry—Hemphillian—Nebraska
Ürgüp (Kayseri)—Late Miocene—Turkey
Urlau (Leutkirch)—Astaracian—Germany
Ursendorf—Astaracian—Germany
Ushuk Mountains—(= Orleanian)—Mongolia
Ust'-Urt—(= Orleanian)—Kazakhstan
Vaison-la-Romaine (Mollans)—Late Miocene—France
Valalto—Astaracian?—Spain
Valdecebro—Turolian—Spain
Valdemoros—Orleanian—Spain
Vale de Chelas—Astaracian—Portugal
Valençay—Orleanian—France
Valentine (St.-Gaudens)—Astaracian—France
Valentine Formation sites—Barstovian and Clarendonian—Nebraska
Valle de Oaxaca—Barstovian—Oaxaca, Mexico
Valles-Penedes district—Vallesian and Turolian—Spain
Valley View Quarry (Frick)—Barstovian—California
Valréas (Les Mistrals)—Vallesian—France
Valtorres—Orleanian—Spain
Valverde—Astaracian—Spain
Vanora Grade—Arikareean—Oregon
Van Tassel—Late Arikareean—Wyoming
Van Tassel South—Hemingfordian—Wyoming
Varennes-sur-Teche—Agenian—France
Varna (Varnitsa?)—Vallesian?—Bulgaria
Vathylakkos—Vallesian and Turolian—Macedonia, Greece
Vedder locality—Hemingfordian—California
Veles (Orisari)—Turolian—Macedonia, Greece
Veltheim—Astaracian—Switzerland
Vendant (See St.-Gérand-le-Puy.)
Vendargues—Late Miocene?—France
Venta del Moro—Turolian—Spain
Verdi—Hemphillian—Nevada
Verdigre Quarry—Clarendonian—Nebraska
Vermes—Astaracian—Switzerland
Version Quarry—Barstovian—Nebraska
Vic-Fezensac—Orleanian—France
Vickers' Pocket—Clarendonian—Nevada
Viehhausen—Astaracian—Germany
Vienna—Sarmatian = Astaracian—Austria
Vieux-Collonges—Orleanian—France

Villadecaballs (Can Bayona)—Vallesian—Spain
Villafeliche IIa, IV—Orleanian—Spain
Villavieja Formation and district—Friasian—Colombia
Villedieu (Point Rouge)—Vallesian—France
Villefranche-d'Astarac—Astaracian—France
Villette (Serpaize)—Late Miocene—France
Vim-Peetz—Barstovian—Colorado
Vingrau—Astaracian—France
Vinogradi—Late? Miocene—Bulgaria
Virgil Clark Gravel Pit—Hemphillian—Oklahoma (Tedford et al. (1983?))
Virgin Valley Formation sites—Barstovian—Nevada
Visan—Late Miocene—France
Vista Quarry—Hemingfordian—Nebraska
Vizirka—Late Miocene—Ukraine
Voitsberg—Astaracian—Austria
Vorderndorf—Karpatian = Late Orleanian—Austria
Vösendort (Vienna)—(= Vallesian)—Austria
Vrabec—marine Miocene—Yugoslavia
Vranje Basin (Rusce)—Astaracian—Yugoslavia
Vrbica (Timok Basin)—Astaracian—Yugoslavia
Vully I, II, III—Orleanian—Switzerland
Wadi-el-Hammam (See Bou Hanifia.)
Wadi Faregh—(= Orleanian) Rusingan—Egypt
Wadi Natrun—(= Turolian) Lothagamian—Egypt
Wafangyingtze (See Changpeihsien.)
WaKeeney—Clarendonian—Kansas
Walhcim—Late Miocene—Germany
Wallerstein—Astaracian—Germany
Walnut Grove (Milk Creek)—Clarendonian—Arizona
Wangen—Astaracian—Switzerland
Warm Springs—Hemingfordian—Oregon
Warren Syncline—Hemphillian—California
Warrior-Douglas—Clarendonian—Nevada
Warwal (Dhok Pathan)—(= Turolian)—Pakistan
Wash Basin—Arikareean—Oregon
Washington County, miscellaneous—Hemingfordian—Texas
Watt I and II—Astaracian—Switzerland
Watze—Late Miocene—Gansu
Weatherford—Arikareean—Oregon
Weaver—Arikareean—Oregon
Weid ob Wipkingen—Astaracian—Switzerland
Weisenau—Agenian—Germany
Weiser North—Clarendonian—Idaho
Weissenburg 6—Agenian—Germany
Weller—Barstovian?—Colorado
Wellton—Late Arikareean—Arizona
Wemmel—marine and Astaracian—Belgium
Wenas Valley South—Clarendonian—Washington
Wenneberg (Nördlingen)—Astaracian—Germany
Wenquen (Zungaria Basin)—Late Miocene—Xinjiang
Werners Place—Barstovian—Oregon
Wesley Gordon—Clarendonian—California
Wessington Springs—Barstovian—South Dakota
West End Blowout—Hemphillian—Oregon
West Fork—Barstovian—California
Westhofen (Worms)—Late Miocene—Germany
West Jenkins Quarry—Barstovian—Nebraska

West Sand Quarry (Frick)—Barstovian—Nebraska
West Surface Quarry (Frick)—Barstovian—Nebraska
Wewala fauna—Arikareean—South Dakota
Whistler Microsite—Barstovian—California
White Cone—Hemphillian—Arizona
Whitehall District—Hemphillian—Montana
White Operation Quarry (Frick)—Barstovian—New Mexico
White Seale—marine and Clarendonian—California
White Sulphur Springs Northwest—Hemingfordian—Montana
Whitten—Clarendonian—California
Wien (Vienna, Heiligenstadt)—Sarmatian = Astaracian—Austria
Wien (Hernals)—Sarmatian = Astaracian—Austria
Wien (Inzerdorf)—Pannonian = Vallesian—Austria
Wien (Ottakring)—Sarmatian = Astaracian—Austria
Wien (Türkenschanze)—Sarmatian = Astaracian—Austria
Wies—Karpatian = Late Orleanian—Austria
Wiesholz—Astaracian—Switzerland
Wikieup—Hemphillian—Arizona
Wildcat Canyon—Clarendonian—California
Wildensbuch—Astaracian—Switzerland
Willmandingen—Astaracian—Germany
Willow Canyon—Clarendonian—Nevada
Wills Pit—Clarendonian—California
Windham Place—Barstovian—Texas
Winnewala (Dhok Pathan)—(= Turolian)—Pakistan
Wintershof-Ost—Orleanian—Germany
Wintershof-West—Orleanian—Germany
Wischberg (Langenthal)—Agenian—Switzerland
Wissberg (Witzberg)—Late Miocene—Germany
Withlacoochee River—Hemphillian—Florida
Wolf Creek—Clarendonian—South Dakota
Wolfsheim—Late Miocene—Germany
Wollerau—Orleanian—Switzerland
Woodburne's—Clarendonian—California
Wood Mountain Gravels sites—Barstovian—Saskatchewan
Woodruff Creek—Barstovian?—Nevada
Woodville East—Barstovian—Texas
Woodville Northwest and West Doucette—Barstovian—Texas
Woody—marine Early Miocene—California
Wosskressensk—Astaracian—USSR
Wounded Knee District—Arikareean and Hemingfordian—South Dakota
Wray (Yuma)—Hemphillian—Colorado
Wright Farm   (See Windham Place.)
Würenlos—Orleanian—Switzerland
Wutherich—Orleanian—Germany
Wynyard (Fossil Bluff)—Early Miocene—Tasmania
Xiakou (Yushe district)—Late Miocene—Shanxi
Xiehu   (See Lantien.)
Xmas Quarry—Clarendonian—Nebraska
Yabichna—Vallesian?—Bulgaria
Yack-Harvey (= Cañadón Jack or Yak-Harvey)   —Marshall (1976)—Santacrucian—Argentina
Yaogou—(= Turolian)—Nei Mongol
Yaylacilar (Afyon-Suzuk)—(= Astaracian?)—Turkey
Yecora—Hemingfordian?—Mexico
Yegua Quemada—Santacrucian—Argentina
Yellow Bear Quarry—Late Arikareean—Nebraska

Yemliha (Kayseri-Himmetdede)—(= Turolian)—Turkey
Yenangyaung—Middle? and Late Miocene—Burma
Yeni Eskihisar (Muğla-Yatağan)—(= Astaracian?)—Turkey
Yepomera—Hemphillian—Mexico
Yerington district—Hemphillian—Nevada
Yermo—Hemingfordian—California
Yevpatoriya l—Late Miocene—Crimea, USSR
Yiğitler Köy (Yozgat-Fakili)—(= Vallesian)—Turkey
Yuan-chuchen Basin—Late Miocene—Gansu?
Yukari Kizilca (Manisa-Kemalpaşa)—(= Astaracian?)—Turkey
Yürükali (Bursa-Mudanya)—(= Astaracian?)—Turkey
Yushe Basin (Xiakou, etc.)—Late Miocene—Shanxi
Zangtal—Karpatian = Late Orleanian—Austria
Zapresič Breg—Astaracian—Yugoslavia
Zaraza—Late Miocene—Venezuela
Zeglingen (Ebnet)—Orleanian?—Switzerland
Zelten   (See Gebel Zelten.)
Zetzwil—Astaracian—Switzerland
Zheltokamenka—Late Miocene—Ukraine
Zhenam River—Late Miocene—Kazakhstan
Zhongxiang—(= Vallesian?)—Hubei
Zillah—Clarendonian—Washington
Zillingdorf—Pontian = Late Miocene—Austria
Zivra—(= Turolian?)—Turkey
Zogelsdorf   (See Eggenburg)—Eggenburgian = Agenian—Austria
Zoyatal—Hemingfordian—Mexico
Zungaria Basin—Miocene—Xinjiang
Zvezdan (Timok Basin)—Astaracian—Yugoslavia
Zvižd Basin (Popovac)—Astaracian—Yugoslavia
ZX Bar—Hemphillian—Nebraska

ENDNOTES

1. We thank Dr. P. Mein for revising the age assignment of many of the Miocene localities of Europe and for revising our nomenclature, especially of many of the Miocene rodents.
2. We thank Dr. K. Rothausen for revising many of the age assignments and the nomenclature of many of the Oligocene and early Miocene cetaceans.
3. We thank Dr. L. Barnes for revising many of the age assignments and the nomenclature of many of the Miocene cetaceans.

BIBLIOGRAPHY

ADDICOTT, W. O. 1969. Tertiary climatic change in the marginal northeast Pacific Ocean. *Science* 165:583–586.

AGUILAR, J.P. 1977. Les gisements continentaux de Plaissan et de la Nouvelle Faculté de Médecine (Hérault). Leur position stratigraphique. *Géobios, Mém. Spéc.*, 10:81–101.

AGUILAR, J.P. 1979–0007. Principaux résultats biostratigraphiques de l'étude des Rongeurs miocènes du Languedoc. *C. R. Acad. Sci., Paris, Sér. D*, 288:473–476.

AGUILAR, J.P., AGUSTI, J., and GIBERT, J. 1979–0008. Rongeurs miocènes dans le Vallés-Penedés. 2. Les rongeurs de Castell de Barbera. *Palaeovertebr.* 9:17–31.

AGUILAR, J.P., and CLAUZON, G. 1979–0009. Un gisement à mammifères dans la formation lacustre d'âge Miocène moyen du Collet Redon près de St.-Cannat (Bouches-du-Rhône); implications stratigraphiques. *Palaeovertebr.* 8:327–341.

AGUILAR, J.-P., and MAGNE, J. 1978. Nouveaux gisements à rongeurs dans des formations marines miocènes du Languedoc méditeranéen. *Bull. Soc. Géol. Fr.* (7) 20:803–805.

AGUSTI, J. 1979. El Vallesiense inferior de la Peninsula Iberica y su fauna de roedores. (Mammalia). *Acta Geol. Hisp.* 13:137–141.

ALBERDI, M.T. 1974A. *El genero Hipparion en España.* Trab. Sobre Neogeno-Cuaternario. Madrid.

ALBERDI, M.T. 1974B. Las faunas de *Hipparion* de los yacimientos españoles. *Estud. Geol.* (Inst. Invest. Geol. "Lucas Mallada"), 30:189–212.

AMEUR, R., BIZON, G., JAEGER, J.-J., MICHAUX, J., and MULLER, C. 1979. A propos de l'immigration des hipparions en Afrique du Nord. *Réun. Ann. Sci. Terre, 7ᵉ, Lyon,* 1979:8.

ANDREWS, P.J. 1978A. A revision of the Miocene Hominoidea of East Africa. *Bull. Geol. Br. Mus.* (*Nat. Hist.*) 30:85–224.

ANDREWS, P.J., and VAN COUVERING, J.H. 1975. Palaeoenvironments in the East African Miocene. In *Approaches to Primate Paleobiology,* ed. F.S. Szalay, pp. 62–103. Vol. 5. Basel: Karger Press.

ANDREWS, P.J., and WALKER, A. 1976. The primate and other fauna from Fort Ternan, Kenya. In *Human Origins, Louis Leakey, and the East African Evidence,* ed. G. Isaac and E. McCown, pp. 279–304. Menlo Park, Calif.: Staples Press.

ARCHER, M. 1979. *Wabularoo naughtoni,* gen. et sp. nov., an enigmatic kangaroo (Marsupialia) from the middle Tertiary Carl Creek Limestone of northwest Queensland. *Queensl. Mus., Mem.,* 19:299–307.

ARCHER, M., and RICH, T.H.V. 1979. *Wakamatha tasselli,* gen. et sp. nov., a fossil dasyurid (Marsupialia) from South Australia convergent on modern *Sminthopsis. Queensl. Mus., Mem.,* 19:309–317.

AZZAROLI, A., and GUAZZONE, G. 1979. Terrestrial mammals and land connections in the Mediterranean before and during the Messinian. *Palaeogeogr., Palaeoclimatol., Palaeoecol.* 29:155–167.

BACHMAYER, F., and WILSON, R.W. 1970A. Die Fauna der altpliozänen Höhlen- und Spaltenfüllungen bei Kohfidisch, Burgenland (Österreich). *Ann. Naturhist. Mus. Wien* 74:533–587.

BACHMAYER, F. and WILSON, R.W. 1979. A second contribution to the small mammal fauna of Kohfidisch, Austria. *Ann. Naturhist. Mus. Wien* 81:129–161.

BAKALOV, P., and NIKOLOV, I. 1962. *Tertiary mammals.* Vol. 10. (*The Fossils of Bulgaria*). Sofiya, Bulgaria: Acad. Sci. Press. 162 pp. (In Bulgarian. Translated by G. Shkurkin, 1970.)

BALLESIO, R., CARBONNEL, G., MEIN, P., and TRUC, G. 1979. Sur un nouveau gisement fossilifère du Miocène supérieur (Tortonien-Turolien moyen) de Cucuron (Vaucluse). *Geobios, Mém. Spéc.,* 12:467–471.

BARNES, L.G. 1976. Outline of eastern North Pacific fossil cetacean assemblages. *Syst. Zool.* 25:321–343.

BARNES, L.G. 1978–0087. A review of *Lophocetus* and *Liolithax* and their relationships to the delphinoid family Kentriodontidae (Cetacea: Odontoceti). *Sci. Bull. Los Ang. Cty. Mus. Nat. Hist.* 28:1–35.

BARNES, L.G. 1979. Fossil enaliarctine pinnipeds (Mammalia: Otariidae) from Pyramid Hill, Kern County, California. *Los Ang. Cty. Mus. Nat. Hist. Contrib. Sci.* 318:1–41.

BASKIN, J.A. 1979. Small mammals of the Hemphillian White Cone local fauna, northeastern Arizona. *J. Paleontol.* 53:695–708.

BASKIN, J.A. 1980. The generic status of *Aelurodon* and *Epicyon* (Carnivora, Canidae). *J. Paleontol.* 54:1349–1351.

BASKIN, J.A. 1981. *Barbourofelis* (Nimravidae) and *Nimravides* (Felidae), with a description of two new species from the late Miocene of Florida. *J. Mammal.* 62:122–139.

BEAUMONT, G. DE. 1975. Recherches sur les Félidés (Mammifères, Carnivores) du Pliocène inférieur des sables à *Dinotherium* des environs d'Eppelsheim (Rheinhessen). *Arch. Sci.* (Soc. Phys. Hist. Nat. Genève) 28:369–405.

BERGGREN, W.A., and VAN COUVERING, J.A. 1974. The Late Neogene. *Palaeogeogr., Palaeoclim., Palaeoecol.* 16 (special issue):1–216.

BERNOR, R.L. 1983. Geochronology and zoogeographic relationships of Miocene Hominoidea. In *New Interpretations of Ape and Human Ancestry,* ed. R.L. Ciochon and R.S. Corruccini. New York: Plenum Press, 1983.

BERNOR, R.L., WOODBURNE, M.O., and VAN COUVERING, J.A. 1980. A contribution to the chronology of some Old World Miocene faunas based on hipparionine horses. *Géobios, Mém. Spéc.,* 13:705–739.

BISHOP, W.W., MILLER, J.A., and FITCH, F.J. 1969. New potassium-argon age determinations relevant to the Miocene fossil mammal sequence in East Africa. *Am. J. Sci.* 267:669–699.

BISHOP, W.W., and PICKFORD, H.H.L. 1975. Geology, fauna, and palaeo-environments of the Ngorora Formation, Kenya Rift Valley. *Nature* 254:185–192.

BLACK, C.C. 1961. Rodents and lagomorphs from the Miocene Fort Logan and Deep River formations of Montana. *Postilla* 48:1–20.

BLACK, C.C. 1963. A review of the North American Tertiary Sciuridae. *Bull. Mus. Comp. Zool.* (Harv. Univ.) 130:111–248.

BLACK, C.C. 1972A. A new species of *Merycopotamus* (Artiodactyla: Anthracotheriidae) from the late Miocene of Tunisia. *Trav. Geol. Tunisienne* 6:5–40.

BLACK, C.C., KRISHTALKA, L., and SOLOUNIAS, N. 1980. Mammalian fossils of Samos and Pikermi. Part 1. The Turolian rodents and insectivores of Samos. *Ann. Carnegie Mus.* 49:359–378.

BOAZ, N.T., GAZIRY, A.W., and EL ARNUATI. 1979. New fossil finds from the Libyan upper Neogene site of Sahabi. *Nature* 280:137–140.

BONIS, H. DE. 1973. Contribution à l'étude des Mammifères de l'Aquitanien de l'Agenais, Rongeurs-Carnivores-Périssodactyles. *Mus. Natl. Hist. Nat.* (*Paris*), *Mém. Sér. C,* 28:1–122.

BOUVRAIN, G. 1979. Un nouveau genre de Bovidé de la fin du Miocène. *Bull. Soc. Géol. Fr.* 21:507–511.

BRUIJN, H. DE. 1966A. Some new Miocene Gliridae (Rodentia, Mammalia) from the Calatayud area (Prov. Zaragosa, Spain) I. *Proc. Ned. Akad. Wet., Ser. B,* 69:1–21.

BRUIJN, H. DE. 1976. Vallesian and Turolian rodents from Biotia, Attica and Rhodes (Greece) I. *Proc. Ned. Akad. Wet., Ser. B,* 79:361–384.

BRUIJN, H. DE, VAN DER MEULEN, A.J., and KATSIKATSOS, G. 1980. The mammals from the Lower Miocene of Aliveri (Island of Evia, Greece). *Proc. Ned. Akad. Wet., Ser. B, Palaeont.,* 83:241–261.

BUTLER, P.M. 1969. Insectivores and bats from the Miocene of East Africa: New material. In *Fossil Mammals of Africa*, ed. L.S.B. Leakey, pp. 1–37. Vol. 1. New York and London: Academic Press.

CAMPBELL, B.G., AMINI, M.H., BERNOR, R.L., DICKINSON, W., DRAKE, R., MORRIS, R., VAN COUVERING, J.A., and VAN COUVERING, J.H. 1980. Maragheh, a classical late Miocene vertebrate locality in northwestern Iran. *Nature* 287:837–841.

CARRANZA-CASTANEDA, O., and FERRUSQUIA-V., I. 1979. El genero *Neohipparion* (Mammalia: Perissodactyla) de la fauna local Rancho El Ocote, (Plioceno medio) de Guanajuato, Mexico. *Revista Instituto de Geol., Univ. Nac. Autonom. Mex.*, 3:29–38.

CHESTERS, K.I.M. 1959. The Miocene flora of Rusinga Island, Lake Victoria, Kenya. *Palaeontogr., Abt. B*, 101:30–71.

CHIU CHAN-SIANG (QIU ZHAN-XIANG), HUANG WEI-LONG, and GAO ZHI-HUI. 1979. Hyaenidae of the Qingyang (K'ingyang) *Hipparion* fauna. *Vertebr. PalAsiatica* 17:200–220.

CHIU CHAN-SIANG, LI CHUAN-KUEI, and CHIU CHU-TING (QIU ZHU-DING). 1979A. The Chinese Neogene—A preliminary review of the mammalian localities and faunas. *Ann. Geol. Hellen.* 1979:263–272.

CHIU CHU-TING (QIU ZHU-DING). 1979B. Some mammalian fossils from the Pliocene of Inner Mongolia and Gansu (Kansu). *Vertebr. PalAsiatica* 17:221–235.

CICHA, I., FAHLBUSCH, V., and FEJFAR, O. 1972. Die biostratigraphische korrelation einiger jungtertiärer Wirbeltierfaunen Mitteleuropas. *Abh. Neuesjahrb. Geol. Palaeont.* 40:129–145.

CIRIC, A., and THENIUS, E. 1959. Über das vorkommen von *Giraffokeryx* (Giraffidae) im europäischen Miozän. *Anz. Akad. Wiss. Wien* 96:153–162.

CHOPRA, S.R.K., and VASISHAT, R.N. 1979. Sivalik fossil tree shrew from Haritalyangar, India. *Nature* 281:214–215.

CLARK, J.B., DAWSON, M.R., and WOOD, A.E. 1964. Fossil mammals from the Lower Pliocene Fish Lake Valley, Nevada. *Bull. Mus. Comp. Zool.* (Harv. Univ.) 131:1–63.

CLEMENS, W.A., and PLANE, M. 1974. Mid-Tertiary Thylacoleonidae (Marsupialia, Mammalia). *J. Paleontol.* 48:652–660.

COLBERT, E.H. 1935A. Siwalik mammals in the American Museum of Natural History. *Trans. Amer. Phil. Soc.*, n.s., 26:1–401.

COLLIER, A., and HUIN, J. 1979. Découverte d'un gisement d'âge burdigalien inférieur dans des sables sous-jacents aux Faluns de la Touraine; étude de la faune de Rongeurs et intérêt biostratigraphique. *C. R. Acad. Sci., Paris, Sér. D*, 289:249–252.

COOK, H.J. 1965. Running Water formation, Middle Miocene of Nebraska. *Am. Mus. Novit.* 2227:1–8.

COOK, H.J., and COOK, M. C. 1933. Faunal lists of the Tertiary Vertebrata of Nebraska and adjacent areas. *Nebr. Geol. Surv., Pap.*, 5:1–58.

COOKE, H.B.S. 1978. Africa: The physical setting. In *Evolution of African Mammals*, ed. V.J. Maglio and H. B. S. Cooke, pp. 17–45. Cambridge: Harvard Univ. Press.

COOMBS, M.C. 1978. Re-evaluation of early Miocene North American *Moropus* (Perissodactyla, Chalicotheriidae, Schizotheriinae). *Bull. Carnegie Mus. Nat. Hist.* 4:1–62.

COOMBS, M. C. 1979. *Tylocephalonyx*, a new genus of North American dome-skulled chalicotheres (Mammalia, Perissodactyla). *Bull. Am. Mus. Nat. Hist.* 164:1–64.

COPPENS, Y. 1978. Le Lothagamien et le Shungurien, étages continentaux du Pliocène est-africain. *Bull. Soc. Géol. Fr.* 20:39–44.

CROCHET, J.-Y. 1975. Diversité des Insectivores soricidés du Miocène inférieur de France. *Fr., Cent. Natl. Rech. Sci., Colloq. Int.*, 218:631–652.

CROUZEL, F. 1957. Le Miocène continental du Bassin d'Aquitaine. *Bull. Serv. Carte Géol. Fr.* 54:1–264.

CROUZEL, F. 1979. Suite à l'étude des gisements fossilifères du Miocène gersois. *Bull. Soc. Hist. Nat. Toulouse.* 114:263–269.

CRUSAFONT, M. 1950E. La cuestión del llamado Méotico español. *Arrahona, Mus. Sabadell*, 1–2.

CRUSAFONT, M. 1958B. Endemism and paneuropeism in Spanish fossil mammalian faunas, with special regard to the Miocene. *Soc. Sci. Fenn., Commentat. Biol.*, 18:1–30.

CRUSAFONT, M. 1965N. Observations à un travail de M. Freudenthal et P.Y. Sondaar sur des nouveaux gisements à *Hipparion* d'Espagne. *Proc. Ned. Akad. Wet., Ser. B*, 68:121–126.

CRUSAFONT, M. 1979. Transicion Vindoboniense-Vallesiense en los alrededores de Sabadell. *Paleontol. Evol.—Barc., Inst. Prov. Paleontol.*, 13:7–9.

CRUSAFONT., M., and GOLPE-POSSE, G.P. 1971. Biozonation des mammifères néogènes d'Espagne. *Bur. Rech. Géol. Min.* (Lyon) 78:121–129.

CRUSAFONT, M., and PETTER, G. 1969. Contribution à l'étude des Hyaenidae. La sous-famille des Ictitheriinae. *Ann. Paléont.* 55:89–127.

DAAMS, R. 1976. Miocene rodents (Mammalia) from Cetina de Aragon (Prov. Zaragosa) and Bunol (Prov. Valencia), Spain. *Proc. Ned. Akad. Wet., Ser. B*, 9:152–182.

DAAMS, R. 1981. The dental pattern of the dormice *Dryomys, Myomimus, Microdyromys* and *Peridyromys*. *Utrecht. Micropaleontol. Bull., Spec. Publ.*, 3:1–115.

DAAMS, R., and FREUDENTHAL, M. 1981. Aragonian: The Stage concept versus Neogene Mammal Zones. *Scripta Geologica* 62:1–17.

DAAMS, R., FREUDENTHAL, M., and VAN DE WEERD, A. 1977. Aragonian, a new stage for continental deposits of Miocene age. *Newsl. Stratigr.* 6:42–55.

DALQUEST, W.W., and MOOSER, O. 1974. Miocene vertebrates from Aguascalientes, Mexico. *Tex. Mem. Mus., Pearce-Sellards Ser.*, 21:1–10.

DALQUEST, W.W., and MOOSER, O. 1980. Late Hemphillian mammals of the Ocote local fauna, Guanajuato, Mexico. *Tex. Mem. Mus., Pearce-Sellards Ser.*, 32:1–25.

DANIS, J.C. 1982. A new mammal *Laventatherium* (Notoungulata: Leontiniidae) from the Miocene of Colombia. *PaleoBios.* (In press.) Univ. Calif., Berkeley.

DAWSON, M.R. 1958. Later Tertiary Leporidae of North America. *Univ. Kans., Paleontol. Contrib.* 22:1–75.

DAWSON, M.R. 1961A. On two ochotonids (Mammalia, Lagomorpha) from the later Tertiary of Inner Mongolia. *Am. Mus. Novit.* 2061:1–15.

DAWSON, M.R. 1965. *Oreolagus* and other Lagomorpha (Mammalia) from the Miocene of Colorado, Wyoming, and Oregon. *Colo. State Univ., Stud. Ser. Earth Sci.* 1:1–36.

DAWSON, M.R. 1967. Lagomorph history and the stratigraphic record. In *Essays in Paleontology and Stratigraphy*, pp. 287–

316. Raymond C. Moore Commemorative Volume. *Univ. Kans., Dept. Geol. Spec. Publ.* 2:287–316.

DAXNER-HOCK, G., and BRUIJN, H. DE. 1981. Gliridae (Rodentia, Mammalia) des Eichkogels bei Mödling (Niederösterreich). *Z. Palaeont.* 55:157–172.

DEHM, R. 1950A. Die Raubtiere aus dem Mittel-Miocän (Burdigalium) von Wintershof-West bei Eichstätt in Bayern. *Abh. Bayer. Akad. Wiss., Math.-Naturwiss. Kl.* 58:1–141.

DEHM, R. 1950B. Die Nagetiere aus dem Mittel-Miocän (Burdigalium) von Wintershof-West bei Eichstätt in Bayern. *Abh. Bayer. Akad. Wiss., Math.-Naturwiss. Kl., B,* 91:321–427.

DELSON, E. 1979. *Prohylobates* (Primates from the early Miocene of Libya; a new species and its implications for cynomorph origins. *Geobios* 12:725–733.

DELSON, E. 1980. Fossil macaques, phyletic relationships and a scenario of deployment. In *The Macaques: Studies in Ecology, Behavior and Evolution,* ed. D.G. Lindburg, chap. 2. New York: Van Nostrand.

DIMITRIEVA, E.L. 1977. Neogene antelopes of Mongolia and adjacent territories. *Trans. Joint Soviet-Mongolian Paleontological Expedition* 6:1–116.

DOMNING, D.P. 1978. Sirenian evolution in the North Pacific Ocean. *Univ. Calif. Publ. Geol. Sci.* 118:1–176.

DOWNS, R. 1956D. The Mascall fauna from the Miocene of Oregon. *Univ. Calif. Publ. Geol. Sci.* 31:199–354.

EDWARDS, S.W. 1982. A new species of *Hipparion* (Mammalia, Equidae) from the Clarendonian (Miocene) of California. *J. Vertebr. Paleont.*

ENGESSER, B. 1972A. Die obermiozäne Säugetierfauna von Anwil (Baselland). *Naturforsch. Ges. Baselland, Tatigkeitsber.,* 28:37–363.

ENGESSER, B. 1975. Revision der europaischen Heterosoricinae (Insectivora, Mammalia). *Eclagae Geol. Helv.* 68:649–671.

ENGESSER, B. 1979. Relationships of some insectivores and rodents from the Miocene of North America and Europe. *Bull. Carnegie Mus. Nat. Hist.* 14:5–68.

ENGESSER, B. 1980. Insectivora und Chiroptera (Mammalia) aus dem Neogen der Türkei. *Abh. Schweiz. Palaeont.* 102:47–149.

EVERNDEN, J.F., SAVAGE, D.E., CURTIS, G.H., and JAMES, G.T. 1964. Potassium-argon dates and the Cenozoic mammalian chronology of North America. *Am. J. Sci.* 262:145–198.

FAHLBUSCH, V. 1976. Report on the International Symposium on mammalian stratigraphy of the European Tertiary. *Newsl. Stratigr.* 5:160–167.

FEJFAR, O. 1974. Die Eomyiden und Cricetiden (Rodentia, Mammalia) des Miozäns der Tschechoslowakei. *Palaeontographica, Abt. A,* 146:100–180.

FICCARELLI, G., and TORRE, D. 1970. Remarks on the taxonomy of hyaenids. *Palaeontogr. Ital.* 66, n.s., 36:13–33.

FISHER, R.V., and RENSBERGER, J.M. 1972. Physical stratigraphy of the John Day Formation. *Univ. Calif. Publ. Geol. Sci.* 101:1–45.

FLEAGLE, J.G., and SIMONS, E.L. 1978. *Micropithecus clarki,* a small ape from the Miocene of Uganda. *Am. J. Phys. Anthropol.* 49:427–440.

FRAILEY, D. 1979. The large mammals of the Buda local fauna (Arikareean) Alachua County, Florida. *Fla. State Mus., Bull. Biol. Sci.* 24:123–173.

FRANZEN, J.L., and STORCH, G. 1975. Die unterpliozäne (Turolische) Wirbeltierfauna von Dorn-Durkheim, Rheinhessen (SW-Deutschland). 1. Entdeckung, Geologie, Mammalia: Carnivora, Proboscidea, Rodentia. *Senckenb. Lethaea* 56:233–303.

FRANZEN, J.L., and STORCH, G. 1976. Die unterpliozäne Fundstelle von Dorn-Durkheim (Rheinhessen). *Rhein-Main. Forsch.* 82:61–72.

FREUDENTHAL, M. 1963. Entwicklungsstufen der Miozänen Cricetodontinae (Mammalia, Rodentia) Mittelspaniens und ihre stratigraphische Bedeutung. *Beaufortia* 10:51–157.

FRICK, C. 1937. Horned ruminants of North America. *Bull. Am. Mus. Nat. Hist.* 69:1–669.

FRICK, C., and TAYLOR, G. 1968. A generic review of the stenomyline camels. *Am. Mus. Novit.* 2353:1–51.

FRICK, C., and TAYLOR, B. 1971. *Michenia,* a new protolabine (Mammalia, Camelidae) and a brief review of the early taxonomic history of the genus *Protolabis. Am. Mus. Novit.* 2444:1–24.

GABUNIA, L.K. 1979. Biostratigraphic correlations between the Neogene land mammal faunas of the East and Central Paratethys. *Ann. Geol. Pays. Hell.* 1979:421–423. (7th Internat. Congr. Medit. Neogene, Athens).

GALUSHA, T. 1975. Stratigraphy of the Box Butte Formation, Nebraska. *Bull. Am. Mus. Nat. Hist.* 156:1–68.

GALUSHA, T., and BLICK, J.C. 1971. Stratigraphy of the Santa Fe Group, New Mexico. *Bull. Am. Mus. Nat. Hist.* 144:1–128.

GAWNE, C.E. 1975. Rodents from the Zia Sand Miocene of New Mexico. *Am. Mus. Novit.* 2608:1–15.

GENTRY, A.W. 1970. The Bovidae (Mammalia) of the Fort Ternan fossil fauna. In *Fossil Vertebrates of Africa,* Vol. 2, ed. L.S.B. Leakey and R.J.G. Savage, pp. 243–323. New York and London: Academic Press.

GENTRY, A.W. 1971A. The earliest goats and other antelopes from the Samos *Hipparion* fauna. *Br. Mus. (Nat. Hist.), Bull., Geol.,* 20:229–296.

GIGOT, P., and MEIN, P. 1973. Découvertes de mammifères aquitaniens dans la molasse burdigalienne du Golfe de Digne. *C. R. Acad. Sci., Paris, Sér. D.,* 276:3293–3294.

GINSBURG, L. 1955. De la subdivision du genre *Hemicyon* Lartet (Carnassier du Miocène). *Bull. Soc. Géol. Fr.* 5:85–99.

GINSBURG, L. 1968C. Les mustelidés piscivores du Miocène français. *Bull. Mus. Natl. Hist. Nat.* 40:228–238.

GINSBURG, L. 1970A. Les Mammifères des faluns helvétiens du Nord de la Loire. *C. R. Soc. Géol. Fr., Sér. D.,* 1970, 6:189–190.

GINSBURG, L. 1972A. Sur l'âge des Mammifères des faluns miocènes du Nord de la Loire. *C. R. Acad. Sci., Paris, Sér. D.,* 274:3345–3347.

GINSBURG, L. 1973. Les tayassuidés des Phosphorites du Quercy. *Palaeovertebr.* 6:55–85.

GINSBURG, L. 1974. Les rhinocerotidés du Miocène de Sansan (Gers). *C. R. Acad. Sci., Paris, Sér. D.,* 278:597–600.

GINSBURG, L. 1975A. Etude paléontologique des vertébrés pliocènes de Pont-de-Gail (Cantal). *Bull. Soc. Géol. Fr., Sér. 7,* 17:752–759.

GINSBURG, L. 1975B. Le pliopithèque des faluns Helvétiens de la Touraine et de l'Anjou. *Fr., Cent. Natl. Rech. Sci., Colloq. Int.,* 218:877–886.

GINSBURG, L. 1975C. Une échelle stratigraphique continentale pour l'Europe occidentale et un nouvel étage: l'Orléanien. *Bull. Trimestr. Assoc. Naturalistes Orléanais* 18:1–11.

GINSBURG, L. 1980. *Hyainailouros sulzeri,* mammifère créodonte du Miocène d'Europe. *Ann. Paléontol., Vertébrés,* 66:19–55.

GINSBURG, L., and ANTUNES, M.T. 1979. Les rhinocérotidés du Miocène inférieur et moyen de Lisbonne (Portugal). Succession stratigraphique et incidences paléogéographiques. *C. R. Acad. Sci., Paris, Sér. D.,* 288:493–495.

GINSBURG, L., and MEIN, P. 1980. *Crouzelia rhodanica,* nouvelle espèce de Primate catarrhinien, et essai sur la position systématique des Pliopithecidae. *Bull. Mus. Nat. Hist. Natur., Paris, C,* 2:57–85.

GINSBURG, L., and TASSY, P. 1977. Les nouveaux gisements à mastodontes du Vindobonien moyen de Simorre (Gers). *C. R. Soc. Géol. Fr.* 1977:24–26.

GOLPE-POSSE, J.M. 1979. Un nuevo Tayasuido en el Vindoboniense terminal de Castell de Barbera (Cuenca del Valles, Espana). *Paleontol. Evol.* (Barc., Inst. Prov. Paleontol.) 13:13–17.

Green, M. 1971. Additions to the Mission vertebrate fauna, lower Pliocene of South Dakota. *J. Paleontol.* 45:486–490.

Green, M. 1972. Lagomorpha from the Rosebud Formation, South Dakota. *J. Paleontol.* 46:377–385.

Green, M. 1977B. Neogene Zapodidae (Mammalia, Rodentia) from South Dakota. *J. Paleontol.* 51:996–1015.

GREEN, M., and HOLMAN, J.A. 1977. A late Tertiary stream channel fauna from South Bijou Hill, South Dakota. *J. Paleontol.* 51:543–547.

GREGORY, J.T. 1942. Pliocene vertebrates from Big Spring Canyon, South Dakota. *Univ. Calif., Publ. Dept. Geol. Sci.* 26:307–446.

GREGORY, J.T., and DOWNS, T. 1951. *Bassariscus* in Miocene faunas and *"Potamotherium lycopotamicum"* Cope. *Postilla* 8:1–10.

GUÉRIN, C. 1976. Les restes de rhinoceros du gisement Miocène de Beni Mellal, Maroc. *Géol. Médit.* 3:105–108.

HAMILTON, W.R. 1973A. North African lower Miocene rhinoceroses. *Bull. Br. Mus. (Nat. Hist.),* 24:351–395.

HAMILTON, W.R. 1973B. The lower Miocene ruminants of Gebel Zelten, Libya. *Bull. Br. Mus. (Nat. Hist.)* 21:75–150.

HAMILTON, W.R., WHYBROW, P.J., and McCLURE, H.A. 1978. Fauna of fossil mammals from the Miocene of Saudi Arabia. *Nature* 274:248.

HARRISON, J.A. 1981. A review of the extinct wolverine, *Plesiogulo* Carnivora: Mustelidae) from North America. *Smithson. Contrib. Paleobiol.* 46:1–27.

HEINTZ, E. 1976. Les Giraffidae (Artiodactyla, Mammalia) du Miocène de Beni Mellal, Maroc. *Géol. Médit.* 3:91–104.

HEINTZ, E., GINSBURG, L., and HARTENBERGER, J.-L. 1978. Mammifères fossiles en Afghanistan: état des connaissances et résultats d'une prospection. *Bull. Mus. Natl. Hist. Nat., 3rd ser.,* 69:101–119.

HEISSIG, K. 1976. Rhinocerotidae (Mammalia) aus der *Anchitherium*-Fauna Anatoliens. *Geol. Jahrb., Series B,* 19:1–121.

HEISSIG, K. 1979. Die hypothetische Rolle Südosteuropas bei dem Säugetier wanderungen in Eozän und Oligozän. *Neues Jahrb. Geol. Palaeontol., H2*:83–96.

HENDEY, Q.B. 1974. The Late Cenozoic Carnivora of the Southwestern Cape Province. *Ann. S. Af. Mus.* 63:1–369.

HESSE, C.J. 1943. A preliminary report on the Miocene vertebrate faunas of southeast Texas. *Proc. and Trans. Tex. Acad. Sci.* 26:157–179.

HIBBARD, C.W., and JAMMOT, D. 1971. The shrews of the WaKeeney local fauna, lower Pliocene of Trego County, Kansas. *Univ. Mich., Mus. Paleontol., Contrib.* 23:377–380.

HIBBARD, C.W., and KEENMAN, K.A. 1950. New evidence of the lower Miocene age of the Blacktail Deer Creek Formation in Montana. *Univ. Mich., Mus. Paleontol., Contrib.* 8:199–204.

HIRSCHFELD, S.E., and MARSHALL, L.G. 1976. Revised faunal list of the La Venta fauna (Friasian-Miocene) of Colombia, South America. *J. Paleontol.* 50:433–436.

HONEY, J.C., and TAYLOR, B.E. 1978. A generic revision of the Protolabidinae (Mammalia, Camelidae), with a description of two new protolabidines. *Bull. Am. Mus. Nat. Hist.* 161:367–426.

HOWELL, F.C. 1980. Zonation of late Miocene and early Pliocene circum-Mediterranean faunas. *Geobios, Mém. Spéc.* 13:653–657.

HSEN HSU HU, and CHANEY, R.W. 1940. A Miocene flora from Shantung Province, China. *Palaeontologia Sinica, n.s., A,* 1:1–147.

HUGUENEY, M. 1974. Gisements de petits mammifères dans la région de Saint-Gérand-le-Puy. *Rev. Scient. Bourbonnais* 1974:52–68.

HUGUENEY, M., and MEIN, P. 1965. Lagomorphes et rongeurs du Néogène de Lissieu (Rhône). *Univ. Lyon, Trav. Lab. Géol.* 12:109–123.

HUIN, J. 1979. Les faunes miocènes du Haut-Armagnac; 2, Les Lagomorphes (Mammalia); Première partie, Le genre *Prolagus. Bull. Soc. Hist. Nat. Toulouse* 114:382–392.

HUNT, R.M., JR. 1972A. Miocene amphicyonids (Mammalia, Carnivora) from the Agate Springs quarries, Sioux County, Nebraska. *Am. Mus. Novit.* 2506:1–39.

HUNT, R.M., JR. 1978. Depositional setting of a Miocene mammalian assemblage, Sioux County, Nebraska (USA) *Palaeogr., Palaeoclimatol., Palaeoecol.* 24:1–52.

HÜRZELER, J. 1944B. Zur Revision der europäischen Hemicyoniden. *Verh. Naturforsch. Ges. Basel.* 55:131–157.

HÜRZELER, J., and ENGESSER, B. 1976. Les faunes de mammifères néogènes du Bassin de Baccinello (Grosseto, Italie). *C. R. Acad. Sci., Paris, Sér. D,* 283:333–336.

HUSSAIN, S.T., BRUIJN, H. DE, AND LEINDERS, J.M. 1978. Middle Eocene rodents from the Kala Chitta Range (Punjab, Pakistan). *Proc. Ned. Akad. Wet., Ser. B,* 81:74–112.

HUSSAIN, S.T., MUNTHE, J., WEST, R.M., and LUKACS, J.R. 1977. The Daud Khel local fauna: A preliminary report on a Neogene vertebrate assemblage from the Trans-Indus Siwaliks, Pakistan. *Milw. Publ. Mus., Contrib. Biol. Geol.* 16:1–17.

HUTCHISON, J.H. 1966A. Notes on some upper Miocene shrews from Oregon. *Univ. Oreg., Bull. Mus. Nat. Hist.* 2:1–23.

HUTCHISON, J.H. 1968. Fossil Talpidae (Insectivora, Mammalia) from the later Tertiary of Oregon. *Univ. Oreg., Bull. Mus. Nat. Hist.* 11:1–117.

HUTCHISON, J.H. 1972A. Review of the Insectivora from the early Miocene Sharps Formation of South Dakota. *Los Ang. Cty. Mus. Nat. Hist., Contrib. Sci.* 235:1–16.

HUTCHISON, J.H., and LINDSAY, E.H. 1974. The Hemingfordian

mammal fauna of the Vedder Locality, Branch Canyon Formation, Santa Barbara County, California. Part 1: Insectivora, Chiroptera, Lagomorpha and Rodentia (Sciuridae). *PaleoBios* 15:1–19.

IZETT, G.A. 1968. Geology of the Hot Sulphur Springs Quadrangle, Grand County, Colorado. *U.S. Geol. Surv. Prof. Pap.* 586:1–79.

JACOBS, L.L. 1977. Rodents of the Hemphillian age Redington local fauna, San Pedro Valley, Arizona. *J. Paleontol.* 51:505–519.

JACOBS, L.L. 1978. Fossil rodents (Rhizomyidae and Muridae) from Neogene Siwalik deposits, Pakistan. *Bull. Mus. North. Ariz.* 52:1–103.

JAEGER, J.-J. 1974. Nouvelles faunes de rongeurs (Mammalia, Rodentia) du Miocène supérieur d'Afrique nord-occidentale. *Ann. Geol. Surv. Egypt* 4:263–268.

JAEGER, J.-J. 1977A. Rongeurs (Mammalia, Rodentia) du Miocène de Beni Mellal. *Palaeovertebr.* 7:91–132.

JAEGER, J.-J. 1977B. Les rongeurs du Miocène moyen et supérieur du Maghreb. *Palaeovertebr.* 8:1–166.

JAEGER, J.-J., and MARTIN, J. 1971. Découverte au Maroc des premiers micromammifères du Pontien d'Afrique. *C. R. Acad. Sci., Paris, Sér. D*, 272:2155–2158.

JAHNS, R.H. 1940. Stratigraphy of the easternmost Ventura basin, California, with a description of a new lower Miocene mammalian fauna from the Tick Canyon Formation. *Carnegie Inst. Wash. Publ.* 514:145–194.

JAMES, G.T. 1963. Paleontology and nonmarine stratigraphy of the Cuyama Valley, Badlands, California. Part 1, Geology, faunal interpretations, and systematic descriptions of Chiroptera, Insectivora, and Rodentia. *Univ. Calif., Publ. Geol. Sci.* 45:1–154.

JANVIER, P., and MONTENAT, C. 1970. La plus ancien Léporidé d'Europe occidentale, *Hispanolagus crusafonti*, nov. gen., nov. sp., du Miocène supérieur de Murcia (Espagne). *Bull. Mus. Natl. Hist. Nat. (Paris)* 42:780–788.

JOHNSON, S.C. 1982. Astrapotheres from the Miocene of Colombia, South America, with special reference to their feeding mechanisms, ecology, and behavior. Ph.D. dissertation, Dept. Paleont., Univ. Calif., Berkeley.

JULLIEN, R., GUÉRIN, C., HUGUENEY, M., and MEIN, P. 1979. Découverte d'un gisement de Mammifères du Miocène moyen à Collet-Redon, près Saint-Cannat (Bouches-du-Rhône, France): liste faunique, implications stratigraphiques et paléogéographiques. *Geobios* 2:297–301.

KLINGENER, D.J. 1966. Dipodoid rodents from the Valentine Formation of Nebraska. *Univ. Mich., Mus. Zool., Occ. Pap.* 644:1–9.

KOENIGSWALD, W. VON. 1980. Schmelzstruktur und Morphologie in den Molaren der Arvicolidae (Rodentia). *Abh. Senckenb. Naturforsch. Gesell.* 539:1–129.

KOERNER, H.E. 1940. The geology and vertebrate paleontology of the Fort Logan and Deep River formations of Montana. *Am. J. Sci.* 238:837–862.

KROHN, I.M. 1978. Functional adaptation in Miocene Glyptodontidae (Edentata, Mammalia) from the Honda Group, Colombia, South America. *Ph.D. dissertation*, Dept. Paleont., Univ. Calif., Berkeley.

KUENZI, W.D., and FIELDS, R.W. 1971. Tertiary stratigraphy, structure, and geologic history, Jefferson Basin, Montana. *Bull. Geol. Soc. Am.* 82:3373–3394.

LANCE, J.F. 1950. Paleontologia y estratigraphia del Plioceno de Yepomera, Estado de Chihuahua. Part 1: Equidos, excepto *Neohipparion*. *Bol. Inst. Geol. Mex.* 54:1–83.

LAVOCAT, R. 1961. Le gisement de vertébrés miocènes de Beni Mellal (Maroc). Part 2. Étude systematique de la faune de mammifères. *Serv. Géol. Maroc, Notes et Mém.*, 155:29–92.

LAVOCAT, R. 1973. Les rongeurs du Miocène d'Afrique orientale. *Inst. Montpellier, Mém. et Trav.*, 1 A–B: 1–284.

LEAKEY, L.S.B., and SAVAGE, R.J.G., eds. 1970. *Fossil Vertebrates of Africa.* Vols. 1 and 2. Edinburgh: Scottish Press.

LI CHUAN-KUEI, and QIU ZHU-DING. 1980. Early Miocene mammalian fossils of Xining Basin, Qinghai. *Vertebr. PalAsiatica* 18:198–214.

LINDSAY, E.H. 1972B. Small mammal fossils from the Barstow Formation, California. *Univ. Calif. Publ. Geol. Sci.* 93:1–104.

LINDSAY, E.H. 1974. The Hemingfordian mammal fauna of the Vedder locality, Branch Canyon Formation, Santa Barbara County, California. Part 2: Rodentia (Eomyidae and Heteromyidae). *PaleoBios* 16:1–19.

LOPEZ-MARTINEZ, N. 1978. Cladistique et paléontologie. Application à la phylogénie des ochotonidés européens (Lagomorpha, Mammalia). *Bull. Soc. Géol. Fr.* (7) 20:821–830.

LOPEZ-MARTINEZ, N., and THALER, L. 1975. Biogéographie, évolution et compléments à la systématique du groupe d'ochotonidés *Piezodus-Prolagus* (Mammalia, Lagomorpha). *Bull. Soc. Géol. Fr.* (7) 17:850–865.

LUGN, A.L. 1939. Classification of the Tertiary System in Nebraska. *Bull. Geol. Soc. Am.* 50:1245–1276.

MACDONALD, J.R. 1970. Review of the Miocene Wounded Knee faunas of southwestern South Dakota. *Bull. Los Ang. Cty. Mus. Nat. Hist.* 8:1–82.

MACDONALD, L. 1972. Monroe Creek (Early Miocene) microfossils from the Wounded Knee area, South Dakota. *S. Dak. Geol. Surv., Rep. Invest.*, 105:1–43.

MACFADDEN, B.J. 1980. The Miocene horse *Hipparion* from North America and from the type locality in southern France. *Palaeontology* 23:617–637.

MACFADDEN, B.J., and BAKR, A. 1979. The horse *Cormohipparion theobaldi* from the Neogene of Pakistan, with comments on Siwalik hipparions. *Palaeontology* 22:439–447.

MACFADDEN, B.J., JOHNSON, N.M., and OPDYKE, N.P. 1979. Magnetic polarity stratigraphy of the Mio-Pliocene mammal-bearing Big Sandy Formation of western Arizona. *Earth Planet. Sci. Lett.* 44:349–364.

MACFADDEN, B.J., and SKINNER, M.F. 1979. Diversification and biogeography of the one-toed horses *Onohippidium* and *Hippidion*. *Postilla* 175:1–10.

MACFADDEN, B.J., and SKINNER, M.F. 1981. Earliest Holarctic hipparion, *Cormohipparion goorisi* n. sp. (Mammalia, Equidae), from the Barstovian (medial Miocene) Texas Gulf Coastal Plain. *J. Paleontol.* 55:619–627.

MARSHALL, L.G. 1976. Fossil localities for Santacrucian (Early Miocene) mammals, Santa Cruz Province, southern Patagonia, Argentina. *J. Paleontol.* 50:1129–1142.

MARSHALL, L.G. 1978–0808. Evolution of the Borhyaenidae, extinct South American predaceous marsupials. *Univ. Calif. Publ. Geol. Sci.* 117:1–89.

MARSHALL, L.G., PASCUAL, R., CURTIS, G.H., and DRAKE, R.E. 1977. South American geochronology: Radiometric time scale for middle to late Tertiary mammal-bearing horizons in Patagonia. *Science* 195:1325–1328.

MARTIN, J.E. 1976. Small mammals from the Miocene Batesland Formation of South Dakota. *Contrib. Geol., Univ. Wyo.,* 14:69–98.

MARTIN, L.D. 1975. Microtine rodents from the Ogallala Pliocene of Nebraska and the early evolution of the Microtinae in North America. *Univ. Mich., Mus. Paleontol., Claude W. Hibbard Memorial Volume* 3:101–110.

MARTIN, L.D. 1980. The early evolution of the Cricetidae in North America. *Univ. Kans., Paleontol. Contrib.* 102:1–42.

MATTHEW, W.D. 1924C. Third contribution to the Snake Creek fauna. *Bull. Am. Mus. Nat. Hist.* 50:59–210.

MAYR, H., and FAHLBUSCH, V. 1975. Eine unterpliozäne Kleinsäugerfauna aus der Oberen Susswasser-Molasse Bayerns. *Bayer. Staatssamml. Palaeont. Hist. Geol. Mitt.* 15:91–111.

McKENNA, M.C. 1965. Stratigraphic nomenclature of the Miocene Hemingford Group, Nebraska. *Am. Mus. Novit.* 2228:1–21.

McKENNA, M.C. 1980. Early history and biogeography of South America's extinct land mammals. In *Evolutionary Biology of the New World Monkeys and Continental Drift,* ed. R.L. Ciochon and A.B. Chiarelli, pp. 43–78. New York and London: Plenum Press.

McKENNA, M.C., and LOVE, J.D. 1972. High-level strata containing early Miocene mammals on the Big Horn Mountains, Wyoming. *Am. Mus. Novit.* 2490:1–31.

MEIN, P. 1958. Les Mammifères de la faune sidérolithique de Vieux-Collonges. *Nouv. Arch. Mus. Hist. Nat. Lyon* 5:1–122.

MEIN, P. 1964. Chiroptera (Miocène) de Lissieu (Rhône). *Soc. Savantes, 89th Congr., Lyon,* 1964:237–253.

MEIN, P. 1970A. Les sciuroptères (Mammalia, Rodentia) néogènes d'Europe occidentale. *Geobios* 3:7–77.

MEIN, P. 1975. Résultats du Groupe de travail des Vertébrés. (Rept. on activities of the R.C.M.N.S. working groups). *Int. Union Géol. Sci., Rég. Comm. Médit. Neógène Stratigr.* Bratislava, Czech. Pp. 77–81.

MEIN, P. 1979. Rapport d'activité du groupe de travail vertébrés mise à jour de la biostratigraphie du Néogène basée sur les mammifères. 7th Internatl. Congr. Médit. Néogène. *Ann. Géol. Pays Hellen.* 1979:1367–1372.

MEIN, P., BIZON, J.-J., MONTENAT, C. 1973. Le gisement de Mammifères de La Alberca (Murcia, Espagne méridionale). Corrélations avec les formations marines du Miocène terminal. *C. R. Acad. Sci., Paris, Sér. D,* 276:3077–3080.

MEIN, P., and FREUDENTHAL, M. 1971A. Une nouvelle classification des Cricetidae (Mammalia, Rodentia) du Tertiaire de l'Europe. *Sér. Géol. (Leiden)* 2:1–37.

MEIN, P., MOISSENET, E., and TRUC, G. 1978. Les formations continentales du Néogène supérieur des vallées du Jucar et du Cabriel au NE d'Albacete (Espagne). Biostratigraphie et environment. *Lyon, Fac. Sci., Lab. Géol., Doc.,* 72:99–147.

MEIN, P., and TRUC, G. 1966. Faciès et association faunique dans le Miocène supérieur continental du Haut-Comtat venaissin. *Lyon, Fac. Sci., Lab. Géol., Trav.,* 13:273–276.

MEIN, P., TRUC, G., and DEMARCQ, G. 1971. Micromammifères et gastéropodes continentaux des biozones de Paulhiac et de La

Romieu dans le Miocène de La Bastidonne et de Mirabeau (Vaucluse, Sud-Est de la France). *C. R. Acad. Sci., Paris, Sér. D,* 273:566–568.

MILLER, W.E., and DOWNS, T. 1974. A Hemphillian local fauna containing a new genus of antilocaprid from southern California. *Los Ang. Cty. Mus. Nat. Hist., Contrib. Sci.,* 258:1–36.

MITCHELL, E., and TEDFORD, R.H. 1973. The Enaliarctinae, a new group of extinct aquatic Carnivora and a consideration of the origin of the Otariidae. *Bull. Am. Mus. Nat. Hist.* 151:201–284.

MOTTL, M. 1970. Die jungtertiären säugetierfaunen der Steiermark, Südöst-Osterreich. Graz, Joanneum, *Mitt. Mus. Berg. Geol. Techn.* 31:79–168.

MUNTHE, J., and COOMBS, M.C. 1979. Miocene dome-skulled chalicotheres (Mammalia, Perissodactyla) from the western United States: A preliminary discussion of a bizarre structure. *J. Paleontol.* 53:79–91.

NICHOLS, R. 1976. Early Miocene mammals from the Lemhi Valley of Idaho. *Tebiwa* 18:9–47.

NICHOLS, R. 1979. Additional early Miocene mammals from the Lemhi Valley of Idaho. *Tebiwa* 17:1–12.

NICOLAS, P.J. 1978. Un nouveau gisement de Vertébrés dans le Chersonien: Khutchuk-Tchekmedjé Ouest (Thrace turque). *C. R., Acad. Sci., Paris, Sér. D,* 287:455–458.

OPDYKE, N.D., LINDSAY, E.H., JOHNSON, G.D., et al. 1979. Magnetic polarity stratigraphy and vertebrate paleontology of the Upper Siwalik Subgroup of northern Pakistan. *Palaeogeogr., Palaeoclimatol., Palaeoecol.* 27:1–34.

OZANSOY, F. 1965. Etude des gisements continentaux et des Mammifères du Cénozoïque de Turquie. *Soc. Géol. Fr., Mém.,* n.s., 44:1–92.

PATTERSON, B. 1975. The fossil aardvark (Mammalia: Tubulidentata). *Bull. Mus. Comp. Zool.* (Harv. Univ.) 147:185–237.

PATTON, T.H. 1969B. Miocene and Pliocene artiodactyls, Texas Gulf coastal plain. *Fla. State Mus., Bull. Biol. Sci.* 14:115–226.

PATTON, T.H., and TAYLOR, B.E. 1971. The Synthetoceratinae (Mammalia, Tylopoda, Protoceratidae)—the systematics of the Protoceratidae. *Bull. Am. Mus. Nat. Hist.* 145:119–218.

PATTON, T.H., and TAYLOR, B.E. 1973. The Protoceratinae (Mammalia, Tylopoda, Protoceratidae) and the systematics of the Protoceratidae. *Bull. Am. Mus. Nat. Hist.* 150:347–414.

PETERSON, O.A. 1906. The Miocene beds of western Nebraska and eastern Wyoming and their vertebrate faunae. *Ann. Carnegie Mus. Nat. Hist.* 4:21–72.

PETRONIJEVIC, Z.M. 1967. (Middle Miocene and Lower Sarmatian (Steier) mammalian fauna of Serbiya.) *Paleontol. Jugoslav.* 7:1–160. Translated by G. Shkurkin. 1980.

PETTER, G. 1976. Etude d'un nouvel ensemble de petits carnivores du Miocène d'Espagne. *Géol. Médit.* 3:135–154.

PICKFORD, M., and ERTURK, C. 1979. Suidae and Tayassuidae from Turkey. *Bul. Turk. Jeol. Kurumu* 22:141–154.

PILBEAM, D., BARRY, J., MEYER, G.E., SHAH, S.M.I., PICKFORD, M.H.L., BISHOP, W.W., THOMAS, H., and JACOBS, L.L. 1977. Geology and paleontology of Neogene strata of Pakistan. *Nature* 270:684–689.

PILBEAM, D.R., BEHRENSMEYER, A.K., BARRY, J.C., and SHAH, S.M.I. 1979. Miocene sediments and faunas of Pakistan. *Postilla* 179:1–45.

PILGRIM, G.E. 1931. *Catalogue of the Pontian Carnivora of Europe.* London: Br. Mus. (Nat. Hist.), Publ. Dept. Geol.

PILGRIM, G.E., and HOPWOOD, A.T. 1928. *Catalogue of the Pon-*

tian Bovidae of Europe. London: Br. Mus. (Nat. Hist.), Publ. Dept. Geol.

QUINN, J.H. 1955. Miocene Equidae of the Texas Gulf Coastal Plain. *Univ. Tex., Bull. Econ. Geol.* 5516:1–102.

RABEDER, G. 1978. 7. Die Säugetiere des Badenian. In *Chronostratigraphie und Neostratotypen Miozän der Zentral Paratethys,* pp. 467–480. Bd. VI, m4–Badenian. Bratislava.

RAY, C. 1976. Geography of phocid evolution. *Syst. Zool.* 25:391–406.

REED, K.M. 1960. Insectivores of the middle Miocene Split Rock local fauna, Wyoming. *Breviora* 116:1–11.

REINHART, R.H. 1976. Fossil sirenians and desmostylids from Florida and elsewhere. *Bull. Fla. St. Mus. Biol. Sci.* 20:187–300.

RENSBERGER, J.M. 1971. Entoptychine pocket gophers (Mammalia, Geomyidae) of the early Miocene John Day Formation, Oregon. *Univ. Calif. Publ. Geol. Sci.* 90:1–209.

RENSBERGER, J.M. 1973A. Pleurolicine rodents (Geomyidae) of the John Day Formation, Oregon, and their relationships to taxa from the early and middle Miocene, South Dakota. *Univ. Calif. Publ. Geol. Sci.* 102:1–95.

RENSBERGER, J.M. 1973B. *Sanctimus* (Mammalia, Rodentia) and the phyletic relationships of the large Arikareean geomyoids. *J. Paleontol.* 47:835–853.

RENSBERGER, J.M. 1979. *Promylagaulus,* progressive aplodontoid rodents of the Early Miocene. *Los Ang. Cty. Mus. Nat. Hist., Contrib. Sci.,* 312:1–18.

REPENNING, C.A. 1967C. Subfamilies and genera of the Soricidae. *U.S. Geol. Surv. Prof. Pap.* 565:1–74.

REPENNING, C.A. 1976. *Enhydra* and *Enhydriodon* from the Pacific Coast of North America. *U.S. Geol. Surv., Jour. Res.,* 4:305–315.

REPENNING, C.A. 1980. Faunal exchanges between Siberia and North America. Proceedings of 5th Biennial Conference, American Quaternary Association, 1978. *Can. J. Anthropol.* 1:37–44.

REPENNING, C.A., RAY, C.E., and GRIGORESCU, D. 1979. Pinniped biogeography. In *Historical Biogeography, Plate Tectonics and the Changing Environment,* ed. J. Gray and A.J. Boucot, pp. 357–369. Corvallis: Oregon State University Press.

REPENNING, C.A., and VEDDER, J.G. 1961. Continental vertebrates and their stratigraphic correlation with marine mollusks, eastern Caliente Range, California. *U.S. Geol. Surv. Prof. Pap.* 424C:235–239.

RICH, P.V., and THOMPSON, E.M., eds. 1982. *The Fossil Vertebrate Record of Australasia.* Monash Univ. (In press.) Clayton, Victoria, Aust.

RICH, T.H.V., and ARCHER M. 1979. *Namilamadeta snideri,* a new diprotodontan (Marsupialia, Vombatoidea) from the medial Miocene of South Australia. *Alcheringa* 3:197–208.

RICH, T.H.V., and RASMUSSEN, D.L. 1973. New World American erinaceine hedgehogs (Mammalia, Insectivora). *Univ. Kans., Mus. Nat. Hist., Occ. Pap.,* 210:1–64.

RICH, T.H.V., and RICH, P.V. 1971. *Brachyerix,* a Miocene hedgehog from North America with a description of the tympanic regions of *Paraechinus* and *Podogymnura. Am. Mus. Novit.* 2477:1–58.

RICHARD, M. 1948. Contribution à l'étude du Bassin d'Aquitaine. Les gisements de mammifères tertiaires. *Soc. Géol. Fr., Mém., Sér. 5,* 52:1–380.

RINGEADE, M. 1978–1021. Micromammifères et biostratigraphie des horizons aquitaniens d'Aquitaine. *Bull. Soc. Géol. Fr.* (7) 20:806–813.

SAHNI, A., and MITRA, H.C. 1980. Neogene palaeobiogeography of the Indian subcontinent with special reference to fossil vertebrates. *Palaeogeogr., Palaeoclim., Palaeoecol.* 31:39–62.

SAVAGE, R.J.G. 1965A. Fossil mammals of Africa: 19. The Miocene Carnivora of East Africa. *Bull. Br. Mus. (Nat. Hist.), Geol.* 10:239–316.

SAVAGE, R.J.G. 1967. Early Miocene mammal faunas of the Tethyan region. *Syst. Assoc. Publ.* 7:247–282.

SAVAGE, R.J.G., and HAMILTON, W.R. 1973. Introduction to the Miocene mammalian fauna of Gebel Zelten, Libya. *Bull. Br. Mus. (Nat. Hist.), Geol.* 22:515–527.

SCHLAIKJER, E.M. 1935C. Contribution to the stratigraphy and paleontology of the Goshen Hole Area, Wyoming. Part 4, New vertebrates and the stratigraphy of the Oligocene and early Miocene. *Bull. Mus. Comp. Zool.* (Harvard Univ.) 86:97–189.

SCHMIDT-KITTLER, N. 1976. Carnivores from the Neogene of Asia Minor. *Palaeontographica* 151:1–131.

SCHULTZ, C.B. 1938B. The Miocene of western Nebraska. *Am. J. Sci.* 35:441–444.

SCHULTZ, C.B., and FALKENBACH, C.H. 1941. Ticholeptinae, a new subfamily of oreodonts. *Bull. Am. Mus. Nat. Hist.* 79:1–105.

SCHULTZ, C.B., and FALKENBACH, C.H. 1954. Desmatochoerinae, a new subfamily of oreodonts. *Bull. Am. Mus. Nat. Hist.* 105:143–256.

SCHULTZ, C.B., and FALKENBACH, C.H. 1968. The phylogeny of the oreodonts. *Bull. Am. Mus. Nat. Hist.* 139:1–498.

SCHULTZ, C.B., MARTIN, L.D., and CORNER, R.G. 1975. Part 1. Middle and late Cenozoic tapirs from Nebraska. *Univ. Nebr., Bull. State Mus.* 10:1–21.

SCHULTZ, C.B., SCHULTZ, M.R., and MARTIN, L.D. 1970. A new tribe of saber-toothed cats (Barbourofelini) from the Pliocene of North America. *Univ. Nebr., Bull. State Mus.* 9:1–31.

SCHULTZ, G.E. 1977. Field conference on Late Cenozoic biostratigraphy of the Texas Panhandle and adjacent Oklahoma. *W. Tex. State Univ., Spec. Publ.* 1:1–160.

SCUDDER, H.I. 1981. Upper Miocene insect deposits in Stewart Valley, Mineral County, Nevada. Unpublished report to U.S. Bureau of Land Management, Carson City Nevada District, on file in Univ. Calif. Mus. Paleontol., Berkeley.

SEN, S., and THOMAS, H. 1979. Découverte de rongeurs dans le Miocène moyen de la formation Hofuf (Province du Hasa, Arabie Saoudite). *C. R. Soc. Géol. Fr.* 1979:34–37.

SEN, S., and UNAY, E. 1978. Cricetodontini (Rodentia, Mammalia) miocènes de Turquie. Evolution et biostratigraphie. *Bull. Soc. Géol. Fr.* (7), 20:837–840.

SHOTWELL, J.A. 1956. Hemphillian mammalian assemblage from northeastern Oregon. *Bull. Geol. Soc. Am.* 67:717–738.

SHOTWELL, J.A. 1967B. Late Tertiary geomyid rodents of Oregon. *Univ. Oreg., Bull. Mus. Nat. Hist.* 9:1–51.

SHOTWELL, J.A. 1968. Miocene mammals of southeast Oregon. *Univ. Oreg., Bull. Mus. Nat. Hist.* 14:1–67.

SHOTWELL, J.A. 1970A. Pliocene mammals of southeast Oregon and adjacent Idaho. *Univ. Oreg., Bull. Mus. Nat. Hist.* 17:1–103.

SHOTWELL, J.A., and RUSSELL, D.E. 1963. 4. Mammalian fauna of the upper Juntura Formation, the Black Butte local fauna. In *The Juntura Basin: Studies in Earth History and Palaeoecology,* ed. J.A. Shotwell, pp. 42–69. *Trans. Am. Philos. Soc.* 53:1–77.

SICKENBERG, O., BECKER-PLATEN, J.D., BENDA, L., BERG, D., ENGESSER, B., GAZIRY, W., HEISSIG, K., HUNERMANN, K.A., SONDAAR, P.Y., SCHMIDT-KITTLER, N., STAESCHE, K., STAESCHE, U., STEFFENS, P., and TOBIEN, H. 1975. Die Gliederung des höheren Jungtertiär und Altquartärs in der Türkei noch Vertebraten und ihre Bedeutung für die internationale Neogen-Stratigraphie. *Geol. Jahrb., Series B*, 15:1–167.

SICKENBERG, O., and TOBIEN, H. 1971. New Neogene and lower Quaternary vertebrate faunas in Turkey. *Newsl. Stratigr.* 1:51–61.

SIGÉ, B. 1968. Les chiroptères du Miocène inférieur de Bouzigues. *Palaeovertebrata* 1:65–133.

SKINNER, M.F. 1968. A Pliocene chalicothere from Nebraska, and the distribution of chalicotheres in the late Tertiary of North America. *Am. Mus. Novit.* 2346:1–24.

SKINNER, M.F., and MacFADDEN, B.J. 1977. *Cormohipparion*, n. gen. (Mammalia, Equidae) from the North American Miocene (Barstovian-Clarendonian). *J. Paleontol.* 51:912–926.

SKINNER, M.F., SKINNER, S.M., and GOERIS, A.J. 1968. Cenozoic rocks and faunas of Turtle Butte, south-central South Dakota. *Bull. Am. Mus. Nat. Hist.* 138:381–436.

SKINNER, M.F., SKINNER, S.M., and GOERIS, A.J. 1977. Stratigraphy and biostratigraphy of late Cenozoic deposits in central Sioux County, western Nebraska. *Bull. Am. Mus. Nat. Hist.* 158:263–270.

SKINNER, M.F., and TAYLOR, B.E. 1967. A revision of the geology and paleontology of the Bijou Hills, South Dakota. *Am. Mus. Novit.* 2300:1–53.

SMART, C.L. 1976. The Lothagam 1 fauna: Its phylogenetic, ecological, and biogeographical significance. In *Earliest Man and Environments in Lake Rudolph Basin*, ed. Y. Coppens et al., pp. 361–369. Chicago: Univ. of Chicago Press.

SOLOUNIAS, N. 1981. Mammalian fossils of Samos and Pikermi. Part 2. Resurrection of a classic Turolian fauna. *Ann. Carnegie Mus.* 50:231–270.

SPENCER, B. 1900A. A description of *Wynyardia bassiana*, a fossil marsupial from the Tertiary beds of Table Cape, Tasmania. *Proc. Zool. Soc. London* 1900:776–795.

STEININGER, F., RÖGL, F., and MARTINI, E. 1976. Current Oligocene/Miocene biostratigraphic concept of the Central Paratethys (Middle Europe). *Newsl. Stratigr.* 4:174–202.

STEVENS, M.S. 1977. Further study of Castolon local fauna (Early Miocene), Big Bend National Park, Texas. *Tex. Mem. Mus., Pearce-Sellards Ser.*, 28:1–69.

STEVENS, M.S., STEVENS, J.B., and DAWSON, M.R. 1969A. New Early Miocene formation and vertebrate fauna, Big Bend National Park, Brewster County, Texas. *Tex. Mem. Mus., Pearce-Sellards Ser.*, 15:1–53.

STIRTON, R.A. 1957B. A new koala from the Pliocene Palankarinna fauna of South Australia. *Rec. S. Austral. Mus.* 13:71–81.

STIRTON, R.A. 1967A. The Diprotodontidae from the Ngapakaldi fauna, South Australia. *Bull. Austral. Bur. Min. Res.* 85:1–44.

STIRTON, R.A. 1967B. A diprotodontid from the Miocene Kutjamarpu fauna. *Bull. Austral. Bur. Min. Res.* 85:45–51.

STIRTON, R.A. 1967C. A new species of *Zygomaturus* and additional observations on *Meniscolophus*, Pliocene Palankarinna fauna, South Australia. *Bull. Austral. Bur. Min. Res.* 85:129–147.

STIRTON, R.A., TEDFORD, R.H., and WOODBURNE, M.O. 1967B. A new Tertiary formation and fauna from the Tirari Desert, South Australia. *Rec. S. Austral. Mus.* 15:427–462.

STIRTON, R.A., WOODBURNE, M.O., and PLANE, M.D. 1967. Tertiary Diprotodontidae from Australia and New Guinea. *Bull. Austral. Bur. Min. Res.* 85:1–160.

STORCH, G. 1978. Die turolische Wirbeltierfauna von Dorn-Dürkheim, Rheinhessen (SW-Deutschland). 2. Mammalia: Insectivora. *Senckenb. Lethaea* 58:421–449.

STORER, J.E. 1970. New rodents and lagomorphs from the upper Miocene Wood Mountain Formation of southern Saskatchewan. *Can. J. Earth Sci.* 9:1125–1129.

STORER, J.E. 1973. The entoptychine geomyid *Lignimus* (Mammalia: Rodentia) from Kansas and Nebraska. *Can. J. Earth Sci.* 10:72–83.

STORER, J.E. 1975. Tertiary mammals of Saskatchewan. Part 3. The Miocene fauna. *Roy. Ont. Mus. Life Sci., Contrib.*, 103:1–133.

SZALAY, F.S., and DELSON, E. 1979. *Evolutionary History of the Primates*. New York: Academic Press, 580 pp.

TASSY, P. 1977. Les mastodontes miocènes du Bassin aquitain: une mise au point taxonomique. *C. R. Acad. Sci., Paris, Sér. D.*, 284:1389–1392.

TASSY, P. 1979. Les proboscidiens (Mammalia) du Miocène d'Afrique orientale—résultats préliminaires. *Bull. Soc. Géol. Fr.* (7) 21:265–269.

TAYLOR, B.E., and WEBB, S.D. 1976. Miocene Leptomerycidae (Artiodactyla, Ruminantia) and their relationships. *Am. Mus. Novit.* 2596:1–22.

TEDFORD, R.H. 1967A. Fossil mammals from the Carl Creek Limestone, northwestern Queensland. *Bull. Austral. Bur. Min. Res.* 92:217–237.

TEDFORD, R.H. 1981. Mammalian biochronology of the late Cenozoic basins of New Mexico. *Bull. Geol. Soc. Am.* 92:1008–1022.

TEDFORD, R.H., and FRAILEY, D. 1976. Review of some Carnivora (Mammalia) from the Thomas Farm local fauna (Hemingfordian), Gilchrist County, Florida. *Am. Mus. Novit.* 2610:1–9.

TEDFORD, R.H., GALUSHA, T., SKINNER, M.F., TAYLOR, B.E., FIELDS, R.F., MACDONALD, J.R., PATTON, T.H., RENSBERGER, J.M., and WHISTLER, D.P. 1983?. [Article on definition and calibration of land-mammal ages of the North American Miocene, in symposium volume on status of vertebrate stratigraphy and geochronology, ed. M.O. Woodburne, Univ. Calif., Publ. Geol. Sci. In press.]

TEDFORD, R.H., and TAYLOR, B.E. 1982?. [Manuscript in preparation on the Canidae (Mammalia, Carnivora).]

THENIUS, E. 1949H. Die Carnivoren von Göriach (Steiermark). *Oesterreich. Akad. Wiss., Math.-Naturwiss. Kl., Sitzungsber., Abt.* 1, 9:695–762.

THOMAS, H. 1979A. *Miotragocerus cyrenaicus* sp. nov. (Bovidae, Artiodactyla, Mammalia) du Miocène supérieur de Sahabi (Libye) et ses rapports avec les autres *Miotragocerus*. *Geobios* 12:267–282.

THOMAS, H. 1979B. Le rôle de barrière écologique de la ceinture Saharo-Arabique au Miocène: arguments paléontologiques. *Bull. Mus. Natl. Hist. Nat.* (Paris), Sér. 4, 1:127–135.

Tobien, H. 1970C. Lagomorpha (Mammalia) im Unter-Miozän des Mainzer Beckens und die Altersstellung der Fundschichten. *Abh. Hess. Landessamt. Bodenforsch.* 56:13–36.

Tobien, H. 1974. Zur Gebiss-struktur, Systematik und Evolution der Genera *Amphilagus* und *Titanomys* (Lagomorpha, Mammalia) aus einingen Vorkommen im jüngeren Tertiär Mittel- und Westeuropas. *Mainzer Geowiss. Mitteil.* 3:95–214.

Tobien, H. 1975. Zur Gebiss-struktur, Systematik und Evolution der Genera *Piezodus, Prolagus* und *Ptycholagus* (Lagomorpha, Mammalia) aus einigen Vorkommen im jüngeren Tertiär Mittel- und Westeuropas. *Notizbl. Hess. Landessamt. Bodenforsch.* 103:103–186.

Tobien, H. 1980. Taxonomic status of some Cenozoic mammalian local faunas from the Mainz Basin. *Mainzer. Geowiss. Mitteil.* 9:203–235.

Toohey, L. 1959. The species of Nimravus (Carnivora, Felidae). *Bull. Am. Mus. Nat. Hist.* 118:71–112.

Topal, G. 1979. Fossil bats of the *Rhinolophus ferrumequinum* group in Hungary. (Mammalia, Chiroptera). *Fragm. Mineral. Palaeontol.* 9.61–101.

Torre, D. 1979. The Ruscinian and Villafranchian dogs of Europe. *Boll. Soc. Paleontol. Ital.* 18:162–165.

Turner, D.L. 1970. Potassium-argon dating of Pacific Coast Miocene foraminiferal stages. *Geol. Soc. Am., Spec. Pap.,* 124:91–129.

Van Couvering, J.A., and Miller, J. A. 1969. Miocene stratigraphy and age determinations, Rusinga Island, Kenya. *Nature* 221:628–632.

Van Couvering, J.H. 1980. Community evolution in Africa during the late Cenozoic. In *Fossils in the Making,* ed. A. K. Behrensmeyer and Hill, pp. 272–298. Chicago: Univ. of Chicago Press.

Van Couvering, J.H., and Van Couvering, J.A. 1976. Early Miocene mammal fossils from East Africa: Aspects of the geology; faunistics and paleoecology. In *Human Origins, Louis Leakey and East African Evidence,* ed. G. Issac and E. McCown, pp. 155–207. Menlo Park, Calif.: Staples Press.

Van de Weerd, A., and Daams, R. 1979. A review of the Neogene rodent succession in Spain. 7th Internat. Congr. Medit. Neogene. *Ann. Geol. Pays Hellen.* 1979:1263–1273.

Vaas, D., and Bagdasarjan, G.P. 1978. A radiometric time scale for the Neogene of the Parathethys region. In *The Geologic Time Scale,* ed. G.V. Cohee, M.F. Glaessner, and H.D. Hedberg, pp. 179–204. Am. Assoc. Petrol. Geol., Studies in Geol. 6. Tulsa.

Vondra, C.F., Schultz, C.B., and Stout, T.M. 1969. New members of the Gering Formation (Miocene) in western Nebraska. *Nebr. Geol. Surv. Pap.* 18:1–18.

Voorhies, M.R. 1969. Taphonomy and population dynamics of an early Pliocene vertebrate fauna, Knox County, Nebraska. *Univ. Wyo., Contrib. Geol., Spec. Pap.,* 1:1–69.

Wagner, H. 1976. A new species of *Pliotaxidea* (Mustelidae: Carnivora) from California. *J. Paleontol.* 50:107–127.

Wahlert, J.H. 1976. *Jimomys labaughi,* a new geomyoid rodent from the early Barstovian of North America. *Am. Mus. Novit.* 2591:1–6.

Walker, A.C. 1969A. True affinities of *Propotto leakeyi* Simpson, 1967. *Nature* 223:647–648.

Webb, S.D. 1969. The Burge and Minnechaduza Clarendonian mammalian faunas of north-central Nebraska. *Univ. Calif. Publ. Geol. Sci.* 78:1–191.

Webb, S.D. 1973. Pliocene pronghorns of Florida. *J. Mammal.* 54:203–221.

Webb, S.D., MacFadden, B.J., and Baskin, J.A. 1981. Geology and paleontology of the Love bone bed, from the late Miocene of Florida. *Am. J. Sci.* 281:513–544.

Webb, S.D., and Taylor, B.E. 1980. The phylogeny of hornless ruminants and a description of the cranium of *Archaeomeryx*. *Bull. Am. Mus. Nat. Hist.* 167:117–158.

West, R.M., Lukacs, J.R., Munthe, J., and Hussain, T. 1978. Vertebrate fauna from Neogene Siwaliks Group, Dang Valley, western Nepal. *J. Paleontol.* 52:1015–1022.

Whitmore, F.C., and Stewart, R.H. 1965. Miocene mammals and Central American seaways. *Science* 148:180–185.

Wilkinson, A.F. 1976. The Lower Miocene Suidae of Africa. In *Fossil Vertebrates of Africa,* ed. R.J.G. Savage and S. Coryndon, pp. 173–282. London and New York: Academic Press.

Wilson, J.A. 1956. Miocene formations and vertebrate biostratigraphic units, Texas Coastal Plain. *Bull. Am Assoc. Petrol. Geol.* 40:2233–2246.

Wilson, J.A. 1957. Early Miocene entelodonts, Texas Coastal Plain. *Am. J. Sci.* 255:641–649.

Wilson, R.L. 1968. Systematics and faunal analysis of a lower Pliocene assemblage from Trego County, Kansas. *Univ. Mich., Mus. Paleontol., Contrib.* 22:75–126.

Wilson, R.W. 1960. Early Miocene rodents and insectivores from northeastern Colorado. *Univ. Kans., Paleontol. Contrib. (Vertebrates)* 7:1–92.

Wilson, R.W. 1968. Insectivores, rodents and intercontinental correlation of the Miocene. Rep. *23 Internat. Geol. Congr.,* Czech., 10:19–25.

Wolfe, J.A. 1964. Miocene floras from Fingerrock Wash, southwestern Nevada. *U.S. Geol. Surv., Prof. Pap.,* 454–N.

Wolfe, J.A. 1978. A paleobotanical interpretation of Tertiary climates in the northern hemisphere. *Amer. Sci.* 66:694–703.

Wolfe, J.A. 1980. Tertiary climates and floristic relationships at high latitudes in the northern hemisphere. *Palaeogeogr., Palaeoclimatol., Palaeoecol.* 30:313–323.

Wood, A.E. 1935C. Evolution and relationships of the heteromyid rodents, with new forms from the Tertiary of western North America. *Ann. Carnegie Mus. Nat. Hist.* 24:73–262.

Wood, H.E., Chaney, R.W., Clark, J., Colbert, E.H., Jepsen, G.L., Reeside, J.B. Jr., and Stock, C. 1941. Nomenclature and correlation of the North American continental Tertiary. *Bull. Geol. Soc. Am.* 52:1–48.

Woodburne, M.P. 1967A. Three new diprotodontids from the Tertiary of the Northern Territory, Australia. *Bull. Austral. Bur. Min. Res.* 85:53–103.

Woodburne, M.O. 1969B. Systematics, biogeography, and evolution of *Cynorca* and *Dyseohyus* (Tayassuidae). *Bull. Am. Mus. Nat. Hist.* 141:271–356.

Woodburne, M.O. 1977. Definition and characterization in mammalian chronostratigraphy. *J. Paleontol.* 51:220–234.

Woodburne, M.O., and Bernor, R.L. 1980. On superspecific groups of some Old World hipparionine horses. *J. Paleontol.* 54:1319–1348.

Woodburne, M.O., and Golz, D.J. 1972. Stratigraphy of the Punchbowl Formation, Cajon Valley, southern California. *Univ. Calif., Publ. Geol. Sci.* 92:1–73.

WOODBURNE, M.O., and ROBINSON, P.T. 1977. A new late Hemingfordian mammal fauna from the John Day Formation, Oregon, and its stratigraphic implications. *J. Paleontol.* 51:750–757.

WOODBURNE, M.O., and TEDFORD, R.H. 1975. The first Tertiary monotreme from Australia. *Am. Mus. Novit.* 2588:1–11.

WOODBURNE, M.O., TEDFORD, R.H., PLANE, M.D., TURNBULL, W., ARCHER, M., and LUNDELIUS, E. 1983?. [Biochronology of the continental mammal record of Australia and New Guinea. In symposium on status of vertebrate stratigraphy and geochronology, ed. M.O. Woodburne, *Univ. Calif. Publ. Geol. Sci.* In press.]

WOODBURNE, M.O., TEDFORD, R.H., STEVENS, M.S., and TAYLOR, B.E. 1974. Early Miocene mammalian faunas, Mojave Desert, California. *J. Paleontol.* 48:6–26.

WOODBURNE, M.O., and WHISTLER, D.P. 1978. An early Miocene oreodont (Merychyinae, Mammalia) from the Orocopia Mountains, southern California. *J. Paleontol.* 47:908–912.

XIE WAN-MING. 1979. First discovery of *Palaeotapirus* in China. *Vertebr. PalAsiatica* 17:146–148.

YAN DE-FA. 1979. Einige der Fossilen Miozänen Säugetiere der Kreis von Fangxian in der provinz Hupei. *Vertebr. PalAsiatica* 17:189–199.

ZAPFE, H. 1969B. *Catalogus Fossilium Austriae,* Primates. Pp. 1–15. Vienna.

ZAPFE, H. 1979. *Chalicotherium grande* (Blainv.) aus der Miozänen Spaltenfuellung von Neudorf an der March (Devinska Nova Ves), Czechoslovakia. *Neue Denkschr. Naturhist. Mus. Wien* 2:1–282.

Chapter Seven

Pliocene Mammalian Faunas

Mammals About 5 to 1.8 Million Years Ago

TYPIFICATION AND CONTENT OF THE PLIOCENE

Sir Charles Lyell (1833) stated that his idea of the *Older Pliocene period* was based on the recentness exemplified by the marine shells taken from the formations termed *Subapennine* in the north of Italy and in Tuscany: "The proportion of *recent* shells [that is, extant species represented in the collections of fossils] usually approaches to one half. . . . Out of 569 species examined from these strata in Italy, 238 were found to be still living, and 331 extinct or unknown."

These "Subapennine formations," now usually referred to the Zanclian (lower) and Piacenzian (upper) stages, overlie the evaporite deposits of the Messinian Stage in the Mediterranean Basin and represent a return to more normal, open-water marine conditions in the ancestral Mediterranean Sea. Berggren and Van Couvering (1974) have given us a comprehensive review of the literature on the stratigraphy and paleontology of the Mediterranean area.

As with the Miocene, we shall not relabor the many problems and controversies that have arisen through the past 150 years about boundaries and content of the Pliocene (= Lyell's Older Pliocene). Rather, we simply accept that this now emaciated series-epoch is a 3-million-year interval from a little more than 5 to a little less than 2 million years ago.

Although the boundaries and the marine-nonmarine correlations are not exact to within less than 300,000 years probably, we are using the *Ruscinian* and *Villafranchian* as the land-mammal ages of the type, western European area, as the nonmarine Pliocene, and as correlatives of the marine Zanclian and Piacenzian intervals. And we add a short, post-Piacenzian and pre-Calabrian (= "late Astian") interval to the marine phenomena of the Pliocene. Ruscinian and Villafranchian, as we are using them, include the Csarnotan, "Rebielce," and Villanyian of Hungary, Czechoslovakia, and Poland. The late Pontian, Dacian, Rumanian, and Cimmerian subages and the Moldavian, Khaprovian, and Asperian faunas of the Ponto-Caspian region are approximately equivalent.

RUSCINIAN

The name *Ruscinian* was concocted from the Roussillon fauna and district of southern France by Kretzoi (1962D), but it was based largely on faunas in eastern Europe that were correlated with Roussillon. (The age was usually called late Pliocene in the literature before about 1977.) See Depéret (1890), Hugueney and Mein (1966), Leinders and Michaux (1969), and Michaux (1976) for paleontological details about Roussillon. Following the lead of Berggren and

Van Couvering (1974), we include Kretzoi's Csarnotan as a later part of the Ruscinian; see Tobien (1970A) and Repenning (1980 and other papers), however, for different ways of subdividing this Pliocene interval.

Roussillon was a poor choice for a standard bearer, for it is a small sample of fossils, and as Michaux (1976) noted, actually consists of two local faunas of significant stratigraphic and age difference. Two of Mein's (1975, 1979) Neogene mammal zones appear to be useful subdivisions of the composite Ruscinian in Europe, and we cite Mein's characterizations of these zones.

MN Zone 15

Localities: Layna, Sète, Perpignan, Gundersheim l, Wölfersheim, Csarnota, Ivanovce, Wèze l, Kvabebi, Salci, Kagul, Odessa.

"Formes caractéristiques de lignées évolutives": Pliopentalagus dietrichi, Mimomys occitanus, Cricetus angustidens, Ruscinomys europaeus, Trilophomys pyrenaicus, Kowalskia intermedia, Pliopetaurista pliocaenica, Parabos boodoṇ, Dicerorhinus miguelcrusafonti, Hipparion fissurae.

Associations: Anancus + Zygolophodon, Mammut + Dicerorhinus megarhinus + Hipparion crassum; complimenting association = Hyaena + Euryboas or Parabos boodon + Sus minor. [Curiously, Mein did not use small-mammal taxa in his associations.]

Appearance: Hypolagus, Pliopentalagus, Therailurus [Dinofelis], Dolichopithecus, Eusyncerus, Toribos, Postschizotherium, Trogontherium, Cseria, Allocricetus, Canis michauxi.

MN Zone 14

Representative localities in Europe and Turkey: Goraffe 1, Montpellier, Terrats, Hautimagne, Hauterives, Osztramos 1, Podlesice, Kossiakino, Kamenskoe, Maritsa, Dinar-Akcaköy.

"Formes caractéristiques de lignées évolutives": Promimomys insuliferus, Cricetus barrieri, Kowalskia magna, K. polonica, Dipoides sigmodus, Rhagapodemus hautimagnensis, Parabos cordieri.

Association: Parabos cordieri + Dicerorhinus megarhinus or Paracervulus australis + Agriotherium insigne.

Appearance: Trilophomys, Promimomys, Mimomys, Baranomys, Rhagapodemus, Nyctereutes, Ursus.

Composite Ruscinian Mammalian Fauna

LAGOMORPHA
  Ochotonidae
    Prolagus michauxi Lopez and Thaler, 1975
    P. figaro Lopez and Thaler, 1975
    P. bilobus Heller, 1936A
    Ochotona sp.   —Bruijn, Dawson, and Mein (1970A)
    O. antiqua Pidoplitshko, 1938
    O.? pusilla? Pallas, 1778   —Kurtén (1968)
    Ochotonoides csarnotanus Kretzoi, 1959A
  Leporidae
    Trischizolagus maritsae Bruijn, Dawson, and Mein, 1970A
    Pliopentalagus dietrichi (Fejfar, 1961C)   —Gureev and Kon'Kova (1967)
    P. moldaviensis Gureev and Kon'Kova, 1967

Hypolagus beremendensis (Petényi, 1864?)
H. brachygnathus Kormos, 1934E
Alilepus
Pliolagus beremendensis Kormos, 1934E
Lepus   —Crusafont and Sondaar (1971)
CARNIVORA
  Ursidae
    Ursus minimus Devèze and Bouillet, 1827
    U. wenzensis Stach, 1953
    Agriotherium insigne (Viret, 1939C)
    A. intermedium Stach, 1957
  Mustelidae
    Pannonictis pliocaenica Kormos, 1931D
    Protarctos?, sp. indet.   —Kretzoi (1962)
    Vormela beremendensis (Petényi, 1864?)
    V. cf. *petenyi* Kretzoi, 1942D
    Baranogale antiqua (Pomel, 1853?)   —Kurtén (1968)
    B. helbingi Kormos, 1934A
    Xenictis pilgrimi (Kormos, 1934A)
    Mustela pliocaenica Stach, 1959
    M. palerminea (Petényi, 1864)   —Kormos (1934A)
    Martes wenzensis Stach, 1959
    Arctomeles pliocaenicus Stach, 1951
  Procyonidae
    Parailurus anglicus Dawkins, 1888
  Canidae
    Canis aff. *etruscus* Forsyth-Major, 1877
    C. michauxi Martin, 1973
    Vulpes praecorsac Kormos, 1932C   —Kurtén (1968)
    V. odessana Odintsev, 1966   (dissertation abstract)
    Nyctereutes donnezani (Crusafont, 1950D)
  Hyaenidae
    Hyaena donnezani Viret, 1954B   —Ficcarelli and Torre (1970A)
    Pachycrocuta perrieri (Croizet and Jobert, 1828)   —Ficcarelli and Torre (1970A)
    Chasmaporthetes borissiaki (Khomenko, 1932)   —Kurtén and Anderson (1980)
  Felidae
    Felis wenzensis Stach, 1961
    "F." christoli (Gervais, 1859)   —Ficcarelli and Torre (1975)
    Leo, sp. indet.   —(Kretzoi, 1962)
    Dinofelis diastemata (Astre, 1929)   —Hemmer (1965)
    "Machairodus cultridens" (Cuvier, 1824)
    cf. *Epimachairodus* sp. indet.   —Kretzoi (1962)
    Lynx brevirostris Croizet and Jobert, 1828
    Lynx   —Crusafont and Sondaar (1971)
  Erinaceidae
    Erinaceus samsonowiczi Sulimski, 1959
  Soricidae
    Blarinoides semseyi Kormos
    Blarinella   —de Bruijn, Dawson, and Mein (1970A)
    Beremendia fissidens (Petényi, 1864)   —Kormos (1934D)
    Zelceina soriculoides (Sulimski, 1959)   —Sulimski (1962)
    Sorex minutus Linnaeus, 1766
    S. subminutus Sulimski, 1962
    S. praearaneus Kormos, 1934D

S. alpinoides Kowalski, 1956

S. runtonensis Hinton, 1911

S. hibbardi Sulimski, 1962

S. fallax Heller, 1936A

S. dehneli Kowalski, 1956

Petenyia hungarica Kormos, 1934D

Petenyiella gracilis (Petényi, 1864)   —Sulimski, Szynkiewicz, and Wolozyn (1979)

P. zelcea (Sulimski, 1959)

P. aff. *repenningi* Bachmayer and R. Wilson (1970A)

Allosorex stenodus Fejfar, 1966B

"Soriculus" kubinyi (Kormos, 1934D)

Episoriculus gibberodon (Petényi, 1864)

Suncus hungaricus

Crocidura kornfeldi Kormos, 1934D

Talpidae

Galemys semseyi Kormos

Talpa praeglacialis Kormos, 1930?   —(syn. of *T. fossilis*, acc. Kretzoi)

T. fossilis Petényi, 1864

T. minor Freudenberg, 1914

Amblycoptus topali Janossy, 1972

Desmana pontica Schreuder, 1940

D. nehringi Kormos, 1913

D. kormosi Schreuder, 1940

CHIROPTERA

Rhinolophidae

Rhinolophus delphinensis Gaillard, 1899

R. grivensis (Depéret, 1892)

R. neglectus Heller, 1936A

R. cf. *variabilis* Topal, 1975

Vespertilionidae

Myotis dasycneme Boie, 1825

M. danutae Kowalski, 1956

M. podlesicensis Kowalski, 1956

M. bechsteini (Kuhl, 1818)

M. kormosi Heller, 1936A   ⎤

M. rapax Heller, 1936D | [Whether or not this great diversity of species from one site is credible, Heller's list does indicate the great abundance of micromammal material from Gundersheim.]

M. cf. *aemulus* Heller, 1936D

M. gundersheimensis Heller, 1936D

M. praevius Heller, 1936D

M. delicatus Heller, 1936D

M. cf. *exilis* Heller, 1936D

M. insignis Heller, 1936D   ⎦

M. helleri 1962A

Plecotus crassidens Kormos, 1930B

P. aff. *auritus* (Linnaeus, 1758)   —Heller (1936A)

Miniopterus schreibersi (Kuhl, 1819)   —Heller (1936A)

PRIMATES

Cercopithecidae   —Szalay and Delson (1979)

Dolichopithecus ruscinensis Depéret, 1889

Macaca cf. *sylvanus prisca* Gervais, 1859

ARTIODACTYLA

Suidae

Propotamochoerus provincialis (Gervais, 1852)

Sus minor Depéret, 1890

Sus? arvernensis Croizet and Jobert, 1828

Camelidae

Paracamelus bessarabiensis (Khomenko, 1912)

Cervidae

Paracervulus australis (Serres, 1832)   —Mein (1979)

Procapreolus wenzensis (Czyzewska, 1960)   —Czyzewska (1968)

Anoglochis ramosus Croizet and Jobert, 1828

Cervus australis (Gervais, 1848–1852)   (*Capreolus?, Euprox?, Paracervulus?*)

C. (*Rusa*)   —Czyzewska (1959)

C. pyrenaicus Depéret

C. cusanus (Croizet and Jobert, 1828)

C. warthae Czyzewska, 1968

C. moldavicum

"C." perrieri (Croizet and Jobert, 1828)

Muntiacus polonicus Czyzewska, 1968

Bovidae

Gazella borbonica Depéret, 1884

Gazella   —Crusafont and Sondaar (1971)

Procamptoceras brivatense Schaub, 1923

Hemitragus stehlini

Eusyncerus

Parabos boodon (Gervais, 1853)

P. cordieri (Christol, 1832)

Toribos

Leptobos?   —Crusafont and Sondaar (1971)

"caprid", indet.   —Crusafont and Sondaar (1971)

PERISSODACTYLA

Equidae

Hipparion crassum Gervais, 1867

H. fissurae Crusafont and Sondaar (1971)

H. rocinantis Hernandez-Pacheco, 1921   —Pirlot (1956)

H. longipes Gromova, 1952   —Sen (1977)

Tapiridae

Tapirus arvernensis Croizet and Jobert, 1828

Rhinocerotidae

Stephanorhinus, sp. indet.   —Kretzoi (1962)

Dicerorhinus megarhinus (Christol, 1835)   ⎫

D. etruscus (Falconer, 1859)   ⎬ *Stephanorhinus?*

D. miguelcrusafonti Guérin and Santafé, 1978   ⎭

HYRACOIDEA

Postschizotherium chardini von Koenigswald, 1932D

PROBOSCIDEA

Gomphotheriidae

Anancus arvernensis (Croizet and Jobert, 1828)

Mammutidae

Mammut borsoni (Hays, 1834)

PHOLIDOTA

Manidae

Manis hungarica Kormos, 1934B

RODENTIA

Eomyidae

Keramidomys carpathicus Schaub and Zapfe, 1953

Sciuridae

Pliopetaurista pliocaenica (Depéret, 1897)

P. dehneli (Sulimski, 1964)   —Mein (1970A)

Pliopetes hungaricus Kretzoi, 1950A

Spermophilinus giganteus de Bruijn, Dawson, and Mein, 1970A

Petinomys?   —Mein (1970A)

Eutamias orlovi Sulimski, 1964

Blackia wolfersheimensis Mein, 1970A

Atlantoxerus rhodius de Bruijn, Dawson, and Mein, 1970A

Cryptopterus thaleri Mein, 1970A

Pliosciuropterus schaubi Sulimski, 1964

Sciurus warthae Sulimski, 1964

Sciurus —David (1967)

Castoridae

Castor praefiber Depéret, 1897

Castor —Michaux (1976)

Boreofiber wenzensis (Sulimski, 1964) —Radulesco and Samson (1972A)

Dipoides sigmodus

Trogontherium minus? Newton, 1890

Cricetidae

Blancomys neglectus Van de Weerd, Adrover, Mein, and Soria, 1977

Cricetus barrierei Mein and Michaux, 1970

C. angustidens Depéret, 1890

C. lophidens de Bruijn, Dawson, and Mein, 1970A

Rhinocricetus —Hartenberger, Michaux, and Thaler (1966)

Mesocricetus primitivus de Bruijn, Dawson, and Mein, 1970A

Kowalskia magna

K. intermedia

K. polonica Fahlbusch, 1969

Cricetulus? —de Bruijn, Dawson, and Mein, 1970A

Microtodon kowalskii (Kretzoi, 1962) —W. von Koenigswald (1980)

M. longidens Kowalski, 1960B

Baranomys kowalskii Kretzoi, 1969

B. loczyi Kormos, 1933

Micromys cf. *praeminutus* Kretzoi, 1959A

M. steffensi Van de Weerd, 1979

M. kozaniensis Van de Weerd, 1979

M. bendai Van de Weerd, 1979

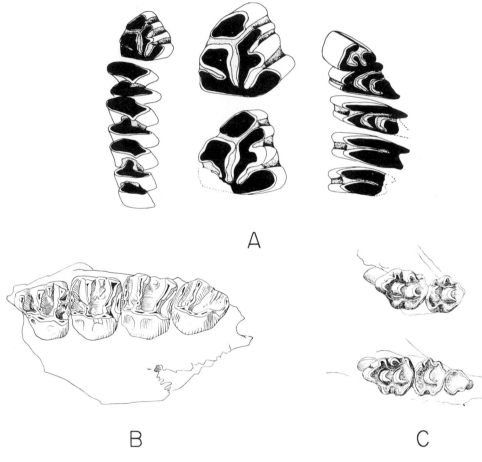

FIGURE 7-1 *Ruscinian Mammals.*

A Prolagus figaro Lopez and Thaler, 1975. Lower P$_3$–M$_3$ and upper P^3–M^2. ×5, approximately. Enlarged views of P$_3$. Figures 1–3 in Lopez and Thaler (1975).

(*Published with permission of the Société géologique de France.*)

B Cryptopterus thaleri Mein, 1970A. P^4–M^3. ×3½, approximately. Figure 35 in Mein (1970A).

(*Published with permission of Geobios, Faculté des Sciences de Lyon.*)

C Stephanomys donnezani (Depéret, 1890). M$^{1-2}$ and M$^{1-3}$. ×3½, approximately. Figures 17 and 18 in Hugueney and Mein (1966).

(*Published with permission of Faculté des Sciences de Lyon.*)

Calomyscus minor de Bruijn, Dawson, and Mein, 1970A
Ruscinomys europaeus Depéret, 1890
Trilophomys pyrenaicus (Depéret, 1890)
T. canterranensis Michaux, 1976
T. depereti Fejfar, 1961B
T. schaubi Fejfar, 1961B
T. vandeweerdi Brandy, 1979
Prosomys insuliferus (Kowalski, 1958A)   —Agad-
   janian and Kowalski (1978)

Ungaromys nanus Kormos, 1932   —Hartenberger,
   Michaux, and Thaler (1966)
Mimomys (Cseria) gracilis (Kretzoi, 1959)
M. (C.) stehlini (Kormos, 1931C)   —Hibbard and
   Zakrzewski (1967)
M. pusillus (Méhely, 1914)

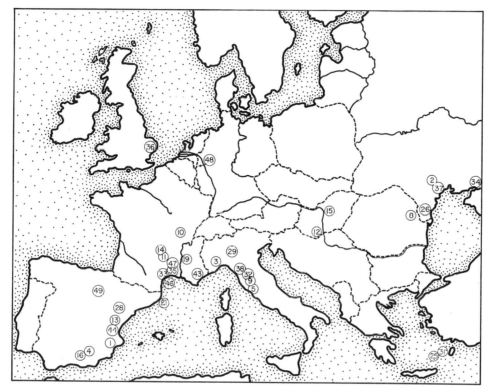

FIGURE 7-2   *Representative Pliocene Localities of Europe (Ruscinian and Villafranchian).*

1 Alcoy—Spain	*21 Iles Medas—Spain*	*38 Olivola—Italy*
2 Anan'yev—Ukraine, USSR	*22 Ivanovce—Poland*	*39 Ostramos 1—Hungary?*
3 Arondelli—Italy	*23 Kadzielna—Poland*	*14 Pardines—France*
4 Arquillo 3—Spain	*24 Kamyk—Poland*	*40 Podlesice—Poland*
5 Baccinello V3—Italy	*25 Karpathos—Greece*	*41 Püspökfürdö (Episcopia)—*
6 Barot—Köpec	*26 Kishinev 2—Ukraine,*	*Hungary?*
7 Beremend—Hungary (in	*USSR*	*3 RDB—Villafranca d'Asti—*
part)	*27 Kislang—Poland*	*Italy*
8 Beresti—Romania	*28 Layna—Spain*	*42 Rebielice Krolewskie—*
9 Casino—Italy	*29 Leffe—Italy*	*Poland*
10 Chagny—France	*30 Malusteni—Romania*	*36 Red Crag—England*
11 Coupet—France	*31 Maritsa—Rhodes, Greece*	*14 Roccaneyra—France*
12 Csarnota—Hungary	*32 Montecarlo—Italy*	*43 St. Vallier—France*
13 Escorihuela—Spain	*32 Montevarchi—Italy*	*44 Sarrion 1—Spain*
14 Etouaires—France	*32 Montopoli—Italy*	*45 Senèze—France*
3 Fornace RDB—Villafranca	*33 Montpellier—France*	*46 Sète—France*
d'Asti—Italy	*4 Moreda 1 and 2—Spain*	*47 Seynes—France*
15 Gödöllö—Hungary	*34 Nikolayevka—Crimea,*	*48 Tegelen—Netherlands*
16 Goraffe 1 and 2—Spain	*USSR*	*32 Val d'Arno—Italy*
17 Gundersheim—Germany	*35 Nîmes—France*	*3 Villafranca d'Asti—Italy*
18 Hajnacka—Czechoslovakia	*36 Norwich Crag—England*	*49 Villaroya—Spain*
19 Hauterives—France	*37 Odessa vicinity—Ukraine,*	*50 Węzé—Poland*
20 Hautimagne—France	*USSR*	*51 Wölfersheim—Germany*

M. occitanus Thaler, 1955B
M. proseki Fejfar, 1961B
M. cappettai Michaux, 1971
M. minor Fejfar, 1961
M. hassiacus Heller, 1936A
M. pliocaenicus (Major, 1902)
M. davakosi Van de Weerd, 1979
Dolomys milleri (Nehring)
D. cf. *nehringi* Kretzoi
D. hungaricus Kormos, 1934D
Germanomys weileri Heller, 1936A
G. trilobodon (Kowalski, 1960B)
G. helleri Fejfar, 1961B
G. parvidens Fejfar, 1961B
} [Identifications disputed; some authors throw these species into *Stachomys* Kowalski or *Ungaromys* Kormos.]
Propliomys hungaricus (Kormos, 1934)
Pliomys kowalskii Schevchenko, 1965B
Leukaristomys vagui Fejfar, 1961B
Clethrionomys? glareolus? —Kurtén (1968)
Laugaritiomys ivanovcensis Fejfar, 1961B (*Stachomys?, Ungaromys?*)
Epimeriones austriacus Daxner-Höck, 1972
E. progressus Kowalski, 1974
Pseudomeriones abbreviatus (Teilhard, 1926)
Lagurus pannonicus Kormos, 1932
Arvicola —Heller (1936A)
Prospalax priscus (Nehring, 1898)
Spalax sotirisi de Bruijn, Dawson, and Mein, 1970A
Pliospalax compositodontus (Topachevski, 1969)
P. macoveii (Simionescu, 1930) —Sen (1977)

Despite the certainty that an international convention of cricetid specialists would be needed to develop a consensus on identification and nomenclature of the Ruscinian cricetids, it is apparent that there was a very great adaptive radiation of members of this family in the European area during (or at the beginning) of Ruscinian time.

Seleviniidae
Plioselevinia gromovi Sulimski, 1962B
Zapodidae
Sminthozapus janossyi Sulimski, 1962B
Muridae
Castillomys crusafonti Michaux, 1969B
C. magnus Sen, 1977
Apodemus sylvaticus (Linnaeus, 1766) —Kurtén (1968)
A. atavus Heller, 1936A
A. cf. *primaevus* Hugueney and Mein, 1965
A. dominans Kretzoi, 1959A
A. jeanteti Michaux, 1967
Anthracomys meini Michaux, 1969B
A. ellenbergeri Thaler, 1966B
Parapodemus lugdunensis Schaub, 1938
P. schaubi Papp, 1947A
P. coronensis Schaub, 1938A
Occitanomys brailloni Michaux, 1969B
O. anomalus de Bruijn, Dawson, and Mein, 1970A
Rhagapodemus hautimagnensis Mein and Michaux, 1970A
R. frequens Kretzoi, 1959A
R. ballesioi Mein and Michaux, 1970A

Paraethomys anomalus de Bruijn, 1970
P. jaegeri Montenat and de Bruijn, 1976
P. meini Montenat and de Bruijn, 1976
Stephanomys donnezani (Depéret, 1890)
S. cf. *minor* Gmelig-Moyling and Michaux, 1973
Pelomys europeus de Bruijn, Dawson, and Mein, 1970A
Orientalomys galaticus Sen, 1977
Hystricidae
Hystrix primigenia (Wagner, 1848) —Sulimski (1960)
Gliridae
Hyponomys —Crusafont and Sondaar (1971)
Glis minor Kowalski, 1956
Muscardinus pliocaenicus Kowalski, 1963
M. aff. *dacicus* Kormos, 1930 —Kowalski (1963)
Glirulus pusillus (Heller, 1936A)
Myomimus maritsensis de Bruijn, Dawson, and Mein, 1970A
Dryomimus —Hartenberger, Michaux, and Thaler (1966)
Eliomys truci Mein and Michaux, 1970
E. intermedius Friant, 1953A

EARLY PLIOCENE OF ASIA

We follow the correlations Chiu, Li, and Chiu (1979) proposed in assigning a few mammalian local faunas known from Shanxi, Yunnan, and Nei Mongol to earlier Pliocene (= Ruscinian, approximately). Also, the Dinar-Akçaköy Faunengruppe of Anatolian Turkey (Becker-Platen, Sickenberg, and Tobien, 1975) is correlative and seems to be composed of species mostly found in the composite Ruscinian mammalian fauna of Europe. Listed here is a composite "Earlier Pliocene" land-mammal fauna for Asia.

Composite "Earlier Pliocene" Mammalian Fauna of Asia

LAGOMORPHA
Ochotonidae
Ochotonoides cf. *complicidens* (Boule and Teilhard, 1928) —Teilhard and Young (1931)
Leporidae
Pliopentalagus cf. *dietrichi* (Fejfar, 1961C)
CARNIVORA
Mustelidae
mustelid, *Mustela*-size
cf. *Enhydrictis* sp.
Felidae
felid, *Lynx*-size
Dinofelis abeli Zdansky, 1926?
INSECTIVORA
Soricidae, spp.
Soriculus (*Episoriculus?*)
Talpidae
Desmanella bifida Engesser, 1980
CHIROPTERA
Vespertilionidae
cf. *Eptesicus*
ARTIODACTYLA
Suidae
Sus minor Depéret, 1890

Cervidae
 cervid 1 (muntjac?; smaller than *Capreolus*)
 cervid 2 (size of *Croizetoceros*)
 cervid 3
Bovidae
 Gazella?
 antilopine
 Kabulicornis ahmadi Heintz and Thomas, 1981
 caprine?
 Leptobos?
 Orchonoceros gromovi Vislobokova, 1979
PERISSODACTYLA
 Equidae
  Hipparion longipes Gromova, 1952
 Rhinocerotidae
  Stephanorhinus?
PROBOSCIDEA, INDET.
RODENTIA
 Eomyidae, gen. indet.
 Cricetidae
  cf. *Rotundomys*
  Promimomys
  Mimomys davakosi Van de Weerd, 1979
  M. steffensi Van de Weerd, 1979
  Micromys bendae Van de Weerd, 1979
  M. kozaniensis Van de Weerd, 1979
  Aratomys mutifidus Zazhigin, 1980?
  Microtoscoptes touvaensis Zazhigin —W. von Koe-
   nigswald (1980)
  Pseudomeriones abbreviatus (Teilhard, 1926) —
   Brunet, Carbonnel, Heintz, and Sen (1980)
  Protatera —Brandy (1979)
 Muridae
  Pelomys orientalus Sen, Brunnet and Heintz, 1979
  Apodemus jeanteti Michaux, 1967
  A. dominans Kretzoi, 1959A
  Parapelomys charkensis Brandy, 1979
  Karnimata afghanensis Brandy, 1979
  Arvicanthis magnus Sen, Brunet and Heintz, 1979
  Mus —Brunet, Carbonnel, Heintz, and Sen (1980)
  Proceromys elongatus Brandy, 1979

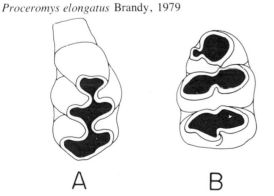

FIGURE 7-3 *Earlier Pliocene Mammals of Asia.*
 A *Proceromys elongatus Brandy, 1979. M¹. ×10½,
  approximately. Figure 7 in Brandy (1979).*
 B *Karnimata afghanensis Brandy, 1979, M². ×10½,
  approximately. Figure 6 in Brandy (1979).*
*(Figures published with permission of Académie des
Sciences, Paris.)*

Rhizomyidae
 Rhizomyoides carbonneli Brandy, 1979
Gliridae
 Myomimus

LANGEBAANIAN

Langebaanian was proposed and characterized by Hendey (1974). Under this designation we list mammalian taxa from miscellaneous local faunas in South Africa, East Africa, and North Africa that seem to be contemporaneous among themselves and are transitional paleontologically between Lothagamian and Makapanian: Langebaanweg 1 and 2, Kaperyon, Lothagam 3, Kubi Algi, Chemeron, Kanapoi, Mursi, Ekora, Aterir, Oued Fouarat, Oued el Akrech. The age name is taken from the outstanding nonmarine-marine fauna of the Varswater Formation at Langebaanweg, southwestern Cape Province (Hendey, 1970A, 1972B, 1974, 1978). Langebaanian faunas show taxonomic intergradation from the preceding *Hipparion* faunas of northern and eastern Africa to the early Makapanian (=*Anancus-Stegodon-Mammuthus-Nyanzachoerus-Hipparion*) faunas. Langebaanian is also the earliest appearance on record of the australopithecine hominids.

Langebaanian Composite Mammalian Fauna

LAGOMORPHA
 Leporidae
  Lepus
MACROSCELIDEA
 Macroscelidae
  Palaeothentoides africanus Stromer, 1932
CARNIVORA
 Ursidae
  Agriotherium africanum Hendey, 1972B (known only from
   Langebaanweg)
 Mustelidae
  Mellivora benfieldi Hendey, 1978–0575
  M. aff. *punjabiensis* Lydekker, 1884
  Enhydriodon africanus Stromer, 1931C —Repen-
   ning (1976)
 Viverridae
  Viverra leakeyi Petter, 1963A
  Genetta
  Herpestes, sp. A —Hendey (1974)
  H., sp. B —Hendey (1974)
 Hyaenidae
  Percrocuta australis Hendey, 1974
  Hyaenictis preforfex Hendey, 1974
  Hyaena namaquensis Stromer, 1931
  H. abronia Hendey, 1974

Felidae
Felis (Sivafelis) obscura Hendey, 1974
Leo
Lynx issiodorensis (Croizet and Jobert, 1828)
Dinofelis diastemata (Astre, 1929)
Machairodus   —R. Savage (1978)
INSECTIVORA   —Butler (1978)
Soricidae, 2 spp.   —Hendey (1974)
Chrysochloridae
Chrysochloris
PRIMATES
Cercopithecidae   —Simons and Delson (1978)
Paracolobus chemeroni R. Leakey, 1969
Macaca
Theropithecus
Parapapio jonesi Patterson, 1968
Papio baringensis R. Leakey, 1969
Hominidae
Australopithecus?
ARTIODACTYLA
Tayassuidae
Pecarichoerus? africanus Hendey, 1976 (known only from Langebaanweg)
Suidae
Nyanzachoerus kanamensis Leakey, 1958
N. pattersoni Cooke and Ewer, 1972
N. jaegeri Coppens, 1971
Hippopotamidae   —Coryndon (1978)
Hexaprotodon harvardi Coryndon, 1977
Hippopotamus?
Giraffidae   —Churcher (1978)
Giraffa cf. *gracilis* Arambourg, 1947
G. cf. *jumae* Leakey, 1965
Sivatherium hendeyi Harris, 1976 (synonym of *S. maurisium?*)
Bovidae   —Gentry (1978)
Gazella aff. *vanhoepeni* Wells and Cooke, 1956
Mesembriportax acrae Gentry, 1974
Kobus aff. *ancystrocera* (Arambourg, 1947)
Raphicerus
Tragelaphus cf. *nukuae* (Arambourg, 1941)
Simatherium?
HYRACOIDEA
cf. *Procavia antiqua* Broom, 1934
PERISSODACTYLA
Equidae   —Churcher (1978)
Hipparion libycum Pomel, 1897 (age?)
H. namaquense (Haughton, 1932A)
H. sitifense Pomel, 1897
H. baardi Boné and Singer, 1965
Rhinocerotidae
Ceratotherium praecox Hooijer and Patterson, 1972
cf. *Brachypotherium*
PROBOSCIDEA   —Coppens, Maglio, Madden, and Beden (1978)
Deinotheriidae
Deinotherium bozasi Arambourg, 1934
Gomphotheriidae

Anancus osiris Arambourg, 1945
A. kenyensis (MacInnes, 1942)
Stegodontidae
Stegodon
Elephantidae
Primelephas gomphotheroides Maglio, 1970B
P. korotorensis (Coppens, 1965)
Elephas ekorensis Maglio, 1970B
Loxodonta adaurora Maglio, 1970B
L. atlantica (Pomel, 1879)
Mammuthus subplanifrons (Osborn, 1928)
Stegotetrabelodon orbus Maglio, 1970B
Stegodibelodon schneideri Coppens, 1972 (syn. of *Stegotetrabelodon?*)
TUBULIDENTATA
Orycteropus
PHOLIDOTA
Cf. *Manis*
RODENTIA   —Lavocat (1978)
Anomaluridae, gen. undet.
Bathyergidae, 2 spp.
Cricetidae
Tatera
Hystricidae
Hystrix
Muridae
Saidomys natrunensis Slaughter and James, 1979

See Figure 7-7 for Langebaanian localities.

NONMARINE PLIOCENE OF NORTH AMERICA

Hibbard (various papers between 1940 and 1974), Hibbard et al. (1965), Lindsay, Johnson, and Opdyke (1975), Lindsay, Opdyke, and Johnson (1980), Repenning (1983 and other papers), Schultz et al. (1978), Gustafson (1978), Bjork (1970A), Taylor (1966), and Dalquest (1978) have been concerned especially with the sequence of Pliocene land-vertebrate and mollusc faunas in North America. But among these workers there has been no exact consensus on intra-American correlation and aging. We follow approximately the chronology Repenning proposed, which he based on species of microtine rodents interpreted to be either immigrant or endemic. And we believe that all the current Pliocene workers in North America can agree on the threefold time subdivision and grouping of North American Pliocene faunas as shown in Figure 7-4.

HEMPHILLO-BLANCAN, OR BLANCAN I

It is difficult to specify mammalian faunas in North America that are equivalent to the Ruscinian of Europe. We lumped all the Hemphillian faunas into Miocene for convenience in organizing this book, but possibly the latest Hemphillian

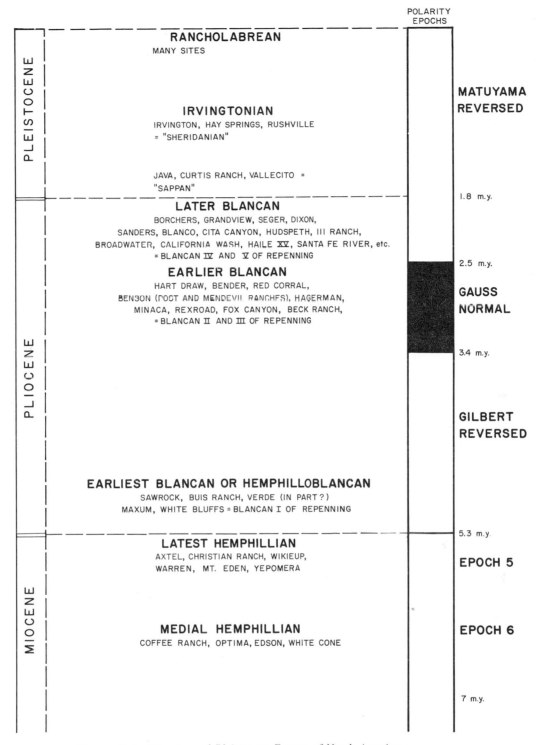

FIGURE 7-4 *Pliocene and Pleistocene Faunas of North America.*

faunas such as Axtel, Wikieup, Mt. Eden, and Yepomera may be equal temporally to part of the Ruscinian. For many years North American mammalian paleontologists believed that there was a significant time hiatus between the Hemphillian and the Blancan faunas; faunal-taxonomic change between the two ages was great. In the 1950s, however, C.W. Hibbard found samples in Kansas and Oklahoma (Saw-rock and Buis Ranch) that are transitional. Unfortunately, there is still no radiometric aging associated with these transitionals, and the array of species is pitifully small. Other samples of land mammals have been found in widely scattered localities in the western United States more recently, and their fossils indicate approximate contemporaneity with Sawrock and Buis Ranch.

Composite Mammalian Fauna of the Hemphillo-Blancan

(= Blancan I)

LAGOMORPHA
  Leporidae
    Hypolagus ringoldensis Gustafson, 1978
    Nekrolagus cf. *progressus* (Hibbard, 1939A)
EDENTATA
  Megalonychidae
    Megalonyx rohrmanni Gustafson, 1978
CARNIVORA
  Ursidae
    Ursus cf. *abstrusus* Bjork (1970A)   —Gustafson (1978)
  Mustelidae
    Buisnictis schoffi Hibbard, 1954A
    Trigonictis cooki (Gazin, 1934C)
  Procyonidae
    Parailurus   —Tedford and Gustafson (1977)
  Canidae
    Osteoborus progressus Hibbard, 1944C   (synonym of *O. hilli* Johnston, 1939A)
    Canis davisi Merriam, 1911B   (from Axtel?)
    Borophagus   —Gustafson (1978)
  Felidae
    Felis   —Gustafson (1978)
    machairodontine, gen. indet.
INSECTIVORA
  Soricidae, gen. unident.
  Talpidae
    Hesperoscalops sewardensis K. Reed, 1962
    Scapanus sp.   —Gustafson (1978)
ARTIODACTYLA
  Tayassuidae
    Platygonus pearcei Gazin, 1938A   —Gustafson (1978)
  Camelidae
    Megatylopus cf. *cochrani* (Hibbard and Riggs, 1949)   —Gustafson (1978)
    Titanotylopus cf. *spatulus* (Cope, 1895A)   —Hibbard
    Hemiauchenia, sp. indet.
  Cervidae
    Bretzia pseudalces Fry and Gustafson, 1974
PERISSODACTYLA
  Equidae
    Equus (*Dolichohippus*) cf. *simplicidens* (Cope, 1892J)   —Gustafson (1978)
    cf. *Nannippus,* sp. unident.   —Hibbard
  Rhinocerotidae, gen. unident.   (Buis Ranch)
PROBOSCIDEA
  Mammutidae
    Mammut americanum (Kerr, 1792)   —Gustafson (1978)
RODENTIA
  Sciuridae
    Spermophilus dotti (Hibbard, 1954A)
    S.? russelli Gustafson, 1978

Marmota sawrockensis Hibbard, 1964
Marmota or *Paenemarmota,* sp. indet.   —Gustafson (1978)
Ammospermophilus hanfordi Gustafson, 1978
Castoridae
  Dipoides wilsoni Hibbard, 1949C
  D. rexroadensis Hibbard and Riggs, 1949   —Gustafson (1978)
  Castor californicus L. Kellogg, 1911A   —Gustafson (1978)
Geomyidae
  Pliogeomys   —Hibbard
  P. buisi Hibbard, 1954A
  Thomomys cf. *gidleyi* R. Wilson, 1933C   —Gustafson (1978)
Heteromyidae
  Perognathus mclaughlini Hibbard, 1949C
  Prodipodomys
Cricetidae
  Baiomys sawrockensis Hibbard, 1953D
  Peromyscus sawrockensis Hibbard, 1964
  P. nosher Gustafson, 1978
  Onychomys larrabeei Hibbard, 1953D
  Cimarronomys stirtoni Hibbard, 1953D
  Mimomys (*Cosomys*) *sawrockensis* (Hibbard, 1957A)   —Repenning (1982)
  M. (*C.*) *primus* R. Wilson, 1932   —Repenning (1982)
  M. (*Ogmodontomys*) *poaphagus* Hibbard, 1941A   —Repenning (1982)
  M. (*Ophiomys*) *mcknighti* Gustafson, 1978
  Pliophenacomys wilsoni Lindsay and Jacobs, 1981   —Repenning (1982)
  P. finneyi Hibbard and Zakrzewski, 1972A   —Repenning (1982)
  Nebraskomys rexroadensis Hibbard, 1970B   —Repenning (1982)
  Pliolemmus antiquus Hibbard, 1937D   —Repenning (1982)
  Neotoma (*Paraneotoma*) *sawrockensis* Hibbard, 1969
  N. cf. *quadriplicatus* (Hibbard, 1941A)   —Gustafson (1978)

Repenning (1983) concludes that 4.8 million years ago *Nebraskomys, Mimomys* (*Ophiomys*), *Mimomys* (*Cosomys*), *Ursus,* certain cervids, *Nekrolagus,* and *Trigonictis* arrived in North America from Eurasia, and that this immigration event marks the beginning of the Blancan mammal age.

See Figure 7-9A,B for Hemphillo-Blancan localities.

MONTEHERMOSAN

The mammal-age name Montehermosan is based on the mammalian fauna from the formation called Monte Hermoso by Argentine stratigraphers. The formation is exposed in sea cliffs near the town of Monte Hermoso, Buenos Aires

Province (Marshall, Hoffstetter, and Pascual, 1982). The Montehermosan age, according to Marshall et al. (1982), is best identified by *Notocynus, Parahyaenodon, Nopachthus, Neocavia, Anchimysops, Neoanchimys, Xenodontomys, Diplasiotherium, Alitoxodon,* and *Trigodon.* Marshall et al. (1982) list a radiometric minimum age for the Montehermosan at Corral Quemado of 3.59 m.y.

The following generic list is taken from Marshall et al. (1982).

Composite Montehermosan Fauna

MARSUPIALIA
  Didelphidae
    Chironectes Illiger, 1811
    Lutreolina Thomas, 1910
    Marmosa Gray, 1821
    Paradidelphys Ameghino, 1904
    Philander Brisson, 1762
    Thylatheridium Reig, 1952D
    Zygolestes Ameghino, 1898
  Sparassocynidae
    Sparassocynus Mercerat, 1899
  Borhyaenidae
    Notocynus Mercerat, 1891
    Parahyaenodon Ameghino, 1904
  Thylacosmilidae
    Achlysictis Ameghino, 1891
    Thylacosmilus Riggs, 1933
  Argyrolagidae
    Argyrolagus Ameghino, 1904
    Microtragulus Ameghino, 1904
EDENTATA
  Dasypodidae
    Chaetophractus? Fitzinger, 1871
    Doellotatus Bordas, 1932
    Macroeuphractus Ameghino, 1887
    Paraeuphractus?
    Ringueletia
    Tolypeutes? Illiger, 1811
    Zaedyus Ameghino, 1889
  Glyptodontidae
    Eleutherocercus Koken, 1888
    Nopachtus Ameghino, 1888
    Palaeodoedicurus Castellanos, 1927
    Urotherium Castellanos, 1926
  Megalonychidae
    Pronothrotherium Ameghino, 1907
  Megatheriidae
    Plesiomegatherium? Roth, 1911
  Mylodontidae
    Proscelidodon Bordas, 1935B
  Myrmecophagidae
    Myrmecophaga Linnaeus, 1758
CARNIVORA
  Procyonidae
    Chapalmalania Kraglievich and Olazabal, 1959
    Cyonasua Ameghino, 1885
LITOPTERNA
  Proterotheriidae
    Brachytherium Ameghino, 1883

    Diplasiotherium Rovereto, 1914
    Eoauchenia Cabrera, 1939D
  Macraucheniidae
    Promacrauchenia Ameghino, 1904
NOTOUNGULATA
  Toxodontidae
    Alitoxodon Rovereto, 1914
    Palaeotoxodon? Ameghino, 1904
    Trigodon Ameghino, 1882
    Xotodon Ameghino, 1887
  Mesotheriidae
    Pseudotypotherium Ameghino, 1904
  Hegetotheriidae
    Paedotherium Burmeister, 1888
    Tremacyllus Ameghino, 1891
RODENTIA
  Cricetidae
    Auliscomys Osgood, 1915
    Bolomys Thomas, 1916
  Octodontidae
    Actenomys Burmeister, 1888
    Eucoleophorus Ameghino, 1909
    Phtoramys Ameghino, 1887
    Pithanotomys Ameghino, 1887
    Pseudoplataeomys   —Rusconi (1938B)
    Xenodontomys Kraglievich, 1927
  Echimyidae
    Cercomys Cuvier, 1832
    Eumysops Ameghino, 1888
  Myocastoridae
    Isomyopotamus Rovereto, 1914
  Dinomyidae
    Telicomys Kraglievich, 1926
  Caviidae
    Cardiomys Ameghino, 1885
    Caviodon? Ameghino, 1885
    Neocavia Kraglievich, 1932
    Orthomyctera Ameghino, 1889
    Palaeocavia Ameghino, 1889
  Chinchillidae
    Lagostomopsis Kraglievich, 1929C
  Hydrochoeridae
    Anchimysops Kraglievich, 1929
    Chapalmatherium Ameghino, 1908
    Neoanchimys Pascual and Bondesio, 1961
    Protohydrochoerus Rovereto, 1914
  Abrocomyidae
    Protabrocoma Kraglievich, 1927

VILLAFRANCHIAN

The nonmarine Villafranchian Stage was proposed by Pareto in 1865; he gave a lithologic description of the stratotype near Villafranca d'Asti, northwestern Italy, and listed four species of proboscideans that had been found in the type area. Hürzeler (1967), Savage and Curtis (1970), Francavilla, Bertolani-Marchetti and Tomadin (1970), Azzaroli and Vialli (1971), and Azzaroli (1977) have supplied further descriptions and summaries of the stratigraphy and paleontology of the type Villafranchian. The stratotype local faunas (Fornace RDB and Arondelli) were grouped as the "Triversa faunal unit" by Azzaroli (1977) and evidently represent early Villafranchian time in an interval about 3.5 million years ago. "Triversa faunal unit" represents an interval predating the Villafranchian that has been applied widely throughout Eurasia and into Africa. This latter, here recognized as later Villafranchian, is the classic "Equus-Elephas-Bos" fauna (=*Equus-Mammuthus-Leptobos*) of the paleontologic-stratigraphic publications of the first half of the Twentieth century and is best-known from the fauna of the upper Val d'Arno, Italy.

Kurtén (1963) and Azzaroli (1977) believed that at least five temporal subdivisions of Villafranchian in its widest sense are recognizable in the European area. Mein (1975, 1979) recognizes three zones in the Villafranchian: MN 16a, MN 16b, and MN 17. Mein's characterizations of these zones are shown below.

MN 17 (late Villafranchian)
Localities: La Puebla de Valverde, Saint Vallier, Villany 3, Kadzielna, Taman, Nogaysk, Psecopus, Yükari-Sögütönü.
Characteristic forms of evolving lineages: *Eucladoceros senezensis vireti, Cervus philisi valliensis, Croizetoceros ramosus medius, Equus stenonis vireti.*
Associations: *Anancus + Mammuthus meridionalis* or *Anancus + Leptobos stenometopon.*
Appearances: *Sciurus, Mimomys pliocaenicus, Mammuthus meridionalis, Eucladoceros, Gallogoral, Gazellospira.*

MN 16b (mid-Villafranchian)
Localities: Montopoli, Etouaires, Roccaneyra.
Characteristic forms of evolving lineages: *"Cervus" pardinensis, Mimomys polonicus, M. (Kislangia) rex.*
Association: *Equus + Hipparion.*
Appearances: *Equus, Mammuthus gromovae, Dicerorhinus etruscus.*

MN 16a (early Villafranchian)
Localities: Triversa = Villafranca d'Asti (Fornace RDB and Arondelli), Escorihuela, Vialette, Seynes, Balaruc II, Khapry.
Characteristic forms of evolving lineages: *Dicerorhinus jeanvireti, Mimomys stehlini, Mimomys capettai.*
Association: *Anancus + Zygolophodon + Tapirus.*
Appearances: *Leptobos, Pliotragus, Arvernoceros, Croizetoceros, Mammuthus gromovae.*

The past history of characterizations such as the above for MN 16 and 17 indicates that there will be many changes made in the near future. For this reason, we have decided to list the complete species ensemble for a well-known fauna in each of these zones.

The type-Villafranchian mammalian fauna (Fornace RDB plus Arondelli)—MN 16a—comprises the following.

LAGOMORPHA
  Ochotonidae
    Prolagus savagei Berzi, 1970
  Leporidae
    Hypolagus cf. *brachygnathus* Kormos, 1934E
CARNIVORA
  Ursidae
    Ursus cf. *minimus* Devèze and Bouillet, 1827
  Procyonidae
    Parailurus cf. *hungaricus* Kormos, 1934 —or *anglicus* Dawkins, 1888
  Mustelidae
    cf. *Baranogale helbingi* Kormos, 1934A
    Enhydrictis, sp. unident.
    mustelid, gen. indet.
  Viverridae
    Viverra cf. *pepraxti* Depéret, 1885
  Hyaenidae
    Chasmaporthetes bielawskyi Schaub, 1942B —Kurtén and Anderson (1980)
    "Hyaena", sp. unident.
  Felidae
    felid, gen. indet.
    Acinonyx, sp. unident.
INSECTIVORA
  Soricidae
    Blarinoides mariae Sulimski, 1959
    Sorex hibbardi Sulimski, 1962A
    Episoriculus gibberodon (Petényi, 1864)
    Petenyia hungarica Kormos, 1934D
    Beremendia fissidens Kormos, 1934D
  Talpidae
    Talpa cf. *minor* Freudenberg, 1914
    T. cf. *fossilis* Petényi, 1864
CHIROPTERA
  Fam. and gen. indet.
PRIMATES
  Cercopithecidae —Szalay and Delson (1979)
    Macaca sylvanus prisca Gervais, 1859
    cf. *Mesopithecus monspessulanus* (Gervais, 1849)
ARTIODACTYLA
  Suidae
    Sus cf. *minor* Depéret, 1890
  Cervid
    cervid, small, cf. *Cervulus* or *Elaphodus* —Hürzeler (1967)
    "Cervus" cf. *cusanus* (Croizet and Jobert, 1828)
    Cervus —Azzaroli (1977)
  Bovidae
    Leptobos stenometopon Rütimeyer, 1866 —Azzaroli (1977)
PERISSODACTYLA
  Tapiridae
    Tapirus arvernensis Croizet and Jobert, 1828
  Rhinocerotidae

Dicerorhinus jeanvireti Guérin, 1972
PROBOSCIDEA
  Mammutidae
    Mammut borsoni (Hays, 1834)
  Gomphotheriidae
    Anancus arvernensis (Croizet and Jobert, 1828)
RODENTIA
  Sciuridae
    sciurids, small
    cf. *Sciuropterus,* sp. unident.
  Castoridae
    Castor, sp. unident.
  Cricetidae
    Mimomys stehlini Kormos, 1931C
    M. gracilis Kretzoi, 1959   —Azzaroli (1977)
  Muridae
    Apodemus cf. *alsomyoides* Schaub, 1938 —Azzaroli (1977)
  Gliridae
    Muscardinus cf. *pliocaenicus* Kowalski, 1963
    Glirulus pusillus (Heller, 1963)
  Hystricidae
    Hystrix, sp. unident.

Early Villafranchian, E. Europe

(Hajňačka, Czechoslovakia —Fejfar, 1964A)
LAGOMORPHA
  Leporidae
    Hypolagus
CARNIVORA
  Procyonidae
    Parailurus hungaricus Kormos, 1934
  Hyaenidae
    Pachycrocuta perrieri (Croizet and Jobert, 1828)
INSECTIVORA
  Soricidae
    Petenyia hungarica Kormos, 1934D
    Beremendia
PRIMATES
  Cercopithecidae
    cf. Colobinae
ARTIODACTYLA
  Suidae
    Sus minor Depéret, 1890
PERISSODACTYLA
  Tapiridae
    Tapirus arvernensis Croizet and Jobert, 1828
  Rhinocerotidae
    Stephanorhinus megarhinus (Christol, 1835)
PROBOSCIDEA
  Gomphotheriidae
    Anancus arvernensis Croizet and Jobert, 1828
  Mammutidae
    Mammut borsoni (Hays, 1834)
RODENTIA
  Sciuridae
    Petauristinae
  Castoridae
    Castor fiber Linnaeus, 1766
    Trogontherium minus Newton, 1890
  Cricetidae

Mimomys hajnackensis Fejfar, 1961B
M. hintoni Fejfar, 1961B
M. kretzoii Fejfar, 1961B
M. pliocaenicus Major, 1902
arvicoline, 2 spp.
prometheomyine
Prospalax priscus (Nehring, 1897)
Baranomys
  Muridae
    Apodemus cf. *atavus* Heller, 1936

Late Villafranchian = MN 17

(Villány 3, —Kretzoi, 1956)
LAGOMORPHA
  Leporidae
    Pliolagus beremendensis (Kormos, 1934E)
    Lagotherium beremendense (Petényi, 1864)   —(includes *Hypolagus brachygnathus* Kormos, acc. to Kretzoi)
CARNIVORA
  Ursidae
    Ursus etruscus Cuvier, 1812
    Ursulus stehlini (Kretzoi, 1954B)   —Kretzoi (1956)
  Mustelidae
    Baranogale beremendensis (Petényi, 1864)
    Vormela petenyii Kretzoi, 1942D
    Mustela palerminea (Petényi, 1864)
    Pannonictis pliocaenica Kormos, 1931D
    Xenictis pilgrimi (Kormos, 1934A)
  Canidae
    Canis mosbachensis Soergel, 1925
    C. lupus Linnaeus, 1758
    Vulpes vulpes? (Linnaeus)
    V. praecorsac Kormos, 1932C
  Felidae
    Felis
    Lynx lynx (Linnaeus, 1766)
    Leo gombaszogensis Kretzoi, 1937
    Epimachairodus hungaricus Kretzoi, 1927
INSECTIVORA
  Erinaceidae
    Erinaceus
  Soricidae
    Sorex runtonensis Hinton, 1911
    S. minutus Linnaeus, 1766
    Beremendia fissidens (Petényi, 1864)
    Petenyia hungarica Kormos, 1934D
    Episoriculus gibberodon (Petényi, 1864)
    Crocidura kornfeldi Kormos, 1934D
  Talpidae
    Talpa fossilis Petényi, 1864   —(includes *T. praeglacialis* Kormos, acc. to Kretzoi)
    T. minor Freudenberg, 1914   —(includes *T. gracilis* Kormos, acc. to Kretzoi)
    Desmana nehringi Kormos, 1913
CHIROPTERA
  Rhinolophidae
    Rhinolophus aff. *ferrumequinum* (Schreber, 1774)

Vespertilionidae
  Vespertilio majori Kormos, 1934D
  Myotis baranensis Kormos, 1934D
  M. steiningeri Kormos, 1934D
  M. schaubi Kormos, 1934D
  M. wüsti Kormos, 1934D
  Eptesicus praeglacialis Kormos, 1934D
ARTIODACTYLA
  Cervidae
    Cervus
    Capreolus, sp. indet.
  Bovidae
    bovid, gen. indet.
    Gazellospira cf. *torticornis* (Aymard, 1855)
    Tragospira cf. *pannonica* Kretzoi
    Procamptoceras cf. *brivatense* Schaub, 1923
    Hemitragus cf. *bonali* Harlé and Schaub, 1913
PERISSODACTYLA
  Equidae
    Equus, sp. indet.
  Rhinocerotidae
    Stephanorhinus etruscus (Falconer, 1859)
PHOLIDOTA
  Manis hungarica Kormos, 1934B   (locality?)
RODENTIA
  Cricetidae
    Rhinocricetus ehiki (Schaub, 1930)
    Mimomys mehelyi Kretzoi
    M. fejervaryi Kormos, 1934D
    M. hungaricus Kormos, 1934D
    M. petenyii Méhely, 1914
    M. pusillus Méhely, 1914
    Kislangia rex (Kormos, 1934D)
    Pliomys episcopalis Méhely
    Clethrionomys (*glareolus* group)   —Kretzoi (1956)
    Lagurodon
    Prospalax priscus (Nehring, 1898)
  Gliridae
    Glis hofmanni Kormos
    Eliomys, sp. indet.
  Muridae
    Apodemus sylvaticus (Linnaeus, 1766)
    A. alsomyoides Schaub, 1938
  Hystricidae
    Hystrix, sp. indet.

Composite Villafranchian Mammalian Fauna

LAGOMORPHA
  Ochotonidae
    Prolagus savagei Berzi, 1970A
    P. ibericus Lopez and Thaler, 1975
    Ochotona? pusilla? Pallas, 1778   —Kurtén (1968)
  Leporidae
    Hypolagus brachygnathus Kormos, 1934E   (a synonym?)
    Lagotherium beremendensis (Petényi, 1864)
    Pliolagus beremendensis (Kormos, 1934E)

Oryctolagus lacosti (Pomel)
Lepus terraerubrae Kretzoi, 1956   (*nomen nudum*)
Trischizolagus dumitrescuae Rădulesco and Samson, 1967C
CARNIVORA
  Ursidae
    Ursus minimus Devèze and Bouillet, 1827
    U. etruscus Cuvier, 1812   —Kurtén (1968)
    Ursulus stehlini (Kretzoi, 1954B)
    Agriotherium insigne (Viret, 1939C)
  Mustelidae
    Baranogale antiqua (Pomel, 1853?)
    B. beremendensis (Petényi, 1864)
    B. helbingi Kormos, 1934
    Vormela beremendensis (Petényi, 1864)
    V. petenyii Kretzoi, 1942D
    Enhydrictis ardea Bravard, 1828
    Enhydriodon? reevi (Newton, 1890)
    Pannonictis pliocaenica Kormos, 1931D
    Xenictis pilgrimi (Kormos, 1934A)
    Mustela stromeri   —Kurtén (1968)
    M. palerminea (Petényi, 1864)   —Heller (1930)
    M. praenivalis Kormos
    Meles thorali Viret, 1954
    Aonyx bravardi (Pomel, 1843)
    Gulo schlosseri Kormos, 1914   —Viret (1939C)
  Procyonidae
    Parailurus anglicus Dawkins, 1888
    P. hungaricus Kormos, 1934
  Canidae
    Canis falconeri Forsyth-Major, 1877
    C. arnensis Del Campana, 1913
    C. lupus Linnaeus, 1758
    C. etruscus Major, 1877   —Kurtén (1968)
    C. mosbachensis Soergel, 1925
    Cuon majori (Del Campana, 1913)
    Vulpes alopecoides Forsyth-Major, 1877
    V. vulpes? (Linnaeus, 1758)
    V. praecorsac Kormos, 1932C
    Nyctereutes megamastoides Pomel, 1854
  Hyaenidae
    Pachycrocuta perrieri (Croizet and Jobert, 1828)   —Ficcarelli and Torre (1970A)
    P. brevirostris (Aymard, 1854?)   —Ficcarelli and Torre (1970A)
    Hyaena hyaena? (Linnaeus, 1766)
    Chasmaporthetes lunensis (Del Campana, 1914B)   —Kurtén and Anderson (1980)
    C. bielawskyi (Schaub, 1942B)   Kurtén and Anderson (1980)
  Felidae
    Viretailurus schaubi (Viret, 1954B)   —Hemmer (1965C)
    Felis sylvestris Schreber   —Ficcarelli and Torre (1974)
    F. pardoides? (Owen, 1846)   —Kurtén (1968)
    Leo gombaszögensis (Kretzoi, 1938)
    Lynx issiodorensis (Croizet and Jobert, 1828)
    L. lynx (Linnaeus, 1766)
    Acinonyx pardinensis (Croizet and Jobert, 1828)
    Homotherium sainzelli Aymard   —Kurtén (1968)
    Megantereon megantereon Croizet and Jobert, 1828
    Epimachairodus hungaricus Kretzoi, 1938

INSECTIVORA
  Erinaceidae
    Erinaceus lechei Kormos, 1934D
  Soricidae
    Sorex hibbardi Sulimski, 1962
    S. cf. *praealpinus* Heller, 1930
    S. minutus Linnaeus, 1766
    S. subminutus Sulimski, 1962
    S. runtonensis Hinton, 1911
    S. praearaneus Kormos, 1934D
    Blarinoides mariae Sulimski, 1959
    Episoriculus gibberodon (Petényi, 1864)
    Petenyia hungarica Kormos, 1934D
    Beremendia fissidens Kormos, 1934D
    "Soriculus" kubinyi (Kormos, 1934D)
    Suncus hungaricus
    S. pannonicus (Kormos, 1934D)
    Crocidura kornfeldi Kormos, 1934D
  Talpidae
    Talpa fossilis Petényi, 1864
    T. minor Freudenberg, 1914
    T. gracilis Kormos   (synonym?)
    T. praeglacialis   (synonym?)
    Desmana kormosi Schreuder, 1940
    D. nehringi Kormos, 1913
    Galemys   —Heintz, Guérin, Martin, and Prat
      (1974)
CHIROPTERA
  Rhinolophidae
    Rhinolophus euryale Blasius, 1857   —Kurtén (1968)
    R. delphinensis Gaillard, 1899
    R. cf. *ferrumequinum* (Schreber, 1774)
  Vespertilionidae
    Myotis dasycneme Boie, 1825
    M. wüsti Kormos, 1934D
    M. steiningeri Kormos, 1934D
    M. schaubi Kormos, 1934D
    M. baranensis Kormos, 1934D
    Plecotus crassidens Kormos, 1930B
    Miniopterus schreibersi Kuhl, 1819
    Vespertilio majori Kormos, 1934D
    Eptesicus praeglacialis Kormos, 1934D
PRIMATES
  Cercopithecidae   —Szalay and Delson (1979)
    Macaca sylvanus prisca Gervais, 1859
    Paradolichopithecus arvernensis (Depéret, 1929A)
      —Necrasov, Samson, and Rădulesco
      (1961)
    Mesopithecus monspessulanus (Gervais, 1849)
ARTIODACTYLA
  Suidae
    Sus arvernensis Croizet and Jobert, 1828
    S. strozzii Meneghini (Major, 1881?)   —Azzaroli
      (1954)
    S. minor Depéret, 1890
  Hippopotamidae
    Hippopotamus amphibius Linnaeus
  Camelidae
    Paracamelus alutensis   —Havesson (1954B)
  Cervidae
    Arvernoceros ardei (Croizet and Jobert, 1828)

Cervus etuerarium (Croizet and Jobert, 1828)
"Cervus" philisi (Schaub, 1942A)   (1941?)
"C." issiodorensis (Croizet and Jobert, 1828)
"C." perolensis (Azzaroli, 1952)
"C." cusanus (Croizet and Jobert, 1828)
"C." pardinensis (Croizet and Jobert, 1828)
"C." perrieri (Croizet and Jobert, 1828)
Anoglochis ramosus (Croizet and Jobert, 1828)
  (*Croizetoceros?*)
Euctenoceros ctenoides (Nesti, 1841)   —Azzaroli
  (1953)
E. senezensis (Depéret, 1910)
Eucladoceros dicranios (Nesti, 1879)   —same as
  Euctenoceros?   —Azzaroli (1953)
E. falconeri (Dawkins, 1868)
E. tetraceros (Dawkins, 1878)
Praemegaceros verticornis Dawkins, 1872
Dama nestii (Forsyth-Major)
Capreolus, sp. indet.
Libralces gallicus Azzaroli, 1952
  Bovidae
    Gazella borbonica Depéret, 1884
    G. julieni Munier-Chalmas, 1889
    Gazellospira torticornis (Aymard, 1855)
    Procamptoceras brivatense Schaub, 1923
    Gallogoral meneghinii (Rütimeyer, 1878)
    Pliotragus ardeus (Depéret, 1884)
    Deperetia ardea (Depéret)   —Kurtén (1968)
    Megalovis latifrons Schaub, 1923
    Ovis   —Schaub (1944C)
    Hesperidoceras merlae Crusafont, 1965E
    Hemitragus stehlini
    H. cf. *bonali* Harlé and Stehlin, 1913
    Leptobos elatus Pomel, 1853
    L. etruscus Falconer, 1868
    L. stenometopon Rütimeyer, 1866
    Syncerus? iselini?   —Kurtén (1968)
    Tragospira cf. *pannonica* Kretzoi
    Myotragus batei Crusafont and Angol, 1966
PERISSODACTYLA
  Equidae
    Equus stenonis Cocchi, 1867
    E. bressanus Viret, 1954
    E. hydruntinus? Regalia, 1904
    Hipparion crusafonti Villalta, 1948
    H. rocinantis Forsten, 1968
  Tapiridae
    Tapirus arvernensis Croizet and Jobert, 1828
  Rhinocerotidae
    Stephanorhinus etruscus (Falconer, 1859)
    S. jeanvireti (Guérin, 1973)   (or *Dicerorhinus*)
PROBOSCIDEA
  Gomphotheriidae
    Anancus arvernensis (Croizet and Jobert, 1828)
  Mammutidae
    Mammut borsoni (Hays, 1834)
  Elephantidae
    Mammuthus meridionalis (Nesti, 1825)

M. gromovi Baĭgusheva, 1971A
PHOLIDOTA
    Manis hungarica Kormos, 1934B
RODENTIA
  Sciuridae
    Spermophilus primigenius (Kormos, 1934D)
    S. nogaici (Topachevsky, 1957B)
    Pliosciuropterus schaubi Sulimski, 1964
  Castoridae
    Castor fiber Linnaeus, 1766
    Trogontherium cuvieri Fischer, 1812
    T. minus Newton, 1890
    Romanocastor filipescui Rădulesco and Samson, 1967
    R.? capiniensis Rădulesco and Samson, 1972
    Zamolxifiber covurluiensis (Simonescu, 1930C)
      —Rădulesco and Samson (1972)
  Cricetidae
    Microtus   —Kretzoi and Vértes (1965)
    Cricetus nanus Schaub, 1930
    Allocricetus bursae Schaub, 1930B
    Baranomys loczyi Kormos, 1933
    Rhinocricetus ehiki (Schaub, 1930)   —W. von Ko-
      enigswald (1980)
    Dolomys milleri Nehring, 1898
    D. hungaricus Kormos, 1934D
    Clethrionomys hintoni Kormos, 1934D
    C. kretzoii (Kowalski, 1958A)
    C. glareolus (Schreber, 1792)
    Mimomys (*Cseria*) *gracilis* Kretzoi, 1959
    M. (C.) stehlini Kormos, 1931C
    M. pusillus (Méhely, 1914)
    M. hintoni Fejfar, 1961B
    M. reidi Hinton, 1910
    M. newtoni (Nehring, 1898)
    M. pliocaenicus Major, 1902
    M. intermedius Newton, 1881
    M. fejervaryi Kormos, 1934D
    M. ostramosensis Jánossy and Van der Meulen, 1976
    M. petenyii Méhely
    M. hungaricus Kormos
    M. polonicus Kowalski, 1960A
    M. cappettai Michaux, 1971
    M. mehelyi Kretzoi
    M. milleri Kretzoi, 1958
    M. savini Hinton, 1910

Lagurus pannonicus Kormos, 1930   —(*Lagurodon*
  sp., acc. to Kretzoi, 1956)
Allophaiomys pliocaenicus Kormos, 1932D
Kislangia rex (Kormos, 1934D)   —Kretzoi (1956)
Trilophomys
Praesynaptomys europaeus Kowalski, 1977
Prospalax priscus (Nehring, 1898)
Pliomys episcopalis Méhely, 1914
Muridae
  Apodemus sylvaticus (Linnaeus, 1766)
  A. flavicollis (Melchior, 1836)
  A. alsomyoides Schaub, 1930
  Parapodemus coronensis Schaub, 1938
  Rhagapodemus frequens Kretzoi, 1959A
Gliridae
  Eliomys sp. indet.   —Kretzoi (1956)
  Glirulus pusillus (Heller, 1963)
  Muscardinus pliocaenicus Kowalski, 1963
  M. avellanarius (Linnaeus, 1776)   —Kowalski (1964)
  Glis minor Kowalski, 1956
  G. hofmanni Kormos
Hystricidae
  Hystrix refossa Gervais, 1853
  H. major Gervais, 1895
  H. etrusca Bosco, 1898
Zapodidae
  Sicista cf. *praeloriger* Kormos, 1930B

See Figure 7-2 for Villafranchian localities.

VILLAFRANCHIAN OF TURKEY

Early Villafranchian Mammalian Fauna of Turkey

(Gülyazi Faunengruppe of Becker-Platen, Sickenberg and
Tobien, 1975.)

LAGOMORPHA
  Leporidae, gen. indet.
CARNIVORA
  Mustelidae
    Enhydrictis or *Pannonictis*
  Canidae
    Vulpes odessana Odintsov, 1967
    V. alopecoides Forsyth-Major, 1877
  Hyaenidae
    Pachycrocuta perrieri (Croizet and Jobert, 1828)

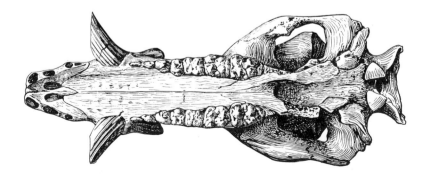

FIGURE 7-5 *Villafranchian Mammal.
Sus strozzii F.-Major, 1881. Skull.
×1/4, approximately. Figure 4b,
Plate XV (VII) in Azzaroli (1954).
(Published with permission of
Palaeontographia Italica, Pisa.)*

Chasmaporthetes lunensis (Del Campana, 1914B)
Felidae
    Lynx issidorensis (Croizet and Jobert, 1828)
    Dinofelis?
    Homotherium?
INSECTIVORA
  Soricidae
    cf. *Petenyia*
  Talpidae
    cf. *Desmana kormosi* Schreuder, 1940
ARTIODACTYLA
  Suidae
    Potamochoerus   —Maglio (1973)
  Camelidae
    Paracamelus cf. *alexejevi* Iatsko, 1962B
  Cervidae
    Eucladocerus? n. sp.
    cervid, size of *Cervus philisi* (Schaub, 1942A)
  Giraffidae
    Macedonitherium martini Sickenberg, 1967A
  Bovidae
    "Gazella" cf. *sinensis*
    Gazella
    Spiroceros?
    Rupicaprinae, gen. indet.
    Caprinae, gen. indet.
    Leptobos?
PERISSODACTYLA
  Equidae
    Hipparion cf. *crusafonti* Villalta, 1948
  Rhinocerotidae
    Stephanorhinus megarhinus (Christol, 1835)
PROBOSCIDEA
  Gomphotheriidae
    Anancus
RODENTIA
  Castoridae
    Trogontherium minus Newton, 1890

FIGURE 7-6 *Pliocene Localities of Turkey.*
(=*Ruscinian and* =*Villafranchian*)
  1 Akcayir (Eskişehir)
  2 Babadat
  3 Beçin (Kotanyaka)
  4 Belekler (Konya)
  5 Datça
  6 Delibayir Siirti
  7 Elbistan (Maras)
  8 Gülyazi
  9 Hasanpaşa
  10 Kalinköy
  11 Kamişli
  12 Kavakdere
  13 Köprübasi
  14 Tekkeköy (Hüyük)
  15 Tekman
  16 Yukari Sögütönü

Pliocene Locality of Israel
(=*Villafranchian*).
  17 Bethlehem

Cricetidae
    cf. *Allocricetus* or *Cricetulus*
    Mimomys polonicus Kowalski, 1960A
    M. septimanus Michaux, 1971
    Pliomys

Late Villafranchian Mammalian Fauna of Turkey

(Yukari-Söğütönü Faunengruppe of Becker-Platen, Sickenberg, and Tobien, 1975.)

LAGOMORPHA
  Leporidae
CARNIVORA
  Canidae
    Canis sp., size of *C. etruscus* Forsyth-Major, 1877
    Vulpes
INSECTIVORA
  Soricidae, gen. indet.
ARTIODACTYLA
  Camelidae
    Paracamelus
  Cervidae
    cf. *Eucladoceros*
  Bovidae
    Gazella borbonica Depéret, 1884
    cf. *Leptobos*
PERISSODACTYLA
  Equidae
    Hipparion
    Equus stenonis Cocchi, 1867
  Rhinocerotidae
    Stephanorhinus sp., small form
PROBOSCIDEA
  Gomphotheriidae
    Anancus arvernensis (Croizet and Jobert, 1828)

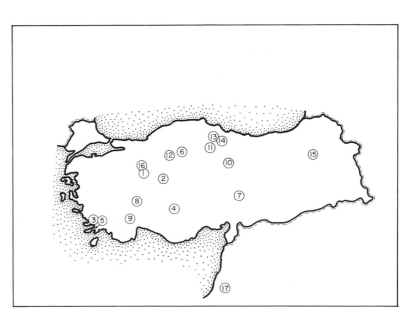

Elephantidae
Elephas planifrons Falconer and Cautley, 1846
Mammuthus meridionalis (Nesti, 1825)
RODENTIA
Cricetidae
Mimomys, sp. 1
Mimomys, sp. 2
Muridae
Orientalomys galaticus Sen, 1975

LATE PLIOCENE (= VILLAFRANCHIAN) OF ASIA

The following composite faunal list of (= Villafranchian) mammals from Asia is taken from Chow and Zhang (1974), Tang et al. (1974), Liu, Tang, and You (1973), Liu and You (1974), Liu, Li, and Zhai (1978), Pope (1982), Colbert (1935A), Pilgrim (1937, 1939A), Hooijer (1958B), and the references cited below. The list comprises genera and species from Yushe, Ling-Yi, Xihoudu, Nihewan, and Xin-She-Chang in China, from the Tatrot and Pinjor "zones" in Pakistan and India, from Afghanistan, and from Bethlehem in Israel. Probably we have missed many proposed species of this age from the USSR. We have included all the species from Nihewan and the Pinjor with the knowledge, following Pope (1982) and others, that these latter may be partly post-Villafranchian.

Late Pliocene Mammalian Fauna of Asia

LAGOMORPHA
Ochotonidae
Ochotona complicidens Boule, Breuil, Licent, and Teilhard, 1928
O. lagreli Bohlin
Leporidae
Caprolagus sivalensis Major, 1899
CARNIVORA
Ursidae
Agriotherium sivalense (Falconer and Cautley, 1836)
Melursus? theobaldi (Lydekker, 1884)
Ursus cf. *etruscus* Cuvier, 1812
Mustelidae
Sinictis lydekkeri Pilgrim, 1932
Mellivora sivalensis (Falconer and Cautley, 1868)
Lutra palaeindica Falconer and Cautley, 1868
L. licenti Teilhard and Piveteau, 1930
Enhydriodon sivalensis Falconer, 1868
Mustela pachygnatha Teilhard and Piveteau, 1930
Meles cf. *leucurus* Hodgson, 1847 —Teilhard and Young (1936A)
Canidae
Canis cautleyi Bose, 1880
Sivacyon curvipalatus (Bose, 1880)
Vulpes
Nyctereutes sinensis (Schlosser, 1903)
N. chihliensis (Zdansky, 1924)

Viverridae
Viverra bakeri Bose, 1880
Vishnuictis durandi (Lydekker, 1884)
Hyaenidae
"Hyaena" sinensis (Owen, 1870)
Hyaenictis bosei Matthew, 1929
Pachycrocuta felina (Bose, 1880) —Ficcarelli and Torre (1970)
Crocuta sivalensis (Falconer and Cautley, 1868)
C. colvini (Lydekker, 1884)
Felidae
Homotherium nihowanensis (Teilhard and Piveteau, 1930)
Megantereon palaeindicus (Bose, 1880)
M. falconeri Pomel, 1853
Felis subhimalayana Brown, 1848
Dinofelis cristata (Falconer and Cautley, 1868) — Kurtén and Anderson (1980)
Sivafelis potens Pilgrim, 1932
S. brachygnathus (Lydekker, 1884)
INSECTIVORA
Erinaceidae
Erinaceus cf. *dealbatus* Swinhoe, 1870 —Pope (1982)
PRIMATES —Szalay and Delson (1979)
Cercopithecidae
Procynocephalus?
Presbytis palaeindicus (Lydekker, 1884)
Papio subhimalayanus (Meyer, 1848)
P. falconeri (Lydekker, 1886)
Hominidae
Sivapithecus? indicus Pilgrim, 1910?
Pongo cf. *satyrus* Linnaeus, 1766)
Ramapithecus punjabicus (Pilgrim, 1910)
ARTIODACTYLA
Suidae
Sivachoerus giganteus (Falconer and Cautley, 1847)
Tetraconodon magnus Falconer, 1868
T. mirabilis Pilgrim, 1926
Potamochoerus theobaldi Pilgrim, 1926
P. palaeindicus Pilgrim, 1926
Dicoryphochoerus titanoides Pilgrim, 1926
D. vagus Pilgrim, 1926
D. durandi Pilgrim, 1926
Hippohyus sivalensis Falconer and Cautley, 1840–1845
Sus hysudricus Falconer and Cautley, 1847
S. peregrinus Pilgrim, 1926
S. bakeri Pilgrim, 1926
S. falconeri Lydekker, 1884
S. cautleyi Pilgrim, 1926
Anthracotheriidae
Merycopotamus dissimilis Falconer and Cautley, 1836
M. nanus Falconer and Cautley, 1847
Hippopotamidae
Hippopotamus sivalensis Falconer and Cautley, 1836
Camelidae
Paracamelus sivalensis (Falconer and Cautley, 1849)
P. gigas —Orlov (1929)
Tragulidae
Dorcatherium
Cervidae
Axis shansius Teilhard and Trassaert, 1937B
A. rugosus Chow M. and Chow B., 1965

Cervus bifurcatus Teilhard and Piveteau, 1930
C. sivalensis Lydekker, 1880
C. chinanensis
C. punjabicus Brown, 1926
C. boulei Teilhard and Piveteau, 1930
C. simplicidens Lydekker, 1876
C. elegans Teilhard and Piveteau, 1930
C. triplidens Lydekker, 1876
Eucladoceros or *Euctenoceros* cf. *tetraceros* Heintz, 1968
Cervulus sinensis Teilhard and Piveteau, 1930
Giraffidae
Sivatherium giganteum Falconer and Cautley, 1836
Indratherium majori Pilgrim, 1910
Giraffa sivalense (Falconer and Cautley, 1843)
G. affinis (Falconer and Cautley, 1843)
Bovidae
Kabulicornis ahmadi Heintz and Thomas, 1981
Antilope subtorta Pilgrim, 1937
Pachyportax latidens (Lydekker, 1876)
Sivaportax dolabella Pilgrim, 1939A
Proamphibos hasticornis Pilgrim, 1939A
P. kasmiricus Pilgrim, 1939A
P. lachrymans Pilgrim, 1939A
Leptobos falconeri Rütimeyer, 1877
L. crassus
Hemibos acuticornis (Rütimeyer, ex. MS Falconer, 1866)
H. occipitalis (Lydekker, 1878)
Bucapra daviesi Rütimeyer, 1877
Platybos platyrhinus Lydekker, 1878
Bos acutifrons Lydekker, 1878
Bison sivalensis (Falconer, 1868)
B. palaeosinensis
Hydaspicobus auritus Pilgrim, 1939A
Vishnucobus patulicornis Pilgrim, 1939A
Indoredunca sterilis Pilgrim, 1939A
I. theobaldi Pilgrim, 1939A
Gangicobus asinalis Pilgrim, 1939A
Sivadenota biforis Pilgrim, 1939A
S. sepulta Pilgrim, 1939A
Sivacobus palaeindicus Pilgrim, 1939A
Sivatragus bohlini Pilgrim, 1939A
S. brevicornis Pilgrim, 1939A
Sivoryx cautleyi Pilgrim, 1939A
S. sivalensis (Lydekker, 1878)   —Pilgrim (1939A)
Hippotragus?
Damalops palaeindicus (Falconer, 1859)   —Pilgrim (1939A)
Antilospira licenti Teilhard and Young, 1931
Gazellospira torticornis (Aymard)
Spirocerus falconeri Boule, Breuil, Licent, and Teilhard, 1928
S. wongi Teilhard and Piveteau, 1930
Dorcadoryx triguetricornis Teilhard and Trassaert, 1938
Gazella cf. *blacki* Teilhard and Young, 1931
G. cf. *sinensis* Teilhard and Piveteau, 1930
G. cf. *subgutturosa* Teilhard and Piveteau, 1930
Sivacapra crassicornis Pilgrim, 1939A
Ovis shantungensis
PERISSODACTYLA
Equidae
Hipparion sinense Teilhard and Piveteau, 1930
Equus sivalensis Falconer and Cautley, 1849

E. namadicus Falconer and Cautley, 1849
E. sanmeniensis Teilhard and Piveteau, 1930
E. huanghoensis Chow-M. and Chow-B., 1965
Rhinocerotidae
Coelodonta platyrhinus (Falconer and Cautley, 1847)
C. antiquitatis Blumenbach, 1803
Rhinoceros sivalensis (Falconer and Cautley, 1847)
R. palaeindicus Falconer and Cautley, 1847
R. cf. *sinensis* Owen, 1870
R. cf. *tichorhinus* Cuvier, 1834
Elasmotherium cf. *inexpectatum* Chow, 1958C
Chalicotheriidae
Nestoritherium sivalense Falconer and Cautley, 1843
HYRACOIDEA
Postschizotherium chardini v. Koenigswald, 1932D
P. intermedium v. Koenigswald, 1966B
PROBOSCIDEA
Gomphotheriidae
Pentalophodon sivalensis (Cautley, 1836)   (included in *Anancus* by some)
P. falconeri Osborn, 1935   (included in *Anancus* by some)
Stegodontidae
Stegolophodon stegodontoides? Pilgrim, 1913
Stegodon bombifrons (Falconer and Cautley, 1847)
S. ganesa (Falconer and Cautley, 1845)
S. insignis (Falconer and Cautley, 1845)
S. pinjorensis Osborn, 1929
S. primitium Liu, Tang, and You, 1973
S. preorientalis
S. zhaotongensis
Elephantidae
Elephas planifrons (Falconer and Cautley, 1845)
E. hysudricus (Falconer and Cautley, 1845)
E. platycephalus Osborn, 1929
RODENTIA
Castoridae
Trogontherium
Cricetidae
Mimomys
Myospalax tingi (Young, 1927)   —Chow and Li (1965B)
M. fontanieri (Milne-Edwards, 1868) —Chow and Li (1965B)
Rhizomyidae
Rhizomys
Hystricidae
Hystrix cf. *leucurus* Sykes, 1831
Dipodidae
Alactaga annulata Milne-Edwards, 1876   —Pope (1982)
Muridae
Nesokia hardwicki (Gray, 1837)
Orientalomys nihowanius Zheng

MAKAPANIAN

Makapanian (Hendey, 1974) was based on South African cave localities. There remains considerable uncertainty about the relative age of various pockets of fossils at the different cave sites, and some of the species we list may be early Pleistocene. We apply Makapanian to all Africa and believe it includes the "late Pliocene Villafranchian" faunas such as Lower Omo and Lower Kaiso, which are without *Equus* and are with *Anancus*, *Stegodon*, *Nyanzachoerus*, and *Mammuthus subplanifrons*. It also includes the "early Pleistocene Villafranchian" faunas that show *Equus*, *Hipparion*, and *Elephas recki* (Cooke and Corydon, 1970). Representative localities are Shungura Formation sites up to Member F, Lower Koobi Fora, Kaiso, Brown and White Sands of the Omo, Laetoli, Kanam, Ouadi Derdemi, Koula, Olduvai I, Ta'ung, Sterkfontein (older), Makapansgat, Bolt's Farm (older?), Ichkeul, Aïn Brimba, St. Arnaud, and Aïn Jourdel.

Composite Makapanian Mammalian Fauna

LAGOMORPHA
  Leporidae
    Lepus   —Leakey (1964)
    Pronolagus cf. *randensis*
MACROSCELIDEA
  Macroscelidae
    Elephantulus fuscus (Peters, 1852)
    E. broomi Corbet and Hanks, 1968
    E. antiquus Broom, 1948
    Macroscelides proboscideus Shaw, 1800
    Mylomygale spiersi Broom, 1948
    Nasilio cf. *brachyrhynchus* (A. Smith, 1838)
CARNIVORA
  Mustelidae
    Aonyx capensis Schinz, 1821
    Enhydriodon
    Lutra
  Canidae
    Canis africanus Pohle, 1928
    C. mesomelas Schreber, 1775
    C. brevirostris Croizet *in* Blainville, 1845
    Vulpes pattisoni Broom, 1948
    Otocyon recki (Pohle, 1928)
  Viverridae
    Viverra leakeyi Petter, 1963A
    Genetta genetta (Linnaeus, 1766)
    Civettictis civetta (Schreber, 1766)
    Pseudocivetta ingens Petter, 1966
    Herpestes primitivus Petter, 1973
    H. delibis Petter, 1973
    Helogale sp.
    Mungos dietrichi Petter, 1973
    M. minutus Petter, 1973
    Crossarchus transvaalensis Ewer
    Cynictis penicillata (Cuvier, 1829)
  Hyaenidae
    Hyaena brunnea Thunberg, 1820
    H. hyaena (Linnaeus, 1766)
    Pachycrocuta brevirostris (Aymard, 1854?)
    Crocuta crocuta Erxleben, 1777
    C. aff. *ultra* Ewer, 1954
    Leecyaena forfex Ewer, 1955
    Chasmaporthetes silberbergi Broom, 1948)

  Felidae
    Felis caracal Schreber, 1776
    Leo leo (Linnaeus, 1766)
    L. pardus (Linnaeus, 1766)
    L. crassidens Broom, 1948
    Dinofelis barlowi (Broom, 1937)
    Acinonyx
    Machairodus transvaalensis Broom, 1939
    Megantereon whitei Broom, 1937
    M. gracile Broom, 1948F
    M. eurynodon Ewer, 1955
    Homotherium ethiopicum Arambourg, 1947
    H. problematicus (Collins, 1972)
INSECTIVORA
  Erinaceidae
    Erinaceus broomi Butler and Greenwood, 1973
  Soricidae
    Myosorex robinsoni Meester, 1955
    Sylvisorex cf. *granti* Thomas
    Suncus varilla (Thomas, 1895)
    S. infinitesimus (Heller)
    Crocidura taungsensis Broom, 1948
    C. cf. *hindei* Thomas
    C. cf. *bicolor* Bocage, 1889
    Diplomesodon fossorius Repenning, 1965
  Chrysochloridae
    Proamblysomus antiquus Broom, 1941
    Chlorotalpa spelea Broom, 1941
    Amblysomus hamiltoni (De Graaff, 1958)
CHIROPTERA   —Butler (1978)
  Megadermatidae
    Cardioderma
  Rhinolophidae
    Rhinolophus cf. *capensis* Lichtenstein, 1823
  Myzopodidae
    Myzopoda
  Vespertilionidae
    Myotis
    cf. *Nycticeius schlieffeni* (Peters, 1859)
    cf. *Pipistrellus rueppelli* (Fischer, 1829)
    Eptesicus cf. *hottentotus* (Smith)
    Miniopterus cf. *schreibersi* (Kuhl, 1819)
  Molossidae
    gen. indet.
PRIMATES
  Lorisidae
    Galago senegalensis Geoffroy-Saint-Hilaire, 1796
  Cercopithecidae   —Simons and Delson (1978)
    Cercopithecoides williamsi Mollett, 1947
    Colobus? flandrini (Arambourg, 1959)
    Cercopithecus
    Papio izodi Gear, 1926
    P. hamadryas (Linnaeus, 1758)
    P. cf. *cynocephalus* (Linnaeus, 1758)
    Parapapio broomi Jones, 1937
    P. jonesi Broom, 1940
    P. whitei Broom, 1940
    P. antiquus (Haughton, 1925)
    Theropithecus oswaldi (Andrews, 1916)
    T. brumpti (Arambourg, 1947)
    T. darti Jolly, 1972

Cercocebus
Hominidae   —Howell (1978)
  Australopithecus africanus Dart, 1925
  A. afarensis Johanson, White, and Coppens, 1978
  A. robustus (Broom, 1938I)
  A. boisei (Leakey, 1959)
  Homo habilis Leakey, Tobias, and Napier, 1964
ARTIODACTYLA
  Suidae   —Cooke and Wilkinson (1978)
    Notochoerus scotti (Leakey, 1943)
    N. capensis Broom, 1925
    N. euilus (Hopwood, 1926)
    N. jaegeri Coppens, 1971
    Metridiochoerus jacksoni (Leakey, 1943)
    M. nyanzae (Leakey, 1958)   (age?)
    M. andrewsi Hopwood, 1926
    Potamochoeroides shawi (Dale, 1948)
    Promesochoerus mukiri Leakey, 1965
    Kolpochoerus limnetes (Hopwood, 1926)
    K. afarensis Cooke, 1978
    K. olduvaiensis (Leakey, 1942)
    Potamochoerus porcus (Linnaeus, 1758)
    Phacochoerus aethiopicus (Pallas, 1767)   (age?)
  Hippopotamidae   —Coryndon (1978)
    Hexaprotodon hipponensis (Gaudry, 1876)
    H. protamphibius (Arambourg, 1947)
    H. karumensis Coryndon, 1977
    H. imagunculus Hopwood, 1926
    Hippopotamus gorgops Dietrich, 1928
    H. aethiopicus Coryndon and Coppens, 1975
    H. kaisensis Hopwood, 1926
    H. cf. amphibius Linnaeus, 1766
  Camelidae
    Camelus
  Cervidae?
    Dama?   (N. Africa; age?)
  Giraffidae   —Churcher (1978)
    Giraffa gracilis Arambourg, 1947
    G. stillei (Dietrich, 1941)
    G. pygmaea Harris, 1976
    G. camelopardalis (Linnaeus, 1766)   (age?)
    Sivatherium maurusium (Pomel, 1892)
    S. olduvaiensis (Hopwood, 1934)
  Bovidae   —Gentry (1978)
    Simatherium kohllarseni Dietrich, 1941
    cf. *Aepyceros melampus* (Lichtenstein, 1814)
    Cephalophus parvus Broom, 1934
    C. caeruleus Wells and Cooke, 1956
    cf. *C. monticola* Bocage, 1890
    Beatragus antiquus Leakey, 1965
    Connochaetes olduvaiensis (Leakey, 1965)
    Parmularius altidens Hopwood, 1934
    Gazella wellsi Cooke, 1949
    G. thomasi (Pomel, 1895)
    G. setifense (Pomel, 1895)
    G. praethompsoni Arambourg, 1947
    G. kohllarseni Dietrich, 1950
    G. vanhoepeni (Wells and Cooke, 1956)
    G. janenschi Dietrich, 1950
    Ugandax gautieri Cooke and Coryndon, 1970
    Kobus sigmoidalis Arambourg, 1941

    K. ancystrocera (Arambourg, 1947)
    Oreotragus major Wells and Cooke, 1956
    cf. *Onotragus*
    Oreonager tournoueri (Thomas, 1884)   (N. Africa)
    Menelikia lyrocera Arambourg, 1941
    Palaeotragus longiceps Broom, 1934
    cf. *Hippotragus gigas* Leakey, 1965
    Hippotragus cf. *equinus* (Geoffroy-Saint-Hilaire, 1816)
    Redunca darti Wells and Cooke, 1956
    cf. *R. fulvorufula* (Afael, 1815)
    Damaliscus antiquus Leakey, 1965
    Antilope cf. *subtorta* Pilgrim, 1937
    Oryx
    Syncerus cf. *caffer* (Sparman, 1779)
    Madoqua avifluminis (Dietrich, 1950)
    cf. *Raphicerus campestris* Thunberg, 1811
    Tragelaphus strepsiceros (Pallas, 1766)
    T. nakuae Arambourg, 1941
    T. cf. *spekei* (Blyth, 1855)
    T. cf. *angasi* Gray, 1848
    T. cf. *buxtoni* Dietrich, 1942
    T. scriptus (Pallas, 1766)
    T. gaudryi (Thomas, 1884)
    Pultiphagonides cf. *africanus* Hopwood, 1934
    Alcelaphus
    Megalotragus kattwinkeli (Schwartz, 1932)
    Capra
    Antidorcas recki (Schwartz, 1932)
    Makapania broomi Well and Cooke, 1956
    Bos? makapaani Broom, 1937
    aff. *Leptobos*
HYRACOIDEA   —G. Meyer (1978)
  Procaviidae
    Gigantohyrax maguirei Kitching, 1965
    Procavia transvaalensis Shaw, 1937
    P. antiqua Broom, 1934
PERISSODACTYLA
  Equidae   —Churcher and Richardson (1978)
    Hipparion baardi Bone and Singer, 1965
    H. libycum Pomel, 1897   (age?)
    H. afarense Eisenmann, 1976
    H. albertense Hopwood, 1926
    Equus (Dolichohippus) numidicus Pomel, 1897
    E. (D.) oldowayensis Hopwood, 1937D
    E. (D.) capensis Broom, 1909
    E. (Hippotigris) burchelli (Gray, 1824)   (this early?)
  Rhinocerotidae
    Dicerorhinus africanus Arambourg, 1970   (Ichkeul)
    Ceratotherium simum (Burchell, 1817)
    Diceros bicornis Linnaeus, 1758
  Chalicotheriidae
    Ancylotherium hennigi (Dietrich, 1942)
PROBOSCIDEA
  Deinotheriidae
    Deinotherium bozasi Arambourg, 1934
  Gomphotheriidae
    Anancus kenyensis (MacInnes, 1942)
  Stegodontidae

FIGURE 7-7 *Representative Pliocene and Pleistocene Sites in Africa. (See opposite page.)*

1 *Afar—Makapanian and Early Pleistocene—Ethiopia*
2 *Aïn Brimba—Makapanian—Tunisia*
3 *Aïn Boucherit—Makapanian—Algeria*
4 *Aïn el Bey—Makapanian—Algeria*
5 *Aïn el Hadj Baba—Makapanian—Algeria*
6 *Aïn Fouarat—Makapanian—Morocco*
7 *Aïn Hanech—Makapanian—Algeria*
8 *Aïn Jourdel—Makapanian—Algeria*
  Argoub el Hafid (See Oued Akrech.)
9 *Aterir beds sites—Langebaanian—Kenya*
  Baard's Quarry (See Langebaanweg.)
  Behanga (See Kaiso Formation sites.)
10 *Bel Hacel—Makapanian—Algeria*
  Beni Foudda (See St. Arnaud.)
11 *Bolt's Farm—Early Pleistocene (in part?)—South Africa Boucheron (See Si Abd el Aziz.)*
12 *Bou Regreg (Salé)—Makapanian—Morocco*
13 *Bourille—Makapanian and? Early Pleistocene?—Ethiopia*
14 *Broken Hill—Late Pleistocene—Zambia*
  Bugoma (See Kaiso Formation sites)—Uganda
15 *Buxton-Norlim—Makapanian—South Africa*
  Cave of Hearths—Late Pleistocene—South Africa
  Chad (See Lake Chad sites.)
16 *Chelmer—Late Pleistocene—South Africa*
9 *Chemeron Beds sites—Langebaanian (in part)—Kenya*
9 *Chesowanja—Makapanian—Kenya*
  Constantine district—Late Pleistocene and Makapanian—Algeria
19 *Cornelia—Late Pleistocene—South Africa*

21 *East Turkana (East Lake Rudolph)—Langebaanian, Makapanian, and Early Pleistocene—Kenya*
22 *Ekora Formation sites—Langebaanian—Kenya*
23 *Elandsfontein—Late Pleistocene—South Africa*
  El Gara (See Si Abd el Aziz.)
24 *Florisbad-Vlakkraal—Late Pleistocene—South Africa*
6 *Fouarat (See Aïn Fouarat.)*
25 *Gambetta—Makapanian—Algeria*
26 *Gamble's Cave—Late Pleistocene—Kenya*
  Goz-Kerki (See Lake Chad.)
  Hadar Formation sites (See Afar.)
27 *Hamada Damous—Makapanian—Tunisia*
28 *Haua Fteah—Late Pleistocene—Libya*
29 *Hopefield (Saldanha)—Late Pleistocene—South Africa*
30 *Ichkeul-Bizerte—Makapanian—Tunisia*
21 *Ileret—Makapanian and Early Pleistocene—Kenya*
32 *Kaiso Formation sites—Makapanian—Uganda*
33 *Kanam—Langebaanian—Kenya*
34 *Kanapoi—Langebaanian—Kenya*
35 *Kanjera—Late Pleistocene—Kenya*
9 *Kaperyon Formation sites—Langebaanian?—Kenya*
37 *Karmosit beds sites—Makapanian—Kenya*
38 *Katorora (Chiwondo beds)—Makapanian—Malawi*
39 *Kazinga Channel—Makapanian—Uganda*
40 *Kebili—Makapanian—Tunisia*
41 *Kikagati—Makapanian—Uganda*
  Kolinga (See Lake Chad.)
42 *Kom Ombo—Late Pleistocene—Egypt*
  Koobi Fora—(Makapanian and Early Pleistocene) (See East Turkana.)
  Koulá (See Lake Chad.)

15 *Kromdraai—Early Pleistocene—South Africa*
48 *Kubi Algi (See East Lake Turkana.)—Langebaanian—Kenya*
44 *Laetoli (Laetolil)—Makapanian—Tanzania*
45 *Lake Chad sites—Makapanian—Chad*
  Lake Turkana (Lake Rudolph) sites (See East Turkana.)
46 *Langebaanweg—Langebaanian—South Africa*
47 *Lothagam 3—Langebaanian—Kenya*
  Maghreb region—Makapanian—Tunisia
15 *Makapansgat—Makapanian—South Africa*
49 *Mascara—Late Pleistocene (in part)—Algeria*
50 *Mena House–Cairo—Langebaanian?—Egypt*
51 *Mumba Hills—Tanzania?—Late Pleistocene*
  Mursi Formation sites (See Omo, lower)—Langebaanian—Ethiopia
38 *Mwenirondo 1—Makapanian—Malawi*
38 *Mwimbi North—Makapanian and Pleistocene—Malawi*
  Nyabrogo (See Kaiso Formation sites.)
52 *Nyamavi district—Langebaanian? Zaire*
  Nyawiega (See Kaiso Formation sites.)
53 *Olduvai (Old way)—Makapanian and Early Pleistocene—Tanzania*
54 *Omo, Omo Basin, Omo beds—Makapanian and Early Pleistocene—Ethiopia*
54 *Omo, lower—Langebaanian (in part)—Ethiopia*
55 *Orlogesaillie—Late Pleistocene—Kenya*
  Ouadi Derdemi (See Lake Chad sites.)
56 *Oued Akrech—Langebaanian—Morocco*
57 *Oued Tessa—Langebaanian?—Tunisia*
  Rudolph Basin (See under East Turkana.)

58 *Saint Arnaud (Constantine, Setif, Beni Foudda)—Makapanian—Algeria*
  St. Charles, St. Donat, St. Eugène (See Constantine or St. Arnaud.)
  Saldanha (See Hopefield.)
  Saragata Deare (See Afar)—Langebaanian to Makapanian—Ethiopia
59 *Semliki—Langebaanian? (or older)—Zaire*
  Shungura Formation sites (See Omo)—Langebaanian and Makapanian
60 *Si Abd el Aziz—Makapanian—Morocco*
61 *Sidi Abdallah—Makapanian—Morocco*
  Sidi Bouknadel (See Sidi Abdallah.)
62 *Sidi Bou Kouffa—Makapanian—Tunisia*
63 *Sidney-on-Vaal—Makapanian—South Africa*
  Sinda-Mohari (See Semliki.)
64 *Springbok—Langebaanian? Makapanian?—South Africa*
11 *Sterkfontein, upper—Early Pleistocene—South Africa*
11 *Sterkfontein, lower—Makapanian—South Africa*
66 *Swartklip—Late Pleistocene—South Africa*
15 *Swartkrans—Early Pleistocene—South Africa*
68 *Tau'ng—Makapanian—South Africa*
69 *Ternifine—Late Pleistocene—Algeria*
70 *Todenyang—Makapanian—Ethiopia*
38 *Uraha Hill—Langebaanian—Malawi*
  Usno Formation sites (See Omo.)
72 *Vaal River Gravels sites—Pleistocene—South Africa*
73 *Wonderwerk Cave—Late Pleistocene—South Africa*
74 *Virginia—Makapanian?—Orange Free State*
  Vlakkraal-Florisbad—Late Pleistocene—South Africa
75 *Vogel River "series"—Langebaanian—Tanzania*
  Yellow Sands–Mursi Formation (See Omo.)

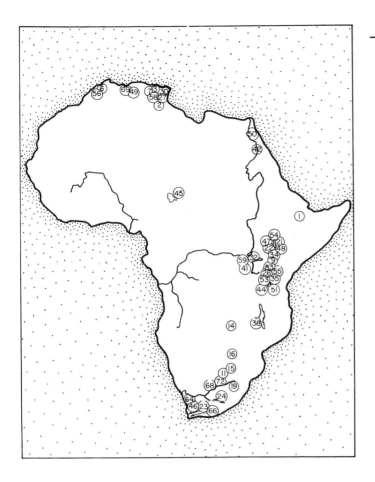

Cricetidae
  Ruscinomys
  Tatera cf. *brantsi*
  Gerbillus sp.
  Prototomys campbelli
  Mystromys darti
  M. antiquus
  M. hausleitneri Broom, 1937
Muridae
  Dendromus antiquus
  D. cf. *mesomelas* (Brants, 1827)
  Malacothrix typica (Smith, 1834)
  M. makapani Graaff, 1961A
  Steatomys cf. *pratensis* Peters, 1846
  Paraethomys anomalus deBruijn, Dawson and Mein, 1970
  Grammomys cf. *dolichurus* (Smuts, 1832)
  Dasymys sp.
  D. cf. *incomptus* Sundevall, 1846
  Arvicanthis
  Acomys cf. *cahirinus* (Geoffroy-Saint-Hilaire, 1812)
  Pelomys cf. *fallax* Peters, 1852
  Limniscomys cf. *griselda*
  Rhabdomys cf. *pumilio* (Wagner, 1820)
  Aethomys cf. *namaquensis* (A. Smith, 1835)
  Rattus debruyni
  Mastomys cf. *natalensis*
  Zelotomys
  Mus cf. *minutoides*
  Otomys gracilis (Broom, 1937E)
  O. campbelli
  Otomys kempi
Ctenodactyloidea
  Irhoudia bohlini Jaeger, 1971
Sciuridae
  Xerus
  sciurid

BLANCAN

(= Blancan II and III of Repenning)

The following composite faunal list is based on the fossils from the Fox Canyon, Rexroad, Beck Ranch, Hagerman and Benson localities of Kansas, Texas, Idaho, and Arizona, and it probably represents the mammals living in various communities of North America during Early Blancan, from about 4.2 to about 3.2 million years ago according to Repenning (1982).

Early Blancan Mammalian Fauna

LAGOMORPHA
Leporidae
  Hypolagus regalis Hibbard, 1939A
  H. limnetus Gazin, 1934A
  H. near *vetus* (L. Kellogg, 1910)   —Gazin (1934A)
  Pratilepus kansasensis Hibbard, 1939A
  P. vagus (Gazin, 1934A)   —Hibbard (1969)

Stegodon kaisensis Hopwood, 1939
Elephantidae
  Mammuthus africanavus (Arambourg, 1952A)
  M. meridionalis (Nesti, 1825)   (N. Africa)
  Elephas recki Dietrich, 1916
  E. ekorensis Maglio, 1970B
  Loxodonta adaurora Maglio, 1970B
  L. atlantica (Pomel, 1879)
TUBULIDENTATA
  Orycteropus
RODENTIA   —Lavocat (1978)
Thryonomyidae
  Petromus minor
Bathyergidae
  Heterocephalus
  Cryptomys robertsi Broom, 1937
  Gypsorhychus darti Broom, 1930
  G. minor Broom, 1939
  G. makapani Broom, 1948
Hystricidae
  Hystrix cf. *africae-australis* Peters, 1852
  H. major Gervais, 1895
  H. cristata Linnaeus, 1758
  H. makapanensis Greenwood, 1958
  Xenohystrix crassidens (Lydekker, 1886)
Pedetidae
  Pedetes gracilis Broom, 1937

Notolagus lepusculus (Hibbard, 1939A)

N. cf. *velox* R. Wilson, 1937D   —Downey (1968A)

Nekrolagus progressus (Hibbard, 1939A)   —Hibbard (1939F)

Aluralagus bensonensis (Gazin, 1942B)   —Downey (1968A)

EDENTATA

Megalonychidae

Megalonyx leptonyx (Marsh, 1874C)?   —Gazin (1935A); *nomen dubium* acc. Hirschfeld and Webb (1968)

CARNIVORA

Ursidae

Ursus abstrusus Bjork, 1970A

tremarctine, gen. indet.

Mustelidae

Brachypsigale dubius Hibbard, 1954B

Mustela rexroadensis Hibbard, 1950

Buisnictis meadensis Hibbard, 1950

B. breviramus (Hibbard, 1941A)

Taxidea taxus (Schreber, 1778)

Spilogale rexroadi Hibbard, 1941A

S.? microdens Dalquest, 1978

Mephitis rexroadensis Hibbard, 1952C

Satherium piscinaria (Leidy, 1873B)

S. sp., large

Ferinestrix vorax Bjork, 1970A

Trigonictis idahoensis (Gazin, 1934C)

T. cooki (Gazin, 1934C)

Sminthosinus bowleri Bjork, 1970A

Procyonidae

Bassariscus rexroadensis Hibbard, 1950

B. casei Hibbard, 1952A

Procyon rexroadensis Hibbard, 1941A

Nasua pronarica Dalquest, 1978

Canidae

Canis lepophagus Johnston, 1938B

Borophagus diversidens Cope, 1892Y

Vulpes near *velox* (Say)

Urocyon, sp. unident.

U. progressus Stevens, 1965

Hyaenidae, gen. unident.

Felidae

Felis lacustris Gazin, 1933A

F. rexroadensis Stephens, 1959B

Homotherium

Megantereon? hesperus (Gazin, 1933A)

INSECTIVORA

Soricidae

Sorex taylori Hibbard, 1937D

S. rexroadensis Hibbard, 1953B

S. hagermanensis Hibbard and Bjork, 1971A

S. meltoni Hibbard and Bjork, 1971A

S. powersi Hibbard and Bjork, 1971A

Beckiasorex hibbardi Dalquest, 1972A

Paracryptotis rex Hibbard, 1950

P. gidleyi (Gazin, 1933B)

Cryptotis adamsi (Hibbard, 1953B)   —Repenning (1967)

C.? meadensis Hibbard, 1953B

Notiosorex jacksoni Hibbard, 1950

Talpidae

Scalopus rexroadi (Hibbard, 1941A)

Scapanus

CHIROPTERA

Vespertilionidae

Lasiurus fossilis Hibbard, 1950

L. borealis (Müller, 1776)

Lasionycteris noctivagus (LeConte)

near *Eptesicus fuscus* (Beauvois)

Antrozous pallidus (LeConte, 1831)

ARTIODACTYLA

Tayassuidae

Platygonus bicalcaratus Cope, 1893A

P. pearcei Gazin, 1938A

Camelidae

Megatylopus cochrani (Hibbard and Riggs, 1949)

Titanotylopus spatulus (Cope, 1893A)

Hemiauchenia blancoensis (Meade, 1945)

Camelops? arenarum Hay, 1927D   —Taylor (1966)

Camelops

Cervidae

Odocoileus brachyodontus Oelrich, 1953B

O. sp., large

Antilocapridae

Ceratomeryx prenticei Gazin, 1935C

antilocaprids, large and small

PERISSODACTYLA

Equidae

Nannippus beckensis Dalquest and Donovan, 1973

N. phlegon (Hay, 1899G)

Equus (*Dolichohippus*) *simplicidens* (Cope, 1892J)

E. (*Asinus?*) *cumminsi* (Cope, 1893A)

PROBOSCIDEA

Gomphotheriidae

Stegomastodon rexroadensis Woodburne, 1961

cf. *Cuvieronius*

cf. *Haplomastodon*

Mammutidae

Mammut adamsi Hibbard and Riggs, 1949

RODENTIA

Sciuridae

Paenemarmota barbouri Hibbard and Schultz, 1948

Spermophilus howelli (Hibbard, 1941A)

S. rexroadensis (Hibbard, 1941A)

S. bensoni (Gidley, 1922B)

Ammospermophilus?   —Zakrzewski (1969A)

Geomyidae

Geomys smithi Hibbard, 1967

G. adamsi Hibbard, 1967

G. minor Gidley, 1922B

G. jacobi Hibbard, 1967

Pliogeomys parvus Zakrzewski, 1969A

Thomomys gidleyi R. Wilson, 1933G

Cratogeomys bensoni Gidley, 1922B

Heteromyidae

Perognathus pearlettensis Hibbard, 1941C

P. rexroadensis Hibbard, 1950

P. carpenteri Dalquest, 1978

P. gidleyi Hibbard, 1941A

P. magnus Zakrzewski, 1969A

P. maldei Zakrzewski, 1969A

Prodipodomys centralis (Hibbard, 1941A)
   P. idahoensis Hibbard, 1962A
   P. minor (Gidley, 1922B)
Castoridae
   Castor californicus Kellogg, 1911
   Dipoides rexroadensis Hibbard and Riggs, 1949
   D. intermedius Zakrzewski, 1969
   Procastoroides sweeti Barbour and Schultz, 1937
Cricetidae
   Baiomys rexroadi Hibbard, 1941A
   B. kolbi Hibbard, 1952G
   B. aquilonius Zakrzewski, 1969A
   Onychomys gidleyi Hibbard, 1941D
   Symmetrodontomys simplicidens Hibbard, 1941A
   Calomys (*Bensonomys*) *arizonae* (Gidley, 1922B) —
      Baskin (1978)
   C. (*B.*) *eliasi* (Hibbard, 1937D)
   Neotoma quadriplicatus (Hibbard, 1941A)
   N. sawrockensis Hibbard, 1967
   N. fossilis Gidley, 1922B
   Pliophenacomys primaevus Hibbard, 1937D
   P. finneyi Hibbard and Zakrzewski, 1972A
   Reithrodontomys rexroadensis Hibbard, 1952C
   R. wetmorei Hibbard, 1952C
   Peromyscus beckensis Dalquest, 1978
   P. kansasensis Hibbard, 1937D
   P. baumgartneri Hibbard, 1954B
   Pliopotamys meadensis Hibbard, 1937D
   P. minor (R. Wilson, 1933C)
   Neondatra kansasensis Hibbard, 1937D
   Nebraskomys rexroadensis Hibbard, 1970A
   Pliolemmus antiquus Hibbard, 1937D
   Mimomys (*Ogmodontomys*) *pouphagus* Hibbard, 1941A
   M. (*Ophiomys*) *taylori* (Hibbard, 1959)
   M. (*Cosomys*) *primus* R. Wilson, 1932
Zapodidae
   Zapus sandersi Hibbard, 1956B
   Z. rinkeri Hibbard, 1951D

Blancan III and IV of Repenning

The following composite faunal list, based on fossils from Blanco, Cita Canyon, Hudspeth, Santa Fe River, Haile XV, Borchers, Dixon, Sanders, Broadwater, Grand View, 111 Ranch (Tusker), California Wash, Coso, Arroyo Seco, and other sites in Texas, Florida, Kansas, Nebraska, Idaho, Arizona, and California, represents mammals living in various communities in North America from about 3.2 (Repenning, 1982) to about 1.8 million years ago.

Composite Blancan Fauna

LAGOMORPHA
   Leporidae
      Hypolagus furlongi Gazin, 1934A
      H. arizonensis Downey, 1962
      H. browni (Hay, 1921)
EDENTATA
   Megalonychidae
      Megalonyx leptostomus Cope, 1893A
      M. leptonyx (Marsh, 1874C)   —(*nomen dubium*, acc.
         Hirschfeld and Webb (1968))

Mylodontidae
   Glossotherium chapadmalense Kraglievich, 1925
Glyptodontidae
   Glyptotherium texanum Osborn, 1903B
Dasypodidae
   Kraglievichia floridanus Robertson, 1976
   Dasypus bellus (Simpson, 1929H)
CARNIVORA
   Ursidae
      Ursus abstrusus Bjork, 1970
      Tremarctos floridanus (Gidley, 1928)
   Mustelidae
      Mustela frenata Lichtenstein, 1831
      Canimartes cumminsi Cope, 1892Y
      Trigonictis idahoensis (Gazin, 1934C)
      T. cooki (Gazin, 1934C)
      Satherium ingens Gazin, 1934C
      S. piscinaria (Leidy, 1873B)
      Taxidea cf. *taxus* (Schreber, 1778)
      Buisnictis burrowsi Hibbard, 1972A
      Mephitis sp. unident.
   Procyonidae
      Procyon cf. *rexroadensis* Hibbard, 1941A
   Canidae
      Canis lepophagus Johnston, 1938B
      Borophagus diversidens Cope, 1892Y
   Hyaenidae
      Chasmaporthetes ossifragus Hay, 1921A
   Felidae
      Felis lacustris Gazin, 1933A
      Lynx cf. *issiodorensis* (Croizet and Jobert, 1828)
      L. rufus (Schreber, 1777)   —(retained
      Dinofelis palaeoonca (Meade, 1945)   in   *Ischyrosmilus*
      Homotherium ischyrus (Merriam, 1905B) Merriam, 1918A,
      H. johnstoni (Mawby, 1965B)   by Mawby, 1965,
      H. idahoensis (Merriam, 1918A)   and Kurtén and An-
      Acinonyx (*Miracinonyx*) *studeri* (Savage, derson, 1980)
         1960H)   —Adams (1979)
INSECTIVORA
   Soricidae
      Sorex sandersi Hibbard, 1956B
      S. leahyi Hibbard, 1956B
      S. cinereus Kerr, 1792   —Eshelman (1975)
      Planisorex dixonensis (Hibbard, 1956B)
         —Hibbard (1972A)
      Blarina, sp. unident.
   Talpidae
      Scalopus aquaticus (Linnaeus, 1758)
      S. blancoensis (Dalquest, 1975)
      Scapanus, sp. unident.
CHIROPTERA
   Vespertilionidae
      Anzanycteris anzensis White, 1969
ARTIODACTYLA
   Tayassuidae
      Platygonus bicalcaratus Cope, 1893A
      P. pearcei Gazin, 1938A
      Mylohyus floridanus Kinsey, 1974
   Camelidae

Titanotylopus spatulus (Cope, 1893A)

Megatylopus, sp. unident.

Camelops traviswhitei Mooser and Dalquest, 1975

C. cf. *kansanus* Leidy, 1854F

Hemiauchenia blancoensis (Meade, 1945)

H. macrocephala (Cope, 1893A)

Blancocamelus meadei Dalquest, 1975

Cervidae

Odocoileus cf. *brachyodontus* Oelrich, 1953

O. virginianus (Zimmermann, 1780)

Cervus, sp. unident.

Antilocapridae

Capromeryx arizonensis Skinner, 1942

Tetrameryx, sp. unident.

PERISSODACTYLA

Equidae

Nannippus phlegon (Hay, 1899G)

Equus (Dolichohippus) simplicidens (Cope, 1892J)

E. (Asinus?) cumminsi (Cope, 1893A)

E. (Hemionus) calobatus Troxell, 1915B

Astrohippus? or very small *Equus,* sp. unident.

Equus giganteus Gidley, 1901

E. scotti Gidley, 1900

Tapiridae

Tapirus copei Simpson, 1945H

PROBOSCIDEA

Gomphotheriidae

Rhynchotherium praecursor (Cope, 1893A)

Cuvieronius, sp. unident.

Haplomastodon?, sp. unident.

Stegomastodon mirificus (Leidy, 1858B)

Mammutidae

Mammut americanum (Kerr, 1791)

RODENTIA

Sciuridae

Paenemarmota barbouri Hibbard and Schultz, 1948

Marmota arizonae Hay, 1921

Spermophilus (Otospermophilus) boothi Hibbard, 1972A

Spermophilus johnsoni Hibbard, 1972A

S. meltoni Hibbard, 1972A

S. howelli (Hibbard, 1941A)

S. magheei (Strain, 1966)

S. finlayensis (Strain, 1966)

S. cragini (Hibbard, 1941A)

S. tuitus (Hay, 1921)

S. meadensis (Hibbard, 1941A)

Cynomys hibbardi Eshelman, 1975

C. vetus Hibbard, 1942A

Cryptopterus webbi Robertson, 1976

Geomyidae

Geomys quinni McGrew, 1944C

G. tobinensis Hibbard, 1944D

Nerterogeomys paenebursarius (Strain, 1966)

Heteromyidae

Prodipodomys centralis (Hibbard, 1944A)

P. idahoensis Hibbard, 1962A

Etadonomys tiheni Hibbard, 1943B   —(*nomen dubium* acc. to Kurtén and Anderson, 1980)

Perognathus gidleyi Hibbard, 1941A

P. pearlettensis Hibbard, 1941C

Castoridae

Dipoides rexroadensis Hibbard and Riggs, 1949

Procastoroides sweeti Barbour and Schultz, 1937

P. idahoensis Shotwell, 1970A

Castoroides ohioensis Foster, 1838

Castor accessor Hay, 1927

C. canadensis Kuhl, 1820

Cricetidae

Reithrodontomys pratincola Hibbard, 1941A

Peromyscus kansasensis Hibbard, 1941A

P. cragini Hibbard, 1944D

Calomys (Bensonomys) meadensis (Hibbard, 1937D)   —Baskin (1978)

Neotoma taylori Hibbard, 1967

Sigmodon medius Gidley, 1922B

S. hudspethensis Strain, 1966

S. curtisi Gidley, 1922B

Nebraskomys mcgrewi Hibbard, 1957A

Mimomys (Ogmodontomys) poaphagus (Hibbard, 1941A)

M. (Ophiomys) magilli (Hibbard, 1972A)

M. (O.) fricki (Hibbard, 1972A)

M. (O.) parvus R. Wilson, 1933C

Pliopotamys meadensis Hibbard, 1937D

Pliophenacomys primaevus (Hibbard, 1937D)

P. osborni L. Martin, 1972

Pliolemmus antiquus Hibbard, 1937D

Synaptomys (Synaptomys) rinkeri Hibbard, 1956B

S. (Metaxyomys) vetus R. Wilson, 1933C

S. (M.) landesi Hibbard, 1954

Onychomys fossilis Hibbard, 1941A

Ondatra idahoensis R. Wilson, 1933C

Zapodidae

Zapus sandersi Hibbard, 1956B

Z. burti Hibbard, 1941C

Erethizontidae

Coendu bathygnathum (R. Wilson, 1935C)   —White (1968, 1970)

Hydrochoeridae

gen. unident.

Of the 106 genera here recognized in the Blancan, 72 percent are new appearances and 36 percent do not appear later in North America. About 10 percent of the 76 new genera were evidently immigrants from South America or Central America or both, whereas as much as 22 percent may have included immigrants from Asia. It seems to us, therefore, that two-thirds of the new genera of the Blancan were produced by species evolution in North America.

Generic extinction (36 percent of the genera) at the end of the Blancan, about 1.8 million years ago, was greater than at the end of the succeeding Irvingtonian (19 percent about 500,000 years ago) or at the end of the Rancholabrean ("end of Pleistocene"; 29 percent of the genera about 10,000 years ago.)

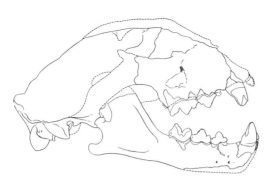

A

D

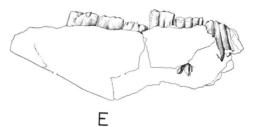

E

B

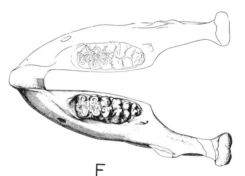

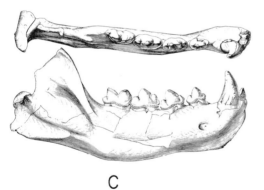

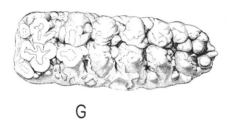

F

C

G

FIGURE 7-8   *Blancan Mammals.*

   A   Acinonyx studeri (Savage, 1960H). Cranium, reconstructed. ×3/8 approx. Figure 3a in Savage (1960H).

   B   Acinonyx studeri (Savage, 1960H). Maxilla. ×4/9 approx. Figure 2 in Savage (1960H).

   C   Chasmaporthetes ossifragus Hay, 1921A (=Ailuraena johnstoni Stirton and Christian, 1940). Lower jaw.
      ×1/3 approx. Figure 1 in Stirton and Christian (1940).

   D   Borophagus diversidens Cope, 1892Y (=Hyaenognathus pachyodon Merriam, 1903B). Lower jaw. ×4/9 ap-
      prox. Figure 1 in Stirton and VanderHoof (1933).

   E   Equus (Dolichohippus) simplicidens (Cope, 1892J). Lower jaw. ×1/4 approx. Figure 1 in Savage (1951C).

   F   Stegomastodon mirificus (Leidy, 1858B). Mandible. ×1/14 approx. Figure 5 in Savage (1955B).

   G   Stegomastodon mirificus (Leidy, 1858B). M³. ×1/5 approx. Figure 2a in Savage (1955B).

(Figures A, B, D, E, F, and G published with permission of the University of California Press. Figure C published with permission of the Journal of Mammalogy.)

FIGURE 7-9 *Representative Pliocene (=Blancan, sensu lato) Localities of North America.*
(More localities from the United States and Mexico are listed by Repenning, 1983?)

1 Alturas (upper)—California
2 Anita Mine—Arizona
3 Arroyo Seco—California
4 Asphalto—California
5 Beck Ranch—Texas
6 Bender—Kansas
7 Benson—Arizona
8 Blanco (Mt. Blanco)—Texas
9 Bonanza—Arizona   (See 7.)
10 Borchers—Kansas   (See 6.)
11 Broadwater—Nebraska
12 Buis Ranch—Oklahoma
13 Buttonwillow—California
14 California Wash—Arizona (See 7.)
15 Cita Canyon—Texas
16 Cochise County sites—Arizona   (See 7.)
17 Comosi Wash—Arizona
18 Coso—California
19 Deer Park—Kansas   (See 6.)
20 Delmont—South Dakota
21 Dixon—Kansas   (See 6.)
22 Donnelly Ranch—Colorado
23 Duncan—Arizona
24 Flat Tire (fauna)—Arizona
25 Fox Canyon—Kansas (See 6.)
26 Grand View—Idaho
27 Hagerman—Idaho
28 Haile XVA—Florida
29 Hammett—Idaho
30 Hart Draw—Kansas
31 Henry Ranch—Arizona
32 Hudspeth—Texas
33 Hungry Valley—California
34 Keam's Canyon—Arizona
35 Keefe Canyon—Kansas (See 6.)
36 Keim Formation sites—Nebraska
37 Las Tunas—Baja California
38 Layer Cake (fauna)—California   (See 3.)
39 Lisco—Nebraska   (See 11.)
40 Maxum—California
41 Mendivil Ranch (Benson)—Arizona   (See 7.)
42 Minaca—Mexico
43 Mullen I (in part?)—Nebraska
44 111 Ranch (Tusker)—Arizona   (See 23.)
45 Panaca—Nevada
46 Port Charlotte—Florida
47 Red Corral (Channing)—Texas
48 Red Knolls—Arizona (See 31.)
49 Red Light—Texas   (See 32.)
50 Rexroad—Kansas   (See 6.)
51 Ringold (White Bluffs)—Washington
52 Sand Draw—Nebraska (See 36.)
53 Sanders—Kansas (See 6.)
54 San Diego Formations sites—California
55 San Joaquin Formations sites—California
56 San Timoteo—California
57 Santa Fe River—Florida
58 Sarasota—Florida
59 Saw Rock—Kansas   (See 6.)
60 Seger—Kansas   (See 6.)
61 Tassajara Formation (upper) sites—California
62 Taunton—Washington
63 Tehama Formation sites—California
64 Tusker (111 Ranch)   (See 23.)
65 Vallecito (lower)—California   (See 3.)
66 Verde Formation sites—Arizona
51 White Bluffs (Ringold)—Washington   (See 51.)
67 White Rock—Kansas
68 Wichman—Nevada
69 Yankton—South Dakota

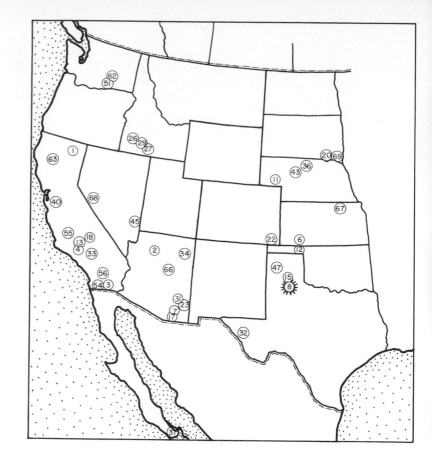

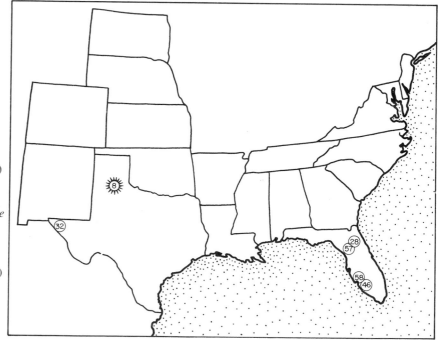

CHAPADMALALAN

Marshall, Hoffstetter, and Pascual (1982) remind us that the Chapadmalal Formation (and its mammalian fauna) is exposed in sea cliffs between Mar del Plata and Miramar, Buenos Aires Province, Argentina. They further note that the Chapadmalalan is distinguished by the first record of abundant mammals of North American origin (*Conepatus, Argyrolagus,* and four genera of cricetid rodents) and by the co-occurrence of *Thylophorops, Scelidotherium, Glossotheridium, Paraglyptodon,* and *Dolicavia.*

The following generic list is taken from Marshall, Hoffstetter, and Pascual (1982).

Composite Chapadmalalan Fauna

MARSUPIALIA
  Didelphidae
    Didelphis Linnaeus, 1758
    Lutreolina Thomas, 1910
    Marmosa Gray, 1821
    Paradidelphys Ameghino, 1904
    Thylatheridium Reig, 1952D
    Thylophorops Reig, 1952D
  Sparassocynidae
    Sparassocynus Mercerat, 1899
  Thylacosmilidae
    Hyaenodonops Ameghino, 1908
    Notosmilus Kraglievich, 1960A
  Argyrolagidae
    Argyrolagus Ameghino, 1904
    Microtragulus Ameghino, 1904
EDENTATA
  Dasypodidae
    Chaetophractus Fitzinger, 1871
    Chorobates
    Doellotatus Bordas, 1932
    Macroeuphractus Ameghino, 1887
    Propraopus? Ameghino, 1881
    Ringueletia
    Tolypeutes? Illiger, 1811
    Zaedyus Ameghino, 1889
  Glyptodontidae
    Palaeodoedicurus? Castellanos, 1927
    Paraglyptodon Castellanos, 1932
    Plohophoroides Castellanos, 1928
    Trachycalyptus Ameghino, 1908
    Urotherium Castellanos, 1926
  Megalonychidae
    Diheterocnus Kraglievich, 1928
  Mylodontidae
    Glossotheridium
    Scelidotherium Owen, 1840
CARNIVORA
  Procyonidae
    Chapalmalania Kraglievich and Olazabal, 1959
    Cyonasua Ameghino, 1885
  Mustelidae
    Conepatus Gray, 1837
ARTIODACTYLA
  Tayassuidae
    Argyrohyus

LITOPTERNA
  Proterotheriidae
    Brachytherium Ameghino, 1883
  Macraucheniidae
    Promacrauchenia Ameghino, 1904
NOTOUNGULATA
  Toxodontidae
    Toxodon Owen, 1840
    Xotodon Ameghino, 1887
  Mesotheriidae
    Pseudotypotherium Ameghino, 1904
  Hegetotheriidae
    Paedotherium Burmeister, 1888
    Tremacyllus Ameghino, 1891
RODENTIA
  Cricetidae
    Akodon Meyen, 1833
    Dankomys
    Graomys Thomas, 1916
    Reithrodon Waterhouse, 1839
  Octodontidae
    Actenomys Burmeister, 1888
    Eucelophorus Ameghino, 1909
    Megactenomys? Rusconi, 1930
    Pithanotomys Ameghino, 1887
    Pseudoplataeomys
  Echimyidae
    Cercomys Cuvier, 1832
    Eumysops Ameghino, 1888
  Myocastoridae
    Isomyopotamus Rovereto, 1914
  Dinomyidae
    Telicomys Kraglievich, 1926
  Caviidae
    Cardiomys Ameghino, 1885
    Caviodon Ameghino, 1885
    Caviops
    Dolicavia Ameghino, 1916
    Orthomyctera Ameghino, 1889
    Palaeocavia Ameghino, 1889
    Pascualia? Ortega Hinojosa, 1963
    Propediolagus?
  Chinchillidae
    Lagostomopsis
  Hydrochoeridae
    Chapalmatherium Ameghino, 1908
    Protohydrochoerus Rovereto, 1914
  Abrocomidae
    Protabrocoma Kraglievich, 1927

PLIOCENE OF AUSTRALASIA

A few land mammal faunas from the Australian area can be assigned to the Pliocene. Representative faunas include

Hamilton in the Grange Burn Formation (northwestern Victoria)—Gill (1957), Turnbull and Lundelius (1970), *Palankarinna* in the Mampuwordu Sands (northeastern South Australia)—Stirton, Tedford and Woodburne (1967), *Chinchilla* in the Chinchilla Sands (southeastern Queensland)—Woods (1962), Bartholomai (1967), *Bow* in the alluvial rocks of New South Wales—Archer (1980), and *Awe* in the Otibanda Formation of New Guinea—Plane (1967B).

A composite roster of taxa from these sites comprises:

Composite Pliocene Fauna of Australasia

MONOTREMATA
    Ornithorhynchus
MARSUPIALIA
  Dasyuridae
    Glaucodon ballaratensis Stirton, 1957A age?
    Antechinus
    Dasyurus dunmalli Bartholomai, 1971
    Planigale
  Thylacinidae
    Thylacinus cynocephalus Harris, 1808
  Peramelidae
    Perameles allinghamensis Archer, 1976
    Isoodon
    Ischnodon australis Stirton, 1955D family?
  Thylacoleonidae
    Thylacoleo crassidentatus Bartholomai, 1962
    T. hilli Pledge, 1977 age?
  Phalangeridae
    Phalanger
    Trichosurus
  Burramyidae
    Burramys
  Phascolarctidae
    Koobor jimbarrati Archer, 1976
    K. notabilis Archer, 1977
  Petauridae
    Pseudokoala erlita Turnbull and Lundelius, 1970
    Pseudocheirus marshalli Turnbull and Lundelius, 1970
    P. stirtoni Turnbull and Lundelius, 1970
    Petaurus
  Potoroidae
    Aepyprimnus
    Hypsiprymnodon
    Propleopus
    cf. *Bettongia*
  Macropodidae
    Petrogale
    Dendrolagus
    Dorcopsis
    Sthenurus antiquus Bartholomai, 1963
    S. notabilis Bartholomai, 1963
    Troposodon minor (Owen, 1877)
    Protemnodon buloloensis Plane, 1967B

    P. snewini Bartholomai, 1978
    P. devisi Bartholomai, 1973
    P. chinchillaensis Bartholomai, 1973
    P. anak Owen, 1874 age?
    P. otibandus Plane, 1967B
    Thylogale
    Wallabia indra De Vis, 1895
    Prionotemnus narada Bartholomai, 1978
    P. palankarinnicus Stirton, 1955D
    P. dryas (De Vis, 1895)
    Macropus giganteus Shaw, 1790
    M. mundjabus Flannery, 1980 age?
    Osphranter pavana Bartholomai, 1978
    O. pan Bartholomai, 1975
    O. woodsi Bartholomai, 1975
  Palorchestidae
    Palorchestes parvus De Vis, 1895
  Vombatidae
    Ramsayia lemleyi Archer, 1976
    Phascolonus
    Vombatus? prior De Vis, 1883
  Ektopodontidae, n. gen.
  Diprotodontidae
    Zygomaturus keanei Stirton, 1967C
    Meniscolophus mawsoni Stirton, 1967C
    Nototherium watutense Anderson, 1937
    Euowenia grata De Vis, 1888
    Euryzygoma dunense De Vis, 1887
    Kolopsis rotundus Plane, 1967B
    Kolopsoides cultridens Plane, 1967B
MICROCHIROPTERA—genus indet.
RODENTIA
  Muridae, genus indet.

PLIOCENE MARINE MAMMALS

The Pliocene record of marine mammals is generally only a hint of the diversity and abundance of mammals that must have been present. You will recall that most of the animals that were formerly dated early to middle Pliocene are now placed in the Miocene. Probably the best specimens of cetaceans and sirenians from the Pliocene are from Italy, where a few skeletons and skulls have been unearthed. The following is a representative list of taxa known.

Representative Pliocene Marine Mammals

CARNIVORA
  Otariidae
    Thalassoleon macnallyae Repenning and Tedford, 1977
    cf. *Callorhinus ursinus* (Linnaeus, 1758)
    Eumetopias
    Arctocephalus
  Odobenidae
    Valenictus
    Prorosmarus alleni Berry and Gregory, 1906
  Phocidae
    Phoca cf. *vitulina* Linnaeus, 1766

Mustelidae
Enhydriodon?
CETACEA —Barnes (1976)
Odontoceti, *Inc. Sed.*
"*Stenodelphis*" *sternbergi* Gregory and Kellogg, 1927
Platanistidae, gen. and sp. indet.
Physeteridae
Scaldicetus Dubus, 1867
Monodontidae, indet.
Delphinidae
aff. *Tursiops* Gervais, 1855
Delphinus or *Stenella*
aff. *Globicephala*
Phocoenidae?
MYSTICETI
Balacnidae
Balaenula astensis Trevisan, 1941
Balaenopteridae
cf. *Plesiocetus* Van Beneden, 1859
Balaenoptera acutirostrata Lacépède, 1804
Idiocetus guicciardinii Capellini, 1876
Fam. *inc. sed.*
"*Eschrichtius*" *davidsonii* (Cope, 1872)
SIRENIA
Felsinotherium gastaldii De Zigno, 1878
F. gervaisii Capellini, 1872

STRATIGRAPHIC DEMONSTRATIONS

A few places scattered around the world give a stratigraphic demonstration of earlier Pliocene mammal-bearing strata superposed on later Miocene mammal-bearing strata. Some of these places, plus a small number of other districts, show a stratigraphic succession of mammal-bearing strata within the Pliocene or through the Pliocene and Pleistocene. Most of the Late Cenozoic land-mammal localities, however, have been arranged chronologically and placed in a correlation system on the basis of their *paleontological* character, that is, on the basis of whether or not their species are interpreted as primitive (= "less derived," usually thought to be older) or advanced (= "more derived," usually thought to be younger). We emphasize, accordingly, that most of the so-called *biostratigraphic* datings and correlations of the past and current literature are nothing more than the usual, old-fashioned *paleontologic dating and correlation* masquerading under a more fashionable (and profitable?) alias. Similarly, the term *biochronology* (=*paleontologic chronology*) is now fashionable, seemingly as paleontologists try to make their siblings in the neontological sciences more sensitive to the fact that paleontology also is concerned with organisms and evolution.

Europe

Districts in Spain and in southeastern Europe, through the Ukraine into the Caucasus, have successions extending from

Turolian or older into Ruscinian and Villafranchian correlatives. To our knowledge, none of these areas have complete faunas of the Pliocene overlying complete faunas of the Miocene. A classic area for demonstration of part of the sequence within the Villafranchian is the Issoire district of central France. Here on one mountainside, the earlier Villafranchian Etouaires locality lies under the "middle" Villafranchian Roccaneyra locality, which in turn lies under the late Villafranchian Pardines locality. The faunas are not outstanding but the stratigraphic succession is clear; moreover, volcaniclastic rocks are intercalated. Etouaires is associated with a potassium-argon date of 3.5 million years before present, and Roccaneyra is associated with a date of 2.6 m.y.

Asia

The Siwaliks succession of northern India and adjacent Pakistan is renowned through the past 150 years as a clear-cut demonstration of the stratigraphic succession of middle Miocene into early Pleistocene land mammal faunas. Here Turolian-equivalent Dhok Pathan faunas lie under Early Villafranchian Tatrot faunas, which lie under Late Villafranchian and Early Pleistocene Pinjor faunas.

Africa

In the Omo district of Ethiopia, and nearby, in the Lake Turkana district, there is a succession of faunas, stratigraphically arranged, which show a sequence from earlier Pliocene, Langebaanian, through later Pliocene, Makapanian, and on into earlier Pleistocene.

North America

Thanks to the work of C.W. Hibbard, the classic district for stratigraphic demonstration of the latest Miocene through Pliocene and Pleistocene mammalian faunas of the North American continent is in southwestern Kansas, in the United States. From Hemphillo-Blancan through Pliocene, these comprise Saw Rock, Fox Canyon, Rexroad, Bender, Deer Park, Sanders, Seger, and Borchers. Other districts—Panhandle of Texas, southeastern Arizona, southeastern California—show a Blancan fauna on top of Hemphillian or a succession of faunas within the Blancan.

PLIOCENE LITHOSTRATIGRAPHIC UNITS

Aterir beds—Langebaanian (in part)—Kenya
Ballard Formation (in part)—Blancan—Kansas
Blanco Formation—Blancan—Texas

Cache Formation—Blancan—California
Chemeron Formation (Aterir beds)—Langebaanian (in part)—
 Kenya
Chinglo Formation—Early Pliocene—Shanxi
Chiwondo beds—Makapanian—Malawi
Coso Formation—Blancan—California
Dacian beds and Stage—approximately Ruscinian—Romania
Denan Dora Member, Hadar Formation—Makapanian—Ethiopia
Djetis beds—Late Pliocene or Early Pleistocene—Java
Ekora Formation—Langebaanian—Kenya
Gila Conglomerate, in part—Blancan—Arizona
Glenns Ferry Formation—Blancan—Idaho
Grange Burn Formation—Pliocene—Victoria, Australia
Hadar Formation—Langebaanian?, Makapanian, and Early Pleis-
 tocene—Ethiopia
Hamada Damous beds, lower—Makapanian—Tunisia
Kada Hadar Member, Hadar Formation—Makapanian—Ethiopia
Kaiso Formation—Makapanian (and later?)—Uganda
Kali-Glagah Formation (marine?)—Late Pliocene—Java
Kanapoi Formation—Langebaanian—Kenya
Kaperyon Formation—Langebaanian, in part—Kenya
Karmosit beds (Chemeron Fm.)—Lothagamian-Langebaanian—
 Kenya
Koobi Fora Formation—Makapanian—Kenya
Kubi Algi beds—Langebaanian—Kenya
Laetoli or Laetolil beds—Langebaanian-Makapanian—Tanzania
"Livermore Gravels", in part—Blancan—California
Lothagam Formation, upper—Langebaanian—Kenya
Makapansgat beds I–V—Makapanian—South Africa
Mampuwordu Sands—Late Pliocene—South Australia
Meade Formation—Blancan—Kansas
Milet Member—Ruscinian—Turkey
Missler Silt Member, Ballard Fm.—Blancan—Kansas
Mono Lake beds, in part—Blancan—California
Mursi Formation—Pliocene—Ethiopia
Olduvai I beds—Makapanian—Tanzania
Otibanda Formation—Pliocene—New Guinea
lll Ranch beds—Blancan—Arizona
Palm Springs Formation, in part—Blancan—California
Panaca Formation (in part?)—Blancan—Nevada
Paso Robles Formation, in part—Blancan—California
Petaluma Formation, upper—Blancan—California
Rexroad Formation—Blancan—Kansas
St. David Formation—Blancan—Arizona
Salt Lake Group, upper, in part—Blancan—Utah-Idaho
San Benito Gravels, in part—Blancan—California
Sand Draw beds—Blancan—Nebraska
San Joaquin Clay—Blancan—California
Santa Clara Formation, in part—Blancan—California
Shagou Formation—Late Pliocene—Yunnan
Shungura Formation—Langebaanian and Makapanian—Ethiopia
Sidi Hakoma Member, Hadar Formation—Makapanian—Ethio-
 pia
Star Valley beds—Blancan—Wyoming
Tassajara Formation, upper, in part—Blancan—California
Tehama Formation—Blancan—California
Truckee Group, in part—Blancan—Nevada
Tulare Formation—Blancan and later—California

Usno Formation—Makapanian—Ethiopia
Vaal River Gravels—Makapanian and later—South Africa
Varswater Formation—Langebaanian—South Africa
Verde Formation—Blancan—Arizona

PLIOCENE LOCALITIES

Afar—Makapanian—Ethiopia
Aïn Brimba—Makapanian—Tunisia
Aïn Boucherit—Makapanian—Algeria
Aïn el Bey—Langebaanian or Makapanian—Algeria
Aïn el Hadj Baba—Makapanian—Algeria
Aïn Fouarat—Makapanian—Morocco
Aïn Hanech—Makapanian—Algeria
Aïn Jourdel—Makapanian—Algeria
Akça assemblage—Villafranchian—Turkey
Akçaköy—Villafranchian—Turkey
Akcayir—Ruscinian or Villafranchian—Turkey
Akkerman—Ukraine, USSR—Ruscinian?
Alcolea de Calatrava—Ruscinian—Spain
Alcoy—Ruscinian—Spain
Aleksandrovskiy fort—Ruscinian?—Kazakhstan
Almoradier—Ruscinian—Spain
Anan'yev 4, 5—Ruscinian?—Ukraine, USSR
Anastas'yevka—Villafranchian?—Ukraine, USSR
Andreyev estuary—Ruscinian—Ukraine, USSR
Andriashevka—Ruscinian—Ukraine, USSR
Anita mine—Blancan—Arizona
Ankara N.E.—Ruscinian?—Turkey
Ankara S.W.—Late Ruscinian—Turkey
Apsheron Range—Ruscinian?—Caucasus, USSR
Argoub el Hafid   (See Oued Akrech.)
Arondelli—Villafranchian—Italy
Arquillo 3—Ruscinian—Spain
Aterir beds sites—Langebaanian—Kenya
Awe—Pliocene—New Guinea
Aydyrlya—Villafranchian—Ural Mts., USSR
Ayman-Kuyu—Ruscinian—Crimea, USSR
Azovstal quarry—Villafranchian—Ukraine, USSR
Baard's Quarry (Langebaanweg)—Langebaanian—South Africa
Babadat—Late Ruscinian—Turkey
Baccinello V3—Ruscinian?—Italy
Balaruc 2—Villafranchian—France
Bamian basin—Pliocene—Afghanistan
Banguo Basin—Pliocene—Yunnan
Bargo—Villafranchian—Italy
Barót-Köpec—Ruscinian—Romania
Baza—Villafranchian—Spain
Beçin-Kolanyaka Tepe—Villafranchian?—Turkey
Beck Ranch—Blancan—Texas
Behanga—Makapanian—Uganda
Balaya River—Pliocene—Caucasus, USSR
Belekler—Villafranchian?—Turkey
Bel Hacel—Makapanian—Algeria
Beni Foudda   (See St. Arnaud.)
Benson—Blancan—Arizona
Berbeşti—Villafranchian—Romania?   (= Beresti?)
Berdyansk—Villafranchian—Ukraine, USSR
Beremend 5—Villafranchian—Hungary
Besymiannae—Villafranchian—USSR
Beteke and Beteke fauna—Villafranchian—USSR

Bethelehem—Villafranchian—Jordan
Bharmar—(= Ruscinian?)—Pakistan
Bhittanni Range—(= Villafranchian)—Pakistan
Binagady (Azerbadzhan, Baku)—Villafranchian—USSR
Bolt's Farm—Makapanian and Early Pleistocene—South Africa
Bone Gulch—Plio-Pleistocene—New South Wales, Australia
Boucheron   (See Si Abd el Aziz.)
Bou Regreg—Makapanian—Morocco
Bourille—Makapanian—Ethiopia
Boz-Dag Range—Villafranchian—Caucasus, USSR
Bra—Villafranchian—Italy
Bugiuleşti—Villafranchian—Romania?
Bugoma (Kaiso)—Makapanian—Uganda
Buxton-Norlim—Makapanian—South Africa
Bystraya River—Ruscinian—Caucasus, USSR
Cairy Khapry—Villafranchian—USSR
Çalta—Ruscinian—Turkey
Çamli Bayir—Villafranchian and Early Pleistocene—Turkey
Çankiri—Late Turolian or Ruscinian—Turkey
Casino—Ruscinian?—Italy
Castelnuovo del Sabbioni—Villafranchian—Italy
Chad   (See Lake Chad.)
Chagny—Villafranchian—France
Chamar—(= Ruscinian?)—Mongolia
Cheleken Island—Villafrachian—USSR
Chermeron—Langebaanian—Kenya
Chiachiashan—Hopei—(= Villafranchian)
Chilhac—Villafranchian—France
Chinchilla fauna—Plio-Pleistocene—Queensland, Australia
Chingchuan—Pliocene—Gansu
Chinglo Formation sites—(= Villafranchian)—Shanxi
Chingning district (Shui-Chuan-Kou.)
Chiwondo beds sites—Makapanian—Malawi
Chlum cave—Villafranchian—Czechoslovakia
Chono Haria section (Hara Usu Nur and Hara Nur)—(= Villafranchian)—Mongolia
Chuagun (Tana Range)—Pliocene—Caucasus, USSR
Chumay—Ruscinian—Moldavia
Cinaglio d'Asti—Villafranchian—Italy
Cochise County Misc.—Blancan—Arizona
Comosi Wash—Blancan—Arizona
Cordoba fissure—Villafranchian—Spain
Coupet—Villafranchian—France
Csarnota—Ruscinian—Hungary
Da-fou-sche (Pien-Hsien)—(= Villafranchian)—Shaanxi
Dalai-nor—(= Villafranchian)—Nei Mongol
Daourat—Makapanian—Morocco
Datça—Pliocene—Turkey
Dawrankhel—(= Ruscinian)—Afghanistan
Delibayir Siirti—Pliocene—Turkey
Delmont—Blancan—South Dakota
Denizli N.E.—Ruscinian—Turkey
Dermedzhi (Moskovey)—Ruscinian—Moldavia
Dinar (Akçaköy)—Ruscinian—Turkey
Djebel Mellah—Makapanian—Tunisia
Domnitsy—Ruscinian—Ukraine, USSR
Duncan—Blancan—Arizona
Dzhenama River—(= Ruscinian?)—Kazakhstan
Dzheyrangoli River valley—Villafranchian—Caucasus, USSR
East Turkana (= E. Lake Rudolph)—Langebaanian into Early Pleistocene—Kenya

Ekora Formation sites—Langebaanian—Kenya
Elbistan (Maras)—Ruscinian—Turkey
El Gara (See Si Abd el Aziz.)
Erpfingen—Villafranchian—Germany
Ertemte—(= Turolian? and = Ruscinian?)—Nei Mongol
Erzurum—Ruscinian—Turkey
Escorihuela—Ruscinian?, Villafranchian?—Spain
Etouaires—Villafranchian—France
Farladany—Ruscinian—Moldavia
Fisherman's Cliff—Plio-Pleistocene—New South Wales, Australia
Flat Tire fauna—Blancan—Arizona
Fornace RDB (Villafranca d'Asti)—Villafranchian—Italy
Forsyth's Bank fauna—Pliocene Victoria, Australia
Fouarat   (See Aïn Fouarat.)
Fox Canyon—Blancan—Kansas
Gambetta—Makapanian—Algeria
Gara Ziad—Pliocene—Morocco
Gaville—Villafranchian—Italy
Gidhniya district—(= Villafranchian)—Nepal
Gödöllö—Ruscinian—Hungary?
Goldashevka—Ruscinian—Ukraine, USSR
Goraffe 1 and 2—Ruscinian—Spain
Goz-Kerki—Chad—Makapanian—Chad
Grange Burn (Forsyth's Bank)—Plio-Pleistocene, Pliocene marine—Australia
Grossulovo 2—Pliocene—Ukraine
Groznyy—Villafranchian—Caucasus
Grushevka—Ruscinian—Caucasus
Gülyazi—Villafranchian—Turkey
Gundersheim 1—Ruscinian—Germany
Guyot Farm (See Sidi Abdallah.)
Hadar Formation sites—Langebaanian into Early Pleistocene—Ethiopia
Hadj Rona—(= Ruscinian)—Afghanistan
Hagerman—Blancan—Idaho
Haile XVa—Blancan—Florida
Hajnáčka—Villafranchian—Czechoslovakia
Hamada Damous—Makapanian—Tunisia
Hamareyama Formation site—(= Ruscinian?)—Japan
Hamilton fauna (Grange Burn)—mid-Pliocene—Victoria, Australia
Hammett—Blancan—Idaho
Hasanpaşa—Ruscinian?—Turkey
Hauterives—Ruscinian—France
Hautimagne—Ruscinian—France
Henry Ranch—Blancan—Arizona
Hieh Sand Pit = Delmont—Blancan—South Dakota
Hirgis-Nur 2—(= Ruscinian)—W. Mongolia
Holu-Fluss—(= Ruscinian)—Tuviskaja ASSR
Homa Mt.—Pliocene?—Kenya
Houfeng (Chinglo)—(= Ruscinian?)—Shanxi
Hsiachaohwan—Villafranchian—China
Ichkeul-Bizerte—Makapanian—Tunisia
Ileret (East Turkana)—Makapanian and Early Pleistocene—Kenya
Iles Medas (Gerona)—Villafranchian—Spain
Incisa Belbo—Villafranchian—Italy

Irimeşti—Villafranchian—Romania?

Irrawaddy ''series'', upper, in part—(= Villafranchian and later)—Burma

Ivanovce—Ruscinian—Poland

Kadzielna—Villafranchian—Poland

Kagula River basin—Ruscinian—Moldavia

Kaiso Village—Makapanian and later—Uganda

Kali Glagah (= Villafranchian)—Java

Kalinköy—Pliocene—Turkey

Kamenskoe—Ruscinian—Romania?, USSR

Kamii—Villafranchian and later—Turkey

Kamyk—Villafranchian—Poland

Kanam—Makapanian—Kenya

Kanapoi Formation sites—Langebaanian—Kenya

Karmosit beds sites—Makapanian—Kenya

Karpathos—Ruscinian—Greece

Katorora (Chiwondo)—Makapanian—Malawi

Kavakdere—Ruscinian—Turkey

Kayseri ESE—Ruscinian—Turkey

Kazinga Channel—Villafranchian—Uganda

Keams Canyon—Blancan—Arizona

Kebili—Makapanian—Tunisia

Keim Formation sites—Blancan—Nebraska

Kemiklitepe—Ruscinian?—Turkey

Kharuzin—Ruscinian—USSR

Khodzhent—Villafranchian—Tadzhikistan, USSR

Kholobolchi Nor 2—(= Villafranchian)—Mongolia

Kholu (= Ruscinian and = Villafranchian)—Mongolia

Kiev suburbs—Pliocene—Ukraine

Kikagati—Makapanian—Africa

Kirnasivka—Ruscinian?—Ukraine

Kisatibi—Pliocene—Caucasus

Kishinev 2—Ruscinian—Ukraine

Kislang—Villafranchian—Poland?

Kizikha—Villafranchian—West Siberia

Klein Zee—Pliocene—South Africa

Klyastitsy—Ruscinian?—Ukraine

Kolinga (Lake Chad)—Makapanian—Chad

Koobi Fora I (East Turkana)—Makapanian—Kenya

Köprübaşi—Villafranchian—Turkey

Kossiakino—Ruscinian—Romania?

Koulá (Lake Chad)—Makapanian—Chad

Koybyn River—(= Villafranchian)—Kazakhstan

Krasnoye—Ruscinian?—Ukraine

Krasnyye Kolodtsy—Ruscinian?—Caucasus

Krements—Ruscinian?—Ukraine

Krivaya ravine—Ruscinian?—Ukraine

Kryzhopol—Ruscinian?—Ukraine

Ksapry (= Cairy Khapry?)—Villafranchian—USSR

Kubi Algi—Langebaanian—Kenya

Kuchurgan—Ruscinian—Ukraine

Kurgloye—Villafranchian—Caucasus

Kvabebi—Ruscinian—Georgian SSR

Laetoli (Laetolil)—Makapanian—Tanzania

Lagman (Jalalabad basin)—Pliocene—Afghanistan

Laiatico—Villafranchian—Italy

La Juliana (Murcia)—Ruscinian—Spain

Lake Chad—Langebaanian and Makapanian and later—Chad

Lake Palankarinna—Pliocene—South Australia

Lake Turkana (Lake Rudolph)—Langebaanian and later—Kenya

Langebaanweg—Langebaanian and Makapanian—South Africa

La Puebla de Valverde—Villafranchian—Spain

Las Tunas—Blancan—Baja California, Mexico

Lataband (Sarobi basin)—(= Ruscinian)—Afghanistan

Layer Cake fauna—Blancan—California

Layna—Ruscinian—Spain

Lazaret cave (Hérault)—Ruscinian—France

Leffe—Villafranchian—Italy

Likhtental—Ruscinian?—Ukraine

Lingyi district—(= Villafranchian)—Shanxi

Lintai—(= Villafranchian)—Gansu

Lipetskoye 1, 2—Ruscinian?—Ukraine

Lok-Batan—Ruscinian?—Caucasus

Lothagam 3 (upper Lothagam Fm.)—Langebaanian—Kenya

Lucheshti—Ruscinian?—Moldavia

Lyapino—Villafranchian—USSR

Maghreb district—Villafranchian—Tunisia   (many sublocalities)

Ma Kai Valley—Villafranchian—Yunnan

Makapansgat—Makapanian—South Africa

Malgobok—Villafranchian—Caucasus

Malusteni—Villafranchian—Romania?

Maritsa—Ruscinian—Rhodes

Mariupol—Villafranchian—Ukraine

Markhal—Ruscinian?—Pakistan

Marwat Range—(= Villafranchian)—Pakistan

Mascara—Langebaanian?—Algeria

Mashuk Mt. (Parekal Cliff)—Villafranchian—Caucasus

Mateesti—Villafranchian—Romania

Maxum—Blancan—California

Médas—Villafranchian—Spain

Mena House (Cairo)—Makapanian—Egypt

Mildura West—Plio-Pleistocene—New South Wales

Mombercelli—Villafranchian—Italy

Montecarlo monastery—Villafranchian—Italy

Montevarchi—Villafranchian—Italy

Montopoli—Villafranchian—Italy

Montpellier—Ruscinian—France

Moreda 1 and 2—Villafranchian—Spain

Morozovka—Ruscinian?—Ukraine

Motru—Ruscinian—Romania

Mt. Zhevaknov—Villafranchian—Ukraine

Mursi Formation sites (Lower Omo)—Ruscinian—Ethiopia

Mwenirondo 1 (Chiwondo)—Makapanian—Malawi

Mwimbi North (Chiwondo)—Makapanian—Malawi

Naftalan—Villafranchian—Caucasus

Nakhichevan—Villafranchian—Caucasus

Nanking—(= Ruscinian?)—China

Navrukho fauna—(= Villafranchian)—(Tadjikstan) USSR and Afghanistan

Nihewan—(= Villafranchian)—Hopei

Nikolayevka—Villafranchian—Crimea, USSR

Nîmes fissures—Ruscinian and later—France

Nogaysk—Villafranchian—USSR?

Norwich Crag—Villafranchian—England

Novaya Yemetovka—Turolian or Ruscinian—Ukraine

Novo-Nikolayevka (See Nikolayevka.)

Novorosiya—Ruscinian?—Ukraine

Novyye Dragusheny—Ruscinian—Moldavia

Nurnus village (Zang River)—Villafranchian—USSR

Nyabrogo (Kaiso)—Makapanian—Uganda
Nyamavi region (Lake Albert)—Pliocene—Zaire
Nyawiega (Kaiso)—Makapanian—Uganda
Odessa 2 and Odessa Catacombs—Ruscinian—Ukraine
Olduvai 1—Makapanian—Tanzania
Olivola—Villafranchian—Italy
Omo sites—Langebaanian, Makapanian, and later—Ethiopia
Orlov Farm—Villafranchian—Ukraine
Orrios—Ruscinian—Spain
Ostramos 1—Ruscinian
Ouadi Derdemi (Lake Chad)—Makapanian—Chad
Oued Akrech—Langebaanian—Morocco
Oued Tessa—Pliocene—Tunisia
Palankarinna   (See Lake Palankarinna.)
Panaca—Blancan—Nevada
Pardines—Villafranchian—France
Peñas—Ruscinian—Spain
Peschany—Ruscinian?—Ukraine
Pichugino 2—Ruscinian—Ukraine
Pieve Fosciana—Villafranchian—Italy
Pinjor—(= Villafranchian)—India and Pakistan
Plešivec—Villafranchian—Czechoslovakia
Podlesice—Ruscinian—Poland
Port Charlotte—Blancan—Florida
Providans Mine—Ruscinian?—Kerch Peninsula, USSR
Pruta River basin—Villafranchian—Moldavia
Psecopus—Villafranchian—Romania
Puchen (Tungyao)—(= Ruscinian and Villafranchian)—Shaanxi
Puebla de Almuradier—Ruscinian—Spain
Puimoisson—Villafranchian—France
Pul-e Charki—(= Ruscinian and Turolian)—Afghanistan
Püspökfürdö—Villafranchian—Hungary?
Rakhanh Lesovyye—Ruscinian?—Ukraine
Ravin de la Chapelle—Villafranchian—France
RDB (See Fornace RDB.)
Rębielice Królewskie—Villafranchian—Poland
Red Crag—Villafranchian—England
Red Knolls—Blancan—Arizona
Rexroad—Blancan—Kansas
Ringold   (See White Bluffs.)
Roccaneyra—Villafranchian—France
St. Arnaud—Makapanian—Algeria
St. Charles—Makapanian—Algeria
St. Donat—Makapanian—Algeria
St. Eugène—Makapanian—Algeria
St. Vallier—Villafranchian—France
Salchi River basin—Villafranchian—Moldavia
San Donato—Villafranchian—Italy
San Joaquin Formation sites—Blancan—California
Santa Fe River—Blancan—Florida
Saragata Deare (Afar)—Langebaanian—Ethiopia
Sarasota—Blancan—Florida
Sarobi—Early Pliocene—Afghanistan
Sarion 1—Villafranchian—Spain
Sarion 2—Ruscinian—Spain
Sarzana—Sarzanello—Ruscinian or Villafranchian—Italy
Saw Rock—Blancan—Kansas
Semliki—Mioc.-Pliocene—Zaire
Semipalatinsk—Ruscinian?—Caucasus
Senèze—Villafranchian—France
Sête—Ruscinian—France

Sétif   (See St. Arnaud.)
Seynes—Villafranchian—France
Shikhovo—Villafranchian—Azerbaydzhan USSR
Shirokaya—Ruscinian?—Ukraine
Shungura Formation sites (Omo)—Makapanian and later—Ethiopia
Si Abd el Aziz—Makapanian—Morocco
Sidi Abdallah—Villafranchian—Morocco
Sidi Bou Kouffa—Villafranchian—Tunisia
Sidney-on-Vaal—Pliocene—South Africa
Sinda–Mohari–Lower Semliki—Pliocene—Zaire
Sivas (Şarkişla)
Springbok—Langebaanian or Makapanian—South Africa
Starovo—Ruscinian?—Ukraine
Stauropol or Stavropol 2—Ruscinian—Caucasus
Stavropol 3 or 4—Villafranchian—Caucasus
Sterkfontein (in part)—Makapanian—South Africa
Stranzendorf—Villafranchian—Austria?
Taganrog—Villafranchian—USSR
Taman peninsula—Villafranchian—Caucasus
Tartul de Sal'chi—Villafranchian—Moldavia
Tash-kala—Villafranchian—Caucasus
Tatrot—(= Villafranchian)—India and Pakistan
Taung—Makapanian—South Africa
Taunton—Blancan—Washington
Tegelen—Villafranchian—Netherlands
Tehama Formation sites—Blancan—California
Tehrenshan district—Pliocene—Xinjiang
Tekkeköy–Hüyük Tepe—Villafranchian?—Turkey
Tekman—Ruscinian—Turkey
Trifeshty—Ruscinian?—Moldavia
Tsagan-Nor—(= Villafranchian?)—Mongolia
Tschono–Choriach 2—(= Ruscinian)—Mongolia
Tsochin—(= Villafranchian)—Taiwan
Tuluceşti—Villafranchian—USSR?, Romania?
Tusker—Blancan—Arizona
Ulanbulan (Dzungaria)—(= Villafranchian)—Xinjiang
Ulhu—Pliocene—Xinjiang
Uraha Hill–Chiwondo—Makapanian—Malawi
Usatovo-Odessa—Ruscinian?—Ukraine
Usno Formation sites–Omo—Makapanian—Ethiopia
Vaal River Gravels sites—Makapanian? and later—South Africa
Val d'Arno localities—Villafranchian—Italy
Val d'Era (Siena)—Ruscinian—Italy   (= Casino?)
Verde Formation sites—Blancan—Arizona
Villaba Alta—Ruscinian—Spain
Villafranca d'Asti   (See Arondelli and Fornace RDB.)
Villaroya—Villafranchian—Spain
Vinodel'noye—Villafranchian—Caucasus
Virginia—Pliocene—South Africa
Vogel River "series"—Pliocene—Tanzania
Voronkovo—Ruscinian?—Ukraine
Weinan—(= Villafranchian?)—Shaanxi
Weybourne Crag (E. Runton)—Villafranchian—England
Weźe—Ruscinian—Poland
White Bluffs—Blancan—Washington
Wichman—Blancan—Nevada
Wölfersheim—Ruscinian—Germany

Yankton—Blancan—South Dakota
Yassiören—Ruscinian?—Turkey
Yellow Sands, Mursi Fm. sites (Omo)—Makapanian—Ethiopia
Yevpatoriya 2—Villafranchian—Crimea
Yuanmo—(= Villafranchian)—Yunnan
Yukari Sögütönü—Villafranchian—Turkey
Zamkowa Dolna Cave—Villafranchian—Poland
Zhmerinka (Yaroshenko)—Ruscinian?—Ukraine
Zhuravlevka—Ruscinian?—Ukraine

BIBLIOGRAPHY

APOSTOL, L., and ENACHE, C. 1979. Étude de l'espèce *Dicerorhinus megarhinus* (de Christol) du bassin carbonifère de Motru (Roumanie). *Trav. Mus. Hist. Nat. "Grigore Antiqua"* 20:533–540.

AZZAROLI, A. 1977. The Villafranchian Stage in Italy and the Plio-Pleistocene boundary. *Gior. Geol. Bologna* 41:61–79.

AZZAROLI, A., and VIALLI, V. 1971. Villafranchian. *Gior. Geol. Bologna* 37:221–232.

BARNES, L.G. 1976. Outline of eastern North Pacific fossil cetacean assemblages. *Syst. Zool.* 23:321–343.

BARTHOLOMAI, A. 1967. *Troposodon*, a new genus of fossil Macropodinae (Marsupialia). *Queensland Mus., Mem.*, 15:21–33.

BECKER-PLATEN, J.D., SICKENBERG, O., and TOBIEN, H. 1975. 2. Die Gliederung der känozoischen Sedimente der Türkei nach Vertebraten-Faunengruppen. In Sickenberg, et al., Der Gliederung des höheren Jungtertiärs und Altquartärs in der Türkei nach Vertebraten und ihre Bedeutung für die internationale Neogen-Stratigraphie, pp. 19–100. *Geol. Jahrb.* (B) 15:1–167.

BJORK, P.R. 1970. The Carnivora of the Hagerman local fauna (Late Pliocene) of southwestern Idaho. *Trans. Am. Philos. Soc.* n.s. 60:1–54.

BRANDY, L.D. 1979–0220. Rongeurs nouveaux du Néogène d'Afghanistan. *C. R. Acad. Sci., Paris*, Sér. D, 289:81–83.

BRUNET, M., CARBONNEL, J.P., HEINTZ, E., and SEN, S. 1980. Première découverte de Vertébrés dans les formations continentales de Pul-e Charkhi, bassin de Kabul, Afghanistan; implications stratigraphiques. *Bull. Mus. Natl. d'Hist. Nat.*, Paris, Sect. C: Sci. de la Terre, 2:277–285.

CHURCHER, C.S., and RICHARDSON, M.L. 1978. Equidae. In *Evolution of African Mammals*, ed. V.J. Maglio and H.B.S. Cooke, pp. 379–422. Cambridge: Harvard Univ. Press.

COOKE, H.B.S. 1974. Plio-Pleistocene deposits and mammalian faunas of Eastern and Southern Africa. Proc. 5 Congr. Neogene Medit., *Bur. Rech. Géol. Minières, Mém.*, 28:99–108.

COOKE, H.B.S. 1978. Africa: The physical setting. In *Evolution of African Mammals*, ed. V.J. Maglio and H.B.S. Cooke, pp. 17–45. Cambridge: Harvard Univ. Press.

COOKE, H.B.S., and CORYNDON, S.C. 1970. Pleistocene mammals from the Kaiso Formation and other related deposits in Uganda. In *Fossil Vertebrates of Africa*, vol. 2, ed. L.S.B. Leakey and R.J.G. Savage, pp. 107–224. New York: Academic Press.

DAAMS, R., and VAN DE WEERD, A. 1980. Early Pliocene small mammals from the Aegean island of Karpathos (Greece) and their palaeogeographic significance. *Geol. Mijnbouw.* 59:327–331.

DAVID, A.I. 1967. On the Roussillon mammalian fauna of Moldavia. *Izv. Akad. Nauk Moldav., SSR* 1967, 4:26–28. (In Russian; translated by G. Shkurkin, 1970.)

ESU, D., and KOTSAKIS, T. 1979. Restes de vertébrés et de mollusques continentaux dans le Villafranchien de Sardaigne. *Géobios* 12:101–106.

FEJFAR, O. 1964A. The lower Villafranchian vertebrates from Hajnáčka near Filakovo in southern Slovakia. *Rozprovy Ústřed. ustavu Geol.* 30:1–115.

FEJFAR, O. 1976. Plio-Pleistocene mammal sequences. In *Quaternary Glaciations in the Northern Hemisphere,* ed. D.J. Easterbrook and V. Sibrava, pp. 351–366. Rept. No. 3. I.G.C.P. Proj. 73/1/24.

FICCARELLI, G. 1979. The Villafranchian machairodonts of Tuscany. *Palaeontogr. Ital.* 71:17–26.

FICCARELLI, G., and TORRE, D. 1974. Nuovi reporti del gatto villafranchiano di Olivola. *Atti. Soc. Tosc., Sci. Nat., Mem.,* 81:312–317.

FICCARELLI, G., and TORRE, D. 1975. "*Felis*" *christoli* Gervais delle sabbie plioceniche di Montpellier. *Boll. Soc. Paleontol. Ital.* 14:217–220.

FORSTEN, A. 1978. *Hipparion* and possible Iberian–North African connections. *Zool. Fenn.* 15:294–297.

FRANCAVILLA, F., BERTOLANI-MARCHETTI, D., and TOMADIN, L. 1970. Ricerche stratigrafiche, sedimentologiche e palinologiche sul Villafranchiano tipo. *Gior. Geol. Bologna* 36:701–742.

FRY, W.E., and GUSTAFSON, E.P. 1974. Cervids from the Pliocene and Pleistocene of central Washington. *J. Paleontol.* 48:375–386.

GHENEA, C. 1970. Stratigraphy of the Upper Pliocene–Lower Pleistocene interval in the Dacic Basin (Romania). *Palaeogeogr., Palaeoclim., Palaeoecol.* 8:165–174.

GUSTAFSON, E.P. 1978. The vertebrate faunas of the Pliocene Ringold Formation, south-central Washington. *Univ. Oreg., Bull. Mus. Nat. Hist.* 23:1–62.

HENDY, Q.B. 1974. The Late Cenozoic Carnivora of the South-western Cape Province. *Ann. S. Af. Mus.* 63:1–369.

HENDY, Q.B. 1978A. Late Tertiary Hyaenidae from Langebaanweg, South Africa, and their relevance to the phylogeny of the family. *Ann. S. Af. Mus.* 76:265–297.

HENDY, Q.B. 1978B. Late Tertiary Mustelidae (Mammalia, Carnivora) from Langebaanweg, South Africa. *Ann. S. Af. Mus.* 76:329–357.

HIRSCHFELD, S.E., and WEBB, S.D. 1968. Pliocene-Pleistocene megalonychid sloths of North America. *Univ. Flor., State Mus., Bull. Biol. Sci.* 12:213–296.

HOOIJER, D.A. 1958B. An early Pleistocene mammalian fauna from Bethlehem. *Bull. Brit. Mus. (Nat. Hist.)* 3:265–292.

HÜRZELER, J. 1967. Nouvelles découvertes de mammifères dans les sédiments fluviolacustres de Villafranca d'Asti. *Colloq. Internat., Fr., Cent. Nat. Rech. Sci.* 163:633–636.

JAEGER, J.-J. 1979. Les faunes de Rongeurs et de Lagomorphes du Pliocène et du Pleistocène d'Afrique orientale. *Soc. Géol. Fr.* 21:301–308.

JOHANSON, D.C., and WHITE, T.D. 1979. A systematic assessment of early African hominids. *Science* 203:321–330.

JOHANSON, D.E., WHITE, T.D., and COPPENS, Y. 1978. A new species of the genus *Australopithecus* (Primates: Hominidae) from the Pliocene of eastern Africa. *Kirtlandia* 28:1–14.

JOHNSON, N.M., OPDYKE, N.D., and LINDSAY, E.H. 1975. Magnetic polarity stratigraphy of Pliocene-Pleistocene terrestrial deposits and vertebrate faunas, San Pedro Valley, Arizona. *Bull. Geol. Soc. Am.* 86:5–12.

KOENIGSWALD, W. VON. 1980. Schmalzstruktur und Morphologie in den Molaren der Arvicolidae (Rodentia) *Abh. Senckenb. Naturforsch. Gesell.* 539:1–129.

KOWALSKI, K. 1964. (Palaeoecology of mammals from the Pliocene and early Pleistocene of Poland.) *Polska. Akad. Nauk, Acta Theriologica* 8:73–88.

KOWALSKI, K. 1974. Remains of Gerbillinae (Rodentia, Mammalia) from the Pliocene of Poland. *Bull. Akad. Polon. Sci.* 22:591–595.

KRETZOI, M. 1956. Die altpleistozänen Wirbeltierfaunen des Villanyer Gebirges. *Geol. Hung., Ser. Paleontol.* 27:1–23.

KURTÉN, B. 1968. *Pleistocene Mammals of Europe.* London: Weidenfeld and Nicolson, Publ.

LINDSAY, E.H., JOHNSON, N.M., and OPDYKE, N.D. 1975. Preliminary correlation of North American land mammal ages and geomagnetic chronology. In *Studies on Cenozoic Paleontology and Stratigraphy in Honor of Claude W. Hibbard. Univ. Mich. Pap. Paleontol.* 12:111–119.

LINDSAY, E.H., OPDYKE, N.D., and JOHNSON, N.M. 1980. Pliocene dispersal of the horse *Equus* and late Cenozoic mammalian dispersal events. *Nature* 287:135–138.

LIU, H., TANG, Y., and YOU, Y. 1973. A new species of *Stegodon* from Upper Pliocene of Yuanmou, Yunnan. *Vertebr. PalAsiatica* 11:192–200.

MAGLIO, V.J. 1973. Origin and evolution of the Elephantidae. *Trans. Am. Philos. Soc.,* n.s., 63 (pt. 3):1–149.

MAGLIO, V.J., 1974. Late Tertiary fossil vertebrate successions in the northern Gregory Rift, East Africa. *Ann. Geol. Surv. Egypt* 4:269–386.

MANKINEN, E.A., and DALRYMPLE, G.B. 1979. Revised geomagnetic polarity time scale for the interval 0 to 5 million years B.P. *J. Geophys. Res.* 84:615–626.

MARSHALL, L.G., HOFFSTETTER, R., and PASCUAL, R. 1982. Geochronology of the continental mammal-bearing Tertiary of South America. Fieldiana. (In press.)

MAY, S.R. 1981. *Repomys* (Mammalia: Rodentia, gen. nov.) from the late Neogene of California and Nevada. *J. Vertebr. Paleontol.* 1:219–230.

MOLINA, E., PEREZ-GONZALEZ, A., and AGUIRRE, E. 1972. Observaciones geologicas en el Campo de Calatrava. *Estud. Geol.* (Inst. Investig. Geol. ''Lucas Mallada'') 28:3–11.

MONTENAT, C., and DE BRUIJN, H. 1976. The Ruscinian rodent faunule from La Juliana (Murcia); its implication for the correlation of continental and marine biozones. *Koninkl. Ned. Akad. Wetensch., B,* 79:245–255.

POPE, G.G., 1982. Hominid evolution in eastern and southeastern Asia. Ph.D. dissertation. Univ. Calif. Berkeley, Dept. Anthropology.

QI TAO. 1979. A note on some species of fossil mammals from the late Pliocene of the Damiao Region, Inner Mongolia. *Vertebr. PalAsiatica* 17:259–260.

QIU ZHAN-XIANG, HUANG WEI-LONG, and GUO ZHI-HUI. 1979. Fossil Hyaenidae from the Qingyang County, Gansu Province. *Vertebr. PalAsiatica* 17:200–221.

QIU, ZHU-DING. 1979. Fossil Pliocene mammals from several localities in North China. *Vertebr. PalAsiatica* 17:222–235.

REIG, O.A. 1979. Roedores cricetides del Plioceno superior de la Provincia de Buenos Aires (Argentina). *Mus. Munic., Cienc. Natur., Mar del Plata,* 2:164–190.

REPENNING, C.A. 1976. *Enhydra* and *Enhydriodon* from the Pacific Coast of North America. *J. Res. U.S. Geol. Surv.* 4:305–315.

REPENNING, C.A. 1980. Faunal exchange between Siberia and North America. *Proc. 5th Biennial Conf. Am. Quat. Assoc., Canad. Jour. Anthropol.* 1:37–44.

REPENNING, C.A. 1983? Biochronology of the microtine rodents of the United States. In symposium on status of vertebrate stratigraphy and geochronology, ed. M.O. Woodburne. Calif. Univ., Publ. Geol. Sci. (In press.)

REPENNING, C.A., and TEDFORD, R.H. 1977. Otarioid seals of the Neogene. *U.S. Geol. Surv. Prof. Pap.* 992:1–93.

SABATIER, M. 1979. Les Rongeurs fossiles de la formation de Hadar et leur intérêt paléoécologique. *Bull. Géol. Soc. Fr.* 21:309–311.

SAVAGE, D.E., and CURTIS, G.H. 1970. The Villafranchian Stage-Age and its radiometric dating. *Geol. Soc. Am. Spec. Pap.* 124:207–231.

SAVAGE, R.J.G. 1978. Carnivora. In *Evolution of African Mammals,* ed. V.J. Maglio and H.B.S. Cooke, pp. 249–267. Cambridge: Harvard Univ. Press.

SCHULTZ, C.B., MARTIN, L.D., TANNER, L.G., and CORNER, R.G. 1978. Provincial land mammal ages for the North American Quaternary. *Trans. Nebr. Acad. Sci.* 5:59–64.

SHOTWELL, J.A. 1970A. Pliocene mammals of southeast Oregon and adjacent Idaho. *Bull. Univ. Oreg., Mus. Nat. Hist.* 17:1–103.

SKILBECK, C.G. 1980. A preliminary report on the late Cainozoic geology and fossil fauna of Bow, New South Wales. *Proc. Linn. Soc. New S. Wales* 104:171–181.

SULIMSKI, A., SAYNKIEWICZ, A., and WOLOSZYN, B. 1979. The middle Pliocene micromammals from central Poland. *Acta Palaeontol. Pol.* 24:377–403.

TANG, Y., YOU, Y., LIU, H., and PAN, Y. 1974. New materials of Pliocene mammals from Banguo Basin of Yuanmou, Yunnan, and their stratigraphical significance. *Vertebr. PalAsiatica* 12:60–68.

TAYLOR, D.W. 1966. Summary of North American Blancan nonmarine mollusks. *Malacologia* 4:1–172.

TEILHARD, P., and YOUNG, C.C. 1931. Fossil mammals from the late Cenozoic of northern China. *Pal. Sinica (C)* 9:1:1–67.

TOBIEN, H. 1970. Biostratigraphy of the mammalian faunas at the Pliocene-Pleistocene boundary in middle and western Europe. *Palaeogeogr., Palaeoclim., Palaeoecol.* 8:77–93.

TURNBULL, W.D., and LUNDELIUS, E.L., JR. 1970A. The Hamilton fauna. A late Pliocene mammalian fauna from the Grange Burn, Victoria, Australia. *Fieldiana, Geol.,* 19:1–163.

VAN DE WEERD, A. 1979. Early Ruscinian rodents and lagomorphs (Mammalia) from the lignites near Ptolemais (Macedonia, Greece). *Proc. Ned. Akad. Wet., Ser. B,* 82:127–170.

VAN DE WEERD, A., ADROVER, R., MEIN, P., and SORIA, P. 1977. A new genus and species of the Cricetidae (Mammalia, Rodentia) from the Pliocene of south-western Europe. *Proc. Koninkl. Ned. Akad. Wetensch., B,* 80:429–439.

WANG, Z., and ZHANG, X. 1979. The first discovery of a fossil *Stegodon* in the early Pliocene beds of Yunnan Province. *Vertebr. PalAsiatica* 16. Inside back cover (in Chinese).

ZAKRZEWSKI, R.J. 1969A. The rodents from the Hagerman local fauna, Upper Pliocene, Idaho. *Univ. Mich. Mus. Paleontol. Contrib.* 23:1–36.

CHAPTER EIGHT

Pleistocene Mammalian Faunas

Mammals About 1.8 to About 0.01 Million Years Ago

Sir Charles Lyell's *Newer Pliocene* (Lyell, 1833) was to be based on a fossil aggregate of many species of marine molluscs from the southern Italian area, of which 90–95 percent are living today. Later he proposed Pleistocene as a replacement name. The name *Pleistocene* remains in our usage, despite Lyell's still later recommendation that the term be forgotten. The beginning of Pleistocene and of the *Quaternary* of Desnoyers (1829) are taken as synchronous and are more or less coincident with the base of the marine Calabrian Stage of Italy and with the beginning of foraminiferal and molluscan occurrences signifying the onset of cooling climate and "Ice Age" in Europe. Many small problems and disputes remain about correlation and provincial dating of this suite of events, but we shall follow Berggren and Van Couvering (1974) in dating the beginning of Pleistocene-Quaternary at approximately 1.8 million years ago; this date is also contemporaneous with the beginning of the Olduvai normal-polarity event in the Matuyama reversed-polarity epoch. We use this year date also as the approximate beginning of the Biharian land-mammal age in Europe and the Irvingtonian land-mammal age in North America, following Repenning (1980 and other papers), although the correlations are tenuous. Berggren and Van Couvering concluded that Biharian and Irvingtonian began later.

We use the date of 10,000 years ago as an arbitrary date for the end of the Pleistocene and the beginning of the Recent (including Holocene), although it is probable that a number of species of land mammals now extinct were living for as much as 2–3 thousand years after this date.

BIHARIAN

The following list is compiled from species in faunas related to Gunz glaciation, Waalian interglacial, Cromer interglacial, Mindel-Elsterian glacial, and Riss-Holsteinian interglacial especially. The principal samples come from Forest Bed and Cromer in England, Mauer, Mosbach, Sussenborn, and Schernfeldt in Germany, Sainzelles, Saint-Prest, and Abbeville in France, Villany Mountains district in Hungary, Episcopia (= Püspökfürdö) in Hungary, Brasso in Romania, Stranska Skala and Gombasek in Czechoslovakia, and Kamyk in Poland.

Biharian Mammalian Fauna

LAGOMORPHA
  Ochotonidae
    Ochotona? pusilla Pallas, 1778
  Leporidae
    Hypolagus brachygnathus Kormos
    Lepus terraerubrae Kretzoi
    L. capensis Linnaeus, 1776

CARNIVORA
  Ursidae
    Ursus etruscus Cuvier, 1812
    U. deningeri Reichenau, 1906
  Mustelidae
    Martes vetus Kretzoi, 1942
    Baranogale antiqua Pomel
    Vormela beremendensis Petenyi, 1864
    Enhydrictis ardea Bravard, 1828
    Pannonictis pliocaenica Kormos, 1931D
    Mustela stromeri Kormos
    M. putorius Linnaeus, 1766
    M. palerminea (Petenyi, 1864)
    M. praenivalis Kormos, 1934
    Meles meles (Linnaeus, 1766)
    Aonyx bravardi Pomel, 1843
    Lutra simplicidens Thenius, 1965
    Gulo schlosseri Kormos, 1914
  Canidae
    Canis arnensis DelCampana, 1913
    C. lupus Linnaeus, 1758
    Cuon majori (DelCampana, 1913)
    Lycaon lycaonoides Kretzoi, 1937
    Vulpes alopecoides Major, 1877
    V. vulpes (Linnaeus, 1758)
    V. praecorsac Kormos, 1932C
  Hyaenidae
    Pachycrocuta perrieri (Croizet and Jobert, 1828)
    P. brevirostris (Aymard, 1854?)
    Crocuta crocuta Erxleben, 1777
  Felidae
    Homotherium sainzelli Aymard (= *"crenatidens"*)
    Dinobastis latidens (Owen, 1846)
    Felis lunensis Martelli, 1905
    F. sylvestris Schreber
    Leo toscana (Schaub, 1949)
    L. leo (Linnaeus, 1766)
    L. pardus (Linnaeus, 1766)
    Lynx pardina (Oken, 1877)
    Acinonyx pardinensis Croizet and Jobert, 1828
INSECTIVORA
  Erinaceidae
    Erinaceus
  Soricidae
    Sorex minutus Linnaeus, 1766
    S. araneus Linnaeus, 1746
    S. praealpinus Heller, 1930
    S. runtonensis Hinton, 1911
    S. margaritodon Kormos
    Neomys fodiens (Pennant)
    Beremendia fissidens (Petenyi, 1864)
    Petenyia hungarica Kormos, 1934D
    Soriculus kubinyi Kormos, 1934D
    Suncus hungaricus
    Crocidura kornfeldi Kormos, 1934D
    Nesiotites hidalgo Bate (age?)
    N. corsicanus Bate (age?)
    N. similis Hensel, 1855 (age?)

  Talpidae
    Desmana moschata (Pallas, 1781)
    D. thermalis Kormos
    Talpa fossilis Petenyi, 1864
    T. gracilis Kormos
    T. episcopalis Kormos
CHIROPTERA
  Rhinolophidae
    Rhinolophus euryale Blasius, 1857
    R. delphinensis Gaillard, 1899
    R. ferrum-equinum (Schreber, 1774)
  Vespertilionidae
    Myotis daubentoni Leisler
    M. dasycneme Boie, 1825
    M. mystacinus Leisler
    M. emarginatus (Geoffroy-Saint-Hilaire, 1806)
    M. nattereri (Kuhl, 1819)
    M. bechsteini (Kuhl, 1818)
    M. baranensis Kormos, 1934D
    M. oxygnathus (Monticelli, 1887)
    Plecotus auritus (Linnaeus, 1758)
    P. crassidens Kormos, 1930B
    Miniopterus schreibersi (Kuhl, 1819)
    Barbastella barbastella (Schreber, 1775)
    B. leucomelas (Rüppel, 1825)
    Vespertilio nilssoni Keyserling and Blasius
    V. praeglacialis Kormos, 1937
PRIMATES
  Cercopithecidae   —Szalay and Delson (1979)
    Macaca sylvanus prisca Gervais, 1859
    Paradolichopithecus arvernensis (Depéret, 1929A) (age?)
  Hominidae
    Homo sapiens heidelbergensis Schoetensack, 1908
ARTIODACTYLA
  Suidae
    Sus strozzii Meneghini
    S. scrofa Linnaeus, 1758
  Hippopotamidae
    Hippopotamus amphibius Linnaeus, 1766
  Cervidae
    "Cervus" perrieri (Croizet and Jobert, 1828)
    Anoglochis ramosus Croizet and Jobert, 1828
    Euctenoceros ctenoides (Nesti, 1841)
    E. senezensis (Depéret, 1910)
    Eucladoceros falconeri (Dawkins, 1868)
    E. sedgwicki Falconer, 1868
    Cervus etuerarium Croizet and Jobert, 1828
    C. elaphus Linnaeus, 1766
    Praemegaceros verticornis Dawkins, 1872
    Capreolus capreolus Linnaeus, 1766
    Alces latifrons (Johnson, 1874)
    Rangifer tarandus (Linnaeus, 1758)
  Bovidae
    Praeovibos priscus Staudinger
    Ovibos moschatus (Zimmermann, 1780)
    Ovis
    Soergelia elisabethae Schaub, 1951
    Hemitragus stehlini Freudenberg
    Leptobos etruscus Falconer, 1868
    Bison priscus (Bojanus, 1827)
    B. schoetensacki Freudenberg

PERISSODACTYLA
  Equidae
    Equus stenonis Cocchi, 1867   (zebrine)
    E. süssenbornensis Wüst, 1901   (zebrine)
    E. (Equus) bressanus Viret, 1954
    E. (E.) mosbachensis Reichenau, 1915
    E. (Asinus?) hydruntinus Regalia, 1904
  Tapiridae
    Tapirus arvernensis Croizet and Jobert, 1828
  Rhinocerotidae
    Stephanorhinus etruscus (Falconer, 1859)
    Elasmotherium sibiricum Fischer
PROBOSCIDEA
  Gomphotheriidae
    Anancus arvernensis (Croizet and Jobert, 1828)
  Elephantidae
    Mammuthus meridionalis (Nesti, 1825)
    M. armeniacus (Falconer, 1857)
    Elephas namadicus Falconer and Cautley, 1845
RODENTIA
  Sciuridae
    Sciurus whitei Hinton, 1917
    Spermophilus primigenius (Kormos, 1934D)
    S. citellus (Linnaeus, 1766)
  Castoridae
    Castor fiber Linnaeus, 1766
    Trogontherium cuvieri Fischer, 1812
  Cricetidae
    Prospalax priscus (Nehring, 1898)
    Spalax leucodon (Nordmann, 1840)
    Cricetus cricetus (Linnaeus, 1766)
    C. praeglacialis Schaub, 1930
    C. nanus Schaub, 1930
    Cricetulus bursae (Schaub, 1930)
    C. migratorius (Pallas, 1794)
    Rhinocricetus ehiki (Schaub, 1930)
    Pliomys episcopalis Méhely
    P. coronensis Méhely
    Clethrionomys glareolus (Schreber, 1792)
    C. esperi Heller
    Mimomys pusillus (Méhely, 1914)
    M. reidi Hinton, 1910
    M. newtoni (Nehring, 1898)
    M. pliocaenicus (Major, 1902)
    M. intermedius Newton, 1882
    M. rex Kormos, 1934
    Arvicola greeni Hinton, 1926
    A. mosbachensis Schmidtgen, 1911
    Lagurus pannonicus Kormos, 1932
    Allophaiomys
    Microtus arvalis Pallas, 1778
    M. arvalinus Hinton, 1923
    M. subarvalis Heller, 1933
    M. coronensis Kormos
    M. raticeps Keyserling and Blasius, 1841
    M. ratticepoides Hinton, 1928
    M. nivalis Martins, 1842
    M. nivalinus Hinton, 1926
    M. nivaloides Major, 1902
    M. subnivalis Pasa, 1949
    M. agrestis (Linnaeus, 1761)

    M. gregalis (Pallas, 1778)
    Pitymys gregaloides Hinton
    Pitymys arvaloides Hinton
    P. dehmi Heller
    P. hintoni Kretzoi
    Tyrrhenicola henseli Major
    Lemmus lemmus (Linnaeus, 1766)
    Dicrostonyx torquatus (Pallas, 1778)
  Zapodidae
    Sicista praeloriger Kormos, 1930
  Gliridae
    Muscardinus avellanarius (Linnaeus, 1766)
    M. dacicus Kormos, 1930
    Glis glis Linnaeus, 1766
  Hystricidae
    Hystrix refossa Gervais, 1853
    H. vinogradovi Argyropulo, 1941
  Muridae
    Parapodemus coronensis Schaub, 1938
    Apodemus sylvaticus (Linnaeus, 1766)
    A. flavicollis (Melchior, 1836)
    A. alsomyoides Schaub, 1930
    A. mystacinus (Danford and Alston, 1877)
    Rhagamys orthodon (Hensel, 1856)

EARLY PLEISTOCENE OF ASIA

We list under this heading the Chinese genera tabulated by Pope (1982) as Middle Pleistocene taxa. Pope noted that most of the recent literature from Chinese colleagues uses a 3-million-year Pleistocene, and he chose to follow their chronological designations in his charts. The badly needed modern synthesis of eastern and northeastern Asian Pleistocene mammalian faunas is being undertaken by Pope and by his Chinese colleagues and others, and we do not presume to be knowledgeable enough to designate "earlier Pleistocene" species. The genera listed below have been found in the Lantian district and at He-Shui of Shaanxi Province, at Zhoukoudian (Choukoutien) in the Beijing area, and at Jin-Niu-Shan in Liaoning Province. The list constitutes a representative array of genera of early Pleistocene land mammals from the "temperate" latitudes of eastern Asia.

Early Pleistocene Mammalian Genera of Asia

LAGOMORPHA
  Ochotonidae
    Ochotona
    Ochotonoides
  Leporidae
    Hypolagus
    Lepus
CARNIVORA
  Ursidae

Ursus
Mustelidae
   Mustela
   Meles
   Lutra
Canidae
   Canis
   Cuon
   Vulpes
   Nyctereutes
Hyaenidae
   Pachycrocuta   ("Hyaena")
   Hyaena?
   Crocuta
Felidae
   Leo ("Panthera" tigris and pardus)
   Lynx
   Dinobastis ("Homotherium")
   Megantereon
   Acinonyx
INSECTIVORA
Soricidae
   Neomys
   Crocidura
Talpidae
   Scaptochirus
PRIMATES
Cercopithecidae
   Macaca
Hominidae
   Homo erectus
ARTIODACTYLA
Suidae
   Sus
Camelidae
   Paracamelus
Cervidae
   Elaphurus
   Cervus
   Megaloceros
   Moschus
   Hydropotes
   Pseudaxis
   Muntiacus
   Capreolus
Bovidae
   Gazella
   Antilope
   Spirocerus
   Ovis
   Leptobos?
   Bison
   Bubalus
PERISSODACTYLA
Equidae
   Equus
Tapiridae
   Tapirus

Rhinocerotidae
   Rhinoceros
   Dicerorhinus
Chalicotheriidae
   Nestoritherium
PROBOSCIDEA
Stegodontidae
   Stegodon
Elephantidae
   Elephas   ("Palaeoloxodon")
RODENTIA
Sciuridae
   Tamias
   Marmota
   Spermophilus   ("Citellus")
Castoridae
   Trogontherium
Cricetidae
   Myospalax
   Gerbillus
   Cricetulus
   Mimomys
   Microtus
Dipodidae
   Sminthoides
Hystricidae
   Hystrix
Muridae
   Rattus
   Mus
   Apodemus

As would be expected, this fauna shows strong generic resemblance to the contemporaneous fauna in Europe and North America. It does maintain, however, a significant number of endemic taxa. Also, it shows reduced family diversity when compared with the preceding faunas of China that we have called (= Villafranchian) Late Pliocene: the viverrids, erinaceids?, anthracotheriids, hippopotamids, tragulids, giraffids, hyracoids, and gomphotheriids of the late Pliocene appear to be extinct in eastern Asia by about 2 million years ago. And there is also apparent reduction in diversity of genera—especially among suids and bovids. We expect, however, that a better knowledge of the land-mammal faunas of this approximate age in southern and southeastern Asia may well show that this earlier Pleistocene reduction in familial and generic diversity was accentuated in the northern regions. See Pope's (1982) extensive analysis of the many remaining uncertainties regarding correlations among the Pleistocene faunas of eastern Asia.

Earlier Pleistocene Mammals in Northern Asia

Vangengeim and Sher (1970) and the paleontologists of the USSR whom Vangengeim and Sher cite have summarized the knowledge on the Tiraspol and Tologoy "faunal complexes" and correlated faunas of the northern USSR. These faunas are generally correlated with Mindel glaciation in Europe and with the ''Sanmenian'' earlier Pleistocene of China. The listed array of genera from this great geographic region, known from many scattered sites, is taken from Vangengeim and Sher.

Representative Early Pleistocene Mammalian Fauna from Northern Asia

LAGOMORPHA
  Ochotonidae
    Ochotona
CARNIVORA
  Ursidae
    Ursus
  Canidae
    Canis
  Hyaenidae
    Hyaena?
ARTIODACTYLA
  Cervidae
    Cervus
    Alces
    Rangifer
  Bovidae
    Spirocerus
    Soergelia?
    Praeovibos
    Bison
PERISSODACTYLA
  Equidae
    Equus
  Rhinocerotidae
    Dicerorhinus
    Coelodonta
    Elasmotherium
PROBOSCIDEA
  Elephantidae
    Elephas
    Mammuthus
RODENTIA
  Castoridae
    Sinocastor
  Cricetidae
    Mimomys
    Clethrionomys
    Ellobius
    Eolagurus
    Lagurus
    Pitymys
    Arvicola
    Microtus
    Myospalax

The list obviously is incomplete but does show in conjunction with the roughly contemporaneous fauna of northern Europe the beginning of the Northern Hemisphere boreal fauna that is well exemplified in Eurasia and North America in later Pleistocene time.

EARLIER PLEISTOCENE OF AFRICA

We group under the designation Earlier Pleistocene the various genera and species of land mammals that have been assigned ''Early Pleistocene,'' ''Plio-Pleistocene,'' and ''Early-to-Mid-Pleistocene'' age by various previous authors. Representative samples come from Swartkrans, Kromdraai, Upper Sterkfontein, Bolt's Farm in South Africa, Members H, J, K, and L of the Shungura Formation (= ''U. Omo'') in Ethiopia, the Upper Member of the Koobi Fora Formation in Kenya, Olduvai II through IV in Tanganyika, and a few sites in North Africa. See Figure 7-7.

Representative Earlier Pleistocene Mammalian Fauna from Africa

MACROSCELIDEA   —Butler (1978)
  Elephantulus antiquus Broom, 1948
  E. broomi Corbet and Hanks, 1968
  E. fuscus
LAGOMORPHA
  Leporidae
    cf. *Lepus capensis* Linnaeus
CARNIVORA   —R. Savage (1978)
  Mustelidae
    Aonyx capensis Schinz, 1821
    Lutra
  Canidae
    Canis mesomelas Schreber, 1775
    C. africanus Pohle, 1928
    C. atrox Broom, 1939
    C. terblanchei Broom, 1948
    Vulpes pulcher Broom, 1937
    V. chama (A. Smith, 1835)
    Lycaon pictus Temminck, 1820
    Otocyon recki Pohle, 1928
  Viverridae
    cf. *Suricata suricatta* (Erxleben, 1777)
    Civettictus civetta (Schreber, 1778)
    Herpestes mesotes Ewer, 1956
    Pseudocivetta ingens Petter, 1966
    Crossarchus transvaalensis Ewer
    Cynictis penicellata (Cuvier, 1829)
    Atilax paludinosus (Cuvier, 1829)
  Hyaenidae
    Chasmaporthetes silberbergi (Broom, 1948)
    Leecyaena forfex Ewer, 1955
    Crocuta crocuta Erxleben, 1777
    Hyaena brunnea Thunderg, 1820
    H. bellax Ewer, 1954
  Felidae
    Felis (Leptailurus) serval Schreber, 1977
    F.? libyca Forster, 1780
    Dinofelis barlowi (Broom, 1937)
    D. piveteaui Ewer, 1955
    Leo leo (Linnaeus, 1766)

L. pardus (Linnaeus, 1766)
L. crassidens Broom, 1948
Megantereon eurynodon Ewer, 1955
M.? whitei Broom, 1937
Acinonyx

INSECTIVORA   —Butler (1978)
  Erinaceidae
    Erinaceus broomi Butler and Greenwood, 1973
  Soricidae
    Crocidura cf. *bicolor* Bocage, 1889
    C. cf. *hindei* Thomas
    Sylvisorex cf. *granti* Thomas
    Myosorex robinsoni Meester, 1955
    Suncus varilla (Thomas, 1895)
    S. infinitesimus (Heller)
  Chrysochloridae
    Proamblysomus antiquus Broom, 1941

CHIROPTERA
  Megadermatidae
    Cardioderma   —Butler and Greenwood
      (1965)
  Myzopodidae
    Myzopoda
  Vespertilionidae
    cf. *Myotis*
    cf. *Nycticeius schlieffeni* (Peters, 1859)
    cf. *Pipistrellus ruppelli* (Fischer, 1829)
    Eptesicus cf. *hottentotus* (Smith)
    Miniopterus cf. *schreibersi* (Kuhl, 1819)
  Molossidae, gen. indet.

PRIMATES
  Cercopithecidae
    Theropithecus danieli Freedman, 1957
    T. oswaldi (Andrews, 1916)
    T. jonathani Leakey and Whitworth, 1958
    Gorgopithecus major (Broom, 1940)
    Parapapio jonesi Jones, 1968
    P. whitei? Broom, 1940
    P. broomi? Jones, 1937
    Papio hamadryas (Linnaeus, 1758)
    P. izodi Gear, 1926
    Cercopithecoides williamsi Mollett, 1947
  Hominidae
    Australopithecus boisei (Leakey, 1959)
    A. crassidens (Broom, 1949)
    A. robustus (Broom, 1938)
    Homo habilis Leakey, Tobias, and Napier, 1964
    H. erectus (Dubois, 1894)

ARTIODACTYLA
  Suidae   —White and Harris (1977); Cooke and Wilkinson (1978)
    Mesochoerus majus (Hopwood, 1934)
    M. limnetes (Hopwood, 1926)
    Notochoerus scotti (Leakey, 1953)
    N. euilus (Hopwood, 1926)
    Kolpochoerus olduvaiensis (Leakey, 1942)
    Metridiochoerus modestus (Van Hoepen and Van Hoepen, 1932)

M. andrewsi Hopwood, 1926
M. nyanzae (Leakey, 1958)
Potamochoerus porcus (Linnaeus, 1758)
Sus scrofa Linnaeus, 1758
  Hippopotamidae
    Hippopotamus gorgops
  Camelidae
    Camelus cf. *thomasi* Pomel, 1893
  Giraffidae   —Churcher (1978)
    Giraffa camelopardalis Brisson, 1756
    G. stillei (Dietrich, 1941)
    G. gracilis Arambourg, 1947
    G. jumae Leakey, 1965
    Sivatherium maurusium (Pomel, 1892)
  Cervidae   —Hamilton (1978)
    Cervus elaphus Linnaeus, 1758
    Dama dama Linnaeus, 1758
  Bovidae   —Gentry (1978)
    Tragelaphus arkelli (Leakey, 1965)
    T. strepsiceros (Pallas, 1766)
    T. cf. *spekei* (Blyth, 1855)
    T. cf. *angasi* Gray, 1848
    T. nakuae (Arambourg, 1941)
    Syncerus
    Pelorovis antiquus (Duvernoy, 1851)
    P. oldowayensis Reck, 1928
    Kobus sigmoidalis Arambourg, 1941
    K. kob Erxleben, 1777
    K. ancystrocera (Arambourg, 1947)
    Menelikia lyrocera Arambourg, 1941
    Thaleroceros radiciformis Reck, 1935
    Hippotragus gigas Leakey, 1965
    Oryx
    Megalotragus kattwinkeli (Schwarz, 1932)
    Connochaetes gnou (Zimmermann, 1777)
    C. africanus (Hopwood, 1934)
    C. taurinus Burchell, 1824
    Beatragus antiquus Leakey, 1965
    Rhynotragus semiticus Reck, 1935B
    Rabaticeras arambourgi Ennouchi, 1953
    R. porrocornutus Vrba, 1971
    Damaliscus
    D. niro (Hopwood, 1936)
    Parmularius angusticornis (Schwarz, 1937)
    P. altidens Hopwood, 1934
    P. rugosus Leakey, 1965
    Aepyceros cf. *melampus* (Lichtenstein, 1814)
    Pelea
    Antidorcas recki (Schwarz, 1932)
    A. bondi (Cooke and Wells, 1951)
    Gazella

PERISSODACTYLA
  Equidae   —Churcher and Richardson (1978)
    Hipparion libycum Pomel, 1897
    Equus (Dolichohippus) capensis Broom, 1909
    E. (D.) oldowayensis Hopwood, 1937
    E. (D.) grevyi Oustalet, 1882
    E. (Hippotigris) burchelli (Gray, 1824)
    E. (H.) quagga Gmelin, 1788
  Rhinocerotidae
    Diceros bicornis (Linnaeus, 1758)

Ceratotherium simum (Burchell, 1817)
TUBULIDENTATA
  Orycteropus?
HYRACOIDEA
  Procaviidae
    Procavia transvaalensis Shaw, 1937
    P. antiqua Broom, 1934
PROBOSCIDEA   —Maglio (1973); Coppens, Maglio,
    Madden, and Beden (1978)
  Deinotheriidae
    Deinotherium bozasi Arambourg, 1934
  Gomphotheriidae
    Anancus kenyensis MacInnes, 1942 (age?)
    A. osiris Arambourg, 1945 (age?)
  Elephantidae
    Elephas recki Dietrich, 1916
    E. iolensis Pomel, 1895 (age?)
    Loxodonta atlantica (Pomel, 1879)
    L. africana (Blumenbach, 1797)
RODENTIA   —Cooke (1964)
  Bathyergidae
    Cryptomys robertsi
  Pedetidae
    Pedetes cf. *caffer* (Pallas, 1778)
  Cricetidae
    Mystromys hausleitneri Broom, 1937
    Tatera cf. *brantsi*
    T. leucogaster
    Desmodillus? auricularis
    Palaeotomys gracilis
  Hystricidae
    cf. *Hystrix africaeaustralis* Peters, 1852
  Muridae
    Pelomys cf. *fallax* Peters, 1852
    Rhabdomys cf. *pumilio* (Wagner, 1820)
    Aethomys cf. *namaquensis* (A. Smith, 1835)
    cf. *Thallomys debruyni*
    Leggada or *Mus* cf. *minutoides* A. Smith, 1835
    Leggada or *Mus* cf. *major* Hofmann, 1887
    Dendromus antiquus
    Malacothrix cf. *typica* (Smith, 1839)
    Steatomys cf. *pratensis* Peters, 1846
    Grammomys cf. *dolichurus* (Smuts, 1832)
    Dasymys bolti
    Arvicanthis
    Rattus debruyni
    Otomys gracilis (Broom, 1937E)
    O. saundersiae

IRVINGTONIAN

D. Savage (1951C) proposed *Irvingtonian* for the "early
Mammuthus–pre-*Bison*" Pleistocene faunas of North
America. The Irvington fauna of the San Francisco Bay re-
gion evidently falls in the mid- to later part of what is here
included in the Irvingtonian land-mammal age. Lindsay,
Johnson, and Opdyke (1975) sampled the strata at Irvington

and concluded that these rocks were deposited during the
later half of the Matuyama reversed-polarity epoch.

The following faunal list is provided by fossils from
Alaska, California, Arizona, South Dakota, Nebraska,
Kansas, Texas, Arkansas, Pennsylvania, and Florida. It is
slightly modified but essentially based on the information
Kurtén and Anderson (1980) have provided so meticulously
in their book on Pleistocene mammals of North America.

Irvingtonian Mammalian Fauna

LAGOMORPHA
  Ochotonidae
    Ochotona whartoni Guthrie and Matthews, 1971
    O. cf. *princeps* (Richardson, 1868)
  Leporidae
    Notolagus cf. *velox* R. Wilson, 1937D
    Sylvilagus floridanus (Allen, 1880)
    Lepus alleni Mearns, 1890
    L. californicus Gray, 1837
    L. townsendi Bachman, 1839
    L. americanus Erxleben, 1777
EDENTATA
  Megalonychidae
    Megalonyx wheatleyi Cope, 1871I
  Megatheriidae
    Nothrotheriops shastense (Sinclair, 1905)
    Eremotherium rusconii (Schaub, 1935)
  Mylodontidae
    Glossotherium chapadmalense Kraglievich, 1925
    G. harlani (Owen, 1840)
  Glyptodontidae   —Gillette and Ray (1981)
    Glyptotherium arizonae Gidley, 1926A
  Dasypodidae
    Kraglievichia floridanus Robertson, 1976
    Holmesina septentrionalis (Leidy, 1889E)
    Dasypus bellus (Simpson, 1929H)
CARNIVORA
  Ursidae
    Tremarctos floridanus (Gidley, 1928)
    Arctodus pristinus Leidy, 1854C
    A. simus (Cope, 1879U)
    Ursus americanus Pallas, 1780
  Mustelidae
    Martes diluviana (Cope, 1889A)
    Mustela frenata Lichtenstein, 1831
    M. erminea Linnaeus, 1758
    M. vison Schreber, 1777
    Tisisthenes parvus R. Martin, 1973
    Gulo cf. *schlosseri* Kormos, 1914
    Taxidea taxus (Schreber, 1778)
    Lutra canadensis (Schreber, 1776)
    Enhydra macrodonta Kilmer, 1972 (marine)
    Spilogale putorius (Linnaeus, 1758)
    Brachyprotoma obtusata Cope, 1899A
    Osmotherium spelaeum Cope, 1896F
    Mephitis mephitis (Schreber, 1776)

Conepatus leuconotus (Lichtenstein, 1832)
Procyonidae
  Procyon lotor (Linnaeus, 1758)
Canidae
  Borophagus diversidens Cope, 1892Y
  Canis lepophagus Johnston, 1938B
  Canis latrans Say, 1823
  C. priscolatrans Cope, 1899A
  C. armbrusteri Gidley, 1913B
  C. lupus Linnaeus, 1758
  C. dirus Leidy, 1858E
  Protocyon texanus (Troxell, 1915A)
  Urocyon cineroargenteus (Schreber, 1775)
  Vulpes velox (Say, 1823)
Hyaenidae
  Chasmaporthetes cf. *ossifragus* Hay, 1921A
Felidae
  Smilodon gracilis Cope, 1880D
  S. fatalis (Leidy, 1868A)
  Dinobastis serus Cope, 1893K   —(synonym of *D. latidens* (Owen, 1846)?; incl. in *Homotherium* by Kurtén and Anderson (1980))
  Leo atrox (Leidy, 1853F)
  L. onca (Linnaeus, 1758)
  Acinonyx studeri (Savage, 1960H)   —Adams (1979)
  Felis lacustris Gazin, 1933A
  F. yagouaroundi Lacepede, 1808   (age?)
  Lynx issiodorensis (Croizet and Jobert, 1828)
INSECTIVORA
  Soricidae
    Sorex cudahyensis Hibbard, 1944D
    S. cinereus Kerr, 1792   —Jammot (1972); Eshelman (1975)
    S. vagrans Baird, 1858
    S. lacustris (Hibbard, 1944D)
    S. megapalustris Paulson, 1961
    S. fumeus Miller, 1895
    S. arcticus Kerr, 1792
    Microsorex pratensis Hibbard, 1944D
    M. minutus Brown, 1908A
    Cryptotis parva (Say, 1823)
    Blarina ozarkensis Brown, 1908A
    B. brevicauda (Say, 1823)
    Notiosorex jacksoni Hibbard, 1950
  Talpidae
    Condylura cristata (Linnaeus, 1758)
    Parascalops breweri (Bachman, 1842)
    Scalopus aquaticus (Linnaeus, 1758)
    Scapanus?
CHIROPTERA
  Vespertilionidae
    Myotis leibi (Audubon and Bachman, 1842)
    M. grisecens Howell, 1909
    M. austroriparius (Rhoads, 1897)
    Pipistrellus subflavus (F. Cuvier, 1832)
    Eptesicus fuscus (Palisot de Beauvois, 1876)
    Histiotus stocki (Stirton, 1931)
    Lasiurus fossilis Hibbard, 1950

Plecotus alleganiensis (Gidley and Gazin, 1933)
P. rafinesquii (Lesson, 1818)

The great number of species of bats should not be regarded as evolutionary first occurrences, for Irvingtonian is the first (earliest) record of fossiliferous cave deposits in the North American Cenozoic.

ARTIODACTYLA
  Tayassuidae
    Mylohyus nasutus (Leidy, 1868I)
    Platygonus bicalcaratus Cope, 1893A
    P. vetus Leidy, 1882A
  Camelidae
    Titanotylopus nebraskensis Barbour and Schultz, 1934
    T. spatulus (Cope, 1893A)
    Camelops kansanus Leidy, 1854F
    C. sulcatus (Cope, 1893A)
    C. minidokae Hay, 1927D
    Hemiauchenia macrocephala (Cope, 1893A)
    Palaeolama mirifica (Simpson, 1929H)
  Cervidae
    Odocoileus virginianus (Zimmermann, 1780)
    O. cf. *hemionus* (Rafinesque, 1817)
    Rangifer tarandus (Linnaeus, 1758)
    Cervus elaphus Linnaeus, 1758
  Antilocapridae
    Capromeryx furcifer Matthew, 1902F
    C. arizonensis Skinner, 1942
    Tetrameryx irvingtonensis Stirton, 1939D
    T. shuleri Lull, 1921B
    T. knoxensis Hibbard and Dalquest, 1960
    Hayoceros falkenbachi Frick, 1937
  Bovidae
    Euceratherium collinum Furlong and Sinclair, 1904
    Soergelia mayfieldi (Troxell, 1915B)
    Symbos cavifrons (Leidy, 1852F)
PERISSODACTYLA
  Equidae
    Equus (Hemionus) calobatus Troxell, 1915B
    E. (H.) tau Owen, 1869
    E. giganteus Gidley, 1901
    E. scotti Gidley, 1900
    E. niobrarensis Hay, 1913B
    E. conversidens Owen, 1869
    E. excelsus Leidy, 1858E
  Tapiridae
    Tapirus merriami Frick, 1921
    T. copei Simpson, 1945H
PROBOSCIDEA
  Gomphotheriidae
    Stegomastodon mirificus (Leidy, 1858B)
  Mammutidae
    Mammut americanum (Kerr, 1791)
  Elephantidae
    Mammuthus meridionalis (Nesti, 1825)
    M. imperator (Leidy, 1858B)
    M. columbi (Falconer, 1857)
RODENTIA
  Sciuridae

Marmota monax Linnaeus, 1758
Spermophilus bensoni (Gidley, 1922B)
S. cf. *meadensis* (Hibbard, 1941A)
S. cochisei (Gidley, 1922A)
S. richardsoni (Sabine, 1822)
S. tridecimlineatus (Mitchell, 1821)
S. spilosoma Bennett, 1833
S. franklini (Sabine, 1822)
Cynomys niobrarius Hay, 1921A
C. ludovicianus (Ord, 1815)
Tamias striatus (Linnaeus, 1758)
Sciurus niger Linnaeus, 1758
S. carolinensis Gmelin, 1788
Tamiasciurus hudsonicus (Erxleben, 1777)
Glaucomys volans (Linnaeus, 1758)

Geomyidae
    Thomomys potomacensis (Gidley and Gazin, 1933)
    Nerterogeomys persimilis (Hay, 1927)
    Geomys garbanii White and Downs, 1961
    G. tobinensis (Hibbard, 1944D)
    G. bursarius (Shaw, 1800)
    G. pinetus Rafinesque, 1817
Heteromyidae
    Prodipodomys idahoensis Hibbard, 1962A
    Etadonomys tiheni Hibbard, 1943B   (taxon questioned)
    Dipodomys gidleyi A. Wood, 1935
    D. ordi Woodhouse, 1853

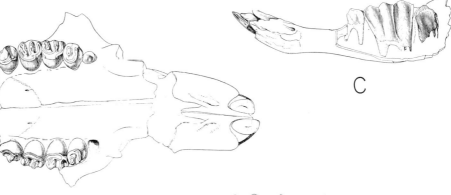

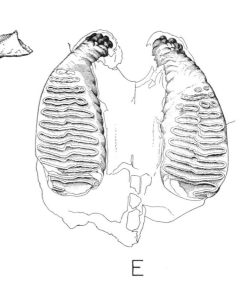

FIGURE 8-1 *Irvingtonian Mammals.*

A *Spermophilus bensoni* (Gidley, 1922B). Palate. ×1¾. Figure 3 in Savage (1951C).

B *Canis "irvingtonensis"* (Savage, 1951C). Lower jaw. ×5/8 approx. Figure 9 in Savage (1951C).

C *Camelops minidokae* Hay, 1927D. Juvenile lower jaw with DI_{1-3}, DP_{3-4} and anterior of M_1. ×3/8 approx. Figure 20 in Savage (1951C).

D *Camelops minidokae* Hay, 1927D. Lower jaw with P_4–M_3. ×3/8 approx. Figure 22 in Savage (1951C).

E *Mammuthus columbi* (Falconer, 1857). Palate with right and left M^3. ×1/5 approx. Figure 12 in Savage (1951C).

(Figures published with permission of the University of California Press.)

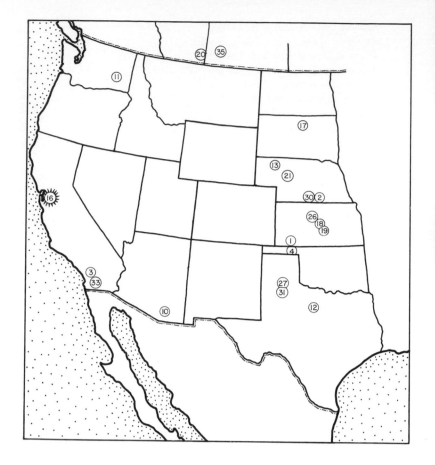

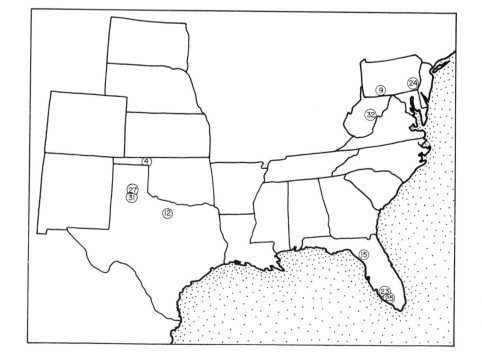

FIGURE 8-2 *Representative Irvingtonian Localities.*

1 *Adams—Butler Spring—Kansas*

2 *Angus—Nebraska*

3 *Bautista—California*

4 *Berends—Oklahoma*

5 *Cape Deceit—Alaska*

6 *Coleman IIA—Florida*

7 *Conard Fissure—Arkansas*

8 *Cudahy—Kansas   (See 1.)*

9 *Cumberland Cave—Pennsylvania*

10 *Curtis Ranch—Arizona*

11 *Delight—Washington*

12 *Gilliland—Texas*

13 *Gordon—Nebraska*

14 *Hay Springs—Nebraska (See 13.)*

15 *Inglis IA—Florida*

16 *Irvington—California*

17 *Java—South Dakota*

18 *Kanopolis—Kansas*

19 *Kentuck—Kansas*

20 *Medicine Hat—Alberta*

21 *Mullen II—Nebraska*

22 *Nash—Kansas*

23 *Pool Branch—Florida*

24 *Port Kennedy Cave—Pennsylvania*

25 *Punta Gorda—Florida*

26 *Rezabek—Kansas*

27 *Rock Creek—Texas*

28 *Rushville—Nebraska (See 13.)*

29 *Sandahl—Kansas (See 19.)*

30 *Sappa—Nebraska*

31 *Slaton—Texas*

32 *Trout Cave—West Virginia*

33 *Vallecito (upper)—California*

34 *Vera—Texas   (See 12.)*

35 *Wellsch Valley—Saskatchewan*

Perognathus hispidus Baird, 1858
Castoridae
　Paradipoides stovalli Rinker and Hibbard, 1952
　Procastoroides sweeti Barbour and Schultz, 1937
　Castoroides ohioensis Foster, 1838
　Castor californicus Kellogg, 1911
　C. accessor Hay, 1927D
　C. canadensis Kuhl, 1820
Cricetidae
　Reithrodontomys moorei (Hibbard, 1944D)
　R. humulis (Audubon and Bachman, 1841)
　R. fulvescens Allen, 1894
　Peromyscus irvingtonensis Savage, 1951C
　P. cragini Hibbard, 1944D
　P. progressus Hibbard and Taylor, 1960
　P. cumberlandensis Guilday and Handley, 1967
　Baiomys brachygnathus (Gidley, 1922B)
　Onychomys pedroensis Gidley, 1922B
　O. jinglebobensis Hibbard, 1955B
　Sigmodon medius Gidley, 1922B
　S. curtisi Gidley, 1922B
　S. bakeri R. Martin, 1974
　Neotoma spelaea (Gidley and Gazin, 1933)
　N. floridana (Ord, 1818)
　N. albigula Hartley, 1894
　Pliophenacomys osborni L. Martin, 1972A
　Pliomys deeringensis Guthrie and Matthews, 1971
　Atopomys texensis Patton, 1965
　A. salvelinus Zakrzewski, 1975
　Mimomys monahani L. Martin, 1972A
　Clethrionomys gapperi (Vigors, 1830)
　Phenacomys intermedius Merriam, 1889
　Microtus pliocaenicus (Kormos, 1933)
　M. deceitensis Guthrie and Matthews, 1971
　M. paropararius Hibbard, 1944D
　M. meadensis (Hibbard, 1944D)
　M. pennsylvanicus (Ord, 1815)
　M. chrotorrhinus (Miller, 1894)
　M. guildayi Van der Meulen, 1978
　M. llanensis Hibbard, 1944D
　M. ochrogaster (Wagner, 1842)
　M. cumberlandensis (Van der Meulen, 1978)
　M. aratai (R. Martin, 1974)
　Proneofiber diluvianus (Cope, 1896F)
　Neofiber alleni True, 1884
　N. leonardi Hibbard, 1943A
　Ondatra idahoensis R. Wilson, 1933C
　O. hiatidens (Cope, 1871I)
　O. annectens (Brown, 1908)
　O. nebrascensis (Hollister, 1911)
　Lemmus sibiricus (Kerr, 1792)
　Predicrostonyx hopkinsi Guthrie and Matthews, 1971
　Synaptomys cooperi Baird, 1858
　S. anzaensis Zakrzewski, 1972
　S. meltoni Paulson, 1961
　S. kansasensis Hibbard, 1952D
Zapodidae
　Zapus sandersi Hibbard, 1956B
　Z. hudsonicus (Zimmermann, 1780)
　Napaeozapus insignis (Miller, 1891)
Erethizontidae

　Coendou stirtoni White, 1968
　C. cumberlandicus White, 1970
　Erethizon dorsatum (Linnaeus, 1758)
Hydrochoeridae
　Hydrochoerus holmesi Simpson, 1928H

Of the approximately 119 genera of land mammals here recognized in the Irvingtonian, 34 percent are new appearances and 19 percent do not appear later. Twenty (20) percent of the 41 first-appearing genera are believed to be immigrants from South or Central America, and at least 27 percent were probably immigrants from Asia. Thus we believe that about half the new genera were immigrants and about half were the product of autochthonous evolution.

PLEISTOCENE OF SOUTH AMERICA

We follow Webb (1976), Marshall, Pascual, Curtis, and Drake (1977), Marshall, Webb, Sepkowski, and Raup (1982) and Marshall and Berta (1982, in press) in recognizing Uquian, Ensenadan, and Lujanian as the Plio-Pleistocene, earlier Pleistocene, and later Pleistocene mammal ages of South America. All these ages are based on land-mammal aggregates from Argentina. As of 1982, there is no radiometric dating associated with the fossils representing these ages, except at the approximate end of Lujanian, about 10,000 years ago. During the 2 million years of these three mammal ages, there was noteworthy flooding of North American land-mammal stocks into South America via the Panamanian Isthmus. And a few taxa from South America became token envoys to North America at the same time. This phenomenon has been termed the Great American Interchange by Webb and the other authors listed above. It was recognized by various South American and North American and European paleontologists before the beginning of the twentieth century.

The best-known Pleistocene faunas in South America include those from the Talara brea in Peru (Lemon and Churcher, 1961), Tarija in Bolivia (Hoffstetter, 1963), Lagoa Santos in Brazil (Paula Couto, 1970), the Andean district in Ecuador (Hoffstetter, 1952), Muaco in Venezuela (Royo y Gomez, 1968) and the type localities of the mammal ages in Argentina.

By mid-Lujanian and mid-Rancholabrean time, about 50,000 years ago, the Great American Interchange had been completed. South and Central American authchthons such as *Didelphis, Megalonyx, Nothrotheriops, Eremotherium, Glossotherium, Dasypus, Kraglievichia, Glyptotherium, Palaeolama?*, hydrochoerids, *Erethizon, Coendu*, and possibly other rodents, and possibly some bats were extant in the United States. And many cricetid rodents, canids, mustelids, felids, tayassuids, camelids, cervids, equids, tapirids, and mastodonts were established in South America. Most of these North American stocks (especially cricetids) evolved into new species and genera after arriving in South America.

Some of the "North American stocks," such as species of *Tremarctos, Tapirus, Haplomastodon, Stegomastodon,* and *Neochoerus,* could have originated in Central America. Many large edentates plus *Macrauchenia* (litoptern) and *Toxodon* lived until about the end of the Pleistocene in South America and maintained the megafaunal autochthon of that continent.

UQUIAN

We follow Marshall, Webb, Sepkowski, and Raup (1982) as to their recent estimates of the duration of the Pleistocene land-mammal ages of South America. Marshall and Berta (1982) review Uquian and note that its beginning is identified by the first appearance in South America of *Scelidodon, Lagostomus, Hydrochoerus, Ctenomys, Myocastor,* Canidae, Ursidae, *Galictis,* Proboscidea, Perissodactyla, and *Palaeolama.* It is based on the fauna from the Uquian Formation, and the Barranca de los Lobos, Vorohue, "Pulchense," and Malacara formations. Our list of Uquian genera follows Pascual et al., 1966.

Composite Uquian Mammalian Fauna

MARSUPIALIA
  Didelphidae
    Didelphis?
    Lutreolina?
    Marmosa?
EDENTATA
  Megatheriidae
    Diheterocnus?
  Mylodontidae
    Scelidodon
    Scelidotherium
    Glossotherium
  Dasypodidae
    Zaedyus?
    Tolypeutes?
    Eutatus?
  Glyptodontidae
    Hoplophorus?
    Panochthus?
    Paraglyptodon?
CARNIVORA
  Ursidae
    Arctodus
  Procyonidae
    Cyonasua

  Mustelidae
    Conepatus
    Galictis
    Stipanicicia
  Felidae
    Smilodontidion
ARTIODACTYLA
  Tayassuidae
    Platygonus
  Camelidae
    Palaeolama
    Lama
  Cervidae
    Hippocamelus?
    Habromeryx?
    Antifer?
PERISSODACTYLA
  Equidae
    Hippidion
    Onohippidium
  Tapiridae
    Tapirus
PROBOSCIDEA
  Gomphotheriidae?
    Stegomastodon?
NOTOUNGULATA
  Toxodontidae
    Toxodon
  Mesotheriidae?
    Mesotherium?
RODENTIA
  Caviidae
    Orthomyctera
    Dolicavia
    Palaeocavia
  Hydrochoeridae
    Hydrochoerus
  Chinchillidae
    Lagostomus
  Octodontidae?
    Pithanotomys?
  Ctenomyidae
    Ctenomys
    Megactenomys
    Actenomys
  Capromyidae
    Eumysops
    Myocastor
    Tramyocastor

ENSENADAN

We follow Marshall, Webb, Sepkowski, and Raup (1982) in their estimation of the duration of the Ensenadan land-mammal age. Marshall and Berta (1982) note that this age was sometimes called Lower Pampean, that it is based on the fauna from the Ensenada and Miramar Formations of Argentina, and that its beginning is best identified by the first appearance of *Lomaphorus*, *Neothoracophorus*, *Plaxhaplous*, *Microcavia*, *Lyncodon*, *Lutra*, *Felis*, *Leo*, *Smilodon*, *Tayassu*, and *Vicugna*. A composite list of Ensenadan genera, following Pascual et al. (1966), comprises the following.

Composite Ensenadan Mammalian Fauna

MARSUPIALIA
  Didelphidae
    Didelphis
    Lutreolina?
    Marmosa?
EDENTATA
  Megalonychidae
    Megalonychops?
  Megatheriidae
    Megatherium
  Mylodontidae
    Scelidotherium
    Scelidodon
    Glossotherium
    Mylodon
  Dasypodidae
    Chaetophractus
    Zaedyus?
    Eutatus
    Tolypeutes
    Propraopus
    Dasypus?
    Pampatherium
  Glyptodontidae
    Hoplophorus
    Sclerocalyptus
    Panochthus
    Lomaphorus
    Neothoracophorus
    Neuryurus
    Plaxhaplous
    Doedicuroides
    Doedicurus
    Glyptodon
CARNIVORA
  Ursidae
    Arctodus
  Procyonidae
    Brachynasua
  Mustelidae
    Conepatus
    Lyncodon
    Galictis?
    Lutra
  Canidae
    Dusicyon
    Cerdocyon
    Theriodictis
  Felidae
    Felis (*Puma*)
    Leo (*Jaguarius*)
    Smilodon
ARTIODACTYLA

  Tayassuidae
    Tayassu?
    Platygonus
    Catagonus
  Camelidae
    Palaeolama
    Lama
    Vicugna
  Cervidae
    Ozotoceros?
    Habromeryx
    Antifer
PERISSODACTYLA
  Equidae
    Hippidion
    Onohippidium
  Tapiridae
    Tapirus
PROBOSCIDEA
  Gomphotheriidae
    Stegomastodon
LITOPTERNA
  Macraucheniidae
    Macraucheniopsis
NOTOUNGULATA
  Toxodontidae
    Toxodon
  Mesotheriidae
    Mesotherium
RODENTIA
  Cricetidae
    Reithrodon
    Necromys
    Ptyssophorus
  Caviidae
    Dolichotis
    Orthomyctera
    Microcavia
  Hydrochoeridae
    Hydrochoerus?
    Neochoerus?
  Chinchillidae
    Lagostomus
  Ctenomyidae
    Ctenomys
  Capromyidae
    Myocastor

STEINHEIMIAN

A multitude of local faunas, including many cave local faunas, represent the fauna of Europe during the interval 400,000 to 5000 years ago. The local faunas here included are usually assigned to subdivisions of the glacial chronology of Europe, and the Steinheimian interval correlates approximately with the Holstein interglacial, Riss glacial, Ilford and Eem interglacials, Würm glacials and interglacial, and the earlier part of the "Holocene," or post-Pleistocene. We list some of the localities in the glossary at the end of this chapter. Many occurrences of *Homo sapiens neanderthalensis* and *Homo sapiens sapiens* fall into this age. The following list is taken from Kurtén (1968) with slight modification. Those entries marked with * are also in the Recent fauna. Those marked with ** are in Recent fauna outside Europe.

Composite Steinheimian Mammalian Fauna

LAGOMORPHA
  Ochotonidae
    Prolagus sardus (Hensel, 1856)
    Ochotona pusilla Pallas, 1778**
  Leporidae
    Oryctolagus cuniculus Linnaeus, 1766*
    Lepus timidus Linnaeus, 1766*
    L. europaeus Pallas, 1778*
    L. capensis Linnaeus, 1766*
CARNIVORA
  Ursidae
    Ursus spelaeus Rosenmüller and Heinroth, 1794
    U. arctos Linnaeus, 1758*
    U. maritimus Desmarest, 1820*
  Mustelidae
    Gulo gulo (Linnaeus, 1758)*
    Martes martes Linnaeus, 1766*
    M. foina Erxleben, 1777*
    Mustela lutreola Linnaeus, 1766*
    M. putorius Linnaeus, 1766*
    M. eversmanni Lesson, 1827*
    M. erminea Linnaeus, 1758*
    M. nivalis Linnaeus, 1766*
    M. rixosa Bangs, 1896*
    Meles meles (Linnaeus, 1766)*
    Aonyx antiqua Serres
    Lutra lutra Linnaeus, 1766*
  Canidae
    Canis lupus Linnaeus, 1758*
    Cuon alpinus (Pallas, 1811)**
    Vulpes vulpes (Linnaeus, 1758)*
    V. corsac Linnaeus, 1768*
    Alopex lagopus (Linnaeus, 1758)*
  Hyaenidae
    Hyaena hyaena (Linnaeus, 1766)**
    Crocuta crocuta Erxleben, 1777**
  Felidae
    Dinobastis latidens (Owen, 1846)
    Felis sylvestris Schreber*
    F. manul Pallas, 1776**
    F. chaus Gueldenstaedt, 1770**
    Lynx pardina (Oken, 1877)*
    L. lynx (Linnaeus, 1766)*
    Leo leo (Linnaeus, 1766)**

    L. pardus (Linnaeus, 1766)**
INSECTIVORA
  Erinaceidae
    Erinaceus*
  Soricidae
    Sorex minutus Linnaeus, 1766*
    S. araneus Linnaeus, 1746*
    S. alpinus Schinz*
    S. kennardi Hinton
    S. tasnadii Kretzoi
    Neomys fodiens (Pennant)*
    N. browni Hinton
    Crocidura leucodon Hermann and Zimmermann, 1780*
    C. russula Hermann and Zimmermann, 1780*
    C. suaveolens Pallas, 1811*
    Nesiotites
  Talpidae
    Desmana moschata (Pallas, 1781)*
    Talpa europaea Linnaeus, 1757*
    T. gracilis Kormos
    T. caeca Savi, 1822*
    T. episcopalis Kormos
    T. romana Thomas*
    T. tyrrhenica Major *in* Lydekker, 1887
CHIROPTERA
  Rhinolophidae
    Rhinolophus ferrum-equinum (Schreber, 1724)*
    R. hipposideros Bechstein*
  Vespertilionidae
    Myotis daubentoni Leisler*
    M. dasycneme Boie, 1825*
    M. mystacinus Leisler*
    M. emarginatus (Geoffroy-Saint-Hilaire, 1806)*
    M. nattereri (Kuhl, 1819)*
    M. bechsteini Leisler*
    M. myotis (Bechstein, 1791)*
    M. oxygnathus (Monticelli, 1887)*
    Plecotus auritus (Linnaeus, 1758)*
    P. austriacus Fischer*
    Miniopterus schreibersi (Kuhl, 1819)*
    Barbastella barbastella (Schreber, 1775)*
    B. leucomelas (Rüppel, 1825)*
    Pipistrellus pipistrellus (Schreber, 1775)*
    Vespertilio serotinus (Schreber, 1775)*
    V. nilssoni Keyserling and Blasius*
    V. murinus Linnaeus, 1758*
    Nyctalus noctula (Schreber, 1775)*
  Molossidae?
    Tadarida? teniotus Rafinesque?*
PRIMATES
  Cercopithecidae
    Macaca sylvana (Linnaeus, 1758)
  Hominidae
    Homo sapiens neanderthalensis King, 1864
    H. sapiens sapiens Linnaeus, 1758*
ARTIODACTYLA
  Suidae
    Sus scrofa Linnaeus, 1758*
  Hippopotamidae
    Hippopotamus amphibius Linnaeus, 1766**
    H. minor Desmarest (Cyprus)

H. pentlandi Meyer, 1832   (Crete, Sicily, Malta)
H. melitensis Major, 1902   (Crete, Malta)
Cervidae
  Cervus elaphus (Linnaeus, 1758)*
  C. tyrrhenicus Azzaroli
  C. siciliae Pohlig, 1892
  Praemegaceros cazioti (Depéret, 1897)
  P. messinae Pohlig
  P. cretensis Simonelli
  Megaloceros giganteus (Blumenbach, 1803)
  Dama clactoniana Falconer, 1868
  D. dama Linnaeus, 1758*
  Capreolus capreolus (Linnaeus, 1766)*
  Alces alces (Linnaeus, 1766)*
  Rangifer tarandus (Linnaeus, 1758)*
Bovidae
  Saiga tatarica (Linnaeus, 1758)*
  Nemorhaedus? melonii (Dehaut, 1911)
  Rupicapra rupicapra (Linnaeus, 1766)*
  Myotragus balearicus Bate
  Ovibos moschatus (Zimmermann, 1780)**
  Ovis musimon Schreber, 1795*
  Capra ibex Linnaeus, 1766*
  Hemitragus bonali Harlé and Stehlin, 1913
  Bison priscus (Bojanus, 1827)
  B. schoetensacki Freudenberg
  Bubalus murrensis Berckhemer
  Bos primigenius Bojanus, 1827*
PERISSODACTYLA
Equidae
  Equus (Equus) germanicus Nehring, 1884
  E. (E.) przewalskii Poliakoff, 1881**
  E. (E.) steinheimensis Reichenau, 1915
  E. (E.) taubachensis Freudenberg, 1911
  E. (Asinus?) hydruntinus Regalia, 1904
  E. (A.) graziosi Azzaroli, 1966
  E. (Hemionus) hemionus Pallas, 1775**
Rhinocerotidae
  Stephanorhinus or *Dicerorhinus kirchbergensis* (Jäger, 1835)
  S. or *D. hemitoechus* (Falconer)
  Coelodonta antiquitatis Blumenbach, 1803
  Elasmotherium sibiricum (Fischer)   (age?)
PROBOSCIDEA
Elephantidae
  Elephas namadicus Falconer and Cautley, 1845   —
    Maglio (1973)
  E. falconeri Busk, 1867, and other dwarfs   —Mag-
    lio (1973)
  Mammuthus armeniacus (Falconer, 1857)
  M. primigenius (Blumenbach, 1803)
RODENTIA
Sciuridae
  Sciurus whitei Hinton, 1917
  S. vulgaris Linnaeus, 1776*
  Marmota marmota (Linnaeus, 1776)*
  M. bobak (Pallas, 1778)*
  Spermophilus citellus (Linnaeus, 1776)*
  S. suslicus (Gueldenstaedt, 1770)*
  S. major (Pallas, 1778)**
Castoridae
  Castor fiber Linnaeus, 1766*

  Trogontherium cuvieri Fischer, 1812
Cricetidae
  Spalax leucodon (Nordmann, 1840)*
  Cricetus cricetus (Linnaeus, 1766)*
  Cricetulus bursae Schaub
  C. migratorius (Pallas, 1794)*
  Pliomys coronensis Méhely
  Clethrionomys glareolus (Schreber, 1792)*
  Arvicola greeni Hinton, 1926
  A. mosbachensis Schmidtgen, 1911
  A. terrestris (Linnaeus, 1766)*
  A. amphibius (Linnaeus, 1766)*
  A. antiquus Pomel
  Lagurus lagurus (Pallas, 1773)*
  Microtus arvalis Pallas, 1778*
  M. agrestis (Linnaeus, 1761)*
  M. ratticeps Keyserling and Blasius, 1841*
  M. nivalis Martins, 1842*
  M. gregalis (Pallas, 1778)*
  Pitymys subterraneus Sélys, 1836*
  Lemmus lemmus (Linnaeus, 1766)*
  Dicrostonyx torquatus (Pallas, 1778)*
  Tyrrhenicola henseli Major
Hystricidae
  Hystrix vinogradovi Argyropulos, 1941
Zapodidae
  Sicista cf. *betulina* Pallas or cf. *subtilis* Pallas*
Dipodidae
  Alactaga jaculus (Pallas, 1778)*
Gliridae
  Muscardinus avellanarius (Linnaeus, 1766)*
  Glis glis Linnaeus, 1766*
  Eliomys quercinus (Linnaeus, 1766)*
  Dryomys nitedula (Pallas, 1778)*
  Leithia melitensis (Leith Adams, 1863)
  Hyponomys mahonensis Bate
  H. morphaeus Bate
Muridae
  Apodemus sylvaticus (Linnaeus, 1766)*
  A. flavicollis (Melchior, 1836)*
  A. agrarius (Pallas, 1778)*
  Micromys minutus (Pallas, 1778)*
  Mus musculus Linnaeus, 1776*

LATER PLEISTOCENE OF ASIA

Using Simpson (1945I) and a variety of more recent references, we have compiled a tentative composite list of mammalian genera from the Pleistocene of the eastern two-thirds of Asia. A list of the known genera from the Recent of Asia might be added to show a complete roster of taxa, portions of which may someday be found in the many late Pleistocene sites of this huge area. No one locality has produced the majority of the entries in our later Pleistocene list, of

course, and we urge you not to infer that this composite
"fauna" closely matches an actual local fauna from any of
the communities that existed in this ecologically and cli-
matically diverse region. A genus marked with an † is extinct.

Composite List of Later Pleistocene Genera of Eastern Asia

LAGOMORPHA
  Ochotonidae
    Ochotona
  Leporidae
    Lepus
    Caprolagus?
CARNIVORA
  Ursidae
    Selenarctos?
    Ursus
    Helarctos?
    Melursus?
    Ailuropoda
  Mustelidae
    Mustela
    Martes?
    Meles
    Arctonyx?
    Lutra
  Canidae
    Canis
    Vulpes
    Nyctereutes
    Cuon
  Viverridae
    Viverricula?
    Viverra
  Hyaenidae
    Crocuta?
    Hyaena?
  Felidae
    Felis
    Leo
    Lynx
INSECTIVORA
  Erinaceidae
    Erinaceus
  Soricidae
    Soriculus
    Neomys
    Blarinella?
    Crocidura
    Anourosorex?
  Talpidae
    Talpa?
    Scaptochirus
    Scapanulus?
    Desmana?
SCANDENTIA
  Tupaiidae

    Tupaia?
PRIMATES
  Cercopithecidae
    Macaca
  Hominidae   —Szalay and Delson (1979)
    Pongo
    Homo
    Gigantopithecus? †
ARTIODACTYLA
  Suidae
    Sus
  Camelidae
    Camelus
  Tragulidae
    Tragulus
  Cervidae
    Moschus
    Muntiacus
    Megaloceros †
    Axis
    Cervus
    Hydropotes
    Capreolus
  Bovidae
    Boselaphus
    Tetracerus
    Bubalus
    Bos
    Bibos
    Bison
    Antilope
    Gazella
    Procapra
    Pantholops
    Saiga
    Budorcas
    Ovibos
    Ovis
PERISSODACTYLA
  Equidae
    Equus
  Tapiridae
    Tapirus
  Rhinocerotidae
    Dicerorhinus
    Rhinoceros?
    Coelodonta †
    Elasmotherium? †
PROBOSCIDEA
  Elephantidae
    Elephas
    Mammuthus †
PHOLIDOTA
  Manidae
    Manis
RODENTIA
  Sciuridae
    Sciurus
    Sciurotamias
    Marmota
    Spermophilus

Tamias
Petaurista
Castoridae
Castor
Trogontherium? †
Cricetidae
Cricetinus?
Cricetulus
Myospalax
Dicrostonyx
Clethrionomys
Eothenomys
Alticola
Arvicola
Pitymys
Microtus
Lagurus
Ellobius
Gerbillus
Rhizomyidae
Rhizomys
Muridae
Micromys
Apodemus
Rattus
Mus
Dipodidae
Dipus
Allactaga
Hystricidae
Hystrix

LATER PLEISTOCENE OF AFRICA

We group under the designation Later Pleistocene representative species of land mammals that have been assigned "Middle Pleistocene," "Late Pleistocene," "Holocene," "Post-Pleistocene," and "Sub-Recent" by various previous authors. The samples come from the "Cornelian" and "Florisian" faunal intervals, Hopefield (Saldanha)–Elandsfontein, Broken Hill, Vaal River Gravels (in part), Cave of Hearths, Wonderwerk Cave, Vlakkraal, Swartklip, Chelmer, Olorgesaillie, Gamble's Cave, Kanjera, Kom Ombo, Mumba Hills, Ternifine, Constantine, Haua Fteah, and hundreds of other sites in South, East, and North Africa.

Representative Later Pleistocene and Holocene Mammalian Fauna from Africa

MACROSCELIDEA   —Cooke (1964)
Elephantulus langi Broom, 1937E
LAGOMORPHA   —Cooke (1964)
Leporidae
Pronolagus randensis
Lepus saxatalis Cuvier, 1823
CARNIVORA   —R. Savage (1978)
Ursidae
Ursus arctos Linnaeus, 1758

Mustelidae
Ictonyx striatus (Shaw, 1800)
Mellivora capensis Desmarest, 1820
Mustela nivalis Linnaeus, 1766
Lutra lutra Linnaeus, 1766
L. maculicollis Lichtenstein, 1835
Aonyx capensis Schinz, 1821
Canidae
Canis mesomelas Schreber, 1775
C. terblanchei Broom, 1948
C?. africanus Pohle, 1928
C. adustus Sundevall, 1846
C. aureu Linnaeus, 1766
Vulpes chama (A. Smith, 1835)
V. vulpes (Linnaeus, 1758)
Fennecus zerda Zimmermann, 1780
Lycaon pictus Temminck, 1820
Otocyon megalotis Desmarest, 1822
Viverridae
Genetta genetta (Linnaeus, 1766)
Civettictis civetta (Linnaeus, 1766)
Herpestes ichneumon (Linnaeus, 1766)
H. cauui Smith, 1835
II. sanguineus Rüppel, 1835
Atilax paludinosus (Cuvier, 1829)
Paracynictis
Cynictis? selousi Winton, 1896
Suricata suricatta (Erxleben, 1777)
S. major Hendey, 1974
Helogale
Hyaenidae
Hyaena hyaena (Linnaeus, 1766)
H. brunnea Thunderg, 1820
Crocuta crocuta Erxleben, 1777
Felidae
Megantereon? gracile (Broom)
Machairodus
Acinonyx jubatus (Erxleben, 1777)
Felis libyca Forster, 1780
F. (Leptailurus) serval Schreber, 1777
F. (L.) hintoni
Lynx lynx (Linnaeus, 1766)
L. caracal Güldenstaedt, 1766
Leo leo (Linnaeus, 1766)
L. pardus (Linnaeus, 1766)
L. crassidens Broom, 1848
INSECTIVORA   —Cooke (1964)
Soricidae
Crocidura cf. bicolor Bocage, 1889
CHIROPTERA   —Cooke (1964)
Rhinolophidae
Rhinolophus cf. geoffroyi Smith, 1829
Vespertilionidae
Miniopterus
PRIMATES
Cercopithecidae   —Cooke (1964)
Simopithecus
Papio ursinus (Kerr, 1792)

Hominidae —Howell (1978)
   Homo erectus (Dubois, 1894)
   H. sapiens rhodesiensis Woodward, 1921
   H. sapiens neanderthalensis King, 1864
   H. sapiens Linnaeus, 1758
ARTIODACTYLA
Suidae —Cooke and Wilkinson (1978)
   Notochoerus capensis Broom, 1925 (age?)
   Sus scrofa Linnaeus 1758
   Potamochoerus porcus (Linnaeus, 1758)
   Kolpochoerus limnetes (Hopwood, 1926)
   K. phacochoeroides (Thomas, 1884) (age?)
   K. paiceae (Broom, 1931) (age?)
   Hylochoerus meinertzhageni (Thomas, 1904)
   Phacochoerus modestus (E.C. and H.E. Van Hoepen, 1932)
   P. kabuae Whitworth, 1965 (age?)
   P. aethiopicus (Pallas, 1767)
   P. africanus (Gmelin, 1788)
   Metridiochoerus andrewsi Hopwood, 1926
   Stylochoerus compactus (E.C. and H.E. Van Hoepen, 1932)
Hippopotamidae
   Hippopotamus amphibius Linnaeus, 1766
   Hexaprotodon liberiensis Morton, 1849
Camelidae
   Camelus thomasi Pomel, 1893
   C. cf. *bactrianus* Linnaeus, 1766
Cervidae
   Cervus elaphus Linnaeus, 1758
   Dama dama Linnaeus, 1758
   Megaloceros algericus (Lydekker, 1890)
Giraffidae
   Giraffa camelopardalis (Linnaeus, 1766)
   G. jumae Leakey, 1965
   G. stillei (Dietrich, 1942)
   G. gracilis Arambourg, 1947A
   Sivatherium maurusium (Pomel, 1892)
Bovidae
   Tragelaphus
   Syncerus cf. *caffer* (Sparman, 1779)
   Pelorovis antiquus (Duvernoy, 1851)
   Bos primigenius Bojanus, 1827
   cf. *Cephalophus*
   Kobus leche Gray, 1850
   K. ellipsiprymnus (Ogilby, 1833)
   K. kob Erxleben, 1777
   Redunca redunca (Pallas, 1767)
   R. arundinum (Boddaert, 1785)
   Hippotragus niger (Harris, 1838)
   H. leucophaeus (Pallas, 1767)
   H. equinus (Geoffroy-Saint-Hilaire, 1816)
   Oryx helmoedi (Van Hoepen, 1932C)
   Megalotragus priscus Broom, 1909
   Connochaetes gnou (Zimmermann, 1777)
   Damaliscus niro (Hopwood, 1936)
   Rabaticeras arambourgi Ennouchi, 1953
   Aepyceros cf. *melampus* (Lichtenstein, 1814)
   Oreotragus major Wells, 1951 (age?)
   Antidorcas australis Hendey and Hendey, 1968

   A. bondi (Cooke and Wells, 1951)
   A. marsupialis (Zimmermann, 1780)
   Gazella atlantica Bourguignat, 1870
   G. tingitana Arambourg, 1957
   G. dorcas Linnaeus, 1766
   G. rufina Thomas, 1894
   G. cuvieri Ogilby, 1840
   Capra (*Ammotragus*) *lervia* (Pallas, 1777)
PERISSODACTYLA
Equidae —Churcher and Richardson (1978)
   Hipparion libycum Pomel, 1897
   Equus (*Dolichohippus*) *oldowayensis* Hopwood, 1937
   E. (*D.*) *capensis* Broom, 1909
   E. (*D.*) *grevyi* Oustalet, 1882
   E. (*Hippotigris*) *burchelli* (Gray, 1824)
   E. (*H.*) *zebra* Linnaeus, 1758
   E. (*H.*) *quagga* Gmelin, 1788
   E. (*Asinus*) *asinus* Linnaeus, 1758
Rhinocerotidae
   Diceros bicornis (Linnaeus, 1758)
   Ceratotherium simum (Burchell, 1817)
TUBULIDENTATA
   Orycteropus
HYRACOIDEA
Procaviidae
   Procavia capensis Storr, 1780
PROBOSCIDEA
Elephantidae
   Loxodonta atlantica (Pomel, 1879)
   L. africana (Blumenbach, 1797)
   Elephas recki Dietrich, 1916
RODENTIA —Cooke (1964)
Sciuridae
   Xerus capensis Kerr, 1792
Pedetidae
   Pedetes hagenstadi Dreyer and Lyle, 1931
Hystricidae
   Hystrix africaeaustralis Peters, 1852
Muridae
   Aethomys cf. *chrysophilus* (Thomas, 1896)
   Thallomys cf. *paedulcus* (Sundevall, 1846)
   Saccostomus campestris (Desmarest, 1822)
   Dendromus cf. *mesomelas* (Brants, 1827)
   Steatomys cf. *pratensis* Peters, 1846
   Myotomys cf. *turneri* Dreyer and Lyle, 1931
   M. cf. *unisulcatus* (Cuvier, 1829)
   Otomys cf. *irroratus* Brants, 1827
   O. gracilis (Broom, 1937E)
Thryonomyidae
   Thryonomys swinderianus (Temminck, 1827)

RANCHOLABREAN

D. Savage (1951C) proposed the Rancholabrean mammal age for the late Pleistocene faunas of North America, characterized by the presence of species of *Bison* and many species also known in the Recent fauna. There are, however, many extinct species and genera of land mammals in this age. The renowned Rancho La Brea fauna from Los Angeles, California—now so beautifully displayed in the Page Museum at Hancock Park, Los Angeles—is one of the world's most complete and voluminous samples of extinct land mammals.

The following faunal list is based on species identified in localities in Alaska, Northwest Territory of Canada, Alberta, and most of the states in the United States exclusive of the Great Lakes and New England regions. It is taken essentially from Kurtén and Anderson (1980). The species designated by * are found also in the Recent fauna.

The Rancholabrean Mammalian Fauna

MARSUPIALIA
  Didelphidae
    Didelphis virginiana Kerr, 1792*
LAGOMORPHA
  Ochotonidae
    Ochotona whartoni Guthrie and Matthews, 1971
    O. princeps (Richardson, 1828)*
    O. collaris (Nelson, 1893)*
  Leporidae
    Brachylagus idahoensis (Merriam, 1891)*
    Sylvilagus floridanus (Allen, 1890)*
    S. nuttalli (Bachman, 1837)*
    S. transitionalis (Bangs, 1895)*
    S. auduboni (Baird, 1858)*
    S. bachmani (Waterhouse, 1839)*
    S. palustris (Bachman, 1837)*
    S. leonensis Cushing, 1945*
    S. aquaticus (Bachman, 1837)*
    Lepus alleni Mearns, 1890*
    L. californicus Gray, 1837*
    L. townsendi Bachman, 1839*
    L. americanus Erxleben, 1777*
    L. arcticus Ross, 1819*
EDENTATA
  Megalonychidae
    Megalonyx wheatleyi Cope, 1871
    M. jeffersoni (Desmarest, 1822)
  Megatheriidae
    Nothrotheriops shastensis (Sinclair, 1905)
    Eremotherium rusconii (Schaub, 1935)
  Mylodontidae
    Glossotherium harlani (Owen, 1840)
  Dasypodidae
    Dasypus
    Holmesina
CARNIVORA
  Ursidae
    Arctodus simus (Cope, 1879U)
    A. pristinus Leidy, 1854C
    Tremarctos floridanus (Gidley, 1928)
    Ursus americanus Pallas, 1780*
    U. arctos Linnaeus, 1758*
  Mustelidae
    Martes pennanti (Erxleben, 1777)*
    M. nobilis Hall, 1926A

Mustela frenata Lichtenstein, 1831*
M. erminea Linnaeus, 1758*
M. rixosa Bangs, 1896*
M. vison Schreber, 1777*
M. macrodon (Prentiss, 1903)*
M. eversmanni Lesson, 1827*
M. nigripes Audubon and Bachman, 1851*
Gulo schlosseri Kormos, 1914
G. gulo (Linnaeus, 1758)*
Taxidea taxus (Schreber, 1778)*
Lutra canadensis (Schreber, 1776)*
Enhydra macrodonta Kilmer, 1972   (marine)
E. lutris Linnaeus, 1758*
Spilogale putorius (Linnaeus, 1758)*
Brachyprotoma obtusata Cope, 1899A
Mephitis mephitis (Schreber, 1776)*
Conepatus leuconotus (Lichtenstein, 1832)*
  Procyonidae
    Procyon lotor (Linnaeus, 1758)*
    Bassariscus sonoitensis Skinner, 1942
    B. astutus (Lichtenstein, 1830)*
  Canidae
    Canis latrans Say, 1823*
    C. rufus Audubon and Bachman, 1851*
    C. lupus Linnaeus, 1758*
    C. cedazoensis Mooser and Dalquest, 1975
    C. familiaris Linnaeus, 1758*
    C. dirus Leidy, 1858E
    Cuon alpinus (Pallas, 1811)*
    Urocyon cinereoargenteus (Schreber, 1775)*
    Vulpes velox (Say, 1823)*
    V. vulpes (Linnaeus, 1758)*
    Alopex lagopus (Linnaeus, 1758)*
  Felidae
    Smilodon fatalis (Leidy, 1868A)
    Dinobastis serus Cope, 1893A
    Leo atrox (Leidy, 1853F)
    L. onca (Linnaeus, 1758)*
    Acinonyx trumani (Orr, 1969)   —Adams (1979)
    Felis concolor Linnaeus, 1758*
    F. pardalis Linnaeus, 1758*
    F. amnicola Gillette, 1976
    F. yagouaroundi Lacépède, 1808*
    F. wiedi Schinz, 1821*   (age?)
    Lynx canadensis Kerr, 1792*
    L. rufus (Schreber, 1777)*
INSECTIVORA
  Soricidae
    Sorex cinereus Kerr, 1792*
    S. scottensis Jammot, 1972
    S. kansasensis McMullen, 1975
    S. longirostris Bachman, 1837*
    S. vagrans Baird, 1858*
    S. ornatus C. Merriam, 1895*
    S. palustris Richardson, 1828*
    S. fumeus Miller, 1895*
    S. arcticus Kerr, 1792*
    S. dispar Batchelder, 1896*

S. trowbridgei Baird, 1858*
S. merriami Dobson, 1890*
S. saussurei Merriam, 1892*
Microsorex hoyi (Baird, 1858)*
Cryptotis parva (Say, 1823)*
C. mexicana (Coues, 1877)*
Blarina ozarkensis Brown, 1908A*
B. brevicauda (Say, 1823)*
Notiosorex crawfordi (Coues, 1877)*
Talpidae
Condylura cristata (Linnaeus, 1758)*
Parascalops breweri (Bachman, 1842)*
Scapanus latimanus (Bachman, 1842)*
Scalopus aquaticus (Linnaeus, 1758)*
CHIROPTERA
Phyllostomatidae
Mormoops megaphylla (Peters, 1864)*
Leptonycteris nivalis (Saussure, 1860)*
Desmodus stocki Jones, 1958
Vespertilionidae
Myotis thysanodes Miller, 1897*
M. evotis (H. Allen, 1864)*
M. keeni (Merriam, 1895)*
M. leibi (Audubon and Bachman, 1842)*
M. sodalis Miller and G. Allen, 1928*
M. grisecens Howell, 1909*
M. rectidentis Choate and Hall, 1967
M. magnamolaris Choate and Hall, 1967
M. velifer (J. Allen, 1890)*
M. lucifugus (LeConte, 1831)*
M. austroriparius (Rhoads, 1897)*
M. volans (H. Allen, 1890)*
Lasionycteris noctivagus (LeConte, 1881)*
Pipistrellus subflavus (F. Cuvier, 1832)*
Pipistrellus hesperus (H. Allen, 1864)*
Eptesicus fuscus (Palisot de Beauvois, 1876)*
Lasiurus borealis (Muller, 1776)*
L. cinereus (Palisot de Beauvois, 1796)*
L. golliheri (Hibbard and Taylor, 1960)
L. intermedius H. Allen, 1862*
Nycticeius humeralis (Rafinesque, 1818)*
Plecotus tetralophodon (Handley, 1955)
P. townsendi Cooper, 1837*
Antrozous pallidus (LeConte, 1856)*
Molossidae
Tadarida brasiliensis (Geoffrey St. Hilaire, 1824)*
Eumops perotis (Schinz, 1821)*
E. glaucinus (Wagner, 1843)*
PRIMATES
Hominidae
Homo sapiens Linnaeus*
ARTIODACTYLA
Tayassuidae
Mylohyus nasutus (Leidy, 1868I)
Platygonus vetus Leidy, 1882A
P. compressus LeConte, 1848
Camelidae
Camelops kansanus Leidy, 1854F

C. traviswhitei Mooser and Dalquest, 1975
C. hesternus (Leidy, 1873F)
Hemiauchenia macrocephala (Cope, 1893A)
Palaeolama mirifica (Simpson, 1929H)
Cervidae
Odocoileus virginianus (Zimmermann, 1780)*
O. hemionus (Rafinesque, 1817)*
Blastocerus extraneus Simpson, 1928H
Navahoceros fricki (Schultz and Howard, 1935)
Sangamona fugitiva Hay, 1920
Rangifer tarandus (Linnaeus, 1758)*
Alces latifrons (Johnston, 1874)   (*Libralces?*)
A. alces (Linnaeus, 1758)*
Cervalces scotti (Lydekker, 1898)
Cervus elaphus (Linnaeus, 1758)*
Antilocapridae
Capromeryx minor Taylor, 1911
C. mexicana Furlong, 1925
Tetrameryx shuleri Lull, 1921B
T. mooseri Dalquest, 1974
T. tacubayensis Mooser and Dalquest, 1975
Stockoceros conklingi (Stock, 1930D)
S. onusrosagris (Roosevelt and Burden, 1934)
Antilocapra americana (Ord, 1815)*
Bovidae
Saiga tatarica (Linnaeus, 1758)*
Oreamnos americanus (Blainville, 1816)
O. harringtoni Stock, 1936
Ovis dalli Nelson, 1884*
O. canadensis Shaw, 1804*
Euceratherium collinum Furlong and Sinclair, 1904
Symbos cavifrons (Leidy, 1852F)
Bootherium bombifrons (Harlan, 1825)
Praeovibos priscus Staudinger, 1908
Ovibos moschatus (Zimmermann, 1780)*
Bison priscus (Bojanus, 1827)
B. latifrons (Harlan, 1825)
B. bison (Linnaeus, 1758)*   (extinct subspecies recognized)
Platycerabos dodsoni (Barbour and Schultz, 1941)
Bos grunniens Linnaeus, 1766*
PERISSODACTYLA
Equidae
Equus (Hippotigris?) parastylidens Mooser, 1959
E. (Hemionus) calobatus Troxell, 1915B
E. (H.) hemionus? Pallas, 1775
E. (H.) tau Owen, 1869
E. occidentalis Leidy, 1865C
E. complicatus Leidy, 1858C
E. niobrarensis Hay, 1913B
E. conversidens Owen, 1869
E. excelsus Leidy, 1858E
Tapiridae
Tapirus californicus Merriam, 1912
T. copei Simpson, 1945H
T. veroensis Sellards, 1918
PROBOSCIDEA
Gomphotheriidae
Cuvieronius, sp. unident.
Mammutidae
Mammut americanum (Kerr, 1791)
Elephantidae

Mammuthus columbi (Falconer, 1857)
M. jeffersoni (Osborn, 1922B)
M. primigenius (Blumenbach, 1803)
RODENTIA
  Aplodontidae
    Aplodontia rufa (Rafinesque, 1817)*
  Sciuridae
    Marmota monax Linnaeus, 1758*
    M. flaviventris (Audubon and Bachman, 1841)*
    Spermophilus richardsoni (Sabine, 1822)*
    S. townsendi Bachman, 1839*
    S. parryi (Richardson, 1825)*
    S. columbianus (Ord, 1815)*
    S. armatus Kennicott, 1863*
    S. tridecemlineatus (Mitchell, 1821)*
    S. spilosoma Bennett, 1833*
    S. mexicanus (Erxleben, 1777)*
    S. beecheyi (Richardson, 1823)*
    S. variegatus (Erxleben, 1777)*
    S. franklini (Sabine, 1822)*
    S. lateralis (Say, 1823)*
    Ammospermophilus leucurus (Merriam, 1889)*
    Cynomys ludovicianus (Ord, 1815)*
    C. leucurus Merriam, 1890*
    C. gunnisoni (Baird, 1858)*
    Tamias aristus Ray, 1965B
    T. striatus (Linnaeus, 1758)*
    Eutamias minimus (Bachman, 1839)*
    Sciurus niger Linnaeus, 1758*
    S. carolinensis Gmelin, 1788*
    S. arizonensis Coues, 1867*
    S. alleni Nelson, 1898*
    Tamiasciurus hudsonicus (Erxleben, 1777)*
    T. douglasi (Bachman, 1839)*
    Glaucomys volans (Linnaeus, 1758)*
  Geomyidae
    Thomomys orientalis Simpson, 1928H
    T. bottae (Eydoux and Gervais, 1836)*
    T. umbrinus (Richardson, 1829)*
    T. townsendi (Bachman, 1839)*
    T. talpoides (Richardson, 1828)*
    T. microdon Sinclair, 1950B
    Geomys bursarius (Shaw, 1800)*
    G. pinetus Rafinesque, 1817*
    Cratogeomys castanops (Baird, 1852)*
    Heterogeomys onerosus Russell, 1960
  Heteromyidae
    Dipodomys ordi Woodhouse, 1853*
    D. ingens (Merriam, 1904)
    D. agilis Gambel, 1848*
    D. spectabilis Merriam, 1890*
    D. merriami Mearns, 1890*
    Perognathus californicus Merriam, 1889*
    P. flavus Baird, 1855*
    P. inornatus Merriam, 1889*
    P. intermedius Merriam, 1889*
    P. parvus (Peale, 1848)*
    P. hispidus Baird, 1858*
    P. merriami J. Allen, 1892*
    P. longimembris (Coues, 1875)*
    P. apache Merriam, 1889*

    Liomys irroratus (Gray)*
  Castoridae
    Castoroides ohioensis Foster, 1838
    Castor accessor Hay, 1927D
    C. canadensis Kuhl, 1820*
  Cricetidae
    Oryzomys palustris (Harlan, 1837)*
    Reithrodontomys montanus (Baird, 1855)*
    R. humulis (Audubon and Bachman, 1841)*
    R. megalotis (Baird, 1858)*
    R. fulvescens Allen, 1894*
    Peromyscus crinitus (Merriam, 1891)*
    P. nesodytes R. Wilson, 1936B
    P. anyapahensis White, 1966
    P. imperfectus Dice, 1925A
    P. eremicus (Baird, 1858)*
    P. progressus Hibbard and Taylor, 1960
    P. berendsensis Starrett, 1956
    P. cumberlandensis Guilday and Handley, 1967
    P. cochrani Hibbard, 1955B
    P. maniculatus (Wagner, 1845)*
    P. polionotus Wagner, 1843)*
    P. leucopus (Rafinesque, 1818)*
    P. gossypinus (LeConte, 1853)*
    P. boylii (Baird, 1855)*
    P. pectoralis Osgood, 1904*
    P. truei (Shufeldt, 1885)*
    P. difficilis (Allen, 1891)*
    P. oklahomensis Stephens, 1960
    P. floridanus (Chapman, 1889)*
    Ochrotomys nuttali (Harlan, 1832)*
    Baiomys taylori (Thomas, 1887)*
    Onychomys jinglebobensis Hibbard, 1955B
    O. leucogaster (Wied-Neuwied, 1841)*
    O. torridus (Coues, 1874)*
    Sigmodon bakeri R. Martin, 1974
    S. ochrognathus Bailey, 1902*
    Neotoma floridana (Ord, 1818)*
    N. micropus Baird, 1855*
    N. albigula Hartley, 1894*
    N. lepida Thomas, 1893*
    N. mexicana Baird, 1855*
    N. fuscipes Baird, 1858*
    N. cinerea (Ord, 1815)*
    Clethrionomys gapperi (Vigors, 1830)*
    Phenacomys intermedius Merriam, 1889*
    Microtus pennsylvanicus (Ord, 1815)*
    M. montanus (Peale, 1848)*
    M. californicus (Peale, 1848)*
    M. longicaudus (Merriam, 1888)*
    M. mexicanus (Saussure, 1861)*
    M. chrotorrhinus (Miller, 1894)*
    M. xanthognathus (Leach, 1815)*
    M. miurus Osgood, 1901*
    M. ochrogaster (Wagner, 1842)*
    M. hibbardi (Holman, 1959)
    M. mcnowni (Hibbard, 1937A)
    M. pinetorum (LeConte, 1830)*

Lagurus curtatus (Cope, 1868)*
Neofiber leonardi Hibbard, 1943A
N. alleni True, 1884*
Ondatra nebrascensis (Hollister, 1911)
O. zibethicus (Linnaeus, 1766)*
Lemmus sibiricus (Kerr, 1792)*
Dicrostonyx torquatus (Pallas, 1799)*
D. hudsonius (Pallas, 1778)*
Synaptomys bunkeri Hibbard, 1939B
S. cooperi Baird, 1858*
S. australis Simpson, 1928H
S. borealis (Richardson, 1828)*
Zapodidae
Zapus hudsonius (Zimmermann, 1780)*
Z. princeps Allen, 1893*

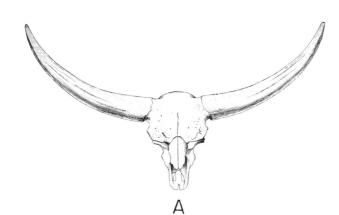

A

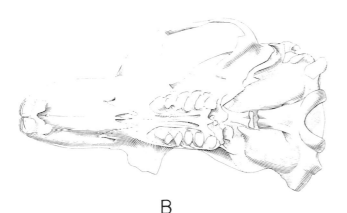

B

FIGURE 8-3 *Rancholabrean Mammals.*
 A *Bison latifrons* (Harlan, 1825). *Skull* ×1/24 approx.
    Figure 2 in VanderHoof (1942B).
 B *Thomomys bottae* (Eydoux and Gervais). *Skull.* ×2
    approx. Figure 8 in W. Miller (1971A).
(*Figure A published with permission of the University of
California Press. Figure B published with permission of the
Natural History Museum of Los Angeles County.*)

Napaeozapus insignis (Miller, 1891)*
Erethizontidae
Erethizon dorsatum (Linnaeus, 1758)*
Hydrochoeridae
Hydrochoerus holmesi Simpson, 1928H
Neochoerus pinckneyi (Hay, 1923)

Of the 133 genera of land mammals here recognized in the Rancholabrean, 27 percent are new appearances. Thirty percent of the 34 first-appearing genera (mostly bats and rodents) may have been immigrants from South America or Central America, and 47 percent (mostly "boreal" mammals) were probably immigrants from Asia. Thus we believe that the new genera in the Rancholabrean were immigrants by as much as 80 percent.

Twenty-seven percent of the 311 species and 29 percent of the genera here recognized in the Rancholabrean are extinct in North America. These extinct taxa were mostly, but not entirely, the larger animals. Among the approximate 110 species of mammals as large as a fox or larger (about 8 kg weight), about 58 percent became extinct. This is the "end-of-Pleistocene megafaunal mass extinction of land mammals" that many authors have discussed previously.

LUJANIAN

Marshall, Webb, Sepkowski, and Raup (1982) estimate the Lujanian age to be about 0.3 million years long. Marshall and Berta (1982) note that it is based on the fauna from the Lujan, Buenos Aires, Arroyo Seco formations and the Conglomerados de Magdalena and Arroyo Sauce Grande in Argentina. Its beginning is best identified by the first appearance of *Euphractus, Cabassous, Chamyphorus, Holochilius,* and *Cavia,* according to Marshall and Berta (1982). The list of genera follows Pascual et al. (1966).

Representative Lujanian Mammalian Fauna

MARSUPIALIA
 Didelphidac
  Didelphis
  Lutreolina?
  Marmosa?
EDENTATA
 Megatheriidae
  Nothrotherium
  Megatherium
  Essonodontherium
 Mylodontidae
  Scelidotherium
  Glossotherium
  Mylodon
  Lestodon
 Dasypodidae
  Chaetophractus
  Euphractus
  Zaedyus

Eutatus
Cabassous
Tolypeutes?
Propraopus
Dasypus
Chlamyphorus
Pampatherium
Glyptodontidae
Hoplophorus
Sclerocalyptus   (syn. of *Hoplophorus?*)
Panochthus
Lomaphorus
Neothoracophorus
Plaxhaplous
Doedicurus
Glyptodon
CARNIVORA
Ursidae
Arctodus
Pararctotherium
Mustelidae
Conepatus
Lyncodon
Galictis
Lutra
Canidae
Dusicyon
Cerdocyon
Felidae
Felis (*Puma*)
Leo (*Jaguarius*)
Smilodon
ARTIODACTYLA
Tayassuidae
Tayassu
Platygonus
Camelidae
Lama
Eulameops
Vicugna
Cervidae
Hippocamelus
Ozotoceros
Blastocerus
Habromeryx
Antifer
PERISSODACTYLA
Equidae
Equus
Hippidion
Onohippidium?
Tapiridae
Tapirus
PROBOSCIDEA
Gomphotheriidae
Stegomastodon
Notiomastodon
LITOPTERNA
Macraucheniidae
Macrauchenia
NOTOUNGULATA

Toxodontidae
Toxodon
RODENTIA
Cricetidae
Reithrodon
Holochilius
Necromys
Ptyssophorus
Caviidae
Dolichotis
Microcavia
Cavia
Hydrochoeridae
Hydrochoerus
Neochoerus
Chinchillidae
Lagostomus
Ctenomyidae
Ctenomys
Capromyidae?
Myocastor?

PLEISTOCENE OF AUSTRALIA

Late Cenozoic (mostly later Pleistocene) vertebrates have been known from Australia since the 1800s. Richard Owen, C.W. DeVis, E.D. Gill, L. Glauert, R. Broom, W. Ride, R. Lydekker and J.T. Woods contributed much of the early literature on these fossils. R.A. Stirton and his students and Australian colleagues continued the collecting and describing, beginning about 1955. Subsequently, many workers have investigated fossiliferous Cenozoic deposits in various parts of the continent: R. Tedford, M.O. Woodburne, M. Plane, M. Archer, T.H. Rich and P.V. Rich, E. Lundelius, W. Turnbull, and others. Most recently, P. Rich and Thompson compiled a review of the fossil-vertebrate record of Australasia.

Pleistocene localities of Australia are found mostly in Queensland, New South Wales, and South Australia; however, important sites are known also in Western Australia and Tasmania. See Stirton, Tedford, and Miller (1961) and P. Rich and Thompson (1982). We shall not enumerate or plot these sites, and we shall not try to distinguish between earlier and later Pleistocene faunas.

Representative Pleistocene Mammalian Fauna of Australasia

MONOTREMATA
Ornithorhynchus
Tachyglossus
Zaglossus
MARSUPIALIA
Dasyuridae
Dasyurus

Phascogale
Sarcophilus
Thylacinidae
Thylacinus
Peramelidae
Perameles
Thylacomys
Thylacis
Phalangeridae
Trichosurus
Cercaertus
Petaurus
Palaeopetaurus
Phascolarctos
Pseudocheirus

Burramyidae
Burramys
Thylacoleonidae
Thylacoleo
Vombatidae
Vombatus
Lasiorhinus
Phascolonus
Ramsayia
Macropodidae
Protemnodon
Macropus
Sthenurus
Procoptodon
Brachalletes
Synaptodon
Petrogale
Thylogale

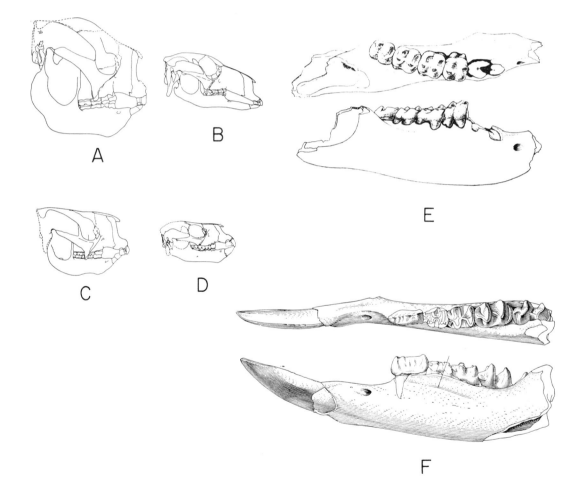

FIGURE 8-4 *Pleistocene Mammals of Australia.*
  A Procoptodon goliah (Owen, 1846). Cranium, reconstructed. ×1/6 approx. Figure 1a in Tedford (1966B).
  B Macropus canguru (Lacépède, 1979). Cranium (Recent age). ×1/6 approx. Figure 1c in Tedford (1966B).
  C Sthenurus occidentalis Glauert, 1910. Cranium, reconstructed. ×1/6 approx. Figure 1b in Tedford (1966B).
  D Sthenurus atlas Owen, 1838. Cranium, juvenile. ×1/6 approx. Figure 1d in Tedford (1966B).
  E Propleopus oscillans (De Vis, 1888). Lower jaw. ×2/3 approx. Figure 5b in Tedford (1967B).
  F Protemnodon. Lower Jaw. ×2/3 approx. Figure 8 in Stirton (1963C).
(Figures published with permission of the University of California Press.)

Bettongia
Aepyprymnus
Potorous
Propleopus
Hypsiprymnodon
Diprotodontidae
  Euowenia
  Palorchestes
  Nototherium
  Zygomaturus
  Diprotodon
  Stenomerus
RODENTIA
  Muridae
    Hydromys
    Mastacomys

REMARKS ON ISLAND FAUNAS

We should be remiss in outlining the succession of mammalian faunas of the world if we did not mention the intriguing phenomena of island dwellers, past and present. Many larger islands are parts of continental plates and have had geologic and biologic histories closely associated with the closest continents. A.R. Wallace (1892, *Island Life*) was one of the first of the ''classic'' naturalists to concentrate on these phenomena. W.D. Matthew (1915, *Climate and Evolution*) debated the island paleogeographic problems and gave proper emphasis to the fossil records involved. Many authors have contributed to the fund of knowledge regarding paleobiogeography and evolution of organisms on islands since the days of Wallace and Matthew.

1. *Madagascar.* Unique Pleistocene and Recent land mammals suggest restricted dispersals from Africa through the Cenozic: early Tertiary (Tenrecidae), early to mid-Tertiary (Lemuridae), middle Tertiary (Malagasy viverrids and cricetid rodents), Pliocene or Pleistocene (*Potamochoerus, Hexaprotodon* (= ''*Choeropsis*''), *Crocidura*.

2. *Celebes* (See Hooijer, 1958A and other papers.) The Tjabenge fauna in southwestern Celebes has an association with a number of Paleolithic artifacts and is ''entirely different from any Pleistocene fauna elsewhere in the Indo-Australian region.'' In it are two proboscideans, a suid and a pigmy buffalo. *Elephas celebensis* Hooijer is represented by adults that stood only 6 ft high at the shoulders. *Stegodon* is the other proboscidean. *Celebochoerus* Hooijer is not related to the two living suids of Celebes, and the buffalo was referred to *Anoa depressicornis*, the living species known only from Celebes. The living fauna includes *Anoa, Cynopithecus* (the crested macaque with some baboonlike characters), *Macrogalidia* (the unique brown palm civet), and *Babirussa* (the strange suid), all endemic. The Pleistocene and Re-

cent mammalian fauna of the Celebes appears to be the result of various fortuitous dispersals (perhaps through the Philippine archipelago as Hooijer believed) during the middle and later Tertiary.

3. The Philippines, Taiwan, the Japanese Archipelago. (Many references.) All these great islands have a Pleistocene land-mammal fauna reflecting some restricted dispersals from mainland Asia. Japan, especially, also produces Eocene and Miocene land mammals and evidently was more closely associated with the mainland in those days.

4. Java-Sumatra. The Pleistocene fauna of these islands has been well publicized, primarily because of the existence of *Homo erectus,* and it is being studied currently. It is well established that this area was a peninsula of southern, mainland Asia in the Pleistocene. See Pope (1982).

5. Sicily, Malta and other Mediterranean islands. Dwarf elephants, hippopotamuses, rodents, and other land mammals have been described from these islands and have been known since the mid-1800s. See Maglio (1973) for some nomenclatural revisions and a beginning guide to the earlier literature.

6. The Channel Islands of California. (Stock, 1943A; Wendorf, 1982.) Dwarf mammoths and rodents of the Pleistocene fauna may not be directly associated with evidences of early human habitation on these islands, as previously concluded.

7. New Caledonia. *Zygomaturus* (diprotodontid) from this island indicates a 600-mile dispersal from New Guinea or Australia proper in Pliocene or perhaps early Pleistocene time.

8. Seymour Island in the Antarctic Archipelago. M.O. Woodburne and colleagues at the University of California, Riverside, have indicated to the news services that they have found *Polydolops* here. This supports the idea that Antarctica shared an Eocene land-mammal fauna with South America and was also a ''rest stop'' for dasyuroids or ''proto-dasyuroids'' en route to colonize and give rise to the great adaptive radiation of marsupials in Australasia.

These islands just mentioned and many other islands not mentioned offer exciting studies for future workers and will undoubtedly be the subject of several books in the next decade.

BIBLIOGRAPHY

CHURCHER, C.S. 1978. Giraffidae. In *Evolution of African Mammals,* ed. V.J. Maglio, pp. 509–535. Cambridge: Harvard Univ. Press.

CHURCHER, C.S., and RICHARDSON, M.L. 1978. Equidae. In *Evolution of African Mammals,* ed. V.J. Maglio, pp. 379–422. Cambridge: Harvard Univ. Press.

COOKE, H.B.S. 1964. The Pleistocene environment in Southern Africa. In *Ecological Studies in Southern Africa,* D.H.S. Davis, pp. 1–23. The Hague: *Junk, Monogr. Biol.* 14.

COOKE, H.B.S. 1978. Faunal evidence for the biotic setting of early African hominids. In *Early Hominids in Africa,* C.J. Jolly, pp. 267–281. New York: Gerald Duckworth and Co.

COOKE, H.B.S., and WILKINSON, A.F. 1978. Suidae and Tayassuidae. In *Evolution of African Mammals,* ed. V.J. Maglio, pp. 435–482. Cambridge: Harvard Univ. Press.

COPPENS, Y., MAGLIO, V.J., MADDEN, C.T., and BEDEN, M. 1978. Proboscidea. In *Evolution of African Mammals,* ed. V.J. Maglio, pp. 336–367. Cambridge: Harvard Univ. Press.

ESHELMAN, R.E. 1975. Geology and paleontology of the early Pleistocene (late Blancan) White Rock fauna from north-central Kansas. Claude W. Hibbard Memorial Volume. *Univ. Mich., Pap. on Paleontol.,* 13:1–60.

GENTRY, A.W. 1978. Bovidae. Tragulidae and Camelidae. In *Evolution of African Mammals,* ed. V.J. Maglio, pp. 536–572. Cambridge: Harvard Univ. Press.

GILLETTE, D.G., and RAY, C.E. 1981. Glyptodonts of North America. *Smithson. Contr. Paleobiol.* 40:1–255.

HAMILTON, W.R. 1978. Cervidae and Palaeomerycidae. In *Evolution of African Mammals,* ed. V.J. Maglio, pp. 496–508. Cambridge: Harvard Univ. Press.

HOWELL, F.C. 1978. Hominidae. In *Evolution of African Mammals,* ed. V.J. Maglio, pp. 154–248. Cambridge: Harvard Univ. Press.

JAMMOT, D. 1972. Relationships between the new species *Sorex scottensis* and the fossil shrews *Sorex cinereus* Kerr. *Mammalia* 3613:449–458.

KURTÉN, B. 1968. *Pleistocene Mammals of Europe.* In *The World Naturalist,* ed. R. Carrington. London: Weidenfeld and Nicolson.

KURTÉN, B., and ANDERSON, E. 1980. *Pleistocene Mammals of North America.* New York: Columbia Univ. Press.

LINDSAY, E.H., JOHNSON, N.M., and OPDYKE, N.D. 1975. Preliminary correlation of North American land mammal ages and geomagnetic chronology. Claude W. Hibbard Memorial Volume. *Univ. Mich., Pap. on Paleontol.* 12:111–119.

MAGLIO, V.J. 1973. Origin and evolution of the Elephantidae. *Trans. Am. Phil. Soc.* 63:1–149.

MARSHALL, L.G., and BERTA, A. 1982?. Biochronology of the continental Quaternary mammal record of South America. *Fieldiana.* (In press.)

MARSHALL, L.G., PASCUAL, R., CURTIS, G.H., and DRAKE, R.E. 1977. South American geochronology: Radiometric time scale for middle to late Tertiary mammal-bearing horizons in Patagonia. *Science* 195:1325–1328.

MARSHALL, L.G., WEBB, S.D., SEPKOSKI, J.H., JR., and RAUP, D.M. 1982. Mammalian evolution and the Great American Interchange. *Science* 215:1351–1357.

PASCUAL, R., ORTEGA-H., E.J., GONDAR, D., and TONNI, E. 1966. *Paleontografia Bonaerense.* Fasc. 4. *Vertebrata.* Com. Investig. Cient. Prov. Buenos Aires, pp. 1–202.

POPE, G.G. 1982. Hominid evolution in East and Southeast Asia. Ph.D. dissertation. Univ. Calif. Berkeley, Dept. Anthropology.

RICH, P.V., and THOMPSON, E.M. 1982. *The Fossil Vertebrate Record of Australasia.* Monash University.

STIRTON, R.A., TEDFORD, R.H., and MILLER, A.H. 1961. Cenozoic stratigraphy and vertebrate paleontology of the Tirari Desert, South Australia. *S. Austral. Mus., Rec.,* 14:19–61.

VANGENGEIM, E.A., and SHER, A.V. 1970. Siberian equivalents of the Tiraspol faunal complex. *Palaeogeogr., Palaeoclim., Palaeoecol.* 8:197–207.

WEBB, S.D. 1976. Mammalian faunal dynamics of the great American interchange. *Paleobiology* 2:220–234.

WENDORF, M. 1982. Re: early people and mammoths on the Channel Islands, California. Ph.D. Dissertation. Univ. Calif., Dept. Anthropology.

CHAPTER NINE

Faunal Turnover

We have compiled data and have formulated simple conclusions about *faunal turnover* as a phenomenon among land mammals during the Cenozoic in Europe and in North America. Faunal turnover is a comparison between appearing and disappearing populations of animals representing certain taxonomic categories. We have selected two aggregates of taxa to represent the fauna of a Cenozoic age:

1. The array of known genera in that age.
2. The array of known families in that age.

Taxonomic lists are a weak expression of the faunas. They tell us little about the actual abundance or the ecological and biogeographical associations of the animals represented. They are, however, an index of sorts for extinction in the area considered and for the appearance and adaptive radiation. So, lacking a better tool, we use such lists for some interpretation and a few speculations.

Our consideration of each mammalian fauna of the succession of land-mammal ages in Europe and North America in preceding chapters included a terse summary of some of the aspects of generic and family turnover. Species in these faunas were not assessed for two principal reasons:

1. The validity and utility of many, if not most, fossil species is hotly debated.

2. Because of the probable biases in recognition of species and for a combination of other reasons (including the high-speed rate of morphologic evolution of mammals, as described in the Introduction), there was an almost complete replacement of species from one mammal age to the next.

We believe that before you make grandiose conclusions about evolutionary turnover, rates of change, faunal equilibria, and the like within the succession of Cenozoic mammalian faunas, as these phenomena relate to the physical history of the earth, you should be reminded of the probable and possible biases and imprecisions in the data.

Question: In a given area, for example Europe or North America, what are the possible causes for a spectacular increase in number of genera or families in Age *B* as contrasted with the preceding Age *A?*

Reply:
1. This phenomenon may be partly the result of accelerated ("explosive") evolution and the resultant greater taxonomic diversification.
2. This phenomenon may be partly the result of sudden immigration of new mammalian groups into the area.

We hope to be able to distinguish the reasons after meticulous comparison of the faunas of Ages *A* and *B* with preceding faunas of the world, but there are problems:

a. Bigger samples tend to produce larger taxonomic lists. More genera and more families will be named from a sample of thousands of specimens than from a sample of only hundreds of specimens.

b. A sample from local faunas that lived and were entombed and preserved in divers taphonomic environments—for example, a combination of fluvial deposits plus lacustrine deposits plus paludal deposits plus fissure deposits in woodlands and in more open country and at different elevations and with different microclimates—should have more taxa and be more diversified than a sample from less diversified environments.

c. Even though the samples representing A and B might be comparable in size and in provenience diversity, if B is from a stratal succession deposited during a 5-million-year interval and A only a 2-million, total B may have many new genera that could have originated in situ or that could have immigrated into the area during the last 3 million years of its duration. All this could be but is not necessarily so!

d. If more people have studied and published on the sample from B, as opposed to A, Age B is likely to have a greater number of proposed taxa.

e. "One worker's species is another worker's genus." Probably no two workers will recognize exactly the same number of genera from a given sample.

f. Other incongruities and imprecisions may result from such factors as differing intensity and meticulousness in the collecting techniques for the sample from A versus the sample from B, or from something as simple as a 20-year lapse between collections.

EUROPEAN AND NORTH AMERICAN CENOZOIC MAMMALIAN GENERA AND FAMILIES

The accompanying graphs and tables show data on genera and families of land mammals in the Cenozoic of Europe and North America. Data for the Pleistocene and Recent of Europe are from Kurtén (1968); data for the Pleistocene and Recent of North America are from Kurtén and Anderson (1980).

North American Cenozoic Mammalian Genera

The table on faunal life of the North American Cenozoic states that the Barstovian land-mammal age, lasting approximately 4 million years, has 146 recognized genera of land mammals; that 29 of these genera are known before and after Barstovian; that 17 genera are known only in this age; that 65 genera (45 percent) appear for the first time in North America at this time; that the genera are appearing at a rate

TABLE 1 *Faunal-Life Table: North American Cenozoic Mammalian Genera*

North American Land-Mammal Ages	Years Duration	Total Genera	Genera Known Before and After This Age	Genera Known Only in This Age	Genera Appearing in This Age	Percentage Genera Appearing	Appearance Rate	Genera Not Known After This Age	Percentage Genera Disappearing	Disappearing Rate	Turnover Rate
Rancholabrean	0.49	133	68	9	34	26	69	36	27	73	71
Irvingtonian	1.3	119	54	10	41	34	31	23	19	18	25
Blancan	3.2	106	25	34	75	71	23	38	36	12	18
Hemphillian	4.0	132	10	33	66	50	17	87	66	22	19
Clarendonian	3.0	134	37	19	54	40	18	70	52	23	21
Barstovian	4.0	146	29	17	65	45	16	65	45	16	16
Hemingfordian	4.0	131	21	20	71	54	18	59	45	15	17
Arikareean 2	4.0	128	17	36	67	52	17	75	59	19	18
Arikareean 1	4.0	82	15	17	42	51	11	25	30	6	9
Whitneyan	2.0	71	22	2	21	30	11	40	56	20	16
Orellan	2.0	75	44	9	26	34	13	25	33	13	13
Chadronian	5.0	127	9	47	96	76	19	78	61	16	18
Duchesnean	2.0	30	5	6	11	37	6	11	37	6	6
Uintan	8.0	132	5	58	95	72	12	106	80	13	13
Bridgerian	2.0	77	20	26	39	51	20	40	52	20	20
Wasatchian	4.0	113	5	42	77	69	19	75	66	19	19
Clarkforkian	2.0	63	18	5	27	43	14	16	25	8	11
Tiffanian	4.0	94	12	14	51	54	13	53	56	13	13
Torrejonian	2.0	88	16	29	54	61	27	45	51	23	25
Puercan	3.0	55	1	22	46	84	15	28	51	9	12
Means of the 17 "good" samples	3.4	107	21	25	57	52	18	54	49	16	17

FAUNAL-LIFE GRAPH - NORTH AMERICAN CENOZOIC MAMMALIAN GENERA

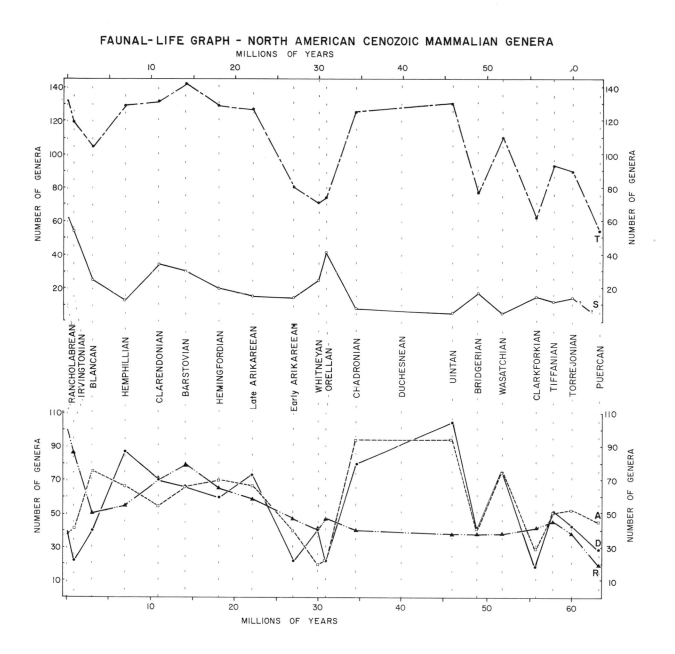

FIGURE 9-1 *Faunal-Life Graph: North American Cenozoic Mammalian Genera. This graph depicts changing numbers of genera through time in some of the categories listed in Table 1. We have plotted an approximate midpoint for each mammal age on the scale of years (for example, Whitneyan is plotted at 30 million years before present).*

Key:

T, *total number of genera recognized in a given age.*

S, *number of genera appearing before, in, and after a given age = standing fauna.*

A, *number of genera appearing for the first time in North America in a given age.*

D, *number of genera disappearing in North America after a given age.*

R, *running mean, or "standing crop," of genera, as used by Marshall et al. (1982) = average of appearances in and disappearances after a given age subtracted from total number of genera recognized in that age.*

403

of 16 per 1 million years; that 65 genera (45 percent) do not appear in North America after Barstovian and are disappearing at a rate of 16 per 1 million years; and that the turnover rate, which equals the average of appearing and disappearing, is also 16 per 1 million years.

For general averaging (establishing the mean) of the phenomena, we exclude the data on *Rancholabrean*. This is done with the rationale that the Recent and some of the future should be included with it for rate determinations. We exclude the *Duchesnean,* which is a miserably small sample. And we exclude the *Puercan,* with the rationale that good taxonomic comparison from the preceding late Cretaceous fauna is not yet available.

Other workers' counts of the total genera, genera known before and after, genera restricted, genera appearing, and genera disappearing are likely to differ from our count by at least 5 percent. Also, remember that a restricted genus is tabulated in the *appearing* column and also the *disappearing*.

We thank Mr. Patrick F. Fields of the Department of Paleontology, University of California, Berkeley, for his criticism of these tables and graphs and for suggesting several categories of data that will be most meaningful to a student of the dynamics of mammalian paleofaunas.

European Cenozoic Mammalian Genera

Table 2 states, for example, that the Turolian land-mammal age, lasting approximately 4 million years, has 178 recognized genera of land mammals; that only 18 of these genera are known before and after Turolian; that 65 genera (37 percent) are known only in this age; that 89 genera (50 percent) appear for the first time in Europe during Turolian; that the genera are appearing at a rate of 22 per million years; that 122 genera (69 percent) do not appear in Europe after the Turolian and are disappearing at a rate of 31 per million years; and that the turnover rate for Turolian is 27 per million years. In the averaging (means), we exclude Steinheimian and Cernaysian.

INTERPRETATIONS AND SPECULATIONS

The data of our graphs and tables will be interpreted differently by various workers. Our conservative belief is that the data show an approximate balance between appearance ("origination") and disappearance ("extinction") of genera and families in North America and Europe ("faunal equilibrium") with the following exceptions:

1. Supported is the often-published conclusion that there was an "explosive" origin (by evolution and? immigra-

TABLE 2   *Faunal-Life Table: European Cenozoic Mammalian Genera*

European Land-Mammal Ages	Years Duration	Total Genera	Genera Known Before and After This Age	Genera Known Only in This Age	Genera Appearing in This Age	Percentage Genera Appearing	Appearance Rate	Genera Not Known After This Age	Percentage Genera Disappearing	Disappearance Rate	Turnover Rate
Steinheimian	0.5	94	52	5	6	6	12	26	28	56	34
Biharian	1.8	98	43	5	46	47	26	28	29	16	21
Villafranchian	1.7	123	48	22	37	30	22	25	20	15	19
Ruscinian	1.5	142	30	34	86	61	57	58	41	39	48
Turolian	4.0	178	18	65	89	50	22	122	69	31	27
Vallesian	3.0	124	31	26	52	42	17	59	48	20	19
Astaracian	3.0	148	47	33	61	41	20	85	57	28	24
Orleanian	5.0	148	24	20	83	56	17	65	44	13	15
Agenian	4.0	95	24	17	37	39	9	45	47	11	10
Late Oligocene	4.0	79	21	9	31	39	8	34	43	9	9
M. Oligocene	4.0	81	35	10	64	79	16	35	43	9	13
Early Oligocene	3.0	114	19	26	78	68	26	49	43	16	21
Headonian	4.0	97	19	23	41	42	10	57	59	14	12
Robiacian	5.0	84	23	7	40	48	8	31	37	6	7
Lutetian	5.0	69	11	18	45	65	9	26	38	5	7
Cuisian	2.0	58	10	4	24	41	12	34	59	17	15
Sparnacian	3.0	61	1	12	55	90	18	22	36	7	27
Cernaysian	4.0	29	20	5	28	96	7	23	79	6	7
Means of 16 samples, Sparnacian to Biharian inclusive	3.38	106	25	21	54	52	19	48	39	16	18

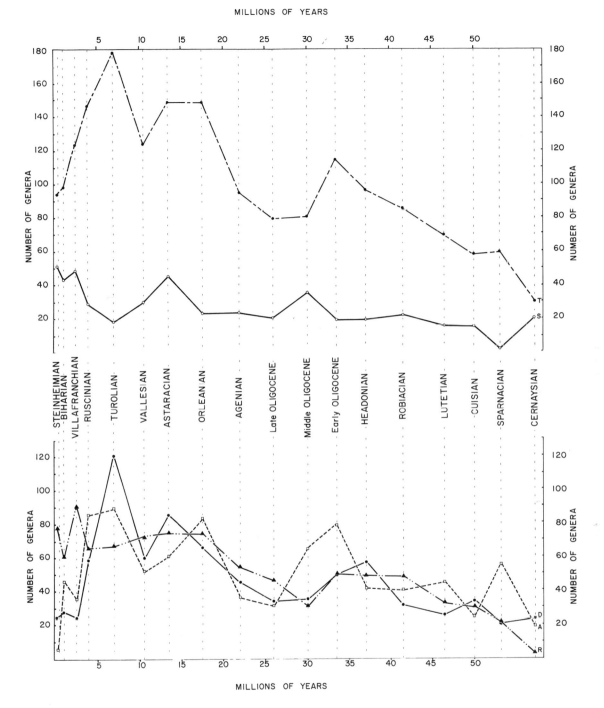

FIGURE 9-2 *Faunal-Life Graph: European Cenozoic Mammalian Genera. This graph depicts changing numbers of genera through time in some of the categories listed in Table 2. Explanations for the plots and the abbreviations are the same as those for Figure 9-1, with* Europe *substituted for* North America.

405

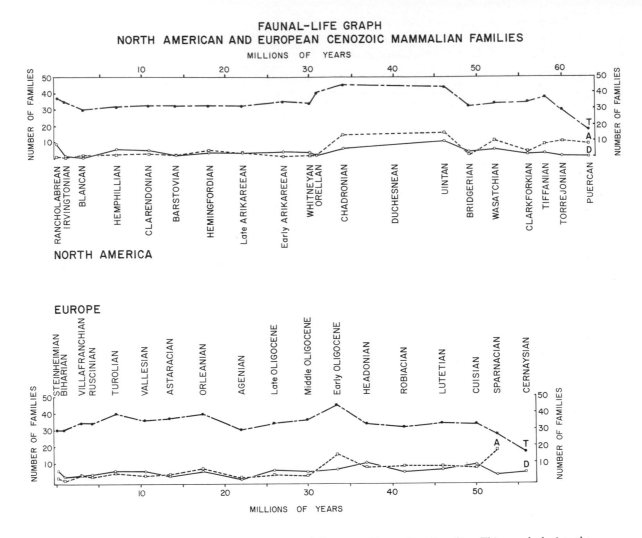

FIGURE 9-3 *Faunal-Life Graph: North American and European Mammalian Families. This graph depicts the changing number of total families and of the first and last appearance of families of land mammals through the Cenozoic of North America and Europe.*

Key:

T, *total number of families recognized in a given age.*

A, *number of families appearing for the first time in a given age.*

D, *number of families disappearing after a given age.*

tion) of families and genera of mammals in earlier Paleocene time (= Puercan plus Torrejonian) following the age of dinosaurs: 50 percent of the Puercan families and 84 percent of the Puercan genera are post-Cretaceous appearances.

2. The noteworthy increase in numbers of genera and families of the Wasatchian reflects probable immigration from Asia of adapids, omomyids, hyaenodontids, meniscotheriids?, perissodactyls?, and artiodactyls?.

3. Sparnacian (early Eocene) in Europe witnessed a mass immigration of land mammals from North America (and Asia?), discussed in our chapter on the Eocene, and this was accompanied by a remarkable decrease in its standing fauna (= genera appearing before, in, and after the

age). The record shows that about 80 percent of the known Cernaysian (late Paleocene) genera of Europe—a small fauna of only 29 genera—became extinct about the time Sparnacian immigrants arrived or just preceding their arrival. If such a percentage applies to the unknown Cernaysian fauna when and if it is found, or if the complete Cernaysian fauna was indeed uniquely small, this *regional* mammalian faunal revolution was the greatest (most complete) in the known history of mammals. It was, to be sure, a *supergrande coupure* for the European area. But Asia may have been a theater of relatively steady evolutionary diversification at that time.

4. The Early Oligocene mammalian fauna of Europe, a large fauna of 114 genera in 45 families, shows many new

genera and families following a 25 percent disappearance of families and a 60 percent disappearance of genera known in the preceding, large Headonian fauna (97 genera in 37 families).

The Chadronian (Eocene-Oligocene) of North America—also a large fauna of 127 genera in 45 families—like the Early Oligocene of Europe, shows many new genera and families following numerous disappearances of genera and families at the end of the long, combined Uintan-Duchesnean interval. Only 20 percent of the Uintan genera are known in later faunas of North America. Lillegraven (1972) concluded that this Eocene-into-Oligocene big faunal turnover "correlated well with" lowering of mean annual temperature and decrease of equability. W. Alvarez, et al. (1982) find an increase in iridium in deep-sea sediments of the Caribbean, accompanied by a turnover in the radiolarian species and an interval of microtektites (Glass and Zwart, 1979). They are considering the possibility that a collision of a giant asteroid or meteorite with the earth somewhere in the equatorial belt triggered profound climatic and organic changes. This is an enticing hypothesis but needs comprehensive testing in the low-latitude rocks of this age, and it needs a thoroughly demonstrated synchroneity of the marine and nonmarine paleontostratigraphic data. We believe that the fossil mammal record *may* or *may not* support Alvarez's hypothesis. Although the Eocene-into-Oligocene turnover was very high, it was not the only interval of great turnover in the Cenozoic history of the Mammalia. All these intervals of turnover should be examined meticulously by the various disciplines.

5. We cannot escape the conclusion that the remarkable similarity of the "curves" for total number of genera, numbers of appearing and disappearing genera, and running mean of genera from the Puercan through the Chadronian is the product of size of the respective samples, and that these "curves" do *not* indicate fluctuating evolutionary or immigrational rates. By contrast, the strong drop in number (and diversity) of genera from the Chadronian (smaller sample) to Orellan (larger sample) does indicate a real reduction in diversity (whatever the reasons are) in our opinion, even though Chadronian samples have a much greater north-south spread and a probable greater ecologic-taphonomic range.

6. Fifty-seven percent of the genera and 23 percent of the families of the Orleanian (Early Miocene) of Europe are new. This is an unusually high percentage in both categories and graphically depicts the "Coupure miocène," an immigration and first appearance in Europe of hyaenids, pliopithecids, giraffids, bovids, equids, and three families of proboscideans. We judge that the earliest representatives of all the above-listed families were outside the European area, with the equids coming from North America and the other groups coming from Africa and Asia. The new genera representing these families, combined with genera that could have evolved from European or African-Asian predecessors give the very high generic-origination number. We refrain from speculating, however, about which of the new genera not in the immigrant families are allochthonous to the European area and which autochthonous.

Faunal turnover during the remainder of the Miocene in Europe seems to have been an approximate balance between appearances and disappearances, with disappearances and the resulting decrease in taxonomic diversity slightly more conspicuous through Astaracian and Vallesian time. Although the Turolian (latest Miocene) had the largest-known array of genera of any recognized land-mammal age, much of this array is found in one family, the Bovidae, and it does not necessarily token greater adaptive radiation and greater taxonomic diversity than several other ages in which more families are represented.

7. The North American Miocene land-mammal ages contained a very stable number of genera (128 to 143) and families (33 to 34). Numbers of new genera (54 to 71) and new families (2 to 3) and of disappearing genera (59 to 76) and disappearing families (1 to 3) are correspondingly stable until the end of the Hemphillian (latest Miocene). At this time (end of Miocene), 74 percent of the genera and 18 percent of the families evidently became extinct. This was another "time of great dying" among North American land mammals.

Scattered emigrants from Eurasia have been identified in the North American Later Arikareean through Clarendonian faunas (see R. Wilson, 1968B, for example) but these seem to have played a minor role in amassing the numbers of first-appearing genera. Thus the North American Miocene scene was basically autochthons evolving from and replacing autochthons. Hemphillian time, however, witnessed a significant immigration of new genera and families from Eurasia via Beringia and from South America via the Panamanian isthmus. This dual immigration gave the Hemphillian and post-Hemphillian faunas a taxonomic composition differing strongly from the dominantly autochthonous ensembles earlier in the Miocene.

8. Corresponding approximately to the end of the Hemphillian and end of Miocene in North America, the end of Turolian in Europe saw a noteworthy extermination of families (13 percent) and of genera (73 percent) of land mammals. So in Europe as well as in North America, an interval of duration unknown but probably much less than a million years, was a time of great faunal turnover. A quick look at the Asian mammalian succession suggests no great faunal turnover there at the same time, and we can find no contemporaneous faunal "revolution" in Africa. South America seems to be a unique

example of faunal change owing to the beginning of infiltration of North American placentals via Panama at about this time.

9. The final mass extinction of land mammals in Europe and North America is the much publicized, widely discussed, and vigorously debated "megafaunal extermination" about the end of the Pleistocene, about 10,000 years before the present. Thirty-one percent of the Rancholabrean (late Pleistocene) families of North America and 23 percent of the European Steinheimian (late Pleistocene) families do not carry on in those areas into the Recent. And 32 percent of the North American genera and 27 percent of the European genera become extinct. The extinction numbers for the families are more impressive than for the genera compared with earlier land-mammal ages. And there was a similar mass extinction in each of Asia, South America, and Australia.

The main point of interest relating to this extinction, well documented by previous writers, is that most of the taxa exterminated were populations of the larger-sized, "game" land mammals. Thus, the role of ancient peoples in effecting this filtered extermination as a result of "overkill" in hunting is pondered. We shall not review the many publications on this question; Kurtén (1967) and Kurtén and Anderson (1980) have compiled detailed reviews and conclusions, and we refer you to their summary: "We believe that changes in vegetation, sudden storms, droughts, loss of habitat, interspecific competition, low reproductive rates, and overspecialization, to name a few factors, reduced or weakened populations, making them vulnerable to environmental pressures, including man, the hunter, who probably delivered the *coup de grace* to some of the megafauna between 12,000 and 9,000 years ago" Kurtén and Anderson (1980, p. 363).

BIBLIOGRAPHY

ALVAREZ, W., ASARO, F., MICHEL, H.V., and ALVAREZ, L.W. 1982. Iridium anomaly approximately synchronous with terminal Eocene extinctions. *Science* 216:886–888.

GLASS, B.P., and ZWART, M.J. 1979. North American microtektites in Deep Sea Drilling Project cores from the Caribbean Sea and Gulf of Mexico. *Bull. Geol. Soc. Am.* 90:595–602.

KURTEN, B. 1968. *Pleistocene Mammals of Europe.* In *The World Naturalist,* ed. R. Carrington. London: Weidenfeld and Nicolson.

KURTEN, B., and ANDERSON, E. 1980. *Pleistocene Mammals of North America.* New York: Columbia Univ. Press.

MARSHALL, L.G., WEBB, S.D., SEPKOSKI, J.J., JR., and RAUP, D.M. 1982. Mammalian evolution and the great American interchange. *Science* 215:1351–1357.

WILSON, R.W. 1968B. Insectivores, rodents, and intercontinental correlation of the Miocene. *Int. Geol. Congr., 23rd, Czech., Rep. Proc.* 10:19–25.

Index of Families and Genera

(Excluding taxa known only from the Mesozoic and only from the Recent)

410

Subject Index

MAMMALIAN PALEOFAUNAS OF THE WORLD

About the book:

Authors Savage and Russell have organized a vast compendium of paleontologic, stratigraphic, and geographic facts on mammalian faunas of the last 200 million years—a synthesis which has not been attempted in over 70 years. Emphasis is on fossil mammals and the application of the fossil-mammal record to studies in evolution, mammalogy, paleogeography, paleoecology, and stratigraphy, particularly with respect to the last 65 million years of the earth's history.

MAMMALIAN PALEOFAUNAS OF THE WORLD is the first book in its field to give a detailed list of species in the recognized major Cenozoic mammalian faunas of the entire world. It provides bibliographic citations for taxa and glossaries of producing localities and rock-stratigraphic units. These data are coupled with succint interpretations of chronologic correlations and the associated paleobiogeography and paleo-environments. Included is a pertinent bibliography, organized under chapter headings and updated through 1982.

Advanced university students and researchers concerned with fossil vertebrates should find this text a valuable source for general background and bibliography.

About the authors:

Senior author **Donald E. Savage** is Professor of Paleontology and Assistant Dean, College of Letters and Science, at the University of California, Berkeley. He received the Ph.D. degree in Paleontology from the University of California. His research publications have focused on the interrelationships of fossil mammals from western North America, France, Italy, Colombia, and Burma and the application of the fossil-mammal record to stratigraphy and historical geology. Dr. Savage has served as Chairman of the Department and Director of the Museum of Paleontology at the University of California. He has been President of the Society of Vertebrate Paleontology, a Fellow of the Geologic Society of America, a member of the Paleontological Society, and a Correspondant of the Muséum National d'Histoire Naturelle of France.

Donald E. Russell received an M.A. degree in vertebrate paleontology from the University of California, Berkeley. He then moved to Paris where he completed the requirements for a Doctorat ès Sciences degree from the University of Paris. His initial field of research concerned the early Tertiary mammals of Europe and more recently he has worked with early Asian faunas. He is currently Maître de Recherche with the Centre National de la Recherche Scientifique in Paris.

Addison-Wesley Publishing Co., Inc.
Advanced Book Program/World Science Division
Reading, Massachusetts